Student's Solutions Manual

Christine S. Verity

Elementary & Intermediate Algebra: Concepts and Applications

Fifth Edition

Marvin Bittinger
Indiana University Purdue University Indianapolis

David Ellenbogen
Community College of Vermont

Barbara Johnson

Addison-Wesley
is an imprint of

Reproduced by Pearson Addison-Wesley from electronic files supplied by the author.

Publishing as Pearson Addison-Wesley, 75 Arlington Street, Boston, MA 02116.

ISBN-13: 978-0-321-58623-0
ISBN-10: 0-321-58623-9

1 2 3 4 5 6 BB 12 11 10 09

Addison-Wesley
is an imprint of

www.pearsonhighered.com

Table of Contents

Chapter 1

Introduction to Algebraic Expressions

Exercise Set 1.1

1. $10n-1$ does not contain an equals sign, so it is an expression.

3. $2x-5=9$ contains an equals sign, so it is an equation.

5. $38=2t$ contains an equals sign, so it is an equation.

7. $4a-5b$ does not contain an equals sign, so it is an expression.

9. $2x-3y=8$ contains an equals sign, so it is an equation.

11. $r(t+7)+5$ does not contain an equals sign, so it is an expression.

13. Substitute 9 for a and multiply.

$$5a=5\cdot 9=45$$

15. Substitute 4 for r and subtract.

$$12-4=8$$

17. $\dfrac{a}{b}=\dfrac{45}{9}=5$

19. $\dfrac{x+y}{4}=\dfrac{2+14}{4}=\dfrac{16}{4}=4$

21. $\dfrac{p-q}{7}=\dfrac{55-20}{7}=\dfrac{35}{7}=5$

23. $\dfrac{5z}{y}=\dfrac{5\cdot 9}{15}=\dfrac{45}{15}=3$

25.
$$\begin{aligned} bh &= (6 \text{ ft})(4 \text{ ft}) \\ &= (6)(4)(\text{ft})(\text{ft}) \\ &= 24 \text{ ft}^2, \text{ or } 24 \text{ square feet} \end{aligned}$$

27.
$$\begin{aligned} A &= \frac{1}{2}bh \\ &= \frac{1}{2}(5 \text{ cm})(6 \text{ cm}) \\ &= \frac{1}{2}(5)(6)(\text{cm})(\text{cm}) \\ &= \frac{5}{2}\cdot 6 \text{ cm}^2 \\ &= 15 \text{ cm}^2, \text{ or } 15 \text{ square centimeters} \end{aligned}$$

29. $\dfrac{h}{a}=\dfrac{10}{29}\approx 0.345$

31. Let r represent Ron's age. Then we have $r+5$, or $5+r$.

33. $6b$, or $b\cdot 6$

35. $c-9$

37. $6+q$, or $q+6$

39. Let m represent Mai's speed. Then we have $8m$, or $m\cdot 8$.

41. $y-x$

43. $x\div w$, or $\dfrac{x}{w}$

45. Let l and h represent the box's length and height, respectively. Then we have $l+h$, or $h+l$.

47. $9\cdot 2m$, or $2m\cdot 9$

49. Let y represent "some number." Then we have $\frac{1}{4}y-13$, or $\frac{y}{4}-13$.

51. Let a and b represent the two numbers. Then we have $5(a-b)$.

53. Let w represent the number of women attending. Then we have 64% of w, or $0.64w$.

55.

$x+17=42$		Writing the equation
$25+17$	42	Substituting 25 for x
$42 \overset{?}{=} 42$		$42=42$ is TRUE.

Since the left-hand and right-hand sides are the same, 25 is a solution.

57.

$a-28=75$		Writing the equation
$93-28$	75	Substituting 93 for a
$65 \overset{?}{=} 75$		$65=75$ is FALSE.

Since the left-hand and right-hand sides are not the same, 93 is not a solution.

59.

$\frac{t}{7}=9$	
$\frac{63}{7}$	9
$9 \overset{?}{=} 9$	$9=9$ is TRUE.

Since the left-hand and right-hand sides are the same, 63 is a solution.

61.

$\frac{108}{x}=36$	
$\frac{108}{3}$	36
$36 \overset{?}{=} 36$	$36=36$ is TRUE.

Since the left-hand and right-hand sides are the same, 3 is a solution.

63. Let x represent the number.

What number added to 73 is 201?

Translating: $x \quad + \quad 73 \; = \; 201$

$$x+73=201$$

65. Let x represent the number.

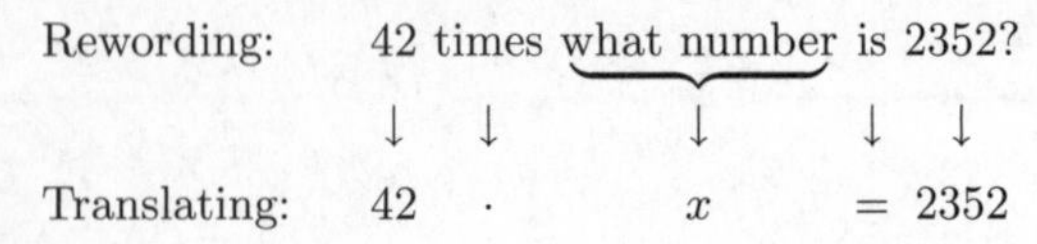

$42x = 2352$

67. Let s represent the number of unoccupied squares.

Rewording: The number of unoccupied squares added to 19 is 64.

Translating: $s + 19 = 64$

$s + 19 = 64$

69. Let x represent the total amount of waste generated, in millions of tons.

Rewording: 32% of the total amount of waste is 79 million tons.

Translating: $32\% \cdot x = 79$

$32\% \cdot x = 79$, or $0.32x = 79$

71. The sum of two numbers m and n is $m + n$, and twice the sum is $2(m + n)$. Choice (f) is the correct answer.

73. Twelve more than a number t is $t + 12$. If this expression is equal to 5, we have the equation $t + 12 = 5$. Choice (d) is the correct answer.

75. The sum of a number t and 5 is $t+5$, and 3 times the sum is $3(t + 5)$. Choice (g) is the correct answer.

77. The product of two numbers a and b is ab, and 1 less than this product is $ab - 1$. If this expression is equal to 48, we have the equation $ab - 1 = 48$. Choice (e) is the correct answer.

79. *Writing Exercise.* A *variable* is a letter that is used to stand for any number chosen from a set of numbers. An *algebraic expression* is an expression that consists of variables, constants, operation signs, and/or grouping symbols. A *variable expression* is an algebraic expression that contains a variable. An *equation* is a number sentence with the verb =. The symbol = is used to indicate that the algebraic expressions on either side of the symbol represent the same number.

81. *Writing Exercise.* No; for a square with side s, the area is given by $A = s \cdot s$. The area of a square with side $2s$ is given by $(2s)(2s) = 4 \cdot s \cdot s = 4A \neq 2A$.

83. Area of sign: $A = \frac{1}{2}(3\text{ ft})(2.5\text{ ft}) = 3.75\text{ ft}^2$

Cost of sign: $\$120(3.75) = \$450$

85. When x is twice y, then y is one-half x, so $y = \frac{12}{2} = 6$.

$\frac{x-y}{3} = \frac{12-6}{3} = \frac{6}{3} = 2$

87. When a is twice b, then b is one-half a, so $b = \frac{16}{2} = 8$.

$\frac{a+b}{4} = \frac{16+8}{4} = \frac{24}{4} = 6$

89. The next whole number is one more than $w + 3$:

$w + 3 + 1 = w + 4$

91. $l + w + l + w$, or $2l + 2w$

93. If t is Molly's race time, then Joe's race time is $t + 3$ and Ellie's race time is

$t + 3 + 5 = t + 8.$

95. *Writing Exercise.* Yes; the area of a triangle with base b and height h is given by $A = \frac{1}{2}bh$. The area of a triangle with base b and height $2h$ is given by $\frac{1}{2}b(2h) = 2\left(\frac{1}{2}bh\right) = 2A$.

Exercise Set 1.2

1. Commutative

3. Associative

5. Distributive

7. Associative

9. Commutative

11. $t + 11$ Changing the order

13. $8x + 4$

15. $3y + 9x$

17. $5(1 + a)$

19. $x \cdot 7$ Changing the order

21. ts

23. $5 + ba$

25. $(a + 1)5$

27. $x + (8 + y)$

29. $(u + v) + 7$

31. $ab + (c + d)$

33. $8(xy)$

35. $(2a)b$

37. $(3 \cdot 2)(a + b)$

39. $s+(t+6)=(s+t)+6$
$=(t+6)+5$

41. $(17a)b=b(17a)$ Using the commutative law
$=b(a17)$ Using the commutative law again

$(17a)b=(a17)b$ Using the commutative law
$=a(17b)$ Using the associative law

Answers may vary.

43. $(1+x)+2=(x+1)+2$ Commutative law
$=x+(1+2)$ Associative law
$=x+3$ Simplifying

45. $(m\cdot 3)7=m(3\cdot 7)$ Associative law
$=m\cdot 21$ Simplifying
$=21m$ Commutative law

47. $2(x+15)=2\cdot x+2\cdot 15=2x+30$

49. $4(1+a)=4\cdot 1+4\cdot a=4+4a$

51. $8(3+y)=8\cdot 3+8\cdot y=24+8y$

53. $10(9x+6)=10\cdot 9x+10\cdot 6=90x+60$

55. $5(r+2+3t)=5\cdot r+5\cdot 2+5\cdot 3t=5r+10+15t$

57. $(a+b)2=a(2)+b(2)=2a+2b$

59. $(x+y+2)5=x(5)+y(5)+2(5)=5x+5y+10$

61. $x+xyz+1$

The terms are separated by plus signs. They are x, xyz, and 1.

63. $2a+\frac{a}{3b}+5b$

The terms are separated by plus signs. They are $2a$, $\frac{a}{3b}$, and $5b$.

65. x, y

67. $4x$, $4y$

69. $2\cdot a+2\cdot b$ Using the distributive law
$=2(a+b)$ The common factor is 2.

Check: $2(a+b)=2\cdot a+2\cdot b=2a+2b$

71. $7+7y=7\cdot 1+7\cdot y$ The common factor is 7.
$=7(1+y)$ Using the distributive law

Check: $7(1+y)=7\cdot 1+7\cdot y=7+7y$

73. $32x+4=4\cdot 8x+4\cdot 1=4(8x+1)$

Check: $4(8x+1)=4\cdot 8x+4\cdot 1=32x+4$

75. $5x+10+15y=5\cdot x+5\cdot 2+5\cdot 3y=5(x+2+3y)$

Check: $5(x+2+3y)=5\cdot x+5\cdot 2+5\cdot 3y=5x+10+15y$

77. $7a+35b=7\cdot a+7\cdot 5b=7(a+5b)$

Check: $7(a+5b)=7\cdot a+7\cdot 5b=7a+35b$

79. $44x+11y+22z=11\cdot 4x+11\cdot y+11\cdot 2z$
$=11(4x+y+2z)$

Check: $11(4x+y+2z)=11\cdot 4x+11\cdot y+11\cdot 2z$
$=44x+11y+22z$

81. $5n=5\cdot n$

The factors are 5 and n.

83. $3(x+y)=3\cdot(x+y)$

The factors are 3 and $(x+y)$.

85. The factors are $7, a$ and b.

87. $(a-b)(x-y)=(a-b)\cdot(x-y)$

The factors are $(a-b)$ and $(x-y)$.

89. *Writing Exercise.* No; in general, when subtracting, the result depends on the order in which the operation is performed.

91. Let k represent Kara's salary. Then we have $\frac{1}{2}k$ or $\frac{k}{2}$.

93. *Writing Exercise.* Answers will vary.

95. The expressions are equivalent by the distributive law.

$8+4(a+b)=8+4a+4b=4(2+a+b)$

97. The expressions are not equivalent.

Let $m=1$. Then we have:

$7\div 3\cdot 1=\frac{7}{3}\cdot 1=\frac{7}{3}$, but

$1\cdot 3\div 7=3\div 7=\frac{3}{7}$.

99. The expressions are not equivalent.

Let $x=1$ and $y=0$. Then we have:

$30\cdot 0+1\cdot 15=0+15=15$, but

$5[2(1+3\cdot 0)]=5[2(1)]=5\cdot 2=10$.

101. *Writing Exercise.* $3(2+x)=3(2+0)=3\cdot 2=6$;

$6+x=6+0=6$

The result indicates that $3(2+x)$ and $6+x$ are equivalent when $x=0$. (By the distributive law, we know they are not equivalent for all values of x.)

Exercise Set 1.3

1. Since $35 = 5 \cdot 7$, choice (b) is correct.

3. Since 65 is an odd number and has more than two different factors, choice (d) is correct.

5. 9 is composite because it has more than two different factors. They are 1, 3, and 9.

7. 41 is prime because it has only two different factors, 41 and 1.

9. 77 is composite because it has more than two different factors. They are 1, 7, 11, and 77.

11. 2 is prime because it has only two different factors, 2 and 1.

13. The terms "prime" and "composite" apply only to natural numbers. Since 0 is not a natural number, it is neither prime nor composite.

15. Factorizations:

$1 \cdot 50,\ 2 \cdot 25,\ 5 \cdot 10$

List all of the factors of 50:

1, 2, 5, 10, 25, 50

17. Factorizations:

$1 \cdot 42,\ 2 \cdot 21,\ 3 \cdot 14,\ 6 \cdot 7$

List all of the factors of 42:

1, 2, 3, 6, 7, 14, 21, 42

19. $39 = 3 \cdot 13$

21. We begin factoring 30 in any way that we can and continue factoring until each factor is prime.

$30 = 2 \cdot 15 = 2 \cdot 3 \cdot 5$

23. We begin by factoring 27 in any way that we can and continue factoring until each factor is prime.

$27 = 3 \cdot 9 = 3 \cdot 3 \cdot 3$

25. We begin by factoring 150 in any way that we can and continue factoring until each factor is prime.

$150 = 2 \cdot 75 = 2 \cdot 3 \cdot 25 = 2 \cdot 3 \cdot 5 \cdot 5$

27. We begin by factoring 40 in any way that we can and continue factoring until each factor is prime.

$40 = 4 \cdot 10 = 2 \cdot 2 \cdot 2 \cdot 5$

29. 31 has exactly two different factors, 31 and 1. Thus, 31 is prime.

31. $210 = 2 \cdot 105 = 2 \cdot 3 \cdot 35 = 2 \cdot 3 \cdot 5 \cdot 7$

33. $115 = 5 \cdot 23$

35. $\dfrac{21}{35} = \dfrac{7 \cdot 3}{7 \cdot 5}$ Factoring numerator and denominator

$= \dfrac{7}{7} \cdot \dfrac{3}{5}$ Rewriting as a product of two fractions

$= 1 \cdot \dfrac{3}{5}$ $\dfrac{7}{7} = 1$

$= \dfrac{3}{5}$ Using the identity property of 1

37. $\dfrac{16}{56} = \dfrac{2 \cdot 8}{7 \cdot 8} = \dfrac{2}{7} \cdot \dfrac{8}{8} = \dfrac{2}{7} \cdot 1 = \dfrac{2}{7}$

39. $\dfrac{12}{48} = \dfrac{1 \cdot 12}{4 \cdot 12}$ Factoring and using the identity property of 1 to write 12 as $1 \cdot 12$

$= \dfrac{1}{4} \cdot \dfrac{12}{12}$

$= \dfrac{1}{4} \cdot 1 = \dfrac{1}{4}$

41. $\dfrac{52}{13} = \dfrac{13 \cdot 4}{13 \cdot 1} = 1 \cdot \dfrac{4}{1} = 4$

43. $\dfrac{19}{76} = \dfrac{1 \cdot 19}{4 \cdot 19}$ Factoring and using the identity property of 1 to write 19 as $1 \cdot 19$

$= \dfrac{1 \cdot \cancel{19}}{4 \cdot \cancel{19}}$ Removing a factor equal to 1: $\dfrac{19}{19} = 1$

$= \dfrac{1}{4}$

45. $\dfrac{150}{25} = \dfrac{6 \cdot 25}{1 \cdot 25}$ Factoring and using the identity property of 1 to write 25 as $1 \cdot 25$

$= \dfrac{6 \cdot \cancel{25}}{1 \cdot \cancel{25}}$ Removing a factor equal to 1: $\dfrac{25}{25} = 1$

$= \dfrac{6}{1}$

$= 6$ Simplifying

47. $\dfrac{42}{50} = \dfrac{2 \cdot 21}{2 \cdot 25}$ Factoring the numerator and the denominator

$= \dfrac{\cancel{2} \cdot 21}{\cancel{2} \cdot 25}$ Removing a factor equal to 1: $\dfrac{2}{2} = 1$

$= \dfrac{21}{25}$

49. $\dfrac{120}{82} = \dfrac{2 \cdot 60}{2 \cdot 41}$ Factoring

$= \dfrac{\cancel{2} \cdot 60}{\cancel{2} \cdot 41}$ Removing a factor equal to 1: $\dfrac{2}{2} = 1$

$= \dfrac{60}{41}$

51. $\dfrac{210}{98} = \dfrac{2 \cdot 7 \cdot 15}{2 \cdot 7 \cdot 7}$ Factoring

$= \dfrac{\cancel{2} \cdot \cancel{7} \cdot 15}{\cancel{2} \cdot \cancel{7} \cdot 7}$ Removing a factor equal to 1: $\dfrac{2 \cdot 7}{2 \cdot 7} = 1$

$= \dfrac{15}{7}$

53. $\dfrac{1}{2} \cdot \dfrac{3}{5} = \dfrac{1 \cdot 3}{2 \cdot 5}$ Multiplying numerators and denominators

$= \dfrac{3}{10}$

55. $\dfrac{9}{2} \cdot \dfrac{4}{3} = \dfrac{9 \cdot 4}{2 \cdot 3} = \dfrac{3 \cdot \cancel{3} \cdot \cancel{2} \cdot 2}{\cancel{2} \cdot \cancel{3}} = 6$

57. $\dfrac{1}{8} + \dfrac{3}{8} = \dfrac{1+3}{8}$ Adding numerators; keeping the common denominator

$= \dfrac{4}{8}$

$= \dfrac{1 \cdot \cancel{4}}{2 \cdot \cancel{4}} = \dfrac{1}{2}$ Simplifying

59. $\dfrac{4}{9} + \dfrac{13}{18} = \dfrac{4}{9} \cdot \dfrac{2}{2} + \dfrac{13}{18}$ Using 18 as the common denominator

$= \dfrac{8}{18} + \dfrac{13}{18}$

$= \dfrac{21}{18}$

$= \dfrac{7 \cdot \cancel{3}}{6 \cdot \cancel{3}} = \dfrac{7}{6}$ Simplifying

61. $\dfrac{3}{a} \cdot \dfrac{b}{7} = \dfrac{3b}{7a}$ Multiplying numerators and denominators

63. $\dfrac{4}{n} + \dfrac{6}{n} = \dfrac{10}{n}$ Adding numerators; keeping the common denominator

65. $\dfrac{3}{10} + \dfrac{8}{15} = \dfrac{3}{10} \cdot \dfrac{3}{3} + \dfrac{8}{15} \cdot \dfrac{2}{2}$ Using 30 as the common denominator

$= \dfrac{9}{30} + \dfrac{16}{30}$

$= \dfrac{25}{30}$

$= \dfrac{5 \cdot \cancel{5}}{6 \cdot \cancel{5}} = \dfrac{5}{6}$ Simplifying

67. $\dfrac{11}{7} - \dfrac{4}{7} = \dfrac{7}{7} = 1$

69. $\dfrac{13}{18} - \dfrac{4}{9} = \dfrac{13}{18} - \dfrac{4}{9} \cdot \dfrac{2}{2}$ Using 18 as the common denominator

$= \dfrac{13}{18} - \dfrac{8}{18}$

$= \dfrac{5}{18}$

71. Note that $\dfrac{20}{30} = \dfrac{2}{3}$. Thus, $\dfrac{20}{30} - \dfrac{2}{3} = 0$.

We can also do this exercise by finding a common denominator:

$\dfrac{20}{30} - \dfrac{2}{3} = \dfrac{20}{30} - \dfrac{20}{30} = 0$

73. $\dfrac{7}{6} \div \dfrac{3}{5} = \dfrac{7}{6} \cdot \dfrac{5}{3}$ Multiplying by the reciprocal of the divisor

$= \dfrac{35}{18}$

75. $\dfrac{8}{9} \div \dfrac{4}{15} = \dfrac{8}{9} \cdot \dfrac{15}{4} = \dfrac{2 \cdot \cancel{4} \cdot \cancel{3} \cdot 5}{\cancel{3} \cdot 3 \cdot \cancel{4}} = \dfrac{10}{3}$

77. $12 \div \dfrac{4}{9} = \dfrac{12}{1} \cdot \dfrac{9}{4} = \dfrac{\cancel{4} \cdot 3 \cdot 9}{1 \cdot \cancel{4}} = 27$

79. Note that we have a number divided by itself. Thus, the result is 1. We can also do this exercise as follows:

$\dfrac{7}{13} \div \dfrac{7}{13} = \dfrac{7}{13} \cdot \dfrac{13}{7} = \dfrac{7 \cdot 13}{7 \cdot 13} = 1$

81. $\dfrac{\frac{2}{7}}{\frac{5}{3}} = \dfrac{2}{7} \div \dfrac{5}{3} = \dfrac{2}{7} \cdot \dfrac{3}{5} = \dfrac{2 \cdot 3}{7 \cdot 5} = \dfrac{6}{35}$

83. $\dfrac{9}{\frac{1}{2}} = 9 \div \dfrac{1}{2} = \dfrac{9}{1} \cdot \dfrac{2}{1} = \dfrac{9 \cdot 2}{1 \cdot 1} = 18$

85. *Writing Exercise.* If the fractions have the same denominator and the numerators and/or denominators are very large numbers, it would probably be easier to compute the sum of the fractions than their product.

87. $5(x+3) = 5(3+x)$ Commutative law of addition

Answers may vary.

89. *Writing Exercise.* Bryce is canceling incorrectly. The number 2 is not a common factor of both terms in the numerator, so it cannot be canceled. For example, let $x = 1$. Then $(2+1)/8 = 3/8$ but $(1+1)/4 = 2/4 = 1/2$. The expressions are not equivalent.

91.

Product	56	63	36	72	140	96	168
Factor	7	7	2	36	14	8	8
Factor	8	9	18	2	10	12	21
Sum	15	16	20	38	24	20	29

93. $\dfrac{16 \cdot 9 \cdot 4}{15 \cdot 8 \cdot 12} = \dfrac{\cancel{4} \cdot \cancel{4} \cdot \cancel{3} \cdot \cancel{3} \cdot \cancel{2} \cdot 2}{\cancel{3} \cdot 5 \cdot \cancel{2} \cdot \cancel{4} \cdot \cancel{3} \cdot \cancel{4}} = \dfrac{2}{5}$

95. $\dfrac{45pqrs}{9prst} = \dfrac{5 \cdot \cancel{9} \cdot \cancel{p} \cdot q \cdot \cancel{r} \cdot \cancel{s}}{\cancel{9} \cdot \cancel{p} \cdot \cancel{r} \cdot \cancel{s} \cdot t} = \dfrac{5q}{t}$

97. $\dfrac{15 \cdot 4xy \cdot 9}{6 \cdot 25x \cdot 15y} = \dfrac{\cancel{15} \cdot \cancel{2} \cdot 2 \cdot \cancel{x} \cdot \cancel{y} \cdot \cancel{3} \cdot 3}{\cancel{2} \cdot \cancel{3} \cdot 25 \cdot \cancel{x} \cdot \cancel{15} \cdot \cancel{y}} = \dfrac{6}{25}$

99. $\dfrac{\frac{27ab}{15mn}}{\frac{18bc}{25np}} = \dfrac{27ab}{15mn} \div \dfrac{18bc}{25np} = \dfrac{27ab}{15mn} \cdot \dfrac{25np}{18bc}$

$= \dfrac{27ab \cdot 25np}{15mn \cdot 18bc} = \dfrac{\cancel{3} \cdot \cancel{9} \cdot a \cdot \cancel{b} \cdot \cancel{5} \cdot 5 \cdot \cancel{n} \cdot p}{\cancel{3} \cdot \cancel{5} \cdot m \cdot \cancel{n} \cdot 2 \cdot \cancel{9} \cdot \cancel{b} \cdot c}$

$= \dfrac{5ap}{2mc}$

101. $\dfrac{5\frac{3}{4}rs}{4\frac{1}{2}st} = \dfrac{\frac{23}{4}rs}{\frac{9}{2}st} = \dfrac{\frac{23rs}{4}}{\frac{9st}{2}} = \dfrac{23rs}{4} \div \dfrac{9st}{2}$

$= \dfrac{23rs}{4} \cdot \dfrac{2}{9st} = \dfrac{23rs \cdot 2}{4 \cdot 9st} = \dfrac{23 \cdot r \cdot \cancel{s} \cdot \cancel{2}}{\cancel{2} \cdot 2 \cdot 9 \cdot \cancel{s} \cdot t} = \dfrac{23r}{18t}$

103. $A = lw = \left(\frac{4}{5}\text{ m}\right)\left(\frac{7}{9}\text{ m}\right)$

$= \left(\frac{4}{5}\right)\left(\frac{7}{9}\right)(\text{m})(\text{m})$

$= \frac{28}{45}\text{ m}^2$, or $\frac{28}{45}$ square meters

105. $P = 4s = 4\left(3\frac{5}{9}\text{ m}\right) = 4 \cdot \frac{32}{9}\text{ m} = \frac{128}{9}\text{ m}$, or

$14\frac{2}{9}$ m

107. There are 12 edges, each with length $2\frac{3}{10}$ cm. We multiply to find the total length of the edges.

$12 \cdot 2\frac{3}{10}\text{ cm} = 12 \cdot \frac{23}{10}\text{ cm}$

$= \frac{12 \cdot 23}{10}\text{ cm}$

$= \frac{\not{2} \cdot 6 \cdot 23}{\not{2} \cdot 5}\text{ cm}$

$= \frac{138}{5}\text{ cm}$, or $27\frac{3}{5}$ cm

Exercise Set 1.4

1. Since $\frac{4}{7} = 0.\overline{571428}$, the correct choice is "repeating."

3. The set of integers consists of all whole numbers along with their opposites, so the correct choice is "integer."

5. A "rational number" has the form described.

7. A "natural number" can be thought of as a counting number.

9. The real number $-10,500$ corresponds to borrowing \$10,500. The real number 27,482 corresponds to the award of \$27,482.

11. The real number 136 corresponds to 136°F. The real number -4 corresponds to 4°F below zero.

13. The real number -554 corresponds to a 554-point fall. The real number 499.19 corresponds to a 499.19-point gain.

15. The real number 650 corresponds to a \$650 deposit, and the real number -180 corresponds to a \$180 withdrawal.

17. The real number 8 corresponds to an 8-yd gain, and the real number -5 corresponds to a 5-yd loss.

19. Since $\frac{10}{3} = 3\frac{1}{3}$, its graph is $\frac{1}{3}$ of a unit to the right of 3.

$\frac{10}{3}$

−5 −4 −3 −2 −1 0 1 2 3 4 5

21. The graph of -4.3 is $\frac{3}{10}$ of a unit to the left of -4.

−4.3

−5 −4 −3 −2 −1 0 1 2 3 4 5

23.

−5 −4 −3 −2 −1 0 1 2 3 4 5

25. $\frac{7}{8}$ means $7 \div 8$, so we divide.

$$\begin{array}{r} 0.8\,7\,5 \\ 8\,\overline{)\,7.0\,0\,0} \\ \underline{6\,4} \\ 6\,0 \\ \underline{5\,6} \\ 4\,0 \\ \underline{4\,0} \\ 0 \end{array}$$

We have $\frac{7}{8} = 0.875$.

27. We first find decimal notation for $\frac{3}{4}$. Since $\frac{3}{4}$ means $3 \div 4$, we divide.

$$\begin{array}{r} 0.7\,5 \\ 4\,\overline{)\,3.0\,0} \\ \underline{2\,8} \\ 2\,0 \\ \underline{2\,0} \\ 0 \end{array}$$

Thus, $\frac{3}{4} = 0.75$, so $-\frac{3}{4} = -0.75$.

29. $\frac{7}{6}$ means $7 \div 6$, so we divide.

$$\begin{array}{r} 1.1\,6\,6 \\ 6\,\overline{)\,7.0\,0\,0} \\ \underline{6} \\ 1\,0 \\ \underline{6} \\ 4\,0 \\ \underline{3\,6} \\ 4\,0 \\ \underline{3\,6} \\ 4 \end{array}$$

Thus $\frac{7}{6} = 1.1\overline{6}$, so $-\frac{7}{6} = -1.1\overline{6}$.

31. $\frac{2}{3}$ means $2 \div 3$, so we divide.

$$\begin{array}{r} 0.6\,6\,6\,\ldots \\ 3\,\overline{)\,2.0\,0\,0} \\ \underline{1\,8} \\ 2\,0 \\ \underline{1\,8} \\ 2\,0 \\ \underline{1\,8} \\ 2 \end{array}$$

We have $\frac{2}{3} = 0.\overline{6}$.

33. We first find decimal notation for $\frac{1}{2}$. Since $\frac{1}{2}$ means $1 \div 2$, we divide.

$$\begin{array}{r} 0.5 \\ 2\,\overline{)\,1.0} \\ \underline{1\,0} \\ 0 \end{array}$$

Thus, $\frac{1}{2} = 0.5$, so $-\frac{1}{2} = -0.5$.

35. Since the denominator is 100, we know that $\frac{13}{100} = 0.13$. We could also divide 13 by 100 to find this result.

37.

$\sqrt{5}$

$-4\ -2\ \ 0\ \ 2\ \ 4$

39.

$-\sqrt{22}$

$-4\ -2\ \ 0\ \ 2\ \ 4$

41. Since 5 is to the right of 0, we have $5 > 0$.

43. Since -9 is to the left of 9, we have $-9 < 0$.

45. Since -8 is to the left of -5, we have $-8 < -5$.

47. Since -5 is to the right of -11, we have $-5 > -11$.

49. Since -12.5 is to the left of -10.2, we have $-12.5 < -10.2$.

51. We convert to decimal notation.
$\frac{5}{12} = 0.41\overline{6}$ and $\frac{11}{25} = 0.44$. Thus, $\frac{5}{12} < \frac{11}{25}$.

53. $-2 > x$ has the same meaning as $x < -2$.

55. $10 \le y$ has the same meaning as $y \ge 10$.

57. $-3 \ge -11$ is true, since $-3 > -11$ is true.

59. $0 \ge 8$ is false, since neither $0 > 8$ nor $0 = 8$ is true.

61. $-8 \le -8$ is true because $-8 = -8$ is true.

63. $|-58| = 58$ since -58 is 58 units from 0.

65. $|-12.2| = 12.2$ since -12.2 is 12.2 units from 0.

67. $|\sqrt{2}| = \sqrt{2}$ since $\sqrt{2}$ is $\sqrt{2}$ units from 0.

69. $\left|-\frac{9}{7}\right| = \frac{9}{7}$ since $-\frac{9}{7}$ is $\frac{9}{7}$ units from 0.

71. $|0| = 0$ since 0 is 0 units from itself.

73. $|x| = |-8| = 8$

75. $-83, -4.7, 0, \frac{5}{9}, 2.1\overline{6}, 62$

77. $-83, 0, 62$

79. All are real numbers.

81. *Writing Exercise.* Yes; every integer can be written as $n/1$, a quotient of the form a/b where $b \ne 0$.

83. $3xy = 3 \cdot 2 \cdot 7 = 42$

85. *Writing Exercise.* No; $|0| = 0$ which is neither positive nor negative.

87. *Writing Exercise.* No; every positive number is nonnegative, but zero is nonnegative and zero is not positive.

89. List the numbers as they occur on the number line, from left to right: $-23, -17, 0, 4$

91. Converting to decimal notation, we can write
$\frac{4}{5}, \frac{4}{3}, \frac{4}{8}, \frac{4}{6}, \frac{4}{9}, \frac{4}{2}, -\frac{4}{3}$ as
$0.8, 1.3\overline{3}, 0.5, 0.6\overline{6}, 0.4\overline{4}, 2, -1.3\overline{3}$, respectively. List the numbers (in fractional form) as they occur on the number line, from left to right:
$-\frac{4}{3}, \frac{4}{9}, \frac{4}{8}, \frac{4}{6}, \frac{4}{5}, \frac{4}{3}, \frac{4}{2}$

93. $|4| = 4$ and $|-7| = 7$, so $|4| < |-7|$.

95. $|23| = 23$ and $|-23| = 23$, so $|23| = |-23|$.

97. $|x| = 19$

x represents a number whose distance from 0 is 19. Thus, $x = 19$ or $x = -19$.

99. $2 < |x| < 5$

x represents an integer whose distance from 0 is greater than 2 and also less than 5. Thus, $x = -4, -3, 3, 4$.

101. $0.9\overline{9} = 3(0.3\overline{3}) = 3 \cdot \frac{1}{3} = \frac{3}{3}$

103. $7.7\overline{7} = 70(0.1\overline{1}) = 70 \cdot \frac{1}{9} = \frac{70}{9}$

(See Exercise 100.)

105. Nonpositive numbers include zero. Thus, $x \le 0$.

107. Distance from zero can go in the positive or negative direction. Thus, $|t| \ge 20$.

109. *Writing Exercise.* The statement $\sqrt{a^2} = |a|$ for any real number a is true. If a is nonnegative, then $\sqrt{a^2} = a$. If a is negative, $\sqrt{a^2} = -a$ since $\sqrt{a^2}$ must be nonnegative. Thus, for a nonnegative number, the result is the number and, for a negative number, the result is the opposite of the number. This describes the absolute value of a number.

Exercise Set 1.5

1. Choice (f), $-3n$, has the same variable factor as $8n$.

3. Choice (e), 9, is a constant as is 43.

5. Choice (b), $5x$, has the same variable factor as $-2x$.

7. Start at 5. Move 8 units to the left.

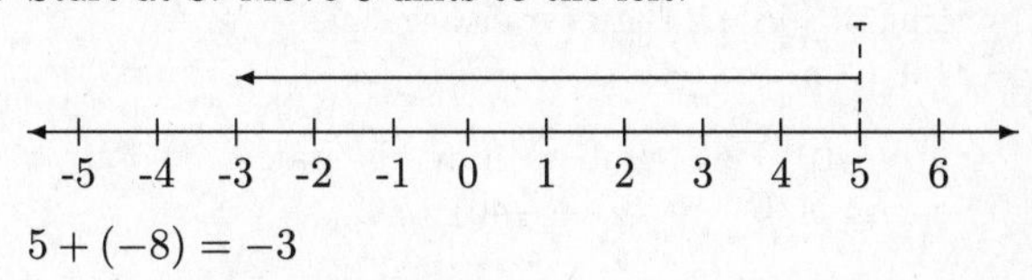

$5 + (-8) = -3$

9. Start at -6. Move 10 units to the right.

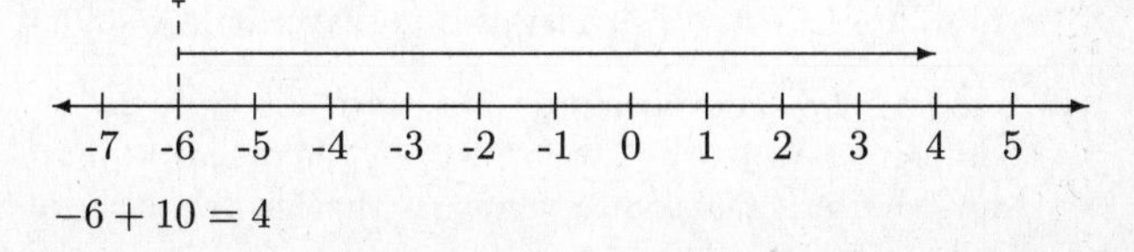

$-6 + 10 = 4$

11. Start at -7. Move 0 units.

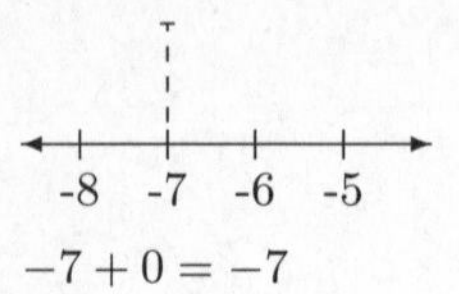

$-7+0=-7$

13. Start at -3. Move 5 units to the left.

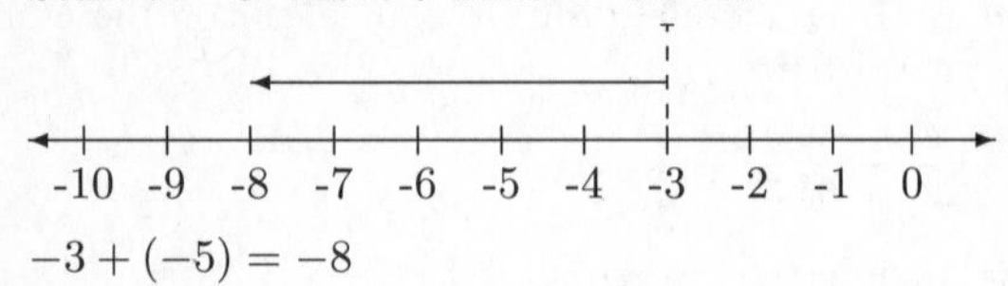

$-3+(-5)=-8$

15. $-35+0$ One number is 0. The answer is the other number. $-35+0=-35$

17. $0+(-8)$ One number is 0. The answer is the other number. $0+(-8)=-8$

19. $12+(-12)$ The numbers have the same absolute value. The sum is 0. $12+(-12)=0$

21. $-24+(-17)$ Two negatives. Add the absolute values, getting 41. Make the answer negative. $-24+(-17)=-41$

23. $-13+13$ The numbers have the same absolute value. The sum is 0. $-13+13=0$

25. $20+(-11)$ The absolute values are 20 and 11. The difference is $20-11$, or 9. The positive number has the larger absolute value, so the answer is positive. $20+(-11)=9$

27. $10+(-12)$ The absolute values are 10 and 12. The difference is $12-10$, or 2. The negative number has the larger absolute value, so the answer is negative. $10+(-12)=-2$

29. $-3+14$ The absolute values are 3 and 14. The difference is $14-3$, or 11. The positive number has the larger absolute value, so the answer is positive. $-3+14=11$

31. $-24+(-19)$ Two negatives. Add the absolute values, getting 43. Make the answer negative. $-24+(-19)=-43$

33. $19+(-19)$ The numbers have the same absolute value. The sum is 0. $19+(-19)=0$

35. $23+(-5)$ The absolute values are 23 and 5. The difference is $23-5$ or 18. The positive number has the larger absolute value, so the answer is positive. $23+(-5)=18$

37. $-31+(-14)$ Two negatives. Add the absolute values, getting 45. Make the answer negative. $-31+(-14)=-45$

39. $40+(-40)$ The numbers have the same absolute value. The sum is 0. $40+(-40)=0$

41. $85+(-69)$ The absolute values are 85 and 69. The difference is $85-69$, or 16. The positive number has the larger absolute value, so the answer is positive. $85+(-69)=16$

43. $-3.6+2.8$ The absolute values are 3.6 and 2.8. The difference is $3.6-2.8$, or 0.8. The negative number has the larger absolute value, so the answer is negative. $-3.6+2.8=-0.8$

45. $-5.4+(-3.7)$ Two negatives. Add the absolute values, getting 9.1. Make the answer negative. $-5.4+(-3.7)=-9.1$

47. $\frac{4}{5}+\left(-\frac{1}{5}\right)$ The absolute values are $\frac{4}{5}$ and $\frac{1}{5}$. The positive number has the larger absolute value, so the answer is positive. $\frac{4}{5}+\left(-\frac{1}{5}\right)=\frac{3}{5}$

49. $\frac{-4}{7}+\frac{-2}{7}$ Two negatives. Add the absolute values, getting $\frac{6}{7}$. Make the answer negative.

$\frac{-4}{7}+\frac{-2}{7}=\frac{-6}{7}$

51. $-\frac{2}{5}+\frac{1}{3}$ The absolute values are $\frac{2}{5}$ and $\frac{1}{3}$. The difference is $\frac{6}{15}-\frac{5}{15}$, or $\frac{1}{15}$. The negative number has the larger absolute value, so the answer is negative.

$-\frac{2}{5}+\frac{1}{3}=-\frac{1}{15}$

53. $\frac{-4}{9}+\frac{2}{3}$ The absolute values are $\frac{4}{9}$ and $\frac{2}{3}$. The difference is $\frac{6}{9}-\frac{4}{9}$, or $\frac{2}{9}$. The positive number has the larger absolute value, so the answer is positive.

$\frac{-4}{9}+\frac{2}{3}=\frac{2}{9}$

55.
$$\begin{aligned}&35+(-14)+(-19)+(-5)\\&=35+[(-14)+(-19)+(-5)] \quad \text{Using the associative law of addition}\\&=35+(-38) \quad \text{Adding the negatives}\\&=-3 \quad \text{Adding a positive and a negative}\end{aligned}$$

57. $-4.9+8.5+4.9+(-8.5)$

Note that we have two pairs of numbers with different signs and the same absolute value: -4.9 and 4.9, 8.5 and -8.5. The sum of each pair is 0, so the result is $0+0$, or 0.

59. Rewording: First increase plus decrease plus second increase is change in price.

Translating: $15¢ + (-3¢) + 17¢ =$ change in price

Since $15+(-3)+17$
$=12+17$
$=29$,

the price rose 29¢ during the given period.

61. Rewording: July bill plus payment plus

Translating: $-82 \quad + \quad 50 \quad +$

$$\underbrace{\text{August charges}} \quad \text{is} \quad \underbrace{\text{new balance.}}$$
$$\downarrow \quad \downarrow \quad \downarrow$$
$$(-63) \quad = \quad \text{new balance}$$

Since $-82 + 50 + (-63) = -32 + (-63)$
$= -95,$

Chloe's new balance was $95.

63. Rewording: $\underbrace{\text{First try yardage}} \quad \text{plus} \quad \underbrace{\text{second try yardage}}$
$$\downarrow \quad \downarrow \quad \downarrow$$
Translating: $(-13) \quad + \quad 12$

$$\text{plus} \quad \underbrace{\text{third try yardage}} \quad \text{is} \quad \underbrace{\text{total gain or loss.}}$$
$$\downarrow \quad \downarrow \quad \downarrow \quad \downarrow$$
$$+ \quad 21 \quad = \quad \text{total gain or loss}$$

Since $(-13) + 12 + 21$
$= -1 + 21$
$= 20,$

the total gain was 20 yd.

65. Rewording: $\underbrace{\text{first drop}} \text{ plus } \underbrace{\text{rise}} \text{ plus}$
$$\downarrow \quad \downarrow \quad \downarrow \quad \downarrow$$
Translating: $\left(-\frac{2}{5}\right) \quad + \quad 1\frac{1}{5} \quad +$

$$\underbrace{\text{drop}} \quad \text{is} \quad \underbrace{\text{total level change.}}$$
$$\downarrow \quad \downarrow \quad \downarrow$$
$$\left(-\frac{1}{2}\right) = \text{total level change}$$

Since $-\frac{2}{5} + 1\frac{1}{5} + \left(-\frac{1}{2}\right) = -\frac{4}{10} + \frac{12}{10} + \left(-\frac{5}{10}\right)$
$= \frac{8}{10} - \frac{5}{10},$
$= \frac{3}{10},$

The lake rose $\frac{3}{10}$ ft.

67. Rewording: $\underbrace{\text{Original balance}} \quad \text{plus} \quad \underbrace{\text{first payment}} \quad \text{plus}$
$$\downarrow \quad \downarrow \quad \downarrow \quad \downarrow$$
Translating: $-470 \quad + \quad 45 \quad +$

$$\underbrace{\text{new charges}} \quad \text{plus} \quad \underbrace{\text{second payment}} \quad \text{is} \quad \underbrace{\text{new balance.}}$$
$$\downarrow \quad \downarrow \quad \downarrow \quad \downarrow \quad \downarrow$$
$$-160 \quad + \quad 500 \quad = \quad \text{new balance}$$

Since $-470 + 45 + (-160) + 500$
$= [-470 + (-160)] + (45 + 500)$
$= -630 + 545$
$= -85,$

Logan owes $85 on his credit card.

69. $7a + 10a = (7 + 10)a$ Using the distributive law
$= 17a$

71. $-3x + 12x = (-3 + 12)x$ Using the distributive law
$= 9x$

73. $4t + 21t = (4 + 21)t = 25t$

75. $7m + (-9m) = [7 + (-9)]m = -2m$

77. $-8y + (-2y) = [-8 + (-2)]y = -10y$

79. $-3 + 8x + 4 + (-10x)$
$= -3 + 4 + 8x + (-10x)$ Using the commutative law of addition
$= (-3 + 4) + [8 + (-10)]x$ Using the distributive law
$= 1 - 2x$ Adding

81. Perimeter $= 8 + 5x + 9 + 7x$
$= 8 + 9 + 5x + 7x$
$= (8 + 9) + (5 + 7)x$
$= 17 + 12x$

83. Perimeter $= 3t + 3r + 7 + 5t + 9 + 4r$
$= 3t + 5t + 3r + 4r + 7 + 9$
$= (3 + 5)t + (3 + 4)r + (7 + 9)$
$= 8t + 7r + 16$

85. Perimeter $= 9 + 6n + 7 + 8n + 4n$
$= 9 + 7 + 6n + 8n + 4n$
$= (9 + 7) + (6 + 8 + 4)n$
$= 16 + 18n$

87. *Writing Exercise.* Answers may vary. One possible explanation follows.

Consider performing the addition on a number line. We start to the left of 0 and then move farther left, so the result must be a negative number.

89. $7(3z + y + 1) = 7 \cdot 3z + 7 \cdot y + 7 \cdot 1 = 21z + 7y + 7$

91. *Writing Exercise.* The sum will be positive when the positive number is greater than the sum of the absolute values of the negative numbers.

93. Starting with the final value, we "undo" the deposit and original amount by adding their opposites. The result is the amount of the check.

Rewording: $\underbrace{\text{Final value}}$ plus $\underbrace{\text{opposite of deposit}}$ plus

Translating: $-\$42.37 + (-\$152) +$

$\underbrace{\text{opposite of original amount}}$ is $\underbrace{\text{check amount.}}$

$-\$257.33$ = check amount.

$$\text{Since } -42.37 + (-152) + (-257.33) = (-194.37) + (-257.33) = -451.70,$$

So the amount of the check was \$451.70.

95. $4x + \_\_ + (-9x) + (-2y)$
$= 4x + (-9x) + \_\_ + (-2y)$
$= [4 + (-9)]x + \_\_ + (-2y)$
$= -5x + \_\_ + (-2y)$

This expression is equivalent to $-5x - 7y$, so the missing term is the term which yields $-7y$ when added to $-2y$. Since $-5y + (-2y) = -7y$, the missing term is $-5y$.

97. $3m + 2n + \_\_ + (-2m)$
$= 2n + \_\_ + (-2m) + 3m$
$= 2n + \_\_ + (-2 + 3)m$
$= 2n + \_\_ + m$

This expression is equivalent to $2n + (-6m)$, so the missing term is the term which yields $-6m$ when added to m. Since $-7m + m = -6m$, the missing term is $-7m$.

99. Note that, in order for the sum to be 0, the two missing terms must be the opposites of the given terms. Thus, the missing terms are $-7t$ and -23.

101. $-3 + (-3) + 2 + (-2) + 1 = -5$

Since the total is 5 under par after the five rounds and $-5 = -1 + (-1) + (-1) + (-1) + (-1)$, the golfer was 1 under par on average.

Exercise Set 1.6

1. $-x$ is read "the opposite of x," so choice (d) is correct.

3. $12 - (-x)$ is read "twelve minus the opposite of x," so choice (f) is correct.

5. $x - (-12)$ is read "x minus negative twelve," so choice (a) is correct.

7. $-x - x$ is read "the opposite of x minus x," so choice (b) is correct.

9. $6 - 10$ is read "six minus ten."

11. $2 - (-12)$ is read "two minus negative twelve."

13. $9 - (-t)$ is read "nine minus the opposite of t."

15. $-x - y$ is read "the opposite of x minus y."

17. $-3 - (-n)$ is read "negative three minus the opposite of n."

19. The opposite of 51 is -51 because $51 + (-51) = 0$.

21. The opposite of $-\frac{11}{3}$ is $\frac{11}{3}$ because $-\frac{11}{3} + \frac{11}{3} = 0$.

23. The opposite of -3.14 is 3.14 because $-3.14 + 3.14 = 0$.

25. If $x = -45$, then $-x = -(-45) = 45$. (The opposite of 45 is -45.)

27. If $x = -\frac{14}{3}$, then $-x = -\left(-\frac{14}{3}\right) = \frac{14}{3}$.

$\left(\text{The opposite of } -\frac{14}{3} \text{ is } \frac{14}{3}.\right)$

29. If $x = 0.101$, then $-x = -(0.101) = -0.101$.
(The opposite of 0.101 is -0.101.)

31. If $x = 37$, then $-(-x) = -(-37) = 37$
(The opposite of the opposite of 37 is 37.)

33. If $x = -\frac{2}{5}$, then $-(-x) = -\left[-\left(-\frac{2}{5}\right)\right] = -\frac{2}{5}$.

$\left(\text{The opposite of the opposite of } -\frac{2}{5} \text{ is } -\frac{2}{5}.\right)$

35. When we change the sign of -1 we obtain 1.

37. When we change the sign we obtain -15.

39. $7 - 10 = 7 + (-10) = -3$

41. $0 - 6 = 0 + (-6) = -6$

43. $2 - 5 = 2 + (-5) = -3$

45. $-4 - 3 = -4 + (-3) = -7$

47. $-9 - (-3) = -9 + 3 = -6$

49. Note that we are subtracting a number from itself. The result is 0. We could also do this exercise as follows:

$$-8 - (-8) = -8 + 8 = 0$$

51. $14 - 19 = 14 + (-19) = -5$

53. $30 - 40 = 30 + (-40) = -10$

55. $0 - 11 = 0 + (-11) = -11$

57. $-9 - (-9) = -9 + 9 = 0$
(See Exercise 49.)

59. $5 - 5 = 5 + (-5) = 0$
(See Exercise 49.)

61. $4 - (-4) = 4 + 4 = 8$

63. $-7 - 4 = -7 + (-4) = -11$

65. $6 - (-10) = 6 + 10 = 16$

67. $-4 - 15 = -4 + (-15) = -19$

69. $-6-(-7)=-6+7=1$

71. $5-(-12)=5+12=17$

73. $0-(-3)=0+3=3$

75. $-5-(-2)=-5+2=-3$

77. $-7-14=-7+(-14)=-21$

79. $0-(-10)=0+10=10$

81. $-8-0=-8+0=-8$

83. $-52-8=-52+(-8)=-60$

85. $2-25=2+(-25)=-23$

87. $-4.2-3.1=-4.2+(-3.1)=-7.3$

89. $-1.3-(-2.4)=-1.3+2.4=1.1$

91. $3.2-8.7=3.2+(-8.7)=-5.5$

93. $0.072-1=0.072+(-1)=-0.928$

95. $\frac{2}{11}-\frac{9}{11}=\frac{2}{11}+\left(-\frac{9}{11}\right)=-\frac{7}{11}$

97. $\frac{-1}{5}-\frac{3}{5}=\frac{-1}{5}+\left(\frac{-3}{5}\right)=\frac{-4}{5}$, or $-\frac{4}{5}$

99. $-\frac{4}{17}-\left(-\frac{9}{17}\right)=-\frac{4}{17}+\frac{9}{17}=\frac{5}{17}$

101. We subtract the smaller number from the larger.

Translate: $3.8-(-5.2)$

Simplify: $3.8-(-5.2)=3.8+5.2=9$

103. We subtract the smaller number from the larger.

Translate: $114-(-79)$

Simplify: $114-(-79)=114+79=193$

105. $-8-32=-8+(-32)=-40$

107. $18-(-25)=18+25=43$

109. $16-(-12)-1-(-2)+3=16+12+(-1)+2+3=32$

111. $-31+(-28)-(-14)-17=(-31)+(-28)+14+(-17)$
$=-62$

113. $-34-28+(-33)-44=(-34)+(-28)+(-33)+(-44)$
$=-139$

115. $-93+(-84)-(-93)-(-84)$

Note that we are subtracting -93 from -93 and -84 from -84. Thus, the result will be 0. We could also do this exercise as follows:

$$-93+(-84)-(-93)-(-84)=-93+(-84)+93+84=0$$

117. $-3y-8x=-3y+(-8x)$, so the terms are $-3y$ and $-8x$.

119. $9-5t-3st=9+(-5t)+(-3st)$, so the terms are 9, $-5t$, and $-3st$.

121. $10x-13x$
$=10x+(-13x)$ Adding the opposite
$=(10+(-13))x$ Using the distributive law
$=-3x$

123. $7a-12a+4$
$=7a+(-12a)+4$ Adding the opposite
$=(7+(-12))a+4$ Using the distributive law
$=-5a+4$

125. $-8n-9+7n$
$=-8n+(-9)+7n$ Adding the opposite
$=-8n+7n+(-9)$ Using the commutative law of addition
$=-n-9$ Adding like terms

127. $5-3x-11$
$=5+(-3x)+(-11)$
$=-3x+5+(-11)$
$=-3x-6$

129. $2-6t-9-2t$
$=2+(-6t)+(-9)+(-2t)$
$=2+(-9)+(-6t)+(-2t)$
$=-7-8t$

131. $5y+(-3x)-9x+1-2y+8$
$=5y+(-3x)+(-9x)+1+(-2y)+8$
$=5y+(-2y)+(-3x)+(-9x)+1+8$
$=3y-12x+9$

133. $13x-(-2x)+45-(-21)-7x$
$=13x+2x+45+21+(-7x)$
$=13x+2x+(-7x)+45+21$
$=8x+66$

135. We subtract the lower temperature from the higher temperature:

$134-(-80)=134+80=214$

The temperature range is 214°F.

137. We subtract the lower elevation from the higher elevation:

$$29,035-(-1312)=29,035+1312=30,347$$

The difference in elevation is 30,347 ft.

139. We subtract the lower elevation from the higher elevation:

$$-40-(-156)=-40+156=116$$

Lake Assal is 116 m lower than the Valdes Peninsula.

141. *Writing Exercise.* Yes; rewrite subtraction as addition of the opposite.

143. Area $= lw =$ (36 ft)(12 ft)=432 ft^2

145. Answers will vary. The symbol "–" can represent the opposite, or a negative, or subtraction.

147. If the clock reads 8:00 A.M. on the day following the blackout when the actual time is 3:00 P.M., then the clock is 7 hr behind the actual time. This indicates that the power outage lasted 7 hr, so power was restored 7 hr after 4:00 P.M., or at 11:00 P.M. on August 14.

149. False. For example, let $m = -3$ and $n = -5$. Then $-3 > -5$, but $-3 + (-5) = -8 \not> 0$.

151. True. For example, for $m = 4$ and $n = -4$, $4 = -(-4)$ and $4 + (-4) = 0$; for $m = -3$ and $n = 3$, $-3 = -3$ and $-3 + 3 = 0$.

153. [(−)] [9] [−] [(−)] [7] [ENTER]

Exercise Set 1.7

1. The product of two reciprocals is 1.

3. The sum of a pair of additive inverses is 0.

5. The number 0 has no reciprocal.

7. The number 1 is the multiplicative identity.

9. A nonzero number divided by itself is 1.

11. $-4 \cdot 10 = -40$ Think: $4 \cdot 10 = 40$, make the answer negative.

13. $-8 \cdot 7 = -56$ Think: $8 \cdot 7 = 56$, make the answer negative.

15. $4 \cdot (-10) = -40$

17. $-9 \cdot (-8) = 72$ Multiplying absolute values; the answer is positive.

19. $-6 \cdot 7 = -42$

21. $-5 \cdot (-9) = 45$ Multiplying absolute values; the answer is positive.

23. $-19 \cdot (-10) = 190$

25. $11 \cdot (-12) = -132$

27. $-25 \cdot (-48) = 1200$

29. $4.5 \cdot (-28) = -126$

31. $-5 \cdot (-2.3) = 11.5$

33. $-(25) \cdot 0 = 0$ The product of 0 and any real number is 0.

35. $\frac{2}{5} \cdot \left(-\frac{5}{7}\right) = -\left(\frac{2 \cdot 5}{5 \cdot 7}\right) = -\left(\frac{2}{7} \cdot \frac{5}{5}\right) = \frac{-2}{7}$

37. $-\frac{3}{8} \cdot \left(-\frac{2}{9}\right) = \frac{\cancel{3} \cdot \cancel{2} \cdot 1}{4 \cdot \cancel{2} \cdot \cancel{3} \cdot 3} = \frac{1}{12}$

39. $(-5.3)(2.1) = -11.13$

41. $-\frac{5}{9} \cdot \frac{3}{4} = -\frac{5 \cdot \cancel{3}}{\cancel{3} \cdot 3 \cdot 4} = -\frac{5}{12}$

43. $3 \cdot (-7) \cdot (-2) \cdot 6$

$= -21 \cdot (-12)$ Multiplying the first two numbers and the last two numbers

$= 252$

45. 0, The product of 0 and any real number is 0.

47. $-\frac{1}{3} \cdot \frac{1}{4} \cdot \left(-\frac{3}{7}\right) = -\frac{1}{12} \cdot \left(-\frac{3}{7}\right) = \frac{3}{12 \cdot 7}$

$= \frac{\cancel{3} \cdot 1}{\cancel{3} \cdot 4 \cdot 7} = \frac{1}{28}$

49. $-2 \cdot (-5) \cdot (-3) \cdot (-5) = 10 \cdot 15 = 150$

51. 0, The product of 0 and any real number is 0.

53. $(-8)(-9)(-10) = 72(-10) = -720$

55. $(-6)(-7)(-8)(-9)(-10) = 42 \cdot 72 \cdot (-10)$
$= 3024 \cdot (-10) = -30,240$

57. $18 \div (-2) = -9$ Check: $-9 \cdot (-2) = 18$

59. $\frac{36}{-9} = -4$ Check: $-4 \cdot (-9) = 36$

61. $\frac{-56}{8} = -7$ Check: $-7 \cdot 8 = -56$

63. $\frac{-48}{-12} = 4$ Check: $4(-12) = -48$

65. $72 \div 8 = -9$ Check: $-9 \cdot 8 = -72$

67. $-10.2 \div (-2) = 5.1$ Check: $5.1(-2) = -10.2$

69. $-100 \div (-11) = \frac{100}{11}$

71. $\frac{400}{-50} = -8$ Check: $-8 \cdot (-50) = 400$

73. Undefined

75. $-4.8 \div 1.2 = -4$ Check: $-4(1.2) = -4.8$

77. $\frac{0}{-9} = 0$

79. $\frac{9.7(-2.8)0}{4.3}$

Since the numerator has a factor of 0, the product in the numerator is 0. The denominator is nonzero, so the quotient is 0.

81. $\frac{-8}{3} = \frac{8}{-3}$ and $\frac{-8}{3} = -\frac{8}{3}$

83. $\frac{29}{-35} = \frac{-29}{35}$ and $\frac{29}{-35} = -\frac{29}{35}$

85. $-\frac{7}{3} = \frac{-7}{3}$ and $-\frac{7}{3} = \frac{7}{-3}$

87. $\frac{-x}{2} = \frac{x}{-2}$ and $\frac{-x}{2} = -\frac{x}{2}$

89. The reciprocal of $\frac{4}{-5}$ is $\frac{-5}{4}$ $\left(\text{or equivalently, } -\frac{5}{4}\right)$ because $\frac{4}{-5} \cdot \frac{-5}{4} = 1$.

91. The reciprocal of $\frac{51}{-10}$ is $-\frac{10}{51}$ because $\frac{51}{-10}\cdot\left(-\frac{10}{51}\right)=1$.

93. The reciprocal of -10 is $\frac{1}{-10}$ $\left(\text{or equivalently, } -\frac{1}{10}\right)$ because $-10\left(\frac{1}{-10}\right)=1$.

95. The reciprocal of 4.3 is $\frac{1}{4.3}$ because $4.3\left(\frac{1}{4.3}\right)=1$.
Since $\frac{1}{4.3}=\frac{1}{4.3}\cdot\frac{10}{10}=\frac{10}{43}$, the reciprocal can also be expressed as $\frac{10}{43}$.

97. The reciprocal of $\frac{-9}{4}$ is $\frac{4}{-9}$ $\left(\text{or equivalently, } -\frac{4}{9}\right)$ because $\frac{-9}{4}\cdot\frac{4}{-9}=1$.

99. The reciprocal of 0 does not exist. (There is no number n for which $0\cdot n=1$.)

101. $\left(\frac{-7}{4}\right)\left(-\frac{3}{5}\right)$

$=\left(-\frac{7}{4}\right)\left(-\frac{3}{5}\right)$ Rewriting $\frac{-7}{4}$ as $-\frac{7}{4}$

$=\frac{21}{20}$

103. $\frac{-3}{8}+\frac{-5}{8}=\frac{-8}{8}=-1$

105. $\left(\frac{-9}{5}\right)\left(\frac{5}{-9}\right)$
Note that this is the product of reciprocals. Thus, the result is 1.

107. $\left(-\frac{3}{11}\right)-\left(-\frac{6}{11}\right)=\frac{-3}{11}+\frac{6}{11}=\frac{3}{11}$

109. $\frac{7}{8}\div\left(-\frac{1}{2}\right)=\frac{7}{8}\cdot\left(-\frac{2}{1}\right)=-\frac{14}{8}=-\frac{7\cdot\cancel{2}}{\cancel{2}\cdot4\cdot1}=-\frac{7}{4}$

111. $-\frac{5}{9}\div\left(-\frac{5}{9}\right)$
Note that we have a number divided by itself. Thus, the result is 1.

113. $\frac{-3}{10}+\frac{2}{5}=\frac{-3}{10}+\frac{2}{5}\cdot\frac{2}{2}=\frac{-3}{10}+\frac{4}{10}=\frac{1}{10}$

115. $\frac{7}{10}\div\left(\frac{-3}{5}\right)=\frac{7}{10}\div\left(-\frac{3}{5}\right)=\frac{7}{10}\cdot\left(-\frac{5}{3}\right)=-\frac{35}{30}$

$=-\frac{7\cdot\cancel{5}}{2\cdot\cancel{5}\cdot3}=-\frac{7}{6}$

117. $\frac{14}{-9}\div\frac{0}{3}=\frac{14}{-9}\cdot-\frac{3}{0}$ Undefined

119. $\frac{-4}{15}+\frac{2}{-3}=\frac{-4}{15}+\frac{-2}{3}=\frac{-4}{15}+\frac{-2}{3}\cdot\frac{5}{5}=\frac{-4}{15}+\frac{-10}{15}$

$=\frac{-14}{15}$, or $-\frac{14}{15}$

121. *Writing Exercise.* You get the original number. The reciprocal of the reciprocal of a number is the original number.

123. $\frac{264}{468}=\frac{\cancel{2}\cdot\cancel{2}\cdot2\cdot\cancel{3}\cdot11}{\cancel{2}\cdot\cancel{2}\cdot\cancel{3}\cdot3\cdot13}=\frac{22}{39}$

125. *Writing Exercise.* Yes; consider n $(n\neq0)$ and its opposite $-n$. The reciprocals of these numbers are $\frac{1}{n}$ and $\frac{1}{-n}$. Now $\frac{1}{n}+\frac{1}{-n}=\frac{1}{n}+\frac{-1}{n}=0$, so the reciprocals are also opposites.

127. Let a and b represent the numbers. The $a+b$ is the sum and $\frac{1}{a+b}$ is the reciprocal of the sum.

129. Let a and b represent the numbers. Then $a+b$ is the sum and $-(a+b)$ is the opposite of the sum.

131. Let x represent a real number. $x=-x$

133. For 2 and 3, the reciprocal of the sum is $\frac{1}{(2+3)}$ or $\frac{1}{5}$. But $\frac{1}{5}\neq\frac{1}{2}+\frac{1}{3}$.

135. The starting temperature is -3°F.

Rewording:	starting temp.	rise	2°	for	3hr	rise	3°	for	6hrs
	↓	↓	↓	↓	↓	↓	↓	↓	↓
Translating:	-3	$+$	2	$\cdot$	3	$+$	3	$\cdot$	6

fall	2°	for	3hr	fall	5°	for	2hr
↓	↓	↓	↓	↓	↓	↓	
$-$	2	$\cdot$	3	$-$	5	$\cdot$	2

Since $-3+2\cdot3+3\cdot6-2\cdot3-5\cdot2$

$=-3+6+18-6-10$

$=5$.

The temperature forecast at 8PM is 5°F

137. $-n$ and $-m$ are both positive, so $\frac{-n}{-m}$ is positive.

139. $-m$ is positive, so $\frac{n}{-m}$ is negative and $-\left(\frac{n}{-m}\right)$ is positive.

141. $-n$ and $-m$ are both positive, so $-n-m$, or $-n+(-m)$ is positive; $\frac{n}{m}$ is also positive, so $(-n-m)\frac{n}{m}$ is positive.

143. $a(-b)+ab=a[-b+b]$ Distributive law

$=a(0)$ Law of opposites

$=0$ Multiplicative property of 0

Therefore, $a(-b)$ is the opposite of ab by the law of opposites.

1.7 Connecting the Concepts

1. $-8+(-2)=-10$

3. $-8\div(-2)=\frac{-8}{-2}=4$

5. $12\cdot(-10)=-120$

7. $-5-18=-23$

9. $\frac{3}{5}-\frac{8}{5}=-\frac{5}{5}=-1$

11. $-5.6+4.8=-0.8$

13. $-44.1\div 6.3=-7$

15. $\frac{9}{5}\cdot\left(-\frac{20}{3}\right)=-\frac{3\cdot 3\cdot 4\cdot 5}{5\cdot 3}=-12$

17. $38-(-62)=38+62=100$

19. $(-15)\cdot(-12)=180$

Exercise Set 1.8

1. a) $4+8\div 2\cdot 2$

There are no grouping symbols or exponential expressions, so we multiply and divide from left to right. This means that we divide first.

b) $7-9+15$

There are no grouping symbols, exponential expressions, multiplications, or divisions, so we add and subtract from left to right. This means that we subtract first.

c) $5-2(3+4)$

We perform the operation in the parentheses first. This means that we add first.

d) $6+7\cdot 3$

There are no grouping symbols or exponential expressions, so we multiply and divide from left to right. This means that we multiply first.

e) $18-2[4+(3-2)]$

We perform the operation in the innermost grouping symbols first. This means that we perform the subtraction in the parentheses first.

f) $\dfrac{5-6\cdot 7}{2}$

Since the denominator does not need to be simplified, we consider the numerator. There are no grouping symbols or exponential expressions, so we multiply and divide from left to right. This means that we multiply first.

3. $\underbrace{x\cdot x\cdot x\cdot x\cdot x\cdot x}_{\text{6 factors}}=x^6$

5. $\underbrace{(-5)(-5)(-5)}_{\text{3 factors}}=(-5)^3$

7. $3t\cdot 3t\cdot 3t\cdot 3t\cdot 3t=(3t)^5$

9. $2\cdot\underbrace{n\cdot n\cdot n\cdot n}_{\text{4 factors}}=2n^4$

11. $4^2=4\cdot 4=16$

13. $(-3)^2=(-3)(-3)=9$

15. $-3^2=-(3\cdot 3)=-9$

17. $4^3=4\cdot 4\cdot 4=16\cdot 4=64$

19. $(-5)^4=(-5)(-5)(-5)(-5)=25\cdot 25=625$

21. $7^1=7$ (1 factor)

23. $(-2)^5=(-2)(-2)(-2)(-2)(-2)=-32$

25. $(3t)^4=(3t)(3t)(3t)(3t)$
$=3\cdot 3\cdot 3\cdot 3\cdot t\cdot t\cdot t\cdot t=81t^4$

27. $(-7x)^3=(-7x)(-7x)(-7x)$
$=(-7)(-7)(-7)(x)(x)(x)=-343x^3$

29. $5+3\cdot 7=5+21$ Multiplying
$=26$ Adding

31. $10\cdot 5+1\cdot 1=50+1$ Multiplying
$=51$ Adding

33. $6-70\div 7-2=6-10-2$ Multiplying
$=-4-2$ Adding
$=-6$

35. $14\cdot 19\div(19\cdot 14)$

Since $14\cdot 19$ and $19\cdot 14$ are equivalent, we are dividing the product $14\cdot 19$ by itself. Thus the result is 1.

37. $3(-10)^2-8\div 2^2$
$=3(100)-8\div 4$ Simplifying the exponential expressions
$=300-8\div 4$ Multiplying and
$=300-2$ dividing from left to right
$=298$ Subtracting

39. $8-(2\cdot 3-9)$
$=8-(6-9)$ Multiplying inside the parentheses
$=8-(-3)$ Subtracting inside the parentheses
$=8+3$ Removing parentheses
$=11$ Adding

41. $(8-2)(3-9)$
$=6(-6)$ Subtracting inside the parentheses
$=-36$ Multiplying

43. $13(-10)^2+45\div(-5)$
$=13(100)+45\div(-5)$ Simplifying the exponential expression
$=1300+45\div(-5)$ Multiplying and
$=1300-9$ dividing from left to right
$=1291$ Subtracting

45. $5+3(2-9)^2=5+3(-7)^2=5+3\cdot 49=5+147=152$

47. $[2\cdot(5-8)]^2=[2\cdot(-3)]^2=(-6)^2=36$

49. $\dfrac{7+2}{5^2-4^2}=\dfrac{9}{25-16}=\dfrac{9}{9}=1$

51. $8(-7)+|3(-4)| = -56+|-12| = -56+12 = -44$

53. $36\div(-2)^2+4[5-3(8-9)^5]$
$= 36\div 4+4[5-3(-1)^5]$
$= 9+4[5+3]$
$= 9+32$
$= 41$

55. $\dfrac{(7)^2-(-1)^7}{5\cdot 7-4\cdot 3^2-2^2} = \dfrac{49-(-1)}{35-4\cdot 9-4} = \dfrac{49+1}{35-36-4} =$
$\dfrac{50}{-5} = -10$

57. $\dfrac{-3^3-2\cdot 3^2}{8\div 2^2-(6-|2-15|)} = \dfrac{-27-2\cdot 9}{8\div 4-(6-|-13|)}$
$= \dfrac{-27-18}{2-(6-13)} = \dfrac{-45}{2-(-7)} = \dfrac{-45}{9} = -5$

59. $9-4x$
$= 9-4\cdot 7$ Substituting 7 for x
$= 9-28$ Multiplying
$= -19$

61. $24\div t^3$
$= 24\div(-2)^3$ Substituting -2 for t
$= 24\div(-8)$ Simplifying the exponential expression
$= -3$ Dividing

63. $45\div a\cdot 5 = 45\div(-3)\cdot 5$ Substituting -3 for a
$= -15\cdot 5$ Multiplying and dividing in order from left to right
$= -75$

65. $5x\div 15x^2$
$= 5\cdot 3\div 15(3)^2$ Substituting 3 for x
$= 5\cdot 3\div 15\cdot 9$ Simplifying the exponential expression
$= 15\div 15\cdot 9$ Multiplying and dividing
$= 1\cdot 9$ in order from
$= 9$ left to right

67. $45\div 3^2x(x-1)$
$= 45\div 3^2\cdot 3(3-1)$ Substituting 3 for x
$= 45\div 3^2\cdot 3(2)$ Subtracting inside the parentheses
$= 45\div 9\cdot 3(2)$ Evaluating the exponential expression
$= 5\cdot 3(2)$ Dividing and
$= 15(2)$ multiplying
$= 30$ from left to right

69. $-x^2-5x = -(-3)^2-5(-3) = -9-5(-3)$
$= -9+15 = 6$

71. $\dfrac{3a-4a^2}{a^2-20} = \dfrac{3\cdot 5-4(5)^2}{(5)^2-20} = \dfrac{3\cdot 5-4\cdot 25}{25-20}$
$= \dfrac{15-100}{5} = \dfrac{-85}{5} = -17$

73. $-(9x+1) = -9x-1$ Removing parentheses and changing the sign of each term

75. $-[7n+8] = 7n-8$ Removing grouping symbols and changing the sign of each term

77. $-(4a-3b+7c) = -4a+3b-7c$

79. $-(3x^2+5x-1) = -3x^2-5x+1$

81. $8x-(6x+7)$
$= 8x-6x-7$ Removing parentheses and changing the sign of each term
$= 2x-7$ Collecting like terms

83. $2x-7x-(4x-6) = 2x-7x-4x+6 = -9x+6$

85. $9t-7r+2(3r+6t) = 9t-7r+6r+12t = 21t-r$

87. $15x-y-5(3x-2y+5z)$
$= 15x-y-15x+10y-25z$ Multiplying each term in parentheses by -5
$= 9y-25z$

89. $3x^2+11-(2x^2+5) = 3x^2+11-2x^2-5$
$= x^2+6$

91. $5t^3+t+3(t-2t^3) = 5t^3+t+3t-6t^3$
$= -t^3+4t$

93. $12a^2-3ab+5b^2-5(-5a^2+4ab-6b^2)$
$= 12a^2-3ab+5b^2+25a^2-20ab+30b^2$
$= 37a^2-23ab+35b^2$

95. $-7t^3-t^2-3(5t^3-3t)$
$= -7t^3-t^2-15t^3+9t$
$= -22t^3-t^2+9t$

97. $5(2x-7)-[4(2x-3)+2]$
$= 5(2x-7)-[8x-12+2]$
$= 5(2x-7)-[8x-10]$
$= 10x-35-8x+10$
$= 2x-25$

99. *Writing Exercise.* Operations should be performed in the following order.

1. **P**arentheses: Perform all calculations within parentheses (and other grouping symbols) first.
2. **E**xponents: Evaluate all exponential expressions.
3. **M**ultiply and **D**ivide in order from left to right.
4. **A**dd and **S**ubtract in order from left to right.

This is a valid approach because it describes the rules for the order of operations.

101. Let n represent "a number." Then we have $2n - 9$.

103. *Writing Exercise.* Finding the opposite of a number and then squaring it is not equivalent to squaring the number and then finding the opposite of the result.
$(-x)^2 = (-1 \cdot x)^2 = (-1 \cdot x)(-1 \cdot x)$
$= (-1)(-1)(x)(x) = x^2 \neq -x^2$ for $x \neq 0$.

105. $5t - \{7t - [4r - 3(t - 7)] + 6r\} - 4r$
$= 5t - \{7t - [4r - 3t + 21] + 6r\} - 4r$
$= 5t - \{7t - 4r + 3t - 21 + 6r\} - 4r$
$= 5t - \{10t + 2r - 21\} - 4r$
$= 5t - 10t - 2r + 21 - 4r$
$= -5t - 6r + 21$

107. $\{x - [f - (f - x)] + [x - f]\} - 3x$
$= \{x - [f - f + x] + [x - f]\} - 3x$
$= \{x - [x] + [x - f]\} - 3x$
$= \{x - x + x - f\} - 3x$
$= x - f - 3x$
$= -2x - f$

109. *Writing Exercise.* No; let $a = 2$, $b = -1$, and $c = 3$. Then $a|b - c| = 2|-1 - 3| = 2|-4| = 2 \cdot 4 = 8$, but $ab - ac = 2(-1) - 2 \cdot 3 = -2 - 6 = -8$.

111. True; $m - n = -n + m = -(n - m)$

113. False; let $m = 2$ and $n = 1$. Then $-2(1-2) = -2(-1) = 2$, but $-(2 \cdot 1 + 2^2) = -(2 + 4) = -6$.

115. $[x + 3(2 - 5x) \div 7 + x](x - 3)$
When $x = 3$, the factor $x - 3$ is 0, so the product is 0.

117. $\dfrac{x^2 + 2^x}{x^2 - 2^x} = \dfrac{3^2 + 2^3}{3^2 - 2^3} = \dfrac{9 + 8}{9 - 8} = \dfrac{17}{1} = 17$

119. $4 \cdot 20^3 + 17 \cdot 20^2 + 10 \cdot 20 + 0 \cdot 1$
$= 4 \cdot 8000 + 17 \cdot 400 + 10 \cdot 20 + 0 \cdot 1$
$= 32,000 + 6800 + 200 + 0$
$= 39,000$

121. The tower is composed of cubes with sides of length x. The volume of each cube is $x \cdot x \cdot x$, or x^3. Now we count the number of cubes in the tower. The two lowest levels each contain 3×3, or 9 cubes. The next level contains one cube less than the two lowest levels, so it has $9 - 1$, or 8 cubes. The fourth level from the bottom contains one cube less than the level below it, so it has $8 - 1$, or 7 cubes. The fifth level from the bottom contains one cube less than the level below it, so it has $7 - 1$, or 6 cubes. Finally, the top level contains one cube less than the level below it, so it has $6 - 1$, or 5 cubes. All together there are $9 + 9 + 8 + 7 + 6 + 5$, or 44 cubes, each with volume x^3, so the volume of the tower is $44x^3$.

Chapter 1 Review

1. True

3. False

5. False

7. True

9. False

11. $8t = 8 \cdot 3$ substitute 3 for t
$= 24$

13. $9 - y^2 = 9 - (-5)^2$ Substitute -5 for y
$= 9 - 25$
$= -16$

15. $y - 7$

17. Let b represent Brandt's speed and w represent the wind speed; $15(b - w)$

19. Let d represent the number of digital prints made in 2006, in billions.

	14.1	is	3.2	more than	d.
Translating:	↓	↓	↓	↓	↓
	14.1	=	3.2	+	d
or	14.1	=	$d + 3.2$		

21. $(2x + y) + z = 2x + (y + z)$

23. $6(3x + 5y) = 6 \cdot 3x + 6 \cdot 5y$
$= 18x + 30y$

25. $21x + 15y = 3 \cdot 7x + 3 \cdot 5y$
$= 3(7x + 5y)$

27. $56 = 7 \cdot 8 = 7 \cdot 2 \cdot 2 \cdot 2$ or $2 \cdot 2 \cdot 2 \cdot 7$

29. $\dfrac{18}{8} = \dfrac{2 \cdot 3 \cdot 3}{2 \cdot 2 \cdot 2} = \dfrac{9}{4}$

31. $\dfrac{9}{16} \div 3 = \dfrac{9}{16} \cdot \dfrac{1}{3}$ Multiply by the reciprocal of the divisor
$= \dfrac{3 \cdot 3 \cdot 1}{16 \cdot 3}$
$= \dfrac{3}{16}$

33. $\dfrac{9}{10} \cdot \dfrac{6}{5} = \dfrac{9 \cdot 6}{10 \cdot 5} = \dfrac{9 \cdot \cancel{2} \cdot 3}{\cancel{2} \cdot 5 \cdot 5} = \dfrac{27}{25}$

35. $\frac{-1}{3}$
−4 −2 0 2 4

37. $2 \geq -8$ True, since 2 is to the right of -8.

38. $0 \leq -1$. False; 0 is not left of -1.

39. $-\frac{4}{9} = -\left(\frac{4}{9}\right) = -(4 \div 9)$, so we divide.

$$\begin{array}{r} 0.444 \\ 9\,\overline{)\,4.000} \\ \underline{36} \\ 40 \\ \underline{36} \\ 40 \\ \underline{36} \\ 4 \end{array}$$

$\frac{4}{9} = 0.\overline{4}$, so $-\frac{4}{9} = -0.\overline{4}$

41. $-(-x) = -(-(-12))$ Subsitute -12 for x

$= -(12) = -12$

43. $-\frac{2}{3} + \frac{1}{12} = \frac{-8}{12} + \frac{1}{12}$ The absolute values are $\frac{8}{12}$ and $\frac{1}{12}$. The difference is $\frac{8}{12} - \frac{1}{12}$, or $\frac{8-1}{12} = \frac{7}{12}$. The negative number has the greater absolute value, so the answer is negative. $-\frac{2}{3} + \frac{1}{12} = -\frac{7}{12}$

45. $-3.8 + 5.1 + (-12) + (-4.3) + 10$

$= (5.1 + 10) + [(-3.8) + (-12) + (-4.3)]$

$= 15.1 + [-20.1]$ Adding positive nos. and adding negative nos.

$= -5$ Adding a positive and a negative no.

47. $\frac{-9}{10} - \frac{1}{2} = \frac{-9}{10} + \left(\frac{-1}{2}\right)$

$= \frac{-9}{10} + \left(\frac{-5}{10}\right) = \frac{-14}{10} = \frac{-7}{5}$

49. $-9 \cdot (-7) = 63$

51. $\frac{2}{3} \cdot \left(-\frac{3}{7}\right) = -\left(\frac{2}{3} \cdot \frac{3}{7}\right) = -\left(\frac{2 \cdot 3}{3 \cdot 7}\right) = -\frac{2}{7}$

53. $35 \div (-5) = -7$

55. $-\frac{3}{5} \div \left(-\frac{4}{15}\right) = \frac{3}{5} \cdot \frac{15}{4} = \frac{3 \cdot 3 \cdot 5}{5 \cdot 4} = \frac{9}{4}$

57. $(120 - 6^2) \div 4 \cdot 8 = (120 - 36) \div 4 \cdot 8$

$= 84 \div 4 \cdot 8$

$= 21 \cdot 8$

$= 168$

59. $16 \div (-2)^3 - 5[3 - 1 + 2(4 - 7)]$

$= 16 \div (-8)^3 - 5[3 - 1 + 2(4 - 7)]$

$= 16 \div (-2)^3 - 5[3 - 1 + 2(4 - 7)]$

$= -2^3 - 5[3 - 1 + 2(-3)]$

$= -2 - 5[-4]$

$= -2 + 20$

$= 18$

61. $\frac{4(18 - 8) + 7 \cdot 9}{9^2 - 8^2}$

$= \frac{4(18 - 8) + 7 \cdot 9}{81 - 64}$

$= \frac{4(10) + 7 \cdot 9}{81 - 64}$

$= \frac{40 + 63}{81 - 64}$

$= \frac{103}{17}$

63. $7x - 3y - 11x + 8y$

$= 7x + (-11x) + (-3y) + 8y$

$= [7 + (-11)]\,x + (-3 + 8)\,y$

$= -4x + 5y$

65. The reciprocal of -7 is $-\frac{1}{7}$, since $-\frac{1}{7} \cdot -7 = 1$

67. $(-5x)^3$

$= -5x \cdot (-5x) \cdot (-5x)$

$= -5 \cdot (-5) \cdot (-5) \cdot x \cdot x \cdot x$

$= -125x^3$

69. $5b + 3(2b - 9)$

$= 5b + 3 \cdot 2b - 3 \cdot 9$

$= 5b + 6b - 27$

$= 11b - 27$

71. $2n^2 - 5(-3n^2 + m^2 - 4mn) + 6m^2$

$= 2n^2 + 15n^2 - 5m^2 + 20mn + 6m^2$

$= (2 + 15)n^2 + (-5 + 6)m^2 + 20mn$

$= 17n^2 + m^2 + 20mn$

73. *Writing Exercise.* The value of a constant never varies. A variable can represent a variety of numbers.

75. *Writing Exercise.* The distributive law is used in factoring algebraic expressions, multiplying algebraic expressions, combining like terms, finding the opposite of a sum, and subtracting algebraic expressions.

77. Substitute 1 for a, 2 for b, and evaluate: $a^{50} - 20a^{25}b^4 + 100b^8$

$= 1^{50} - 20 \cdot 1^{25}2^4 + 100 \cdot 2^8$

$= 1 - 20 \cdot 1 \cdot 16 + 100 \cdot 256$

$= 1 - 320 + 25,600$

$= 25,281$

79. $-\left|\frac{7}{8} - \left(-\frac{1}{2}\right) - \frac{3}{4}\right|$ Use 8: the common denominator

$-\left|\frac{7}{8} - \left(-\frac{4}{8}\right) - \frac{6}{8}\right| = -\left|\frac{7}{8} + \frac{4}{8} + \frac{-6}{8}\right|$

$= -\left|\frac{5}{8}\right| = -\frac{5}{8}$

81. i

83. a

85. k

87. c

89. d

91. g

Chapter 1 Test

1. Substitute 10 for x and 5 for y.

$\frac{2x}{y} = \frac{2 \cdot 10}{5} = \frac{2 \cdot 2 \cdot \cancel{5}}{\cancel{5}} = 4$

3. $A = \frac{1}{2} \cdot b \cdot h$
$= \frac{1}{2} \cdot (16 \text{ ft}) \cdot (30 \text{ ft})$
$= \frac{1}{2} \cdot 16 \cdot 30 \cdot \text{ft} \cdot \text{ft}$
$= 8 \cdot 30 \text{ ft} \cdot \text{ft}$
$= 240 \text{ ft}^2$, or 240 square feet

5. $x \cdot (4 \cdot y) = (x \cdot 4) \cdot y$

7. 45,950 is 4250 less than the maximum

Rewording: $\underbrace{45,950}$ is $\underbrace{\text{maximum}}$ less $\underbrace{4250}$

Translating: $45,950 = p - 4250$

where p represents the maximum production capability.

9. $-5(y-2) = -5 \cdot y - 5(-2) = -5y + 10$

11. $7x + 7 + 49y$
$= 7 \cdot x + 7 \cdot 1 + 7 \cdot 2y$
$= 7(x + 1 + 7y)$

13. $\frac{24}{56} = \frac{3 \cdot \cancel{8}}{7 \cdot \cancel{8}} = \frac{3}{7}$

15. $-3 > -8$, since -3 is right of -8.

17. $|-3.8| = 3.8$, since -3.8 is 3.8 units from 0.

19. $\frac{-7}{4}$, since $\frac{-7}{4} \cdot \frac{-4}{7} = 1$

21. $-5 \geq x$ has the same meaning as $x \leq -5$.

23. $-8 + 4 + (-7) + 3$
$= [-8 + (-7)] + [4 + 3]$
$= -15 + 7 = -8$

25. $-\frac{1}{8} - \frac{3}{4}$
$= -\frac{1}{8} - \frac{2}{2} \cdot \frac{3}{4}$
$= -\frac{1}{8} - \frac{6}{8} = \frac{-1-6}{8} = -\frac{7}{8}$

27. $-\frac{1}{2} \cdot \left(-\frac{4}{9}\right) = \frac{1}{2} \cdot \frac{4}{9} = \frac{1 \cdot \cancel{2} \cdot 2}{\cancel{2} \cdot 9} = \frac{2}{9}$

29. $-\frac{3}{5} \div \left(-\frac{4}{5}\right) = \frac{3}{5} \cdot \frac{5}{4} = \frac{3 \cdot \cancel{5}}{\cancel{5} \cdot 4} = \frac{3}{4}$

31. $-2(16) - |2(-8) - 5^3|$
$= -2(16) - |2(-8) - 125|$
$= -2(16) - |-16 - 125|$
$= -2(16) - |-141|$
$= -2(16) - 141$
$= -32 - 141$
$= -32 + (-141) = -173$

33. $256 \cdot (-16) \cdot 4 = -16 \cdot 4 = -64$

35. $18y + 30a - 9a + 4y$
$= 30a + (-9a) + 18y + 4y$
$= [30 + (-9)]a + (18 + 4)y$
$= 21a + 22y$

37. $4x - (3x - 7)$
$= 4x - 3x + 7$
$= (4 - 3)x + 7 = x + 7$

39. $3[5(y-3) + 9] - 2(8y - 1)$
$= 3[5y - 15 + 9] - 16y + 2$
$= 3[5y - 6] - 16y + 2$
$= 15y - 18 - 16y + 2$
$= -y - 16$

41. $9 - (3 - 4) + 5 = 15$

43. $a - \{3a - [4a - (2a - 4a)]\}$
$= a - \{3a - [4a - (-2a)]\}$
$= a - \{3a - [4a + 2a]\}$
$= a - \{3a - 6a\}$
$= a - \{-3a\}$
$= a + 3a = 4a$

Chapter 2

Equations, Inequalities, and Problem Solving

Exercise Set 2.1

1. The equations $x+3=7$ and $6x=24$ are equivalent equations.

3. A solution is a replacement that makes an equation true.

5. The multiplication principle is used to solve $\frac{2}{3}\cdot x=-4$.

7. For $6x=30$, the next step is (d) divide both sides by 6.

9. For $\frac{1}{6}x=30$, the next step is (c) multiply both sides by 6.

11.
$$\begin{aligned} x+10&=21\\ x+10-10&=21-10\\ x&=11 \end{aligned}$$

Check:
$$\begin{array}{c|c} \multicolumn{2}{c}{x+10=21} \\ \hline 11+10 & 21 \\ 21 & \stackrel{?}{=}\ 21 \end{array} \quad \text{TRUE}$$

The solution is 11.

13.
$$\begin{aligned} y+7&=-18\\ y+7-7&=-18-7\\ y&=-25 \end{aligned}$$

Check:
$$\begin{array}{c|c} \multicolumn{2}{c}{y+7=-18} \\ \hline -25+7 & 18 \\ -18 & \stackrel{?}{=}\ -18 \end{array} \quad \text{TRUE}$$

The solution is -25.

15.
$$\begin{aligned} -6&=y+25\\ -6-25&=y+25-25\\ -31&=y \end{aligned}$$

Check:
$$\begin{array}{c|c} \multicolumn{2}{c}{-6=y+25} \\ \hline -6 & -31+25 \\ -6 & \stackrel{?}{=}\ -6 \end{array} \quad \text{TRUE}$$

The solution is -31.

17.
$$\begin{aligned} x-18&=23\\ x-18+18&=23+18\\ x&=41 \end{aligned}$$

Check:
$$\begin{array}{c|c} \multicolumn{2}{c}{x-18=23} \\ \hline 41-18 & 23 \\ 23 & \stackrel{?}{=}\ 23 \end{array} \quad \text{TRUE}$$

The solution is 41.

19.
$$\begin{aligned} 12&=-7+y\\ 7+12&=7+(-7)+y\\ 19&=y \end{aligned}$$

Check:
$$\begin{array}{c|c} \multicolumn{2}{c}{12=-7+y} \\ \hline 12 & -7+19 \\ 12 & \stackrel{?}{=}\ 12 \end{array} \quad \text{TRUE}$$

The solution is 19.

21.
$$\begin{aligned} -5+t&=-11\\ 5+(-5)+t&=5+(-11)\\ t&=-6 \end{aligned}$$

Check:
$$\begin{array}{c|c} \multicolumn{2}{c}{-5+t=-11} \\ \hline -5+(-6) & -11 \\ -11 & \stackrel{?}{=}\ -11 \end{array} \quad \text{TRUE}$$

The solution is -6.

23.
$$\begin{aligned} r+\frac{1}{3}&=\frac{8}{3}\\ r+\frac{1}{3}-\frac{1}{3}&=\frac{8}{3}-\frac{1}{3}\\ r&=\frac{7}{3} \end{aligned}$$

Check:
$$\begin{array}{c|c} \multicolumn{2}{c}{r+\frac{1}{3}=\frac{8}{3}} \\ \hline \frac{7}{3}+\frac{1}{3} & \frac{8}{3} \\ \frac{8}{3} & \stackrel{?}{=}\ \frac{8}{3} \end{array} \quad \text{TRUE}$$

The solution is $\frac{7}{3}$.

25.
$$\begin{aligned} x-\frac{3}{5}&=-\frac{7}{10}\\ x-\frac{3}{5}+\frac{3}{5}&=-\frac{7}{10}+\frac{3}{5}\\ x&=-\frac{7}{10}+\frac{3}{5}\cdot\frac{2}{2}\\ x&=-\frac{7}{10}+\frac{6}{10}\\ x&=-\frac{1}{10} \end{aligned}$$

Check:
$$\begin{array}{c|c} \multicolumn{2}{c}{x-\frac{3}{5}=-\frac{7}{10}} \\ \hline -\frac{1}{10}-\frac{3}{5} & -\frac{7}{10} \\ -\frac{1}{10}-\frac{6}{10} & \\ -\frac{7}{10} & \stackrel{?}{=}\ -\frac{7}{10} \end{array} \quad \text{TRUE}$$

The solution is $-\frac{1}{10}$.

27.
$$x - \frac{5}{6} = \frac{7}{8}$$
$$x - \frac{5}{6} + \frac{5}{6} = \frac{7}{8} + \frac{5}{6}$$
$$x = \frac{7}{8}\cdot\frac{3}{3} + \frac{5}{6}\cdot\frac{4}{4}$$
$$x = \frac{21}{24} + \frac{20}{24}$$
$$x = \frac{41}{24}$$

Check:
$$\begin{array}{c|c} x - \frac{5}{6} & \frac{7}{8} \\ \hline \frac{41}{24} - \frac{5}{6} & \frac{7}{8} \\ \frac{41}{24} - \frac{20}{24} & \frac{21}{24} \\ \frac{21}{24} & \stackrel{?}{=} \frac{21}{24} \end{array} \quad \text{TRUE}$$

The solution is $\frac{41}{24}$.

29.
$$-\frac{1}{5} + z = -\frac{1}{4}$$
$$\frac{1}{5} - \frac{1}{5} + z = \frac{1}{5} - \frac{1}{4}$$
$$z = \frac{1}{5}\cdot\frac{4}{4} - \frac{1}{4}\cdot\frac{5}{5}$$
$$z = \frac{4}{20} - \frac{5}{20}$$
$$z = -\frac{1}{20}$$

Check:
$$\begin{array}{c|c} -\frac{1}{5} + z & -\frac{1}{4} \\ \hline -\frac{1}{5} + \left(-\frac{1}{20}\right) & -\frac{1}{4} \\ -\frac{4}{20} + \left(-\frac{1}{20}\right) & -\frac{5}{20} \\ -\frac{5}{20} & \stackrel{?}{=} -\frac{5}{20} \end{array} \quad \text{TRUE}$$

The solution is $-\frac{1}{20}$.

31.
$$m - 2.8 = 6.3$$
$$m - 2.8 + 2.8 = 6.3 + 2.8$$
$$m = 9.1$$

Check:
$$\begin{array}{c|c} m - 2.8 & 6.3 \\ \hline 9.1 - 2.8 & 6.3 \\ 6.3 & \stackrel{?}{=} 6.3 \end{array} \quad \text{TRUE}$$

The solution is 9.1.

33.
$$-9.7 = -4.7 + y$$
$$4.7 + (-9.7) = 4.7 + (-4.7) + y$$
$$-5 = y$$

Check:
$$\begin{array}{c|c} -9.7 & -4.7 + y \\ \hline -9.7 & -4.7 + (-5) \\ -9.7 & \stackrel{?}{=} -9.7 \end{array} \quad \text{TRUE}$$

The solution is -5.

35.
$$8a = 56$$
$$\frac{8a}{8} = \frac{56}{8} \quad \text{Dividing both sides by 8}$$
$$1 \cdot a = 7 \quad \text{Simplifying}$$
$$a = 7 \quad \text{Identity property of 1}$$

Check:
$$\begin{array}{c|c} 8a & 56 \\ \hline 8 \cdot 7 & 56 \\ 56 & \stackrel{?}{=} 56 \end{array} \quad \text{TRUE}$$

The solution is 7.

37.
$$84 = 7x$$
$$\frac{84}{7} = \frac{7x}{7} \quad \text{Dividing both sides by 7}$$
$$12 = 1 \cdot x$$
$$12 = x$$

Check:
$$\begin{array}{c|c} 84 & 7x \\ \hline 84 & 7 \cdot 12 \\ 84 & \stackrel{?}{=} 84 \end{array} \quad \text{TRUE}$$

The solution is 12.

39.
$$-x = 38$$
$$-1 \cdot x = 38$$
$$-1 \cdot (-1 \cdot x) = -1 \cdot 38$$
$$1 \cdot x = -38$$
$$x = -38$$

Check:
$$\begin{array}{c|c} -x & 38 \\ \hline -(-38) & 38 \\ 38 & \stackrel{?}{=} 38 \end{array} \quad \text{TRUE}$$

The solution is -38.

41. $-t = -8$

The equation states that the opposite of t is the opposite of 8. Thus, $t = 8$. We could also do this exercise as follows.

$$-t = -8$$
$$-1(-t) = -1(-8) \quad \text{Multiplying both sides by } -1$$
$$t = 8$$

Check:
$$\begin{array}{c|c} -t & -8 \\ \hline -(8) & -8 \\ -8 & \stackrel{?}{=} -8 \end{array} \quad \text{TRUE}$$

The solution is 8.

43.
$$-7x = 49$$
$$\frac{-7x}{-7} = \frac{49}{-7}$$
$$1 \cdot x = -7$$
$$x = -7$$

Check:
$$\begin{array}{c|c} -7x & 49 \\ \hline -7(-7) & 49 \\ 49 & \stackrel{?}{=} 49 \end{array} \quad \text{TRUE}$$

The solution is -7.

45.
$$-1.3a = -10.4$$
$$\frac{-1.3a}{-1.3} = \frac{-10.4}{-1.3}$$
$$a = 8$$

Check:
$$\begin{array}{c|c} \multicolumn{2}{c}{-1.3a = -10.4} \\ \hline -1.3(8) & -10.4 \\ -10.4 & \overset{?}{=} -10.4 \end{array} \quad \text{TRUE}$$

The solution is 8.

47.
$$\frac{y}{8} = 11$$
$$\frac{1}{8}\cdot y = 11$$
$$8\left(\frac{1}{8}\right)\cdot y = 8\cdot 11$$
$$y = 88$$

Check:
$$\begin{array}{c|c} \multicolumn{2}{c}{\frac{y}{8} = 11} \\ \hline \frac{88}{8} & 11 \\ 11 & \overset{?}{=} 11 \end{array} \quad \text{TRUE}$$

The solution is 88.

49.
$$\frac{4}{5}x = 16$$
$$\frac{5}{4}\cdot\frac{4}{5}x = \frac{5}{4}\cdot 16$$
$$x = \frac{5\cdot \not{4}\cdot 4}{\not{4}\cdot 1}$$
$$x = 20$$

Check:
$$\begin{array}{c|c} \multicolumn{2}{c}{\frac{4}{5}x = 16} \\ \hline \frac{4}{5}\cdot 20 & 16 \\ 16 & \overset{?}{=} 16 \end{array} \quad \text{TRUE}$$

The solution is 20.

51.
$$\frac{-x}{6} = 9$$
$$-\frac{1}{6}\cdot x = 9$$
$$-6\left(-\frac{1}{6}\right)\cdot x = -6\cdot 9$$
$$x = -54$$

Check:
$$\begin{array}{c|c} \multicolumn{2}{c}{\frac{-x}{6} = 9} \\ \hline \frac{-(-54)}{6} & 9 \\ \frac{54}{6} & \\ 9 & \overset{?}{=} 9 \end{array} \quad \text{TRUE}$$

The solution is -54.

53.
$$\frac{1}{9} = \frac{z}{-5}$$
$$\frac{1}{9} = -\frac{1}{5}\cdot z$$
$$-5\cdot\frac{1}{9} = -5\cdot\left(-\frac{1}{5}\cdot z\right)$$
$$-\frac{5}{9} = z$$

Check:
$$\begin{array}{c|c} \multicolumn{2}{c}{\frac{1}{9} = \frac{z}{-5}} \\ \hline \frac{1}{9} & \frac{-5/9}{-5} \\ & -\frac{5}{9}\cdot\frac{1}{-5} \\ \frac{1}{9} & \overset{?}{=} \frac{1}{9} \end{array} \quad \text{TRUE}$$

The solution is $\frac{-5}{9}$.

55. $-\frac{3}{5}r = -\frac{3}{5}$

The solution of the equation is the number that is multiplied by $-\frac{3}{5}$ to get $-\frac{3}{5}$. That number is 1. We could also do this exercise as follows:
$$-\frac{3}{5}r = -\frac{3}{5}$$
$$-\frac{5}{3}\cdot\left(-\frac{3}{5}r\right) = -\frac{5}{3}\left(-\frac{3}{5}\right)$$
$$r = 1$$

Check:
$$\begin{array}{c|c} \multicolumn{2}{c}{-\frac{3}{5}r = -\frac{3}{5}} \\ \hline -\frac{3}{5}\cdot 1 & -\frac{3}{5} \\ -\frac{3}{5} & \overset{?}{=} -\frac{3}{5} \end{array} \quad \text{TRUE}$$

The solution is 1.

57.
$$\frac{-3r}{2} = -\frac{27}{4}$$
$$-\frac{3}{2}r = -\frac{27}{4}$$
$$-\frac{2}{3}\cdot\left(-\frac{3}{2}r\right) = -\frac{2}{3}\cdot\left(-\frac{27}{4}\right)$$
$$r = \frac{\not{2}\cdot\not{3}\cdot 3\cdot 3}{\not{3}\cdot\not{2}\cdot 2}$$
$$r = \frac{9}{2}$$

Check:
$$\begin{array}{c|c} \multicolumn{2}{c}{\frac{-3r}{2} = -\frac{27}{4}} \\ \hline -\frac{3}{2}\cdot\frac{9}{2} & -\frac{27}{4} \\ -\frac{27}{4} & \overset{?}{=} -\frac{27}{4} \end{array} \quad \text{TRUE}$$

The solution is $\frac{9}{2}$.

59.
$$\begin{aligned} 4.5+t &= -3.1 \\ 4.5+t-4.5 &= -3.1-4.5 \\ t &= -7.6 \end{aligned}$$
The solution is -7.6.

61.
$$\begin{aligned} -8.2x &= 20.5 \\ \frac{-8.2x}{-8.2} &= \frac{20.5}{-8.2} \\ x &= -2.5 \end{aligned}$$
The solution is -2.5.

63.
$$\begin{aligned} x-4 &= -19 \\ x-4+4 &= -19+4 \\ x &= -15 \end{aligned}$$
The solution is -15.

65.
$$\begin{aligned} t-3 &= 8 \\ t-3+3 &= -8+3 \\ t &= -5 \end{aligned}$$
The solution is -5.

67.
$$\begin{aligned} -12x &= 14 \\ \frac{-12x}{-12} &= \frac{14}{-12} \\ 1\cdot x &= -\frac{7}{6} \\ x &= -\frac{7}{6} \end{aligned}$$
The solution is $-\frac{7}{6}$.

69.
$$\begin{aligned} 48 &= -\frac{3}{8}y \\ -\frac{8}{3}\cdot 48 &= -\frac{8}{3}\left(-\frac{3}{8}y\right) \\ -\frac{8\cdot \cancel{3}\cdot 16}{\cancel{3}} &= y \\ -128 &= y \end{aligned}$$
The solution is -128.

71.
$$\begin{aligned} a-\frac{1}{6} &= -\frac{2}{3} \\ a-\frac{1}{6}+\frac{1}{6} &= -\frac{2}{3}+\frac{1}{6} \\ a &= -\frac{4}{6}+\frac{1}{6} \\ a &= -\frac{3}{6} \\ a &= -\frac{1}{2} \end{aligned}$$
The solution is $-\frac{1}{2}$.

73.
$$\begin{aligned} -24 &= \frac{8x}{5} \\ -24 &= \frac{8}{5}x \\ \frac{5}{8}(-24) &= \frac{5}{8}\cdot\frac{8}{5}x \\ -\frac{5\cdot \cancel{8}\cdot 3}{\cancel{8}\cdot 1} &= x \\ -15 &= x \end{aligned}$$
The solution is -15.

75.
$$\begin{aligned} -\frac{4}{3}t &= -12 \\ -\frac{3}{4}\left(-\frac{4}{3}t\right) &= -\frac{3}{4}(-12) \\ t &= \frac{3\cdot \cancel{4}\cdot 3}{\cancel{4}} \\ t &= 9 \end{aligned}$$
The solution is 9.

77.
$$\begin{aligned} -483.297 &= -794.053+t \\ -483.297+794.053 &= -794.053+t+794.053 \\ 310.756 &= t \quad \text{Using a calculator} \end{aligned}$$
The solution is 310.756.

79. *Writing Exercise.* For an equation $x+a=b$, add the opposite of a (or subtract a) on both sides of the equation. For an equation $ax=b$, multiply by $1/a$ (or divide by a) on both sides of the equation.

81.
$$\begin{aligned} &3\cdot 4-18 \\ &= 12-18 \quad \text{Multiplying} \\ &= -6 \quad \text{Subtracting} \end{aligned}$$

83.
$$\begin{aligned} &16\div(2-3\cdot 2)+5 \\ &= 16\div(2-6)+5 \quad \text{Simplifying inside} \\ &= 16\div(-4)+5 \quad \text{the parentheses} \\ &= -4+5 \quad \text{Dividing} \\ &= 1 \quad \text{Adding} \end{aligned}$$

85. *Writing Exercise.* Yes, it will form an equivalent equation by the addition principle. It will not help to solve the equation, however. The multiplication principle should be used to solve the equation.

87.
$$\begin{aligned} mx &= 11.6m \\ \frac{mx}{m} &= \frac{11.6m}{m} \\ x &= 11.6 \end{aligned}$$
The solution is 11.6.

89.
$$\begin{aligned} cx+5c &= 7c \\ cx+5c-5c &= 7c-5c \\ cx &= 2c \\ \frac{cx}{c} &= \frac{2c}{c} \\ x &= 2 \end{aligned}$$
The solution is 2.

91.
$$7+|x| = 30$$
$$-7+7+|x| = -7+30$$
$$|x| = 23$$

x represents a number whose distance from 0 is 23. Thus $x=-23$ or $x=23$.

93.
$$t-3590 = 1820$$
$$t-3590+3590 = 1820+3590$$
$$t = 5410$$
$$t+3590 = 5410+3590$$
$$t+3590 = 9000$$

95. To "undo" the last step, divide 22.5 by 0.3.

$22.5 \div 0.3 = 75$

Now divide 75 by 0.3.

$75 \div 0.3 = 250$

The answer should be 250 not 22.5.

Exercise Set 2.2

1.
$$3x-1 = 7$$
$$3x-1+1 = 7+1 \quad \text{Adding 1 to both sides}$$
$$3x = 7+1$$

Choice (c) is correct.

3. $6(x-1)=2$

$6x-6=2$ Using the distributive law

Choice (a) is correct.

5.
$$4x = 3-2x$$
$$4x+2x = 3-2x+2x \quad \text{Adding } 2x \text{ to both sides}$$
$$4x+2x = 3$$

Choice (b) is correct.

7.
$$2x+9 = 25$$
$$2x+9-9 = 25-9 \quad \text{Subtracting 9 from both sides}$$
$$2x = 16 \quad \text{Simplifying}$$
$$\frac{2x}{2} = \frac{16}{2} \quad \text{Dividing both sides by 2}$$
$$x = 8 \quad \text{Simplifying}$$

Check:
$$\begin{array}{r|l} \multicolumn{2}{c}{2x+9 = 25} \\ \hline 2\cdot 8+9 & 25 \\ 16+9 & \\ 25 \stackrel{?}{=} 25 & \text{TRUE} \end{array}$$

The solution is 8.

9.
$$6z+5 = 47$$
$$6z+5-5 = 47-5 \quad \text{Subtracting 5 from both sides}$$
$$6z = 42 \quad \text{Simplifying}$$
$$\frac{6z}{6} = \frac{42}{6} \quad \text{Dividing both sides by 6}$$
$$z = 7 \quad \text{Simplifying}$$

Check:
$$\begin{array}{r|l} \multicolumn{2}{c}{6z+5 = 47} \\ \hline 6\cdot 7+5 & 47 \\ 42+5 & \\ 47 \stackrel{?}{=} 47 & \text{TRUE} \end{array}$$

The solution is 7.

11.
$$7t-8 = 27$$
$$7t-8+8 = 27+8 \quad \text{Adding 8 to both sides}$$
$$7t = 35$$
$$\frac{7t}{7} = \frac{35}{7} \quad \text{Dividing both sides by 7}$$
$$t = 5$$

Check:
$$\begin{array}{r|l} \multicolumn{2}{c}{7t-8 = 27} \\ \hline 7\cdot 5-8 & 27 \\ 35-8 & \\ 27 \stackrel{?}{=} 27 & \text{TRUE} \end{array}$$

The solution is 5.

13.
$$3x-9 = 1$$
$$3x-9+9 = 1+9$$
$$3x = 10$$
$$\frac{3x}{3} = \frac{10}{3}$$
$$x = \frac{10}{3}$$

Check:
$$\begin{array}{r|l} \multicolumn{2}{c}{3x-9 = 1} \\ \hline 3\cdot \frac{10}{3}-9 & 1 \\ 10-9 & \\ 1 \stackrel{?}{=} 1 & \text{TRUE} \end{array}$$

The solution is $\frac{10}{3}$.

15.
$$8z+2 = -54$$
$$8z+2-2 = -54-2$$
$$8z = -56$$
$$\frac{8z}{8} = \frac{-56}{8}$$
$$z = -7$$

Check:
$$\begin{array}{r|l} \multicolumn{2}{c}{8z+2 = -54} \\ \hline 8(-7)+2 & -54 \\ -56+2 & \\ -54 \stackrel{?}{=} -54 & \text{TRUE} \end{array}$$

The solution is -7.

17.
$$-37 = 9t+8$$
$$-37-8 = 9t+8-8$$
$$-45 = 9t$$
$$\frac{-45}{9} = \frac{9t}{9}$$
$$-5 = t$$

Check:
$$\begin{array}{r|l} \multicolumn{2}{c}{-37 = 9t+8} \\ \hline -37 & 9\cdot(-5)+8 \\ & -45+8 \\ -37 \stackrel{?}{=} -37 & \text{TRUE} \end{array}$$

The solution is -5.

19.
$$12 - t = 16$$
$$-12 + 12 - t = -12 + 16$$
$$-t = 4$$
$$\frac{-t}{-1} = \frac{4}{-1}$$
$$t = -4$$

Check:
$$\begin{array}{c|c} \multicolumn{2}{c}{12 - t = 16} \\ \hline 12 - (-4) & 16 \\ 12 + 4 & \end{array}$$
$$16 \stackrel{?}{=} 16 \quad \text{TRUE}$$

The solution is -4.

21.
$$-6z - 18 = -132$$
$$-6z - 18 + 18 = -132 + 18$$
$$-6z = -114$$
$$\frac{-6z}{-6} = \frac{-114}{-6}$$
$$z = 19$$

Check:
$$\begin{array}{c|c} \multicolumn{2}{c}{-6z - 18 = -132} \\ \hline -6 \cdot 19 - 18 & -132 \\ -114 - 18 & \end{array}$$
$$-132 \stackrel{?}{=} -132 \quad \text{TRUE}$$

The solution is 19.

23.
$$5.3 + 1.2n = 1.94$$
$$1.2n = -3.36$$
$$\frac{1.2n}{1.2} = \frac{-3.36}{1.2}$$
$$n = -2.8$$

Check:
$$\begin{array}{c|c} \multicolumn{2}{c}{5.31 + 1.2n = 1.94} \\ \hline 5.3 + 1.2(-2.8) & 1.94 \\ 5.3 + (-3.36) & \end{array}$$
$$1.94 \stackrel{?}{=} 1.94 \quad \text{TRUE}$$

The solution is -2.8.

25.
$$32 - 7x = 11$$
$$-32 + 32 - 7x = -32 + 11$$
$$-7x = -21$$
$$\frac{-7x}{-7} = \frac{-21}{-7}$$
$$x = 3$$

Check:
$$\begin{array}{c|c} \multicolumn{2}{c}{32 - 7x = 11} \\ \hline 32 - 7 \cdot 3 & 11 \\ 32 - 21 & \end{array}$$
$$11 \stackrel{?}{=} 11 \quad \text{TRUE}$$

The solution is 3.

27.
$$\frac{3}{5}t - 1 = 8$$
$$\frac{3}{5}t - 1 + 1 = 8 + 1$$
$$\frac{3}{5}t = 9$$
$$\frac{5}{3} \cdot \frac{3}{5}t = \frac{5}{3} \cdot 9$$
$$t = \frac{5 \cdot \cancel{3} \cdot 3}{\cancel{3} \cdot 1}$$
$$t = 15$$

Check:
$$\begin{array}{c|c} \multicolumn{2}{c}{\frac{3}{5}t - 1 = 8} \\ \hline \frac{3}{5} \cdot 15 - 1 & 8 \\ 9 - 1 & \end{array}$$
$$8 \stackrel{?}{=} 8 \quad \text{TRUE}$$

The solution is 15.

29.
$$6 + \frac{7}{2}x = -15$$
$$-6 + 6 + \frac{7}{2}x = -6 - 15$$
$$\frac{7}{2}x = -21$$
$$\frac{2}{7} \cdot \frac{7}{2}x = \frac{2}{7}(-21)$$
$$x = -\frac{2 \cdot 3 \cdot \cancel{7}}{\cancel{7} \cdot 1}$$
$$x = -6$$

Check:
$$\begin{array}{c|c} \multicolumn{2}{c}{6 + \frac{7}{2}x = -15} \\ \hline 6 + \frac{7}{2}(-6) & -15 \\ 6 + (-21) & \end{array}$$
$$-15 \stackrel{?}{=} -15 \quad \text{TRUE}$$

The solution is -6.

31.
$$-\frac{4a}{5} - 8 = 2$$
$$-\frac{4a}{5} - 8 + 8 = 2 + 8$$
$$-\frac{4a}{5} = 10$$
$$-\frac{5}{4}\left(-\frac{4a}{5}\right) = -\frac{5}{4} \cdot 10$$
$$a = -\frac{5 \cdot 5 \cdot \cancel{2}}{2 \cdot \cancel{2}}$$
$$a = -\frac{25}{2}$$

Check:
$$\begin{array}{c|c} \multicolumn{2}{c}{-\frac{4a}{5} - 8 = 2} \\ \hline -\frac{4}{5}\left(-\frac{25}{2}\right) - 8 & 2 \\ 10 - 8 & \end{array}$$
$$2 \stackrel{?}{=} 2 \quad \text{TRUE}$$

The solution is $-\frac{25}{2}$.

33. $4x = x + 3x$
$4x = 4x$

All real numbers are solutions and the equation is an identity.

35. $4x - 6 = 6x$

$-6 = 6x - 4x$ Subtracting $4x$ from both sides

$-6 = 2x$ Simplifying

$\frac{-6}{2} = \frac{2x}{2}$ Dividing both sides by 2

$-3 = x$

Check: $4x - 6 = 6x$

$4(-3) - 6$	$6(-3)$
$-12 - 6$	-18

$-18 \stackrel{?}{=} -18$ TRUE

The solution is -3.

37. $2 - 5y = 26 - y$

$2 - 5y + y = 26 - y + y$ Adding y to both sides

$2 - 4y = 26$ Simplifying

$-2 + 2 - 4y = -2 + 26$ Adding -2 to both sides

$-4y = 24$ Simplifying

$\frac{-4y}{-4} = \frac{24}{-4}$ Dividing both sides by -4

$y = -6$

Check: $2 - 5y = 26 - y$

$2 - 5(-6)$	$26 - (-6)$
$2 + 30$	$26 + 6$

$32 \stackrel{?}{=} 32$ TRUE

The solution is -6.

39. $7(2a - 1) = 21$

$14a - 7 = 21$ Using the distributive law

$14a = 21 + 7$ Adding 7

$14a = 28$

$a = 2$ Dividing by 14

Check: $7(2a - 1) = 21$

$7(2 \cdot 2 - 1)$	21
$7(4 - 1)$	
$7 \cdot 3$	

$21 \stackrel{?}{=} 21$ TRUE

The solution is 2.

41. We can write $11 = 11(x + 1)$ as $11 \cdot 1 = 11(x + 1)$. Then $1 = x + 1$, or $x = 0$. The solution is 0.

43. $2(3 + 4m) - 6 = 48$

$6 + 8m - 6 = 48$

$8m = 48$ Combining like terms

$m = 6$

Check: $2(3 + 4m) - 6 = 48$

$2(3 + 4 \cdot 6) - 6$	48
$2(3 + 24) - 6$	
$2 \cdot 27 - 6$	
$54 - 6$	

$48 \stackrel{?}{=} 48$ TRUE

The solution is 6.

45. $3(x + 4) = 3(x - 1)$

$3x + 12 = 3x - 3$ Using the distributive law

$12 = -3$ Subtracting $3x$ from both sides

Since $12 \neq -3$, there is no solution and the equation is a contradiction.

47. $2r + 8 = 6r + 10$

$2r + 8 - 10 = 6r + 10 - 10$

$2r - 2 = 6r$ Combining like terms

$-2r + 2r - 2 = -2r + 6r$

$-2 = 4r$

$\frac{-2}{4} = \frac{4r}{4}$

$-\frac{1}{2} = r$

Check: $2r + 8 = 6r + 10$

$2\left(-\frac{1}{2}\right) + 8$	$6\left(-\frac{1}{2}\right) + 10$
$-1 + 8$	$-3 + 10$

$7 \stackrel{?}{=} 7$ TRUE

The solution is $-\frac{1}{2}$.

49. $6x + 3 = 2x + 3$

$6x - 2x = 3 - 3$

$4x = 0$

$\frac{4x}{4} = \frac{0}{4}$

$x = 0$

Check: $6x + 3 = 2x + 3$

$6 \cdot 0 + 3$	$2 \cdot 0 + 3$
$0 + 3$	$0 + 3$

$3 \stackrel{?}{=} 3$ TRUE

The solution is 0.

51. $5 - 2x = 3x - 7x + 25$

$5 - 2x = -4x + 25$

$4x - 2x = 25 - 5$

$2x = 20$

$\frac{2x}{2} = \frac{20}{2}$

$x = 10$

Check:

$5 - 2x = 3x - 7x + 25$

$5 - 2 \cdot 10$	$3 \cdot 10 - 7 \cdot 10 + 25$
$5 - 20$	$30 - 70 + 25$
-15	$-40 + 25$

$-15 \stackrel{?}{=} -15$ TRUE

The solution is 10.

53. $7+3x-6=3x+5-x$

$3x+1=2x+5$ Combining like terms on each side

$3x-2x=5-1$

$x=4$

Check: $7+3x-6=3x+5-x$

$7+3\cdot4-6$	$3\cdot4+5-4$
$7+12-6$	$12+5-4$
$19-6$	$17-4$

$13 \stackrel{?}{=} 13$ TRUE

The solution is 4.

55. $4y-4+y+24=6y+20-4y$

$5y+20=2y+20$

$5y-2y=20-20$

$3y=0$

$y=0$

Check:

$4y-4+y+24=6y+20-4y$

$4\cdot0-4+0+24$	$6\cdot0+20-4\cdot0$
$0-4+0+24$	$0+20-0$

$20 \stackrel{?}{=} 20$ TRUE

The solution is 0.

57. $4+7x=7(x+1)$

$4+7x=7x+7$

$4=7$

Since $4 \neq 7$, there is no solution and the equation is a contradiction.

59. $19-3(2x-1)=7$

$19-6x+3=7$

$22-6x=7$

$-6x=7-22$

$-6x=-15$

$x=\frac{15}{6}$

$x=\frac{5}{2}$

Check: $19-3(2x-1)=7$

$19-3(2\cdot\frac{5}{2}-1)$	7
$19-3(5-1)$	
$19-3(4)$	
$19-12$	

$7 \stackrel{?}{=} 7$ TRUE

The solution is $\frac{5}{2}$.

61. $7(5x-2)=6(6x-1)$

$35x-14=36x-6$

$-14+6=36x-35x$

$-8=x$

Check:

$7(5x-2)=6(6x-1)$

$7(5(-8)-2)$	$6(6(-8)-1)$
$7(-40-2)$	$6(-48-1)$
$7(-42)$	$6(-49)$

$-294 \stackrel{?}{=} -294$ TRUE

The solution is -8.

63. $2(3t+1)-5=t-(t+2)$

$6t+2-5=t-t-2$

$6t-3=-2$

$6t=-2+3$

$6t=1$

$t=\frac{1}{6}$

Check: $2(3t+1)-5=t-(t+2)$

$2\left(3\cdot\frac{1}{6}+1\right)-5$	$\frac{1}{6}-\left(\frac{1}{6}+2\right)$
$2\left(\frac{1}{2}+1\right)-5$	$\frac{1}{6}-2\frac{1}{6}$
$2\cdot\frac{3}{2}-5$	-2

$-2 \stackrel{?}{=} -2$ TRUE

The solution is $\frac{1}{6}$.

65. $2(7-x)-20=7x-3(2+3x)$

$14-2x-20=7x-6-9x$

$-2x-6=-2x-6$

All real numbers are solutions and the equation is an identity.

67. $19-(2x+3)=2(x+3)+x$

$19-2x-3=2x+6+x$

$16-2x=3x+6$

$16-6=3x+2x$

$10=5x$

$2=x$

Check: $19-(2x+3)=2(x+3)+x$

$19-(2\cdot2+3)$	$2(2+3)+2$
$19-(4+3)$	$2\cdot5+2$
$19-7$	$10+2$

$12 \stackrel{?}{=} 12$ TRUE

The solution is 2.

69. $\frac{2}{3}+\frac{1}{4}t=2$

The number 12 is the least common denominator, so we multiply by 12 on both sides.

$12\left(\frac{2}{3}+\frac{1}{4}t\right)=12\cdot2$

$12\cdot\frac{2}{3}+12\cdot\frac{1}{4}t=24$

$8+3t=24$

$3t=24-8$

$3t=16$

$t=\frac{16}{3}$

Check: $\frac{2}{3}+\frac{1}{4}t=2$

$\frac{2}{3}+\frac{1}{4}\left(\frac{16}{3}\right)$	2
$\frac{2}{3}+\frac{4}{3}$	

$2 \stackrel{?}{=} 2$ TRUE

The solution is $\frac{16}{3}$.

71. $\frac{2}{3}+4t=6t-\frac{2}{15}$

The number 15 is the least common denominator, so we multiply by 15 on both sides.

$$\begin{aligned}15\left(\frac{2}{3}+4t\right)&=15\left(6t-\frac{2}{15}\right)\\ 15\cdot\frac{2}{3}+15\cdot 4t&=15\cdot 6t-15\cdot\frac{2}{15}\\ 10+60t&=90t-2\\ 10+2&=90t-60t\\ 12&=30t\\ \frac{12}{30}&=t\\ \frac{2}{5}&=t\end{aligned}$$

Check: $\frac{2}{3}+4t=6t-\frac{2}{15}$

$\frac{2}{3}+4\cdot\frac{2}{5}$	$6\cdot\frac{2}{5}-\frac{2}{15}$
$\frac{2}{3}+\frac{8}{5}$	$\frac{12}{5}-\frac{2}{15}$
$\frac{10}{15}+\frac{24}{15}$	$\frac{36}{15}-\frac{2}{15}$

$\frac{34}{15}\stackrel{?}{=}\frac{34}{15}$ TRUE

The solution is $\frac{2}{5}$.

73. $\frac{1}{3}x+\frac{2}{5}=\frac{4}{5}+\frac{3}{5}x-\frac{2}{3}$

The number 15 is the least common denominator, so we multiply by 15 on both sides.

$$\begin{aligned}15\left(\frac{1}{3}x+\frac{2}{5}\right)&=15\left(\frac{4}{5}+\frac{3}{5}x-\frac{2}{3}\right)\\ 15\cdot\frac{1}{3}x+15\cdot\frac{2}{5}&=15\cdot\frac{4}{5}+15\cdot\frac{3}{5}x-15\cdot\frac{2}{3}\\ 5x+6&=12+9x-10\\ 5x+6&=2+9x\\ 5x-9x&=2-6\\ -4x&=-4\\ \frac{-4x}{-4}&=\frac{-4}{-4}\\ x&=1\end{aligned}$$

Check: $\frac{1}{3}x+\frac{2}{5}=\frac{4}{5}+\frac{3}{5}x-\frac{2}{3}$

$\frac{1}{3}\cdot 1+\frac{2}{5}$	$\frac{4}{5}+\frac{3}{5}\cdot 1-\frac{2}{3}$
$\frac{1}{3}+\frac{2}{5}$	$\frac{4}{5}+\frac{3}{5}-\frac{2}{3}$
$\frac{5}{15}+\frac{6}{15}$	$\frac{12}{15}+\frac{9}{15}-\frac{10}{15}$
$\frac{11}{15}$	$\frac{11}{15}$

$\frac{11}{15}\stackrel{?}{=}\frac{11}{15}$ TRUE

The solution is 1.

75. $2.1x+45.2=3.2-8.4x$

Greatest number of decimal places is 1

$10(2.1x+45.2)=10(3.2-8.4x)$

Multiplying by 10 to clear decimals

$$\begin{aligned}10(2.1x)+10(45.2)&=10(3.2)-10(8.4x)\\ 21x+452&=32-84x\\ 21x+84x&=32-452\\ 105x&=-420\\ x&=\frac{-420}{105}\\ x&=-4\end{aligned}$$

Check: $2.1x+45.2=3.2-8.4x$

$2.1(-4)+45.2$	$3.2-8.4(-4)$
$-8.4+45.2$	$3.2+33.6$

$36.8\stackrel{?}{=}36.8$ TRUE

The solution is -4.

77. $0.76+0.21t=0.96t-0.49$

Greatest number of decimal places is 2

$100(0.76+0.21t)=100(0.96t-0.49)$

Multiplying by 100 to clear decimals

$$\begin{aligned}100(0.76)+100(0.21t)&=100(0.96t)-100(0.49)\\ 76+21t&=96t-49\\ 76+49&=96t-21t\\ 125&=75t\\ \frac{125}{75}&=t\\ \frac{5}{3}&=t,\text{ or}\\ 1.\overline{6}&=t\end{aligned}$$

The answer checks. The solution is $\frac{5}{3}$, or $1.\overline{6}$.

79. $\frac{2}{5}x-\frac{3}{2}x=\frac{3}{4}x+3$

The least common denominator is 20.

$$\begin{aligned}20\left(\frac{2}{5}x-\frac{3}{2}x\right)&=20\left(\frac{3}{4}x+3\right)\\ 20\cdot\frac{2}{5}x-20\cdot\frac{3}{2}x&=20\cdot\frac{3}{4}x+20\cdot 3\\ 8x-30x&=15x+60\\ -22x&=15x+60\\ -22x-15x&=60\\ -37x&=60\\ x&=-\frac{60}{37}\end{aligned}$$

Check:

$\frac{2}{5}x-\frac{3}{2}x=\frac{3}{4}x+3$

$\frac{2}{5}\left(-\frac{60}{37}\right)-\frac{3}{2}\left(-\frac{60}{37}\right)$	$\frac{3}{4}\left(-\frac{60}{37}\right)+3$
$-\frac{24}{37}+\frac{90}{37}$	$-\frac{45}{37}+\frac{111}{37}$

$\frac{66}{37}\stackrel{?}{=}\frac{66}{37}$ TRUE

The solution is $-\frac{60}{37}$.

81.
$$\frac{1}{3}(2x-1)=7$$
$$3\cdot\frac{1}{3}(2x-1)=3\cdot 7$$
$$2x-1=21$$
$$2x=22$$
$$x=11$$

Check:
$$\frac{1}{3}(2x-1)=7$$
$$\frac{1}{3}(2\cdot 11-1) \mid 7$$
$$\frac{1}{3}\cdot 21$$
$$7\stackrel{?}{=}7 \quad \text{TRUE}$$

The solution is 11.

83.
$$\frac{3}{4}(3t-4)=15$$
$$\frac{4}{3}\cdot\frac{3}{4}(3t-4)=\frac{4}{3}\cdot 15$$
$$3t-4=20$$
$$3t=24$$
$$t=8$$

Check:
$$\frac{3}{4}(3t-4)=15$$
$$\frac{3}{4}(3\cdot 8-4) \mid 15$$
$$\frac{3}{4}\cdot(24-4)$$
$$\frac{3}{4}\cdot 20$$
$$15\stackrel{?}{=}15 \quad \text{TRUE}$$

The solution is 8.

85.
$$\frac{1}{6}\left(\frac{3}{4}x-2\right)=-\frac{1}{5}$$
$$30\cdot\frac{1}{6}\left(\frac{3}{4}x-2\right)=30\left(-\frac{1}{5}\right)$$
$$5\left(\frac{3}{4}x-2\right)=-6$$
$$\frac{15}{4}x-10=-6$$
$$\frac{15}{4}x=4$$
$$4\cdot\frac{15}{4}x=4\cdot 4$$
$$15x=16$$
$$x=\frac{16}{15}$$

Check:
$$\frac{1}{6}\left(\frac{3}{4}x-2\right)=-\frac{1}{5}$$
$$\frac{1}{6}\left(\frac{3}{4}\cdot\frac{16}{15}-2\right) \mid -\frac{1}{5}$$
$$\frac{1}{6}\left(\frac{4}{5}-2\right)$$
$$\frac{1}{6}\left(-\frac{6}{5}\right)$$
$$-\frac{1}{5}\stackrel{?}{=}-\frac{1}{5} \quad \text{TRUE}$$

The solution is $\frac{16}{15}$.

87.
$$0.7(3x+6)=1.1-(x-3)$$
$$2.1x+4.2=1.1-x+3$$
$$2.1x+4.2=-x+4.1$$
$$10(2.1x+4.2)=10(-x+4.1) \quad \text{Clearing decimals}$$
$$21x+42=-10x+41$$
$$21x=-10x+41-42$$
$$21x=-10x-1$$
$$31x=-1$$
$$x=-\frac{1}{31}$$

The check is left to the student. The solution is $-\frac{1}{31}$.

89.
$$a+(a-3)=(a+2)-(a+1)$$
$$a+a-3=a+2-a-1$$
$$2a-3=1$$
$$2a=1+3$$
$$2a=4$$
$$a=2$$

Check:
$$a+(a-3)=(a+2)-(a+1)$$
$$2+(2-3) \mid (2+2)-(2+1)$$
$$2-1 \mid 4-3$$
$$1\stackrel{?}{=}1 \quad \text{TRUE}$$

The solution is 2.

91. *Writing Exercise.* By adding $t-13$ to both sides of $45-t=13$ we have $32=t$. This approach is preferable since we found the solution in just one step.

93. $3-5a=3-5\cdot 2=3-10=-7$

95. $7x-2x=7(-3)-2(-3)=-21+6=-15$

97. *Writing Exercise.* Multiply by 100 to clear decimals. Next multiply by 12 to clear fractions. (These steps could be reversed.) Then proceed as usual. The procedure could be streamlined by multiplying by 1200 to clear decimals and fractions in one step.

99.
$$8.43x-2.5(3.2-0.7x)=-3.455x+9.04$$
$$8.43x-8+1.75x=-3.455x+9.04$$
$$10.18x-8=-3.455x+9.04$$
$$10.18x+3.455x=9.04+8$$
$$13.635x=17.04$$
$$x=1.\overline{2497},\text{ or }\frac{1136}{909}$$

The solution is $1.\overline{2497}$, or $\frac{1136}{909}$.

101. $-2[3(x-2)+4] = 4(5-x)-2x$
$-2[3x-6+4] = 20-4x-2x$
$-2[3x-2] = 20-6x$
$-6x+4 = 20-6x$
$4 = 20$ Adding $6x$ to both sides

This is a contradiction.

103. $2x(x+5)-3(x^2+2x-1) = 9-5x-x^2$
$2x^2+10x-3x^2-6x+3 = 9-5x-x^2$
$-x^2+4x+3 = 9-5x-x^2$
$4x+3 = 9-5x$ Adding x^2
$4x+5x = 9-3$
$9x = 6$
$x = \frac{2}{3}$

The solution is $\frac{2}{3}$.

105. $9-3x = 2(5-2x)-(1-5x)$
$9-3x = 10-4x-1+5x$
$9-3x = 9+x$
$9-9 = x+3x$
$0 = 4x$
$0 = x$

The solution is 0.

107. $\frac{x}{14}-\frac{5x+2}{49} = \frac{3x-4}{7}$
$98\left(\frac{x}{14}-\frac{5x+2}{49}\right) = 98\left(\frac{3x-4}{7}\right)$
$98\cdot\frac{x}{14}-98\left(\frac{5x+2}{49}\right) = 42x-56$
$7x-10x-4 = 42x-56$
$-3x-4 = 42x-56$
$-4+56 = 42x+3x$
$52 = 45x$
$\frac{52}{45} = x$

109. $2\{9-3[-2x-4]\} = 12x+42$
$2\{9+6x+12\} = 12x+42$
$2\{6x+21\} = 12x+42$
$12x+42 = 12x+42$

All real numbers are solutions and the equation is an identity.

Exercise Set 2.3

1. We substitute 0.9 for t and calculate d.

$d = 344t = 344\cdot 0.9 = 309.6$

The fans were 309.6m from the stage.

3. We substitute 21,345 for n and calculate f.

$f = \frac{n}{15} = \frac{21,345}{15} = 1423$

There are 1423 full-time equivalent students.

5. We substitute 0.025 for I and 0.044 for U and calculate f.
$f = 8.5+1.4(I-U)$
$= 8.5+1.4(0.025-0.044)$
$= 8.5+1.4(-0.019)$
$= 8.5-0.0266$
$= 8.4734$

The federal funds rate should be 8.4734.

7. Substitute 1 for t and calculate n.
$n = 0.5t^4+3.45t^3-96.65t^2+347.7t$
$= 0.5(1)^4+3.45(1)^3-96.65(1)^2+347.7(1)$
$= 0.5+3.45-96.65+347.7$
$= 255$

255 mg of ibuprofen remains in the bloodstream.

9. $A = bh$
$\frac{A}{h} = \frac{bh}{h}$ Dividing both sides by h
$\frac{A}{h} = b$

11. $d = rt$
$\frac{d}{t} = \frac{rt}{t}$ Dividing both sides by t
$\frac{d}{t} = r$

13. $I = Prt$
$\frac{I}{rt} = \frac{Prt}{rt}$ Dividing both sides by rt
$\frac{I}{rt} = P$

15. $H = 65-m$
$H+m = 65$ Adding m to both sides
$m = 65-H$ Subtracting H from both sides

17. $P = 2l+2w$
$P-2w = 2l+2w-2w$ Subtracting $2w$ from both sides
$P-2w = 2l$
$\frac{P-2w}{2} = \frac{2l}{2}$ Dividing both sides by 2
$\frac{P-2w}{2} = l$, or
$\frac{P}{2}-w = l$

19. $A = \pi r^2$
$\frac{A}{r^2} = \frac{\pi r^2}{r^2}$
$\frac{A}{r^2} = \pi$

21. $A = \frac{1}{2}bh$

$2A = 2 \cdot \frac{1}{2}bh$ Multiplying both sides by 2

$2A = bh$

$\frac{2A}{b} = \frac{bh}{b}$ Dividing both sides by h

$\frac{2A}{b} = h$

23. $E = mc^2$

$\frac{E}{m} = \frac{mc^2}{m}$ Dividing both sides by m

$\frac{E}{m} = c^2$

25. $Q = \frac{c+d}{2}$

$2Q = 2 \cdot \frac{c+d}{2}$ Multiplying both sides by 2

$2Q = c+d$

$2Q - c = c + d - c$ Subtracting c from both sides

$2Q - c = d$

27. $A = \frac{a+b+c}{3}$

$3A = 3 \cdot \frac{a+b+c}{3}$ Multiplying both sides by 3

$3A = a+b+c$

$3A - a - c = a + b + c - a - c$ Subtracting a and c from both sides

$3A - a - c = b$

29. $w = \frac{r}{f}$

$f \cdot w = f \cdot \frac{r}{f}$ Multiplying both sides by f

$fw = r$

31. $F = \frac{9}{5}C + 32$

$F - 32 = \frac{9}{5}C$

$\frac{5}{9}(F - 32) = \frac{5}{9} \cdot \frac{9}{5}C$

$\frac{5}{9}(F - 32) = C$

33. $2x - y = 1$

$2x - y + y - 1 = 1 + y - 1$ Adding $y - 1$ to both sides

$2x - 1 = y$

35. $2x + 5y = 10$

$5y = -2x + 10$

$y = \frac{-2x+10}{5}$

$y = -\frac{2}{5}x + 2$

37. $4x - 3y = 6$

$-3y = -4x + 6$

$y = \frac{-4x+6}{-3}$

$y = \frac{4}{3}x - 2$

39. $9x + 8y = 4$

$8y = -9x + 4$

$y = \frac{-9x+4}{8}$

$y = -\frac{9}{8}x + \frac{1}{2}$

41. $3x - 5y = 8$

$-5y = -3x + 8$

$y = \frac{-3x+8}{-5}$

$y = \frac{3}{5}x - \frac{8}{5}$

43. $z = 13 + 2(x + y)$

$z - 13 = 2(x + y)$

$z - 13 = 2x + 2y$

$z - 13 - 2y = 2x$

$\frac{z - 13 - 2y}{2} = x$

$\frac{1}{2}z - \frac{13}{2} - y = x$

45. $t = 27 - \frac{1}{4}(w - l)$

$t - 27 = -\frac{1}{4}(w - l)$

$-4(t - 27) = w - l$ Multiplying by -4

$-4t + 108 = w - l$

$-4t + 108 - w = -l$

$4t - 108 + w = l$ Multiplying by -1

47. $A = at + bt$

$A = t(a + b)$ Factoring

$\frac{A}{a+b} = t$ Dividing both sides by $a+b$

49. $A = \frac{1}{2}ah + \frac{1}{2}bh$

$2A = 2\left(\frac{1}{2}ah + \frac{1}{2}bh\right)$

$2A = ah + bh$

$2A = h(a + b)$

$\frac{2A}{a+b} = h$

51. $R = r + \frac{400(W-L)}{N}$

$N \cdot R = N\left(r + \frac{400(W-L)}{N}\right)$

Multiplying both sides by N

$NR = Nr + 400(W-L)$

$NR = Nr + 400W - 400L$

$NR + 400L = Nr + 400W$ Adding $400L$ to both sides

$400L = Nr + 400W - NR$ Adding $-NR$ to both sides

$L = \frac{Nr + 400W - NR}{400}$

53. *Writing Exercise.* Given the formula for converting Celsius temperature C to Fahrenheit temperature F, solve for C. This yields a formula for converting Fahrenheit temperature to Celsius temperature.

55. $-2+5-(-4)-17$
$= -2+5+4-17$
$= 3+4-17$
$= 7-17$
$= -10$

57. $4.2(-11.75)(0) = 0$

59. $20 \div (-4) \cdot 2 - 3$

$= -5 \cdot 2 - 3$ Dividing and

$= -10 - 3$ multiplying from left to right

$= -13$ Subtracting

61. *Writing Exercise.* Answers may vary. A decorator wants to have a carpet cut for a bedroom. The perimeter of the room is 54 ft and its length is 15 ft. How wide should the carpet be?

63. $K = 21.235w + 7.75h - 10.54a + 102.3$

$2852 = 21.235(80) + 7.75(190) - 10.54a + 102.3$

$2852 = 1698.8 + 1472.5 - 10.54a + 102.3$

$2852 = 3273.6 - 10.54a$

$-421.6 = -10.54a$

$40 = a$

The man is 40 years old.

65. First we substitute 54 for A and solve for s to find the length of a side of the cube.

$A = 6s^2$

$54 = 6s^2$

$9 = s^2$

$3 = s$ Taking the positive square root

Now we substitute 3 for s in the formula for the volume of a cube and compute the volume.

$V = s^3 = 3^3 = 27$

The volume of the cube is 27 in^3.

67. $c = \frac{w}{a} \cdot d$

$ac = a \cdot \frac{w}{a} \cdot d$

$ac = wd$

$a = \frac{wd}{c}$

69. $ac = bc + d$

$ac - bc = d$

$c(a-b) = d$

$c = \frac{d}{a-b}$

71. $3a = c - a(b+d)$

$3a = c - ab - ad$

$3a + ab + ad = c$

$a(3+b+d) = c$

$a = \frac{c}{3+b+d}$

73. $K = 21.235w + 7.75h - 10.54a + 102.3$
$K = 21.235\left(\frac{w}{2.2046}\right) + 7.75\left(\frac{h}{0.3937}\right) - 10.54a + 102.3$
$K = 9.632w + 19.685h - 10.54a + 102.3$

Exercise Set 2.4

1. "What percent of 57 is 23?" can be translated as $n \cdot 57 = 23$, so choice (d) is correct.

3. "23 is 57% of what number?" can be translated as $23 = 0.57y$, so choice (e) is correct.

5. "57 is what percent of 23?" can be translated as $n \cdot 23 = 57$, so choice (c) is correct.

7. "What is 23% of 57?" can be translated as $a = (0.23)57$, so choice (f) is correct.

9. "23% of what number is 57?" can be translated as $57 = 0.23y$, so choice (b) is correct.

11. $49\% = 49.0\%$

49% 0.49.0

Move the decimal point 2 places to the left.

$49\% = 0.49$

13. $1\% = 1.0\%$

1% 0.01.0

Move the decimal point 2 places to the left.

$1\% = 0.01$

15. $4.1\% = 4.10\%$

4.1% 0.04.10

Move the decimal point 2 places to the left.

$4.1\% = 0.041$

17. $20\% = 20.0\%$

20% 0.20.0

Move the decimal point 2 places to the left.

$20\% = 0.20$, or 0.2

19. 62.5% 0.62.5

Move the decimal point 2 places to the left.

$62.5\% = 0.625$

21. 0.2% 0.00.2

Move the decimal point 2 places to the left.

$0.2\% = 0.002$

23. $175\% = 175.0\%$ 1.75.0

Move the decimal point 2 places to the left.

$175\% = 1.75$

25. 0.38

First move the decimal point two places to the right; 0.38.

then write a % symbol: 38%

27. 0.039

First move the decimal point two places to the right; 0.03.9

then write a % symbol: 3.9%

29. 0.45

First move the decimal point two places to the right; 0.45.

then write a % symbol: 45%

31. 0.7

First move the decimal point two places to the right; 0.70.

then write a % symbol: 70%

33. 0.0009

First move the decimal point two places to the right; 0.00. 09

then write a % symbol: 0.09%

35. 1.06

First move the decimal point two places to the right; 1.06.

then write a % symbol: 106%

37. 1.8

First move the decimal point two places to the right; 1.80.

then write a % symbol: 180%

39. $\frac{3}{5}$ $\left(\text{Note: } \frac{3}{5} = 0.6\right)$

Move the decimal point two places to the right; 0.60.

then write a % symbol: 60%

41. $\frac{8}{25}$ $\left(\text{Note: } \frac{8}{25} = 0.32\right)$

First move the decimal point two places to the right; 0.32.

then write a % symbol: 32%

43. ***Translate.***

$$\underbrace{\text{What percent}}_{y} \text{ of } 76 \text{ is } 19?$$

$$y \cdot 76 = 19$$

We solve the equation and then convert to percent notation.

$$y \cdot 76 = 19$$
$$y = \frac{19}{76}$$
$$y = 0.25 = 25\%$$

The answer is 25%.

45. ***Translate.***

$$\underbrace{\text{What percent}}_{y} \text{ of } 150 \text{ is } 39?$$

$$y \cdot 150 = 39$$

We solve the equation and then convert to percent notation.

$$y \cdot 150 = 39$$
$$y = \frac{39}{150}$$
$$y = 0.26 = 26\%$$

The answer is 26%.

47. ***Translate.***

$$14 \text{ is } 30\% \text{ of } \underbrace{\text{what number}}_{y}?$$

$$14 = 30\% \cdot y$$

We solve the equation.

$$14 = 0.3y \qquad (30\% = 0.3)$$
$$\frac{14}{0.3} = y$$
$$46.\overline{6} = y$$

The answer is $46.\overline{6}$, or $46\frac{2}{3}$, or $\frac{140}{3}$.

49. ***Translate.***

$$0.3 \text{ is } 12\% \text{ of } \underbrace{\text{what number}}_{y}?$$

$$0.3 = 12\% \cdot y$$

We solve the equation.

$$0.3 = 0.12y \qquad (12\% = 0.12)$$
$$\frac{0.3}{0.12} = y$$
$$2.5 = y$$

The answer is 2.5.

51. ***Translate.***

What number is 1% of one million?

$$y = 1\% \cdot 1{,}000{,}000$$

We solve the equation.

$y = 0.01 \cdot 1{,}000{,}000 \quad (1\% = 0.01)$

$y = 10{,}000 \quad$ Multiplying

The answer is 10,000.

53. ***Translate.***

What percent of 60 is 75?

$$y \cdot 60 = 75$$

We solve the equation and then convert to percent notation.

$y \cdot 60 = 75$

$y = \frac{75}{60}$

$y = 1.25 = 125\%$

The answer is 125%.

55. ***Translate.***

What is 2% of 40?

$$x = 2\% \cdot 40$$

We solve the equation.

$x = 0.02 \cdot 40 \quad (2\% = 0.02)$

$x = 0.8 \quad$ Multiplying

The answer is 0.8.

57. Observe that 25 is half of 50. Thus, the answer is 0.5, or 50%. We could also do this exercise by translating to an equation.

Translate.

25 is what percent of 50?

$$25 = y \cdot 50$$

We solve the equation and convert to percent notation.

$25 = y \cdot 50$

$\frac{25}{50} = y$

$0.5 = y$, or $50\% = y$

The answer is 50%.

59. ***Translate.***

What percent of 69 is \$23?

$$y \cdot 69 = 23$$

We solve the equation and convert to percent notation.

$$y \cdot 69 = 23y = \frac{23}{69}y = 0.33\overline{3} = 33.\overline{3}\% \text{ or } 33\frac{1}{3}\%$$

The answer is $33.\overline{3}\%$ or 33 1/3%.

61. First we reword and translate, letting c represent Americans who commute to work, in millions.

What is 5% of 57?

$$c = 0.05 \cdot 57$$

$c = 0.05 \cdot 57 = 2.85$

There are 2.85 million Americans who bicycle to commute to school or work.

63. First we reword and translate, letting h represent Americans who bicycle to exercise for health.

What is 41% of 57%

$$h = 0.41 \cdot 57$$

$h = 0.41 \cdot 57 = 23.37$

There are 23.37 million Americans who bicycle to exercise for health.

65. First we reword and translate, letting c represent the number of credits Cody has completed.

What is 60% of 125?

$$c = 0.6 \cdot 125$$

$c = 0.6 \cdot 125 = 75$

Cody has completed 75 credits.

67. First we reword and translate, letting b represent the number of at-bats.

216 is 36.3% of what number?

$$216 = 0.363 \cdot b$$

$\frac{216}{0.363} = b$

$595 \approx b$

Magglio Ordonez had 595 at-bats.

69. a) First we reword and translate, letting p represent the unknown percent.

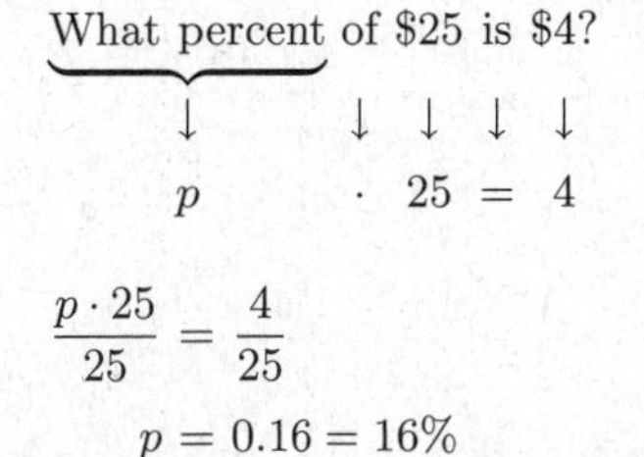

What percent of \$25 is \$4?

$$p \cdot 25 = 4$$

$\frac{p \cdot 25}{25} = \frac{4}{25}$

$p = 0.16 = 16\%$

The tip was 16% of the cost of the meal.

b) We add to find the total cost of the meal, including tip:

$\$25 + \$4 = \$29$

71. To find the percent of crude oil came from Canada and Mexico, we first reword and translate, letting p represent the unknown percent.

3.4 million is what percent of 10.2 million?

$$3.4 = p \cdot 10.2$$

$\frac{3.4}{10.2} = p$

$0.33\overline{3} = p$

$33.\overline{3}\% = p$ or 33 1/3%

About $33.\overline{3}\%$ or 33 1/3% of crude oil came from Canada and Mexico.

To find the percent of crude oil that came from the rest of the world, we subtract:

$$100\% - 33\ 1/3\% = 66\ 2/3\% \text{ or } 66.\overline{6}\%.$$

About 66 2/3% or $66.\overline{6}\%$ of crude oil came from the rest of the world.

73. Let I = the amount of interest Glenn will pay. Then we have:

I is 7% of \$2400.

$\downarrow \downarrow \quad \downarrow \quad \downarrow \quad \downarrow$

$I = 0.07 \cdot \$2400$

$I = \$168$

Glenn will pay \$168 interest.

75. If n = the number of women who had babies in good or excellent health, we have:

n is 95% of 300.

$\downarrow \downarrow \quad \downarrow \quad \downarrow \quad \downarrow$

$n = 0.95 \cdot 300$

$n = 285$

285 women had babies in good or excellent health.

77. A self-employed person must earn 120% as much as a non-self-employed person. Let a = the amount Tia would need to earn, in dollars per hour, on her own for a comparable income.

a is 120% of \$16.

$\downarrow \downarrow \quad \downarrow \quad \downarrow \quad \downarrow$

$a = 1.2 \cdot 16$

$a = 19.20$

Tia would need to earn \$19.20 per hour on her own.

79. We reword and translate.

$\underbrace{\text{What percent}}$ of 103 is 45?

$\downarrow \quad \downarrow \downarrow \downarrow \downarrow$

$p \cdot 103 = 45$

$p \cdot 103 = 45$

$p \approx 0.437 = 43.7\%$

The actual cost exceeds initial estimate by about 43.7%.

81. When the sales tax is 6%, the total amount paid is 106% of the cost of the merchandise. Let c = the cost of the merchandise. Then we have:

\$47.70 is 106% of c.

$\downarrow \quad \downarrow \quad \downarrow \quad \downarrow \downarrow$

$47.70 = 1.06 \cdot c$

$\frac{47.70}{1.06} = c$

$45 = c$

The price of the merchandise was \$45.

83. When the sales tax is 6%, the total amount paid is 106% of the cost of the merchandise. Let c = the amount the school group owes, or the cost of the software without tax. Then we have:

\$157.41 is 106% of c.

$\downarrow \quad \downarrow \quad \downarrow \quad \downarrow \downarrow$

$157.41 = 1.06 \cdot c$

$\frac{157.41}{1.06} = c$

$148.5 = c$

The school group owes \$148.50.

85. First we reword and translate.

What is 16.5% of 191?

$\downarrow \quad \downarrow \quad \downarrow \quad \downarrow \quad \downarrow$

$a = 0.165 \cdot 191$

Solve. We convert 16.5% to decimal notation and multiply.

$a = 0.165 \cdot 191$

$a = 31.515 \approx 31.5$

About 31.5 lb of the author's body weight is fat.

87. Let m = the number of mailed ads that led to a sale or response from customers. Then we have:

m is 2.15% of 114.

$\downarrow \downarrow \quad \downarrow \quad \downarrow \quad \downarrow$

$m = 0.0215 \cdot 114$

$m \approx 2.45$

About 2.45 billion pieces of mail led to a response.

89. The number of calories in a serving of Light Style Bread is 85% of the number of calories in a serving of regular bread. Let c = the number of calories in a serving of regular bread. Then we have:

$\underbrace{\text{140 calories}}$ is 85% of c.

$\downarrow \quad \downarrow \downarrow \downarrow \downarrow$

$140 = 0.85 \cdot c$

$\frac{140}{0.85} = c$

$165 \approx c$

There are about 165 calories in a serving of regular bread.

91. *Writing Exercise.* The book is marked up \$30. Since Campus Bookbuyers paid \$30 for the book, this is a 100% markup.

93. Let l represent represent the length and w the width. Then twice the length plus twice the width is $2l + 2w$.

95. Let p represent the number of points Tino scored. Then $p - 5$ is five fewer than p.

97. Half of a is $\frac{1}{2}a$. So the product of 10 and half of a is $10\left(\frac{1}{2}a\right)$.

99. Let l represent the length and w the width. Then, the width is 2 in. less than the length which is $w = l - 2$.

101. (a) In the survey report, 40% of all sick days on Monday or Friday sounds excessive. However, for a traditional 5-day business week, 40% is the same as $\frac{2}{5}$. That is, just 2 days out of 5.

(b)In the FBI statistics, 26% of home burglaries occurring between Memorial Day and Labor Day sounds excessive. However, 26% of a 365-day year is 73 days, For the months of June, July, and August there are at least 90 days. So 26% is less than one home burglary per day.

103. Let $p =$ the population of Bardville. Then we have:

$$\underbrace{1332}\ \text{is}\ 15\%\ \text{of}\ 48\%\ \text{of}\ \underbrace{\text{the population.}}$$

$$1332 = 0.15 \cdot 0.48 \cdot p$$

$$\frac{1332}{0.15(0.48)} = p$$

$$18,500 = p$$

The population of Bardville is 18,500.

105. Since 6 ft $= 6 \times 1$ ft $= 6 \times 12$ in. $= 72$ in., we can express 6 ft 4 in. as 72 in.+4 in., or 76 in. We reword and translate. Let $a =$ Jaraan's final adult height.

$$\underbrace{\text{76in.}}\ \text{is}\ 96.1\%\ \text{of}\ \underbrace{\text{adult height}}$$

$$76 = 0.961 \cdot a$$

$$\frac{76}{0.961} = a$$

$$79 \approx a$$

Note that 79 in. $= 72$ in. $+ 7$ in. $= 6$ ft 7 in.

Jaraan's final adult height will be about 6 ft 7 in.

107. Using the formula for the area A of a rectangle with length l and width w, $A = l \cdot w$, we first find the area of the photo.

$$A = 8\text{ in.} \times 6\text{ in.} = 48\text{ in}^2$$

Next we find the area of the photo that will be visible using a mat intended for a 5-in. by 7-in. photo.

$$A = 7\text{ in.} \times 5\text{ in.} = 35\text{ in}^2$$

Then the area of the photo that will be hidden by the mat is $48\text{ in}^2 - 35\text{ in}^2$, or 13 in^2.

We find what percentage of the area of the photo this represents.

$$\underbrace{\text{What percent}}\ \text{of}\ \underbrace{48\text{ in}^2}\ \text{is}\ \underbrace{13\text{ in}^2}?$$

$$p \cdot 48 = 13$$

$$\frac{p \cdot 48}{48} = \frac{13}{48}$$

$$p \approx 0.27$$

$$p \approx 27\%$$

The mat will hide about 27% of the photo.

109. *Writing Exercise.* Suppose Jorge has x dollars of taxable income. If he makes a \$50 tax-deductible contribution, then he pays tax of $0.3(x - \$50)$, or $0.3x - \$15$ and his assets are reduced by $0.3x - \$15 + \$50$, or $0.3x + \$35$. If he makes a \$40 non-tax-deductible contribution, he pays tax of $0.3x$ and his assets are reduced by $0.3x + \$40$. Thus, it costs him less to make a \$50 tax-deductible contribution.

Exercise Set 2.5

1. ***Familiarize.*** Let $n =$ the number. Then three less than two times the number is $2n - 3$.

Translate.

$$\underbrace{\text{Three less than twice a number}}\ \text{is}\ 19.$$

$$2n - 3 = 19$$

Carry out. We solve the equation.

$$2n - 3 = 19$$

$$2n = 22 \quad \text{Adding 3}$$

$$n = 11 \quad \text{Dividing by 2}$$

Check. Twice 11 is 22 and three fewer than 19. The answer checks.

State. The number is 11.

3. ***Familiarize.*** Let $a =$ the number. Then "five times the sum of 3 and twice some number" translates to $5(2a + 3)$.

Translate.

$$\underbrace{\text{Five times the sum of 3 and twice some number}}\ \text{is}\ 70.$$

$$5(2a + 3) = 70$$

Carry out. We solve the equation.

$$5(2a + 3) = 70$$

$$10a + 15 = 70 \quad \text{Using the distributive law}$$

$$10a = 55 \quad \text{Subtracting 15}$$

$$a = \tfrac{11}{2} \quad \text{Dividing by 10}$$

Check. The sum of $2 \cdot \frac{11}{2}$ and 3 is 14, and $5 \cdot 14 = 70$. The answer checks.

State. The number is $\frac{11}{2}$.

5. ***Familiarize.*** Let $p =$ the regular price of the iPod. At 20% off, Kyle paid $(100-20)\%$, or 80% of the regular price.

Translate.

$$\$120\ \text{is}\ 80\%\ \text{of}\ \underbrace{\text{the regular price.}}$$

$$120 = 0.80 \cdot p$$

Carry out. We solve the equation.

$$120 = 0.80p$$

$$150 = p$$

Check. 80% of \$150, or 0.80(\$150), is \$120. The answer checks.

State. The regular price was \$150.

7. ***Familiarize.*** Let $c =$ the price of the graphing calculator itself. When the sales tax rate is 6%, the tax paid on the calculator is 6% of c, or $0.06c$.

Translate.

$$\underbrace{\text{Price of calculator}}\ \text{plus}\ \underbrace{\text{sales tax}}\ \text{is}\ \$137.80.$$

$$c + 0.06c = 137.80$$

Carry out. We solve the equation.

$$c + 0.06c = 137.80$$
$$1.06c = 137.80$$
$$c = 130$$

Check. 6% of \$130, or 0.06(\$130), is \$7.80 and \$7.80 + \$130 is \$137.80, the total cost. The answer checks.

State. The graphing calculator itself cost \$130.

9. ***Familiarize***. Let $d =$ Looi's distance, in miles, from the start after 8 hr. Then the distance from the finish line is $2d$.

Translate.

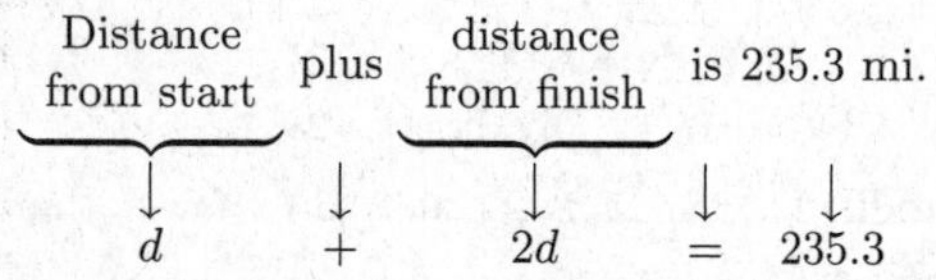

Carry out. We solve the equation.

$$d + 2d = 235.3$$
$$3d = 235.3$$
$$d \approx 78.4$$

Check. If Looi is 78.4 mi from the start, then he is $2 \cdot (78.4)$, or 156.8 mi from the finish. Since 78.4 + 156.8 is approximately 235.3, the total distance, the answer checks.

State. Looi had traveled approximately 78.4 mi.

11. ***Familiarize***.Let $d =$ the distance, in miles, that Danica had traveled to the given point after the start. Then the distance from the finish line was $300d$ miles.

Translate.

Distance to finish | plus | 20 mi more | was | distance to start.
$$300 - d \quad + \quad 20 \quad = \quad d$$

Carry out. We solve the equation.

$$300 - d + 20 = d$$
$$320 - d = d$$
$$320 = 2d$$
$$160 = d$$

Check. If Danica was 160 mi from the start, she was 300160, or 140 mi from the finish. Since 160 is 20 more than 140, the answer checks.

State. Danica had traveled 160 mi at the given point.

13. ***Familiarize***. Let $n =$ the number of Erica's apartment. Then $n+1 =$ the number of her next-door neighbor's apartment.

Translate.

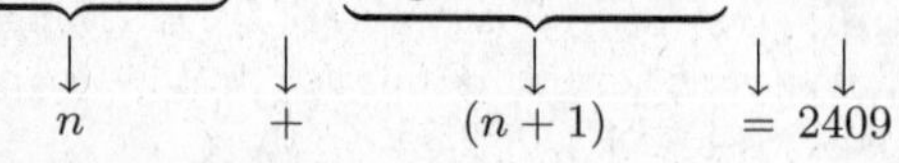

Carry out. We solve the equation.

$$n + (n + 1) = 2409$$
$$2n + 1 = 2409$$
$$2n = 2408$$
$$n = 1204$$

If Erica's apartment number is 1204, then her next-door neighbor's number is 1204 + 1, or 1205.

Check. 1204 and 1205 are consecutive numbers whose sum is 2409. The answer checks.

State. The apartment numbers are 1204 and 1205.

15. ***Familiarize***. Let $n =$ the smaller house number. Then $n + 2 =$ the larger number.

Translate.

Smaller number plus larger number is 572.
$$n \quad + \quad (n + 2) \quad = 572$$

Carry out. We solve the equation.

$$n + (n + 2) = 572$$
$$2n + 2 = 572$$
$$2n = 570$$
$$n = 285$$

If the smaller number is 285, then the larger number is 285 + 2, or 287.

Check. 285 and 287 are consecutive odd numbers and 285 + 287 = 572. The answer checks.

State. The house numbers are 285 and 287.

17. ***Familiarize***. Let $x =$ the first page number. Then $x + 1 =$ the second page number, and $x + 2 =$ the third page number.

Translate.

The sum of three consecutive page numbers is 99.
$$x + (x + 1) + (x + 2) \quad = 99$$

Carry out. We solve the equation.

$$x + (x + 1) + (x + 2) = 99$$
$$3x + 3 = 99$$
$$3x = 96$$
$$x = 32$$

If x is 32, then $x + 1$ is 33 and $x + 2 = 34$.

Check. 32, 33, and 34 are consecutive integers, and 32 + 33 + 34 = 99. The result checks.

State. The page numbers are 32, 33, and 34.

19. ***Familiarize***. Let $m =$ the man's age. Then $m - 2 =$ the woman's age.

Translate.

Man's age plus Woman's age is 204.
$$m \quad + \quad (m - 2) \quad = 204$$

Carry out. We solve the equation.

$$m + (m - 2) = 204$$
$$2m - 2 = 204$$
$$2m = 206$$
$$m = 103$$

If m is 103, then $m - 2$ is 101.

Check. 103 is 2 more than 101, and $103 + 101 = 204$. The answer checks.

State. The man was 103 yr old, and the woman was 101 yr old.

21. ***Familiarize***. Familiarize. Let m = the number non-spam messages, in billions. Then $4m$ is the number of spam messages.

Translate.

$$\underbrace{\text{spam}}_{\downarrow\ 4m}\ \underset{\downarrow\ +}{\text{plus}}\ \underbrace{\text{non-spam}}_{\downarrow\ m}\ \underset{\downarrow\ =}{\text{is}}\ \underset{\downarrow\ 125}{125}$$

Carry out. We solve the equation.

$$4m + m = 125$$
$$5m = 125$$
$$m = 25$$

If m is 25, then $4m$ is 100.

Check. 100 is four times 25, and $25 + 100 = 125$. The answer checks.

State.There were 100 billion spam messages and 25 billion non-spam messages sent each day in 2006.

23. ***Familiarize.*** The page numbers are consecutive integers. If we let x = the smaller number, then $x + 1$ = the larger number.

Translate. We reword the problem.

$$\underbrace{\text{First integer}}_{\downarrow\ x}\ \underset{\downarrow\ +}{+}\ \underbrace{\text{Second integer}}_{\downarrow\ (x+1)}\ \underset{\downarrow\ =}{=}\ \underset{\downarrow\ 281}{281}$$

Carry out. We solve the equation.

$$x + (x + 1) = 281$$
$$2x + 1 = 281 \quad \text{Combining like terms}$$
$$2x = 280 \quad \text{Adding } -1 \text{ on both sides}$$
$$x = 140 \quad \text{Dividing on both sides by 2}$$

Check. If $x = 140$, then $x + 1 = 141$. These are consecutive integers, and $140 + 141 = 281$. The answer checks.

State. The page numbers are 140 and 141.

25. ***Familiarize***. We draw a picture. Let w = the width of the rectangle, in feet. Then $w + 60$ = the length.

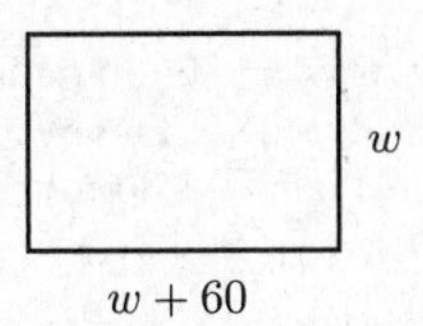

The perimeter is twice the length plus twice the width, and the area is the product of the length and the width.

Translate.

$$\underbrace{\text{Twice the length}}_{\downarrow\ 2(w+60)}\ \underset{\downarrow\ +}{\text{plus}}\ \underbrace{\text{twice the width}}_{\downarrow\ 2w}\ \underset{\downarrow\ =}{\text{is}}\ \underbrace{520\text{ ft}}_{\downarrow\ 520}.$$

Carry out. We solve the equation.

$$2(w + 60) + 2w = 520$$
$$2w + 120 + 2w = 520$$
$$4w + 120 = 520$$
$$4w = 400$$
$$w = 100$$

Then $w + 60 = 100 + 60 = 160$, and the area is $160 \text{ ft} \cdot 100 \text{ ft} = 16,000 \text{ ft}^2$.

Check. The length, 160 ft, is 60 ft more than the width, 100 ft. The perimeter is $2 \cdot 160 \text{ ft} + 2 \cdot 100 \text{ ft}$, or $320 \text{ ft} + 200 \text{ ft}$, or 520 ft. We can check the area by doing the calculation again. The answer checks.

State. The length is 160 ft, the width is 100 ft, and the area is 16,000 ft^2.

27. ***Familiarize***. Let w = the width, in meters. Then $w + 4$ is the length. The perimeter is twice the length plus twice the width.

Translate.

$$\underbrace{\text{Twice the width}}_{\downarrow\ 2w}\ \underset{\downarrow\ +}{\text{plus}}\ \underbrace{\text{twice the length}}_{\downarrow\ 2(w+4)}\ \underset{\downarrow\ =}{\text{is}}\ \underset{\downarrow\ 92}{92}.$$

Carry out. We solve the equation.

$$2w + 2(w + 4) = 92$$
$$2w + 2w + 8 = 92$$
$$4w = 84$$
$$w = 21$$

Then $w + 4 = 21 + 4 = 25$.

Check. The length, 25 m is 4 more than the width, 21 m. The perimeter is $2 \cdot 21 \text{ m} + 2 \cdot 25 \text{ m} = 42 \text{ m} + 50 \text{ m} = 92 \text{ m}$. The answer checks.

State. The length of the garden is 25 m and the width is 21 m.

29. ***Familiarize***. Let w = the width, in inches. Then $2w$ = the length. The perimeter is twice the length plus twice the width. We express $10\frac{1}{2}$ as 10.5.

Translate.

$$\underbrace{\text{Twice the length}}_{\downarrow\ 2\cdot 2w}\ \underset{\downarrow\ +}{\text{plus}}\ \underbrace{\text{twice the width}}_{\downarrow\ 2w}\ \underset{\downarrow\ =}{\text{is}}\ \underbrace{10.5\text{ in.}}_{\downarrow\ 10.5}$$

Carry out. We solve the equation.

$$2 \cdot 2w + 2w = 10.5$$
$$4w + 2w = 10.5$$
$$6w = 10.5$$
$$w = 1.75, \text{ or } 1\frac{3}{4}$$

Then $2w = 2(1.75) = 3.5$, or $3\frac{1}{2}$.

Check. The length, $3\frac{1}{2}$ in., is twice the width, $1\frac{3}{4}$ in. The perimeter is $2\left(3\frac{1}{2} \text{ in.}\right) + 2\left(1\frac{3}{4} \text{ in.}\right) =$

7 in. $+ 3\frac{1}{2}$ in. $= 10\frac{1}{2}$ in. The answer checks.

State. The actual dimensions are $3\frac{1}{2}$ in. by $1\frac{3}{4}$ in.

31. ***Familiarize.*** We draw a picture. We let $x =$ the measure of the first angle. Then $3x =$ the measure of the second angle, and $x + 30 =$ the measure of the third angle.

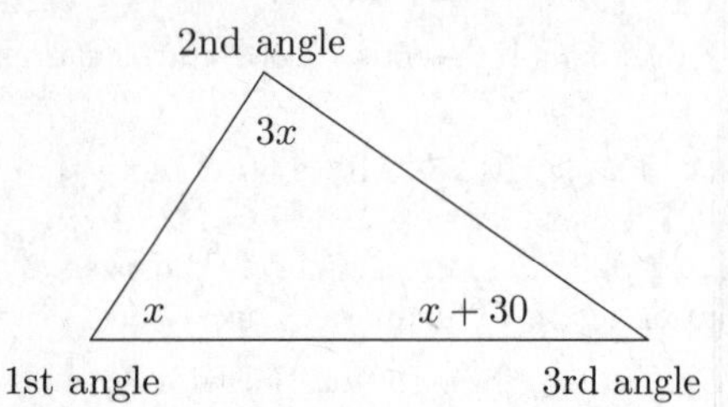

Recall that the measures of the angles of any triangle add up to 180°.

Translate.

Measure of first angle + measure of second angle + measure of third angle is 180°.

$$x + 3x + (x + 30) = 180$$

Carry out. We solve the equation.

$$x + 3x + (x + 30) = 180$$
$$5x + 30 = 180$$
$$5x = 150$$
$$x = 30$$

Possible answers for the angle measures are as follows:

First angle: $x = 30°$

Second angle: $3x = 3(30)° = 90°$

Third angle: $x + 30° = 30° + 30° = 60°$

Check. Consider 30°, 90°, and 60°. The second angle is three times the first, and the third is 30° more than the first. The sum of the measures of the angles is 180°. These numbers check.

State. The measure of the first angle is 30°, the measure of the second angle is 90°, and the measure of the third angle is 60°.

33. ***Familiarize.*** Let $x =$ the measure of the first angle. Then $4x =$ the measure of the second angle, and $x + 4x + 5 = 5x + 5$ is the measure of the third angle.

Translate.

Measure of first angle + measure of second angle + measure of third angle is 180°.

$$x + 4x + (5x + 5) = 180$$

Carry out. We solve the equation.

$$x + 4x + (5x + 5) = 180$$
$$10x + 5 = 180$$
$$10x = 175$$
$$x = 17.5$$

If $x = 17.5$, then $4x = 4(17.5) = 70$, and $5x+5 = 5(17.5)+5 = 87.5 + 5 = 92.5$.

Check. Consider 17.5°, 70°, and 92.5°. The second is four times the first, and the third is 5°more than the sum of the other two. The sum of the measures of the angles is 180°. These numbers check.

State. The measure of the second angle is 70°.

35. ***Familiarize.*** Let $b =$ the length of the bottom section of the rocket, in feet. Then $\frac{1}{6}b =$ the length of the top section, and $\frac{1}{2}b =$ the length of the middle section.

Translate.

Length of top section + length of middle section + length of bottom section is 240 ft.

$$\frac{1}{6}b + \frac{1}{2}b + b = 240$$

Carry out. We solve the equation. First we multiply by 6 on both sides to clear the fractions.

$$\frac{1}{6}b + \frac{1}{2}b + b = 240$$
$$6\left(\frac{1}{6}b + \frac{1}{2}b + b\right) = 6 \cdot 240$$
$$6 \cdot \frac{1}{6}b + 6 \cdot \frac{1}{2}b + 6 \cdot b = 1440$$
$$b + 3b + 6b = 1440$$
$$10b = 1440$$
$$b = 144$$

Then $\frac{1}{6}b = \frac{1}{6} \cdot 144 = 24$ and $\frac{1}{2}b = \frac{1}{2} \cdot 144 = 72$.

Check. 24 ft is $\frac{1}{6}$ of 144 ft, and 72 ft is $\frac{1}{2}$ of 144 ft. The sum of the lengths of the sections is
24 ft + 72 ft + 144 ft = 240 ft. The answer checks.

State. The length of the top section is 24 ft, the length of the middle section is 72 ft, and the length of the bottom section is 144 ft.

37. ***Familiarize.*** Let $m =$ the number of miles that can be traveled on a \$18 budget. Then the total cost of the taxi ride, in dollars, is $2.250 + 1.80m$, or $2.25 + 1.8m$.

Translate.

Cost of taxi ride is \$18.

$$2.25 + 1.8m = 18$$

Carry out. We solve the equation.

$$2.25 + 1.8m = 18$$
$$1.8m = 15.75$$
$$m = \frac{15.75}{1.8} = 8.75 = 8\frac{3}{4}$$

Check. The mileage charge is \$1.80(8.75), or \$15.75, and the total cost of the ride is \$2.25 + \$15.75 = \$18. The answer checks.

State. Debbie can travel 8.75 mi on her budget.

39. ***Familiarize***. The total cost is the daily charge plus the mileage charge. Let d = the distance that can be traveled, in miles, in one day for \$100. The mileage charge is the cost per mile times the number of miles traveled, or $0.39d$.

Translate.

Daily rate plus mileage charge is \$100.

$$\underbrace{49.95}_{\text{Daily rate}} \; + \; \underbrace{0.39d}_{\text{mileage charge}} \; = \; 100$$

Carry out. We solve the equation.

$$49.95 + 0.39d = 100$$
$$0.39d = 50.05$$
$$d = 128.\overline{3}, or 128\frac{1}{3}$$

Check. For a trip of $128\frac{1}{3}$ mi, the mileage charge is $\$0.39\left(128\frac{1}{3}\right)$, or \$50.05, and \$49.95 + \$50.05 = \$100. The answer checks.

State. Concert Productions can travel $128\frac{1}{3}$ mi in one day and stay within their budget.

41. ***Familiarize***. Let x = the measure of one angle. Then $90 - x$ = the measure of its complement.

Translate.

Measure of one angle is 15° more than twice the measure of its complement.

$$x = 15 + 2(90 - x)$$

Carry out. We solve the equation.

$$x = 15 + 2(90 - x)$$
$$x = 15 + 180 - 2x$$
$$x = 195 - 2x$$
$$3x = 195$$
$$x = 65$$

If x is 65, then $90 - x$ is 25.

Check. The sum of the angle measures is 90°. Also, 65° is 15° more than twice its complement, 25°. The answer checks.

State. The angle measures are 65° and 25°.

43. Familiarize. Let x = the measure of one angle. Then $180 - x$ = the measure of its supplement.

Translate.

Measure of one angle is $3\frac{1}{2}$ times measure of second angle.

$$x = 3\tfrac{1}{2} \cdot (180 - x)$$

Carry out. We solve the equation.

$$x = 3\tfrac{1}{2}(180 - x)$$
$$x = 630 - 3.5x$$
$$4.5x = 630$$
$$x = 140$$

If $x = 140$, then $180 - 140 = 40°$.

Check. The sum of the angles is 180°. Also 140°is three and a half times 40°. The answer checks.

State. The angles are 40°and 140°.

45. ***Familiarize***. Let l = the length of the paper, in cm. Then $l - 6.3$ = the width. The perimeter is twice the length plus twice the width.

Translate.

Twice the length plus twice the width is 99 cm.

$$2l + 2(l - 6.3) = 99$$

Carry out. We solve the equation.

$$2l + 2(l - 6.3) = 99$$
$$2l + 2l - 12.6 = 99$$
$$4l - 12.6 = 99$$
$$4l = 111.6$$
$$l = 27.9$$

Then $l - 6.3 = 27.9 - 6.3 = 21.6$.

Check. The width, 21.6 cm, is 6.3 cm less than the length, 27.9 cm. The perimeter is 2(27.9 cm) + 2(21.6 cm) = 55.8 cm + 43.2 cm = 99 cm. The answer checks.

State. The length of the paper is 27.9 cm, and the width is 21.6 cm.

47. ***Familiarize***. Let a = the amount Janeka invested. Then the simple interest for one year is $6\% \cdot a$, or $0.06a$.

Translate.

Amount invested plus interest is \$6996.

$$a + 0.06a = 6996$$

Carry out. We solve the equation.

$$a + 0.06a = 6996$$
$$1.06a = 6996$$
$$a = 6600$$

Check. An investment of \$6600 at 6% simple interest earns 0.06(\$6600), or \$396, in one year. Since \$6600 + \$396 = \$6996, the answer checks.

State. Janeka invested \$6600.

49. ***Familiarize***. Let w = the winning score. Then $w - 340$ = the losing score.

Translate.

Winning score plus losing score was 1320 points.

$$w + w - 340 = 1320$$

Carry out. We solve the equation.

$$w + (w - 340) = 1320$$
$$2w - 340 = 1320$$
$$2w = 1660$$
$$w = 830$$

Then $w - 340 = 830 - 340 = 490$.

Check. The winning score, 830, is 340 points more than the losing score, 490. The total of the two scores is $830 + 490 = 1320$ points. The answer checks.

State. The winning score was 830 points.

51. ***Familiarize.*** Let $a =$ the selling price of the house. Then the commission on the selling price is 6% times a, or $0.06a$.

Translate.

Selling price minus commission is \$ 117,500.

$$a \quad - \quad 0.06a \quad = \quad 117,500$$

Carry out. We solve the equation.

$$a - 0.06a = 117,500$$
$$0.94a = 117,500$$
$$a = 125,000$$

Check. A selling price of \$125,000 gives a commission of \$7500. Since $\$125,000 - \$7500 = \$117,500$, the answer checks.

State. They must sell the house for \$125,000.

53. ***Familiarize.*** We will use the equation

$$T = \frac{1}{4}N + 40.$$

Translate. We substitute 80 for T.

$$80 = \frac{1}{4}N + 40$$

Carry out. We solve the equation.

$$80 = \frac{1}{4}N + 40$$
$$40 = \frac{1}{4}N$$
$$160 = N \quad \text{Multiplying by 4 on both sides}$$

Check. When $N = 160$, we have $T = \frac{1}{4} \cdot 160 + 40 = 40 + 40 = 80$. The answer checks.

State. A cricket chirps 160 times per minute when the temperature is 80°F.

55. *Writing Exercise.* Although many of the problems in this section might be solved by guessing, using the five-step problem-solving process to solve them would give the student practice is using a technique that can be used to solve other problems whose answers are not so readily guessed.

57. Since -8 is to the left of 1, $-8 < 1$.

59. Since $\frac{1}{2}$ is to the right of 0, $\frac{1}{2} > 0$.

61. $-4 \leq x$

63. $y < 5$

65. *Writing Exercise.* Answers may vary.

The sum of three consecutive odd integers is 375. What are the integers?

67. ***Familiarize.*** Let $c =$ the amount the meal originally cost. The 15% tip is calculated on the original cost of the meal, so the tip is $0.15c$.

Translate.

Original cost plus tip less \$10 is \$32.55.

$$c \quad + \quad 0.15c \quad - \quad 10 \quad = \quad 32.55$$

Carry out. We solve the equation.

$$c + 0.15c - 10 = 32.55$$
$$1.15c - 10 = 32.55$$
$$1.15c = 42.55$$
$$c = 37$$

Check. If the meal originally cost \$37, the tip was 15% of \$37, or 0.15(\$37), or \$5.55. Since $\$37 + \$5.55 - \$10 = \$32.55$, the answer checks.

State. The meal originally cost \$37.

69. ***Familiarize.*** Let $s =$ one score. Then four score $= 4s$ and four score and seven $= 4s + 7$.

Translate. We reword .

1776 plus four score and seven is 1863

$$1776 \quad + \quad (4s + 7) \quad = \quad 1863$$

Carry out. We solve the equation.

$$1776 + (4s + 7) = 1863$$
$$4s + 1783 = 1863$$
$$4s = 80$$
$$s = 20$$

Check. If a score is 20 years, then four score and seven represents 87 years. Adding 87 to 1776 we get 1863. This checks.

State. A score is 20.

71. ***Familiarize.*** Let $n =$ the number of half dollars. Then the number of quarters is $2n$; the number of dimes is $2 \cdot 2n$, or $4n$; and the number of nickels is $3 \cdot 4n$, or $12n$. The total value of each type of coin, in dollars, is as follows.

Half dollars: $0.5n$

Quarters: $0.25(2n)$, or $0.5n$

Dimes: $0.1(4n)$, or $0.4n$

Nickels: $0.05(12n)$, or $0.6n$

Then the sum of these amounts is $0.5n + 0.5n + 0.4n + 0.6n$, or $2n$.

Translate.

Total amount of change is \$10.

$$2n \quad = \quad 10$$

Carry out. We solve the equation.

$$2n = 10$$
$$n = 5$$

Then $2n = 2\cdot 5 = 10$, $4n = 4\cdot 5 = 20$, and $12n = 12\cdot 5 = 60$.

Check. If there are 5 half dollars, 10 quarters, 20 dimes, and 60 nickels, then there are twice as many quarters as half dollars, twice as many dimes as quarters, and 3 times as many nickels as dimes. The total value of the coins is $\$0.5(5)+\$0.25(10)+\$0.1(20)+\$0.05(60) = \$2.50+\$2.50+\$2+\$3 = \$10$. The answer checks.

State. The shopkeeper got 5 half dollars, 10 quarters, 20 dimes, and 60 nickels.

73. Familiarize. Let p = the price before the two discounts. With the first 10% discount, the price becomes 90% of p, or $0.9p$. With the second 10% discount, the final price is 90% of $0.9p$, or $0.9(0.9p)$.

Translate.

10% discount	and	10% discount of price	is	$ 77.75.
↓	↓	↓	↓	↓
0.9	·	$0.9p$	=	77.75

Carry out. We solve the equation.

$$0.9(0.9p) = 77.75$$
$$0.81p = 77.75$$
$$p = 95.99$$

Check. Since 90% of \$95.99 is \$86.39, and 90% of \$86.39 is \$77.75, the answer checks.

State. The original price before discounts was \$95.99.

75. ***Familiarize.*** Let n = the number of DVDs purchased. Assume that two more DVDs were purchased. Then the first DVD costs \$9.99 and the total cost of the remaining $(n-1)$ DVDs is$6.99(n-1)$. The shipping and handling costs are \$3 for the first DVD, \$1.50 for the second (half of \$3), and a total of $\$1(n-2)$ for the remaining $n-2$ DVDs.

Translate.

1st DVD	plus	remaining DVDs	plus	1stS&H charges
↓	↓	↓	↓	↓
9.99	+	$6.99(n-1)$	+	3

... plus	2ndS&H charges	plus	remaining S&Hcharges	is	$45.45.
↓	↓	↓	↓	↓	↓
+	1.50	+	$1(n-2)$	=	45.45

Carry out. We solve the equation.

$$9.99 + 6.99(n-1) + 3 + 1.5 + (n-2) = 45.45$$
$$9.99 + 6.99n - 6.99 + 4.5 + n - 2 = 45.45$$
$$7.99n + 5.5 = 45.45$$
$$7.99n = 39.95$$
$$n = 5$$

Check. If there are 5 DVDs, the cost of the DVDs is $\$9.99 + \$6.99(5-1)$, or \$9.99 + \$27.96, or \$37.95. The cost for shipping and handling is $\$3 + \$1.50 + \$1(5-2) = \$7.50$. The total cost is \$37.95 + \$7.50, or \$45.45. The answer checks.

State. There were 5 DVDs in the shipment.

77. ***Familiarize.*** Let d = the distance, in miles, that Glenda traveled. At \$0.40 per $\frac{1}{5}$ mile, the mileage charge can also be given as 5(\$0.40), or \$2 per mile. Since it took 20 min to complete what is usually a 10-min drive, the taxi was stopped in traffic for 20 – 10, or 10 min.

Translate.

Initial charge	plus	$ 2 per mile	plus	stopped in traffic charge	is	$18.50.
↓	↓	↓	↓	↓	↓	↓
2.50	+	$2d$	+	0.40(10)	=	18.50

Carry out. We solve the equation.

$$2.5 + 2d + 0.4(10) = 18.5$$
$$2.5 + 2d + 4 = 18.5$$
$$2d + 6.5 = 18.5$$
$$2d = 12$$
$$d = 6$$

Check. Since \$2(6) = \$12, and \$0.40(10) = \$4, and \$12 + \$4 + \$2.50 = \$18.50, the answer checks.

State. Glenda traveled 6 mi.

79. *Writing Exercise.* If the school can invest the \$2000 so that it earns at least 7.5% and thus grows to at least \$2150 by the end of the year, the second option should be selected. If not, the first option is preferable.

81. ***Familiarize***. Let w = the width of the rectangle, in cm. Then $w + 4.25$ = the length.

Translate.

The perimeter	is	101.74 cm.
↓	↓	↓
$2(w+4.25)+2w$	=	101.74

Carry out. We solve the equation.

$$2(w+4.25) + 2w = 101.74$$
$$2w + 8.5 + 2w = 101.74$$
$$4w + 8.5 = 101.74$$
$$4w = 93.24$$
$$w = 23.31$$

Then $w + 4.25 = 23.31 + 4.25 = 27.56$.

Check. The length, 27.56 cm, is 4.25 cm more than the width, 23.31 cm. The perimeter is 2(27.56) cm + 2(23.31 cm) = 55.12 cm + 46.62 cm = 101.74 cm. The answer checks.

State. The length of the rectangle is 27.56 cm, and the width is 23.31 cm.

Exercise Set 2.6

1. $-5x \le 30$

$x \ge -6$ Dividing by -5 and reversing the inequality symbol

3. $-2t > -14$

$t < 7$ Dividing by -2 and reversing the inequality symbol

5. $x < -2$ and $-2 > x$ are equivalent.

7. If we add 1 to both sides of $-4x-1 \le 15$, we get $-4x \le 16$. The two given inequalities are equivalent.

9. $x > -4$

a) Since $4 > -4$ is true, 4 is a solution.

b) Since $-6 > -4$ is false, -6 is not a solution.

c) Since $-4 > -4$ is false, -4 is not a solution.

11. $y \le 19$

a) Since $18.99 \le 19$ is true, 18.99 is a solution.

b) Since $19.07 \le 19$ is false, 19.01 is not a solution.

c) Since $19 \le 19$ is true, 19 is a solution.

13. $c \ge -7$

a) Since $0 \ge -7$ is true, 0 is a solution.

b) Since $-5.4 \ge -7$ is true, -5.4 is a solution.

c) Since $7.1 \ge -7$ is true, 7.1 is a solution.

15. $z < -3$

a) Since $0 < -3$ is false, 0 is not a solution.

b) Since $-3\frac{1}{3} < -3$ is true, $-3\frac{1}{3}$ is a solution.

c) Since $1 < -3$ is false, 1 is not a solution.

17. The solutions of $y < 2$ are those numbers less than 2. They are shown on the graph by shading all points to the left of 2. The parenthesis at 2 indicates that 2 is not part of the graph.

$y < 2$

−5 −4 −3 −2 −1 0 1 2 3 4 5

19. The solutions of $x \ge -1$ are those numbers greater than or equal to -1. They are shown on the graph by shading all points to the right of -1. The bracket at -1 indicates that the point -1 is part of the graph.

$x \ge -1$

−5 −4 −3 −2 −1 0 1 2 3 4 5

21. The solutions of $0 \le t$, or $t \ge 0$, are those numbers greater than or equal to zero. They are shown on the graph by shading the point 0 and all points to the right of 0. The bracket at 0 indicates that 0 is part of the graph.

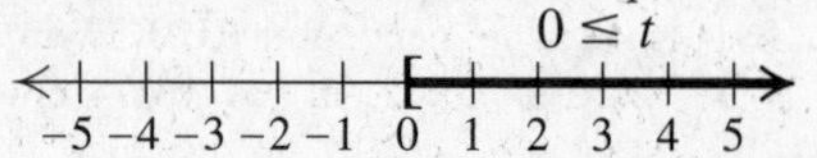

23. In order to be solution of the inequality $-5 \le x < 2$, a number must be a solution of both $-5 \le x$ and $x < 2$. The solution set is graphed as follows:

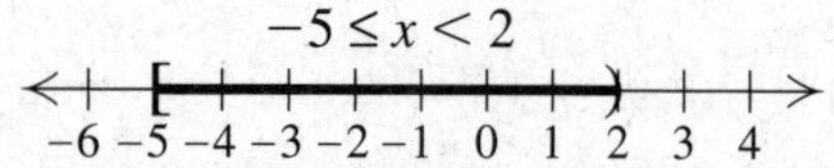

The bracket at -5 means that -5 is part of the graph. The parenthesis at 2 means that 2 is not part of the graph.

25. In order to be a solution of the inequality $-4 < x < 0$, a number must be a solution of both $-4 < x$ and $x < 0$. The solution set is graphed as follows:

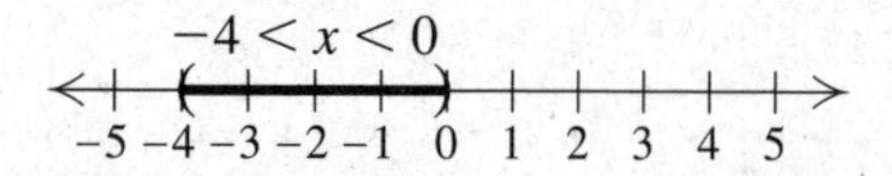

The parenthesis at -4 and 0 mean that -4 and 0 are not part of the graph.

27. $y < 6$
Graph: The solutions consist of all real numbers less than 6, so we shade all numbers to the left of 6 and use a parenthesis at 6 to indicate that it is not a solution.

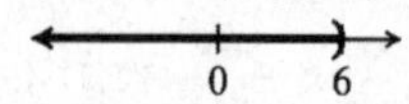

Set builder notation: $\{y|y < 6\}$
Interval notation: $(-\infty, 6)$

29. $x \ge -4$
Graph: We shade all numbers to the right of -4 and use a bracket at -4 to indicate that it is also a solution.

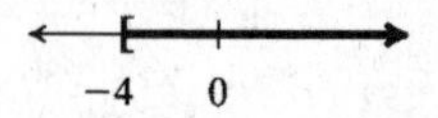

Set builder notation: $\{x|x \ge -4\}$
Interval notation: $[-4, \infty)$

31. $t > -3$
Graph: We shade all numbers to the right of -3 and use a parenthesis at -3 to indicate that it is not a solution.

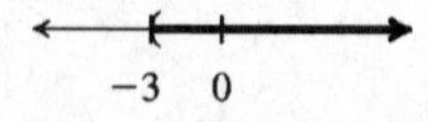

Set builder notation: $\{t|t > -3\}$
Interval notation: $(-3, \infty)$

33. $x \le -7$
Graph: We shade all numbers to the left of -7 and use a bracket at -7 to indicate that it is also a solution.

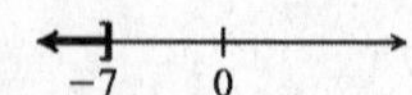

Set builder notation: $\{x|x \le -7\}$
Interval notation: $(-\infty, -7]$

35. All points to the right of -4 are shaded. The parenthesis at -4 indicates that -4 is not part of the graph. Using set-builder notation we have $\{x|x > -4\}$, or $(-4, \infty)$.

37. The point 2 and all points to the left of 2 are shaded. We have $\{x|x \leq 2\}$, or $(-\infty, 2]$.

39. All points to the left of -1 are shaded. The parenthesis at -1 indicates that -1 is not part of the graph. We have $\{x|x < -1\}$, or $(-\infty, -1)$.

41. The point 0 and all points to the right of 0 are shaded. We have $\{x|x \geq 0\}$, or $[0, \infty)$.

43.

$$\begin{aligned} y+6 &> 9 \\ y+6-6 &> 9-6 \quad \text{Adding } -6 \text{ to both sides} \\ y &> 3 \quad \text{Simplifying} \end{aligned}$$

The solution set is $\{y|y > 3\}$, or $(3, \infty)$.

45.

$$\begin{aligned} n-6 &< 11 \\ n-6+6 &< 11+6 \quad \text{Adding 6 to both sides} \\ n &< 17 \quad \text{Simplifying} \end{aligned}$$

The solution set is $\{n|n < 17\}$, or $(-\infty, 17)$.

47.

$$\begin{aligned} 2x &\leq x-9 \\ 2x-x &\leq x-9-x \\ x &\leq -9 \end{aligned}$$

The solution set is $\{x|x \leq -9\}$, or $(-\infty, -9]$.

49.

$$\begin{aligned} y+\frac{1}{3} &\leq \frac{5}{6} \\ y+\frac{1}{3}-\frac{1}{3} &\leq \frac{5}{6}-\frac{1}{3} \\ y &\leq \frac{5}{6}-\frac{2}{6} \\ y &\leq \frac{3}{6} \\ y &\leq \frac{1}{2} \end{aligned}$$

The solution set is $\left\{y\middle|y \leq \frac{1}{2}\right\}$, or $(-\infty, \frac{1}{2}]$.

51.

$$\begin{aligned} t-\frac{1}{8} &> \frac{1}{2} \\ t-\frac{1}{8}+\frac{1}{8} &> \frac{1}{2}+\frac{1}{8} \\ t &> \frac{4}{8}+\frac{1}{8} \\ t &> \frac{5}{8} \end{aligned}$$

The solution set is $\left\{t\middle|t > \frac{5}{8}\right\}$, or $(\frac{5}{8}, \infty)$. The graph is as follows:

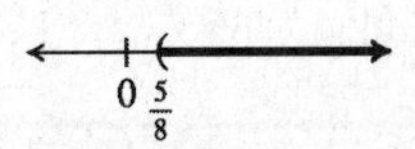

53.

$$\begin{aligned} -9x+17 &> 17-8x \\ -9x+17-17 &> 17-8x-17 \quad \text{Adding } -17 \\ -9x &> -8x \\ -9x+9x &> -8x+9x \quad \text{Adding } 9x \\ 0 &> x \end{aligned}$$

The solution set is $\{x|x < 0\}$, or $(-\infty, 0)$.

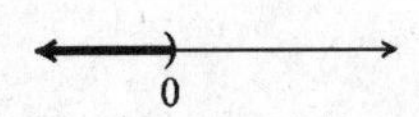

55. $-23 < -t$

The inequality states that the opposite of 23 is less than the opposite of t. Thus, t must be less than 23, so the solution set is $\{t|t < 23\}$. To solve this inequality using the addition principle, we would proceed as follows:

$$\begin{aligned} -23 &< -t \\ t-23 &< 0 \quad \text{Adding } t \text{ to both sides} \\ t &< 23 \quad \text{Adding 23 to both sides} \end{aligned}$$

The solution set is $\{t|t < 23\}$, or $(-\infty, 23)$.

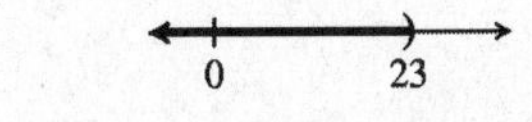

57.

$$\begin{aligned} 10-y &\leq -12 \\ -10+10-y &\leq -10-12 \quad \text{Adding -10} \\ -y &\leq -22 \end{aligned}$$

The symbol has to be reversed.

$$\begin{aligned} -1(-y) &\geq -1(-22) \\ y &\geq 22 \end{aligned}$$

The solution set is $\{y|y \geq 22\}$, or $[22, \infty)$.

59.

$$\begin{aligned} 4x &< 28 \\ \frac{1}{4}\cdot 4x &< \frac{1}{4}\cdot 28 \quad \text{Multiplying by } \frac{1}{4} \\ x &< 7 \end{aligned}$$

The solution set is $\{x|x < 7\}$, or $(-\infty, 7)$.

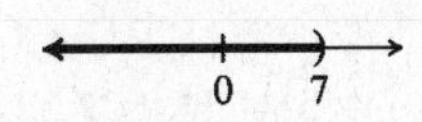

61.

$$\begin{aligned} -24 &> 8t \\ -3 &> t \end{aligned}$$

The solution set is $\{t|t < -3\}$, or $(-\infty, -3)$.

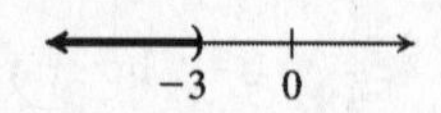

63.

$$1.8 \geq -1.2n$$

$$\frac{-1}{1.2} \cdot 1.8 \leq \frac{-1}{1.2}(-1.2n) \quad \text{Multiplying by } \frac{1}{7}$$

↑ The symbol has to be reversed.

$$-1.5 \leq n$$

The solution set is $\{n|n \geq -1.5\}$, or $[-1.5, \infty)$.

−1.5 0

65.

$$-2y \leq \frac{1}{5}$$

$$-\frac{1}{2} \cdot (-2y) \geq -\frac{1}{2} \cdot \frac{1}{5}$$

↑ The symbol has to be reversed.

$$y \geq -\frac{1}{10}$$

The solution set is $\left\{y \middle| y \geq -\frac{1}{10}\right\}$, or $[-\frac{1}{10}, \infty)$.

$-\frac{1}{10}$ 0

67.

$$-\frac{8}{5} > 2x$$

$$\frac{1}{2} \cdot -\frac{8}{5} > \frac{1}{2} \cdot 2x$$

$$-\frac{4}{5} > x$$

$$\text{or } x < -\frac{4}{5}$$

The solution set is $\left\{-\frac{4}{5} > x\right\}$, or $\left\{x < -\frac{4}{5}\right\}$, or $(-\infty, -\frac{4}{5})$.

$-\frac{4}{5}$ 0

69.

$$2 + 3x < 20$$

$$2 + 3x - 2 < 20 - 2 \quad \text{Adding } -2 \text{ to both sides}$$

$$3x < 18 \quad \text{Simplifying}$$

$$x < 6 \quad \text{Multiplying both sides by } \frac{1}{3}$$

The solution set is $\{x|x < 6\}$, or $(-\infty, 6)$.

71.

$$4t - 5 \leq 23$$

$$4t - 5 + 5 \leq 23 + 5 \quad \text{Adding 5 to both sides}$$

$$4t \leq 28$$

$$\frac{1}{4} \cdot 4t \leq \frac{1}{4} \cdot 28 \quad \text{Multiplying both sides by } \frac{1}{4}$$

$$t \leq 7$$

The solution set is $\{t|t \leq 7\}$, or $(-\infty, 7]$.

73.

$$39 > 3 - 9x$$

$$39 - 3 > 3 - 9x - 3 \quad \text{Adding } -3$$

$$36 > -9x$$

$$-\frac{1}{9} \cdot 36 < -\frac{1}{9} \cdot (-9x) \quad \text{Multiplying by } -\frac{1}{9}$$

↑ The symbol has to be reversed.

$$-4 < x$$

The solution set is $\{x|-4 < x\}$, or $\{x|x > -4\}$, or $(-4, \infty)$.

75.

$$5 - 6y > 25$$

$$-5 + 5 - 6y > -5 + 25$$

$$-6y > 20$$

$$-\frac{1}{6} \cdot (-6y) < -\frac{1}{6} \cdot 20$$

↑ The symbol has to be reversed.

$$y < -\frac{20}{6}$$

$$y < -\frac{10}{3}$$

The solution set is $\left\{y \middle| y < -\frac{10}{3}\right\}$, or $(-\infty, -\frac{10}{3})$.

77.

$$-3 < 8x + 7 - 7x$$

$$-3 < x + 7 \quad \text{Collecting like terms}$$

$$-3 - 7 < x + 7 - 7$$

$$-10 < x$$

The solution set is $\{x|x > -10\}$, or $(-10, \infty)$.

79.

$$6 - 4y > 6 - 3y$$

$$6 - 4y + 4y > 6 - 3y + 4y \quad \text{Adding } 4y$$

$$6 > 6 + y$$

$$-6 + 6 > -6 + 6 + y \quad \text{Adding } -6$$

$$0 > y, \text{ or } y < 0$$

The solution set is $\{y|0 > y\}$, or $\{y|y < 0\}$, or $(-\infty, 0)$.

81.

$$7 - 9y \leq 4 - 7y$$

$$7 - 9y + 9y \leq 4 - 7y + 9y$$

$$7 \leq 4 + 2y$$

$$-4 + 7 \leq -4 + 4 + 2y$$

$$3 \leq 2y$$

$$\frac{3}{2} \leq y, \text{or y} \geq \frac{3}{2}$$

The solution set is $\{y|y \geq \frac{3}{2}\}$, or $[\frac{3}{2}, \infty)$.

83.

$$2.1x + 43.2 > 1.2 - 8.4x$$

$$10(2.1x + 43.2) > 10(1.2 - 8.4x) \quad \text{Multiplying by 10 to clear decimals}$$

$$21x + 432 > 12 - 84x$$

$$21x + 84x > 12 - 432 \quad \text{Adding } 84x \text{ and } -432$$

$$105x > -420$$

$$x > -4 \quad \text{Multiplying by } \frac{1}{105}$$

The solution set is $\{x|x > -4\}$, or $(-4, \infty)$.

85.
$$1.7t + 8 - 1.62t < 0.4t - 0.32 + 8$$
$$0.08t + 8 < 0.4t + 7.68 \quad \text{Collecting like terms}$$
$$100(0.08t + 8) < 100(0.4t + 7.68) \quad \text{Multiplying by 100}$$
$$8t + 800 < 40t + 768$$
$$-8t - 768 + 8t + 800 < 40t + 768 - 8t - 768$$
$$32 < 32t$$
$$1 < t$$
The solution set is $\{t|t > 1\}$, or $(1, \infty)$.

87.
$$\frac{x}{3} + 4 \leq 1$$
$$3\left(\frac{x}{3} + 4\right) \leq 3 \cdot 1 \quad \text{Multiplying by 3 to to clear the fraction}$$
$$x + 12 \leq 3$$
$$x \leq -9$$
The solution set is $\{x|x \leq -9\}$, or $(-\infty, -9]$.

89.
$$3 < 5 - \frac{t}{7}$$
$$-2 < -\frac{t}{7}$$
$$-7(-2) > -7\left(-\frac{t}{7}\right)$$
$$14 > t$$
The solution set is $\{t|t < 14\}$, or $(-\infty, 14)$.

91.
$$4(2y - 3) \leq -44$$
$$8y - 12 \leq -44 \quad \text{Removing parentheses}$$
$$8y \leq -32 \quad \text{Adding 12}$$
$$y \leq -4 \quad \text{Multiplying by } \frac{1}{8}$$
The solution set is $\{y|y \leq -4\}$, or $(-\infty, -4]$.

93.
$$8(2t + 1) > 4(7t + 7)$$
$$16t + 8 > 28t + 28$$
$$-12t + 8 > 28$$
$$-12t > 20$$
$$t < -\frac{5}{3} \quad \text{Multiplying by } -\tfrac{1}{12} \text{ and reversing the symbol}$$
The solution set is $\left\{t\middle|t < -\frac{5}{3}\right\}$, or $(-\infty, -\frac{5}{3})$.

95.
$$3(r - 6) + 2 < 4(r + 2) - 21$$
$$3r - 18 + 2 < 4r + 8 - 21$$
$$3r - 16 < 4r - 13$$
$$-16 + 13 < 4r - 3r$$
$$-3 < r, \text{ or } r > -3$$
The solution set is $\{r|r > -3\}$, or $(-3, \infty)$.

97.
$$\frac{4}{5}(3x - 4) \leq 20$$
$$\frac{5}{4} \cdot \frac{4}{5}(3x + 4) \leq \frac{5}{4} \cdot 20$$
$$3x + 4 \leq 25$$
$$3x \leq 21$$
$$x \leq 7$$
The solution set is $\{x|x \leq 7\}$, or $(-\infty, 7]$.

99.
$$\frac{2}{3}\left(\frac{7}{8} - 4x\right) - \frac{5}{8} < \frac{3}{8}$$
$$\frac{2}{3}\left(\frac{7}{8} - 4x\right) < 1 \quad \text{Adding } \tfrac{5}{8}$$
$$\frac{7}{12} - \frac{8}{3}x < 1 \quad \text{Removing parentheses}$$
$$12\left(\frac{7}{12} - \frac{8}{3}x\right) < 12 \cdot 1 \quad \text{Clearing fractions}$$
$$7 - 32x < 12$$
$$-32x < 5$$
$$x > -\frac{5}{32}$$
The solution is $\left\{x\middle|x > -\frac{5}{32}\right\}$, or $(-\frac{5}{32}, \infty)$.

101. *Writing Exercise.* The inequalities $x > -3$ and $x \geq -2$ are not equivalent because they do not have the same solution set. For example, -2.5 is a solution of $x > -3$, but it is not a solution of $x \geq -2$.

103. $5x - 2(3 - 6x) = 5x - 6 + 12x = 17x - 6$

105.
$$x - 2[4y + 3(8 - x) - 1]$$
$$= x - 2[4y + 24 - 3x - 1]$$
$$= x - 2[4y - 3x + 23]$$
$$= x - 8y + 6x - 46$$
$$= 7x - 8y - 46$$

107.
$$3[5(2a - b) + 1] - 5[4 - (a - b)]$$
$$= 3[10a - 5b + 1] - 5[4 - a + b]$$
$$= 30a - 15b + 3 - 20 + 5a - 5b$$
$$= 35a - 20b - 17$$

109. *Writing Exercise.* The graph of an inequality of the form $a \leq x \leq a$ consists of just one number, a.

111. $x < x + 1$

When any real number is increased by 1, the result is greater than the original number. Thus the solution set is $\{x|x \text{ is a real number}\}$, or $(-\infty, \infty)$.

113.
$$27 - 4[2(4x - 3) + 7] \geq 2[4 - 2(3 - x)] - 3$$
$$27 - 4[8x - 6 + 7] \geq 2[4 - 6 + 2x] - 3$$
$$27 - 4[8x + 1] \geq 2[-2 + 2x] - 3$$
$$27 - 32x - 4 \geq -4 + 4x - 3$$
$$23 - 32x \geq -7 + 4x$$
$$23 + 7 = 4x + 32x$$
$$30 \geq 36x$$
$$\frac{5}{6} \geq x$$
The solution set is $\left\{x\middle|x \leq \frac{5}{6}\right\}$, or $(-\infty, \frac{5}{6}]$.

115.
$$-(x + 5) \geq 4a - 5$$
$$-x - 5 \geq 4a - 5$$
$$-x \geq 4a - 5 + 5$$
$$-x \geq 4a$$
$$-1(-x) \leq -1 \cdot 4a$$
$$x \leq -4a$$
The solution set is $\{x|x \leq -4a\}$, or $(-\infty, -4a]$.

117. $y < ax + b$ Assume $a > 0$.

$y - b < ax$

$\frac{y-b}{a} < x$ Since $a > 0$, the inequality symbol stays the same.

The solution set is $\left\{x \middle| x > \frac{y-b}{a}\right\}$, or $(\frac{y-b}{a}, \infty)$.

119. $|x| > -3$

Since absolute value is always nonnegative, the absolute value of any real number will be greater than -3. Thus, the solution set is $\{x|x \text{ is a real number}\}$.

Chapter 2 Connecting the Concepts

1. $x - 6 = 15$

$x = 21$ Adding 6 to both sides

The solution is 21.

3. $3x = -18$

$x = -6$ Dividing both sides by 3

The solution is -6.

5. $-3x > -18$

$x < 6$ Dividing both sides by-3 and reversing the direction of the inequality symbol

The solution is $\{x < 6\}$.

7. $7 - 3x = 8$

$-3x = 1$ Subtracting 7 from both sides

$x = \frac{-1}{3}$ Dividing both sides by -3

9. $3 - t \geq 19$

$-t \geq 16$ Subtracting 3 from both sides

$t \leq -16$ Dividing both sides by -1 and reversing the direction of the inequality symblol

The solution is $\{t|t \leq -16\}$.

11. $3 - 5a > a + 9$

$-5a > a + 6$ Subtracting 3 from both sides

$-6a > 6$ Subtracting a from both sides

$a < -1$ Dividing both sides by -6 and reversing the direction of the inequality symbol

The solution is $\{a|a < -1\}$.

13. $\frac{2}{3}(x + 5) \geq -4$

$x + 5 \geq -6$ Multiplying both sides by $\frac{3}{2}$

$x \geq -11$ Subracting 5 from both sides

The solution is $\{x|x \geq -11\}$.

15. $0.5x - 2.7 = 3x + 7.9$

$0.5x = 3x + 10.6$ Adding 2.7 to both sides

$-2.5x = 10.6$ Subtracting $3x$ from both sides

$x = -4.24$ Dividing both sides by -2.5

The solution is -4.24.

17. $8 - \frac{y}{3} \leq 7$

$\frac{-y}{3} \leq -1$ Subtracting 8 from both sides

$y \geq 3$ Multiplying both sides by -3 and reversing the direction of the inequality symbol

The solution is $\{y|y \geq 3\}$.

19. $-15 > 7 - 5x$

$-22 > -5x$ Subtracting 7 from both sides

$\frac{22}{5} < x$, or $x > \frac{22}{5}$ Dividing both sides by -5 and reversing the direction of the inequality symbol.

The solution is $\left\{x \middle| x > \frac{22}{5}\right\}$

Exercise Set 2.7

1. a is at least b can be translated as $b \leq a$.

3. a is at most b can be translated as $a \leq b$.

5. b is no more than a can be translated as $b \leq a$.

7. b is less than a can be translated as $b < a$.

9. Let n represent the number. Then we have

$n < 10$.

11. Let t represent the temperature. Then we have

$t \leq -3$.

13. Let d represent the number of years of driving experience. Then we have

$d \geq 5$.

15. Let a represent the age of the Mayan altar. Then we have

$a > 1200$.

17. Let h represent Tania's hourly wage. Then we have

$12 < h < 15$.

19. Let w represent the wind speed. Then we have

$w > 50$.

21. Let c represent the cost of a room at Pine Tree Bed and Breakfast. Then we have

$c \leq 120$.

23. ***Familiarize.*** Let s = the length of the service call, in hours. The total charge is \$55 plus \$40 times the number of hours RJ's was there.

Translate.

$$\underbrace{\text{\$55 charge}}_{\downarrow\ 55} \ \underset{\downarrow\ +}{\text{plus}} \ \underbrace{\text{hourly rate}}_{\downarrow\ 40} \ \underset{\downarrow\ \cdot}{\text{times}} \ \underbrace{\text{number of hours}}_{\downarrow\ s} \ \underbrace{\text{is greater than}}_{\downarrow\ >} \ \underbrace{\$150}_{\downarrow\ 150}.$$

Carry out. We solve the inequality.

$$55 + 40s > 150$$
$$40s > 95$$
$$s > 2.375$$

Check. As a partial check, we show that the cost of a 2.375 hour service call is \$150.

$$\$55 + \$30(2.375) = \$55 + \$95 = \$150$$

State. The length of the service call was more than 2.375 hr.

25. ***Familiarize.*** Let q = Robbin's undergraduate grade point average. Unconditional acceptance is 500 plus 200 times the grade point average.

Translate.

$$\underbrace{\text{GMAT score of 500}}_{\downarrow\ 500} \ \underset{\downarrow\ +}{\text{plus}} \ \underbrace{200}_{\downarrow\ 200} \ \underset{\downarrow\ \cdot}{\text{times}} \ \underbrace{\text{grade point average}}_{\downarrow\ q} \ \underbrace{\text{is at least}}_{\downarrow\ \geq} \ \underbrace{950}_{\downarrow\ 950}.$$

Carry out. We solve the inequality.

$$500 + 200q \geq 950$$
$$200q \geq 450$$
$$q \geq 2.25$$

Check. As a partial check we show that the acceptance score is 950.
$500 + 200(2.25) = 500 + 450 = 950.$

State.For unconditional acceptance, Robbin must have a gpa of at least 2.25.

27. ***Familiarize.*** The average of the five scores is their sum divided by the number of tests, 5. We let s represent Rod's score on the last test.

Translate. The average of the five scores is given by

$$\frac{73 + 75 + 89 + 91 + s}{5}.$$

Since this average must be at least 85, this means that it must be greater than or equal to 85. Thus, we can translate the problem to the inequality

$$\frac{73 + 75 + 89 + 91 + s}{5} \geq 85.$$

Carry out. We first multiply by 5 to clear the fraction.

$$5\Big(\frac{73 + 75 + 89 + 91 + s}{5}\Big) \geq 5 \cdot 85$$
$$73 + 75 + 89 + 91 + s \geq 425$$
$$328 + s \geq 425$$
$$s \geq 97$$

Check. As a partial check, we show that Rod can get a score of 97 on the fifth test and have an average of at least 85:

$$\frac{73 + 75 + 89 + 91 + 97}{5} = \frac{425}{5} = 85.$$

State. Scores of 97 and higher will earn Rod an average quiz grade of at least 85.

29. ***Familiarize.*** Let c = the number of credits Millie must complete in the fourth quarter.

Translate.

$$\underbrace{\text{Average number of credits}}_{\downarrow\ \frac{5+7+8+c}{4}} \ \underbrace{\text{is at least}}_{\downarrow\ \geq} \ \underset{\downarrow\ 7}{7}.$$

Carry out. We solve the inequality.

$$\frac{5 + 7 + 8 + c}{4} \geq 7$$
$$4\Big(\frac{5 + 7 + 8 + c}{4}\Big) \geq 4 \cdot 7$$
$$5 + 7 + 8 + c \geq 28$$
$$20 + c \geq 28$$
$$c \geq 8$$

Check. As a partial check, we show that Millie can complete 8 credits in the fourth quarter and average 7 credits per quarter.

$$\frac{5 + 7 + 8 + 8}{4} = \frac{28}{4} = 7$$

State. Millie must complete 8 credits or more in the fourth quarter.

31. ***Familiarize.*** The average number of plate appearances for 10 days is the sum of the number of appearance per day divided by the number of days, 10. We let p represent the number of plate appearances on the tenth day.

Translate. The average for 10 days is given by

$$\frac{5 + 1 + 4 + 2 + 3 + 4 + 4 + 3 + 2 + p}{10}.$$

Since the average must be at least 3.1, this means that it must be greater than or equal to 3.1. Thus, we can translate the problem to the inequality

$$\frac{5 + 1 + 4 + 2 + 3 + 4 + 4 + 3 + 2 + p}{10} \geq 3.1.$$

Carry out. We first multiply by 10 to clear the fraction.

$$10\Big(\frac{5 + 1 + 4 + 2 + 3 + 4 + 4 + 3 + 2 + p}{10}\Big) \geq 10 \cdot 3.1$$
$$5 + 1 + 4 + 2 + 3 + 4 + 4 + 3 + 2 + p \geq 31$$
$$28 + p \geq 31$$
$$p \geq 3$$

Check. As a partial check, we show that 3 plate appearances in the 10th game will average 3.1

$$\frac{5 + 1 + 4 + 2 + 3 + 4 + 4 + 3 + 2 + 3}{10} = \frac{31}{10} = 3.1$$

State. On the tenth day, 3 or more plate appearances will give an average of at least 3.1.

33. ***Familiarize***. We first make a drawing. We let b represent the length of the base. Then the lengths of the other sides are $b-2$ and $b+3$.

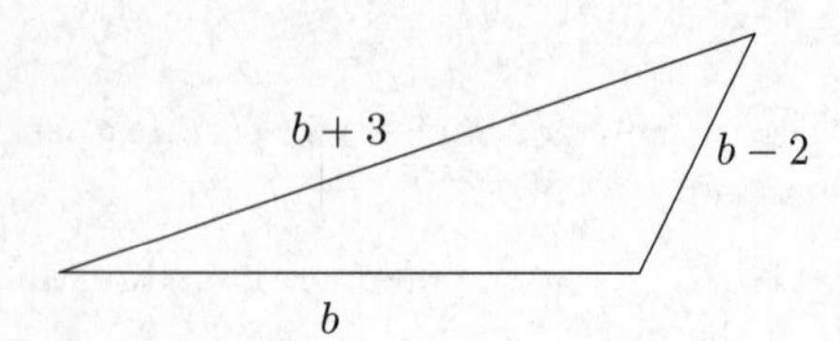

The perimeter is the sum of the lengths of the sides or $b+b-2+b+3$, or $3b+1$.

Translate.

The perimeter | is greater than | 19 cm.
↓ ↓ ↓
$3b+1 \quad > \quad 19$

Carry out.

$$3b+1>19$$
$$3b>18$$
$$b>6$$

Check. We check to see if the solution seems reasonable.

When $b=5$, the perimeter is $3\cdot 5+1$, or 16 cm.

When $b=6$, the perimeter is $3\cdot 6+1$, or 19 cm.

When $b=7$, the perimeter is $3\cdot 7+1$, or 22 cm.

From these calculations, it would appear that the solution is correct.

State. For lengths of the base greater than 6 cm the perimeter will be greater than 19 cm.

35. ***Familiarize***. Let $d=$ the depth of the well, in feet. Then the cost on the pay-as-you-go plan is $\$500+\$8d$. The cost of the guaranteed-water plan is \$4000. We want to find the values of d for which the pay-as-you-go plan costs less than the guaranteed-water plan.

Translate.

Cost of pay-as-you-go plan | is less than | cost of guaranteed-water plan
↓ ↓ ↓
$500+8d \quad < \quad 4000$

Carry out.

$$500+8d<4000$$
$$8d<3500$$
$$d<437.5$$

Check. We check to see that the solution is reasonable.

When $d=437$, $\$500+\$8\cdot 437=\$3996<\$4000$

When $d=437.5$, $\$500+\$8(437.5)=\$4000$

When $d=438$, $\$500+\$8(438)=\$4004>\$4000$

From these calculations, it appears that the solution is correct.

State. It would save a customer money to use the pay-as-you-go plan for a well of less than 437.5 ft.

37. ***Familiarize***. Let $v=$ the blue book value of the car. Since the car was repaired, we know that \$8500 does not exceed $0.8v$ or, in other words, $0.8v$ is at least \$8500.

Translate.

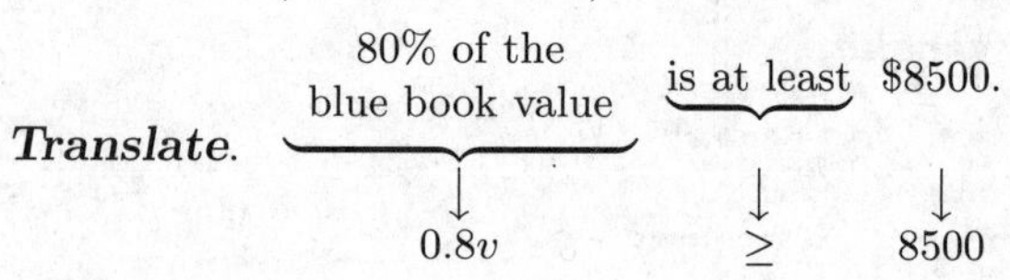

80% of the blue book value | is at least | \$8500.
↓ ↓ ↓
$0.8v \quad \geq \quad 8500$

Carry out.

$$0.8v\geq 8500$$
$$v\geq \frac{8500}{0.8}$$
$$v\geq 10,625$$

Check. As a partial check, we show that 80% of \$10,625 is at least \$8500:

$$0.8(\$10,625)=\$8500$$

State. The blue book value of the car was at least \$10,625.

39. ***Familiarize***. Let $L=$ the length of the package.

Translate.

Length | and | girth | is less than | 84 in
↓ ↓ ↓ ↓ ↓
$L \quad + \quad 29 \quad < \quad 84$

Carry out.

$$L+29<84$$
$$L<55$$

Check. We check to see if the solution seems reasonable.

When $L=60$ $\quad 60+29=89$ in.
When $L=55$ $\quad 55+29=84$ in.
When $L=50$ $\quad 50+29=79$ in.

From these calculations, it would appear that the solution is correct.

State. For lengths less than 55 in, the box is considered a "package."

41. ***Familiarize.*** We will use the formula $F=\frac{9}{5}C+32$.

Translate.

Fahrenheit temperature | is above | 98.6°.
↓ ↓ ↓
$F \quad > \quad 98.6$

Substituting $\frac{9}{5}C+32$ for F, we have

$$\frac{9}{5}C+32>98.6.$$

Carry out. We solve the inequality.

$$\frac{9}{5}C+32>98.6$$
$$\frac{9}{5}C>66.6$$
$$C>\frac{333}{9}$$
$$C>37$$

Check. We check to see if the solution seems reasonable.

When $C=36$, $\frac{9}{5}\cdot 36+32=96.8$.

When $C = 37$, $\frac{9}{5} \cdot 37 + 32 = 98.6$.

When $C = 38$, $\frac{9}{5} \cdot 38 + 32 = 100.4$.

It would appear that the solution is correct, considering that rounding occurred.

State. The human body is feverish for Celsius temperatures greater than 37°.

43. ***Familiarize.*** Let $h =$ the height of the triangle, in ft. Recall that the formula for the area of a triangle with base b and height h is $A = \frac{1}{2}bh$.

Translate.

Area | less than or equal to | 12 ft^2.

$$\frac{1}{2}(8)h \quad \leq \quad 12$$

Carry out. We solve the inequality.

$$\frac{1}{2}(8)h \leq 12$$

$$4h \leq 12$$

$$h \leq 3$$

Check. As a partial check, we show that a length of 3 ft will result in an area of 12 ft^2.

$$\frac{1}{2}(8)(3) = 12$$

State. The height should be no more than 3 ft.

45. ***Familiarize.*** Let $r =$ the amount of fat in a serving of the regular peanut butter, in grams. If reduced fat peanut butter has at least 25% less fat than regular peanut butter, then it has at most 75% as much fat as the regular peanut butter.

Translate.

12 g of fat | is at most | 75% | of | the amount of fat in regular peanut butter.

$$12 \quad \leq \quad 0.75 \quad \cdot \quad r$$

Carry out.

$$12 \leq 0.75r$$

$$16 \leq r$$

Check. As a partial check, we show that 12 g of fat does not exceed 75% of 16 g of fat:

$$0.75(16) = 12$$

State. Regular peanut butter contains at least 16 g of fat per serving.

47. ***Familiarize.*** Let $d =$ the number of days after September 5.

Translate.

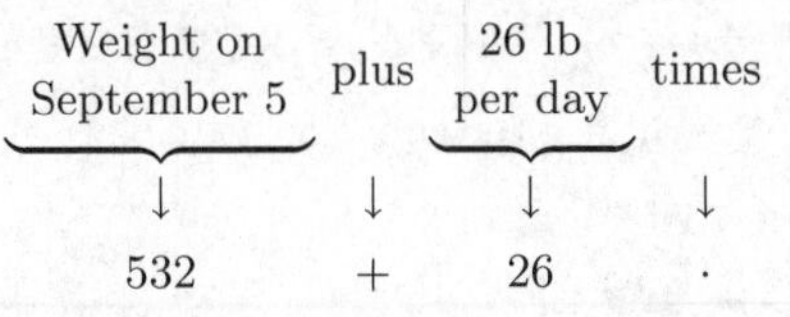

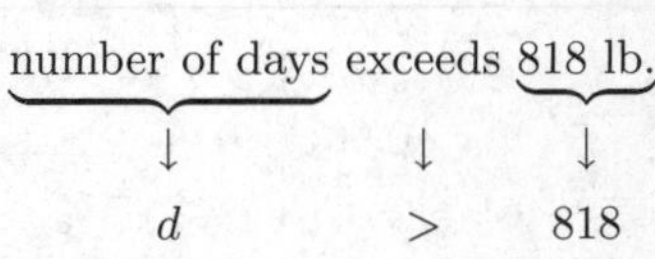

Carry out. We solve the inequality.

$$532 + 26d > 818$$

$$26d > 286$$

$$d > 11$$

Check. As a partial check, we can show that the weight of the pumpkin is 818 lb 11 days after September 5.

$$532 + 26 \cdot 11 = 532 + 286 = 818 \text{ lb}$$

State. The pumpkin's weight will exceed 818 lb more than 11 days after September 5, or on dates after September 16.

49. ***Familiarize.*** Let $n =$ the number of text messages. The total cost is the monthly fee of \$39.95 plus taxes of \$6.65 plus .10 times the number of text messages, or $.10n$.

Translate.

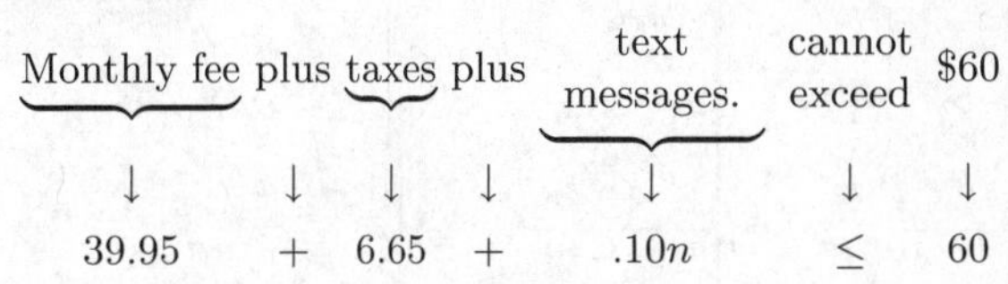

Carry out. We solve the inequality.

$$39.95 + 6.65 + .10n \leq 60$$

$$46.60 + .10n \leq 60$$

$$.10n \leq 13.4$$

$$n \leq 134$$

Check. As a partial check, if the number of text messages is 134, the budget of \$60 will not be exceeded.

State. Liam can send or receive 134 text messages and stay within his budget.

51. ***Familiarize.*** We will use the formula $R = -0.0065t + 4.3259$.

Translate.

The world record | is less than | 3.6 minutes.

$$-0.0065t + 4.3259 \quad < \quad 3.6$$

Carry out. We solve the inequality.

$$-0.0065t + 4.3259 < 3.6$$

$$-0.0065t < -0.7259$$

$$t > 111.68$$

Check. As a partial check, we can show that the record is more than 3.6 min 111 yr after 1900 and is less than 3.6 min 112 yr after 1900.

For $t = 111$, $R = -0.0065(111) + 4.3259 = 3.7709$.

For $t = 112$, $R = -0.0065(112) + 4.3259 = 3.5979$.

State. The world record in the mile run is less than 3.6 min more than 112 yr after 1900, or in years after 2012.

53. ***Familiarize.*** We will use the equation $y = 0.06x + 0.50$.

Translate.

$$\underbrace{\text{The cost}}_{\downarrow\; 0.06x+0.50} \quad \underbrace{\text{is at most}}_{\downarrow\; \leq} \quad \underset{\downarrow\; 14}{\$14.}$$

Carry out. We solve the inequality.

$$0.06x + 0.50 \leq 14$$
$$0.06x \leq 13.50$$
$$x \leq 225$$

Check. As a partial check, we show that the cost for driving 225 mi is $14.

$$0.06(225) + 0.50 = 14$$

State. The cost will be at most $14 for mileages less than or equal to 225 mi.

55. *Writing Exercise.* Answers may vary. Fran is more than 3 years older than Todd.

57. $-2 + (-5) - 7 = -2 + (-5) + (-7) = -14$

59. $3 \cdot (-10) \cdot (-1) \cdot (-2) = (-30) \cdot (-1) \cdot (-2)$
$= (30) \cdot (-2) = -60$

61. $(3 - 7) - (4 - 8) = (-4) - (-4) = (-4) + (4) = 0$

63. $\dfrac{-2 - (-6)}{8 - 10} = \dfrac{-2 + 6}{8 + (-10)} = \dfrac{4}{-2} = -2$

65. *Writing Exercise.* Answers may vary.

A boat has a capacity of 2800 lb. How many passengers can go on the boat if each passenger is considered to weigh 150 lb.

67. ***Familiarize.*** We use the formula $F = \frac{9}{5}C + 32$.

Translate. We are interested in temperatures such that $5° < F < 15°$. Substituting for F, we have:

$$5 < \frac{9}{5}C + 32 < 15$$

Solve.

$$5 < \frac{9}{5}C + 32 < 15$$
$$5 \cdot 5 < 5\left(\frac{9}{5}C + 32\right) < 5 \cdot 15$$
$$25 < 9C + 160 < 75$$
$$-135 < 9C < -85$$
$$-15 < C < -9\frac{4}{9}$$

Check. The check is left to the student.

State. Green ski wax works best for temperatures between $-15°$C and $-9\frac{4}{9}°$C.

69. Since $8^2 = 64$, the length of a side must be less than or equal to 8 cm (and greater than 0 cm, of course). We can also use the five-step problem-solving procedure.

Familiarize. Let s represent the length of a side of the square. The area s is the square of the length of a side, or s^2.

Translate.

$$\underbrace{\text{The area}}_{\downarrow\; s^2} \quad \underbrace{\text{is no more than}}_{\downarrow\; \leq} \quad \underset{\downarrow\; 64}{64 \text{ cm}^2.}$$

Carry out.

$$s^2 \leq 64$$
$$s^2 - 64 \leq 0$$
$$(s + 8)(s - 8) \leq 0$$

We know that $(s + 8)(s - 8) = 0$ for $s = -8$ or $s = 8$. Now $(s+8)(s-8) < 0$ when the two factors have opposite signs. That is:

$s+8>0$ *and* $s-8<0$ *or* $s+8<0$ *and* $s-8>0$

$s>-8$ *and* $s<8$ *or* $s<-8$ *and* $s>8$

This can be expressed as $-8 < s < 8$. This is not possible.

Then $(s + 8)(s - 8) \leq 0$ for $-8 \leq s \leq 8$.

Check. Since the length of a side cannot be negative we only consider positive values of s, or $0 < s \leq 8$. We check to see if this solution seems reasonable.

When $s = 7$, the area is 7^2, or 49 cm^2.

When $s = 8$, the area is 8^2, or 64 cm^2.

When $s = 9$, the area is 9^2, or 81 cm^2.

From these calculations, it appears that the solution is correct.

State. Sides of length 8 cm or less will allow an area of no more than 64 cm^2. (Of course, the length of a side must be greater than 0 also.)

71. ***Familiarize.*** Let f = the fat content of a serving of regular tortilla chips, in grams. A product that contains 60% less fat than another product has 40% of the fat content of that product. If Reduced Fat Tortilla Pops cannot be labeled lowfat, then they contain at least 3 g of fat.

Translate.

40%	of	the fat content of regular tortilla chips	is at least	3 grams of fat
↓	↓	↓	↓	↓
0.4	·	f	$\geq$	3

Carry out.

$$0.4f \geq 3$$
$$f \geq 7.5$$

Check. As a partial check, we show that 40% of 7.5 g is not less than 3 g.

$$0.4(7.5) = 3$$

State. A serving of regular tortilla chips contains at least 7.5 g of fat.

73. ***Familiarize***. Let $p =$ the price of Neoma's tenth book. If the average price of each of the first 9 books is \$12, then the total price of the 9 books is $9 \cdot \$12$, or \$108. The average price of the first 10 books will be $\frac{\$108+p}{10}$.

Translate.

$$\underbrace{\text{The average price of 10 books}}_{\frac{108+p}{10}} \; \underbrace{\text{is at least}}_{\geq} \; \underbrace{\$15.}_{15}$$

Carry out. We solve the inequality.

$$\frac{108+p}{10} \geq 15$$
$$108+p \geq 150$$
$$p \geq 42$$

Check. As a partial check, we show that the average price of the 10 books is \$15 when the price of the tenth book is \$42.

$$\frac{\$108+\$42}{10} = \frac{\$150}{10} = \$15$$

State. Neoma's tenth book should cost at least \$42 if she wants to select a \$15 book for her free book.

75. Let $b =$ the total purchases of bestsellers, $h =$ the total purchases of hardcovers, $p =$ the total purchases of other items at Barnes & Noble.

(1) Solving $0.40b > 25$, we get \$62.50
(2) Solving $0.20h > 25$, we get \$125
(3) Solving $0.10p > 25$, we get \$250

Thus when a customer's bestseller purchases are more than \$62.50, or hardcover purchases are more than \$125, or other purchases are more than \$250, the customer saves money by purchasing the card.

Chapter 2 Review

1. True

3. True

5. True

7. True

9.
$$\begin{aligned} x+9 &= -16 \\ x+9-9 &= -16-9 \quad \text{Adding } -9 \\ x &= -25 \quad \text{Simplifying} \end{aligned}$$

The solution is -25.

11.
$$\begin{aligned} -\tfrac{x}{5} &= 13 \\ -5\left(-\tfrac{x}{5}\right) &= -5\,(13) \quad \text{Multiplying by } -5 \\ x &= -65 \quad \text{Simplifying} \end{aligned}$$

The solution is -65.

13.
$$\begin{aligned} \frac{2}{5}t &= -8 \\ \frac{5}{2}\cdot\frac{2}{5}t &= \frac{5}{2}(-8) \quad \text{Multiplying by } \frac{5}{2} \\ t &= -20 \end{aligned}$$

The solution is -20.

15.
$$\begin{aligned} -\tfrac{2}{3}+x &= -\tfrac{1}{6} \\ 6\left(-\tfrac{2}{3}+x\right) &= 6\left(-\tfrac{1}{6}\right) \quad \text{Multiplying by 6} \\ -4+6x &= -1 \quad \text{Simplifying} \\ -4+6x+4 &= -1+4 \quad \text{Adding 4} \\ 6x &= 3 \quad \text{Simplifying} \\ x &= \tfrac{1}{2} \quad \text{Multiplying by } \tfrac{1}{6} \end{aligned}$$

The solution is $\frac{1}{2}$.

17.
$$\begin{aligned} 5-x &= 13 \\ 5-x-5 &= 13-5 \quad \text{Adding } -5 \\ -x &= 8 \quad \text{Simplifying} \\ x &= -8 \quad \text{Multiplying by } -1 \end{aligned}$$

The solution is -8.

19.
$$\begin{aligned} 7x-6 &= 25x \\ 7x-6-7x &= 25x-7x \quad \text{Adding } -7x \\ -6 &= 18x \quad \text{Simplifying} \\ -\tfrac{1}{3} &= x \quad \text{Multiplying by } \tfrac{1}{18} \end{aligned}$$

The solution is $-\frac{1}{3}$.

21.
$$\begin{aligned} 14y &= 23y-17-10 \\ 14y &= 23y-27 \quad \text{Simplifying} \\ 14y-14y &= 23y-27-14y \quad \text{Adding } -14y \\ 0 &= 9y-27 \quad \text{Simplifying} \\ 0+27 &= 9y-27+27 \quad \text{Adding 27} \\ 27 &= 9y \quad \text{Simplifying} \\ 3 &= y \quad \text{Multiplying by } \tfrac{1}{9} \end{aligned}$$

The solution is 3.

23.
$$\begin{aligned} \tfrac{1}{4}x-\tfrac{1}{8}x &= 3-\tfrac{1}{16}x \\ 16\left(\tfrac{1}{4}x-\tfrac{1}{8}x\right) &= 16\left(3-\tfrac{1}{16}x\right) \quad \text{Multiplying by 16} \\ 4x-2x &= 48-x \quad \text{Distributive Law} \\ 2x &= 48-x \quad \text{Simplifying} \\ 2x+x &= 48-x+x \quad \text{Adding } x \\ 3x &= 48 \quad \text{Simplifying} \\ x &= 16 \quad \text{Multiplying by } \tfrac{1}{3} \end{aligned}$$

The solution is 16.

25.
$$\begin{aligned} 4\,(5x-7) &= -56 \\ 20x-28 &= -56 \quad \text{Distributive Law} \\ 20x-28+28 &= -56+28 \quad \text{Adding 28} \\ 20x &= -28 \quad \text{Simplifying} \\ x &= -\tfrac{28}{20} \quad \text{Multiplying by } \tfrac{1}{20} \\ x &= -\tfrac{7}{5} \quad \text{Simplifying} \end{aligned}$$

The solution is $-\frac{7}{5}$.

27.
$$\begin{aligned} 3(x-4)+2 &= x+2+2(x-5) \\ 3x-12+2 &= x+2x-10 \\ 3x-10 &= 3x-10 \end{aligned}$$

All real numbers are solutions and the equation is an identity.

29.
$$V = \tfrac{1}{3}Bh$$
$$3 \cdot V = 3\left(\tfrac{1}{3}Bh\right) \quad \text{Multiplying by 3}$$
$$3V = Bh \quad \text{Simplifying}$$
$$\tfrac{1}{h}(3V) = \tfrac{1}{h}(Bh) \quad \text{Multiplying by } \tfrac{1}{h}$$
$$\tfrac{3V}{h} = B \quad \text{Simplifying}$$

31.
$$tx = ax + b$$
$$tx - ax = ax + b - ax \quad \text{Adding } -ax$$
$$tx - ax = b \quad \text{Simplifying}$$
$$x(t-a) = b \quad \text{Factoring } x$$
$$x = \frac{b}{t-a} \quad \text{Multiplying by } \frac{1}{t-a}$$

33. $\frac{11}{25} = \frac{4}{4} \cdot \frac{11}{25} = \frac{44}{100} = 0.44$

0.44.

44%

First, move the decimal point two places to the right; then write a % symbol: The answer is 44%.

35. ***Translate.***

49	is	35%	of	What number?
↓	↓	↓	↓	↓
49	=	0.35	·	y

We solve the equation and then convert to percent notation.

$$49 = 0.35y$$
$$\frac{49}{0.35} = y$$
$$140 = y$$

The answer is 140.

37. $x \le -5$

We substitute -7 for x giving $-7 \le -5$, which is a true statement since -7 is to the left of -5 on the number line, so -7 is a solution of the inequality $x \le -5$.

39.
$$5x - 6 < 2x + 3$$
$$5x - 6 + 6 < 2x + 3 + 6 \quad \text{Adding 6}$$
$$5x < 2x + 9 \quad \text{Simplifying}$$
$$5x - 2x < 2x + 9 - 2x \quad \text{Adding } -2x$$
$$3x < 9 \quad \text{Simplifying}$$
$$x < 3 \quad \text{Multiplying by } \tfrac{1}{3}$$

The solution set is $\{x|x < 3\}$. The graph is as follows:

5x − 6 < 2x + 3

−4 −2 0 2 4

41. $t > 0$

The solution set is $\{t|t > 0\}$. The graph is as follows:

$t > 0$

−5 −4 −3 −2 −1 0 1 2 3 4 5

43.
$$9x \ge 63$$
$$\tfrac{1}{9}(9x) \ge \tfrac{1}{9} \cdot 63 \quad \text{Multiplying by } \tfrac{1}{9}$$
$$x \ge 7 \quad \text{Simplifying}$$

The solution set is $\{x|x \ge 7\}$.

45.
$$7 - 3y \ge 27 + 2y$$
$$7 - 3y - 7 \ge 27 + 2y - 7 \quad \text{Adding } -7$$
$$-3y \ge 20 + 2y \quad \text{Simplifying}$$
$$-3y - 2y \ge 20 + 2y - 2y \quad \text{Adding } -2y$$
$$-5y \ge 20 \quad \text{Simplifying}$$
$$y \le -4 \quad \text{Multiplying by } -\tfrac{1}{5} \text{ and reversing the inequality symbol}$$

The solution set is $\{y|y \le -4\}$.

47.
$$-4y < 28$$
$$-\tfrac{1}{4}(-4y) > -\tfrac{1}{4} \cdot 28 \quad \text{Multiplying by } -\tfrac{1}{4} \text{ and reversing the inequality symbol}$$
$$y > -7 \quad \text{Simplifying}$$

The solution set is $\{y|y > -7\}$.

49.
$$4 - 8x < 13 + 3x$$
$$4 - 8x - 4 < 13 + 3x - 4 \quad \text{Adding } -4$$
$$-8x < 9 + 3x \quad \text{Simplifying}$$
$$-8x - 3x < 9 + 3x - 3x \quad \text{Adding } -3x$$
$$-11x < 9 \quad \text{Simplifying}$$
$$-\tfrac{1}{11}(-11x) > -\tfrac{1}{11} \cdot 9 \quad \text{Multiplying by } -\tfrac{1}{11}$$
$$x > -\tfrac{9}{11} \quad \text{Simplifying}$$

The solution set is $\left\{x \,|x > -\frac{9}{11}\right\}$.

51.
$$7 \le 1 - \tfrac{3}{4}x$$
$$7 - 1 \le 1 - \tfrac{3}{4}x - 1 \quad \text{Adding - 1}$$
$$6 \le -\tfrac{3}{4}x \quad \text{Simplifying}$$
$$-\tfrac{4}{3} \cdot 6 \ge -\tfrac{4}{3}\left(-\tfrac{3}{4}x\right) \quad \text{Multiplying by } -\tfrac{4}{3}$$
$$-8 \ge x \quad \text{Simplifying}$$

The solution set is $\{x| - 8 \ge x\}$, or $\{x|x \le -8\}$.

53. ***Familiarize.*** Let x = the length of the first piece, in ft. Since the second piece is 2 ft longer than the first piece, it must be $x + 2$ ft.

Translate.

The sum of the lengths of the two pieces	is	18 ft.
↓	↓	↓
$x + (x + 2)$	=	18

Carry out. We solve the equation.

$$x + (x + 2) = 18$$
$$2x + 2 = 18$$
$$2x = 16$$
$$x = 8$$

Check. If the first piece is 8 ft long, then the second piece must be 8+2, or 10 ft long. The sum of the lengths of the two pieces is 8 ft+10 ft, or 18 ft. The answer checks.

State. The lengths of the two pieces are 8 ft and 10 ft.

55. ***Familiarize.*** Let x = the first odd integer and let $x + 2$ = the next consecutive odd integer.

Translate.

$$\underbrace{\text{The sum of the two consecutive odd integers}}_{x+(x+2)} \quad \text{is} \quad 116 \quad \downarrow\ \downarrow \quad = 116$$

Carry out. We solve the equation.

$$\begin{aligned} x+(x+2) &= 116 \\ 2x+2 &= 116 \\ 2x &= 114 \\ x &= 57 \end{aligned}$$

Check. If the first odd integer is 57, then the next consecutive odd integer would be 57+2, or 59. The sum of these two integers is 57+59, or 116. This result checks.

State. The integers are 57 and 59.

57. **Familiarize.** Let $x =$ the regular price of the picnic table. Since the picnic table was reduced by 25%, it actually sold for 75% of its original price.

Translate.

$$\begin{array}{ccccc} 75\% & \text{of} & \underbrace{\text{the original price}} & \text{is} & \$120? \\ \downarrow & \downarrow & \downarrow & \downarrow & \downarrow \\ 0.75 & \cdot & x & = & 120 \end{array}$$

Carry out. We solve the equation.

$$\begin{aligned} 0.75x &= 120 \\ x &= \frac{120}{0.75} \\ x &= 160 \end{aligned}$$

Check. If the original price was \$160 with a 25% discount, then the purchaser would have paid 75% of \$160, or $0.75 \cdot$ \$160, or \$120. This result checks.

State. The original price was \$160.

59. **Familiarize.** Let $x =$ the measure of the first angle. The measure of the second angle is $x+50^\circ$, and the measure of the third angle is $2x-10^\circ$. The sum of the measures of the angles of a triangle is 180°.

Translate.

$$\begin{array}{ccc} \underbrace{\text{The sum of the measures of the angles}} & \text{is} & 180^\circ \\ \downarrow & \downarrow & \downarrow \\ x+(x+50)+(2x-10) & = & 180 \end{array}$$

Carry out. We solve the equation.

$$\begin{aligned} x+(x+50)+(2x-10) &= 180 \\ 4x+40 &= 180 \\ 4x &= 140 \\ x &= 35 \end{aligned}$$

Check. If the measure of the first angle is 35°, then the measure of the second angle is $35^\circ+50^\circ$, or 85°, and the measure of the third angle is $2 \cdot 35^\circ - 10^\circ$, or 60°. The sum of the measures of the first, second, and third angles is $35^\circ+85^\circ+60^\circ$, or 180°. These results check.

State. The measures of the angles are 35°, 85°, and 60°.

61. **Familiarize.** Let $n =$ the number of copies. The total cost is the setuup fee of \$6 plus \$4 per copy, or $4n$.

Translate.

$$\begin{array}{ccccc} \underbrace{\text{Set up fee}} & \text{plus} & \underbrace{\text{cost per copy}} & \underbrace{\text{cannot exceed}} & \$65 \\ \downarrow & \downarrow & \downarrow & & \\ 6 & + & 4n & \leq & 65 \end{array}$$

Carry out. We solve the inequality.

$$\begin{aligned} 6+4n &\leq 65 \\ 4n &\leq 59 \\ n &\leq \frac{59}{4} \\ n &\leq 14.75 \end{aligned}$$

Check. As a partial check, if the number of copies is 14, the total cost $\$6+\$4 \cdot 14$, ir \$62 does not exceed the budget of \$65. **State.** Myra can make 14 or fewer copies.

63. *Writing Exercise.* The solutions of an equation can usually each be checked. The solutions of an inequality are normally too numerous to check. Checking a few numbers from the solution set found cannot guarantee that the answer is correct, although if any number does not check, the answer found is incorrect.

65. **Familiarize.** Let $x =$ the length of the Nile River, in mi. Let $x+65$ represent the length of the Amazon River, in mi.

Translate.

$$\begin{array}{ccc} \underbrace{\text{The combined length of both rivers}} & \text{is} & 8385 \text{ mi} \\ \downarrow & \downarrow & \downarrow \\ x+(x+65) & = & 8385 \end{array}$$

Carry out. We solve the equation.

$$\begin{aligned} x+(x+65) &= 8385 \\ 2x+65 &= 8385 \\ 2x &= 8320 \\ x &= 4160 \end{aligned}$$

Check. If the Nile River is 4160 mi long, then the Amazon River is 4160 mi+65 mi, or 4225 mi. The combined length of both rivers is then 4160 mi+4225 mi, or 8385 mi. These results check..

State. The Amazon River is 4225 mi long, and the Nile River is 4160 mi long.

67.
$$\begin{aligned} 2|n|+4 &= 50 \\ 2|n| &= 46 \\ |n| &= 23 \end{aligned}$$

The distance from some number n and the origin is 23 units. The solution is $n = 23$, or $n = -23$.

69.
$$\begin{aligned} y &= 2a-ab+3 \\ y &= a(2-b)+3 \\ y-3 &= a(2-b) \\ \tfrac{y-3}{2-b} &= a \end{aligned}$$

The solution is $a = \frac{y-3}{2-b}$.

Chapter 2 Test

1. $t + 7 = 16$
$t + 7 - 7 = 16 - 7$ Adding -7
$t = 9$ Simplifying
The solution is 9.

3. $6x = -18$
$\frac{1}{6}(6x) = \frac{1}{6}(-18)$ Multiplying by $\frac{1}{6}$
$x = -3$ Simplifying
The solution is -3.

5. $3t + 7 = 2t - 5$
$3t + 7 - 7 = 2t - 5 - 7$ Adding -7
$3t = 2t - 12$ Simplifying
$3t - 2t = 2t - 12 - 2t$ Adding -2
$t = -12$ Simplifying
The solution is -12.

7. $8 - y = 16$
$8 - y - 8 = 16 - 8$ Adding -8
$-y = 8$ Simplifying
$y = -8$ Multiply by -1
The solution is -8.

9. $4(x + 2) = 36$
$4x + 8 = 36$ Distributive Law
$4x + 8 - 8 = 36 - 8$ Adding -8
$4x = 28$ Simplifying
$\frac{1}{4}(4x) = \frac{1}{4}(28)$ Multiplying by $\frac{1}{4}$
$x = 7$ Simplifying
The solution is 7.

11. $13t - (5 - 2t) = 5(3t - 1)$
$13t - 5 + 2t = 15t - 5$
$15t - 5 = 15t - 5$
All real numbers are solutions and the equation is an identity.

13. $14x + 9 > 13x - 4$
$14x + 9 - 9 > 13x - 4 - 9$ Adding -9
$14x > 13x - 13$ Simplifying
$14x - 13x > 13x - 13 - 13x$ Adding $-13x$
$x > -13$ Simplifying
The solution set is $\{x|x > -13\}$.

15. $4y \le -30$
$\frac{1}{4}(4y) \le \frac{1}{4}(-30)$ Multiplying by $\frac{1}{4}$
$y \le -\frac{15}{2}$ Simplifying
The solution set is $\{y|y \le -\frac{15}{2}\}$.

17. $3 - 5x > 38$
$3 - 5x - 3 > 38 - 3$ Adding -3
$-5x > 35$ Simplifying
$-\frac{1}{5}(-5x) < -\frac{1}{5}(35)$ Multiplying by $-\frac{1}{5}$ and reversing the inequality symbol
$x < -7$ Simplifying
The solution set is $\{x|x < -7\}$.

19. $5 - 9x \ge 19 + 5x$
$5 - 9x - 5 \ge 19 + 5x - 5$ Adding -5
$-9x \ge 14 + 5x$ Simplifying
$-9x - 5x \ge 14 + 5x - 5x$ Adding $-5x$
$-14x \ge 14$ Simplifying
$-\frac{1}{14}(-14x) \le -\frac{1}{14}(14)$ Multiplying by $-\frac{1}{14}$ and reversing the inequality symbol
$x \le -1$ Simplifying
The solution set is $\{x|x \le -1\}$.

21. $w = \frac{P+l}{2}$
$2 \cdot w = 2(\frac{P+l}{2})$ Multiplying by 2
$2w = P + l$ Simplifying
$2w - P = P + l - P$ Adding $-P$
$2w - P = l$ Simplifying
The solution is $l = 2w - P$.

23. 0.003 First move the decimal point two places to the right; then write a % symbol. The answer is 0.3%.

25. ***Translate.***

What percent	of	75	is	33?
↓	↓	↓	↓	↓
y	$\cdot$	75	=	33

We solve the equation and then convert to percent notation.
$y \cdot 75 = 33$
$y = \frac{33}{75}$
$y = 0.44 = 44\%$
The solution is 44%.

27. $-2 \le x \le 2$

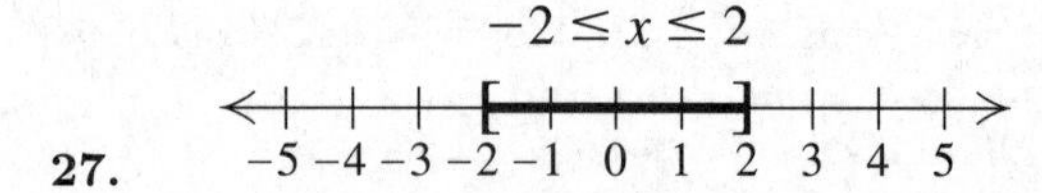

29. ***Familiarize.*** Let $x =$ the distance from Springer Mountain in miles. then $3 \times$ mi is the distance from Mt. Katahdin.

Translate.

Southern Distance	and	northern distance	is	Appalachian trail.
↓	↓	↓	↓	↓
x	+	$3x$	$\cdot$	2100

Carry out. We solve the equation.
$x + 3x = 2100$
$4x = 2100$
$x = 525$
$3x = 1575$

Check.
$525 + 1575 = 2100$.

State. The distance is 525 mi from Springer Moutain and 1575 mi from Mt. Katahdin.

31. ***Familiarize.*** Let $x =$ the electric bill before the temperature of the water heater was lowered. If the bill dropped by 7%, then the Kellys paid 93% of their original bill.

Translate.

93%	of	the original bill	is	\$60.45.
↓	↓	↓	↓	↓
0.93	$\cdot$	x	=	60.45

Carry out. We solve the equation.

$$0.93x = 60.45$$
$$x = \frac{60.45}{0.93}$$
$$x = 65$$

Check. If the original bill was \$65, and the bill was reduced by 7%, or $0.07 \cdot \$65$, or \$4.55, the new bill would be \$65−\$4.55, or \$60.45. This result checks. ***State.*** The original bill was \$65.

33.

$c = \frac{2cd}{a-d}$	
$(a-d)c = (a-d)\left(\frac{2cd}{a-d}\right)$	Multiplying by $a-d$
$ac - dc = 2cd$	Simplifying
$ac - dc + dc = 2cd + dc$	Adding dc
$ac = 3cd$	Simplifying
$\frac{1}{3c}(ac) = \frac{1}{3c}(3cd)$	Multiplying by $\frac{1}{3c}$
$\frac{a}{3} = d$	Simplifying

The solution is $d = \frac{a}{3}$.

35. Let h = the number of hours of sun each day. Then we have $4 \leq h \leq 6$.

Chapter 3

Introduction to Graphing

Exercise Set 3.1

1. The x-values extend from -9 to 4 and the y-values range from -1 to 5, so (a) is the best choice.

3. The x-values extend from -2 to 4 and the y-values range from -9 to 1, so (b) is the best choice.

5. We go to the top of the bar that is above the body weight 100 lb. Then we move horizontally from the top of the bar to the vertical scale listing numbers of drinks. It appears that consuming approximately 2 drinks in one hour will give a 100 lb person a blood-alcohol level of 0.08%.

7. For 3 drinks in one hour, we use the horizontal line at 3. For persons weighing 140 lb or less, their blood-alcohol level is 0.08% or more For persons weighing more than 140 lbs. their blood-alcohol level is under 0.08%. Therefore the person weighs more than 140 lbs.

9. ***Familiarize***. From the pie chart we see that 51% of student aid is Federal loans. The average aid is the total aid distributed of \$134.8 billion divided by the total number of full-time students, 13,334,170, or

$$\frac{\$134.8\text{billion}}{13,334,170}$$

Let $f=$ the average federal loan per full-time student.

Translate. We reword and translate the problem.

What is 51% of $\frac{\$134.8}{13{,}334{,}170}$ billion?

$$\downarrow \quad \downarrow \quad \downarrow \quad \downarrow \qquad \downarrow$$

$$f \quad = \quad 51\% \quad \cdot \quad \frac{134.8}{13,334,170}$$

Carry out. We solve the equation.

$$f = 0.51 \cdot \frac{134.8}{13,334,170} = \$5156$$

Check. We repeat the calculations. The answer checks.

State. The average federal loan per full-time equivalent student is \$5156.

11. ***Familiarize***. From the pie chart we see that 2% of the total aid is Federal campus-based, or 2% of \$134.8 billion = \$2.696 billion. Let $t=$ the amount given to students in two-year public institutions.

Translate. We reword the problem.

What is 8.6% of 2.696?

$$\downarrow \quad \downarrow \quad \downarrow \quad \downarrow \quad \downarrow$$

$$t \quad = \quad 8.6\% \quad \cdot \quad 2.696$$

Carry out.

$$t = 0.086 \cdot 2.696 = 0.231856 \text{ billion}$$

Check. We repeat the calculations.

State. In 2006 the campus-based aid given to students at two-year institutions is \$231,856,000.

13. ***Familiarize***. From the pie chart we see that 11.9% of solid waste is plastic. We let $p=$ the amount of plastic, in millions of tons, in the waste generated in 2005.

Translate. We reword the problem.

What is 11.9% of 245?

$$\downarrow \quad \downarrow \quad \downarrow \quad \downarrow \quad \downarrow$$

$$x \quad = \quad 11.9\% \quad \cdot \quad 245$$

Carry out.

$$p = 0.119 \cdot 245 \approx 29.2$$

Check. We can repeat the calculations.

State. In 2005, about 29.2 million tons of waste was plastic.

15. ***Familiarize***. From the pie chart we see that 5.2% of solid waste is glass. From Exercise 13 we know that Americans generated 245 million tons of waste in 2005. Then the amount of this that is glass is

$$0.052(245), \text{ or } 12.74 \text{ million tons}$$

We let $g=$ the amount of glass, in millions of tons, that Americans recycled in 2005.

Translate. We reword the problem.

What is 25.3% of 12.74 million tons?

$$\downarrow \quad \downarrow \quad \downarrow \quad \downarrow \qquad \downarrow$$

$$g \quad = \quad 25.3\% \quad \cdot \qquad 12.74$$

Carry out.

$$x = 0.253(12.74) \approx 3.2$$

Check. The result checks.

State. Americans recycled about 3.2 million tons of glass in 2005.

17. We locate 2002 on the horizontal axis and then move up to the line. From there we move left to the vertical axis and read the value of home videos, in billions. We estimate that about \$12 billion was spent on home videos in 2002.

19. We locate 10.5 on the vertical axis and move right to the line. From there we move down to the horizontal scale and read the year. We see that approximately \$10.5 billion was spent on home videos in 2001.

21. Starting at the origin:

(1,2) is 1 unit right and 2 units up;

$(-2,3)$ is 2 units left and 3 units up;

$(4,-1)$ is 4 units right and 1 unit down;

$(-5,-3)$ is 5 units left and 3 units down;

(4,0) is 4 units right and 0 units up or down;

$(0, -2)$ is 0 units right or left and 2 units down.

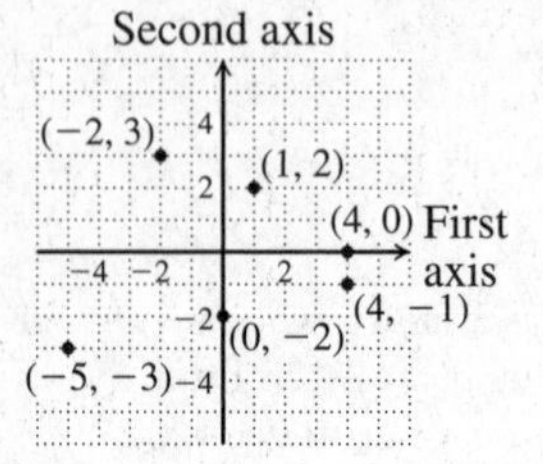

23. Starting at the origin:

(4,4) is 4 units right and 4 units up;

$(-2, 4)$ is 2 units left and 4 units up;

$(5, -3)$ is 5 units right and 3 units down;

$(-5, -5)$ is 5 units left and 5 units down;

(0,4) is 0 units right or left and 4 units up;

$(0, -4)$ is 0 units right or left and 4 units down;

(0,0) is 0 units right and 0 units up or down;

$(-4, 0)$ is 4 units left and 0 units up or down.

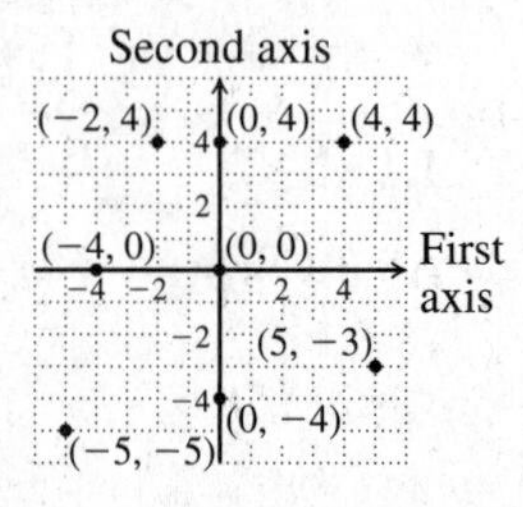

25. We plot the points $(2000, 12)$, $(2001, 34)$, $(2002, 931)$, $(2003, 1221)$, $(2004, 2862)$, $(2005, 7253)$ and $(2006, 8000)$and connect adjacent points with line segments.

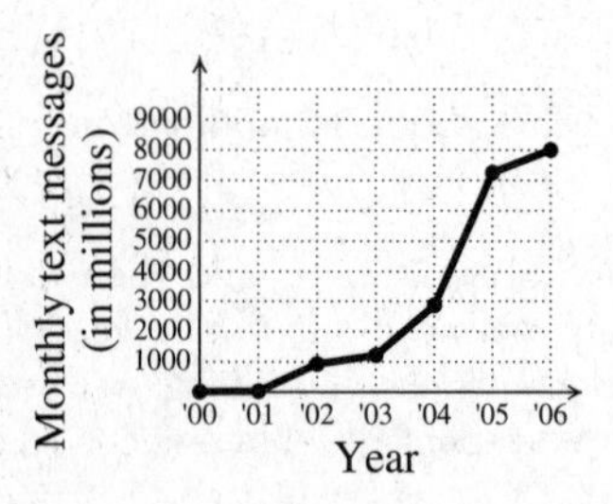

27.

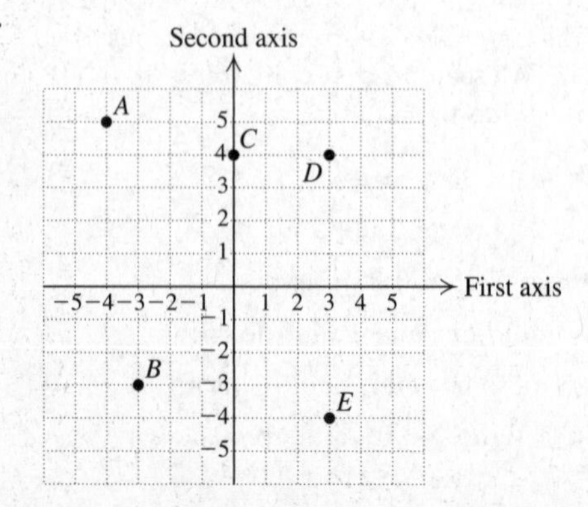

Point A is 4 units left and 5 units up. The coordinates of A are $(-4, 5)$.

Point B is 3 units left and 3 units down. The coordinates of B are $(-3, -3)$.

Point C is 0 units right or left and 4 units up. The coordinates of C are (0,4).

Point D is 3 units right and 4 units up. The coordinates of D are (3,4).

Point E is 3 units right and 4 units down. The coordinates of E are $(3, -4)$.

29.

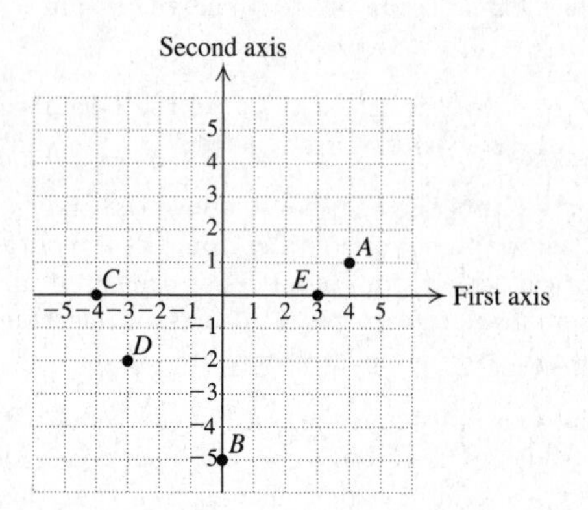

Point A is 4 units right and 1 unit up. The coordinates of A are (4,1).

Point B is 0 units right or left and 5 units down. The coordinates of B are $(0, -5)$.

Point C is 4 units left and 0 units up or down. The coordinates of C are $(-4, 0)$.

Point D is 3 units left and 2 units down. The coordinates of D are $(-3, -2)$.

Point E is 3 units right and 0 units up or down. The coordinates of E are (3,0).

31. Since the x-values range from -75 to 9, the 10 horizontal squares must span $9 - (-75)$, or 84 units. Since 84 is close to 100 and it is convenient to count by 10's, we can count backward from 0 eight squares to -80 and forward from 0 two squares to 20 for a total of $8 + 2$, or 10 squares.

Since the y-values range from -4 to 5, the 10 vertical squares must span $5 - (-4)$, or 9 units. It will be convenient to count by 2's in this case. We count down from 0 five squares to -10 and up from 0 five squares to 10 for a total of $5 + 5$, or 10 squares. (Instead, we might have chosen to count by 1's from -5 to 5.)

Then we plot the points $(-75, 5)$, $(-18, -2)$, and $(9, -4)$.

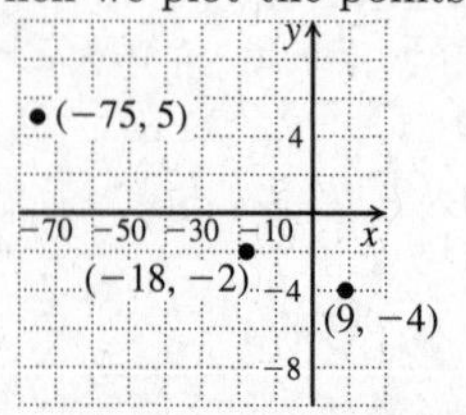

33. Since the x-values range from -5 to 5, the 10 horizontal squares must span $5 - (-5)$, or 10 units. It will be convenient to count by 2's in this case. We count backward from 0 five squares to -10 and forward from 0 five squares to 10 for a total of $5 + 5$, or 10 squares.

Since the y-values range from -14 to 83, the 10 vertical squares must span $83-(-14)$, or 97 units. To include both -14 and 83, the squares should extend from about -20 to 90, or $90-(-20)$, or 110 units. We cannot do this counting by 10's, so we use 20's instead. We count down from 0 four units to -80 and up from 0 six units to 120 for a total of $4+6$, or 10 units. There are other ways to cover the values from -14 to 83 as well.

Then we plot the points $(-1,83)$, $(-5,-14)$, and $(5,37)$.

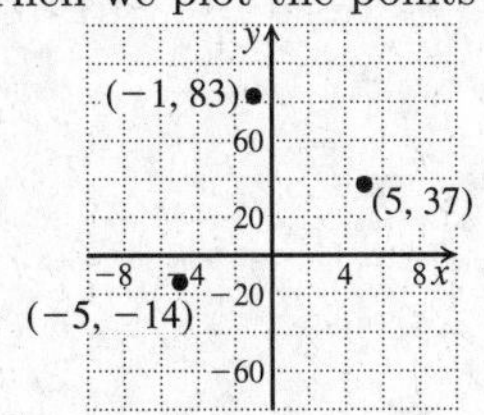

35. Since the x-values range from -16 to 3, the 10 horizontal squares must span $3-(-16)$, or 19 units. We could number by 2's or 3's. We number by 3's, going backward from 0 eight squares to -24 and forward from 0 two squares to 6 for a total of $8+2$, or 10 squares.

Since the y-values range from -4 to 15, the 10 vertical squares must span $15-(-4)$, or 19 units. We will number the vertical axis by 3's as we did the horizontal axis. We go down from 0 four squares to -12 and up from 0 six squares to 18 for a total of $4+6$, or 10 squares.

Then we plot the points $(-10,-4)$, $(-16,7)$, and $(3,15)$.

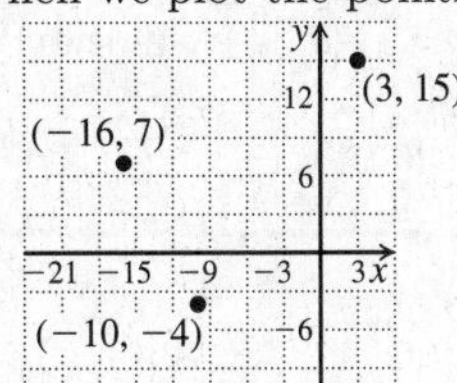

37. Since the x-values range from -100 and 800, the 10 horizontal squares must span $800-(-100)$, or 900 units. Since 900 is close to 1000 we can number by 100's. We go backward from 0 two squares to -200 and forward from 0 eight squares to 800 for a total of $2+8$, or 10 squares. (We could have numbered from -100 to 900 instead.)

Since the y-values range from -5 to 37, the 10 vertical squares must span $37-(-5)$, or 42 units. Since 42 is close to 50, we can count by 5's. We go down from 0 two squares to -10 and up from 0 eight squares to 40 for a total of $2+8$, or 10 squares.

Then we plot the points $(-100,-5)$, $(350,20)$, and $(800,37)$.

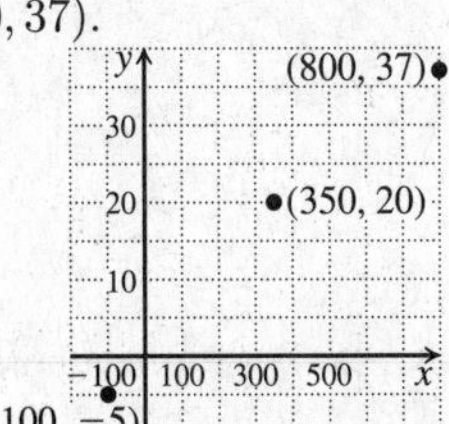

39. Since the x-values range from -124 to 54, the 10 horizontal squares must span $54-(-124)$, or 178 units. We can number by 25's. We go backward from 0 six squares to -150 and forward from 0 four squares to 100 for a total of $6+4$, or 10 squares.

Since the y-values range from -238 to 491, the 10 vertical squares must span $491-(-238)$, or 729 units. We can number by 100's. We go down from 0 four squares to -400 and up from 0 six squares to 600 for a total of $4+6$, or 10 squares.

Then we plot the points $(-83,491)$, $(-124,-95)$, and $(54,-238)$.

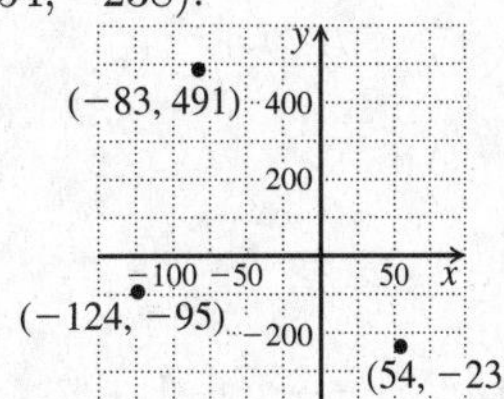

41. Since the first coordinate is positive and the second coordinate negative, the point $(7,-2)$ is located in quadrant IV.

43. Since both coordinates are negative, the point $(-4,-3)$ is in quadrant III.

45. Since both coordinates are positive, the point $(2,1)$ is in quadrant I.

47. Since the first coordinate is negative and the second coordinate is positive, the point $(-4.9,8.3)$ is in quadrant II.

49. First coordinates are positive in the quadrants that lie to the right of the origin, or in quadrants I and IV.

51. Points for which both coordinates are positive lie in quadrant I, and points for which both coordinates are negative life in quadrant III. Thus, both coordinates have the same sign in quadrants I and III.

53. *Writing Exercise.* The vertical scale above 80¢ is not labeled. The actual years in question are not given either.

55.
$$5y = 2x$$
$$y = \tfrac{2}{5}x \quad \text{Divide both sides by 5}$$

57.
$$x - y = 8$$
$$-y = -x + 8 \quad \text{Add } -x \text{ to both sides}$$
$$y = x - 8$$

59.
$$2x + 3y = 5$$
$$3y = -2x + 5$$
$$y = \tfrac{-2}{3}x + \tfrac{5}{3}$$

61. *Writing Exercise.* As time passes from 2004-6, the line graph is almost horizontal. The indicates that there is no new business involving home videos.

63. The coordinates have opposite signs, so the point could be in quadrant II or quadrant IV.

65.

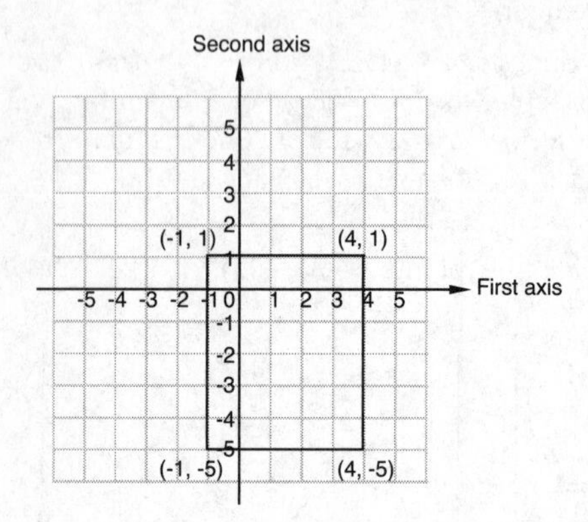

The coordinates of the fourth vertex are $(-1, -5)$.

67. Answers may vary.

We select eight points such that the sum of the coordinates for each point is 7.

$$\begin{array}{ll} (0,7) & 0+7=7 \\ (1,6) & 1+6=7 \\ (2,5) & 2+5=7 \\ (3,4) & 3+4=7 \\ (4,3) & 4+3=7 \\ (5,2) & 5+2=7 \\ (6,1) & 6+1=7 \\ (7,0) & 7+0=7 \end{array}$$

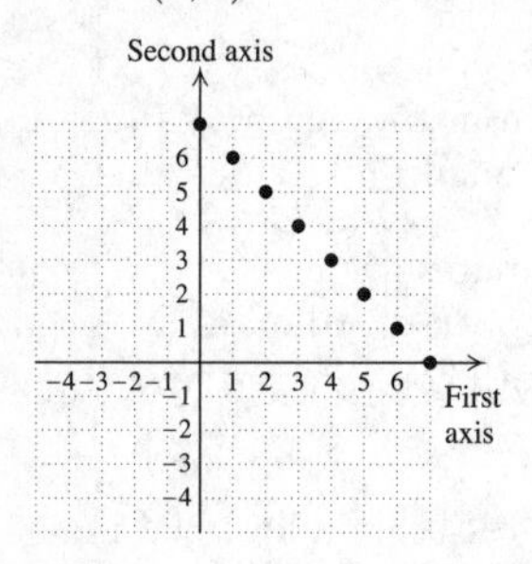

69.

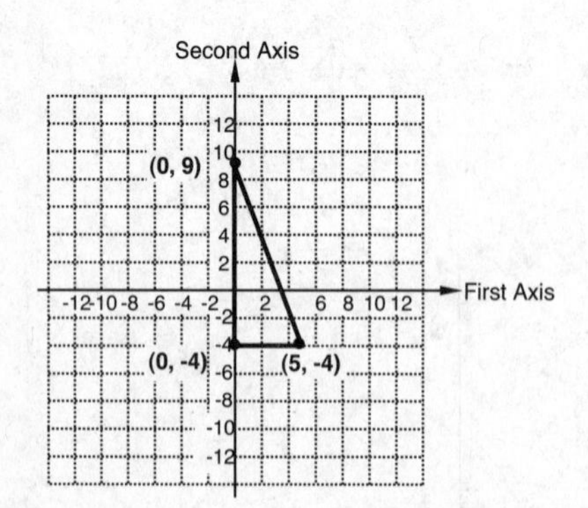

The base is 5 units and the height is 13 units.

$A = \frac{1}{2}bh = \frac{1}{2} \cdot 5 \cdot 13 = \frac{65}{2}$ sq units, or $32\frac{1}{2}$ sq units

71. Latitude 27° North,

Longitude 81° West

73. *Writing Exercise.* Eight "quadrants" will exist. Think of the coordinate system being formed by the intersection of one coordinate plane with another plane perpendicular to one of its axes such that the origins of the two planes coincide. Then there are four quadrants "above" the x,y-plane and four "below" it.

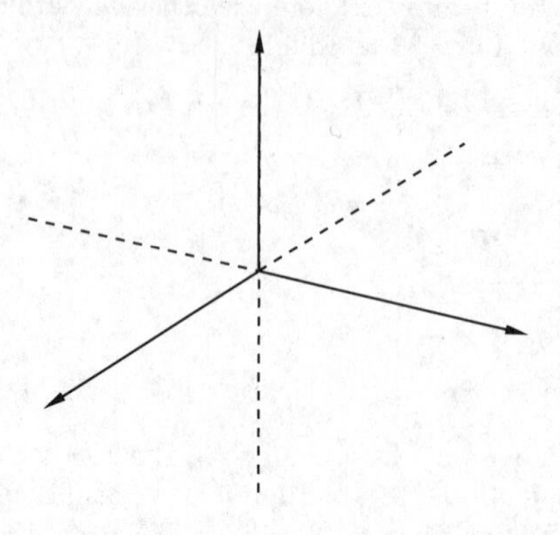

Exercise Set 3.2

1. False. A linear equation in two variables has infinitely many ordered pairs that are solutions.

3. True. All of the points on the graph of the line are solutions to the equation.

5. True. A solution may be found by selecting a value for x and solving for y. The ordered pair is a solution to the equation.

7. We substitute 2 for x and 1 for y.

$$\begin{array}{c|c} y & 4x-7 \\ \hline 1 & 4(2)-7 \\ & 8-7 \\ \end{array}$$

$1 \stackrel{?}{=} 1$ TRUE

Since $1 = 1$ is true, the pair (2,1) is a solution.

9. We substitute 5 for x and 1 for y.

$$\begin{array}{c|c} 3y+4x & 19 \\ \hline 3(1)+4(5) & 19 \\ 3+20 & \\ \end{array}$$

$23 \stackrel{?}{=} 19$ FALSE

Since $23 = 19$ is false, the pair $(5, 1)$ is not a solution.

11. We substitute 3 for m and -1 for n.

$$\begin{array}{c|c} 4m-5n & 7 \\ \hline 4(3)-5(-1) & 7 \\ 12+5 & \\ \end{array}$$

$17 \stackrel{?}{=} 7$ FALSE

Since $17 = 7$ is false, the pair $(3, -1)$ is not a solution.

13. To show that a pair is a solution, we substitute, replacing x with the first coordinate and y with the second coordinate in each pair.

$$\begin{array}{c|c} y = x + 3 & \\ \hline 2 & -1+3 \\ 2 \stackrel{?}{=} 2 & \text{TRUE} \end{array} \qquad \begin{array}{c|c} y = x + 3 & \\ \hline 7 & 4+3 \\ 7 \stackrel{?}{=} 7 & \text{TRUE} \end{array}$$

In each case the substitution results in a true equation. Thus, $(-1,2)$ and $(4,7)$ are both solutions of $y = x + 3$. We graph these points and sketch the line passing through them.

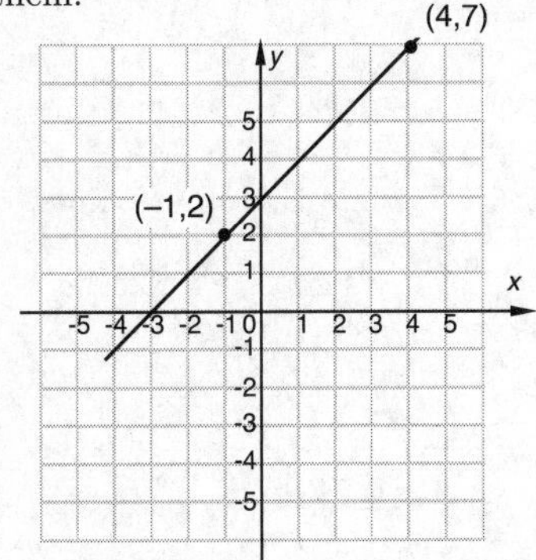

The line appears to pass through $(0,3)$ also. We check to determine if $(0,3)$ is a solution of $y = x + 3$.

$$\begin{array}{c|c} y = x + 3 & \\ \hline 3 & 0+3 \\ 3 \stackrel{?}{=} 3 & \text{TRUE} \end{array}$$

Thus, $(0,3)$ is another solution. There are other correct answers, including $(-5,-2)$, $(-4,-1)$, $(-3,0)$, $(-2,1)$, $(1,4)$, $(2,5)$, and $(3,6)$.

15. To show that a pair is a solution, we substitute, replacing x with the first coordinate and y with the second coordinate in each pair.

$$\begin{array}{c|c} y = \frac{1}{2}x + 3 & \\ \hline 5 & \frac{1}{2}\cdot 4+3 \\ & 2+3 \\ 5 \stackrel{?}{=} 5 & \text{TRUE} \end{array} \qquad \begin{array}{c|c} y = \frac{1}{2}x + 3 & \\ \hline 2 & \frac{1}{2}(-2)+3 \\ & -1+3 \\ 2 \stackrel{?}{=} 2 & \text{TRUE} \end{array}$$

In each case the substitution results in a true equation. Thus, $(4,5)$ and $(-2,2)$ are both solutions of $y = \frac{1}{2}x + 3$. We graph these points and sketch the line passing through them.

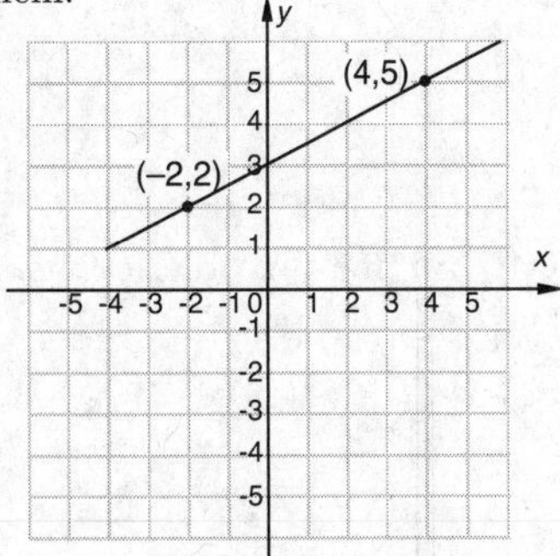

The line appears to pass through $(0,3)$ also. We check to determine if $(0,3)$ is a solution of $y = \frac{1}{2}x + 3$.

$$\begin{array}{c|c} y = \frac{1}{2}x + 3 & \\ \hline 3 & \frac{1}{2}\cdot 0+3 \\ 3 \stackrel{?}{=} 3 & \text{TRUE} \end{array}$$

Thus, $(0,3)$ is another solution. There are other correct answers, including $(-6,0)$, $(-4,1)$, $(2,4)$, and $(6,6)$.

17. To show that a pair is a solution, we substitute, replacing x with the first coordinate and y with the second coordinate in each pair.

$$\begin{array}{c|c} y + 3x = 7 & \\ \hline 1+3\cdot 2 & 7 \\ 1+6 & \\ 7 \stackrel{?}{=} 7 & \text{TRUE} \end{array} \qquad \begin{array}{c|c} y + 3x = 7 & \\ \hline -5+3\cdot 4 & 7 \\ -5+12 & \\ 7 \stackrel{?}{=} 7 & \text{TRUE} \end{array}$$

In each case the substitution results in a true equation. Thus, $(2,1)$ and $(4,-5)$ are both solutions of $y + 3x = 7$. We graph these points and sketch the line passing through them.

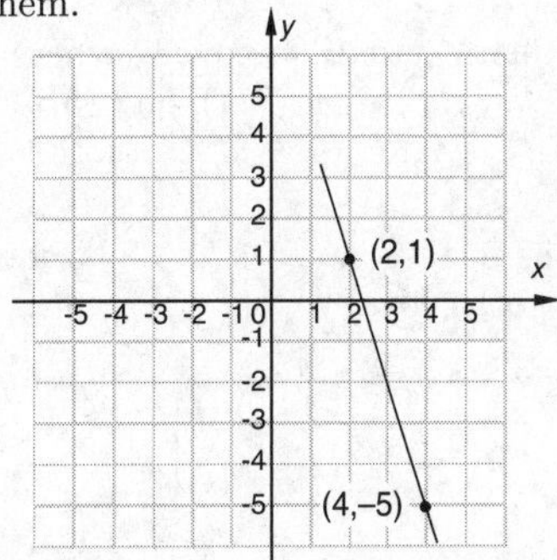

The line appears to pass through $(1,4)$ also. We check to determine if $(1,4)$ is a solution of $y + 3x = 7$.

$$\begin{array}{c|c} y + 3x = 7 & \\ \hline 4+3\cdot 1 & 7 \\ 4+3 & \\ 7 \stackrel{?}{=} 7 & \text{TRUE} \end{array}$$

Thus, $(1,4)$ is another solution. There are other correct answers, including $(3,-2)$.

19. To show that a pair is a solution, we substitute, replacing x with the first coordinate and y with the second coordinate in each pair.

$$\begin{array}{c|c} 4x - 2y = 10 & \\ \hline 4\cdot 0 - 2(-5) & 10 \\ 10 \stackrel{?}{=} 10 & \text{TRUE} \end{array}$$

$$\begin{array}{c|c} 4x - 2y = 10 & \\ \hline 4\cdot 4 - 2\cdot 3 & 10 \\ 16-6 & \\ 10 \stackrel{?}{=} 10 & \text{TRUE} \end{array}$$

In each case the substitution results in a true equation. Thus, $(0,-5)$ and $(4,3)$ are both solutions of $4x - 2y = 10$. We graph these points and sketch the line passing through them.

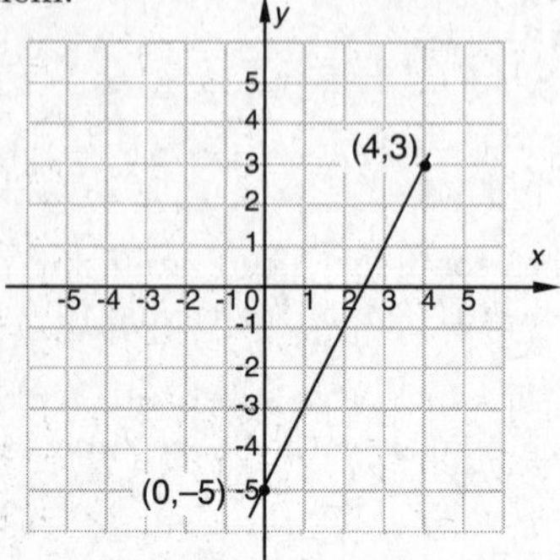

The line appears to pass through $(2,-1)$ also. We check to determine if $(2,-1)$ is a solution of $4x-2y=10$.

$$\begin{array}{c|c} 4x-2y & = 10 \\ \hline 4\cdot 2-2(-1) & 10 \\ 8+2 & \end{array}$$

$$10 \stackrel{?}{=} 10 \quad \text{TRUE}$$

Thus, $(2,-1)$ is another solution. There are other correct answers, including $(1,-3)$, $(2,-1)$, $(3,1)$, and $(5,5)$.

21. $y=x+1$

The equation is in the form $y=mx+b$. The y-intercept is $(0,1)$. We find two other pairs.

When $x=3$, $y=3+1=4$.
When $x=-5$, $y=-5+1=-4$.

x	y
0	1
3	4
−5	−4

Plot these points, draw the line they determine, and label the graph $y=x+1$.

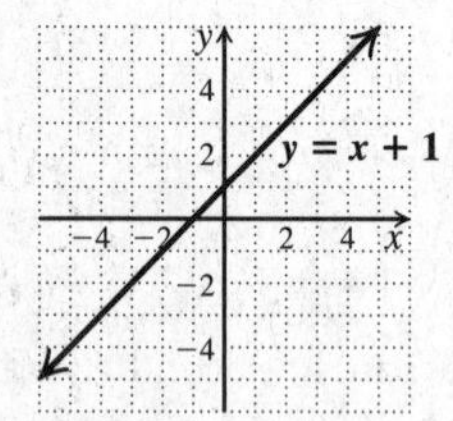

23. $y=-x$

The equation is equivalent to $y=-x+0$. The y-intercept is $(0,0)$. We find two other points.

When $x=-2$, $y=-(-2)=2$.
When $x=3$, $y=-3$.

x	y
0	0
−2	2
3	−3

Plot these points, draw the line they determine, and label the graph $y=-x$.

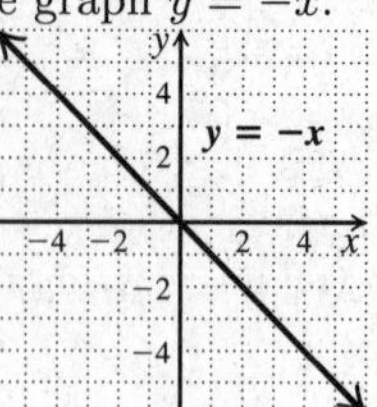

25. $y=2x$

The y-intercept is $(0,0)$. We find two other points.

When $x=1$, $y=2(1)=2$.

When $x=-1$, $y=2(-1)=-2$.

x	y
0	0
1	2
−1	−2

Plot these points, draw the line they determine, and label the graph $y=2x$.

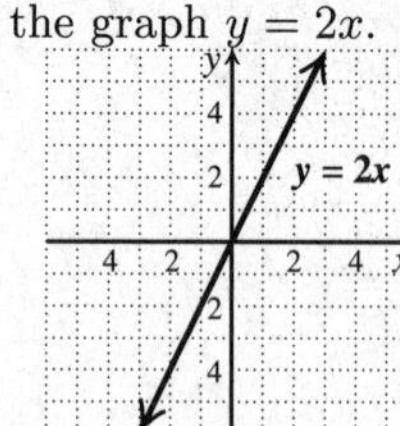

27. $y=2x+2$

The y-intercept is $(0,2)$. We find two other points.

When $x=-3$, $y=2(-3)+2=-6+2=-4$.

When $x=1$, $y=2\cdot 1+2=2+2=4$.

x	y
0	2
−3	−4
1	4

Plot these points, draw the line they determine, and label the graph $y=2x+2$.

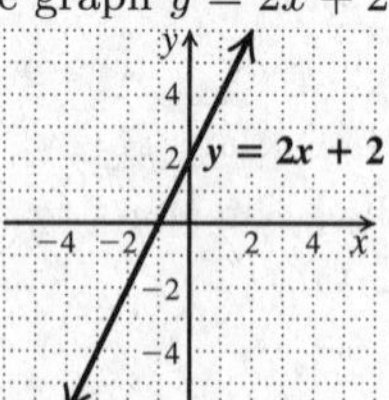

29. $y=-\frac{1}{2}x=-\frac{1}{2}x+0$

The y-intercept is $(0,0)$. We find two other points.

When $x=2$, $y=-\frac{1}{2}(2)=-1$.

When $x=-2$, $y=-\frac{1}{2}(-2)=1$.

x	y
0	0
2	−1
−2	1

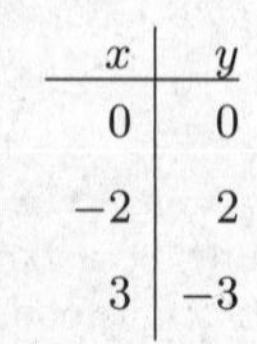

Plot these points, draw the line they determine, and label the graph $y = -\frac{1}{2}x$.

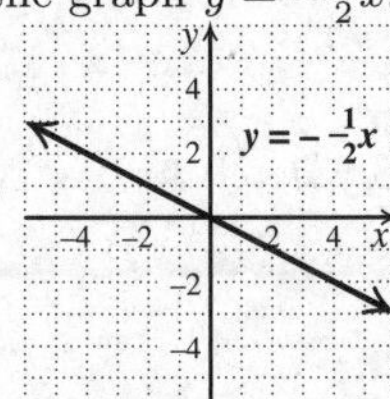

31. $y = \dfrac{1}{3}x - 4 = \dfrac{1}{3}x + (-4)$

The y-intercept is $(0, -4)$. We find two other points, using multiples of 3 for x to avoid fractions.

When $x = -3$, $y = \dfrac{1}{3}(-3) - 4 = -1 - 4 = -5$.

When $x = 3$, $y = \dfrac{1}{3} \cdot 3 - 4 = 1 - 4 = -3$.

x	y
0	-4
-3	-5
3	-3

Plot these points, draw the line they determine, and label the graph $y = \dfrac{1}{3}x - 4$.

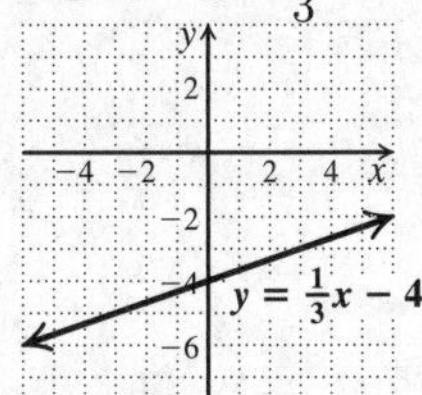

33. $x + y = 4$

$y = -x + 4$

The y-intercept is $(0, 4)$. We find two other points.

When $x = -1$, $y = -(-1) + 4 = 1 + 4 = 5$.

When $x = 2$, $y = -2 + 4 = 2$.

x	y
0	4
-1	5
2	2

Plot these points, draw the line they determine, and label the graph $x + y = 4$.

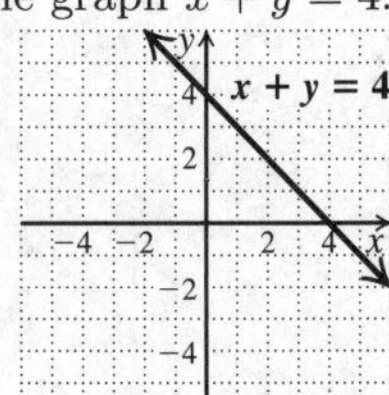

35. $x - y = -2$

$y = x + 2$

The y-intercept is $(0, 2)$. We find two other points.

When $x = 1$, $y = 1 + 2 = 3$.

When $x = -1$, $y = -1 + 2 = 1$.

x	y
0	2
1	3
-1	1

Plot these points, draw the line they determine, and label the graph $x - y = -2$.

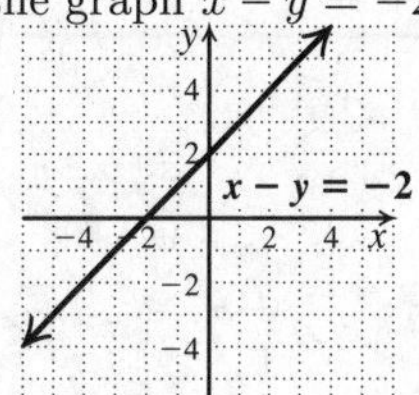

37. $x + 2y = -6$

$2y = -x - 6$

$y = -\dfrac{1}{2}x - 3$

$y = -\dfrac{1}{2}x + (-3)$

The y-intercept is $(0, -3)$. We find two other points, using multiples of 2 for x to avoid fractions.

When $x = -4$, $y = -\dfrac{1}{2}(-4) - 3 = 2 - 3 = -1$.

When $x = 2$, $y = -\dfrac{1}{2} \cdot 2 - 3 = -1 - 3 = -4$.

x	y
0	-3
-4	-1
2	-4

Plot these points, draw the line they determine, and label the graph $x + 2y = -6$.

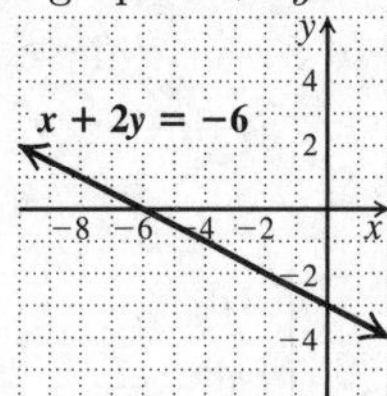

39. $y = -\dfrac{2}{3}x + 4$

The y-intercept is $(0, 4)$. We find two other points, using multiples of 3 for x to avoid fractions.

When $x = 3$, $y = -\dfrac{2}{3} \cdot 3 + 4 = -2 + 4 = 2$.

When $x = 6$, $y = -\dfrac{2}{3} \cdot 6 + 4 = -4 + 4 = 0$.

x	y
0	4
3	2
6	0

Plot these points, draw the line they determine, and label

the graph $y = -\frac{2}{3}x + 4$.

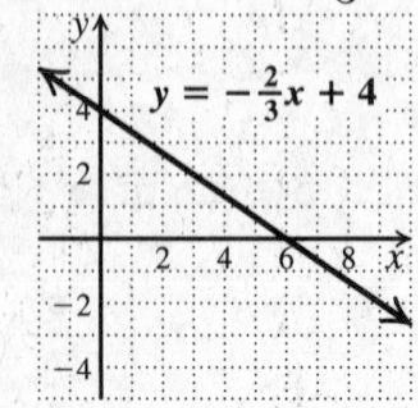

41. $4x = 3y$

$y = \frac{4}{3}x = \frac{4}{3}x + 0$

The y-intercept is $(0, 0)$. We find two other points.

When $x = 3$, $y = \frac{4}{3}(3) = 4$.

When $x = -3$, $y = \frac{4}{3}(-3) = -4$.

x	y
0	0
3	4
-3	-4

Plot these points, draw the line they determine, and label the graph $4x = 3y$.

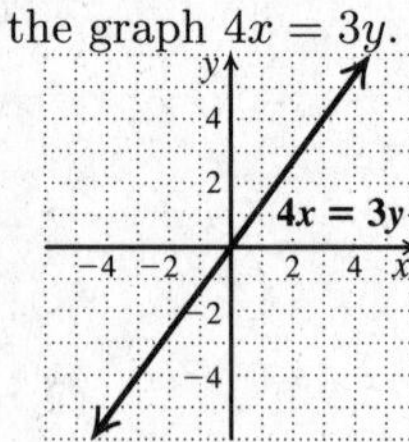

43. $5x - y = 0$

$y = 5x = 5x + 0$

The y-intercept is $(0, 0)$. We find two other points.

When $x = 1$, $y = 5(1) = 5$.

When $x = -1$, $y = 5(-1) = -5$.

x	y
0	0
1	5
-1	-5

Plot these points, draw the line they determine, and label the graph $5x - y = 0$.

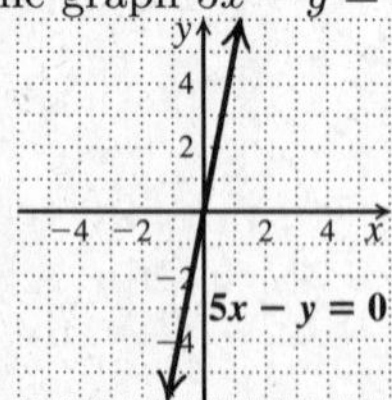

45. $6x - 3y = 9$

$-3y = -6x + 9$

$y = 2x - 3$

$y = 2x + (-3)$

The y-intercept is $(0, -3)$. We find two other points.

When $x = -1$, $y = 2(-1) - 3 = -2 - 3 = -5$.

When $x = 3$, $y = 2 \cdot 3 - 3 = 6 - 3 = 3$.

x	y
0	-3
-1	-5
3	3

Plot these points, draw the line they determine, and label the graph $6x - 3y = 9$.

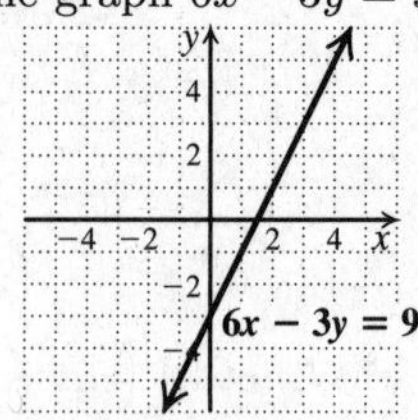

47. $6y + 2x = 8$

$6y = -2x + 8$

$y = -\frac{1}{3}x + \frac{4}{3}$

The y-intercept is $\left(0, \frac{4}{3}\right)$. We find two other points.

When $x = -2$, $y = -\frac{1}{3}(-2) + \frac{4}{3} = \frac{2}{3} + \frac{4}{3} = 2$.

When $x = 1$, $y = -\frac{1}{3} \cdot 1 + \frac{4}{3} = -\frac{1}{3} + \frac{4}{3} = 1$.

x	y
0	$\frac{4}{3}$
-2	2
1	1

Plot these points, draw the line they determine, and label the graph $6y + 2x = 8$.

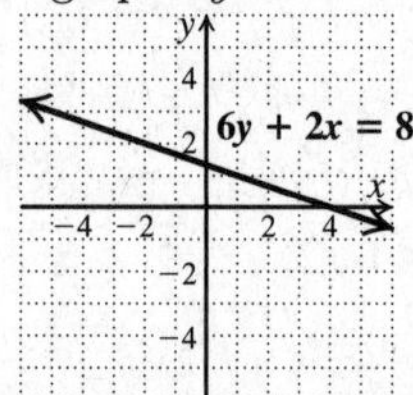

49. We graph $a = 0.08t + 2.5$ by selecting values for t and then calculating the associated values for a.

If $t = 0$, $a = 0.08(0) + 2.5 = 2.5$.

If $t = 10$, $a = 0.08(10) + 2.5 = 3.3$.

If $t = 20$, $a = 0.08(20) + 2.5 = 4.1$.

t	a
0	2.5
10	3.3
20	4.1

We plot the points and draw the graph.

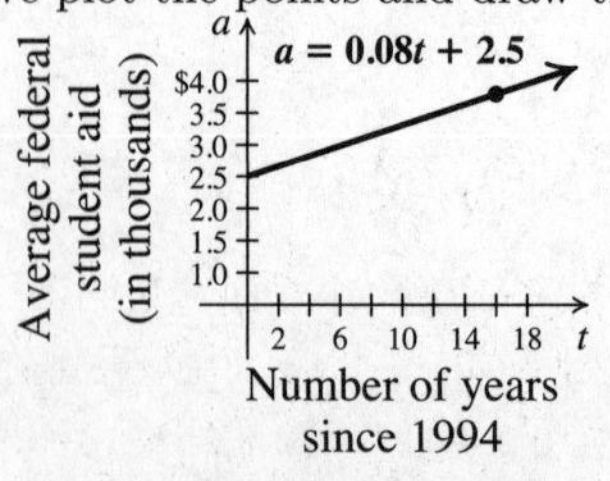

Since $2010 - 1994 = 16$, the year 2010 is 16 years after 1994. Thus, to estimate the average amount of federal student aid per student in 2010, we find the second coordinate associated with 16. Locate the point on the line that is above 16 and then find a value on the vertical axes that corresponds to that point. That value is about 3.8, so we estimate that the average amount of federal student aid per student in 2010 is \$3.8 thousand or \$3,800.

51. We graph $c = 3.1w + 29.07$ by selecting values for w and then calculating the associated values for c.

If $w = 1$, $c = 3.1(1) + 29.07 = 32.17$.

If $w = 5$, $c = 3.1(5) + 29.07 = 44.57$.

If $w = 10$, $c = 3.1(10) + 29.07 = 60.07$.

w	c
1	32.17
5	44.57
10	60.07

We plot the points and draw the graph.

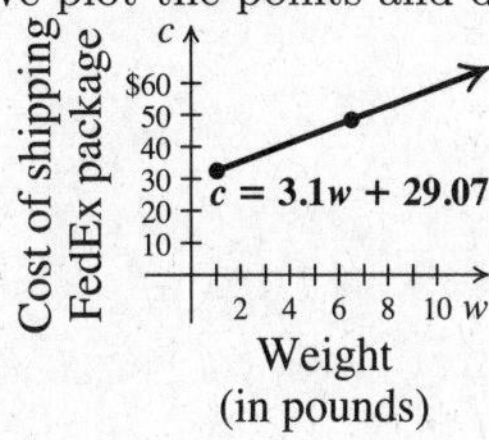

Locate the point on the line that is above $6\frac{1}{2}$ and then find the value on the vertical axes that corresponds to that point. That value is 49, so we estimate the cost of shipping a $6\frac{1}{2}$-lb package to be \$49.

53. We graph $p = 3.5n + 9$ by selecting values for n and then calculating the associated values for p.

If $n = 10$, $p = 3.5(10) + 9 = 44$.

If $n = 20$, $p = 3.5(20) + 9 = 79$.

If $n = 30$, $p = 3.5(30) + 9 = 114$.

n	p
10	44
20	79
30	114

We plot the points and draw the graph.

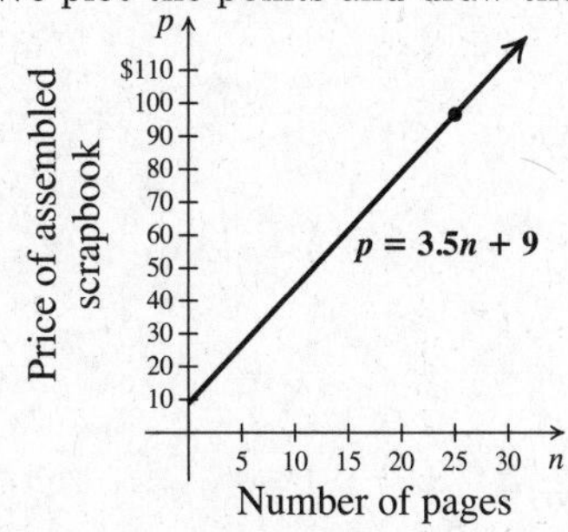

Locate the point on the line that is above 25 and then find the value on the vertical axes that corresponds to that point. That value is 97, so we estimate the price of a scrapbook containing 25 pages as \$97.

55. We graph $w = 1.6t + 16.7$ by selecting values for t and then calculating the associated values for w.

If $t = 5$, $w = 1.6(5) + 16.7 = 24.7$.

If $t = 7$, $w = 1.6(7) + 16.7 = 27.9$.

If $t = 10$, $w = 1.6(10) + 16.7 = 32.7$.

We plot the points and draw the graph.

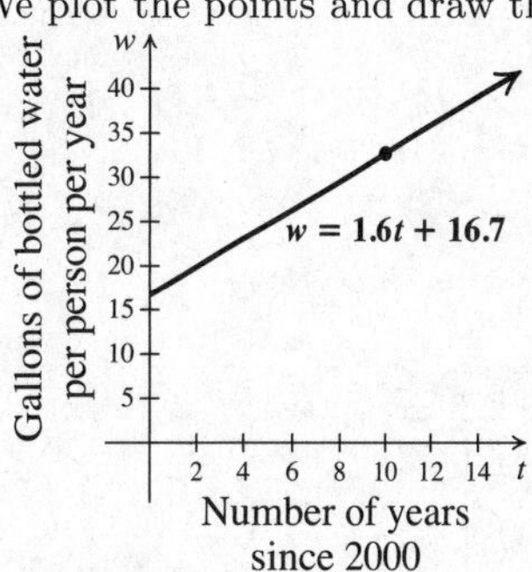

To predict the number of gallons consumed per person in 2010 we find the second coordinate associated with 10. (2010 is 10 years after 2000.) The value is 32.7, so we predict that about 33 gallons of bottled water will be consumed per person in 2010.

57. We graph $T = \frac{5}{4}c + 2$. Since number of credits cannot be negative, we select only nonnegative values for c.

If $c = 4$, $T = \frac{5}{4}(4) + 2 = 7$.

If $c = 8$, $T = \frac{5}{4}(8) + 2 = 12$.

If $c = 12$, $T = \frac{5}{4}(12) + 2 = 17$.

We plot the points and draw the graph.

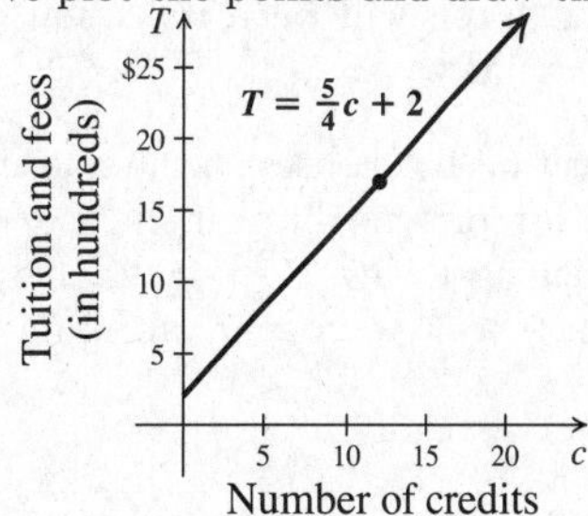

Four three-credit courses total $4 \cdot 3$ or 12, credits. Locate the point in the graph, the value is 17. So tuition and fees will cost \$17 hundred, or \$1700.

59. *Writing Exercise.* Most would probably say that the second equation would be easier to graph because it has been solved for y. This makes it more efficient to find the y-value that corresponds to a given x-value.

61.
$$5x + 3 \cdot 0 = 12$$
$$5x + 0 = 12$$
$$5x = 12$$
$$x = \frac{12}{5}$$

Check:
$$\begin{array}{c|c} 5x+3\cdot 0 & 12 \\ \hline 5\cdot\frac{12}{5}+3\cdot 0 & 12 \\ 12+0 & \\ 12 & \stackrel{?}{=} 12 \quad \text{TRUE} \end{array}$$

The solution is $\frac{12}{5}$.

63.
$$\begin{aligned} 5x+3(2-x) &= 12 \\ 5x+6-3x &= 12 \\ 2x+6 &= 12 \\ 2x &= 6 \\ x &= 3 \end{aligned}$$

Check:
$$\begin{array}{c|c} 5x+3(2-x) & 12 \\ \hline 5(3)+3(2-3) & 12 \\ 15+3(-1) & \\ 15-3 & \\ 12 & \stackrel{?}{=} 12 \quad \text{TRUE} \end{array}$$

The solution is 3.

65.
$$\begin{aligned} A &= \frac{T+Q}{2} \\ 2A &= T+Q \\ 2A-T &= Q \end{aligned}$$

67.
$$\begin{aligned} Ax+By &= C \\ By &= C-Ax \quad \text{Subtracting } Ax \\ y &= \frac{C-Ax}{B} \quad \text{Dividing by } B \end{aligned}$$

69. *Writing Exercise.* Her graph will be a reflection of the correct graph across the line $y = x$.

71. Let s represent the gear that Laura uses on the southbound portion of her ride and n represent the gear she uses on the northbound portion. Then we have $s + n = 24$. We graph this equation, using only positive integer values for s and n.

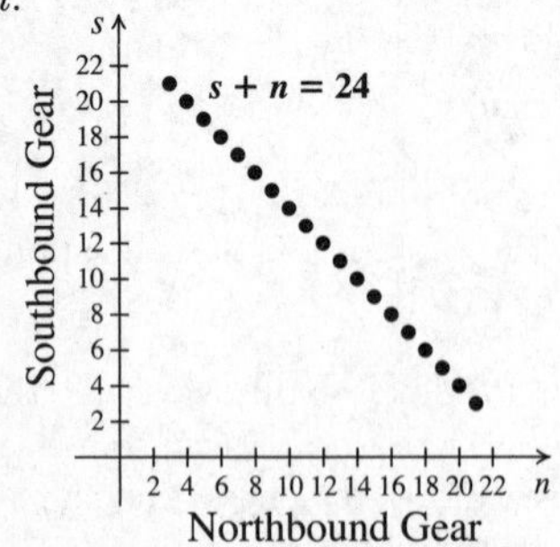

73. Note that the sum of the coordinates of each point on the graph is 5. Thus, we have $x + y = 5$, or $y = -x + 5$.

75. Note that each y-coordinate is 2 more than the corresponding x-coordinate. Thus, we have $y = x + 2$.

77. The equation is $25d + 5l = 225$.

Since the number of dinners cannot be negative, we choose only nonnegative values of d when graphing the equation. The graph stops at the horizontal axis since the number of lunches cannot be negative.

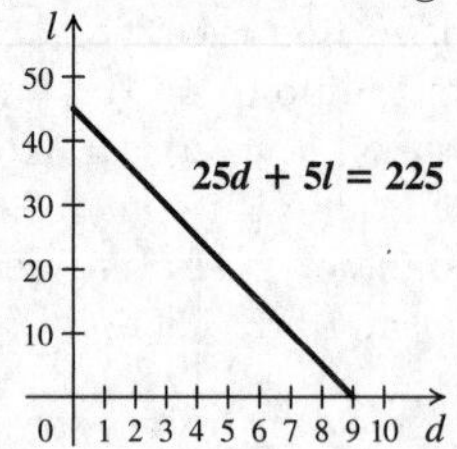

We see that three points on the graph are $(1, 40)$, $(5, 20)$, and $(8, 5)$. Thus, three combinations of dinners and lunches that total $225 are

1 dinner, 40 lunches,

5 dinners, 20 lunches,

8 dinners, 5 lunches.

79. $y = -|x|$

x	y
-3	-3
-2	-2
-1	-1
0	0
1	-1
2	-2
3	-3

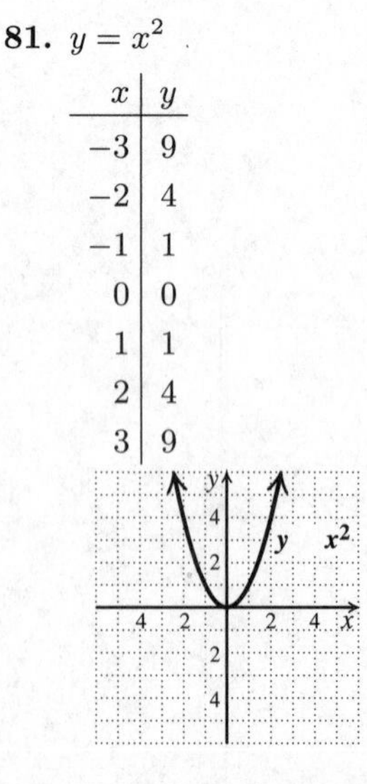

81. $y = x^2$

x	y
-3	9
-2	4
-1	1
0	0
1	1
2	4
3	9

83.

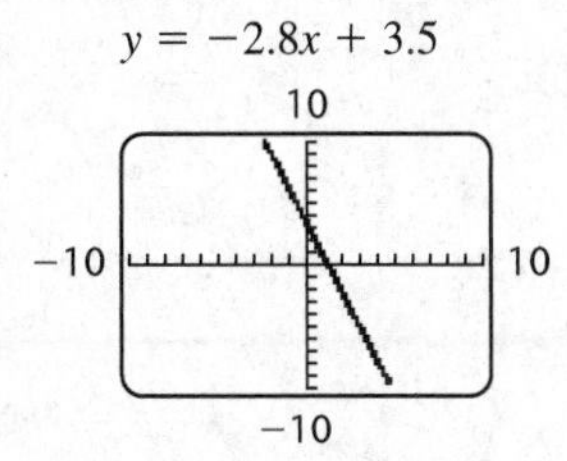

85.

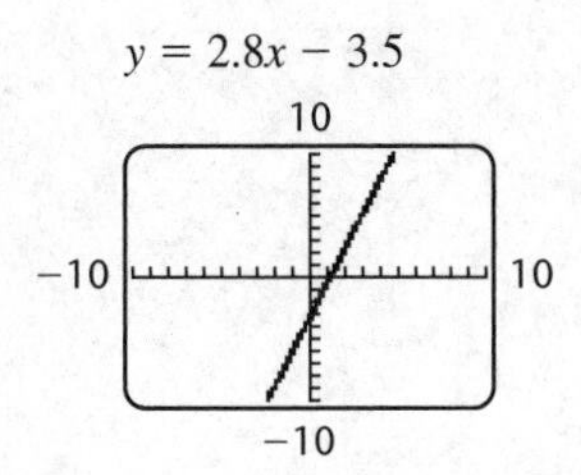

87.

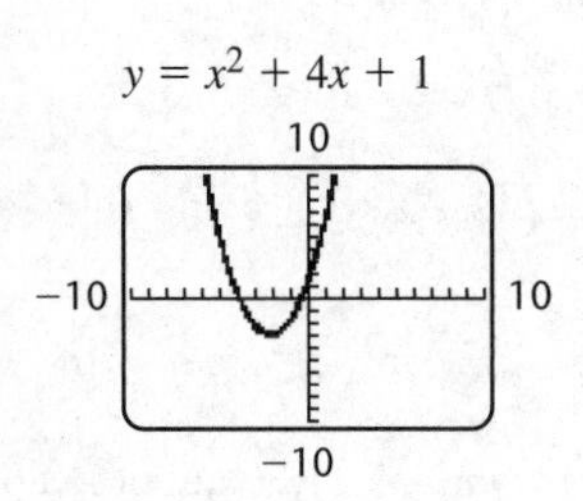

89. $t = -0.1s + 13.1$

s	t
55	7.6
70	6.1

At 55 mph, the efficiency is 7.6 mpg, so a 500 mile trip will use $\frac{500}{7.6} \approx 65.79$ gal at a cost of \$230.27. At 70 mph, the efficiency is 6.1 mpg, so a 500 mile trip will use $\frac{500}{6.1} \approx 81.97$ gal at a cost of \$286.89. Driving at 55 mph instead of 70 mph will save \$56.62 and 16.2 gal.

Exercise Set 3.3

1. The graph of $x = -4$ is a vertical line, so (f) is the most appropriate choice.

3. The point $(0, 2)$ lies on the y-axis, so (d) is the most appropriate choice.

5. The point $(3, -2)$ does not lie on an axis, so it could be used as a check when we graph using intercepts. Thus (b) is the most appropriate choice.

7. (a) The graph crosses the y-axis at $(0, 5)$, so the y-intercept is $(0, 5)$.

(b) The graph crosses the x-axis at $(2, 0)$, so the x-intercept is $(2, 0)$.

9. (a) The graph crosses the y-axis at $(0, -4)$, so the y-intercept is $(0, -4)$.

(b) The graph crosses the x-axis at $(3, 0)$, so the x-intercept is $(3, 0)$.

11. (a) The graph crosses the y-axis at $(0, -2)$, so the y-intercept is $(0, -2)$.

(b) The graph crosses the x-axis at $(-3, 0)$ and $(3, 0)$, so the x-intercepts are $(-3, 0)$ and $(3, 0)$.

13. (a) The graph crosses the y-axis at $(0, 0)$, so the y-intercept is $(0, 0)$.

(b) The graph crosses the x-axis at $(-2, 0), (0, 0)$ and $(5, 0)$, so the x-intercepts are $(-2, 0), (0, 0)$ and $(5, 0)$.

15. $3x + 5y = 15$

(a) To find the y-intercept, let $x = 0$. This is the same as temporarily ignoring the x-term and then solving.

$$5y = 15$$
$$y = 3$$

The y–intercept is $(0, 3)$.

(b) To find the x-intercept, let $y = 0$. This is the same as temporarily ignoring the y-term and then solving.

$$3x = 15$$
$$x = 5$$

The x-intercept is $(5, 0)$.

17. $9x - 2y = 36$

(a) To find the y-intercept, let $x = 0$. This is the same as temporarily ignoring the x-term and then solving.

$$-2y = 36$$
$$y = -18$$

The y-intercept is $(0, -18)$.

(b) To find the x-intercept, let $y = 0$. This is the same as temporarily ignoring the y-term and then solving.

$$9x = 36$$
$$x = 4$$

The x-intercept is $(4, 0)$.

19. $-4x + 5y = 80$

(a) To find the y-intercept, let $x = 0$. This is the same as temporarily ignoring the x-term and then solving.

$$5y = 80$$
$$y = 16$$

The y-intercept is $(0, 16)$.

(b) To find the x-intercept, let $y = 0$. This is the same as temporarily ignoring the y-term and then solving.

$$-4x = 80$$
$$x = -20$$

The x-intercept is $(-20, 0)$.

21. $x = 12$

Observe that this is the equation of a vertical line 12 units to the right of the y-axis. Thus, (a) there is no y-intercept and (b) the x-intercept is $(12, 0)$.

23. $y = -9$

Observe that this is the equation of a horizontal line 9 units below the x-axis. Thus, (a) the y-intercept is $(0, -9)$ and (b) there is no x-intercept.

25. $3x + 5y = 15$

Find the y-intercept:

$$\begin{aligned} 5y &= 15 \quad \text{Ignoring the } x\text{-term} \\ y &= 3 \end{aligned}$$

The y-intercept is $(0, 3)$.

Find the x-intercept:

$$\begin{aligned} 3x &= 15 \quad \text{Ignoring the } y\text{-term} \\ x &= 5 \end{aligned}$$

The x-intercept is $(5, 0)$.

To find a third point we replace x with -5 and solve for y.

$$\begin{aligned} 3(-5) + 5y &= 15 \\ -15 + 5y &= 15 \\ 5y &= 30 \\ y &= 6 \end{aligned}$$

The point $(-5, 6)$ appears to line up with the intercepts, so we draw the graph.

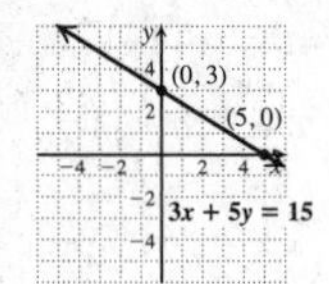

27. $x + 2y = 4$

Find the y-intercept:

$$\begin{aligned} 2y &= 4 \quad \text{Ignoring the } x\text{-term} \\ y &= 2 \end{aligned}$$

The y-intercept is $(0, 2)$.

Find the x-intercept:

$$x = 4 \quad \text{Ignoring the } y\text{-term}$$

The x-intercept is $(4, 0)$.

To find a third point we replace x with 2 and solve for y.

$$\begin{aligned} 2 + 2y &= 4 \\ 2y &= 2 \\ y &= 1 \end{aligned}$$

The point $(2, 1)$ appears to line up with the intercepts, so we draw the graph.

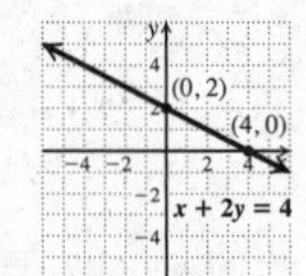

29. $-x + 2y = 8$

Find the y-intercept:

$$\begin{aligned} 2y &= 8 \quad \text{Ignoring the } x\text{-term} \\ y &= 4 \end{aligned}$$

The y-intercept is $(0, 4)$.

Find the x-intercept:

$$\begin{aligned} -x &= 8 \quad \text{Ignoring the } y\text{-term} \\ x &= -8 \end{aligned}$$

The x-intercept is $(-8, 0)$.

To find a third point we replace x with 4 and solve for y.

$$\begin{aligned} -4 + 2y &= 8 \\ 2y &= 12 \\ y &= 6 \end{aligned}$$

The point $(4, 6)$ appears to line up with the intercepts, so we draw the graph.

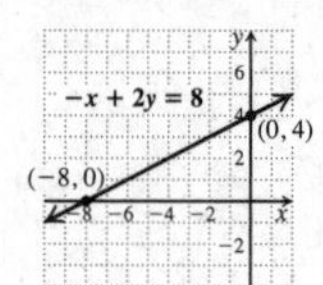

31. $3x + y = 9$

Find the y-intercept:

$$y = 9 \qquad \text{Ignoring the } x\text{-term}$$

The y-intercept is $(0, 9)$.

Find the $x-$intercept:

$$\begin{aligned} 3x &= 9 \quad \text{Ignoring the } y\text{-term} \\ x &= 3 \end{aligned}$$

The x-intercept is $(3, 0)$.

To find a third point we replace x with 2 and solve for y.

$$\begin{aligned} 3 \cdot 2 + y &= 9 \\ 6 + y &= 9 \\ y &= 3 \end{aligned}$$

The point $(2, 3)$ appears to line up with the intercepts, so we draw the graph.

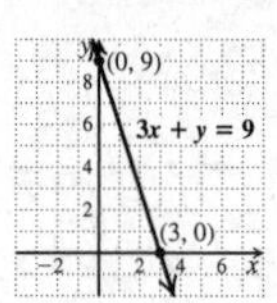

33. $y = 2x - 6$

Find the y-intercept:

$$y = -6 \qquad \text{Ignoring the } x\text{-term}$$

The y-intercept is $(0, -6)$.

Find the $x-$intercept:

$$\begin{aligned} 0 &= 2x - 6 \quad \text{Replacing } y \text{ with } 0 \\ 6 &= 2x \\ 3 &= x \end{aligned}$$

The x-intercept is $(3, 0)$.

To find a third point we replace x with 2 and find y.

$$y = 2 \cdot 2 - 6 = 4 - 6 = -2$$

The point $(2, -2)$ appears to line up with the intercepts, so we draw the graph.

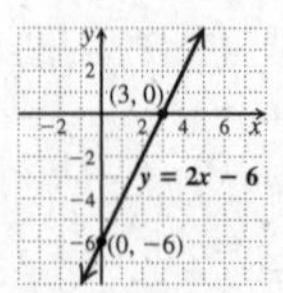

35. $5x - 10 = 5y$

We can leave the equation in the given form or rewrite it in the form $Ax + By = C$. We will use the given form.

Find the y-intercept:

$$\begin{aligned} -10 &= 5y \quad \text{Ignoring the } x\text{-term} \\ -2 &= y \end{aligned}$$

The y-intercept is $(0,-2)$.

To find the x-intercept, let $y = 0$.

$$\begin{aligned} 5x - 10 &= 5 \cdot 0 \\ 5x - 10 &= 0 \\ 5x &= 10 \\ x &= 2 \end{aligned}$$

The x-intercept is $(2,0)$.

To find a third point we replace x with 5 and solve for y.

$$\begin{aligned} 5 \cdot 5 - 10 &= 5y \\ 25 - 10 &= 5y \\ 15 &= 5y \\ 3 &= y \end{aligned}$$

The point $(5,3)$ appears to line up with the intercepts, so we draw the graph.

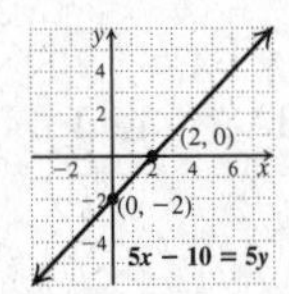

37. $2x - 5y = 10$

Find the y-intercept:

$$\begin{aligned} -5y &= 10 \quad \text{Ignoring the } x\text{-term} \\ y &= -2 \end{aligned}$$

The y-intercept is $(0,-2)$.

Find the x-intercept:

$$\begin{aligned} 2x &= 10 \quad \text{Ignoring the } y\text{-term} \\ x &= 5 \end{aligned}$$

The x-intercept is $(5,0)$.

To find a third point we replace x with -5 and solve for y.

$$\begin{aligned} 2(-5) - 5y &= 10 \\ -10 - 5y &= 10 \\ -5y &= 20 \\ y &= -4 \end{aligned}$$

The point $(-5,-4)$ appears to line up with the intercepts, so we draw the graph.

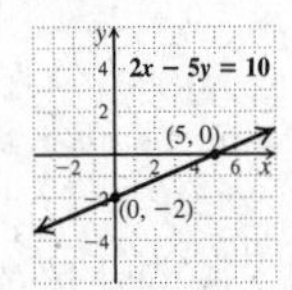

39. $6x + 2y = 12$

Find the y-intercept:

$$\begin{aligned} 2y &= 12 \quad \text{Ignoring the } x\text{-term} \\ y &= 6 \end{aligned}$$

The y-intercept is $(0,6)$.

Find the x-intercept:

$$\begin{aligned} 6x &= 12 \quad \text{Ignoring the } y\text{-term} \\ x &= 2 \end{aligned}$$

The x-intercept is $(2,0)$.

To find a third point we replace x with 3 and solve for y.

$$\begin{aligned} 6 \cdot 3 + 2y &= 12 \\ 18 + 2y &= 12 \\ 2y &= -6 \\ y &= -3 \end{aligned}$$

The point $(3,-3)$ appears to line up with the intercepts, so we draw the graph.

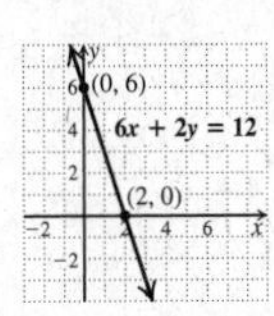

41. $4x + 3y = 16$

Find the y-intercept:

$$\begin{aligned} 3y &= 16 \quad \text{Ignoring the } x\text{-term} \\ y &= \tfrac{16}{3} \end{aligned}$$

The y-intercept is $(0, \frac{16}{3})$.

Find the x-intercept:

$$\begin{aligned} 4x &= 16 \quad \text{Ignoring the } y\text{-term} \\ x &= 4 \end{aligned}$$

The x-intercept is $(4,0)$.

To find a third point we replace x with -2 and solve for y.

$$\begin{aligned} 4(-2) + 3y &= 16 \\ -8 + 3y &= 16 \\ 3y &= 24 \\ y &= 8 \end{aligned}$$

The point $(-2,8)$ appears to line up with the intercepts, so we draw the graph.

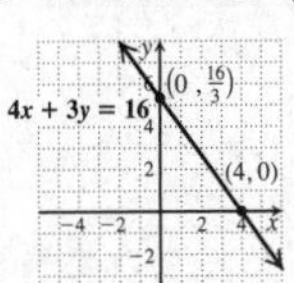

43. $2x + 4y = 1$

Find the y-intercept:

$$\begin{aligned} 4y &= 1 \quad \text{Ignoring the } x\text{-term} \\ y &= \frac{1}{4} \end{aligned}$$

The y-intercept is $\left(0, \frac{1}{4}\right)$.

Find the x-intercept:

$$\begin{aligned} 2x &= 1 \quad \text{Ignoring the } y\text{-term} \\ x &= \tfrac{1}{2} \end{aligned}$$

The x-intercept is $\left(\frac{1}{2}, 0\right)$.

To find a third point we replace x with $-\frac{3}{2}$ and solve for y.

$$\begin{aligned} 2\left(-\tfrac{3}{2}\right) + 4y &= 1 \\ -3 + 4y &= 1 \\ 4y &= 4 \\ y &= 1 \end{aligned}$$

The point $\left(-\frac{3}{2}, 1\right)$ appears to line up with the intercepts, so we draw the graph.

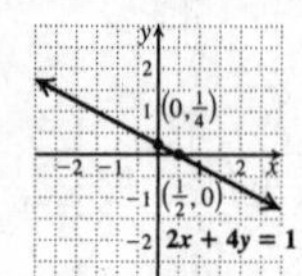

45. $5x - 3y = 180$

Find the y-intercept:

$$-3y = 180 \quad \text{Ignoring the } x\text{-term}$$
$$y = -60$$

The y-intercept is $(0, -60)$.

Find the x-intercept:

$$5x = 180 \quad \text{Ignoring the } y\text{-term}$$
$$x = 36$$

The x-intercept is $(36, 0)$.

To find a third point we replace x with 6 and solve for y.

$$5 \cdot 6 - 3y = 180$$
$$30 - 3y = 180$$
$$-3y = 150$$
$$y = -50$$

This means that $(6, -50)$ is on the graph.

To graph all three points, the y-axis must go to at least 60 and the x-axis must go to at least 36. Using a scale of 10 units per square allows us to display both intercepts and $(6, -50)$ as well as the origin.

The point $(6, -50)$ appears to line up with the intercepts, so we draw the graph.

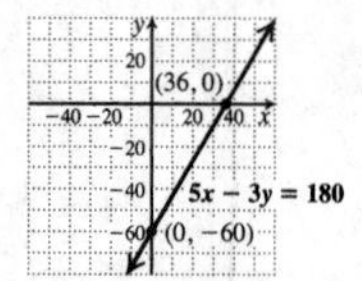

47. $y = -30 + 3x$

Find the y-intercept:

$$y = -30 \quad \text{Ignoring the } x\text{-term}$$

The y-intercept is $(0, -30)$.

To find the x-intercept, let $y = 0$.

$$0 = -30 + 3x$$
$$30 = 3x$$
$$10 = x$$

The x-intercept is $(10, 0)$.

To find a third point we replace x with 5 and solve for y.

$$y = -30 + 3 \cdot 5$$
$$y = -30 + 15$$
$$y = -15$$

This means that $(5, -15)$ is on the graph.

To graph all three points, the y-axis must go to at least -30 and the x-axis must go to at least 10. Using a scale of 5 units per square allows us to display both intercepts and $(5, -15)$ as well as the origin.

The point $(5, -15)$ appears to line up with the intercepts, so we draw the graph.

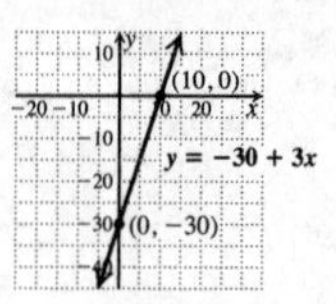

49. $-4x = 20y + 80$

To find the y-intercept, we let $x = 0$.

$$-4 \cdot 0 = 20y + 80$$
$$0 = 20y + 80$$
$$-80 = 20y$$
$$-4 = y$$

The y-intercept is $(0, -4)$.

Find the x-intercept:

$$-4x = 80 \quad \text{Ignoring the } y\text{-term}$$
$$x = -20$$

The x-intercept is $(-20, 0)$.

To find a third point we replace x with -40 and solve for y.

$$-4(-40) = 20y + 80$$
$$160 = 20y + 80$$
$$80 = 20y$$
$$4 = y$$

This means that $(-40, 4)$ is on the graph.

To graph all three points, the y-axis must go at least from -4 to 4 and the x-axis must go at least from -40 to -20. Since we also want to include the origin we can use a scale of 10 units per square on the x-axis and 1 unit per square on the y-axis.

The point $(-40, 4)$ appears to line up with the intercepts, so we draw the graph.

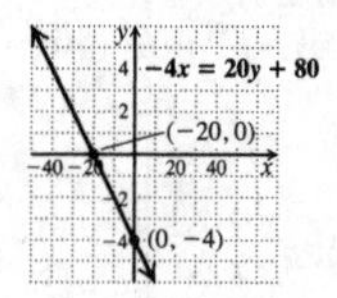

51. $y - 3x = 0$

Find the y-intercept:

$$y = 0 \quad \text{Ignoring the } x\text{-term}$$

The y-intercept is $(0, 0)$. Note that this is also the x-intercept.

In order to graph the line, we will find a second point.

When $x = 1$, $y - 3 \cdot 1 = 0$
$$y - 3 = 0$$
$$y = 3$$

To find a third point we replace $x = -1$ and solve for y.

$$y - 3(-1) = 0$$
$$y + 3 = 0$$
$$y = -3$$

The point $(-1, -3)$ appears to line up with the other two points, so we draw the graph.

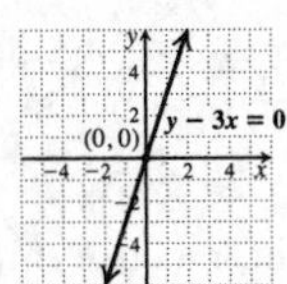

53. $y = 1$

Any ordered pair $(x, 1)$ is a solution. The variable y must be 1, but the x variable can be any number we choose. A few solutions are listed below. Plot these points and draw the line.

x	y
-3	1
0	1
2	1

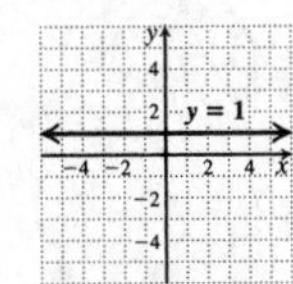

55. $x = 3$

Any ordered pair $(3, y)$ is a solution. The variable x must be 3, but the y variable can be any number we choose. A few solutions are listed below. Plot these points and draw the line.

x	y
3	-2
3	0
3	4

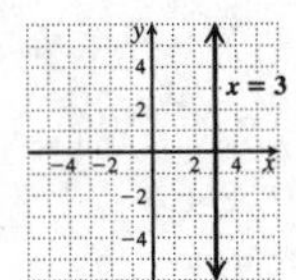

57. $y = -2$

Any ordered pair $(x, -2)$ is a solution. The variable y must be -2, but the x variable can be any number we choose. A few solutions are listed below. Plot these points and draw the line.

x	y
-3	-2
0	-2
4	-2

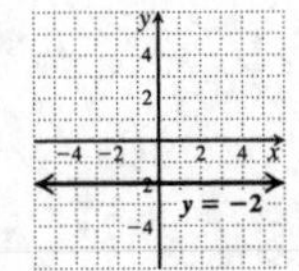

59. $x = -1$

Any ordered pair $(-1, y)$ is a solution. The variable x must be -1, but the y variable can be any number we choose. A few solutions are listed below. Plot these points and draw the line.

x	y
-1	-3
-1	0
-1	2

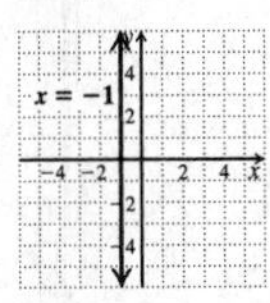

61. $y = -15$

Any ordered pair $(x, -15)$ is a solution. The variable y must be -15, but the x variable can be any number we choose. A few solutions are listed below. Plot these points and draw the line.

x	y
-1	-15
0	-15
3	-15

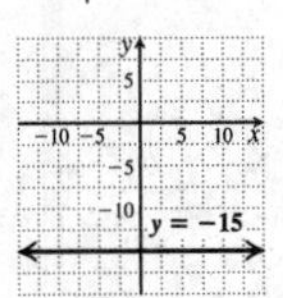

63. $y = 0$

Any ordered pair $(x, 0)$ is a solution. A few solutions are listed below. Plot these points and draw the line.

x	y
-4	0
0	0
2	0

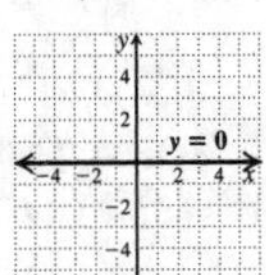

65. $x = -\dfrac{5}{2}$

Any ordered pair $\left(-\dfrac{5}{2}, y\right)$ is a solution. A few solutions are listed below. Plot these points and draw the line.

x	y
$-\frac{5}{2}$	-3
$-\frac{5}{2}$	0
$-\frac{5}{2}$	5

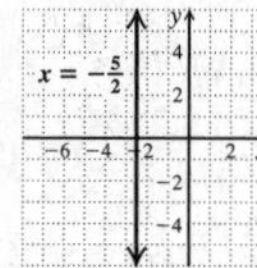

67. $-4x = -100$

$x = 25$ Dividing by -4

The graph is a vertical line 25 units to the right of the y-axis.

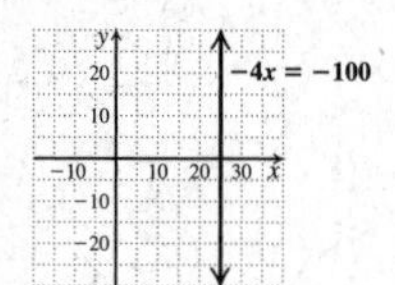

69. $35 + 7y = 0$

$7y = -35$

$y = -5$

The graph is a horizontal line 5 units below the x-axis.

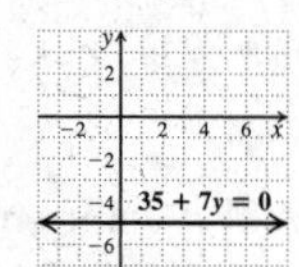

71. Note that every point on the horizontal line passing through $(0, -1)$ has -1 as the y-coordinate. Thus, the equation of the line is $y = -1$.

73. Note that every point on the vertical line passing through $(4, 0)$ has 4 as the x-coordinate. Thus, the equation of the line is $x = 4$.

75. Note that every point on the vertical line passing through $(0, 0)$ has 0 as the x-coordinate. Thus, the equation of the line is $x = 0$.

77. *Writing Exercise.* Any solution of $y = 8$ is an ordered pair $(x, 8)$. Thus, all points on the graph of $y = 8$ are 8 units above the x-axis, so they lie on a horizontal line.

79. $d - 7$

81. Let n represent the number. Then we have $7 + 4n$.

83. Let x and y represent the numbers. Then we have $2(x+y)$.

85. *Writing Exercise.* The graph will be a line with y-intercept $(0, C)$ and x-intercept $(C, 0)$.

87. The x-axis is a horizontal line, so it is of the form $y = b$. All points on the x-axis are of the form $(x, 0)$, so b must be 0 and the equation is $y = 0$.

89. A line parallel to the y-axis has an equation of the form $x = a$. Since the x-coordinate of one point on the line is -2, then $a = -2$ and the equation is $x = -2$.

91. Since the x-coordinate of the point of intersection must be -3 and y must equal 4, the point of intersection is $(-3, 4)$.

93. The y-intercept is $(0, 5)$, so we have $y = mx + 5$. Another point on the line is $(-3, 0)$ so we have

$$0 = m(-3) + 5$$
$$-5 = -3m$$
$$\frac{5}{3} = m$$

The equation is $y = \frac{5}{3}x + 5$, or $5x - 3y = -15$, or $-5x + 3y = 15$.

95. Substitute 0 for x and -8 for y.

$$4 \cdot 0 = C - 3(-8)$$
$$0 = C + 24$$
$$-24 = C$$

97.

$$Ax + D = C$$
$$Ax = C - D$$
$$x = \frac{C - D}{A}$$

The x-intercept is $\left(\frac{C-D}{A}, 0\right)$

99. Find the y-intercept:

$-7y = 80$ Covering the x-term

$y = -\frac{80}{7} = -11.\overline{428571}$

The y-intercept is $\left(0, -\frac{80}{7}\right)$, or $(0, -11.\overline{428571})$.

Find the x-intercept:

$2x = 80$ Covering the y-term

$x = 40$

The x-intercept is $(40, 0)$.

101. From the equation we see that the y-intercept is $(0, -9)$.

To find the x-intercept, let $y = 0$.

$$0 = 0.2x - 9$$
$$9 = 0.2x$$
$$45 = x$$

The x-intercept is $(45, 0)$.

103. Find the y-intercept.

$25y = 1$ Covering the x-term

$y = \frac{1}{25}$, or 0.04

The y-intercept is $\left(0, \frac{1}{25}\right)$, or $(0, 0.04)$.

Find the x-intercept:

$50x = 1$ Covering the y-term

$x = \frac{1}{50}$, or 0.02

The x-intercept is $\left(\frac{1}{50}, 0\right)$, or $(0.02, 0)$.

Exercise Set 3.4

1. $\dfrac{100 \text{ miles}}{5 \text{ hours}} = 20$ miles per hour, or miles/hour

3. $\dfrac{300 \text{ dollars}}{150 \text{ miles}} = 2$ dollars per mile, or dollars/mile

5. $\dfrac{40 \text{ minutes}}{8 \text{ errands}} = 5$ minutes per errand, or minutes/errand

7. a) We divide the number of miles traveled by the number of gallons of gas used for distance traveled.

Rate, in miles per gallon

$$= \frac{14,131 \text{ mi} - 13,741 \text{ mi}}{13 \text{ gal}}$$
$$= \frac{390 \text{ mi}}{13 \text{ gal}}$$
$$= 30 \text{ mi/gal}$$
$$= 30 \text{ miles per gallon}$$

b) We divide the cost of the rental by the number of days. From June 5 to June 8 is 8 – 5, or 3 days.

Average cost, in dollars per day

$$= \frac{118 \text{ dollars}}{3 \text{ days}}$$
$$\approx 39.33 \text{ dollars/day}$$
$$\approx \$39.33 \text{ per day}$$

c) We divide the number of miles traveled by the number of days. The car was driven 390 miles, and was rented for 3 days.

Rate, in miles per day

$$= \frac{390 \text{ mi}}{3 \text{ days}}$$
$$= 130 \text{ mi/day}$$
$$= 130 \text{ mi per day}$$

d) Note that $\$118 = 11,800¢$. The car was driven 390 miles.

$$\text{Rate, in cents per mile} = \frac{11,800¢}{390 \text{ mi}}$$
$$\approx 30¢ \text{ per mi}$$

9. a) From 9:00 to 11:00 is 11 – 9, or 2 hr.

$$\text{Average speed, in miles per hour} = \frac{14 \text{ mi}}{2 \text{ hr}}$$
$$= 7 \text{ mph}$$

b) From part (a) we know that the bike was rented for 2 hr.

$$\text{Rate, in dollars per hour} = \frac{\$15}{2 \text{ hr}}$$
$$= \$7.50 \text{ per hr}$$

c) $$\text{Rate, in dollars per mile} = \frac{\$15}{14 \text{ mi}}$$
$$\approx \$1.07 \text{ per mi}$$

11. a) It is 3 hr from 9:00 A.M. to noon and 2 more hours from noon to 2:00 P.M., so the proofreader worked 3 + 2, or 5 hr.

$$\text{Rate, in dollars per hour} = \frac{\$110}{5 \text{ hr}}$$
$$= \$22 \text{ per hr}$$

b) The number of pages proofread is 195 – 92, or 103.

$$\text{Rate, in pages per hour} = \frac{103 \text{ pages}}{5 \text{ hr}}$$
$$= 20.6 \text{ pages per hr}$$

c) $$\text{Rate, in dollars per page} = \frac{\$110}{103 \text{ pages}}$$
$$\approx \$1.07 \text{ per page}$$

13. Increase in debt: 8612 billion –5770 billion or 2842 billion.

Change in time 2006 – 2001 = 5 year

$$\text{Rate of increase} = \frac{\text{Change in debt}}{\text{Change in time}}$$
$$= \frac{\$2842 \text{ billion}}{5 \text{ yr}}$$
$$\approx \$568.4 \text{ billion/yr}$$

15. a) The elevator traveled 34 – 5, or 29 floors in 2:40 – 2:38, or 2 min.

$$\text{Average rate of travel} = \frac{29 \text{ floors}}{2 \text{ min}}$$
$$= 14.5 \text{ floors per min}$$

b) In part (a) we found that the elevator traveled 29 floors in 2 min. Note that $2 \text{ min} = 2 \times 1 \text{ min} = 2 \times 60 \text{ sec} = 120 \text{ sec}$.

$$\text{Average rate of travel} = \frac{120 \text{ sec}}{29 \text{ floors}}$$
$$\approx 4.14 \text{ sec per floor}$$

17. Ascended $29,028 \text{ ft} - 17,552 \text{ ft} = 11,476 \text{ ft}$. The time of ascent: 8 hr, 10 min, or $8 \text{ hr} + 10 \text{ min} = 480 \text{ min} + 10 \text{ min} = 490 \text{ min}$. a)

$$\text{Rate, in feet per minute} = \frac{11,476 \text{ ft}}{490 \text{ min}}$$
$$\approx 23.42 \text{ ft/min}$$

b) $$\text{Rate, in minutes per foot} = \frac{490 \text{ min}}{11,476 \text{ ft}}$$
$$\approx 0.04 \text{ min/ft}$$

19. The rate of decrease is given in tons per year, so we list the number of tons in the vertical axis and the year on the horizontal axis. If we count by increments of 10 million on the vertical axis we can easily reach 35.7 million and beyond. We label the units on the vertical axis in millions of tons. We list the years on the horizontal axis, beginning with 2006. We plot the point (2006, 35.7 million). Then, to display the decreased rate we move from that point to a point that represents a decrease of 700,000 tons one year later. The coordinates of this point are (2006 + 1, 35.7 –

0.7 million), or (2007, 35 million). Finally, we draw a line through the two points.

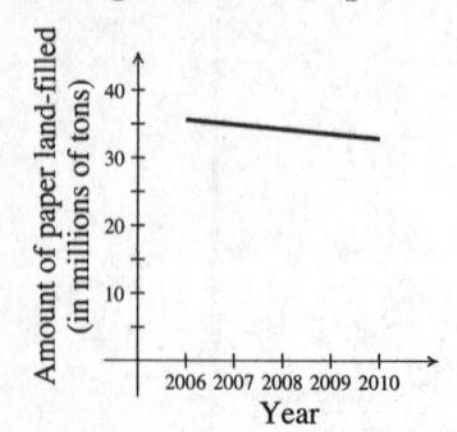

21. The rate is given in dollars per year, so we list the amount in sales of asthma drug products on the vertical axis and year on the horizontal axis. We can count by increments of 2 billion on the vertical axis. We plot the point (2006, \$11 billion). Then to display the rate of growth, we move from that point to a point that represents \$1.2 billion more a year later. The coordinates of this point are (2006+1, \$11+1.2 billion) or (2007, \$12.2 billion). Finally, we draw a line through the two points.

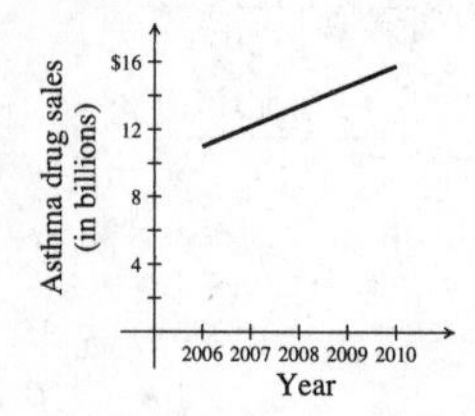

23. The rate is given in miles per hour, so we list the number of miles traveled on the vertical axis and the time of day on the horizontal axis. If we count by 100's of miles on the vertical axis we can easily reach 230 without needing a terribly large graph. We plot the point (3:00, 230). Then to display the rate of travel, we move from that point to a point that represents 90 more miles traveled 1 hour later. The coordinates of this point are (3:00 + 1 hr, 230 + 90), or (4:00, 320). Finally, we draw a line through the two points.

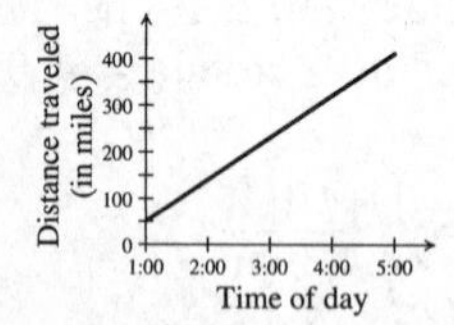

25. The rate is given in dollars per hour so we list money earned on the vertical axis and the time of day on the horizontal axis. We can count by \$20 on the vertical axis and reach \$50 without needing a terribly large graph. Next we plot the point (2:00 P.M., \$50). To display the rate we move from that point to a point that represents \$15 more 1 hour later. The coordinates of this point are (2+1, \$50+\$15), or (3:00 P.M., \$65). Finally, we draw a line through the two points.

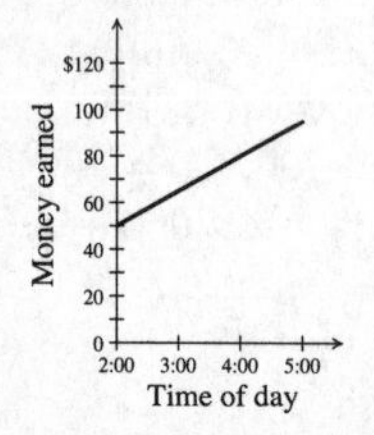

27. The rate is given in cost per minute so we list the amount of the telephone bill on the vertical axis and the number of additional minutes on the horizontal axis. We begin with \$7.50 on the vertical axis and count by \$0.50. A jagged line at the base of the axis indicates that we are not showing amounts smaller than \$7.50. We begin with 0 additional minutes on the horizontal axis and plot the point (0, \$7.50). We move from there to a point that represents \$0.10 more 1 minute later. The coordinates of this point are (0 + 1 min, \$7.50 + \$0.10), or (1 min, \$7.60). Then we draw a line through the two points.

29. The points (10:00, 30 calls) and (1:00, 90 calls) are on the graph. This tells us that in the 3 hr between 10:00 and 1:00 there were $90 - 30 = 60$ calls completed. The rate is

$$\frac{60 \text{ calls}}{3 \text{ hr}} = 20 \text{ calls/hour.}$$

31. The points (12:00, 100 mi) and (2:00, 250 mi) are on the graph. This tells us that in the 2 hr between 12:00 and 2:00 the train traveled $250 - 100 = 150$ mi. The rate is

$$\frac{150 \text{ mi}}{2\text{hr}} = 75 \text{ mi per hr.}$$

33. The points (5 min, 60¢) and (10 min, 120¢) are on the graph. This tells us that in $10 - 5 = 5$ min the cost of the call increased 120¢ − 60¢ = 60¢. The rate is

$$\frac{60¢}{5 \text{ min}} = 12¢ \text{ per min.}$$

35. The points (1970, 140 thousand) and (2000, 80 thousand) are on the graph. This tells us that in $2000 - 1970 = 30$yrs the population decreases $80 - 140 = -60$thousand. The rate is

$$\frac{-60 \text{ thousand}}{30 \text{ yr}} = -2000 \text{ people/year.}$$

37. The points (90 mi, 2 gal) and (225 mi, 5 gal) are on the graph. This tells us that when driven $225 - 90 = 135$ mi the vehicle consumed $5 - 2 = 3$ gal of gas. The rate is

$$\frac{3 \text{ gal}}{135 \text{ mi}} = 0.02 \text{ gal/mi.}$$

39. Since swimming is the slowest of the three sports and biking is the fastest, the slope of the line representing swimming speed will be the least steep of the three and that representing biking speed will be the steepest. The second segment of graph (e) rises most steeply and the third segment is the least steep of the three segments. Thus this graph represents running followed by biking and then swimming.

41. Since swimming is the slowest of the three sports and biking is the fastest, the slope of the line representing swimming speed will be the least steep of the three and that representing biking speed will be the steepest. The first segment of graph (d) is the least steep and the second segment is the steepest of the three segments. Thus this graph represents swimming followed by biking and then running.

43. Since swimming is the slowest of the three sports and biking is the fastest, the slope of the line representing swimming speed will be the least steep of the three and that representing biking speed will be the steepest. The first segment of graph (b) is the steepest and the second segment is the least steep of the three segments. Thus this graph represents biking followed by swimming and then running.

45. *Writing Exercise.* A negative rate of travel indicates that an object is moving backwards.

47. $-2-(-7) = -2+7 = 5$

49. $\dfrac{5-(-4)}{-2-7} = \dfrac{9}{-9} = -1$

51. $\dfrac{-4-8}{11-2} = \dfrac{-12}{9} = \dfrac{-4}{3}$

53. $\dfrac{-6-(-6)}{-2-7} = \dfrac{-6+6}{-2-7} = \dfrac{0}{9} = 0$

55. *Writing Exercise.* a) The graph of Jon's total earnings would be above Jenny's total earnings, with Jon's rate or slope steeper than Jenny's.

b) Jenny's graph is above Jon's but the slope or rate is the same.

c) The final result (total earnings) can be compared, but not the rate.

57. Let t = flight time and a = altitude. While the plane is climbing at a rate of 6300 ft/min, the equation $a = 6300t$ describes the situation. Solving $31,500 = 6300t$, we find that the cruising altitude of 31,500 ft is reached after about 5 min. Thus we graph $a = 6300t$ for $0 \le t \le 5$.

The plane cruises at 31,500 ft for 3 min, so we graph $a = 31,500$ for $5 < t \le 8$. After 8 min the plane descends at a rate of 3500 ft/min and lands. The equation $a = 31,500 - 3500(t-8)$, or $a = -3500t + 59,500$, describes this situation. Solving $0 = -3500t + 59,500$, we find that the plane lands after about 17 min. Thus we graph $a = -3500t + 59,500$ for $8 < t \le 17$. The entire graph is show below.

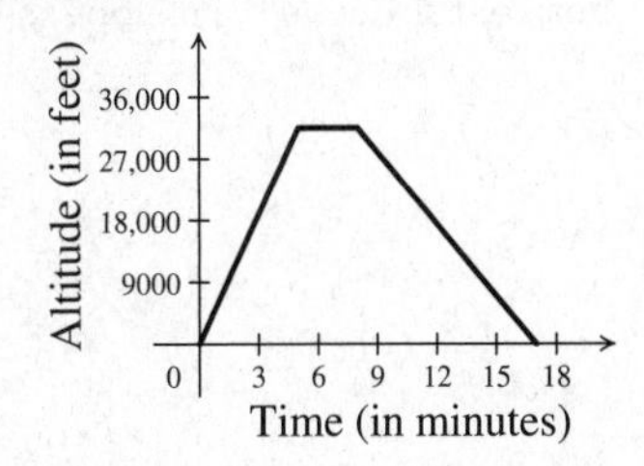

59. Let the horizontal axis represent the distance traveled, in miles, and let the vertical axis represent the fare, in dollars. Use increments of 1/5, or 0.2 mi, on the horizontal axis and of \$1 on the vertical axis. The fare for traveling 0.2 mi is \$2 + \$0.50 · 1, or \$2.50 and for 0.4 mi, or 0.2 mi × 2, we have \$2 + \$0.50(2), or \$3. Plot the points (0.2 mi, \$2.50) and (0.4 mi, \$3) and draw the line through them.

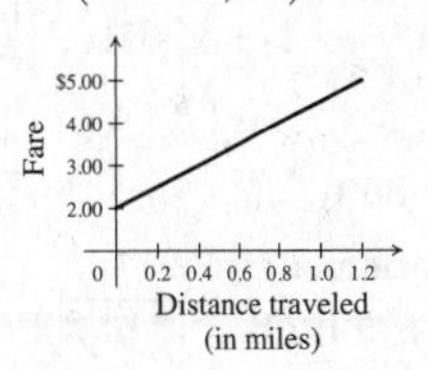

61. 95 mph + 39 mph = 134 mph

$\dfrac{134 \text{ mi}}{1 \text{ hr}}$ gives us $\dfrac{1 \text{ hr}}{134 \text{ mi}}$.

$\dfrac{1 \text{ hr}}{134 \text{ min}} = \dfrac{1 \cancel{\text{hr}}}{134 \text{ mi}} \cdot \dfrac{60 \text{ min}}{1 \cancel{\text{hr}}} \approx 0.45 \text{ min per mi}$

63. First we find Anne's speed in minutes per kilometer.

$\text{Speed} = \dfrac{15.5 \text{ min}}{7 \text{ km} - 4 \text{ km}} = \dfrac{15.5}{3} \dfrac{\text{min}}{\text{km}}$

Now we convert min/km to min/mi.

$\dfrac{15.5}{3} \dfrac{\text{min}}{\text{km}} \approx \dfrac{15.5}{3} \dfrac{\text{min}}{\cancel{\text{km}}} \cdot \dfrac{1 \cancel{\text{km}}}{0.621 \text{ mi}} \approx \dfrac{15.5}{1.863} \dfrac{\text{min}}{\text{mi}}$

At a rate of $\dfrac{15.5}{1.863} \dfrac{\text{min}}{\text{mi}}$, to run a 5-mi race it would take $\dfrac{15.5}{1.863} \dfrac{\text{min}}{\cancel{\text{mi}}} \cdot 5 \cancel{\text{mi}} \approx 41.6$ min.

(Answers may vary slightly depending on the conversion factor used.)

65. First we find Ryan's rate. Then we double it to find Alex's rate. Note that 50 minutes = $\dfrac{50}{60}$ hr = $\dfrac{5}{6}$ hr.

$$\begin{aligned}\text{Ryan's rate} &= \frac{\text{change in number of bushels picked}}{\text{corresponding change in time}}\\ &= \frac{5\frac{1}{2} - 4 \text{ bushels}}{\frac{5}{6} \text{ hr}}\\ &= \frac{1\frac{1}{2} \text{ bushels}}{\frac{5}{6} \text{ hr}}\\ &= \frac{3}{2} \cdot \frac{6}{5} \frac{\text{bushels}}{\text{hr}}\\ &= \frac{9}{5} \text{ bushels per hour, or}\\ &\quad 1.8 \text{ bushels per hour}\end{aligned}$$

Then Alex's rate is 2(1.8) = 3.6 bushels per hour.

Exercise Set 3.5

1. A teenager's height increases over time, so the rate is positive.

3. The water level decreases during a drought, so the rate is negative.

5. The distance from the starting point increases during a race, so the rate is positive.

7. The number of U.S. senators does not change, so the rate is zero.

9. The number of people present decreases in the moments following the final buzzer, so the rate is negative.

11. The rate can be found using the coordinates of any two points on the line. We use (10, \$600) and (25, \$1500).

$$\begin{aligned}\text{Rate} &= \frac{\text{change in compensation}}{\text{corresponding change in number of blogs}}\\ &= \frac{\$1500 - \$600}{25 - 10}\\ &= \frac{\$900}{15 \text{ blogs}}\\ &= \$60 \text{ per blog}\end{aligned}$$

13. The rate can be found using the coordinates of any two points on the line. We use (May2005, \$380) and (Mar2006, \$320).

$$\begin{aligned}\text{Rate} &= \frac{\text{change in price}}{\text{corresponding change in time}}\\ &= \frac{\$320 - \$380}{\text{Mar2006} - \text{May2005}}\\ &= \frac{-\$60}{15 - 5 \text{ months}}\\ &= \frac{-60}{10}\\ &= -\$6 \text{ per month}\end{aligned}$$

15. The rate can be found using the coordinates of any two points on the line. We use (35, 480) and (65, 510), where 35 and 65 are in \$1000's.

$$\begin{aligned}\text{Rate} &= \frac{\text{change in score}}{\text{corresponding change in income}}\\ &= \frac{510 - 480 \text{ points}}{65 - 35}\\ &= \frac{30 \text{ points}}{30}\\ &= 1 \text{ point per } \$1000 \text{ income}\end{aligned}$$

17. The rate can be found using the coordinates of any two points on the line. We use (0 min, 54°) and (27 min, −4°).

$$\begin{aligned}\text{Rate} &= \frac{\text{change in temperature}}{\text{corresponding change in time}}\\ &= \frac{-4^\circ - 54^\circ}{27 \text{ min} - 0 \text{ min}}\\ &= \frac{-58^\circ}{27 \text{ min}}\\ &\approx -2.1^\circ \text{per min}\end{aligned}$$

19. We can use any two points on the line, such as $(0, 1)$ and $(3, 5)$.

$$\begin{aligned}m &= \frac{\text{change in } y}{\text{change in } x}\\ &= \frac{5-1}{3-0} = \frac{4}{3}\end{aligned}$$

21. We can use any two points on the line, such as $(1, 0)$ and $(3, 3)$.

$$\begin{aligned}m &= \frac{\text{change in } y}{\text{change in } x}\\ &= \frac{3-0}{3-1} = \frac{3}{2}\end{aligned}$$

23. We can use any two points on the line, such as $(2, 2)$ and $(4, 6)$.

$$\begin{aligned}m &= \frac{\text{change in } y}{\text{change in } x}\\ &= \frac{6-2}{4-2}\\ &= \frac{4}{2} = 2\end{aligned}$$

25. We can use any two points on the line, such as $(0, 2)$ and $(2, 0)$.

$$\begin{aligned}m &= \frac{\text{change in } y}{\text{change in } x}\\ &= \frac{2-0}{0-2} = \frac{2}{-2} = -1\end{aligned}$$

27. This is the graph of a horizontal line. Thus, the slope is 0.

29. We can use any two points on the line, such as $(0, 2)$ and $(3, 1)$.

$$\begin{aligned}m &= \frac{\text{change in } y}{\text{change in } x}\\ &= \frac{1-2}{3-0} = -\frac{1}{3}\end{aligned}$$

31. This is the graph of a vertical line. Thus, the slope is undefined.

33. We can use any two points on the line, such as $(-2, 1)$ and $(2, -2)$.

$$\begin{aligned}m &= \frac{\text{change in } y}{\text{change in } x}\\ &= \frac{-2-1}{2-(-2)} = -\frac{3}{4}\end{aligned}$$

35. We can use any two points on the line, such as $(-2, 0)$ and $(2, 1)$.

$$\begin{aligned}m &= \frac{\text{change in } y}{\text{change in } x}\\ &= \frac{1-0}{2-(-2)} = \frac{1}{4}\end{aligned}$$

37. This is the graph of a horizontal line, so the slope is 0.

39. $(1, 3)$ and $(5, 8)$

$$m = \frac{8-3}{5-1} = \frac{5}{4}$$

41. $(-2, 4)$ and $(3, 0)$

$$m = \frac{4-0}{-2-3} = \frac{4}{-5} = -\frac{4}{5}$$

43. $(-4, 0)$ and $(5, 6)$

$$m = \frac{6-0}{5-(-4)} = \frac{6}{9} = \frac{2}{3}$$

45. $(0, 7)$ and $(-3, 10)$

$$m = \frac{10-7}{-3-0} = \frac{3}{0-3} = -1$$

47. $(-2, 3)$ and $(-6, 5)$

$$m = \frac{5-3}{-6-(-2)} = \frac{2}{-6+2} = \frac{2}{-4} = -\frac{1}{2}$$

49. $\left(-2, \frac{1}{2}\right)$ and $\left(-5, \frac{1}{2}\right)$

Observe that the points have the same y-coordinate. Thus, they lie on a horizontal line and its slope is 0. We could also compute the slope.

$$m = \frac{\frac{1}{2} - \frac{1}{2}}{-2-(-5)} = \frac{\frac{1}{2} - \frac{1}{2}}{-2+5} = \frac{0}{3} = 0$$

51. $(5, -4)$ and $(2, -7)$

$$m = \frac{-7-(-4)}{2-5} = \frac{-3}{-3} = 1$$

53. $(6, -4)$ and $(6, 5)$

Observe that the points have the same x-coordinate. Thus, they lie on a vertical line and its slope is undefined. We could also compute the slope.

$$m = \frac{-4-5}{6-6} = \frac{-9}{0}, \text{ undefined}$$

55. The line $y = 5$ is a horizontal line. A horizontal line has slope 0.

57. The line $x = -8$ is a vertical line. Slope is undefined.

59. The line $x = 9$ is a vertical line. The slope is undefined.

61. The line $y = -10$ is a horizontal line. A horizontal line has slope 0.

63. The grade is expressed as a percent.

$$m = \frac{792}{5280} = 0.15 = 15\%$$

65. The slope is expressed as a percent.

$$m = \frac{28}{80} = 0.35 = 35\%.$$

67. 2 ft 5 in. $= 2 \cdot 12$ in. $+ 5$ in. $= 24$ in. $+ 5$ in. $= 29$ in.

8 ft 2 in. $= 8 \cdot 12$ in. $+ 2$ in. $= 96$ in. $+ 2$ in. $= 98$ in.

$$m = \frac{29}{98}, \text{ or about } 30\%$$

69. Dooley Mountain rises $5400 - 3500 = 1900$ ft.

$$m = \frac{1900}{37000} \approx 0.051 \approx 5.1\%$$

Yes, it qualifies as part of the Tour de France.

71. *Writing Exercise.*

$$\frac{y_2 - y_1}{x_2 - x_1} = \frac{-(y_1 - y_2)}{-(x_1 - x_2)} = \frac{y_1 - y_2}{x_1 - x_2}$$

73. $ax + by = c$

$by = c - ax$ Adding $-ax$ to both sides

$y = \frac{c - ax}{b}$ Dividing both sides by b

75. $ax - by = c$

$-by = c - ax$ Adding $-ax$ to both sides

$y = \frac{c - ax}{-b}$ Dividing both sides by $-b$

We could also express this result as $y = \frac{ax - c}{b}$.

77. $8x + 6y = 24$

$6y = -8x + 24$

$y = \frac{-4}{3}x + 4$

x	y
0	4
3	0
6	−4

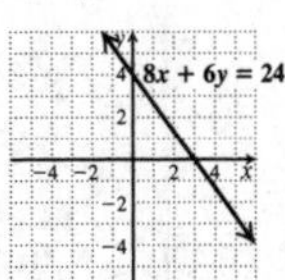

79. *Writing Exercise.*

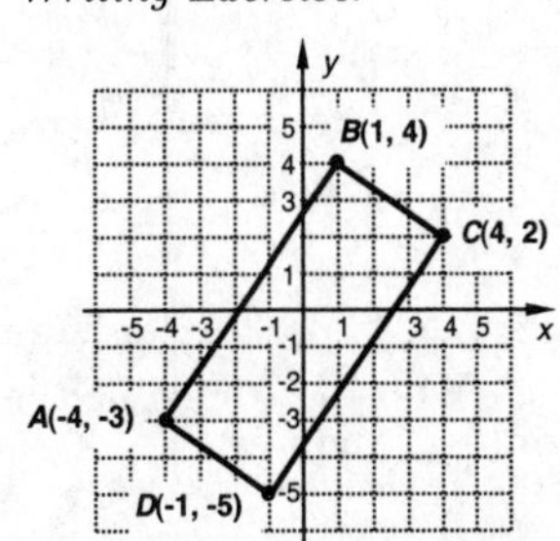

We find the slope of each side of the quadrilateral.

For side $\overline{AB}$, $m = \frac{4-(-3)}{1-(-4)} = \frac{7}{5}$.

For side $\overline{BC}$, $m = \frac{2-4}{4-1} = -\frac{2}{3}$.

For side $\overline{CD}$, $m = \frac{2-(-5)}{4-(-1)} = \frac{7}{5}$.

For side $\overline{DA}$, $m = \frac{-5-(-3)}{-1-(-4)} = -\frac{2}{3}$.

Since the opposite sides of the quadrilateral have the same slopes but lie on different lines, the lines on which they lie never intersect so they are parallel. Thus the quadrilateral is a parallelogram.

81. From the dimensions on the drawing, we see that the ramps labeled A have a rise of 61 cm and a run of 167.6 cm.

$$m = \frac{61 \text{ cm}}{167.6 \text{ cm}} \approx 0.364, \text{ or } 36.4\%$$

83. If the line passes through $(2, 5)$ and never enters the second quadrant, then it slants up from left to right or is vertical. This means that its slope is positive. The line slants least steeply if it passes through $(0, 0)$. In this case, $m = \frac{5-0}{2-0} = \frac{5}{2}$. Thus, the numbers the line could have for it slope are $\left\{m \middle| m \geq \frac{5}{2}\right\}$.

85. Let $t =$ the number of units each tick mark on the vertical axis represents. Note that the graph drops 4 units for every 3 units of horizontal change. Then we have:

$$\frac{-4t}{3} = -\frac{2}{3}$$
$$-4t = -2 \quad \text{Multiplying by 3}$$
$$t = \frac{1}{2} \quad \text{Dividing by } -4$$

Each tick mark on the vertical axis represents $\frac{1}{2}$ unit.

Exercise Set 3.6

1. We can read the slope, 3, directly from the equation. Choice (f) is correct.

3. We can read the slope, $\frac{2}{3}$, directly from the equation. Choice (d) is correct.

5. $y = 3x - 2 = 3x + (-2)$

The y-intercept is $(0, -2)$, so choice (e) is correct.

7. Slope $\frac{2}{3}$; y-intercept $(0, 1)$

We plot $(0, 1)$ and from there move up 2 units and right 3 units. This locates the point $(3, 3)$. We plot $(3, 3)$ and draw a line passing through $(0, 1)$ and $(3, 3)$.

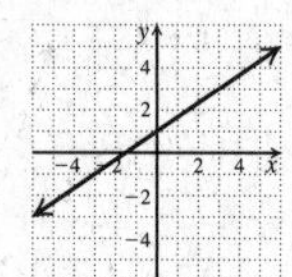

9. Slope $\frac{5}{3}$; y-intercept $(0, -2)$

We plot $(0, -2)$ and from there move up 5 units and right 3 units. This locates the point $(3, 3)$. We plot $(3, 3)$ and draw a line passing through $(0, -2)$ and $(3, 3)$.

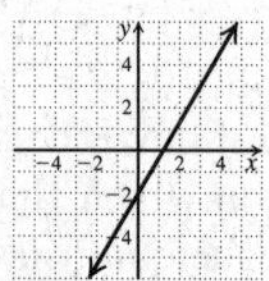

11. Slope $-\frac{1}{3}$; y-intercept $(0, 5)$

We plot $(0, 5)$. We can think of the slope as $\frac{-1}{3}$, so from $(0, 5)$ we move down 1 unit and right 3 units. This locates the point $(3, 4)$. We plot $(3, 4)$ and draw a line passing through $(0, 5)$ and $(3, 4)$.

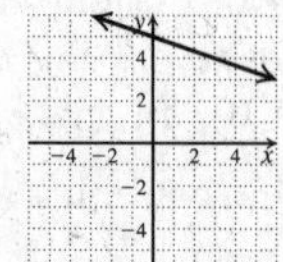

13. Slope 2; y-intercept $(0, 0)$

We plot $(0, 0)$. We can think of the slope as $\frac{2}{1}$, so from $(0, 0)$ we move up 2 units and right 1 unit. This locates the point $(1, 2)$. We plot $(1, 2)$ and draw a line passing through $(0, 0)$ and $(1, 2)$.

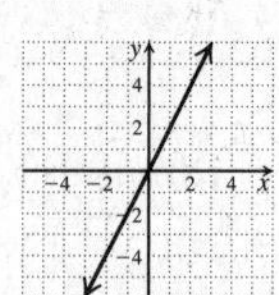

15. Slope -3; y-intercept $(0, 2)$

We plot $(0, 2)$. We can think of the slope as $\frac{-3}{1}$, so from $(0, 2)$ we move down 3 units and right 1 unit. This locates the point $(1, -1)$. We plot $(1, -1)$ and draw a line passing through $(0, 2)$ and $(1, -1)$.

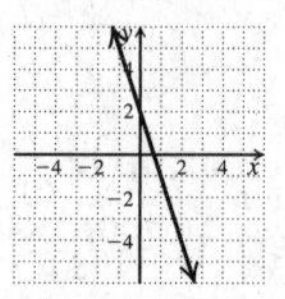

17. Slope 0; y-intercept $(0, -5)$

Since the slope is 0, we know the line is horizontal, so from $(0, -5)$ we move right 1 unit. This locates the point $(1, -5)$. We plot $(1, -5)$ and draw a line passing through $(0, -5)$ and $(1, -5)$.

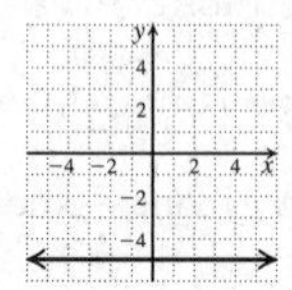

19. We read the slope and y-intercept from the equation.

$$y = -\frac{2}{7}x + 5$$

The slope is $-\frac{2}{7}$. The y-intercept is $(0, 5)$.

21. We read the slope and y-intercept from the equation.

$$y = \frac{1}{3}x + 7$$

The slope is $\frac{1}{3}$. The y-intercept is $(0, 7)$.

23. $y = \frac{9}{5}x - 4$

$y = \frac{9}{5}x + (-4)$

The slope is $\frac{9}{5}$, and the y-intercept is $(0, -4)$.

25. We solve for y to rewrite the equation in the form $y = mx + b$.

$-3x + y = 7$

$y = 3x + 7$

The slope is 3, and the y-intercept is $(0, 7)$.

27. $4x + 2y = 8$

$2y = -4x + 8$

$y = \frac{1}{2}(-4x + 8)$

$y = -2x + 4$

The slope is -2, and the y-intercept is $(0, 4)$.

29. Observe that this is the equation of a horizontal line that lies 3 units above the x-axis. Thus, the slope is 0, and the y-intercept is $(0, 3)$. We could also write the equation in slope-intercept form.

$$y = 3$$
$$y = 0x + 3$$

The slope is 0, and the y-intercept is $(0, 3)$.

31.
$$2x - 5y = -8$$
$$-5y = -2x - 8$$
$$y = -\frac{1}{5}(-2x - 8)$$
$$y = \frac{2}{5}x + \frac{8}{5}$$

The slope is $\frac{2}{5}$, and the y-intercept is $\left(0, \frac{8}{5}\right)$.

33.
$$9x - 8y = 0$$
$$-8y = -9x$$
$$y = \frac{9}{8}x \text{ or } y = \frac{9}{8}x + 0$$

Slope: $\frac{9}{8}$, y-intercept: $(0, 0)$

35. We use the slope-intercept equation, substituting 5 for m and 7 for b:

$$y = mx + b$$
$$y = 5x + 7$$

37. We use the slope-intercept equation, substituting $\frac{7}{8}$ for m and -1 for b:

$$y = mx + b$$
$$y = \frac{7}{8}x - 1$$

39. We use the slope-intercept equation, substituting $-\frac{5}{3}$ for m and -8 for b:

$$y = mx + b$$
$$y = -\frac{5}{3}x - 8$$

41. We use the slope-intercept equation, substituting 0 for m and $\frac{1}{3}$ for b.

$$y = mx + b$$
$$y = 0x + \frac{1}{3}$$
$$y = \frac{1}{3}$$

43. From the graph we see that the y-intercept is $(0, 17)$. We also see that the point $(4, 23)$ is on the graph. We find the slope:

$$m = \frac{23 - 17}{4 - 0} = \frac{6}{4} = \frac{3}{2}$$

Substituting $\frac{3}{2}$ for m and 17 for b in the slope-intercept equation $y = mx + b$, we have

$$y = \frac{3}{2}x + 17,$$

where y is the number of gallons of bottled water consumed per person and x is the number of years since 2000.

45. From the graph we see that the y-intercept is $(0, 15)$. We also see that the point $(5, 17)$ is on the graph. We find the slope:

$$m = \frac{17 - 15}{5 - 0} = \frac{2}{5}$$

Substituting $\frac{2}{5}$ for m and 15 for b in the slope-intercept equation $y = mx + b$, we have

$$y = \frac{2}{5}x + 15,$$

where y is the number of jobs in millions, and x is the number of years since 2000.

47. $y = \frac{2}{3}x + 2$

First we plot the y-intercept $(0, 2)$. We can start at the y-intercept and use the slope, $\frac{2}{3}$, to find another point. We move up 2 units and right 3 units to get a new point $(3, 4)$. Thinking of the slope as $\frac{-2}{-3}$ we can start at $(0, 2)$ and move down 2 units and left 3 units to get another point $(-3, 0)$.

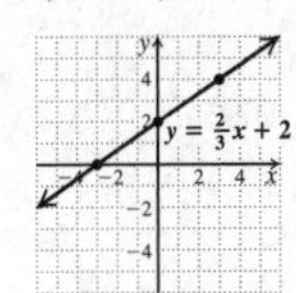

49. $y = -\frac{2}{3}x + 3$

First we plot the y-intercept $(0, 3)$. We can start at the y-intercept and, thinking of the slope as $\frac{-2}{3}$, find another point by moving down 2 units and right 3 units to the point $(3, 1)$. Thinking of the slope as $\frac{2}{-3}$ we can start at $(0, 3)$ and move up 2 units and left 3 units to get another point $(-3, 5)$.

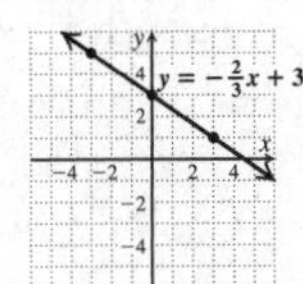

51. $y = \frac{3}{2}x + 3$

First we plot the y-intercept $(0, 3)$. We can start at the y-intercept and use the slope, $\frac{3}{2}$, to find another point. We move up 3 units and right 2 units to get a new point $(2, 6)$. Thinking of the slope as $\frac{-3}{-2}$ we can start at $(0, 3)$ and move down 3 units and left 2 units to get another point $(-2, 0)$.

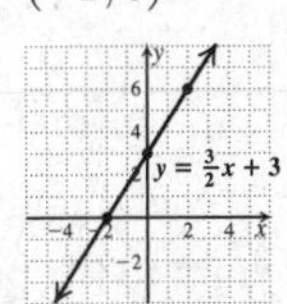

53. $y = \frac{-4}{3}x + 3$

First we plot the y-intercept $(0, 3)$. We can start at the y-intercept and, thinking of the slope as $\frac{-4}{3}$, find another point by moving down 4 units and right 3 units to the point $(3, -1)$. Thinking of the slope as $\frac{4}{-3}$ we can start at $(0, 3)$ and move up 4 units and left 3 units to get another point $(-3, 7)$.

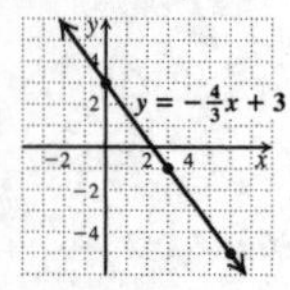

55. We first rewrite the equation in slope-intercept form.

$$2x + y = 1$$
$$y = -2x + 1$$

Now we plot the y-intercept $(0, 1)$. We can start at the y-intercept and, thinking of the slope as $\frac{-2}{1}$, find another point by moving down 2 units and right 1 unit to the point $(1, -1)$. In a similar manner, we can move from the point $(1, -1)$ to find a third point $(2, -3)$.

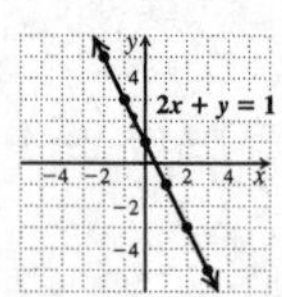

57. We first rewrite the equation in slope-intercept form.

$$3x + y = 0$$
$$y = -3x, \text{ or } y = -3x + 0$$

Now we plot the y-intercept $(0, 0)$. We can start at the y-intercept and, thinking of the slope as $\frac{-3}{1}$, find another point by moving down 3 units and right 1 unit to the point $(1, -3)$. Thinking of the slope as $\frac{3}{-1}$ we can start at $(0, 0)$ and move up 3 units and left 1 unit to get another point $(-1, 3)$.

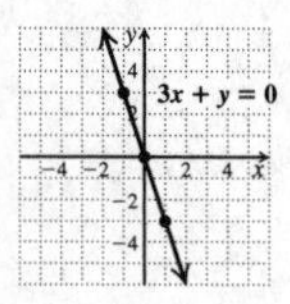

59. We first rewrite the equation in slope-intercept form.

$$4x + 5y = 15$$
$$5y = -4x + 15$$
$$y = \frac{1}{5}(-4x + 15)$$
$$y = -\frac{4}{5}x + 3$$

Now we plot the y-intercept $(0, 3)$. We can start at the y-intercept and, thinking of the slope as $\frac{-4}{5}$, find another point by moving down 4 units and right 5 units to the point $(5, -1)$. Thinking of the slope as $\frac{4}{-5}$ we can start at $(0, 3)$ and move up 4 units and left 5 units to get another point $(-5, 7)$.

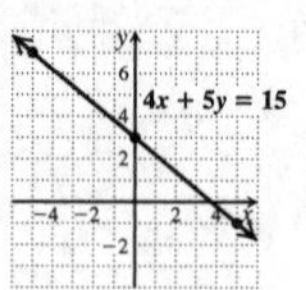

61. We first rewrite the equation in slope-intercept form.

$$x - 4y = 12$$
$$-4y = -x + 12$$
$$y = -\frac{1}{4}(-x + 12)$$
$$y = \frac{1}{4}x - 3$$

Now we plot the y-intercept $(0, -3)$. We can start at the y-intercept and use the slope, $\frac{1}{4}$, to find another point. We move up 1 unit and right 4 units to the point $(4, -2)$. Thinking of the slope as $\frac{-1}{-4}$ we can start at $(0, -3)$ and move down 1 unit and left 4 units to get another point $(-4, -4)$.

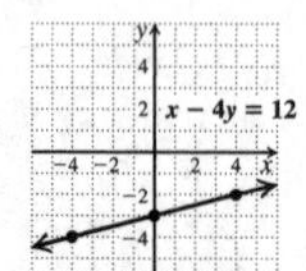

63. The equation $y = \frac{3}{4}x + 6$ represents a line with slope $\frac{3}{4}$, and the y-intercept is $(0, 6)$.

The equation $y = \frac{3}{4}x - 2$ represents a line with slope $\frac{3}{4}$, and the y-intercept is $(0, -2)$.

Since both lines have slope $\frac{3}{4}$ but different y-intercepts, their graphs are parallel.

65. The equation $y = 2x - 5$ represents a line with slope 2 and y-intercept $(0, -5)$. We rewrite the second equation in slope-intercept form.

$$4x + 2y = 9$$
$$2y = -4x + 9$$
$$y = \frac{1}{2}(-4x + 9)$$
$$y = -2x + \frac{9}{2}$$

The slope is -2 and the y-intercept is $\left(0, \frac{9}{2}\right)$. Since the lines have different slopes, their graphs are not parallel.

67. Rewrite each equation in slope-intercept form.

$$3x + 4y = 8$$
$$4y = -3x + 8$$
$$y = \frac{1}{4}(-3x + 8)$$
$$y = -\frac{3}{4}x + 2$$

The slope is $-\frac{3}{4}$, and the y-intercept is $(0, 2)$.

$$7 - 12y = 9x$$
$$-12y = 9x - 7$$
$$y = -\frac{1}{12}(9x - 7)$$
$$y = -\frac{3}{4}x + \frac{7}{12}$$

The slope is $-\frac{3}{4}$, and the y-intercept is $\left(0, \frac{7}{12}\right)$.

Since both lines have slope $-\frac{3}{4}$ but different y-intercepts, their graphs are parallel.

69. $y = 4x - 5$,
$4y = 8 - x$

The first equation is in slope-intercept form. It represents a line with slope 4. Now we rewrite the second equation in slope-intercept form.

$$4y = 8 - x$$
$$y = \frac{1}{4}(8 - x)$$
$$y = 2 - \frac{1}{4}x$$
$$y = -\frac{1}{4}x + 2$$

The slope of the line is $-\frac{1}{4}$.

Since $4\left(-\frac{1}{4}\right) = -1$, the equations represent perpendicular lines.

71. $x - 2y = 5$,
$2x + 4y = 8$

We write each equation in slope-intercept form.

$$x - 2y = 5$$
$$-2y = -x + 5$$
$$y = -\frac{1}{2}(-x + 5)$$
$$y = \frac{1}{2}x - \frac{5}{2}$$

The slope is $\frac{1}{2}$.

$$2x + 4y = 8$$
$$4y = -2x + 8$$
$$y = \frac{1}{4}(-2x + 8)$$
$$y = -\frac{1}{2}x + 2$$

The slope is $-\frac{1}{2}$.

Since $\frac{1}{2}\left(-\frac{1}{2}\right) = -\frac{1}{4} \neq -1$, the equations do not represent perpendicular lines.

73. $2x + 3y = 1$,
$3x - 2y = 1$

We write each equation in slope-intercept form.

$$2x + 3y = 1$$
$$3y = -2x + 1$$
$$y = \frac{1}{3}(-2x + 1)$$
$$y = -\frac{2}{3}x + \frac{1}{3}$$

The slope is $-\frac{2}{3}$.

$$3x - 2y = 1$$
$$-2y = -3x + 1$$
$$y = -\frac{1}{2}(-3x + 1)$$
$$y = \frac{3}{2}x - \frac{1}{2}$$

The slope is $\frac{3}{2}$.

Since $-\frac{2}{3}\left(\frac{3}{2}\right) = -1$, the equations represent perpendicular lines.

75. The slope of the line represented by $y = 5x - 7$ is 5. Then a line parallel to the graph of $y = 5x - 7$ has slope 5 also. Since the y-intercept is $(0, 11)$, the desired equation is $y = 5x + 11$.

77. First find the slope of the line represented by $2x + y = 0$.

$$2x + y = 0$$
$$y = -2x$$

The slope is -2. Then the slope of a line perpendicular to the graph of $2x + y = 0$ is the negative reciprocal of -2, or $\frac{1}{2}$. Since the y-intercept is $(0, 0)$, the desired equation is $y = \frac{1}{2}x + 0$, or $y = \frac{1}{2}x$.

79. The slope of the line represented by $y = x$ is 1. Then a line parallel to this line also has slope 1. Since the y-intercept is $(0, 3)$, the desired equation is
$y = 1 \cdot x + 3$, or $y = x + 3$.

81. First find the slope of the line represented by
$x + y = 3$.

$$x + y = 3$$
$$y = -x + 3, \text{ or } y = -1 \cdot x + 3$$

The slope is -1. Then the slope of a line perpendicular to this line is the negative reciprocal is -1, or 1. Since the y-intercept is -4, the desired equation is $y = 1 \cdot x - 4$, or $y = x - 4$.

83. *Writing Exercise.* Yes; think of the slope as $\frac{0}{a}$ for any nonzero value of a.

85. $y - k = m(x - h)$
$y = m(x - h) + k$ Adding k to both sides

87. $-10 - (-3) = -10 + 3 = -7$

89. $-4 - 5 = -4 + (-5) = -9$

91. *Writing Exercise.* Some such circumstances include using an incorrect slope and/or y-intercept when drawing the graph.

93. When $x = 0$, $y = b$, so $(0, b)$ is on the line. When $x = 1$, $y = m + b$, so $(1, m + b)$ is on the line. Then,

slope$= \dfrac{(m+b)-b}{1-0} = m.$

95. Rewrite each equation in slope-intercept form.

$$2x - 6y = 10$$
$$-6y = -2x + 10$$
$$y = \frac{1}{3}x - \frac{5}{3}$$

The slope of the line is $\frac{1}{3}$.

$$9x + 6y = 18$$
$$6y = -9x + 18$$
$$y = -\frac{3}{2}x + 3$$

The y-intercept of the line is $(0, 3)$.

The equation of the line is $y = \frac{1}{3}x + 3$.

97. Rewrite the first equation in slope-intercept form.

$$3x - 5y = 8$$
$$-5y = -3x + 8$$
$$y = -\frac{1}{5}(-3x + 8)$$
$$y = \frac{3}{5}x - \frac{8}{5}$$

The slope is $\frac{3}{5}$.

The slope of a line perpendicular to this line is a number m such that

$$\frac{3}{5}m = -1, \text{ or}$$
$$m = -\frac{5}{3}.$$

Now rewrite the second equation in slope-intercept form.

$$2x + 4y = 12$$
$$4y = -2x + 12$$
$$y = \frac{1}{4}(-2x + 12)$$
$$y = -\frac{1}{2}x + 3$$

The y-intercept of the line is $(0, 3)$.

The equation of the line is $y = -\frac{5}{3}x + 3$.

99. Rewrite the first equation in slope-intercept form.

$$3x - 2y = 9$$
$$-2y = -3x + 9$$
$$y = -\frac{1}{2}(-3x + 9)$$
$$y = \frac{3}{2}x - \frac{9}{2}$$

The slope of the line is $\frac{3}{2}$.

The slope of a line perpendicular to this line is a number m such that

$$\frac{3}{2}m = -1, \text{ or}$$
$$m = -\frac{2}{3}.$$

Now rewrite the second equation in slope-intercept form.

$$2x + 5y = 0$$
$$5y = -2x$$
$$y = -\frac{2}{5}x, \text{ or } y = -\frac{2}{5}x + 0$$

The y-intercept is $(0, 0)$.

The equation of the line is $y = -\frac{2}{3}x + 0$, or $y = -\frac{2}{3}x$.

Connecting the Concepts 3.6

1. a) $x = 3$ is linear.
b)Draw a vertical line through $(3, 0)$.

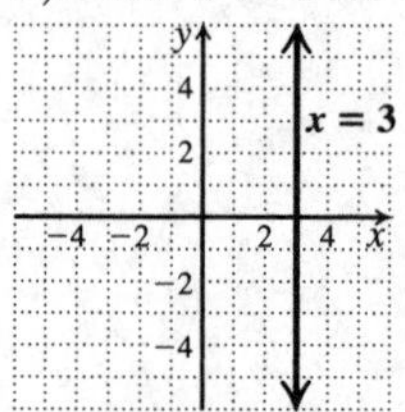

3. a) $y = \frac{1}{2}x + 3$ is linear.
b)The y-intercept is $(0, 3)$. Another point is $(2, 4)$.

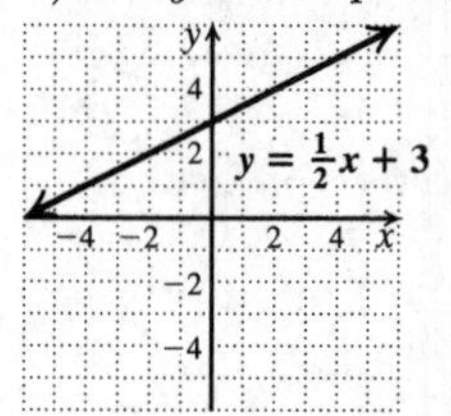

5. **a)** $y - 5 = x$ is linear.

b) Rewriting in point-slope form $y = x + 5$. The y-intercept is $(0, 5)$. Another point is $(1, 6)$.

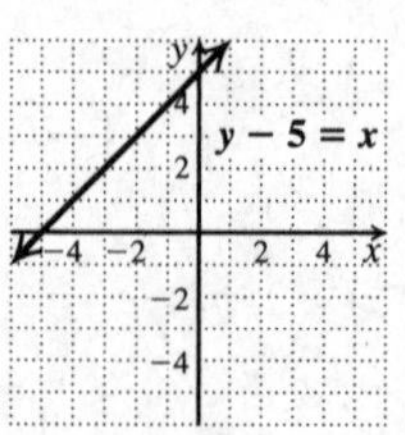

7. $3xy = 6$ is not linear.

9. **a)** $3 - y = 4$ is linear.

b) Solving for y,
$3 - y = 4$
$-1 = y$
Draw a horizontal line through $(0, -1)$.

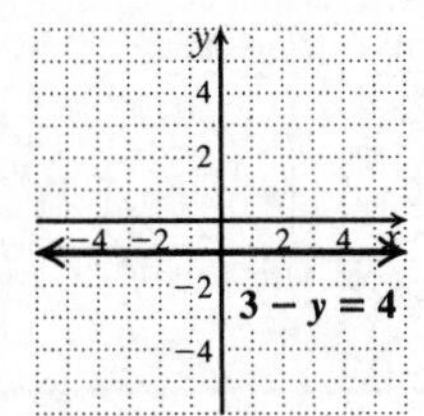

11. **a)** $2y = 9x - 10$ is linear.

b) Rewriting in slope-intercept form $y = \frac{9}{2}x - 5$. The y-intercept is $(0, -5)$. Another point is $(2, 4)$.

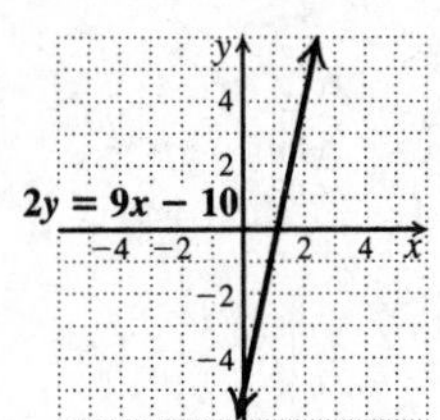

13. **a)** $2x - 6 = 3y$ is linear.

b) Solving for y,
$3y = 2x - 6$
$y = \frac{2}{3}x - 2$
The y-intercept is $(0, -2)$. Another point is $(3, 0)$.

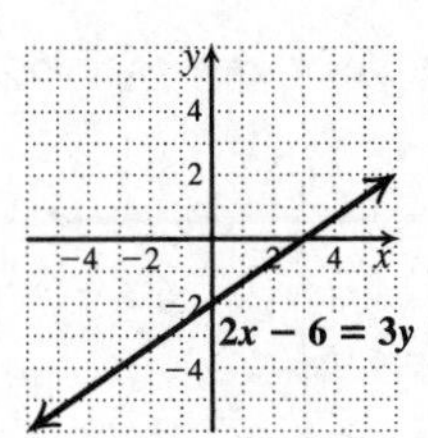

15. **a)** $2y - x = 4$ is linear.

b) When $x = 0$,
$2y = 4$
$y = 2$
the y-intercept is $(0, 2)$.
When $y = 0$,
$-x = 4$
$x = -4$
the x-intercept is $(-4, 0)$.

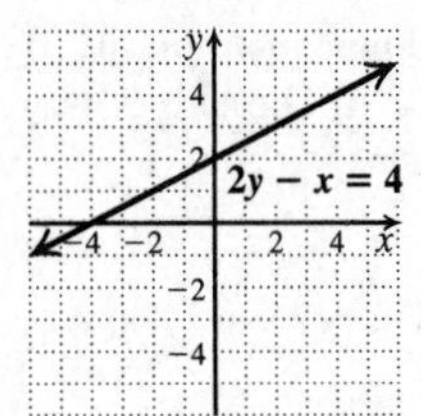

17. **a)** $x - 2y = 0$ is linear.

b) When $x = 0$,
$-2y = 0$
$y = 0$
the y-intercept is $(0, 0)$, also the x-intercept.

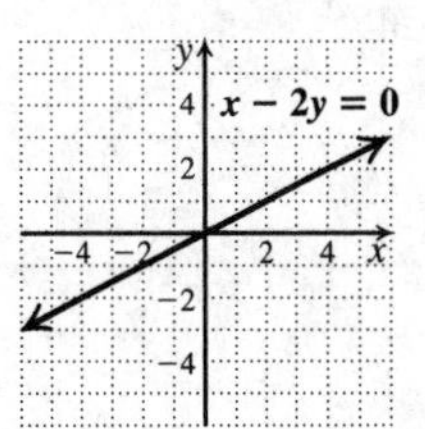

19. **a)** $y = 4 + x$ is linear.

b) The y-intercept is $(0, 4)$. Another point is $(1, 5)$.

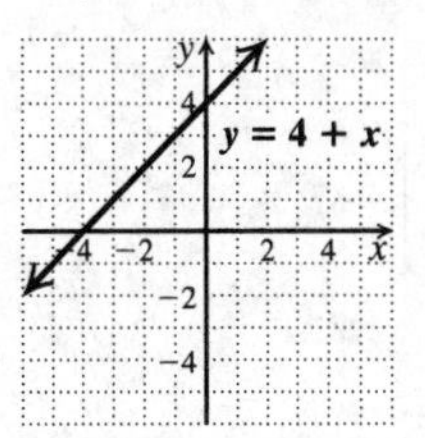

Exercise Set 3.7

1. Substituting 5 for m, 2 for x_1, and 3 for y_1 in the point-slope equation $y - y_1 = m(x - x_1)$, we have $y - 3 = 5(x - 2)$. Choice (g) is correct.

3. Substituting -5 for m, 2 for x_1, and 3 for y_1 in the point-slope equation $y - y_1 = m(x - x_1)$, we have $y - 3 = -5(x - 2)$. Choice (d) is correct.

5. Substituting -5 for m, -2 for x_1, and -3 for y_1 in the point-slope equation $y - y_1 = m(x - x_1)$, we have $y - (-3) = -5(x - (-2))$, or $y + 3 = -5(x + 2)$. Choice (e) is correct.

7. Substituting -5 for m, -3 for x_1, and -2 for y_1 in the point-slope equation $y - y_1 = m(x - x_1)$, we have $y - (-2) = -5(x - (-3))$, or $y + 2 = -5(x + 3)$. Choice (f) is correct.

9. We see that the points $(1, -4)$ and $(-3, 2)$ are on the line. To go from $(1, -4)$ to $(-3, 2)$ we go up 6 units and left 4 units so the slope of the line is $\frac{6}{-4}$, or $-\frac{3}{2}$. Then, substituting $-\frac{3}{2}$ for m, 1 for x_1, and -4 for y_1 in the point-slope equation $y - y_1 = m(x - x_1)$, we have $y - (-4) = -\frac{3}{2}(x - 1)$, or $y + 4 = -\frac{3}{2}(x - 1)$. Choice (c) is correct.

11. We see that the points $(1, -4)$ and $(5, 2)$ are on the line. To go from $(1, -4)$ to $(5, 2)$ we go up 6 units and right 4 units so the slope of the line is $\frac{6}{4}$, or $\frac{3}{2}$. Then, substituting $\frac{3}{2}$ for m, 1 for x_1, and -4 for y_1 in the point-slope equation $y - y_1 = m(x - x_1)$, we have $y - (-4) = \frac{3}{2}(x - 1)$, or $y + 4 = \frac{3}{2}(x - 1)$. Choice (d) is correct.

13. $y - y_1 = m(x - x_1)$
We substitute 3 for m, 1 for x_1, and 6 for y_1.
$y - 6 = 3(x - 1)$

15. $y - y_1 = m(x - x_1)$

We substitute $\frac{3}{5}$ for m, 2 for x_1, and 8 for y_1.

$y - 8 = \frac{3}{5}(x - 2)$

17. $y - y_1 = m(x - x_1)$

We substitute -4 for m, 3 for x_1, and 1 for y_1.

$y - 1 = -4(x - 3)$

19. $y - y_1 = m(x - x_1)$

We substitute $\frac{3}{2}$ for m, 5 for x_1, and -4 for y_1.

$y - (-4) = \frac{3}{2}(x - 5)$

21. $y - y_1 = m(x - x_1)$

We substitute $\frac{-5}{4}$ for m, -2 for x_1, and 6 for y_1.

$y - 6 = \frac{-5}{4}(x - (-2))$

23. $y - y_1 = m(x - x_1)$

We substitute -2 for m, -4 for x_1, and -1 for y_1.

$y - (-1) = -2(x - (-4))$

25. $y - y_1 = m(x - x_1)$

We substitute 1 for m, -2 for x_1, and 8 for y_1.

$y - 8 = 1(x - (-2))$

27. First we write the equation in point-slope form.

$y - y_1 = m(x - x_1)$

$y - 5 = 4(x - 3)$ Substituting

Next we find an equivalent equation of the form $y = mx + b$.

$y - 5 = 4(x - 3)$

$y - 5 = 4x - 12$

$y = 4x - 7$

29. First we write the equation in point-slope form.

$y - y_1 = m(x - x_1)$

$y - (-2) = \frac{7}{4}(x - 4)$ Substituting

Next we find an equivalent equation of the form $y = mx + b$.

$y - (-2) = \frac{7}{4}(x - 4)$

$y + 2 = \frac{7}{4}x - 7$

$y = \frac{7}{4}x - 9$

31. First we write the equation in point-slope form.

$y - y_1 = m(x - x_1)$

$y - 7 = -2(x - (-3))$

Next we find an equivalent equation of the form $y = mx + b$.

$y - 7 = -2(x - (-3))$

$y - 7 = -2(x + 3)$

$y - 7 = -2x - 6$

$y = -2x + 1$

33. First we write the equation in point-slope form.

$y - y_1 = m(x - x_1)$

$y - (-1) = -4(x - (-2))$

Next we find an equivalent equation of the form $y = mx + b$.

$y - (-1) = -4(x - (-2))$

$y + 1 = -4(x + 2)$

$y + 1 = -4x - 8$

$y = -4x - 9$

35. First we write the equation in point-slope form.

$y - y_1 = m(x - x_1)$

$y - 6 = \frac{2}{3}(x - 5)$

Next we find an equivalent equation of the form $y = mx + b$.

$y - 6 = \frac{2}{3}x - \frac{10}{3}$

$y = \frac{2}{3}x + \frac{8}{3}$

37. The slope is $-\frac{5}{6}$ and the y-intercept is $(0, 4)$. Substituting $-\frac{5}{6}$ for m and 4 for b in the slope-intercept equation $y = mx + b$, we have $y = -\frac{5}{6}x + 4$.

39. First solve the equation for y and determine the slope of the given line.

$x - 2y = 3$ Given line

$-2y = -x + 3$

$y = \frac{1}{2}x - \frac{3}{2}$

The slope of the given line is $\frac{1}{2}$.

The slope of every line parallel to the given line must also be $\frac{1}{2}$. We find the equation of the line with slope $\frac{1}{2}$ and containing the point (2, 5).

$y - y_1 = m\,(x - x_1)$ Point-slope equation

$y - 5 = \frac{1}{2}\,(x - 2)$ Substituting

$y - 5 = \frac{1}{2}x - 1$

$y = \frac{1}{2}x + 4$

41. The slope of $y = 4x + 3$ is 4. The given point $(0, -5)$ is the y-intercept, so we substitute in the slope-intercept equation.

$y = 4x - 5$.

43. First solve the equation for y and determine the slope of the given line.

$2x + 3y = -7$ Given line

$3y = -2x - 7$

$y = -\frac{2}{3}x - \frac{7}{3}$

The slope of the given line is $-\frac{2}{3}$.

The slope of every line parallel to the given line must also be $-\frac{2}{3}$. We find the equation of the line with slope $-\frac{2}{3}$ and containing the point $(-2, -3)$.

$y - y_1 = m(x - x_1)$ Point-slope equation

$y - (-3) = -\frac{2}{3}[x - (-2)]$ Substituting

$y + 3 = -\frac{2}{3}(x + 2)$

$y + 3 = -\frac{2}{3}x - \frac{4}{3}$

$y = -\frac{2}{3}x - \frac{13}{3}$

45. $x = 2$ is a vertical line. A line parallel to it that passes through $(5, -4)$ is the vertical line 5 units to the right of the y-axis, or $x = 5$.

47. First solve the equation for y and determine the slope of the given line.

$2x - 3y = 4$ Given line

$-3y = -2x + 4$

$y = \frac{2}{3}x - \frac{4}{3}$

The slope of the given line is $\frac{2}{3}$.

The slope of perpendicular line is given by the opposite of the reciprocal of $\frac{2}{3}$, $-\frac{3}{2}$. We find the equation of the line with slope $-\frac{3}{2}$ and containing the point (3, 1).

$y - y_1 = m(x - x_1)$ Point-slope equation

$y - 1 = -\frac{3}{2}(x - 3)$ Substituting

$y - 1 = -\frac{3}{2}x + \frac{9}{2}$

$y = -\frac{3}{2}x + \frac{11}{2}$

49. First solve the equation for y and determine the slope of the given line.

$x + y = 6$ Given line

$y = -x + 6$

The slope of the given line is -1.

The slope of perpendicular line is given by the opposite of the reciprocal of -1, 1. We find the equation of the line with slope 1 and containing the point (-4, 2).

$y - y_1 = m(x - x_1)$ Point-slope equation

$y - 2 = 1(x - (-4))$ Substituting

$y - 2 = x + 4$

$y = x + 6$

51. The slope of a line perpendicular to $2x - 5 = y$ is $-\frac{1}{2}$ and we are given the y-intercept of the desired line, (0, 6). Then we have $y = -\frac{1}{2}x + 6$.

53. $y = 5$ is a horizontal line, so a line perpendicular to it must be vertical. The equation of the vertical line containing (-3, 7) is $x = -3$.

55. We plot $(1, 2)$, move up 4 and to the right 3 to $(4, 6)$ and draw the line.

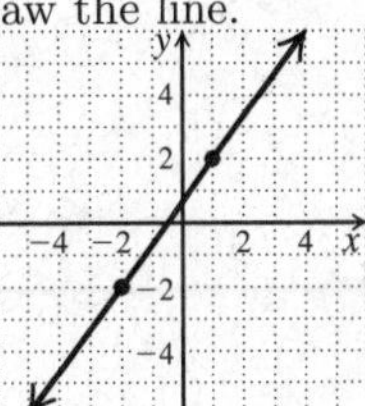

57. We plot $(2, 5)$, move down 3 and to the right 4 to $(6, 2)$ $\left(\text{since } -\frac{3}{4} = \frac{-3}{4}\right)$, and draw the line. We could also think of $-\frac{3}{4}$ and $\frac{3}{-4}$ and move up 3 and to the left 4 from the point $(2, 5)$ to $(-2, 8)$.

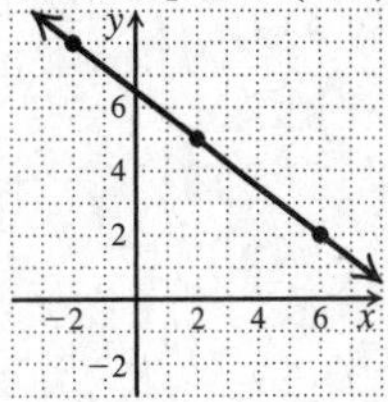

59. $y - 5 = \frac{1}{3}(x - 2)$ Point-slope form

The line has slope $\frac{1}{3}$ and passes through $(2, 5)$. We plot $(2, 5)$ and then find a second point by moving up 1 unit and right 3 units to $(5, 6)$. We draw the line through these points.

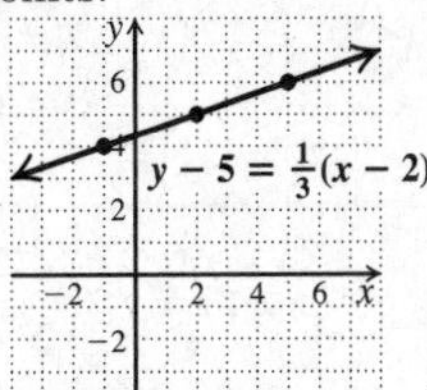

61. $y - 1 = -\frac{1}{4}(x - 3)$ Point-slope form

The line has slope $-\frac{1}{4}$, or $\frac{1}{-4}$ passes through $(3, 1)$. We plot $(3, 1)$ and then find a second point by moving up 1 unit and left 4 units to $(-1, 2)$. We draw the line through these points.

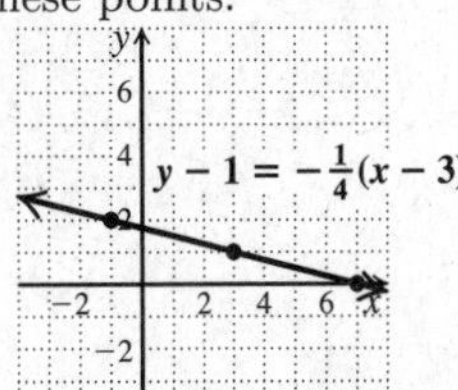

63. $y + 2 = \frac{2}{3}(x - 1)$, or $y - (-2) = \frac{2}{3}(x - 1)$

The line has slope $\frac{2}{3}$ and passes through $(1, -2)$. We plot $(1, -2)$ and then find a second point by moving up 2 units and right 3 units to $(4, 0)$. We draw the line through these points.

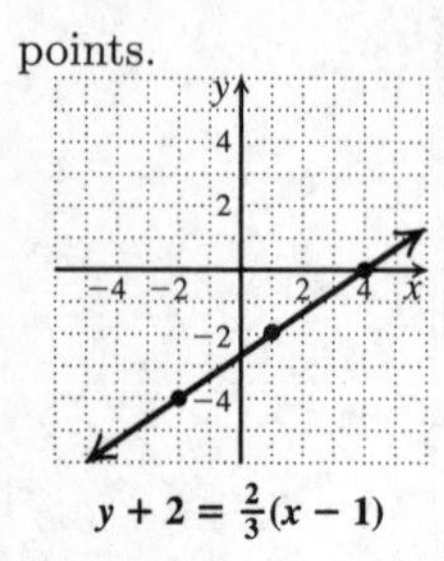

$y+2=\frac{2}{3}(x-1)$

65. $y+4=3(x+1)$, or $y-(-4)=3(x-(-1))$

The line has slope 3, or $\dfrac{3}{1}$, and passes through $(-1,-4)$. We plot $(-1,-4)$ and then find a second point by moving up 3 units and right 1 unit to $(0,-1)$. We draw the line through these points.

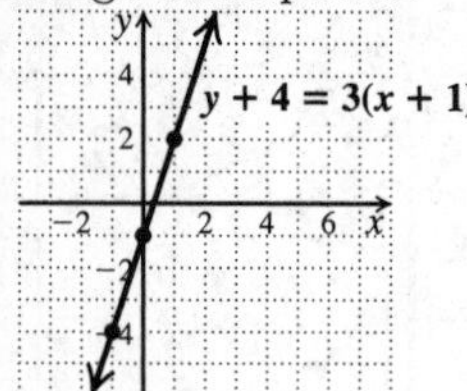

67. $y-4=-2(x+1)$, or $y-4=-2(x-(-1))$

The line has slope -2, or $\dfrac{-2}{1}$, and passes through $(-1,4)$. We plot $(-1,4)$ and then find a second point by moving down 2 units and right 1 unit to $(0,2)$. We draw the line through these points.

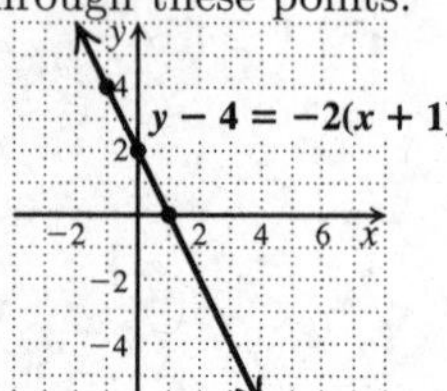

69. $y+1=-\dfrac{3}{5}(x-2)$, or $y-(-1)=-\dfrac{3}{5}(x-(-2))$

The line has slope $-\dfrac{3}{5}$, or $\dfrac{-3}{5}$ and passes through $(-2,-1)$. We plot $(2,-1)$ and then find a second point by moving up 3 units and left 5 units to $(-3,2)$, and draw the line.

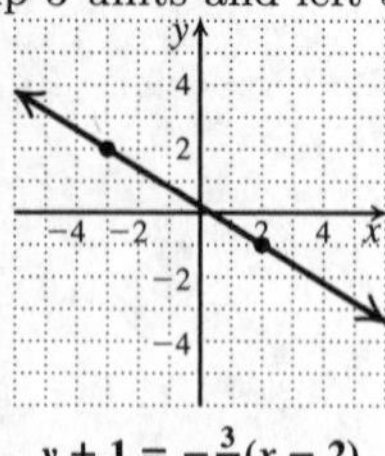

$y+1=-\frac{3}{5}(x-2)$

71. First find the slope of the line passing through the points $(1,62.1)$ and $(17,41.1)$.

$$m=\frac{41.1-62.1}{17-1}=\frac{-21}{16}=-1.3125$$

Now write an equation of the line. We use $(1,62.1)$ in the point-slope equation and then write an equivalent slope-intercept equation.

$$y-y_1=m(x-x_1)$$
$$y-62.1=-1.3125(x-1)$$
$$y-62.1=-1.3125x+1.3125$$
$$y=-1.3125x+63.4125$$

a) Since 1999 is 9 yr after 1990, we substitute 9 for x to calculate the birth rate in 1999.

$y=-1.3125(9)+63.4125=-11.8125+63.4125=51.6$

In 1999, there were 51.6 births per 1000 females age 15 to 19.

b) 2008 is 18 yr after 1990 ($2008-1990=18$), so we substitute 18 for x.

$y=-1.3125(18)+63.4125=-23.625+63.4125=39.7875\approx 39.8$

We predict that the birth rate among teenagers will be 39.8 births per 1000 females in 2008.

73. First find the slope of the line passing through the points $(0,14.2)$ and $(3,10.8)$. In each case, we let the first coordinate represent the number of years after 2000.

$$m=\frac{10.8-14.2}{3-0}=\frac{-3.4}{3}\approx -1.13$$

The y-intercept of the line is $(0,14.2)$. We write the slope-intercept equation: $y=-1.13x+14.2$.

a) Since 2002–2001 = 1, we substitute 1 for x to calculate the percentage in 2002.

$y=-1.13(1)+14.2=-1.131+14.2\approx 13.1\%$

b) Since $2008-2001=7$, we substitute 7 for x to find the percentage in 2008.

$y=-1.13(7)+14.2=-10.17+14.2\approx 4\%$

(Answers will vary depending on how the slope was rounded in part (a).)

75. First find the slope of the line passing through $(0,14.3)$ and $(10,17.4)$. In each case, we let the first coordinate represent the number of years after 1995.

$$m=\frac{17.4-14.3}{10-0}=\frac{3.1}{10}=0.31$$

The y-intercept is $(0,14.3)$. We write the slope-intercept equation: $y=0.31x+14.3$.

a) Since 2002 is 7 yr after 1995, we substitute 7 for x to calculate the college enrollment in 2002.

$y=0.31(7)+14.3=2.17+14.3=16.47$ million students

b) Since $2010-1995=15$ yr after 1990, we substitute 15 for x to find the enrollment in 2010.

$y=0.31(15)+14.3=4.65+14.3=18.9$ million students

77. First find the slope of the line through $(0,31)$ and $(12,36.3)$. In each case, we let the first coordinate represent the number of years after 1990 and the second millions of residents.

$$m=\frac{36.3-31}{14-0}=\frac{5.3}{14}\approx 0.38$$

The y-intercept is $(0, 31)$. We write the slope-intercept equation: $y = 0.38x + 31$.

a) Since 1997 is 7 yr after 1990, we substitute 7 for x to find the number of U.S. residents over the age of 65 in 1997.

$y = 0.38(7) + 31 = 33.6$ million residents

b) Since 2010 is 20 yr after 1990, we substitute 20 for x to find the number of U.S. residents over the age of 65 in 2010.

$y = 0.38(20) + 31 = 38.6$ million residents

(Answers will vary depending on how the slope is rounded.)

79. $(2, 3)$ and $(4, 1)$

First we find the slope.

$$m = \frac{1-3}{4-2} = \frac{-2}{2} = -1$$

Then we write an equation of the line in point-slope form using either of the points above.

$$y - 3 = -1(x - 2)$$

Finally, we find an equivalent equation in slope-intercept form.

$$y - 3 = -1(x - 2)$$
$$y - 3 = -x + 2$$
$$y = -x + 5$$

81. $(-3, 1)$ and $(3, 5)$

First we find the slope.

$$m = \frac{1-5}{-3-3} = \frac{-4}{-6} = \frac{2}{3}$$

Then we write an equation of the line in point-slope form using either of the points above.

$$y - 5 = \frac{2}{3}(x - 3)$$

Finally, we find an equivalent equation in slope-intercept form.

$$y - 5 = \frac{2}{3}(x - 3)$$
$$y - 5 = \frac{2}{3}x - 2$$
$$y = \frac{2}{3}x + 3$$

83. $(5, 0)$ and $(0, -2)$

First we find the slope.

$$m = \frac{0-(-2)}{5-0} = \frac{2}{5}$$

Then we write an equation of the line in point-slope form using either of the points above.

$$y - 0 = \frac{2}{5}(x - 5)$$

Finally, we find an equivalent equation in slope-intercept form.

$$y - 0 = \frac{2}{5}(x - 5)$$
$$y = \frac{2}{5}x - 2$$

85. $(-4, -1)$ and $(1, 9)$

First we find the slope.

$$m = \frac{9-(-1)}{1-(-4)} = \frac{9+1}{1+4} = 2$$

Then we write an equation of the line in point-slope form using either of the points above.

$$y - 9 = 2(x - 1)$$

Finally, we find an equivalent equation in slope-intercept form.

$$y - 9 = 2(x - 1)$$
$$y - 9 = 2x - 2$$
$$y = 2x + 7$$

87. *Writing Exercise.* The equation of a horizontal line $y = b$ can be written in point-slope form:

$$y - b = 0(x - x_1)$$

The equation of a vertical line cannot be written in point-slope form because the slope of a vertical line is undefined.

89. $(-5)^3 = (-5)(-5)(-5) = -125$

91. $-2^6 = -2 \cdot 2 \cdot 2 \cdot 2 \cdot 2 \cdot 2 = -64$

93. $2 - (3 - 2^2) + 10 \div 2 \cdot 5 = 2 - (3 - 4) + 10 \div 2 \cdot 5$
$= 2 - (-1) + 10 \div 2 \cdot 5 = 2 + 1 + 10 \div 2 \cdot 5$
$= 2 + 1 + 5 \cdot 5 = 2 + 1 + 25 = 28$

95. *Writing Exercise.*

(1) Find the slope of the line using $m = \dfrac{y_2 - y_1}{x_2 - x_1}$.

(2) Substitute in the point-slope equation, $y - y_1 = m(x - x_1)$.

(3) Solve for y.

97. $y - 3 = 0(x - 52)$

Observe that the slope is 0. Then this is the equation of a horizontal line that passes through $(52, 3)$. Thus, its graph is a horizontal line 3 units above the x-axis.

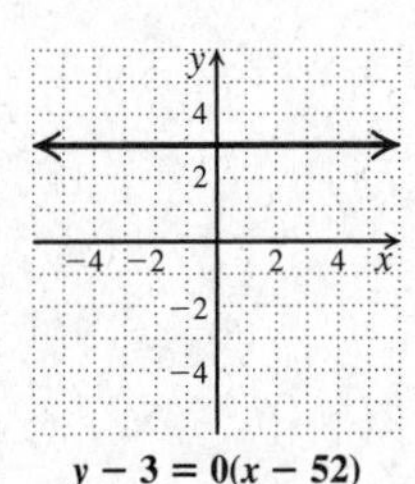

$y - 3 = 0(x - 52)$

99. First we find the slope of the line using any two points on the line. We will use $(3, -3)$ and $(4, -1)$.

$$m = \frac{-3-(-1)}{3-4} = \frac{-2}{-1} = 2$$

Then we write an equation of the line in point-slope form using either of the points above.

$$y - (-3) = 2(x - 3)$$

Finally, we find an equivalent equation in slope-intercept form.

$$y-(-3)=2(x-3)$$
$$y+3=2x-6$$
$$y=2x-9$$

101. First we find the slope of the line using any two points on the line. We will use $(2,5)$ and $(5,1)$.

$$m=\frac{5-1}{2-5}=\frac{4}{-3}=-\frac{4}{3}$$

Then we write an equation of the line in point-slope form using either of the points above.

$$y-5=-\frac{4}{3}(x-2)$$

Finally, we find an equivalent equation in slope-intercept form.

$$y-5=-\frac{4}{3}(x-2)$$
$$y-5=-\frac{4}{3}x+\frac{8}{3}$$
$$y=-\frac{4}{3}x+\frac{23}{3}$$

103. The slope of $y=3-4x$ is -4. We are given the y-intercept of the line, so we use slope-intercept form. The equation is $y=-4x+7$.

105. First find the slope of the line passing through $(2,7)$ and $(-1,-3)$.

$$m=\frac{-3-7}{-1-2}=\frac{-10}{-3}=\frac{10}{3}$$

Now find an equation of the line containing the point $(-1,5)$ and having slope $\frac{10}{3}$.

$$y-5=\frac{10}{3}(x-(-1))$$
$$y-5=\frac{10}{3}(x+1)$$
$$y-5=\frac{10}{3}x+\frac{10}{3}$$
$$y=\frac{10}{3}x+\frac{25}{3}$$

107. $\frac{x}{2}+\frac{y}{5}=1$

Using the form $\frac{x}{a}+\frac{y}{b}=1$

The x-intercept is $(2,0)$.
The y-intercept is $(0,5)$.

109.

$$4y-3x=12$$
$$\frac{1}{12}(4y-3x)=12\cdot\frac{1}{12}$$
$$\frac{y}{3}-\frac{x}{4}=1$$
$$\frac{x}{-4}+\frac{y}{3}=1$$

The x-intercept is $(-4,0)$.
The y-intercept is $(0,3)$.

111. *Writing Exercise.* Equations are entered on most graphing calculators in slope-intercept form. Writing point-slope form in the modified form $y=m(x-x_1)+y_1$ better accommodates graphing calculators.

Connecting the Concepts 3.7

1. $y=-\frac{1}{2}x-7$ is in slope-intercept form.

3. $x=y+2$ is none of these.

5. $y-2=5(x+1)$ is in point-slope form.

7.

$2x=5y+10$
$2x-5y=10$ Subtracting $5y$ from both sides

9.

$y=2x+7$
$-2x+y=7$ Subtracting $2x$ from both sides
$2x-y=-7$ Multiplying -1 to both sides

11.

$y-2=3(x+7)$
$y-2=3x+21$ Using the distributive law
$y=3x+23$ Adding 2
$-3x+y=23$ Subtracting $3x$
$3x-y=-23$ Multiplying by -1

13.

$2x-7y=8$
$-7y=-2x+8$ Subtracting $2x$
$-\frac{1}{7}(-7y)=-\frac{1}{7}(-2x+8)$ Multiplying $-\frac{1}{7}$
$y=\frac{2}{7}x-\frac{8}{7}$ Using distributive law

15.

$8x=y+3$
$8x-3=y$ Subtracting 3
$y=8x-3$ rewriting

17.

$$9y=5-8x$$
$$\frac{1}{9}(9y)=\frac{1}{9}(-8x+5)$$
$$y=-\frac{8}{9}x+\frac{5}{9}$$

19.

$2-3y=5y+6$
$-4-3y=5y$ Subtracting 6
$-4=8y$ Adding $3y$
$-\frac{4}{8}=y$ Multiplying $\frac{1}{8}$
$-\frac{1}{2}=y$ Simplifying
$y=-\frac{1}{2}$

Chapter 3 Review

1. True, see page 153 of the text.

3. False, slope-intercept form is $y=mx+b$.

5. True, see page 185 of the text.

7. True, see page 187 of the text.

9. True, see page 207 in the text.

11. ***Familiarize.*** From the pie chart we see that 23.8% of the searches using Yahoo. We let $x =$ the number of searches using Yahoo, in billions in July 2006.

Translate. We reword the problem.

What	is	23.8%	of	5.6
↓	↓	↓	↓	↓
x	$=$	23.8%	$\cdot$	5.6

Carry out.

$$x = 0.238 \cdot 5 \cdot 6 \approx 1.3$$

Check. We can repeat the calculations.

State. About 1.3 billion searches were done using Yahoo.

13.-15. We plot the points $(5, -1)$, $(2, 3)$ and $(-4, 0)$.

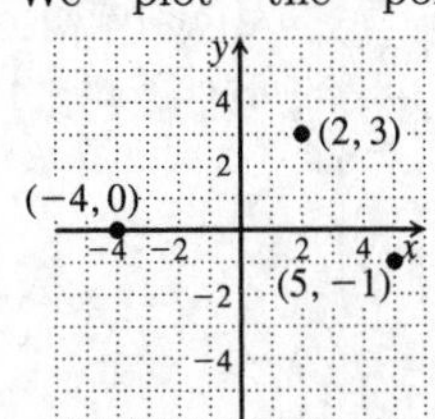

17. Since the first coordinate is positive and the second point is negative, the point $(15.3, -13.8)$ is in quadrant IV.

19. Point A is 5 units left and 1 unit down. The coordinates of A $(-5, -1)$.

21. Point C is 3 units right and 0 units up or down. The coordinates of C are $(3, 0)$

23. a) We substitute 3 for x and 1 for y.

$$\begin{array}{c|l} y & = 2x + 7 \\ \hline 1 & 2(3) + 7 \\ & 6 + 7 \end{array}$$

$1 \stackrel{?}{=} 13$ FALSE

No, the pair $(3, 1)$ is not a solution.

b) We substitute -3 for x and 1 for y.

$$\begin{array}{c|l} y & = 2x + 7 \\ \hline 1 & 2(-3) + 7 \\ & -6 + 7 \end{array}$$

$1 \stackrel{?}{=} 1$ TRUE

Yes, the pair $(-3, 1)$ is a solution.

25. $y = x - 5$

The y-intercept is $(0, -5)$. We find two other points.

When $x = 5$, $y = 5 - 5 = 0$.

When $x = 3$, $y = 3 - 5 = -2$.

x	y
0	−5
5	0
3	−2

We plot these points, draw the line and label the graph $y = x - 5$.

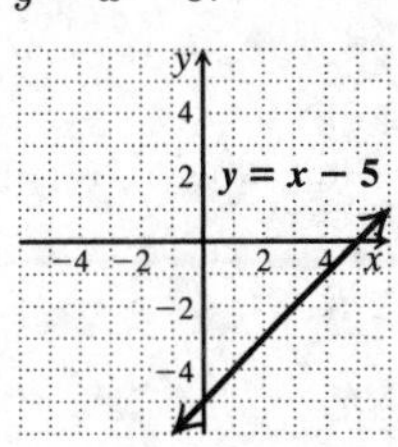

27. $y = -x + 4$

The y-intercept is $(0, 4)$. We find two other points.

When $x = 4$, $y = -4 + 4 = 0$.

When $x = 2$, $y = -2 + 4 = 2$.

x	y
0	4
4	0
2	2

We plot these points, draw the line and label the graph $y = -x + 4$.

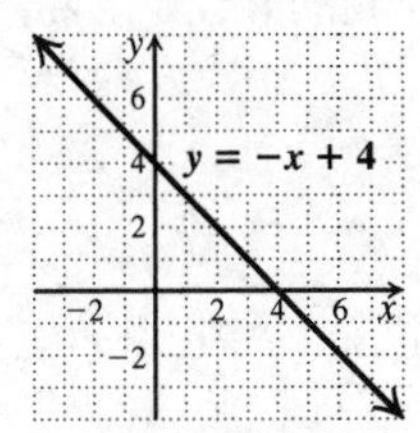

29. $4x + 5 = 3$

$x = -\dfrac{1}{2}$

Any order pair $(-\frac{1}{2}, y)$ is a solution. The variable x must be $-\frac{1}{2}$, but the y variable can be any number. A few are listed below.

When $x = 5$, $y = 5 - 5 = 0$.

When $x = 3$, $y = 3 - 5 = -2$.

x	y
$-\frac{1}{2}$	0
$-\frac{1}{2}$	2
$-\frac{1}{2}$	−2

We plot these points, draw the line and label the graph $4x + 5 = 3$.

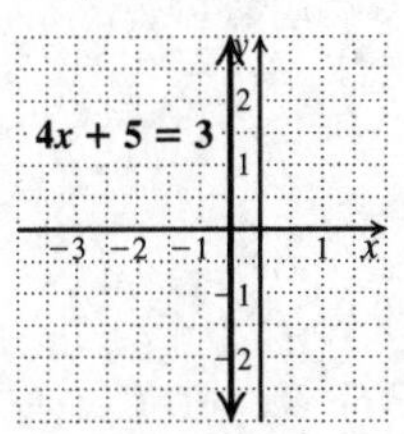

31. We graph $v = -\frac{1}{4}t + 9$ by selecting values of t and calculating the values for v.

If $t = 0$, $v = -\frac{1}{4}(0) + 9 = 9$.

If $t = 4$, $v = -\frac{1}{4}(4) + 9 = 8$.

If $t = 8$, $v = -\frac{1}{4}(8) + 9 = 7$.

t	v
0	9
4	8
8	7

We plot these points, draw the line and label the graph.

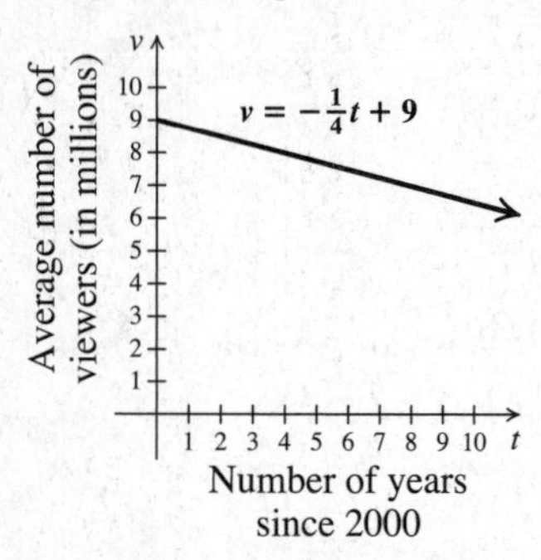

Since $2008 - 2000 = 8$. Locate the point on the line above 8 and find the corresponding value on the vertical axis. The value is 7, so we estimate about 7 million daily viewers in 2008.

33. The points (60 mi, 5 gal) and (120 mi, 10 gal) are on the graph. This tells $120 - 60$ mi and $10 - 5 = 5$gal. The rate is

$$\frac{60 \text{ mi}}{5 \text{ gal}} = 12 \text{ mpg}$$

35. We can use any two points on the line, such as $(-1, -2)$ and $(2, 5)$.

$$m = \frac{\text{change in } y}{\text{change in } x}$$
$$= \frac{5 - (-2)}{2 - (-1)} = \frac{7}{3}$$

37. $(-2, 5)$ and $(3, -1)$

$$m = \frac{-1 - 5}{3 - (-2)} = \frac{-6}{5}$$

39. $(-3, 0)$ and $(-3, 5)$

$$m = \frac{5 - 0}{-3 - (-3)} = \frac{5}{0}, \text{ undefined}$$

41. The grade is expressed as a percent

$$m = \frac{1}{12} \approx 0.08\bar{3} \approx 8.\bar{3}\%$$

43. Rewrite the equation in slope-intercept form.
$3x + 5y = 45$:

$$5y = -3x + 45$$
$$y = -\tfrac{3}{5}x + 9$$

The slope is $-\frac{3}{5}$ and the $y-$intercept is $(0, 9)$.

45. First solve for y and determine the slope of each line.
$3x - 5 = 7y$

$$y = \tfrac{3}{7}x - \tfrac{5}{7}$$

The slope of $3x - 5 = 7y$ is $\frac{3}{7}$.

$$7y - 3x = 7$$

$$y = \tfrac{3}{7}x + 1$$

The slope of $7y - 3x = 7$ is $\frac{3}{7}$.

The slopes are the same, so the lines are parallel.

47. $y - y_1 = m(x - x_1)$
We substitute $-\frac{1}{3}$ for m, -2 for x_1, and 9 for y_1.
$y - 9 = -\frac{1}{3}(x - (-2))$

49. $y - y_1 = m(x - x_1)$
We substitute 5 for m, 3 for x_1, and -10 for y_1.

$$y - (-10) = 5(x - 3)$$
$$y + 10 = 5x - 15$$
$$y = 5x - 25$$

51. $y = \dfrac{2}{3}x - 5$
First we plot the y-intercept $(0, -5)$. We can start at the y-intercept and use the slope, $\frac{2}{3}$, to find another point. We move up 2 units and right 3 units to get a new point $(3, -3)$. Thinking of the slope as $\dfrac{-2}{-3}$ we move down 2 units and left 3 units to get another point $(-3, -7)$.

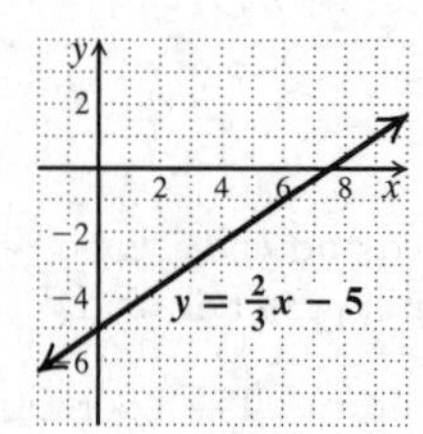

53. $y = 6$
Any ordered pair $(x, 6)$ is a solution. The variable y must be 6, but the x variable can be any number. A few are listed below. Plot these points and draw the graph.

x	y
0	6
1	6
2	6

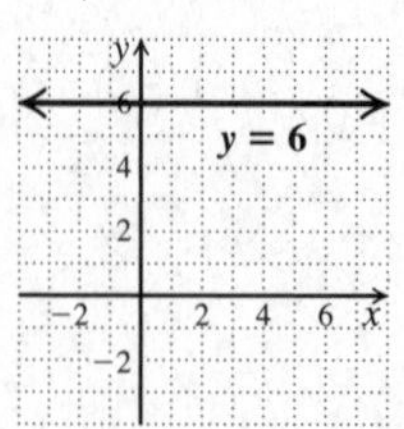

55. $y + 2 = -\dfrac{1}{2}(x - 3)$
We plot the $(3, -2)$. move down 1 unit and right 2 units to the point $(5, -3)$ and draw the line.

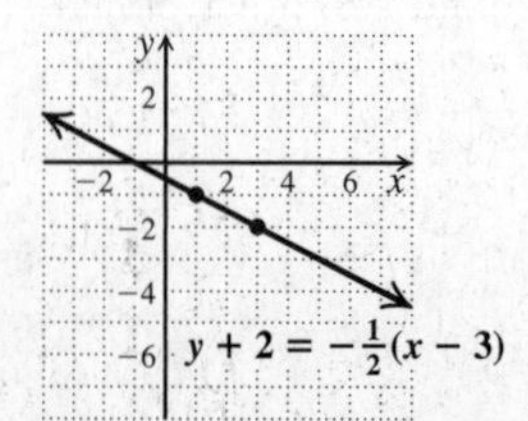

57. *Writing Exercise.* The graph of a vertical line has only an x-intercept. The graph of a horizontal line has only a y-intercept. The graph of a nonvertical, nonhorizontal line will have only one intercept if it passes through the origin: (0,0) is both the x-intercept and the y-intercept.

59. $y = -5x + b$, we substitute $(3, 4)$.

$$4 = -5(3) + b$$
$$4 = -15 + b$$
$$19 = b$$

61. $y = 4 - |x|$

x	y
0	4
1	3
−1	3

Answers may vary.

Chapter 3 Test

1. First determine the number of student volunteers:

$$25\% \times 1200 = 0.25 \times 1200 = 300$$

From the chart, 31.6% of the students will volunteer in education or youth services, so

$$31.6\% \times 300 = 0.316 \times 300 = 94.8$$

Therefore, about 95 students will volunteer in education or youth services.

3. The point having coordinates $(-2, -10)$ is located in quadrant III.

5. Point A has coordinates $(3, 4)$.

7. Point C has coordinates $(-5, 2)$.

9. $2x - 4y = -8$

We rewrite the equation in slope-intercept form.

$$2x - 4y = -8$$
$$-4y = -2x - 8$$
$$4y = 2x + 8$$
$$y = \tfrac{1}{2}x + 2$$

Slope: $\frac{1}{2}$; y-intercept: $(0, 2)$. First we plot the y-intercept $(0, 2)$. We can start at the y-intercept and use the slope $\frac{1}{2}$ to find another point. We move up 1 unit and right 2 units to get a new point $(2, 3)$. Thinking of the slope as $\frac{-1}{-2}$ we can start at $(0, 2)$ and move down 1 unit and left 2 units to get another point $(-2, 1)$. To finish, we draw and label the line.

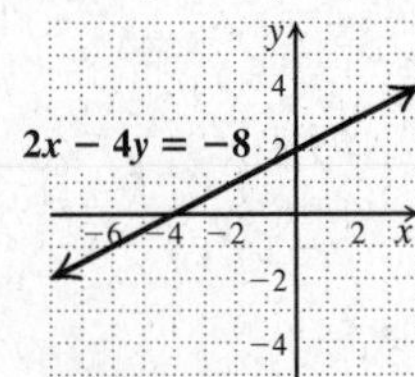

11. $y = \frac{3}{4}x$

We rewrite this equation in slope-intercept form.

$y = \frac{3}{4}x + 0$

Slope: $\frac{3}{4}$; y-intercept: $(0, 0)$

First we plot the y-intercept $(0, 0)$. We can start at the y-intercept and using the slope as $\frac{3}{4}$ we find another point. We move up 3 units and right 4 units to get a new point $(4, 3)$. Thinking of the slope as $\frac{-3}{-4}$ we can start at $(0, 0)$ and move down 3 units and left 4 units to get another point $(-4, -3)$. To finish, we draw and label the line.

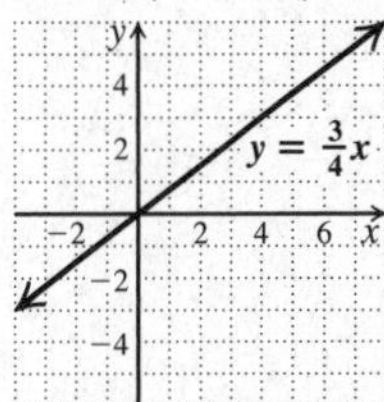

13. $x = -1$ This is a vertical line with x-intercept $(-1, 0)$.

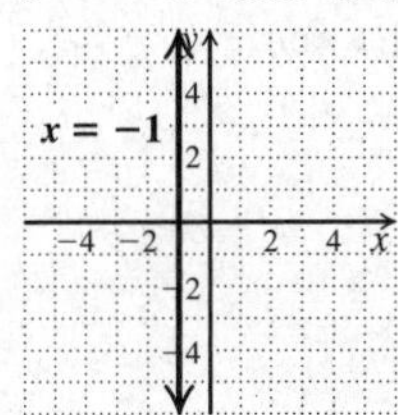

15. $(-5, 6)$ and $(-1, -3)$

$$m = \frac{-3 - 6}{-1 - (-5)} = \frac{-3 - 6}{-1 + 5} = \frac{-9}{4}$$

17.

$$\text{rate} = \frac{\text{change in distance}}{\text{change in time}} = \frac{6\text{ km} - 3\text{ km}}{2{:}24\text{ P.M.} - 2{:}15\text{ P.M.}} = \frac{3\text{ km}}{9\text{min}} = \frac{1}{3}\text{ km/min}$$

19. $5x - y = 30$

To find the x-intercept, we let $y = 0$ and solve for x.

$$5x - 0 \cdot 0 = 30$$
$$5x - 0 = 30$$
$$5x = 30$$
$$x = 6$$

The x-intercept is $(6, 0)$.

To find the y-intercept, we let $x = 0$ and solve for y.

$$5 \cdot 0 - y = 30$$
$$-y = 30$$
$$y = -30$$

The y-intercept is $(0, -30)$.

21. Slope: $-\dfrac{1}{3}$; y-intercept: $(0, -11)$.

The slope-intercept equation is $y = -\dfrac{1}{3}x - 11$.

23. Write both equations in slope-intercept form.

$$2y - x = 6$$

$$y = -2x + 5 \qquad y = \tfrac{1}{2}x + 3$$
$$m = -2$$

$$m = \tfrac{1}{2}$$

The product of their slopes is $(-2)\left(\frac{1}{2}\right)$, or -1; the lines are perpendicular.

25.

a. Plot $(20, 150)$
$(60, 120)$

$$m = \frac{150 - 120}{20 - 60} = \frac{30}{-40} = \frac{-3}{4}$$

Using point-slope form:

$$r - 120 = \tfrac{-3}{4}(a - 60)$$
$$r - 120 = \tfrac{-3}{4}(a - 60)$$
$$r = \tfrac{-3}{4}a + 165$$

To finish we draw and label the line.

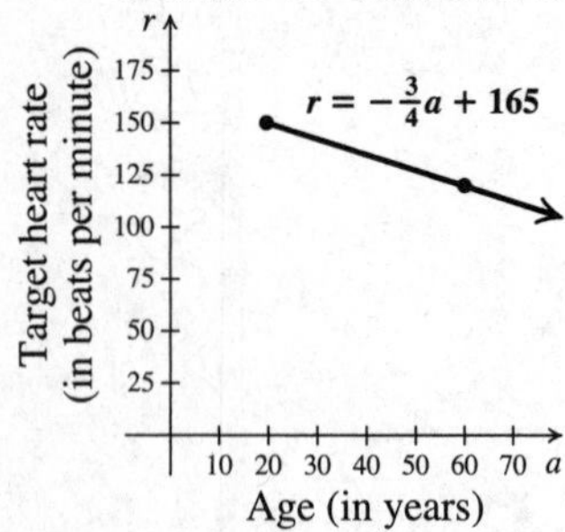

b. Using equation and let $a = 36$

$$r = -\tfrac{3}{4}(36) + 165$$
$$= -27 + 165$$
$$= 138$$

The target heart rate for a 36-year-old is 138 beats per minute.

27. $y + 4 = -\frac{1}{2}(x - 3)$

Rewriting this equation in point-slope form, we have:
$y - (-4) = -\frac{1}{2}(x - 3)$
We have slope of $-\frac{1}{2}$ and a point having coordinates $(3, -4)$. Thinking of the slope as $\frac{-1}{2}$, we start at $(3, -4)$ and move down 1 unit and right 2 units to get a new point $(5, -5)$. Thinking of the slope as $\frac{1}{-2}$, we move up 1 unit and left 2 units to get the point $(1, -3)$. To finish, we draw and label the line.

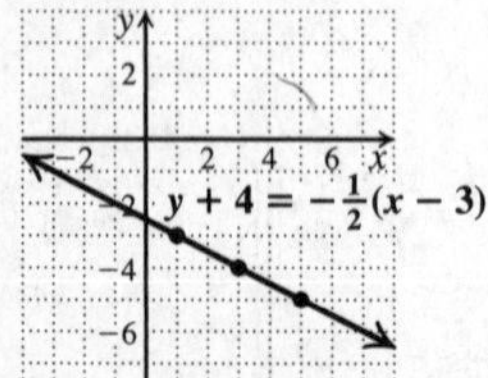

29. First make a sketch.

The height of the square is $4 - (-1) = 4 + 1 = 5$ and the width is $2 - (-3) = 2 + 3 = 5$. Therefore, the area of the square is $5 \times 5 = 25$ square units and the perimeter is $4 \times 5 = 20$ units.

Chapter 4

Polynomials

Exercise Set 4.1

1. By the rule for raising a product to a power on page 232, choice (e) is correct.

3. By the power rule on page 231, choice (b) is correct.

5. By the definition of 0 as an exponent on page 230, choice (g) is correct.

7. By the rule for raising a quotient to a power on page 229, choice (c) is correct.

9. The base is $2x$. The exponent is 5.

11. The base is x. The exponent is 3.

13. The base is $\frac{4}{y}$. The exponent is 7.

15. $d^3 \cdot d^{10} = d^{3+10} = d^{13}$

17. $a^6 \cdot a = a^6 \cdot a^1 = a^{6+1} = a^7$

19. $6^5 \cdot 6^{10} = 6^{5+10} = 6^{15}$

21. $(3y)^4(3y)^8 = (3y)^{4+8} = (3y)^{12}$

23. $(8n)(8n)^9 = (8n)^1(8n)^9 = (8n)^{1+9} = (8n)^{10}$

25. $(a^2b^7)(a^3b^2) = a^2b^7a^3b^2$ Using an associative law

$= a^2a^3b^7b^2$ Using a commutative law

$= a^5b^9$ Adding exponents

27. $(x+3)^5(x+3)^8 = (x+3)^{5+8} = (x+3)^{13}$

29. $r^3 \cdot r^7 \cdot r^0 = r^{3+7+0} = r^{10}$

31.
$$\begin{aligned}(mn^5)(m^3n^4) &= mn^5m^3n^4\\ &= m^1m^3n^5n^4\\ &= m^{1+3}n^{5+4}\\ &= m^4n^9\end{aligned}$$

33. $\frac{7^5}{7^2} = 7^{5-2} = 7^3$ Subtracting exponents

35. $\frac{t^8}{t} = t^{8-1} = t^7$ Subtracting exponents

37. $\frac{(5a)^7}{(5a)^6} = (5a)^{7-6} = (5a)^1 = 5a$

39. $\frac{(x+y)^8}{(x+y)^8}$

Observe that we have an expression divided by itself. Thus, the result is 1.

We could also do this exercise as follows:

$\frac{(x+y)^8}{(x+y)^8} = (x+y)^{8-8} = (x+y)^0 = 1$

41. $\frac{(r+s)^{12}}{(r+s)^4} = (r+s)^{12-4} = (r+s)^8$

43. $\frac{12d^9}{15d^2} = \frac{12}{15}d^{9-2} = \frac{4}{5}d^7$

45. $\frac{8a^9b^7}{2a^2b} = \frac{8}{2} \cdot \frac{a^9}{a^2} \cdot \frac{b^7}{b^1} = 4a^{9-2}b^{7-1} = 4a^7b^6$

47. $\frac{x^{12}y^9}{x^0y^2} = x^{12-0}y^{9-2} = x^{12}y^7$

49. When $t = 15$, $t^0 = 15^0 = 1$. (Any nonzero number raised to the 0 power is 1.)

51. When $x = -22$, $5x^0 = 5(-22)^0 = 5 \cdot 1 = 5$.

53. $7^0 + 4^0 = 1 + 1 = 2$

55. $(-3)^1 - (-3)^0 = -3 - 1 = -4$

57. $(x^3)^{11} = x^{3 \cdot 11} = x^{33}$ Multiplying exponents

59. $(5^8)^4 = 5^{8 \cdot 4} = 5^{32}$ Multiplying exponents

61. $(t^{20})^4 = t^{20 \cdot 4} = t^{80}$

63. $(10x)^2 = 10^2x^2 = 100x^2$

65. $(-2a)^3 = (-2)^3a^3 = -8a^3$

67. $(-5n^7)^2 = (-5)^2(n^7)^2 = 25n^{7 \cdot 2} = 25n^{14}$

69. $(a^2b)^7 = (a^2)^7(b^7) = a^{14}b^7$

71. $(r^5t)^3(r^2t^8) = (r^5)^3(t)^3r^2t^8 = r^{15}t^3r^2t^8 = r^{17}t^{11}$

73. $(2x^5)^3(3x^4) = 2^3(x^5)^3(3x^4) = 8x^{15} \cdot 3x^4 = 24x^{19}$

75. $\left(\frac{x}{5}\right)^3 = \frac{x^3}{5^3} = \frac{x^3}{125}$

77. $\left(\frac{7}{6n}\right)^2 = \frac{7^2}{(6n)^2} = \frac{49}{6^2n^2} = \frac{49}{36n^2}$

79. $\left(\frac{a^3}{b^8}\right)^6 = \frac{(a^3)^6}{(b^8)^6} = \frac{a^{18}}{b^{48}}$

81. $\left(\frac{x^2y}{z^3}\right)^4 = \frac{(x^2y)^4}{(z^3)^4} = \frac{(x^2)^4(y^4)}{z^{12}} = \frac{x^8y^4}{z^{12}}$

83. $\left(\frac{a^3}{-2b^5}\right)^4 = \frac{(a^3)^4}{(-2b^5)^4} = \frac{a^{12}}{(-2)^4(b^5)^4} = \frac{a^{12}}{16b^{20}}$

85. $\left(\frac{5x^7y}{-2z^4}\right)^3 = \frac{(5x^7y)^3}{(-2z^4)^3} = \frac{5^3(x^7)^3y^3}{-2^3(z^4)^3} = \frac{125x^{21}y^3}{-8z^{12}}$

87. $\left(\frac{4x^3y^5}{3z^7}\right)^0$

Observe that for $x \neq 0$, $y \neq 0$, and $z \neq 0$, we have a nonzero number raised to the 0 power. Thus, the result is 1.

89. *Writing Exercise.* -5^2 is the opposite of the square of 5, or the opposite of 25, so it is -25; $(-5)^2$ is the square of the opposite of 5, or $-5(-5)$, so it is 25.

91. $-10 - 14 = -10 + (-14) = -24$

93. $-16 + 5 = -11$

95. $-3 + (-11) = -14$

97. *Writing Exercise.* Any number raised to an even power is nonnegative. Any nonnegative number raised to an odd power is nonnegative. Any negative number raised to an odd power is negative. Thus, a must be a negative number, and n must be an odd number.

99. *Writing Exercise.* Let $s =$ the length of a side of the smaller square. Then $3s =$ the length of a side of the larger square. The area of the smaller square is s^2, and the area of the larger square is $(3s)^2$, or $9s^2$, so the area of the larger square is 9 times the area of the smaller square.

101. Choose any number except 0. For example, let $x = 1$.

$3x^2 = 3 \cdot 1^2 = 3 \cdot 1 = 3$, but

$(3x)^2 = (3 \cdot 1)^2 = 3^2 = 9$.

103. Choose any number except 0 or 1. For example, let $t = -1$. Then $\dfrac{t^6}{t^2} = \dfrac{(-1)^6}{(-1)^2} = \dfrac{1}{1} = 1$, but $t^3 = (-1)^3 = -1$.

105. $y^{4x} \cdot y^{2x} = y^{4x+2x} = y^{6x}$

107. $\dfrac{x^{5t}(x^t)^2}{(x^{3t})^2} = \dfrac{x^{5t}x^{2t}}{x^{6t}} = \dfrac{x^{7t}}{x^{6t}} = x^t$

109.
$$\begin{aligned} \frac{t^{26}}{t^x} &= t^x \\ t^{26-x} &= t^x \\ 26 - x &= x \quad \text{Equating exponents} \\ 26 &= 2x \\ 13 &= x \end{aligned}$$

The solution is 13.

111. Since the bases are the same, the expression with the larger exponent is larger. Thus, $4^2 < 4^3$.

113. $4^3 = 64$, $3^4 = 81$, so $4^3 < 3^4$.

115.
$$\begin{aligned} 25^8 &= (5^2)^8 = 5^{16} \\ 125^5 &= (5^3)^5 = 5^{15} \end{aligned}$$
$5^{16} > 5^{15}$, or $25^8 > 125^5$.

117. $2^{22} = 2^{10} \cdot 2^{10} \cdot 2^2 \approx 10^3 \cdot 10^3 \cdot 4 \approx 1000 \cdot 1000 \cdot 4 \approx 4,000,000$

Using a calculator, we find that $2^{22} = 4,194,304$. The difference between the exact value and the approximation is $4,194,304 - 4,000,000$, or 194,304.

119. $2^{31} = 2^{10} \cdot 2^{10} \cdot 2^{10} \cdot 2 \approx 10^3 \cdot 10^3 \cdot 10^3 \cdot 2$
$\approx 1000 \cdot 1000 \cdot 1000 \cdot 2 = 2,000,000,000$

Using a calculator, we find that $2^{31} = 2,147,483,648$. The difference between the exact value and the approximation is $2,147,483,648 - 2,000,000,000 = 147,483,648$.

121.
$$\begin{aligned} 1.5 \text{ MB} &= 1.5 \times 1000 \text{ KB} \\ &= 1.5 \times 1000 \times 1 \times 2^{10} \text{ bytes} \\ &= 1,536,000 \text{ bytes} \\ &\approx 1,500,000 \text{ bytes} \end{aligned}$$

Exercise Set 4.2

1. $\left(\dfrac{x^3}{y^2}\right)^{-2} = \left(\dfrac{y^2}{x^3}\right)^2 = \dfrac{(y^2)^2}{(x^3)^2} = \dfrac{y^4}{x^6} \Rightarrow$ (c)

3. $\left(\dfrac{y^{-2}}{x^{-3}}\right)^{-3} = \dfrac{(y^{-2})^{-3}}{(x^{-3})^{-3}} = \dfrac{y^6}{x^9} \Rightarrow$ (a)

5. $2^{-3} = \dfrac{1}{2^3} = \dfrac{1}{8}$

7. $(-2)^{-6} = \dfrac{1}{(-2)^6} = \dfrac{1}{64}$

9. $t^{-9} = \dfrac{1}{t^9}$

11. $xy^{-2} = x \cdot \dfrac{1}{y^2} = \dfrac{x}{y^2}$

13. $r^{-5}t = \dfrac{1}{r^5} \cdot t = \dfrac{t}{r^5}$

15. $\dfrac{1}{a^{-8}} = a^8$

17. $7^{-1} = \dfrac{1}{7^1} = \dfrac{1}{7}$

19. $\left(\dfrac{3}{5}\right)^{-2} = \left(\dfrac{5}{3}\right)^2 = \dfrac{5^2}{3^2} = \dfrac{25}{9}$

21. $\left(\dfrac{x}{2}\right)^{-5} = \left(\dfrac{2}{x}\right)^5 = \dfrac{2^5}{x^5} = \dfrac{32}{x^5}$

23. $\left(\dfrac{s}{t}\right)^{-7} = \left(\dfrac{t}{s}\right)^7 = \dfrac{t^7}{s^7}$

25. $\dfrac{1}{9^2} = 9^{-2}$

27. $\dfrac{1}{y^3} = y^{-3}$

29. $\dfrac{1}{5} = \dfrac{1}{5^1} = 5^{-1}$

31. $\dfrac{1}{t} = \dfrac{1}{t^1} = t^{-1}$

33. $2^{-5} \cdot 2^8 = 2^{-5+8} = 2^3$, or 8

35. $x^{-3} \cdot x^{-9} = x^{-12} = \dfrac{1}{x^{12}}$

37. $t^{-3} \cdot t = t^{-3} \cdot t^1 = t^{-3+1} = t^{-2} = \dfrac{1}{t^2}$

39. $(n^{-5})^3 = n^{-5 \cdot 3} = n^{-15} = \dfrac{1}{n^{15}}$

41. $(t^{-3})^{-6} = t^{-3(-6)} = t^{18}$

43. $(t^4)^{-3} = t^{4(-3)} = t^{-12} = \frac{1}{t^{12}}$

45. $(mn)^{-7} = \frac{1}{(mn)^7} = \frac{1}{m^7n^7}$

47. $(3x^{-4})^2 = 3^2(x^{-4})^2 = 9x^{-8} = \frac{9}{x^8}$

49. $(5r^{-4}t^3)^2 = 5^2(r^{-4})^2(t^3)^2 = 25r^{-8}t^6 = \frac{25t^6}{r^8}$

51. $\frac{t^{12}}{t^{-2}} = t^{12-(-2)} = t^{14}$

53. $\frac{y^{-7}}{y^{-3}} = y^{-7-(-3)} = y^{-4} = \frac{1}{y^4}$

55. $\frac{15y^{-7}}{3y^{-10}} = 5y^{-7-(-10)} = 5y^3$

57. $\frac{2x^6}{x} = 2\frac{x^6}{x^1} = 2x^{6-1} = 2x^5$

59. $-\frac{15a^{-7}}{10b^{-9}} = -\frac{3b^9}{2a^7}$

61. $\frac{t^{-7}}{t^{-7}}$

Note that we have an expression divided by itself. Thus, the result is 1. We could also find this result as follows:

$$\frac{t^{-7}}{t^{-7}} = t^{-7-(-7)} = t^0 = 1.$$

63. $\frac{8x^{-3}}{y^{-7}z^{-1}} = \frac{8y^7z}{x^3}$

65. $\frac{3t^4}{s^{-2}u^{-4}} = 3s^2t^4u^4$

67. $(x^4y^5)^{-3} = (x^4)^{-3}(y^5)^{-3} = x^{-12}y^{-15} = \frac{1}{x^{12}y^{15}}$

69. $(3m^{-5}n^{-3})^{-2} = 3^{-2}m^{-5(-2)}n^{-3(-2)} = 3^{-2}m^{10}n^6$

$= \frac{m^{10}n^6}{9}$

71. $(a^{-5}b^7c^{-2})(a^{-3}b^{-2}c^6) = a^{-5+(-3)}b^{7+(-2)}c^{-2+6}$

$= a^{-8}b^5c^4 = \frac{b^5c^4}{a^8}$

73. $\left(\frac{a^4}{3}\right)^{-2} = \left(\frac{3}{a^4}\right)^2 = \frac{3^2}{(a^4)^2} = \frac{9}{a^8}$

75. $\left(\frac{m^{-1}}{n^{-4}}\right)^3 = \frac{(m^{-1})^3}{(n^{-4})^3} = \frac{m^{-3}}{n^{-12}} = \frac{n^{12}}{m^3}$

77. $\left(\frac{2a^2}{3b^4}\right)^{-3} = \left(\frac{3b^4}{2a^2}\right)^3 = \frac{(3b^4)^3}{(2a^2)^3} = \frac{3^3(b^4)^3}{2^3(a^2)^3} = \frac{27b^{12}}{8a^6}$

79. $\left(\frac{5x^{-2}}{3y^{-2}z}\right)^0$

Any nonzero expression raised to the 0 power is equal to 1. Thus, the answer is 1.

81. $\frac{-6a^3b^{-5}}{-3a^7b^{-8}} = \frac{-6}{-3}\cdot\frac{a^3b^{-5}}{a^7b^{-8}}$

$= 2\cdot\frac{b^{8-5}}{a^{7-3}}$

$= \frac{2b^3}{a^4}$

83. $\frac{10x^{-4}yz^7}{8x^7y^{-3}z^{-3}} = \frac{10}{8}\cdot\frac{y^{1+3}z^{7+3}}{x^{7+4}} = \frac{5y^4z^{10}}{2x^{11}}$

85. $4.92 \times 10^3 = 4.920 \times 10^3$ The decimal point moves right three places

$= 4920$

87. 8.92×10^{-3}

Since the exponent is negative, the decimal point will move to the left.

.008.92 The decimal point moves left 3 places.

$8.92 \times 10^{-3} = 0.00892$

89. 9.04×10^8

Since the exponent is positive, the decimal point will move to the right.

9.04000000.

8 places

$9.04 \times 10^8 = 904,000,000$

91. $3.497 \times 10^{-6} = 0000003.497 \times 10^{-6}$ The decimal point left six places

$= 0.000003497$

93. $36,000,000 = 3.6 \times 10^m$ We move the decimal point 7 places to the right. Thus m is 7.

$= 3.6 \times 10^7$

95. $0.00583 = 5.83 \times 10^m$

To write 5.83 as 0.00583 we move the decimal point 3 places to the left. Thus, m is -3 and

$0.00583 = 5.83 \times 10^{-3}$.

97. $78,000,000,000 = 7.8 \times 10^m$

To write 7.8 as 78,000,000,000 we move the decimal point 10 places to the right. Thus, m is 10 and

$78,000,000,000 = 7.8 \times 10^{10}$.

99. $0.000000527 = 5.27 \times 10^m$

To write 5.27 as 0.000000527 we move the decimal point 7 places to the left. Thus, m is -7 and

$0.000000527 = 5.27 \times 10^{-7}$.

101. $0.000001032 = 1.032 \times 10^m$ We move the decimal point 6 places to the left. Thus $m = -6$.

$= 1.032 \times 10^{-6}$

103. $(3\times10^5)(2\times10^8) = (3\cdot2)\times(10^5\cdot10^8)$
$= 6\times10^{5+8}$
$= 6\times10^{13}$

105. $(3.8\times10^9)(6.5\times10^{-2}) = (3.8\cdot6.5)\times(10^9\cdot10^{-2})$
$= 24.7\times10^7$

The answer is not yet in scientific notation since 24.7 is not a number between 1 and 10. We convert to scientific notation.

$24.7\times10^7 = (2.47\times10)\times10^7 = 2.47\times10^8$

107. $(8.7\times10^{-12})(4.5\times10^{-5})$
$= (8.7\cdot4.5)\times(10^{-12}\cdot10^{-5})$
$= 39.15\times10^{-17}$

The answer is not yet in scientific notation since 39.15 is not a number between 1 and 10. We convert to scientific notation.

$39.15\times10^{-17} = (3.915\times10)\times10^{-17} = 3.915\times10^{-16}$

109. $\dfrac{8.5\times10^8}{3.4\times10^{-5}} = \dfrac{8.5}{3.4}\times\dfrac{10^8}{10^{-5}}$
$= 2.5\times10^{8-(-5)}$
$= 2.5\times10^{13}$

111. $(4.0\times10^3)\div(8.0\times10^8) = \dfrac{4.0}{8.0}\times\dfrac{10^3}{10^8}$
$= 0.5\times10^{3-8}$
$= 0.5\times10^{-5}$
$= 5.0\times10^{-6}$

113. $\dfrac{7.5\times10^{-9}}{2.5\times10^{12}} = \dfrac{7.5}{2.5}\times\dfrac{10^{-9}}{10^{12}}$
$= 3.0\times10^{-9-12}$
$= 3.0\times10^{-21}$

115. *Writing Exercise.* $3^{-29} = \dfrac{1}{3^{29}}$ and $2^{-29} = \dfrac{1}{2^{29}}$. Since $3^{29} > 2^{29}$, we have $\dfrac{1}{3^{29}} < \dfrac{1}{2^{29}}$.

117. $9x + 2y - x - 2y = 9x - x + 2y - 2y = 8x$

119. $-3x + (-2) - 5 - (-x)$
$= -3x + x - 2 - 5$
$= -2x - 7$

121. $4 + x^3 = 4 + 10^3 = 4 + 1000 = 1004$

123. *Writing Exercise.* x^{-n} represents a negative integer when x is negative, $\dfrac{1}{x}$ is an integer, and n is an odd number.

125. $\dfrac{1}{1.25\times10^{-6}} = \dfrac{1}{1.25}\times\dfrac{1}{10^{-6}} = 0.8\times10^6$
$= (8\times10^{-1})\times10^6 = 8\times10^5$

127. $8^{-3}\cdot32\div16^2 = (2^3)^{-3}\cdot2^5\div(2^4)^2$
$= 2^{-9}\cdot2^5\div2^8 = 2^{-4}\div2^8 = 2^{-12}$

129. $\dfrac{125^{-4}(25^2)^4}{125} = \dfrac{(5^3)^{-4}((5^2)^2)^4}{5^3}$
$= \dfrac{5^{-12}(5^4)^4}{5^3} = \dfrac{5^{-12}\cdot5^{16}}{5^3} = \dfrac{5^4}{5^3} = 5^1 = 5$

131. $\left[\left(5^{-3}\right)^2\right]^{-1} = 5^{(-3)(2)(-1)} = 5^6$

133. $3^{-1} + 4^{-1} = \dfrac{1}{3} + \dfrac{1}{4} = \dfrac{4}{12} + \dfrac{3}{12} = \dfrac{7}{12}$

135. $\dfrac{27^{-2}(81^2)^3}{9^8} = \dfrac{(3^3)^{-2}((3^4)^2)^3}{(3^2)^8}$
$= \dfrac{3^{-6}\cdot3^{24}}{3^{16}} = \dfrac{3^{18}}{3^{16}} = 3^2 = 9$

137. $\dfrac{5.8\times10^{17}}{(4.0\times10^{-13})(2.3\times10^4)}$
$= \dfrac{5.8}{(4.0\cdot2.3)}\times\dfrac{10^{17}}{(10^{-13}\cdot10^4)}$
$\approx 0.6304347826\times10^{17-(-13)-4}$
$\approx (6.304347826\times10^{-1})\times10^{26}$
$\approx 6.304347826\times10^{25}$

139. $\dfrac{(2.5\times10^{-8})(6.1\times10^{-11})}{1.28\times10^{-3}}$
$= \dfrac{(2.5\cdot6.1)}{1.28}\times\dfrac{(10^{-8}\cdot10^{-11})}{10^{-3}}$
$= 11.9140625\times10^{-8+(-11)-(-3)}$
$= 11.9140625\times10^{-16}$
$= (1.19140625\times10)\times10^{-16}$
$= 1.19140625\times10^{-15}$

141. ***Familiarize.*** Let n = the number of car miles.

Translate. We reword the problem.

Miles for one tree	times	the number of trees	is	n
↓	↓	↓	↓	↓
500	×	600,000	=	n

Carry out. We solve the equation.

$500\times600,000 = n$
$300,000,000 = n$
$n = 3\times10^8$

Check. Review the computation, The answer is reasonable.

State. Trees can clean 3×10^8 miles of car traffic in a year.

143. ***Familiarize.*** Let c = the total cost of the condominiums.

Translate. We reword the problem.

What	is	price $2100 per ft²	times	space 110,000 ft²
↓	↓	↓	↓	↓
c	=	(2100)	×	(110,000)

Carry out.

$c = 2100\times1.1\times10^5$
$c = 231,000,000$

Check. We review calculations.

State. $c = \$2.31\times10^8$ is the total cost.

145. ***Familiarize.*** Let m = the number of minutes.

Convert. 3.3 hours $= 3.3 \times 60 = 198$ minutes

Translate.

$$\underbrace{\text{What}}\ \text{is}\ \underbrace{\text{number of patients}}\ \text{times}\ \underbrace{\text{average time in minutes}}$$

$$m = 115{,}000{,}000 \times 198$$

$m = 2.277 \times 10^{10}$

Check. We review calculations.

State. In 2005 patients spent 2.277×10^{10} minutes in emergency rooms.

4.2 Connecting the Concepts

1. $x^4 x^{10} = x^{4+10} = x^{14}$

3. $\frac{x^{-4}}{x^{10}} = \frac{1}{x^{10+4}} = \frac{1}{x^{14}}$

5. $(x^{-4})^{-10} = x^{40}$

7. $\frac{1}{c^{-8}} = c^8$

9. $(2x^3y)^4 = 2^4x^{3\cdot4}y^4 = 16x^{12}y^4$

11. $(3xy^{-1}z^5)^0 = 1$

13. $\left(\frac{a^3}{b^4}\right)^5 = \frac{a^{15}}{b^{20}}$

15. $\frac{30x^4y^3}{12xy^7} = \frac{30}{12}x^{4-1}y^{3-7} = \frac{5x^3}{2y^4}$

17. $\frac{7p^{-5}}{xt^{-6}} = \frac{7t^6}{xp^5}$

19. $(2p^2q^4)(3pq^5)^2 = 2p^2q^4(9p^2q^{10}) = 18p^4q^{14}$

Exercise Set 4.3

1. The only expression with 4 terms is (b).

3. Expression (h) has three terms and they are written in descending order.

5. Expression (g) has two terms, and the degree of the leading term is 7.

7. Expression (a) has two terms, but it is not a binomial because $\frac{2}{x^2}$ is not a monomial.

9. $8x^3 - 11x^2 + 6x + 1$

The terms are $8x^3$, $-11x^2$, $6x$, and 1.

11. $-t^6 - 3t^3 + 9t - 4$

The terms are $-t^6$, $-3t^3$, $9t$, and -4.

13. $8x^4 + 2x$

Term	Coefficient	Degree
$8x^4$	8	4
$2x$	2	1

15. $9t^2 - 3t + 4$

Term	Coefficient	Degree
$9t^2$	9	2
$-3t$	-3	1
4	4	0

17. $6a^5 + 9a + a^3$

Term	Coefficient	Degree
$6a^5$	6	5
$9a$	9	1
a^3	1	3

19. $x^4 - x^3 + 4x - 3$

Term	Coefficient	Degree
x^4	1	4
$-x^3$	-1	3
$4x$	4	1
-3	-3	0

21. $5t + t^3 + 8t^4$

a)

Term	$5t$	t^3	$8t^4$
Degree	1	3	4

b) The term of highest degree is $8t^4$. This is the leading term. Then the leading coefficient is 8.

c) Since the term of highest degree is $8t^4$, the degree of the polynomial is 4.

23. $3a^2 - 7 + 2a^4$

a)

Term	$3a^2$	-7	$2a^4$
Degree	2	0	4

b) The term of highest degree is $2a^4$. This is the leading term. Then the leading coefficient is 2.

c) Since the term of highest degree is $2a^4$, the degree of the polynomial is 4.

25. $8 + 6x^2 - 3x - x^5$

a)

Term	8	$6x^2$	$-3x$	$-x^5$
Degree	0	2	1	5

b) The term of highest degree is $-x^5$. This is the leading term. Then the leading coefficient is -1 since $-x^5 = -1 \cdot x^5$.

c) Since the term of highest degree is $-x^5$, the degree of the polynomial is 5.

27. $7x^2 + 8x^5 - 4x^3 + 6 - \frac{1}{2}x^4$

Term	Coefficient	Degree of Term	Degree of Polynomial
$8x^5$	8	5	
$-\frac{1}{2}x^4$	$-\frac{1}{2}$	4	
$-4x^3$	-4	3	5
$7x^2$	7	2	
6	6	0	

29. Three monomials are added, so $x^2-23x+17$ is a trinomial.

31. The polynomial $x^3 - 7x + 2x^2 - 4$ is a polynomial with no special name because it is composed of four monomials.

33. Two monomials are added, so $y + 8$ is a binomial.

35. The polynomial 17 is a monomial because it is the product of a constant and a variable raised to a whole number power. (In this case the variable is raised to the power 0.)

37. $5n^2 + n + 6n^2 = (5+6)n^2 + n = 11n^2 + n$

39. $3a^4 - 2a + 2a + a^4 = (3+1)a^4 + (-2+2)a$
$= 4a^4 + 0a = 4a^4$

41. $7x^3 - 11x + 5x + x^2 = 7x^3 + x^2 + (-11+5)x$
$= 7x^3 + x^2 - 6x$

43. $4b^3 + 5b + 7b^3 + b^2 - 6b$
$= (4+7)b^3 + b^2 + (5-6)b$
$= 11b^3 + b^2 - b$

45. $10x^2 + 2x^3 - 3x^3 - 4x^2 - 6x^2 - x^4$
$= -x^4 + (2-3)x^3 + (10-4-6)x^2 = -x^4 - x^3$

47. $\frac{1}{5}x^4 + 7 - 2x^2 + 3 - \frac{2}{15}x^4 + 2x^2$
$= \left(\frac{1}{5} - \frac{2}{15}\right)x^4 + (-2+2)x^2 + (7+3)$
$= \left(\frac{3}{15} - \frac{2}{15}\right)x^4 + 0x^2 + 10 = \frac{1}{15}x^4 + 10$

49. $8.3a^2 + 3.7a - 8 - 9.4a^2 + 1.6a + 0.5$
$= (8.3-9.4)a^2 + (3.7+1.6)a - 8 + 0.5$
$= -1.1a^2 + 5.3a - 7.5$

51. For $x = 3$: $-4x + 9 = -4 \cdot 3 + 9$
$= -12 + 9$
$= -3$

For $x = -3$: $-4x + 9 = -4(-3) + 9$
$= 12 + 9$
$= 21$

53. For $x = 3$: $2x^2 - 3x + 7 = 2 \cdot 3^2 - 3 \cdot 3 + 7$
$= 2 \cdot 9 - 3 \cdot 3 + 7$
$= 18 - 9 + 7$
$= 16$

For $x = -3$: $2x^2 - 3x + 7 = 2(-3)^2 - 3(-3) + 7$
$= 2 \cdot 9 - 3(-3) + 7$
$= 18 + 9 + 7$
$= 34$

55. For $x = 3$:
$-3x^3 + 7x^2 - 4x - 8 = -3 \cdot 3^3 + 7 \cdot 3^2 - 4 \cdot 3 - 8$
$= -3 \cdot 27 + 7 \cdot 9 - 12 - 8$
$= -81 + 63 - 12 - 8$
$= -38$

For $x = -3$:
$-3x^3 + 7x^2 - 4x - 8 = -3(-3)^3 + (-3)^2 - 4(-3) - 8$
$= -3(-27) + 7 \cdot 9 + 12 - 8$
$= 148$

57. For $x = 3$: $2x^4 - \frac{1}{9}x^3 = 2 \cdot 3^4 - \frac{1}{9} \cdot 3^3$
$= 2 \cdot 81 - \frac{1}{9} \cdot 27 = 162 - 3 = 159$

For $x = -3$: $2x^4 - \frac{1}{9}x^3 = 2(-3)^4 - \frac{1}{9}(-3)^3$
$= 2 \cdot 81 - \frac{1}{9}(-27)$
$= 162 + 3$
$= 165$

59. For $x = 3$: $-x-x^2-x^3 = -3-3^2-3^3 = -3-9-27 = -39$

For $x = -3$: $-x - x^2 - x^3 = -(-3) - (-3)^2 - (-3)^3$
$= 3 - 9 + 27 = 21$

61. Since 2006 is 2 years after 2004, we evaluate the polynomial for $t = 2$.
$0.4t + 1.13 = 0.4(2) + 1.13$
$= 0.8 + 1.13$
$= 1.93$

The amount spent on shoes for college in 2006 is about \$1.93 billion.

63. $11.12t^2 = 11.12(10)^2 = 11.12(100) = 1112$

A skydiver has fallen approximately 1112 ft 10 seconds after jumping from a plane.

65. $2\pi r = 2(3.14)(10)$ Substituting 3.14 for π and 10 for r
$= 62.8$

The circumference is 62.8 cm.

67. $\pi r^2 = 3.14(7)^2$ Substituting 3.14 for π and 7 for r
$= 3.14(49)$
$= 153.86$

The area is 153.86 m^2.

69. Since 2006 is 3 years after 2003, we first locate 3 on the horizontal axis. From there we move vertically to the graph and then horizontally to the K-axis. This locates an K-value of about 75. Thus the number of kidney-paired donations in 2006 is about 75 donations.

71. Locate 10 on the horizontal axis. From there move vertically to the graph and then horizontally to the M-axis. This locates an M-value of about 9. Thus, about 9 words were memorized in 10 minutes.

73. Locate 8 on the horizontal axis. From there move vertically to the graph and then horizontally to the M-axis. This locates an M-value of about 6. Thus, the value of $-0.001t^3 + 0.1t^2$ for $t = 8$ is approximately 6.

75. Locate 4 on the horizontal axis. From there move vertically to the graph and then horizontally to the B-axis. This locates an BMI-value of about 16.

Locate 14 on the horizontal axis. From there move vertically to the graph and then horizontally to the B-axis. This locates an BMI-value of about 19.

77. *Writing Exercise.* A term is a number, a variable, or a product of numbers and variables which may be raised to powers whereas a monomial is a number, a variable, or a product of numbers and variables raised to *whole number* powers. For example, the term $5x^{-2}y^4$ is not a monomial.

79. $2x + 5 - (x + 8) = 2x + 5 - x - 8 = x - 3$

81. $4a + 3 - (-2a + 6) = 4a + 3 + 2a - 6 = 6a - 3$

83. $4t^4 + 8t - (5t^4 - 9t) = 4t^4 + 8t - 5t^4 + 9t = -t^4 + 17t$

85. *Writing Exercise.* Yes; the evaluation will yield a sum of products of integers which must be an integer.

87. Answers may vary. Choose an ax^5-term where a is an even integer. Then choose three other terms with different degrees, each less than degree 5, and coefficients $a+2$, $a+4$, and $a+6$, respectively, when the polynomial is written in descending order. One such polynomial is $2x^5 + 4x^4 + 6x^3 + 8$.

89. Find the total revenue from the sale of 30 monitors:

$$\begin{aligned} 250x - 0.5x^2 &= 250(30) - 0.5(30)^2 \\ &= 250(30) - 0.5(900) \\ &= 7500 - 450 \\ &= \$7050 \end{aligned}$$

Find the total cost of producing 30 monitors:

$$\begin{aligned} 4000 + 0.6x^2 &= 4000 + 0.6(30)^2 \\ &= 4000 + 0.6(900) \\ &= 4000 + 540 \\ &= \$4540 \end{aligned}$$

Subtract the cost from the revenue to find the profit: $\$7050 - \$4540 = \$2510$

91.

$$\begin{aligned} &(3x^2)^3 + 4x^2 \cdot 4x^4 - x^4(2x)^2 + [(2x)^2]^3 - \\ &\quad 100x^2(x^2)^2 \\ &= 27x^6 + 4x^2 \cdot 4x^4 - x^4 \cdot 4x^2 + (2x)^6 - 100x^2 \cdot x^4 \\ &= 27x^6 + 16x^6 - 4x^6 + 64x^6 - 100x^6 \\ &= 3x^6 \end{aligned}$$

93. First locate 16 on the vertical axis. Then move horizontally to the graph. We meet the curve at 2 places. At each place move down vertically to the horizontal axis and read the corresponding x-value. We see that the ages for a 16 BMI are 3 and 8.

95. We first find q, the quiz average, and t, the test average.

$$q = \frac{60 + 85 + 72 + 91}{4} = \frac{308}{4} = 77$$

$$t = \frac{89 + 93 + 90}{3} = \frac{272}{3} \approx 90.7$$

Now we substitute in the polynomial.

$$\begin{aligned} A &= 0.3q + 0.4t + 0.2f + 0.1h \\ &= 0.3(77) + 0.4(90.7) + 0.2(84) + 0.1(88) \\ &= 23.1 + 36.28 + 16.8 + 8.8 \\ &= 84.98 \\ &\approx 85.0 \end{aligned}$$

97. When $t = 3$, $-t^2 + 10t - 18 = -3^2 + 10 \cdot 3 - 18$
$= -9 + 30 - 18 = 3.$
When $t = 4$, $-t^2 + 10t - 18 = -4^2 + 10 \cdot 4 - 18$
$= -16 + 40 - 18 = 6.$
When $t = 5$, $-t^2 + 10t - 18 = -5^2 + 10 \cdot 5 - 18$
$= -25 + 50 - 18 = 7.$
When $t = 6$, $-t^2 + 10t - 18 = -6^2 + 10 \cdot 6 - 18$
$= -36 + 60 - 18 = 6.$
When $t = 7$, $-t^2 + 10t - 18 = -7^2 + 10 \cdot 7 - 18$
$= -49 + 70 - 18 = 3.$

We complete the table. Then we plot the points and connect them with a smooth curve.

t	$-t^2 + 10t - 18$
3	3
4	6
5	7
6	6
7	3

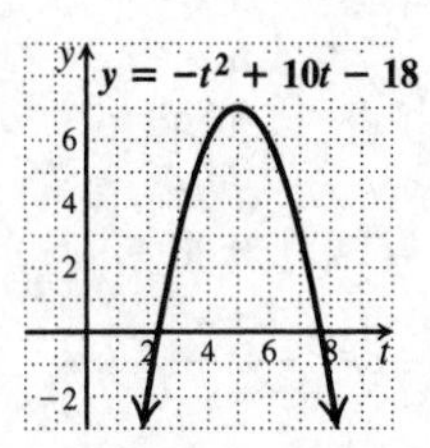

Exercise Set 4.4

1. Since the right-hand side has collected like terms, the correct expression is x^2 to make

$$(3x^2 + 2) + (6x^2 + 7) = (3 + 6)x^2 + (2 + 7).$$

3. Since the right-hand side is the result of using subtraction (the distributive law), the correct operation is $-$ to make

$$(9x^3 - x^2) - (3x^2 + x^2) = 9x^3 - x^2 - 3x^2 - x^2.$$

5. $(3x+2)+(x+7) = (3+1)x+(2+7) = 4x+9$

7. $(2t+7)+(-8t+1) = (2-8)t+(7+1) = -6t+8$

9. $(x^2+6x+3)+(-4x^2-5) = (1-4)x^2+6x+(3-5)$
$= -3x^2+6x-2$

11. $(7t^2-3t-6)+(2t^2+4t+9)$
$= (7+2)t^2+(-3+4)t+(-6+9) = 9t^2+t+3$

13. $(4m^3-7m^2+m-5)+(4m^3+7m^2-4m-2)$
$= (4+4)m^3+(-7+7)m^2+(1-4)m+(-5-2)$
$= 8m^3-3m-7$

15. $(3+6a+7a^2+a^3)+(4+7a-8a^2+6a^3)$
$= (1+6)a^3+(7-8)a^2+(6+7)a+(3+4)$
$= 7a^3-a^2+13a+7$

17. $(3x^6+2x^4-x^3+5x)+(-x^6+3x^3-4x^2+7x^4)$
$= (3-1)x^6+(2+7)x^4+(-1+3)x^3-4x^2+5x$
$= 2x^6+9x^4+2x^3-4x^2+5x$

19. $\left(\frac{3}{5}x^4+\frac{1}{2}x^3-\frac{2}{3}x+3\right)+\left(\frac{2}{5}x^4-\frac{1}{4}x^3-\frac{3}{4}x^2-\frac{1}{6}x\right)$
$= \left(\frac{3}{5}+\frac{2}{5}\right)x^4+\left(\frac{1}{2}-\frac{1}{4}\right)x^3-\frac{3}{4}x^2+\left(-\frac{2}{3}-\frac{1}{6}\right)x+3$
$= x^4+\left(\frac{2}{4}-\frac{1}{4}\right)x^3-\frac{3}{4}x^2+\left(\frac{-4}{6}-\frac{1}{6}\right)x+3$
$= x^4+\frac{1}{4}x^3-\frac{3}{4}x^2-\frac{5}{6}x+3$

21. $(5.3t^2-6.4t-9.1)+(4.2t^3-1.8t^2+7.3)$
$= 4.2t^3+(5.3-1.8)t^2-6.4t+(-9.1+7.3)$
$= 4.2t^3+3.5t^2-6.4t-1.8$

23.
$$\begin{array}{r} -4x^3+8x^2+3x-2 \\ -4x^2+3x+2 \\ \hline -4x^3+4x^2+6x+0 \\ -4x^3+4x^2+6x \end{array}$$

25.
$$\begin{array}{rrrrr} 0.05x^4 & +0.12x^3 & -0.5x^2 & & \\ & -0.02x^3 & +0.02x^2 & +2x & \\ 1.5x^4 & & +0.01x^2 & & +0.15 \\ & 0.25x^3 & & & +0.85 \\ -0.25x^4 & & +10x^2 & & -0.04 \\ \hline 1.3x^4 & +0.35x^3 & +9.53x^2 & +2x & +0.96 \end{array}$$

27. Two forms of the opposite of $-3t^3+4t-7$ are
i) $-(-3t^3+4t-7)$ and
ii) $3t^3-4t+7$. (Changing the sign of every term.)

29. Two forms for the opposite of x^4-8x^3+6x are
i) $-(x^4-8x^3+6x)$ and
ii) $-x^4+8x^3-6x$. (Changing the sign of every term)

31. We change the sign of every term inside parentheses.
$-(9x-10) = -9x+10$

33. We change the sign of every term inside parentheses.
$-(3a^4-5a^2+1.2) = -3a^4+5a^2-1.2$

35. We change the sign of every term inside parentheses.
$-\left(-4x^4+6x^2+\frac{3}{4}x-8\right) = 4x^4-6x^2-\frac{3}{4}x+8$

37. $(3x+1)-(5x+8)$
$= 3x+1-5x-8$ Changing the sign of every term inside parentheses
$= -2x-7$

39. $(-9t+12)-(t^2+3t-1)$
$= -9t+12-t^2-3t+1$
$= -t^2-12t+13$

41. $(4a^2+a-7)-(3-8a^3-4a^2) = 4a^2+a-7-3+8a^3+4a^2$
$= 8a^3+8a^2+a-10$

43. $(1.2x^3+4.5x^2-3.8x)-(-3.4x^3-4.7x^2+23)$
$= 1.2x^3+4.5x^2-3.8x+3.4x^3+4.7x^2-23$
$= 4.6x^3+9.2x^2-3.8x-23$

45. $(7x^3-2x^2+6)-(6-2x^2+7x^3)$
Observe that we are subtracting the polynomial $7x^3-2x^2+6$ from itself. The result is 0.

47. $(3+5a+3a^2-a^3)-(2+4a-9a^2+2a^3)$
$= 3+5a+3a^2-a^3-2-4a+9a^2-2a^3$
$= 1+a+12a^2-3a^3$

49. $\left(\frac{5}{8}x^3-\frac{1}{4}x-\frac{1}{3}\right)-\left(-\frac{1}{2}x^3+\frac{1}{4}x-\frac{1}{3}\right)$
$= \frac{5}{8}x^3-\frac{1}{4}x-\frac{1}{3}+\frac{1}{2}x^3-\frac{1}{4}x+\frac{1}{3}$
$= \frac{9}{8}x^3-\frac{2}{4}x$
$= \frac{9}{8}x^3-\frac{1}{2}x$

51. $(0.07t^3-0.03t^2+0.01t)-(0.02t^3+0.04t^2-1)$
$= 0.07t^3-0.03t^2+0.01t-0.02t^3-0.04t^2+1$
$= 0.05t^3-0.07t^2+0.01t+1$

53.
$$\begin{array}{r} x^3+3x^2+1 \\ -(x^3+x^2-5) \\ \hline \end{array}$$
$$\begin{array}{r} x^3+3x^2+1 \\ -x^3-x^2+5 \\ \hline 2x^2+6 \end{array}$$

55.
$$\begin{array}{r} 4x^4-2x^3 \\ -(7x^4+6x^3+7x^2) \\ \hline \end{array}$$
$$\begin{array}{r} 4x^4-2x^3 \\ -7x^4-6x^3-7x^2 \\ \hline -3x^4-8x^3-7x^2 \end{array}$$

57. a)

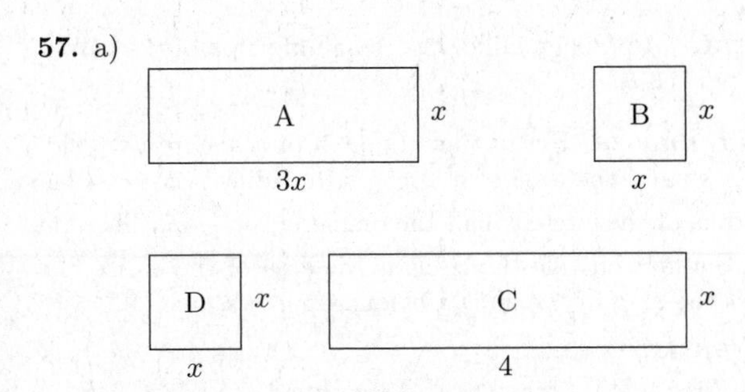

Familiarize. The area of a rectangle is the product of the length and the width.

Translate. The sum of the areas is found as follows:

$$\begin{array}{ccccccc} \text{Area} & + & \text{Area} & + & \text{Area} & + & \text{Area} \\ \text{of } A & & \text{of } B & & \text{of } C & & \text{of } D \end{array}$$
$$= 3x \cdot x \; + \; x \cdot x \; + \; 4 \cdot x \; + \; x \cdot x$$

Carry out. We collect like terms.

$$3x^2 + x^2 + 4x + x^2 = 5x^2 + 4x$$

Check. We can go over our calculations. We can also assign some value to x, say 2, and carry out the computation of the area in two ways.

Sum of areas: $3 \cdot 2 \cdot 2 + 2 \cdot 2 + 4 \cdot 2 + 2 \cdot 2$
$= 12 + 4 + 8 + 4 = 28$

Substituting in the polynomial:
$5(2)^2 + 4 \cdot 2 = 20 + 8 = 28$

Since the results are the same, our solution is probably correct.

State. A polynomial for the sum of the areas is $5x^2 + 4x$.

b) For $x = 5$: $5x^2 + 4x = 5 \cdot 5^2 + 4 \cdot 5$
$= 5 \cdot 25 + 4 \cdot 5 = 125 + 20 = 145$

When $x = 5$, the sum of the areas is 145 square units.

For $x = 7$: $5x^2 + 4x = 5 \cdot 7^2 + 4 \cdot 7$
$= 5 \cdot 49 + 4 \cdot 7 = 245 + 28 = 273$

When $x = 7$, the sum of the areas is 273 square units.

59. The perimeter is the sum of the lengths of the sides.

$4y + 4 + 7 + 2y + 7 + 6 + (3y + 2) + 7y$

$= (4 + 2 + 3 + 7)y + (4 + 7 + 7 + 6 + 2)$

$= 16y + 26$

61.

r 11
9
r + 9
r
r + 11

The length and width of the figure can be expressed as $r + 11$ and $r + 9$, respectively. The area of this figure (a rectangle) is the product of the length and width. An algebraic expression for the area is $(r + 11) \cdot (r + 9)$.

The algebraic expressions $9r + 99 + r^2 + 11r$ and $(r + 11) \cdot (r + 9)$ represent the same area.

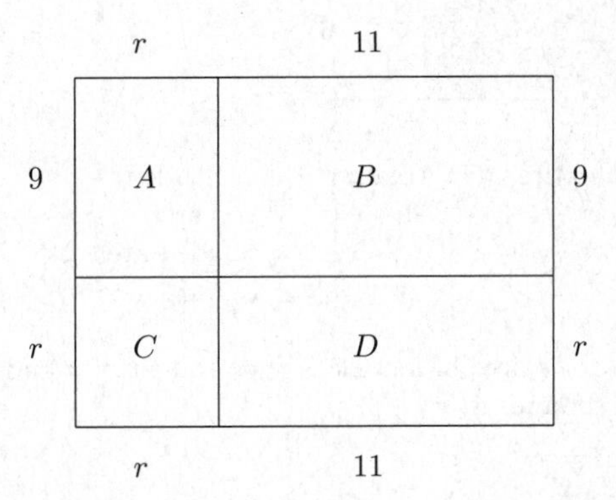

The area of the figure can be found by adding the areas of the four rectangles A, B, C, and D. The area of a rectangle is the product of the length and the width.

$$\begin{array}{ccccccc} \text{Area} & + & \text{Area} & + & \text{Area} & + & \text{Area} \\ \text{of } A & & \text{of } B & & \text{of } C & & \text{of } D \\ = 9 \cdot r & + & 11 \cdot 9 & + & r \cdot r & + & 11 \cdot r \\ = 9r & + & 99 & + & r^2 & + & 11r \end{array}$$

An algebraic expression for the area of the figure is $9r + 99 + r^2 + 11r$.

63.

x + 3
x 3
x A B x
x + 3
3 C D 3
x 3

The length and width of the figure can each be expressed as $x + 3$. The area can be expressed as $(x + 3) \cdot (x + 3)$, or $(x + 3)^2$. Another way to express the area is to find an expression for the sum of the areas of the four rectangles A, B, C, and D. The area of each rectangle is the product of its length and width.

$$\begin{array}{ccccccc} \text{Area} & + & \text{Area} & + & \text{Area} & + & \text{Area} \\ \text{of } A & & \text{of } B & & \text{of } C & & \text{of } D \\ = x \cdot x & + & 3 \cdot x & + & 3 \cdot x & + & 3 \cdot 3 \\ = x^2 & + & 3x & + & 3x & + & 9 \end{array}$$

The algebraic expressions $(x + 3)^2$ and $x^2 + 3x + 3x + 9$ represent the same area.

$$(x + 3)^2 = x^2 + 3x + 3x + 9$$

65. Recall that the area of a rectangle is length times width.

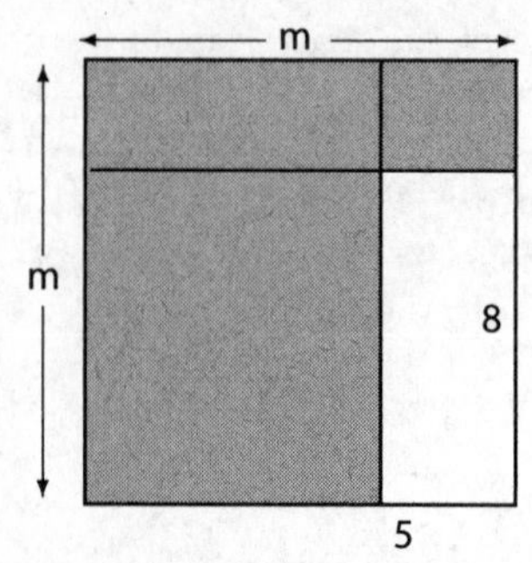

Area of entire square	$-$	Area not shaded	$=$	Shaded area
$m \cdot m$	$-$	$8 \cdot 5$	$=$	Shaded area
		$m^2 - 40$	$=$	Shaded area

67. Recall that the area of a circle is the product of π and the square of the radius, r^2.

$A = \pi r^2$.

The area of a square is the length of on side, s, squared.

$A = s^2$.

Area of circle	$-$	Area of square	$=$	Shaded area
πr^2	$-$	$7 \cdot 7$	$=$	Shaded area
πr^2	$-$	49	$=$	Shaded area

69. ***Familiarize***. Recall that the area of a rectangle is the product of the length and the width and that, consequently, the area of a square with side s is s^2. The remaining floor area is the area of the entire floor less the area of the bath enclosure, in square feet.

Translate.

Area of entire floor	$-$	Area of bath enclosure	$=$	Remaining floor area
x^2	$-$	$2 \cdot 6$	$=$	Remaining floor area

Carry out. We simplify the expression.

$x^2 - 2 \cdot 6 = x^2 - 12$

Check. We go over the calculations. The answer checks.

State. A polynomial for the remaining floor area is $(x^2 - 12)$ ft^2.

71. ***Familiarize***. Recall that the area of a square with side z is z^2. Recall that the area of a circle with radius 6, or half the diameter is πr^2 or $\pi \cdot 6^2$ The remaining area of the garden is the area of the garden less the area of the patio, in square feet.

Translate.

Area of garden	$-$	Area of patio	is	Remaining garden area
z^2	$-$	$\pi \cdot 6^2$	$=$	Remaining garden area

Carry out. We simplify the expression.

$(z^2 - 36\pi)$ft^2.

Check. We go over the calculations. The answer checks.

State. A polynomial for the remaining area of the garden is $(z^2 - 36\pi)$ ft^2.

73. ***Familiarize***. Recall that the area of a square with side s is s^2 and the area of a circle with radius r is πr^2. The radius of the circle is half the diameter, or $\frac{d}{2}$ m. The area of the mat outside the circle is the area of the entire mat less the area of the circle, in square meters.

Translate.

Area of mat	$-$	Area of circle	is	Area outside the circle
12^2	$-$	$\pi \cdot \left(\frac{d}{2}\right)^2$	$=$	Area outside the circle

Carry out. We simplify the expression.

$$12^2 - \pi \cdot \left(\frac{d}{2}\right)^2 = 144 - \pi \cdot \frac{d^2}{4} = 144 - \frac{d^2}{4}\pi$$

Check. We go over the calculations. The answer checks.

State. A polynomial for the area of the mat outside the wrestling circle is $\left(144 - \frac{d^2}{4}\pi\right)$ m^2.

75. *Writing Exercise*. We use the parentheses in $(x^2 - 64\pi)$ft^2 to indicate that the units, square feet, apply to the entire quantity in the expression $x^2 - 64\pi$.

77. $2\left(x^2 - x + 3\right) = 2x^2 - 2x + 6$

79. $x^2 \cdot x^6 = x^{2+6} = x^8$

81. $2n \cdot n^2 = 2n^{1+2} = 2n^3$.

83. *Writing Exercise*. The polynomials are opposites.

85.
$$\begin{aligned} &(6t^2 - 7t) + (3t^2 - 4t + 5) - (9t - 6) \\ &= 6t^2 - 7t + 3t^2 - 4t + 5 - 9t + 6 \\ &= 9t^2 - 20t + 11 \end{aligned}$$

87.
$$\begin{aligned} &4\left(x^2 - x + 3\right) - 2\left(2x^2 + x - 1\right) \\ &= 4x^2 - 4x + 12 - 4x^2 - 2x + 2 \\ &= (4 - 4)x^2 + (-4 - 2)x + (12 + 2) \\ &= -6x + 14 \end{aligned}$$

89.
$$\begin{aligned} &(345.099x^3 - 6.178x) - (94.508x^3 - 8.99x) \\ &= 345.099x^3 - 6.178x - 94.508x^3 + 8.99x \\ &= 250.591x^3 + 2.812x \end{aligned}$$

91. ***Familiarize***. The surface area is $2lw + 2lh + 2wh$, where l = length, w = width, and h = height of the rectangular solid. Here we have $l = 3$, $w = w$, and $h = 7$.

Translate. We substitute in the formula above.

$2 \cdot 3 \cdot w + 2 \cdot 3 \cdot 7 + 2 \cdot w \cdot 7$

Carry out. We simplify the expression.

$$\begin{aligned} &2 \cdot 3 \cdot w + 2 \cdot 3 \cdot 7 + 2 \cdot w \cdot 7 \\ &= 6w + 42 + 14w \\ &= 20w + 42 \end{aligned}$$

Check. We can go over the calculations. We can also assign some value to w, say 6, and carry out the computation in two ways.

Using the formula: $2 \cdot 3 \cdot 6 + 2 \cdot 3 \cdot 7 + 2 \cdot 6 \cdot 7 = 36 + 42 + 84 = 162$

Substituting in the polynomial: $20 \cdot 6 + 42 = 120 + 42 = 162$

Since the results are the same, our solution is probably correct.

State. A polynomial for the surface area is $20w + 42$.

93. ***Familiarize***. The surface area is $2lw + 2lh + 2wh$, where l = length, w = width, and h = height of the rectangular solid. Here we have $l = x$, $w = x$, and $h = 5$.

Translate. We substitute in the formula above.

$$2 \cdot x \cdot x + 2 \cdot x \cdot 5 + 2 \cdot x \cdot 5$$

Carry out. We simplify the expression.

$$\begin{aligned} &2 \cdot x \cdot x + 2 \cdot x \cdot 5 + 2 \cdot x \cdot 5 \\ &= 2x^2 + 10x + 10x \\ &= 2x^2 + 20x \end{aligned}$$

Check. We can go over the calculations. We can also assign some value to x, say 3, and carry out the computation in two ways.

Using the formula: $2 \cdot 3 \cdot 3 + 2 \cdot 3 \cdot 5 + 2 \cdot 3 \cdot 5 = 18 + 30 + 30 = 78$

Substituting in the polynomial: $2 \cdot 3^2 + 20 \cdot 3 = 2 \cdot 9 + 60 = 18 + 60 = 78$

Since the results are the same, our solution is probably correct.

State. A polynomial for the surface area is $2x^2 + 20x$.

95. Length of top edges: $x + 6 + x + 6$, or $2x + 12$

Length of bottom edges: $x + 6 + x + 6$, or $2x + 12$

Length of vertical edges: $4 \cdot x$, or $4x$

Total length of edges: $2x + 12 + 2x + 12 + 4x = 8x + 24$

97. *Writing Exercise*. Yes; $4(-x)^7 - 6(-x)^3 + 2(-x) = -4x^7 + 6x^3 - 2x = -(4x^7 - 6x^3 + 2x)$.

Exercise Set 4.5

1. $3x^2 \cdot 2x^4 = (3 \cdot 2)(x^2 \cdot x^4) = 6x^6$

Choice (c) is correct.

3. $4x^3 \cdot 2x^5 = (4 \cdot 2)(x^3 \cdot x^5) = 8x^8$

Choice (d) is correct.

5. $4x^6 + 2x^6 = (4 + 2)x^6 = 6x^6$

Choice (c) is correct.

7. $(3x^5)7 = (3 \cdot 7)x^5 = 21x^5$

9. $(-x^3)(x^4) = (-1 \cdot x^3)(x^4) = -1(x^3 \cdot x^4) = -1 \cdot x^7 = -x^7$

11. $(-x^6)(-x^2) = (-1 \cdot x^6)(-1 \cdot x^2) = (-1)(-1)(x^6 \cdot x^2) = x^8$

13. $4t^2(9t^2) = (4 \cdot 9)(t^2 \cdot t^2) = 36t^4$

15. $(0.3x^3)(-0.4x^6) = 0.3(-0.4)(x^3 \cdot x^6) = -0.12x^9$

17. $\left(-\frac{1}{4}x^4\right)\left(\frac{1}{5}x^8\right) = \left(-\frac{1}{4} \cdot \frac{1}{5}\right)(x^4 \cdot x^8) = -\frac{1}{20}x^{12}$

19. $(-5n^3)(-1) = (-5)(-1)n^3 = 5n^3$

21. $11x^5(-4x^5) = (-11 \cdot 4)(x^5 \cdot x^5) = -44x^{10}$

23. $(-4y^5)(6y^2)(-3y^3) = -4(6)(-3)(y^5 \cdot y^2 \cdot y^3) = 72y^{10}$

25. $5x(4x + 1) = 5x(4x) + 5x(1) = 20x^2 + 5x$

27. $(a - 9)3a = a \cdot 3a - 9 \cdot 3a = 3a^2 - 27a$

29.
$$\begin{aligned} x^2(x^3 + 1) &= x^2(x^3) + x^2(1) \\ &= x^5 + x^2 \end{aligned}$$

31.
$$\begin{aligned} &-3n(2n^2 - 8n + 1) \\ &= (-3n)(2n^2) + (-3n)(-8n) + (-3n)(1) \\ &= -6n^3 + 24n^2 - 3n \end{aligned}$$

33. $-5t^2(3t + 6) = -5t^2(3t) - 5t^2(6) = -15t^3 - 30t^2$

35.
$$\begin{aligned} &\frac{2}{3}a^4\left(6a^5 - 12a^3 - \frac{5}{8}\right) \\ &= \frac{2}{3}a^4(6a^5) - \frac{2}{3}a^4(12a^3) - \frac{2}{3}a^4\left(\frac{5}{8}\right) \\ &= \frac{12}{3}a^9 - \frac{24}{3}a^7 - \frac{10}{24}a^4 \\ &= 4a^9 - 8a^7 - \frac{5}{12}a^4 \end{aligned}$$

37.
$$\begin{aligned} (x + 3)(x + 4) &= (x + 3)x + (x + 3)4 \\ &= x \cdot x + 3 \cdot x + x \cdot 4 + 3 \cdot 4 \\ &= x^2 + 3x + 4x + 12 \\ &= x^2 + 7x + 12 \end{aligned}$$

39.
$$\begin{aligned} (t + 7)(t - 3) &= (t + 7)t + (t + 7)(-3) \\ &= t \cdot t + 7 \cdot t + t(-3) + 7(-3) \\ &= t^2 + 7t - 3t - 21 \\ &= t^2 + 4t - 21 \end{aligned}$$

41.
$$\begin{aligned} (a - 0.6)(a - 0.7) &= (a - 0.6)a + (a - 6)(-0.7) \\ &= a \cdot a - 0.6 \cdot a + a(-0.7) + (-0.6)(-0.7) \\ &= a^2 - 0.6a - 0.7a + 0.42 \\ &= a^2 - 1.3a + 0.42 \end{aligned}$$

43.
$$\begin{aligned} (x + 3)(x - 3) &= (x + 3)x + (x + 3)(-3) \\ &= x \cdot x + 3 \cdot x + x(-3) + 3(-3) \\ &= x^2 + 3x - 3x - 9 \\ &= x^2 - 9 \end{aligned}$$

45.
$$\begin{aligned} (4 - x)(7 - 2x) &= (4 - x)7 + (4 - x)(-2x) \\ &= 4 \cdot 7 - x \cdot 7 + 4(-2x) - x(-2x) \\ &= 28 - 7x - 8x + 2x^2 \\ &= 28 - 15x + 2x^2 \end{aligned}$$

47.
$$\begin{aligned} \left(t + \frac{3}{2}\right)\left(t + \frac{4}{3}\right) &= \left(t + \frac{3}{2}\right)t + \left(t + \frac{3}{2}\right)\left(\frac{4}{3}\right) \\ &= t \cdot t + \frac{3}{2} \cdot t + t \cdot \frac{4}{3} + \frac{3}{2} \cdot \frac{4}{3} \\ &= t^2 + \frac{3}{2}t + \frac{4}{3}t + 2 \\ &= t^2 + \frac{9}{6}t + \frac{8}{6}t + 2 \\ &= t^2 + \frac{17}{6}t + 2 \end{aligned}$$

49. $\left(\frac{1}{4}a+2\right)\left(\frac{3}{4}a-1\right)$

$=\left(\frac{1}{4}a+2\right)\left(\frac{3}{4}a\right)+\left(\frac{1}{4}a+2\right)(-1)$

$=\frac{1}{4}a\left(\frac{3}{4}a\right)+2\cdot\frac{3}{4}a+\frac{1}{4}a(-1)+2(-1)$

$=\frac{3}{16}a^2+\frac{3}{2}a-\frac{1}{4}a-2$

$=\frac{3}{16}a^2+\frac{6}{4}a-\frac{1}{4}a-2$

$=\frac{3}{16}a^2+\frac{5}{4}a-2$

51. Illustrate $x(x+5)$ as the area of a rectangle with width x and length $x+5$.

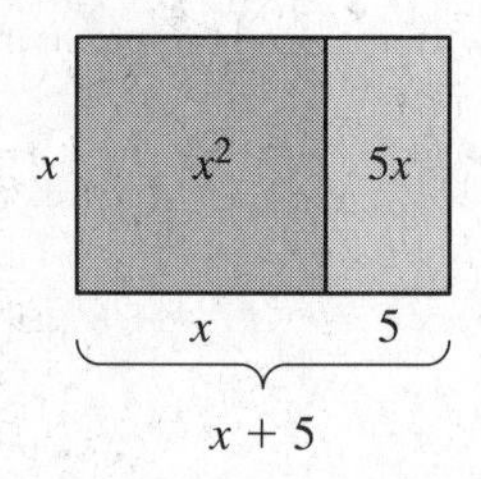

53. Illustrate $(x+1)(x+2)$ as the area of a rectangle with width $x+1$ and length $x+2$.

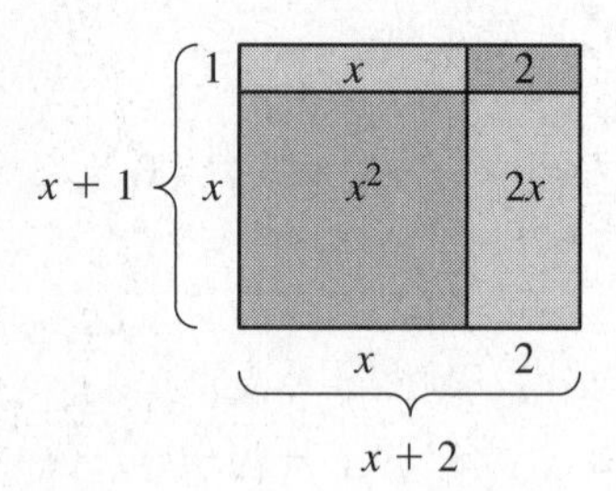

55. Illustrate $(x+5)(x+3)$ as the area of a rectangle with length $x+5$ and width $x+3$.

57. $(x^2-x+3)(x+1)$

$=(x^2-x+3)x+(x^2-x+3)1$

$=x^3-x^2+3x+x^2-x+3$

$=x^3+2x+3$

A partial check can be made by selecting a convenient replacement for x, say 1, and comparing the values of the original expression and the result.

$(1^2-1+3)(1+1)$	$1^3+6\cdot1+3$
$=(1-1+3)(1+1)$	$=1+6+3$
$=3\cdot2$	$=6$
$=6$	

Since the value of both expressions is 6, the multiplication is very likely correct.

59. $(2a+5)(a^2-3a+2)$

$=(2a+5)a^2-(2a+5)(3a)+(2a+5)2$

$=2a\cdot a^2+5\cdot a^2-2a\cdot3a-5\cdot3a+2a\cdot2+5\cdot2$

$=2a^3+5a^2-6a^2-15a+4a+10$

$=2a^3-a^2-11a+10$

A partial check can be made as in Exercise 57.

61. $(y^2-7)(3y^4+y+2)$

$=(y^2-7)(3y^4)+(y^2-7)y+(y^2-7)(2)$

$=y^2\cdot3y^4-7\cdot3y^4+y^2\cdot y-7\cdot y+y^2\cdot2-7\cdot2$

$=3y^6-21y^4+y^3-7y+2y^2-14$

$=3y^6-21y^4+y^3+2y^2-7y-14$

A partial check can be made as in Exercise 57.

63. $(3x+2)(7x+4x+1)=(3x+2)(11x+1)$

$=(3x+2)(11x)+(3x+2)(1)$

$=3x\cdot11x+2\cdot11x+3x\cdot1+2\cdot1$

$=33x^2+22x+3x+2$

$=33x^2+25x+2$

65.

$$\begin{array}{rl} x^2+5x-1 & \text{Line up like terms} \\ x^2-x+3 & \text{in columns} \\ \hline 3x^2+15x-3 & \text{Multiplying by 3} \\ -x^3-5x^2+x \quad & \text{Multiplying by } -x \\ x^4+5x^3-x^2 \qquad\quad & \text{Multiplying by } x^2 \\ \hline x^4+4x^3-3x^2+16x-3 & \end{array}$$

A partial check can be made as in Exercise 57.

67.

$$\begin{array}{rl} 5t^2-t+\frac{1}{2} & \\ 2t^2+t-4 & \\ \hline -20t^2+4t-2 & \text{Multiplying by } -4 \\ 5t^3-t^2+\frac{1}{2}t \quad & \text{Multiplying by } t \\ 10t^4-2t^3+t^2 \qquad\quad & \text{Multiplying by } 2t^2 \\ \hline 10t^4+3t^3-20t^2+\frac{9}{2}t-2 & \end{array}$$

A partial check can be made as in Exercise 57.

69. We will multiply horizontally while still aligning like terms.

$(x+1)(x^3+7x^2+5x+4)$

$=x^4+7x^3+5x^2+4x$ Multiplying by x

$\quad+x^3+7x^2+5x+4$ Multiplying by 1

$=x^4+8x^3+12x^2+9x+4$

A partial check can be made as in Exercise 57.

71. *Writing Exercise.* No; the distributive law is the basis for polynomial multiplication.

73. $(9-3)(9+3)+3^2-9^2=(6)(12)+3^2-9^2$
$=(6)(12)+9-81=72+9-81=0$

75.
$$5+\frac{7+4+2\cdot 5}{7}$$
$$=5+\frac{7+4+10}{7}$$
$$=5+\frac{21}{7}$$
$$=5+3$$
$$=8$$

77.
$$(4+3\cdot 5+5)\div 3\cdot 4$$
$$=(4+15+5)\div 3\cdot 4$$
$$=24\div 3\cdot 4$$
$$=8\cdot 4$$
$$=32$$

79. *Writing Exercise.* $(A+B)(C+D)$ will be a trinomial when there is exactly one pair of like terms among AC, AD, BC, and BD.

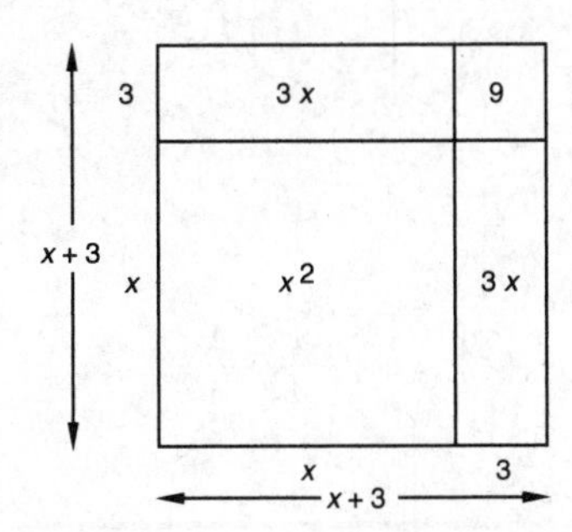

Then we see that the area of the figure is $(x+3)^2$, or $x^2+3x+3x+9 \neq x^2+9$.

81. The shaded area is the area of the large rectangle, $6y(14y-5)$ less the area of the unshaded rectangle, $3y(3y+5)$. We have:

$$6y(14y-5)-3y(3y+5)$$
$$=84y^2-30y-9y^2-15y$$
$$=75y^2-45y$$

83. Let $n=$ the missing number.

The area of the figure is $x^2+3x+nx+3n$. This is equivalent to $x^2+8x+15$, so we have $3x+nx=8x$ and $3n=15$. Solving either equation for n, we find that the missing number is 5.

85.

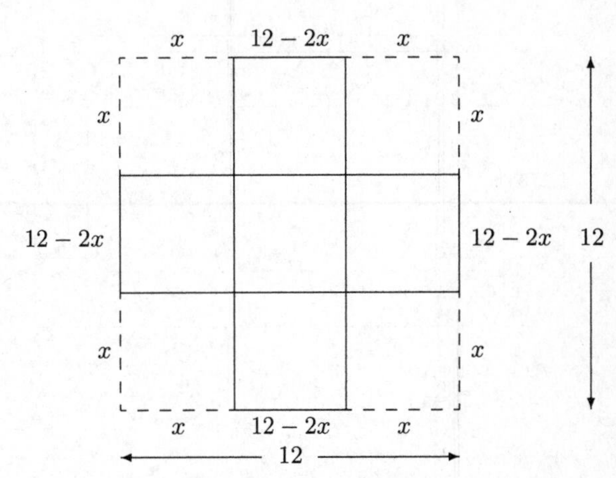

The dimensions, in inches, of the box are $12-2x$ by $12-2x$ by x. The volume is the product of the dimensions (volume = length × width × height):

$$\text{Volume} = (12-2x)(12-2x)x$$
$$=(144-48x+4x^2)x$$
$$=(144x-48x^2+4x^3)\text{ in}^3\text{, or}$$
$$(4x^3-48x^2+144x)\text{ in}^3$$

The outside surface area is the sum of the area of the bottom and the areas of the four sides. The dimensions, in inches, of the bottom are $12-2x$ by $12-2x$, and the dimensions, in inches, of each side are x by $12-2x$.

$$\text{Surface area} = \text{Area of bottom} + 4\cdot\text{Area of each side}$$
$$=(12-2x)(12-2x)+4\cdot x(12-2x)$$
$$=144-24x-24x+4x^2+48x-8x^2$$
$$=144-48x+4x^2+48x-8x^2$$
$$=(144-4x^2)\text{ in}^2\text{, or }(-4x^2+144)\text{ in}^2$$

87.

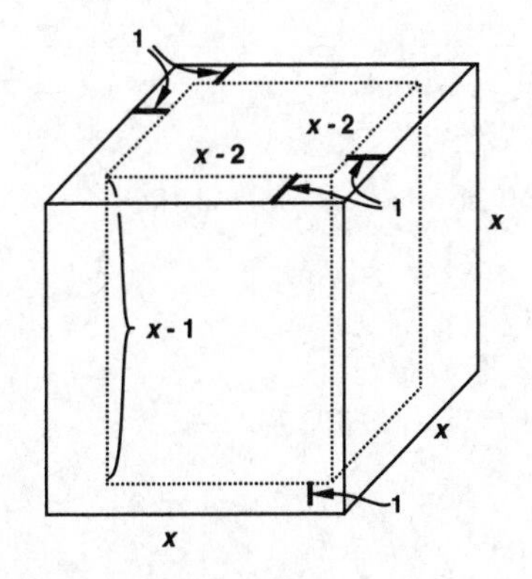

The interior dimensions of the open box are $x-2$ cm by $x-2$ cm by $x-1$ cm.

$$\text{Interior volume} = (x-2)(x-2)(x-1)$$
$$=(x^2-4x+4)(x-1)$$
$$=(x^3-5x^2+8x-4)\text{ cm}^3$$

89. Let $x=$ the width of the garden. Then $2x=$ the length of the garden.

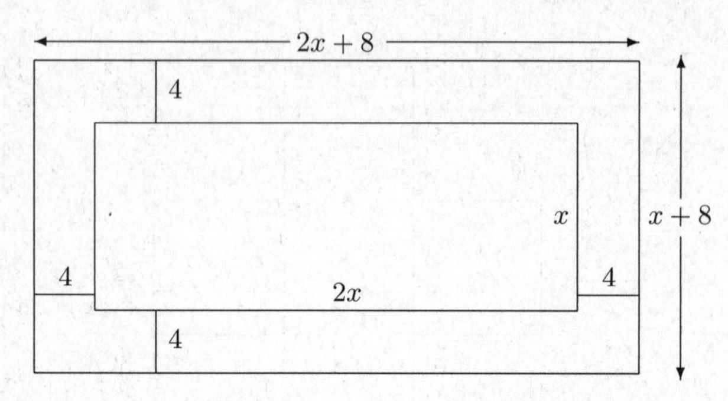

Area of garden and sidewalk together is Area of garden alone + 256 ft²

$$(2x+8)(x+8) = 2x \cdot x + 256$$

$2x^2 + 24x + 64 = 2x^2 + 256$
$24x = 192$
$x = 8$

The dimensions are 8 ft by 16 ft.

91. $(x-2)(x-7)-(x-7)(x-2)$

First observe that, by the commutative law of multiplication, $(x-2)(x-7)$ and $(x-7)(x-2)$ are equivalent expressions. Then when we subtract $(x-7)(x-2)$ from $(x-2)(x-7)$, the result is 0.

93. $(x+2)(x+4)(x-5)$
$= (x^2+2x+4x+8)(x-5)$
$= (x^2+6x+8)(x-5)$
$= x^3+6x^2+8x-5x^2-30x-40$
$= x^3+x^2-22x-40$

95. $(x-a)(x-b)\cdots(x-x)(x-y)(x-z)$
$= (x-a)(x-b)\cdots 0\cdot(x-y)(x-z)$
$= 0$

Exercise Set 4.6

1. It is true that FOIL is simply a memory device for finding the product of two binomials.

3. This statement is false. See Example 2(d).

5. $(x^2+2)(x+3)$
F O I L
$= x^2\cdot x + x^2\cdot 3 + 2\cdot x + 2\cdot 3$
$= x^3+2x+3x^2+6$, or x^3+3x^2+2x+6

7. $(t^4-2)(t+7)$
F O I L
$= t^4\cdot t + t^4\cdot 7 - 2\cdot t - 2\cdot 7$
$= t^5+7t^4-2t-14$

9. $(y+2)(y-3)$
F O I L
$= y\cdot y + y\cdot(-3) + 2\cdot y + 2\cdot(-3)$
$= y^2-3y+2y-6$
$= y^2-y-6$

11. $(3x+2)(3x+5)$
F O I L
$= 3x\cdot 3x + 3x\cdot 5 + 2\cdot 3x + 2\cdot 5$
$= 9x^2+15x+6x+10$
$= 9x^2+21x+10$

13. $(5x-3)(x+4)$
F O I L
$= 5x\cdot x + 5x\cdot 4 - 3\cdot x - 3\cdot 4$
$= 5x^2+20x-3x-12$
$= 5x^2+17x-12$

15. $(3-2t)(5-t)$
F O I L
$= 3\cdot 5 + 3\cdot t - 2t\cdot 5 + (-2t)(-t)$
$= 15-3t-10t+2t^2$
$= 15-13t+2t^2$

17. $(x^2+3)(x^2-7)$
F O I L
$= x^2\cdot x^2 - x^2\cdot 7 + 3\cdot x^2 - 3\cdot 7$
$= x^4-7x^2+3x^2-21$
$= x^4-4x^2-21$

19. $\left(p-\frac{1}{4}\right)\left(p+\frac{1}{4}\right)$
F O I L
$= p\cdot p + p\cdot\frac{1}{4} + \left(-\frac{1}{4}\right)\cdot p + \left(-\frac{1}{4}\right)\cdot\frac{1}{4}$
$= p^2+\frac{1}{4}p-\frac{1}{4}p-\frac{1}{16}$
$= p^2-\frac{1}{16}$

21. $(x-0.3)(x-0.3)$
F O I L
$= x\cdot x - x\cdot 0.3 + -0.3\cdot x + (-0.3)(-0.3)$
$= x^2-0.3x-0.3x+0.09$
$= x^2-0.6x+0.09$

23. $(-3n+2)(n+7)$
F O I L
$= -3n\cdot n - 3n\cdot 7 + 2\cdot n + 2\cdot 7$
$= -3n^2-21n+2n+14$
$= -3n^2-19n+14$

25. $(x+10)(x+10)$
F O I L
$= x^2 + 10x + 10x + 100$
$= x^2+10x+10x+100$
$= x^2+20x+100$

27. $(1-3t)(1+5t^2)$
F O I L
$= 1 + 5t^2 - 3t - 15t^3$
$= 1-3t-5t^2-15t^3$

29. $(x^2+3)(x^3-1)$
F O I L
$= x^5 - x^2 + 3x^3 - 3$, or $x^5+3x^3-x^2-3$

31. $(3x^2-2)(x^4-2)$
F O I L
$= 3x^6 - 6x^2 - 2x^4 + 4$, or $3x^6 - 2x^4 - 6x^2 + 4$

33. $(2t^3+5)(2t^3+5)$
F O I L
$= 4t^6 + 10t^3 + 10t^3 + 25$
$= 4t^6 + 20t^3 + 25$

35. $(8x^3+5)(x^2+2)$
F O I L
$= 8x^5 + 16x^3 + 5x^2 + 10$

37. $(10x^2+3)(10x^2-3)$
F O I L
$= 100x^4 - 30x^2 + 30x^2 - 9$
$= 100x^4 - 9$

39. $(x+8)(x-8)$ Product of sum and difference of the same two terms
$= x^2 - 8^2$
$= x^2 - 64$

41. $(2x+1)(2x-1)$ Product of sum and difference of the same two terms
$= (2x)^2 - 1^2$
$= 4x^2 - 1$

43. $(5m^2+4)(5m^2-4)$ Product of sum and difference of the same two terms
$= (5m)^2 - 4^2$
$= 25m^4 - 16$

45. $(9a^3+1)(9a^3-1)$
$= (9a^3)^2 - 1^2$
$= 81a^6 - 1$

47. $(x^4+0.1)(x^4-.01)$
$= (x^4)^2 - 0.1^2$
$= x^8 - 0.01$

49. $\left(t-\frac{3}{4}\right)\left(t+\frac{3}{4}\right)$
$= t^2 - \left(\frac{3}{4}\right)^2$
$= t^2 - \frac{9}{16}$

51. $(x+3)^2$
$= x^2 + 2 \cdot x \cdot 3 + 3^2$ Square of a binomial
$= x^2 + 6x + 9$

53. $(7x^3-1)^2$ Square of a binomial
$= (7x^3)^2 - 2 \cdot 7x^3 \cdot 1 + (-1)^2$
$= 49x^6 - 14x^3 + 1$

55. $\left(a-\frac{2}{5}\right)^2$ Square of a binomial
$= a^2 - 2 \cdot a \cdot \frac{2}{5} + \left(\frac{2}{5}\right)^2$
$= a^2 - \frac{4}{5}a + \frac{4}{25}$

57. $(t^4+3)^2$ Square of a binomial
$= (t^4)^2 + 2 \cdot t^4 \cdot 3 + 3^2$
$= t^8 + 6t^4 + 9$

59. $(2-3x^4)^2 = 2^2 - 2 \cdot 2 \cdot 3x^4 + (3x^4)^2$
$= 4 - 12x^4 + 9x^8$

61. $(5+6t^2)^2 = 5^2 + 2 \cdot 5 \cdot 6t^2 + (6t^2)^2$
$= 25 + 60t^2 + 36t^4$

63. $(7x-0.3)^2 = (7x)^2 - 2(7x)(0.3) + (0.3)^2$
$= 49x^2 - 4.2x + 0.09$

65. $7n^3(2n^2-1)$
$= 7n^3 \cdot 2n^2 - 7n^3 \cdot 1$ Multiplying each term of
$= 14n^5 - 7n^3$ the binomial by the monomial

67. $(a-3)(a^2+2a-4)$
$= a^3 + 2a^2 - 4a$ Multiplying horizontally
$- 3a^2 - 6a + 12$ and aligning like terms
$= a^3 - a^2 - 10a + 12$

69. $(7-3x^4)(7-3x^4)$
$= 7^2 - 2 \cdot 7 \cdot 3x^4 + (-3x^4)^2$ Squaring a binomial
$= 49 - 42x^4 + 9x^8$

71. $5x(x^2+6x-2)$
$= 5x \cdot x^2 + 5x \cdot 6x + 5x(-2)$ Multiplying each term of the trinomial by the monomial
$= 5x^3 + 30x^2 - 10x$

73. $(q^5+1)(q^5-1)$
$= (q^5)^2 - 1^2$
$= q^{10} - 1$

75. $3t^2(5t^3-t^2+t)$
$= 3t^2 \cdot 5t^3 + 3t^2(-t^2) + 3t^2 \cdot t$ Multiplying each term of the trinomial by the monomial
$= 15t^5 - 3t^4 + 3t^3$

77. $(6x^4-3x)^2$ Squaring a binomial
$= (6x^4)^2 - 2 \cdot 6x^4 \cdot 3x + (-3x^2)$
$= 36x^8 - 36x^5 + 9x^2$

79. $(9a+0.4)(2a^3+0.5)$ Product of two binomials; use FOIL
$= 9a \cdot 2a^3 + 9a \cdot 0.5 + 0.4 \cdot 2a^3 + 0.4 \cdot 0.5$
$= 18a^4 + 4.5a + 0.8a^3 + 0.2$, or
$18a^4 + 0.8a^3 + 4.5a + 0.2$

81. $\left(\frac{1}{5}-6x^4\right)\left(\frac{1}{5}+6x^4\right)$
$= \left(\frac{1}{5}\right)^2 - (6x^4)^2$
$= \frac{1}{25} - 36x^8$

83. $(a+1)(a^2-a+1)$

$$\begin{aligned} &= a^3 - a^2 + a && \text{Multiplying horizontally} \\ & \quad a^2 - a + 1 && \text{and aligning like terms} \\ &= a^3 \qquad\qquad + 1 \end{aligned}$$

85.

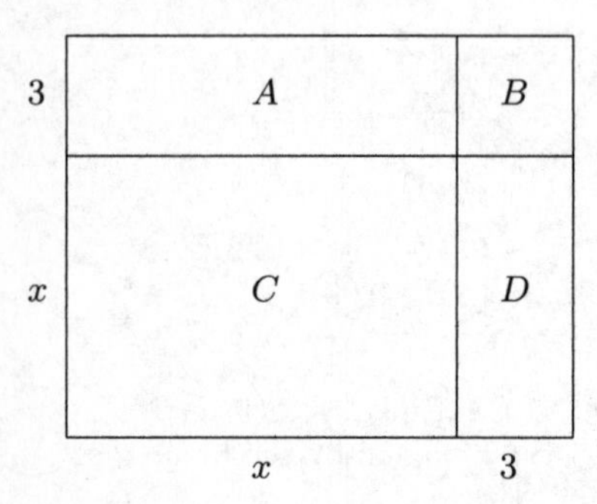

We can find the shaded area in two ways.

Method 1: The figure is a square with side $x+3$, so the area is $(x+3)^2 = x^2+6x+9$.

Method 2: We add the areas of A, B, C, and D.

$3 \cdot x + 3 \cdot 3 + x \cdot x + x \cdot 3 = 3x + 9 + x^2 + 3x$
$= x^2 + 6x + 9$.

Either way we find that the total shaded area is $x^2 + 6x + 9$.

87.

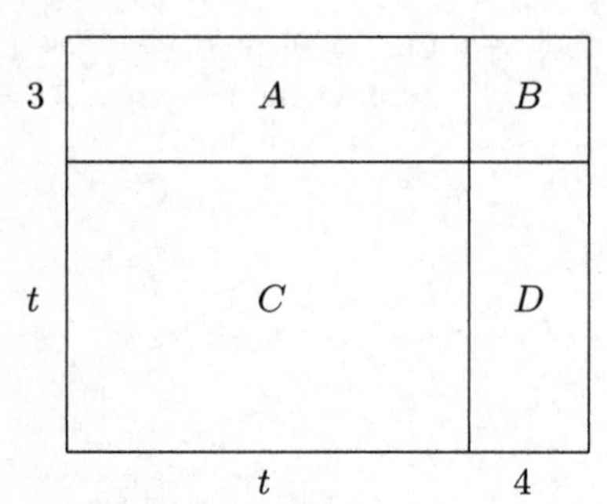

We can find the shaded area in two ways.

Method 1: The figure is a rectangle with dimensions $t+3$ by $t+4$, so the area is

$(t+3)(t+4) = t^2 + 4t + 3t + 12 = t^2 + 7t + 12$.

Method 2: We add the areas of A, B, C, and D.

$3 \cdot t + 3 \cdot 4 + t \cdot t + t \cdot 4 = 3t + 12 + t^2 + 4t = t^2 + 7t + 12$.

Either way, we find that the area is $t^2 + 7t + 12$.

89.

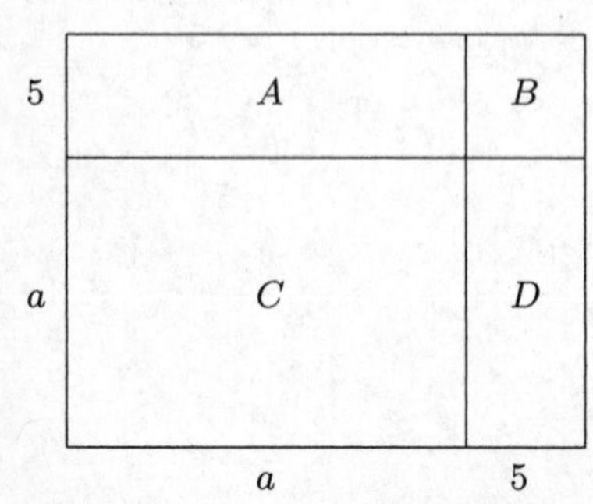

We can find the shaded area in two ways.

Method 1: The figure is a square with side $a+5$, so the area is $(a+5)^2 = a^2 + 10a + 25$.

Method 2: We add the areas of A, B, C, and D.

$5 \cdot a + 5 \cdot 5 + a \cdot a + 5 \cdot a = 5a + 25 + a^2 + 5a = a^2 + 10a + 25$.

Either way, we find that the total shaded area is $a^2 + 10a + 25$.

91.

We can find the shaded area in two ways.

Method 1: The figure is a rectangle with dimensions $x+7$ by $x+3$, so the area is $(x+7)(x+3)$
$= x^2 + 3x + 7x + 21 = x^2 + 10x + 21$.

Method 2: We add the areas of A, B, C, and D.

$x \cdot x + x \cdot 7 + 3 \cdot x + 3 \cdot 7 = x^2 + 10x + 21$.

Either way, we find that the total shaded area is $x^2 + 10x + 21$.

93.

We can find the shaded area in two ways.

Method 1: The figure is a rectangle with dimensions $a+1$ by $a+7$, so the area is

$(a+1)(a+7) = a^2 + 7a + a + 7 = a^2 + 8a + 7$

Method 2: We add the areas of A, B, C, and D.

$a \cdot a + a \cdot 1 + 7 \cdot a + 7 \cdot 1 = a^2 + a + 7a + 7 = a^2 + 8a + 7$.

Either way, we find that the total shaded area is $a^2 + 8a + 7$.

95.

We can find the shaded area in two ways.

Method 1: The figure is a square with side $5t + 2$, so the area is $(5t + 2)^2 = 25t^2 + 20t + 4$.

Method 2: We add the areas of A, B, C, and D.

$5t \cdot 5t + 5t \cdot 2 + 2 \cdot 5t + 2 \cdot 2 = 25t^2 + 10t + 10t + 4 = 25t^2 + 20t + 4.$

Either way, we find that the total shaded area is $25t^2 + 20t + 4$.

97. We draw a square with side $x + 5$.

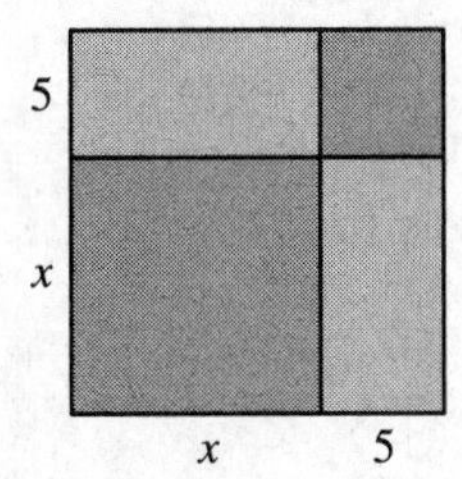

99. We draw a square with side $t + 9$.

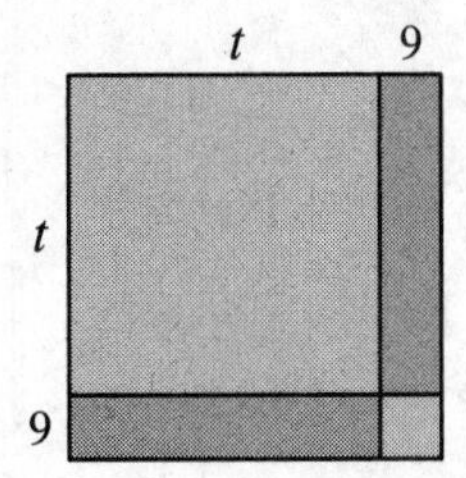

101. We draw a square with side $3 + x$.

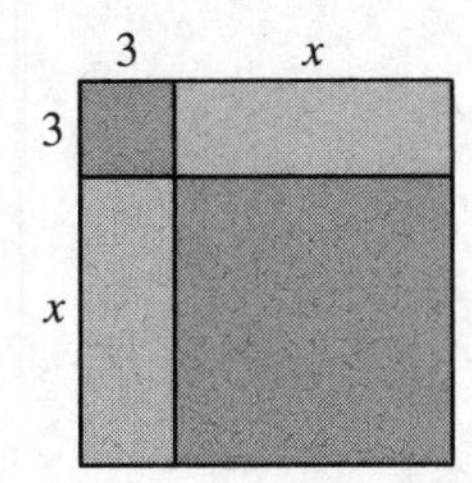

103. *Writing Exercise.* It's a good idea to study the other special products, because they allow for faster computations than the FOIL method.

105. ***Familiarize***. Let w = the energy, in kilowatt-hours per month, used by the washing machine. Then $21w$ = the amount of energy used by the refrigerator, and $11w$ = the amount of energy used by the freezer.

Translate.

Washing Machine and refrigerator and freezer is Total energy

$$w \quad + \quad 21w \quad + \quad 11w \quad = \quad 189$$

Solve. We solve the equation.

$$w + 21t + 11w = 297$$
$$33w = 297$$
$$w = 9$$

Then $21w = 21 \cdot 9 = 189$

and $11w = 11 \cdot 9 = 99$

Check. The energy used by the refrigerator, 189 kWh, is 21 times the energy used by the washing machine. The energy used by the freezer, 99 kWh is 11 times the energy used by the washing machine. Also, $9 + 189 + 99 = 297$, the total energy used.

State. The washing machine used 9 kWh, the refrigerator used 189 kWh, and the freezer used 99 kWh.

107. $5xy = 8$

$y = \dfrac{8}{5x}$ Dividing both sides by $5x$

109. $ax - by = c$

$ax = by + c$ Adding by to both sides

$x = \dfrac{by + c}{a}$ Dividing both sides by a

111. *Writing Exercise.* The computation $(20 - 1)(20 + 1) = 400 - 1 = 399$ is easily performed mentally and is equivalent to the computation $19 \cdot 21$.

113. $(4x^2 + 9)(2x + 3)(2x - 3)$

$= (4x^2 + 9)(4x^2 - 9)$

$= 16x^4 - 81$

115. $(3t - 2)^2(3t + 2)^2$

$= [(3t - 2)(3t + 2)]^2$

$= (9t^2 - 4)^2$

$= 81t^4 - 72t^2 + 16$

117. $(t^3 - 1)^4(t^3 + 1)^4$

$= [(t^3 - 1)(t^3 + 1)]^4$

$= (t^6 - 1)^4$

$= [(t^6 - 1)^2]^2$

$= (t^{12} - 2t^6 + 1)^2$

$= (t^{12} - 2t^6 + 1)(t^{12} - 2t^6 + 1)$

$= t^{24} - 2t^{18} + t^{12} - 2t^{18} + 4t^{12} - 2t^6 +$
$t^{12} - 2t^6 + 1$

$= t^{24} - 4t^{18} + 6t^{12} - 4t^6 + 1$

119. $18 \times 22 = (20 - 2)(20 + 2) = 20^2 - 2^2$

$= 400 - 4 = 396$

121. $(x + 2)(x - 5) = (x + 1)(x - 3)$

$x^2 - 5x + 2x - 10 = x^2 - 3x + x - 3$

$x^2 - 3x - 10 = x^2 - 2x - 3$

$-3x - 10 = -2x - 3$ Adding $-x^2$

$-3x + 2x = 10 - 3$ Adding $2x$ and 10

$-x = 7$

$x = -7$

The solution is -7.

123.

The area of the entire figure is F^2. The area of the unshaded region, C, is $(F-7)(F-17)$. Then one expression for the area of the shaded region is
$F^2-(F-7)(F-17)$.

To find a second expression we add the areas of regions A, B, and D. We have:

$17\cdot 7+7(F-17)+17(F-7)$

$=119+7F-119+17F-119$

$=24F-119$

It is possible to find other equivalent expressions also.

125. The dimensions of the shaded area, regions A and D together, are $y+1$ by $y-1$ so the area is $(y+1)(y-1)$.

To find another expression we add the areas of regions A and D. The dimensions of region A are y by $y-1$, and the dimensions of region D are $y-1$ by 1, so the sum of the areas is $y(y-1)+(y-1)(1)$, or $y(y-1)+y-1$.

It is possible to find other equivalent expressions also.

127.

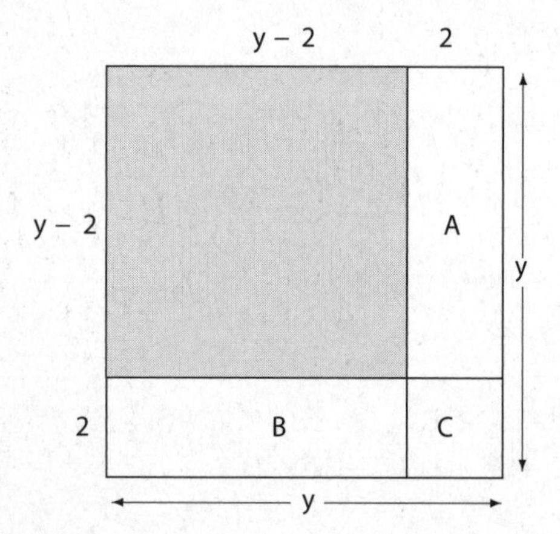

The shaded area is $(y-2)^2$. We find it as follows:

Shaded area	=	Area of square	−	Area of A	−	Area of B	−	Area of C
$(y-2)^2$	=	y^2	−	$2(y-2)$	−	$2(y-2)$	−	$2\cdot 2$

$(y-2)^2=y^2-2y+4-2y+4-4$

$(y-2)^2=y^2-4y+4$

129.

4.6 Connecting the Concepts

1. $(3x^2-2x+6)+(5x-3)$ Addition
$=3x^2-2x+5x+6-3$
$=3x^2+3x+3$

3. $6x^3(8x^2-7)$ Multiplication
$=48x^5-42x^3$

5. $(9x^3-7x+3)-(5x^2-10)$ Subtraction
$=9x^3-7x+3-5x^2+10$
$=9x^3-5x^2-7x+13$

7. $(9x+1)(9x-1)$ Multiplication
$=(9x)^2-1^2$
$=81x^2-1$ $(A+B)(A-B)=A^2-B^2$

9. $(4x^2-x-7)-(10x^2-3x+5)$
$=4x^2-x-7-10x^2+3x-5$
$=-6x^2+2x-12$

11. $8x^5(5x^4-6x^3+2)=40x^9-48x^8+16x^5$

13. $(2m-1)^2=(2m)^2-2(2m)(1)+1^2$
$=4m^2-4m+1$

15. $(5x^3-6x^2-2x)+(6x^2+2x+3)$
$=5x^2-6x^2-2x+6x^2+2x+3$
$=5x^3+3$

17. $(4y^3+7)^2=(4y^3)^2+2\cdot 4y^3\cdot 7+7^2$
$=16y^6+56y^3+49$

19. $(4t^2-5)(4t^2+5)$
$=(4t^2)^2-5^2$ $(A+B)(A-B)=A^2-B^2$
$=16t^4-25$

Exercise Set 4.7

1. $(3x+5y)^2$ is the square of a binomial, choice (a).

3. $(5a+6b)(-6b+5a)$, or $(5a+6b)(5a-6b)$ is the product of the sum and difference of the same two terms, choice (b).

5. $(r-3s)(5r+3s)$ is neither the square of a binomial nor the product of the sum and difference of the same two terms, so choice (c) is appropriate.

7. $(4x-9y)(4x-9y)$, or $(4x-9y)^2$ is the square of a binomial, choice (a).

9. We replace x by 5 and y by -2.

$x^2-2y^2+3xy=5^2-2(-2)^2+3\cdot 5(-2)$

$=25-8-30.$

$=-13.$

11. We replace x by 2, y by -3, and z by -4.

$xy^2z - z = 2(-3)^2(-4) - (-4) = -72 + 4 = -68$

13. Evaluate the polynomial for $h = 160$ and $A = 20$.

$$\begin{aligned}&0.041h - 0.018A - 2.69\\ &= 0.041(160) - 0.018(20) - 2.69\\ &= 6.56 - 0.36 - 2.69\\ &= 3.51\end{aligned}$$

The woman's lung capacity is 3.51 liters.

15. Evaluate the polynomial for $w = 125$, $h = 64$, and $a = 27$.

$$\begin{aligned}&917 + 6w + 6h - 6a\\ &= 917 + 6(125) + 6(64) - 6(27)\\ &= 917 + 750 + 384 - 162\\ &= 1889\end{aligned}$$

The daily caloric needs are 1889 calories.

17. Evaluate the polynomial for $h = 7$, $r = 1\frac{1}{2} = \frac{3}{2}$, and $\pi \approx 3.14$.

$$\begin{aligned}2\pi rh + \pi r^2 &\approx 2(3.14)\left(\frac{3}{2}\right)(7) + 3.14\left(\frac{3}{2}\right)^2\\ &\approx 65.94 + 7.065\\ &\approx 73.005\end{aligned}$$

The surface area is about 73.005 in^2.

19. Evaluate the polynomial for $h = 50$, $v = 18$, and $t = 2$.

$$\begin{aligned}&h + vt - 4.9t^2\\ &= 50 + 18 \cdot 2 - 4.9(2)^2\\ &= 50 + 36 - 19.6\\ &= 66.4\end{aligned}$$

The ball will be 66.4 m above the ground 2 seconds after it is thrown.

21. $3x^2y + 5xy + 2y^2 - 11$

Term	Coefficient	Degree	
$3x^2y$	3	3	
$-5xy$	-5	2	
$2y^2$	2	2	
-11	-11	0	(Think: $-11 = -11x^0$)

The degree of the polynomial is the degree of the term of highest degree. The term of highest degree is $3x^2y$. Its degree is 3, so the degree of the polynomial is 3.

23. $7 - abc + a^2b + 9ab^2$

Term	Coefficient	Degree
7	7	0
$-abc$	-1	3
a^2b	1	3
$9ab^2$	9	3

The terms of highest degree are $-abc$, a^2b and $9ab^2$. Each has degree 3. The degree of the polynomial is 3.

25. $3r + s - r - 7s = (3 - 1)r + (1 - 7)s = 2r - 6s$

27. $5xy^2 - 2x^2y + x + 3x^2$

There are no like terms, so none of the terms can be combined.

29. $$\begin{aligned}&6u^2v - 9uv^2 + 3vu^2 - 2v^2u + 11u^2\\ &= (6 + 3)u^2v + (-9 - 2)uv^2 + 11u^2\\ &= 9u^2v - 11uv^2 + 11u^2\end{aligned}$$

31. $$\begin{aligned}&5a^2c - 2ab^2 + a^2b - 3ab^2 + a^2c - 2ab^2\\ &= (5 + 1)a^2c + (-2 - 3 - 2)ab^2 + a^2b\\ &= 6a^2c - 7ab^2 + a^2b\end{aligned}$$

33. $$\begin{aligned}&(6x^2 - 2xy + y^2) + (5x^2 - 8xy - 2y^2)\\ &= (6 + 5)x^2 + (-2 - 8)xy + (1 - 2)y^2\\ &= 11x^2 - 10xy - y^2\end{aligned}$$

35. $$\begin{aligned}&(3a^4 - 5ab + 6ab^2) - (9a^4 + 3ab - ab^2)\\ &= 3a^4 - 5ab + 6ab^2 - 9a^4 - 3ab + ab^2 \quad \text{Adding the opposite}\\ &= (3 - 9)a^4 + (-5 - 3)ab + (6 + 1)ab^2\\ &= -6a^4 - 8ab + 7ab^2\end{aligned}$$

37. $(5r^2 - 4rt + t^2) + (-6r^2 - 5rt - t^2) + (-5r^2 + 4rt - t^2)$

Observe that the polynomials $5r^2 - 4rt + t^2$ and $-5r^2 + 4rt - t^2$ are opposites. Thus, their sum is 0 and the sum in the exercise is the remaining polynomial, $-6r^2 - 5rt - t^2$.

39. $$\begin{aligned}&(x^3 - y^3) - (-2x^3 + x^2y - xy^2 + 2y^3)\\ &= x^3 - y^3 + 2x^3 - x^2y + xy^2 - 2y^3\\ &= 3x^3 - 3y^3 - x^2y + xy^2, \text{ or}\\ &\quad 3x^3 - x^2y + xy^2 - 3y^3\end{aligned}$$

41. $$\begin{aligned}&(2y^4x^3 - 3y^3x) + (5y^4x^3 - y^3x) - (9y^4x^3 - y^3x)\\ &= (2 + 5 - 9)y^4x^3 + (-3 - 1 + 1)y^3x\\ &= -2y^4x^3 - 3y^3x\end{aligned}$$

43. $$\begin{aligned}&(4x + 5y) + (-5x + 6y) - (7x + 3y)\\ &= 4x + 5y - 5x + 6y - 7x - 3y\\ &= (4 - 5 - 7)x + (5 + 6 - 3)y\\ &= -8x + 8y\end{aligned}$$

45. $$\begin{aligned}(4c - d)(3c + 2d) &= \overset{F}{12c^2} + \overset{O}{8cd} - \overset{I}{3cd} - \overset{L}{2d^2}\\ &= 12c^2 + 5cd - 2d^2\end{aligned}$$

47. $$\begin{aligned}(xy - 1)(xy + 5) &= \overset{F}{x^2y^2} + \overset{O}{5xy} - \overset{I}{xy} - \overset{L}{5}\\ &= x^2y^2 + 4xy - 5\end{aligned}$$

49. $(2a - b)(2a + b)$ $[(A + B)(A - B) = A^2 - B^2]$
$= 4a^2 - b^2$

51. $$\begin{aligned}(5rt - 2)(4rt - 3) &= \overset{F}{20r^2t^2} - \overset{O}{15rt} - \overset{I}{8rt} + \overset{L}{6}\\ &= 20r^2t^2 - 23rt + 6\end{aligned}$$

53. $$\begin{aligned}&(m^3n + 8)(m^3n - 6)\\ &= \overset{F}{m^6n^2} - \overset{O}{6m^3n} + \overset{I}{8m^3n} - \overset{L}{48}\\ &= m^6n^2 + 2m^3n - 48\end{aligned}$$

55. $$\begin{aligned}&(6x - 2y)(5x - 3y)\\ &= \overset{F}{30x^2} - \overset{O}{18xy} - \overset{I}{10xy} + \overset{L}{6y^2}\\ &= 30x^2 - 28xy + 6y^2\end{aligned}$$

57. $(pq+0.1)(-pq+0.1)$
$= (0.1+pq)(0.1-pq)$ $[(A+B)(A-B)=A^2-B^2]$
$= 0.01 - p^2q^2$

59. $(x+h)^2$
$= x^2+2xh+h^2$ $[(A+B)^2 = A^2+2AB+B^2]$

61. $(4a-5b)^2$
$= 16a^2-40ab+25b^2$ $[(A-B)^2=A^2-2AB+B^2]$

63. $(ab+cd^2)(ab-cd^2) = (ab)^2-(cd^2)^2$
$= a^2b^2-c^2d^4$

65. $(2xy+x^2y+3)(xy+y^2)$
$= (2xy+x^2y+3)(xy)+(2xy+x^2y+3)(y^2)$
$= 2x^2y^2+x^3y^2+3xy+2xy^3+x^2y^3+3y^2$

67. $(a+b-c)(a+b+c)$
$= [(a+b)-c][(a+b)+c]$
$= (a+b)^2-c^2$
$= a^2+2ab+b^2-c^2$

69. $[a+b+c][a-(b+c)]$
$= [a+(b+c)][a-(b+c)]$
$= a^2-(b+c)^2$
$= a^2-(b^2+2bc+c^2)$
$= a^2-b^2-2bc-c^2$

71. The figure is a square with side $x+y$. Thus the area is $(x+y)^2 = x^2+2xy+y^2$.

73. The figure is a triangle with base $ab+2$ and height $ab-2$. Its area is $\frac{1}{2}(ab+2)(ab-2) = \frac{1}{2}(a^2b^2-4) = \frac{1}{2}a^2b^2-2$.

75. The figure is a rectangle with dimensions $a+b+c$ by $a+d+c$. Its area is
$(a+b+c)(a+d+c)$
$= [(a+c)+b][(a+c)+d]$
$= (a+c)^2+(a+c)d+b(a+c)+bd$
$= a^2+2ac+c^2+ad+cd+ab+bc+bd$

77. The figure is a parallelogram with base $m-n$ and height $m+n$. Its area is $(m-n)(m+n) = m^2-n^2$.

79. We draw a rectangle with dimensions $r+s$ by $u+v$.

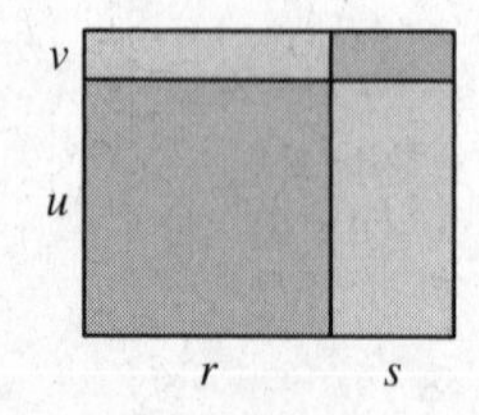

81. We draw a rectangle with dimensions $a+b+c$ by $a+d+f$.

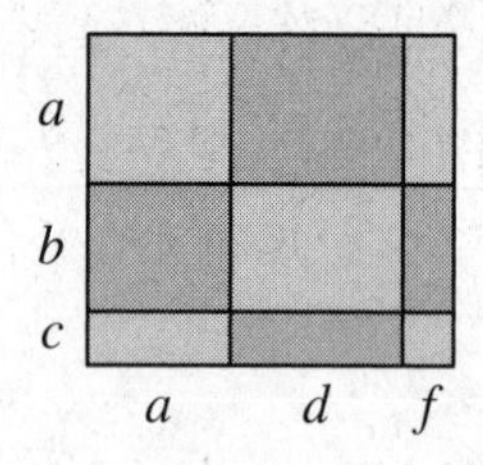

83. *Writing Exercise.* Yes; consider $a+b+c+d$. This is a polynomial in 4 variables but it has degree 1.

85.
$$\begin{array}{r} x^2-3x-7 \\ -(\quad +5x-3) \\ \hline x^2-8x-4 \end{array}$$

87.
$$\begin{array}{r} 3x^2+x+5 \\ -(3x^2+3x) \\ \hline -2x+5 \end{array}$$

89.
$$\begin{array}{r} 5x^3-2x^2+1 \\ -(5x^3-15x^2) \\ \hline 13x^2+1 \end{array}$$

91. *Writing Exercise.* The leading term of a polynomial is the term of highest degree. When a polynomial has several variables it is possible that more than one term has the highest degree as in Exercise 23.

93. It is helpful to add additional labels to the figure.

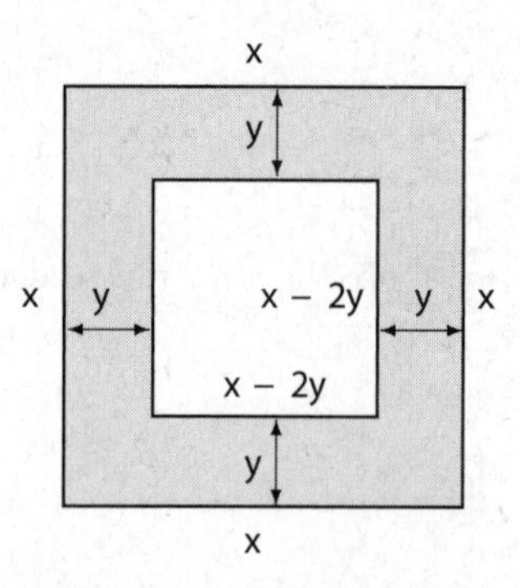

The area of the large square is $x \cdot x$, or x^2. The area of the small square is $(x-2y)(x-2y)$, or $(x-2y)^2$.

Area of shaded region = Area of large square − Area of small square

Area of shaded region = x^2 − $(x-2y)^2$

$= x^2-(x^2-4xy+4y^2)$
$= x^2-x^2+4xy-4y^2$
$= 4xy-4y^2$

95. The unshaded region is a circle with radius $a-b$. Then the shaded area is the area of a circle with radius a less the area of a circle with radius $a-b$. Thus, we have:

$$\begin{aligned}\text{Shaded area} &= \pi a^2 - \pi(a-b)^2\\ &= \pi a^2 - \pi(a^2 - 2ab + b^2)\\ &= \pi a^2 - \pi a^2 + 2\pi ab - \pi b^2\\ &= 2\pi ab - \pi b^2\end{aligned}$$

97. The figure can be thought of as a cube with side x, a rectangular solid with dimensions x by x by y, a rectangular solid with dimensions x by y by y, and a rectangular solid with dimensions y by y by $2y$. Thus the volume is

$x^3 + x \cdot x \cdot y + x \cdot y \cdot y + y \cdot y \cdot 2y$, or

$x^3 + x^2y + xy^2 + 2y^3$.

99. The surface area of the solid consists of the surface area of a rectangular solid with dimensions x by x by h less the areas of 2 circles with radius r plus the lateral surface area of a right circular cylinder with radius r and height h. Thus, we have

$2x^2 + 2xh + 2xh - 2\pi r^2 + 2\pi rh$, or

$2x^2 + 4xh - 2\pi r^2 + 2\pi rh$.

101. *Writing Exercise.* The height of the observatory is 40 ft and its radius is 30/2, or 15 ft, so the surface area is $2\pi rh + \pi r^2 \approx 2(3.14)(15)(40) + (3.14)(15)^2 \approx 4474.5\text{ ft}^2$. Since $4474.5\text{ ft}^2/250\text{ ft}^2 = 17.898$, 18 gallons of paint should be purchased.

103. For the formula $= 2 * A4 + 3 * B4$, we substitute 5 for $A4$ and 10 for $B4$.

$$\begin{aligned}= 2 * A4 + 3 * B4 &= 2 \cdot 5 + 3 \cdot 10\\ &= 10 + 30\\ &= 40\end{aligned}$$

The value of $D4$ is 40.

105. Replace t with 2 and multiply.

$$\begin{aligned}&P(1+r)^2\\ &= P(1 + 2r + r^2)\\ &= P + 2Pr + Pr^2\end{aligned}$$

107. Substitute \$10,400 for P, 8.5%, or 0.085 for r, and 5 for t.

$$\begin{aligned}&P(1+r)^t\\ &= \$10,400(1+0.085)^5\\ &\approx \$15,638.03\end{aligned}$$

Exercise Set 4.8

1. $$\begin{aligned}\frac{40x^6 - 25x^3}{5} &= \frac{40x^6}{5} - \frac{25x^3}{5}\\ &= \frac{40}{5}x^6 - \frac{25}{5}x^3 \quad \text{Dividing coefficients}\\ &= 8x^6 - 5x^3\end{aligned}$$

To check, we multiply the quotient by 5:

$(8x^6 - 5x^3)5 = 40x^6 - 25x^3$.

The answer checks.

3. $$\begin{aligned}&\frac{u - 2u^2 + u^7}{u}\\ &= \frac{u}{u} - \frac{2u^2}{u} + \frac{u^7}{u}\\ &= 1 - 2u + u^6\end{aligned}$$

Check: We multiply.

$u(1 - 2u + u^6) = u - 2u^2 + u^7$

5. $$\begin{aligned}&(18t^3 - 24t^2 + 6t) \div (3t)\\ &= \frac{18t^3 - 24t^2 + 6t}{3t}\\ &= \frac{18t^3}{3t} - \frac{24t^2}{3t} + \frac{6t}{3t}\\ &= 6t^2 - 8t + 2\end{aligned}$$

Check: We multiply.

$3t(6t^2 - 8t + 2) = 18t^3 - 24t^2 + 6t$

7. $$\begin{aligned}&(42x^5 - 36x^3 + 9x^2) \div (6x^2)\\ &= \frac{42x^5 - 36x^3 + 9x^2}{6x^2}\\ &= \frac{42x^5}{6x^2} - \frac{36x^3}{6x^2} + \frac{9x^2}{6x^2}\\ &= 7x^3 - 6x + \frac{3}{2}\end{aligned}$$

Check: We multiply.

$6x^2\left(7x^3 - 6x + \frac{3}{2}\right) = 42x^5 - 36x^3 + 9x^2$

9. $$\begin{aligned}&(32t^5 + 16t^4 - 8t^3) \div (-8t^3)\\ &= \frac{32t^5 + 16t^4 - 8t^3}{-8t^3}\\ &= \frac{32t^5}{-8t^3} + \frac{16t^4}{-8t^3} - \frac{8t^3}{-8t^3}\\ &= -4t^2 - 2t + 1\end{aligned}$$

Check: We multiply.

$-8t^3(-4t^2 - 2t + 1) = 32t^5 + 16t^4 - 8t^3$

11. $$\begin{aligned}&\frac{8x^2 - 10x + 1}{2x}\\ &= \frac{8x^2}{2x} - \frac{10x}{2x} + \frac{1}{2x}\\ &= 4x - 5 + \frac{1}{2x}\end{aligned}$$

Check: We multiply.

$2x\left(4x - 5 + \frac{1}{2x}\right) = 8x^2 - 10x + 1$

13. $$\begin{aligned}&\frac{5x^3y + 10x^5y^2 + 15x^2y}{5x^2y}\\ &= \frac{5x^3y}{5x^2y} + \frac{10x^5y^2}{5x^2y} + \frac{15x^2y}{5x^2y}\\ &= x + 2x^3y + 3\end{aligned}$$

Check: We multiply.

$5x^2y(x + 2x^3y + 3) = 5x^3y + 10x^5y^2 + 15x^2y$

15. $\dfrac{9r^2s^2 + 3r^2s - 6rs^2}{-3rs}$

$= \dfrac{9r^2s^2}{-3rs} + \dfrac{3r^2s}{-3rs} - \dfrac{6rs^2}{-3rs}$

$= -3rs - r + 2s$

Check: We multiply.

$-3rs(-3rs - r + 2s) = 9r^2s^2 + 3r^2s - 6rs^2$

17.

$$\begin{array}{r l}
 & x - 6 \\
x - 2 \,\big)\!\! & \overline{x^2 + 8x + 12} \\
 & \underline{x^2 - 2x} \\
 & -6x + 12 \leftarrow (x^2 + 8x) - (x^2 - 2x) = -6x \\
 & \underline{-6x + 12} \\
 & 0 \leftarrow (-6x + 12) - (-6x + 12) = 0
\end{array}$$

The answer is $x - 6$.

19.

$$\begin{array}{r l}
 & t - 5 \\
t - 5 \,\big)\!\! & \overline{t^2 - 10t - 20} \\
 & \underline{t^2 - 5t} \\
 & -5t - 20 \leftarrow (t^2 - 10t) - (t^2 - 5t) = -5t \\
 & \underline{-5t + 25} \\
 & -45 \leftarrow (-5t - 20) - (-5t + 25) = -45
\end{array}$$

The answer is $t - 5 + \dfrac{-45}{t - 5}$.

21.

$$\begin{array}{r l}
 & 2x - 1 \\
x + 6 \,\big)\!\! & \overline{2x^2 + 11x - 5} \\
 & \underline{2x^2 + 12x} \\
 & -x - 5 \leftarrow (2x^2 + 11x) - (2x^2 + 12x) = -x \\
 & \underline{-x - 6} \\
 & 1 \leftarrow (-x - 5) - (-x - 6) = 1
\end{array}$$

The answer is $2x - 1 + \dfrac{1}{x + 6}$.

23.

$$\begin{array}{r l}
 & t^2 - 3t + 9 \\
t + 3 \,\big)\!\! & \overline{t^3 + 0t^2 + 0t + 27} \leftarrow \text{Writing in the missing terms} \\
 & \underline{t^3 + 3t^2} \\
 & -3t^2 + 0t \leftarrow t^3 - (t^3 + 3t^2) = -3t^2 \\
 & \underline{-3t^2 - 9t} \\
 & 9t + 27 \leftarrow -3t^2 - (-3t^2 - 9t) = 9t \\
 & \underline{9t + 27} \\
 & 0 \leftarrow (9t + 27) = -(9t + 27) = 0
\end{array}$$

The answer is $t^2 - 3t + 9$.

25.

$$\begin{array}{r l}
 & a + 5 \\
a - 5 \,\big)\!\! & \overline{a^2 + 0a - 21} \leftarrow \text{Writing in the missing term} \\
 & \underline{a^2 - 5a} \\
 & 5a - 21 \leftarrow a^2 - (a^2 - 5a) = 5a \\
 & \underline{5a - 25} \\
 & 4 \leftarrow (5a^2 - 21) - (5a - 25) = 4
\end{array}$$

The answer is $a + 5 + \dfrac{4}{a - 5}$.

27.

$$\begin{array}{r l}
 & x - 3 \\
5x - 1 \,\big)\!\! & \overline{5x^2 - 16x + 0} \leftarrow \text{Writing in the missing term} \\
 & \underline{5x^2 - x} \\
 & -15x + 0 \leftarrow (5x^2 - 16x) - (5x^2 - x) = -15x \\
 & \underline{-15x + 3} \\
 & -3 \leftarrow (-15x + 0) - (-15x + 3) = -3
\end{array}$$

The answer is $x - 3 - \dfrac{3}{5x - 1}$.

29.

$$\begin{array}{r l}
 & 3a + 1 \\
2a + 5 \,\big)\!\! & \overline{6a^2 + 17a + 8} \\
 & \underline{6a^2 + 15a} \\
 & 2a + 8 \leftarrow (6a^2 + 17a) - (6a^2 + 15a) = 2a \\
 & \underline{2a + 5} \\
 & 3 \leftarrow (2a + 8) - (2a + 5) = 3
\end{array}$$

The answer is $3a + 1 + \dfrac{3}{2a + 5}$.

31.

$$\begin{array}{r l}
 & t^2 - 3t + 1 \\
2t - 3 \,\big)\!\! & \overline{2t^3 - 9t^2 + 11t - 3} \\
 & \underline{2t^3 - 3t^2} \\
 & -6t^2 + 11t \leftarrow (2t^3 - 9t^2) - (2t^3 - 3t^2) = -6t^2 \\
 & \underline{-6t^2 + 9t} \\
 & 2t - 3 \leftarrow (-6t^2 + 11t) - (-6t^2 + 9t) = 2t \\
 & \underline{2t - 3} \\
 & 0 \leftarrow (2t - 3) - (2t - 3) = 0
\end{array}$$

The answer is $t^2 - 3t + 1$.

33.

$$\begin{array}{r l}
 & x^2 \quad + 1 \\
x - 1 \,\big)\!\! & \overline{x^3 - x^2 + x - 1} \\
 & \underline{x^3 - x^2} \\
 & x - 1 \leftarrow (x^3 - x^2) - (x^3 - x^2) = 0 \\
 & \underline{x - 1} \\
 & 0 \leftarrow (x - 1) - (x - 1) = 0
\end{array}$$

The answer is $x^2 + 1$.

35.

$$\begin{array}{r l}
 & t^2 \quad - 1 \\
t^2 + 5 \,\big)\!\! & \overline{t^4 + 0t^3 + 4t^2 + 3t - 6} \leftarrow \text{Writing in the missing term} \\
 & \underline{t^4 \quad + 5t^2} \\
 & -t^2 + 3t - 6 \leftarrow (t^4 - 4t^2) - (t^4 + 5t^2) = -t^2 \\
 & \underline{-t^2 \quad - 5} \\
 & 3t - 1 \leftarrow (-t^2 + 3t - 6) - (-t^2 - 5) = 3t - 1
\end{array}$$

The answer is $t^2 - 1 + \dfrac{3t - 1}{t^2 + 5}$.

37.

$$\begin{array}{r l}
 & 3x^2 \quad - 3 \\
2x^2 + 1 \,\big)\!\! & \overline{6x^4 + 0x^3 - 3x^2 + x - 4} \leftarrow \text{Writing in the missing term} \\
 & \underline{6x^4 \quad + 3x^2} \\
 & -6x^2 + x - 4 \leftarrow (6x^4 - 3x^2) - (6x^4 + 3x^2) = -6x^2 \\
 & \underline{-6x^2 \quad - 3} \\
 & x - 1 \leftarrow (-6x^2 + x - 4) - (-6x^2 - 3) = x - 1
\end{array}$$

The answer is $3x^2 - 3 + \dfrac{x - 1}{2x^2 + 1}$.

39. *Writing Exercise.* The distributive law is used in each step when a term of the quotient is multiplied by the divisor and when the subtraction is performed. The distributive law is also used when the quotient is checked.

41. $3x - 4y = 12$

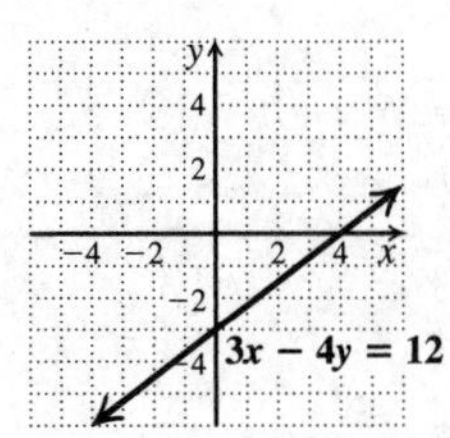

43. $3y - 2 = 7$

$y = 3$

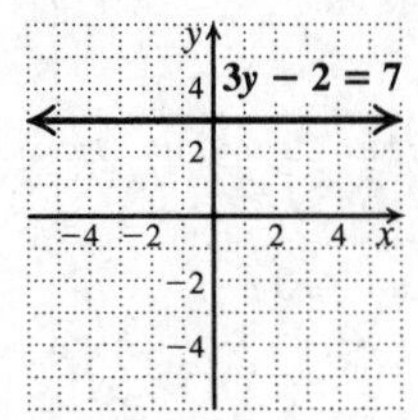

45. $m = \dfrac{5-2}{-7-3} = \dfrac{-3}{10}$

47. $y = -5x - 10$

49. *Writing Exercise.* Find the product of $x - 5$ and a binomial of the form $ax + b$. Then add 3 to this result.

51.
$$\begin{aligned}
&(10x^{9k} - 32x^{6k} + 28x^{3k}) \div (2x^{3k}) \\
&= \frac{10x^{9k} - 32x^{6k} + 28x^{3k}}{2x^{3k}} \\
&= \frac{10x^{9k}}{2x^{3k}} - \frac{32x^{6k}}{2x^{3k}} + \frac{28x^{3k}}{2x^{3k}} \\
&= 5x^{9k-3k} - 16x^{6k-3k} + 14x^{3k-3k} \\
&= 5x^{6k} - 16x^{3k} + 14
\end{aligned}$$

53.
$$\begin{array}{r|l}
 & 3t^{2h} + 2t^{h} - 5 \\
\hline
2t^{h} + 3 & 6t^{3h} + 13t^{2h} - 4t^{h} - 15 \\
 & \underline{6t^{3h} + 9t^{2h}} \\
 & \quad 4t^{2h} - 4t^{h} \\
 & \quad \underline{4t^{2h} + 6t^{h}} \\
 & \qquad -10t^{h} - 15 \\
 & \qquad \underline{-10t^{h} - 15} \\
 & \qquad\qquad 0
\end{array}$$

The answer is $3t^{2h} + 2t^{h} - 5$.

55.
$$\begin{array}{r|l}
 & a + 3 \\
\hline
5a^2 - 7a - 2 & 5a^3 + 8a^2 - 23a - 1 \\
 & \underline{5a^3 - 7a^2 - 2a} \\
 & \quad 15a^2 - 21a - 1 \\
 & \quad \underline{15a^2 - 21a - 6} \\
 & \qquad\qquad 5
\end{array}$$

The answer is $a + 3 + \dfrac{5}{5a^2 - 7a - 2}$.

57.
$$\begin{aligned}
&(4x^5 - 14x^3 - x^2 + 3) + \\
&\quad (2x^5 + 3x^4 + x^3 - 3x^2 + 5x) \\
&= 6x^5 + 3x^4 - 13x^3 - 4x^2 + 5x + 3
\end{aligned}$$

$$\begin{array}{r|l}
 & 2x^2 + x - 3 \\
\hline
3x^3 - 2x - 1 & 6x^5 + 3x^4 - 13x^3 - 4x^2 + 5x + 3 \\
 & \underline{6x^5 \qquad - 4x^3 - 2x^2} \\
 & \quad 3x^4 - 9x^3 - 2x^2 + 5x \\
 & \quad \underline{3x^4 \qquad - 2x^2 - x} \\
 & \qquad -9x^3 \qquad + 6x + 3 \\
 & \qquad \underline{-9x^3 \qquad + 6x + 3} \\
 & \qquad\qquad 0
\end{array}$$

The answer is $2x^2 + x - 3$.

59.
$$\begin{array}{r|l}
 & x - 3 \\
\hline
x - 1 & x^2 - 4x + c \\
 & \underline{x^2 - x} \\
 & \quad -3x + c \\
 & \quad \underline{-3x + 3} \\
 & \qquad c - 3
\end{array}$$

We set the remainder equal to 0.

$c - 3 = 0$

$c = 3$

Thus, c must be 3.

61.
$$\begin{array}{r|l}
 & c^2x + (2c + c^2) \\
\hline
x - 1 & c^2x^2 + 2cx + 1 \\
 & \underline{c^2x^2 - c^2x} \\
 & \quad (2c + c^2)x + 1 \\
 & \quad \underline{(2c + c^2)x - (2c + c^2)} \\
 & \qquad 1 + (2c + c^2)
\end{array}$$

We set the remainder equal to 0.

$c^2 + 2c + 1 = 0$

$(c + 1)^2 = 0$

$c + 1 = 0 \quad \textit{or} \quad c + 1 = 0$

$c = -1 \quad \textit{or} \quad c = -1$

Thus, c must be -1.

Chapter 4 Review

1. True; see page 255 in the text.

3. True; see page 272 in the text.

5. False; the degree of the polynomial is the degree of the leading term.

7. True; see page 283 in the text.

9. $n^3 \cdot n^8 \cdot n = n^{3+8+1} = n^{12}$

11. $t^6 \cdot t^0 = t^{6+0} = t^6$

13. $\dfrac{(a+b)^4}{(a+b)^4} = 1$

15. $(-2xy^2)^3 = (-2)^3 x^{1\cdot3} y^{2\cdot3} = -8x^3y^6$

17. $(a^2b)(ab)^5 = a^2b \cdot a^5b^5 = a^{2+5}b^{1+5} = a^7b^6$

19. $8^{-6} = \dfrac{1}{8^6}$

21. $4^5 \cdot 4^{-7} = 4^{5-7} = 4^{-2} = \dfrac{1}{4^2}$ or $\dfrac{1}{16}$

23. $\left(w^3\right)^{-5} = w^{3(-5)} = w^{-15} = \dfrac{1}{w^{15}}$

25. $\left(\dfrac{2x}{y}\right)^{-3} = \dfrac{2^{-3}x^{-3}}{y^{-3}} = \dfrac{y^3}{8x^3}$

27. $0.0000109 = 1.09 \times 10^m$
To write 1.09 as 0.0000109, we move the decimal point 5 places to the left. Thus, m is -5 and
$0.0000109 = 1.09 \times 10^{-5}$

29.
$$\begin{aligned}\frac{1.28 \times 10^{-8}}{2.5 \times 10^{-4}} &= \frac{1.28}{2.5} \times \frac{10^{-8}}{10^{-4}} \\ &= \frac{1.28}{2.5} \times 10^{-8+4} \\ &= 0.512 \times 10^{-4} \\ &= 5.12 \times 10^{-5}\end{aligned}$$

31. $-4y^5 + 7y^2 - 3y - 2$
The terms are $-4y^5$, $7y^2$, $-3y$, and -2.

33. $7n^4 - \dfrac{5}{6}n^2 - 4n + 10$
The coefficients are 7, $-\dfrac{5}{6}$, -4, and 10.

35. $-2x^5 + 7 + 3x^2 + x$

a)

Term	$-2x^5$	7	$-3x^2$	x
Degree	5	0	2	1

b) the term of highest degree is $-2x^5$. This is the leading term. Then the leading coefficient is -2 since $-2x^2 = -2 \cdot x^5$.

c) Since the term of highest degree is $-2x^5$, the degree of the polynomial is 5.

37. The polynomial $4 - 9t^3 - 7t^4 + 10t^2$ has four monomials so it is a polynomial with no special name.

39. $3x - x^2 + 4x = -x^2 + (3+4)\,x$
$= -x^2 + 7x$

41. $-4t^3 + 2t + 4t^3 + 8 - t - 9$
$= (-4+4)\,t^3 + (2-1)\,t + (8-9)$
$= t - 1$

43. For $x = -2:$
$$\begin{aligned}&9x - 6 \\ &= 9(-2) - 6 \\ &= -18 - 6 \\ &= -24\end{aligned}$$

45. $\left(8x^4 - x^3 + x - 4\right) + \left(x^5 + 7x^3 - 3x - 5\right)$
$= 8x^4 - x^3 + x - 4 + x^5 + 7x^3 - 3x - 5$
$= x^5 + 8x^4 + (-1+7)\,x^3 + (1-3)\,x + (-4-5)$
$= x^5 + 8x^4 + 6x^3 - 2x - 9$

47. $\left(y^2 + 8y - 7\right) - \left(4y^2 - 10\right)$
$= y^2 + 8y - 7 - 4y^2 + 10$
$= (1-4)\,y^2 + 8y + (-7+10)$
$= -3y^2 + 8y + 3$

49.
$$\begin{array}{rrrrr} -\frac{3}{4}x^4 & +\frac{1}{2}x^3 & & & +\frac{7}{8} \\ & -\frac{1}{4}x^3 & -\;x^2 & -\frac{7}{4}x & \\ +\frac{3}{2}x^4 & & +\frac{2}{3}x^2 & & -\frac{1}{2} \\ \hline +\frac{3}{4}x^4 & +\frac{1}{4}x^3 & -\frac{1}{3}x^2 & -\frac{7}{4}x & +\frac{3}{8} \end{array}$$

51. Let $w =$ the width, then $w + 3$ is the length, in meters.

a) Recall that the perimeter of a rectangle is the sum of all sides, so
$$\begin{aligned}\text{perimeter} &= 2w + 2(w+3) = 2w + 2w + 6 \\ &= 4w + 6.\end{aligned}$$

b) Recall that the area of a rectangle is the product of the length and the width. So,
$$\text{area} = w(w+3) = w^2 + 3w.$$

53. $(7x+1)^2 = (7x)^2 + 2 \cdot 7x \cdot 1 + 1^2$ $\quad (A+B)^2 = A + 2AB + B^2$
$= 49x^2 + 14x + 1$

55. $(d-8)(d+8) = d^2 - 8^2$ $\quad (A+B)(A-B) = A^2 - B^2$
$= d^2 - 64$

57. $(x-8)^2 = x^2 - 2 \cdot x \cdot 8 + 8^2$ $\quad (A-B)^2 = A - 2AB + B^2$
$= x^2 - 16x + 64$

59. $(2a+9)(2a-9) = (2a)^2 - 9^2$ $\quad (A+B)(A-B) = A^2 - B^2$
$= 4a^2 - 81$

61. $\left(x^4 - 2x + 3\right)\left(x^3 + x - 1\right)$
$= x^4\left(x^3 + x - 1\right) - 2x\left(x^3 + x - 1\right) + 3\left(x^3 + x - 1\right)$
$= x^7 + x^5 - x^4 - 2x^4 - 2x^2 + 2x + 3x^3 + 3x - 3$
$= x^7 + x^5 - 3x^4 + 3x^3 - 2x^2 + 5x - 3$

63. $(2t^2 + 3)(t^2 - 7)$
F O I L
$= 2t^4 - 14t^2 + 3t^2 - 21$
$= 2t^4 - 11t^2 - 21$

65. $(-7 + 2n)(7 + 2n) = (2n-7)(2n+7)$
$= (2n)^2 - 7^2$ $\quad (A+B)(A-B) = A^2 - B^2$
$= 4n^2 - 49$

67. $x^5y - 7xy + 9x^2 - 8$

Term	Coefficient	Degree
x^5y	1	6
$-7xy$	-7	2
$9x^2$	9	2
-8	-8	0

The term of highest degree is x^5y. Its degree is 6, so the degree of the polynomial is 6.

69. $u+3v-5u+v-7$
$=(1-5)\,u+(3+1)\,v-7$
$=-4u+4v-7$

71. $(4a^2-10ab-b^2)+(-2a^2-6ab+b^2)$
$=4a^2-10ab-b^2-2a^2-6ab+b^2$
$=(4-2)\,a^2+(-10-6)\,ab+(-1+1)\,b^2$
$=2a^2-16ab$

73. $(2x+5y)(x-3y)$
F O I L
$=2x^2-6xy+5xy-15y^2$
$=2x^2-xy-15y^2$

75. The figure is a triangle with base $x+y$ and height $x-y$. Its area is
$\frac{1}{2}(x+y)(x-y)=\frac{1}{2}(x^2-y^2)=\frac{1}{2}x^2-\frac{1}{2}y^2.$

77.

$$\begin{array}{r} 3x^2-7x+4 \\ 2x+3\,\overline{)\,6x^3-5x^2-13x+13} \\ \underline{6x^3+9x^2} \\ -14x^2-13x \\ \underline{-14x^2-21x} \\ 8x+13 \\ \underline{8x+12} \\ 1 \end{array}$$

The answer is $3x^2-7x+4+\frac{1}{2x+3}$.

79. *Writing Exercise.* In the expression $5x^3$, the exponent refers only to the x. In the expression $(5x)^3$, the entire expression $5x$ is the base.

81. a) For $(x^5-6x^2+3)(x^4+3x^3+7)$, the highest terms of each factor are x^5 and x^4. Their product is x^9, which is degree 9. Thus, the degree of the product is 9.
b) For $(x^7-4)^4$, the term of highest degree is x^7, which is degree 7. Taking that term to the fourth power results in a term of degree 28.

83. Let $c=$ the coefficient of x^3.
Let $2c=$ the coefficient of x^4.
Let $2c-3=$ the coefficient of x.
Let $2c-3-7=$ the constant $\cdot$ (the remaining term)
The coefficient of x^2 is 0.

Solve:

$$\begin{aligned} 2c+c+0+2c-3+2c-3-7&=15 \\ 7c-13&=15 \\ 7c&=28 \\ c&=4 \end{aligned}$$

Coefficient of x^3, $c=4$
Coefficient of x^4, $2c=2\cdot 4=8$
Coefficient of x^2, 0
Coefficient of x, $2c-3=2\cdot 4-3=8-3=5$
Constant, $2c-3-7=2\cdot 4-10=8-10=-2$
The polynomial is $8x^4+4x^3+5x-2$.

85.

$$\begin{aligned} (x-7)(x+10)&=(x-4)(x-6) \\ x^2+10x-7x-70&=x^2-6x-4x+24 \\ x^2+3x-70&=x^2-10x+24 \\ 3x-70&=-10x+24 && \text{Subtracting } x^2 \\ 3x&=-10x+94 && \text{Adding 70} \\ 13x&=94 && \text{Adding } 10x \\ x&=\frac{94}{13} \end{aligned}$$

The solution is $\frac{94}{13}$

Chapter 4 Test

1. $x^7\cdot x\cdot x^5=x^{7+1+5}=x^{13}$

3. $\frac{(3m)^4}{(3m)^4}=1$

5.
$$\begin{aligned} (5x^4y)(-2x^5y^3)^3&=(5x^4y)(-2)^3(x^5y^3)^3 \\ &=5x^4y\cdot(-8)x^{5\cdot 3}y^{3\cdot 3} \\ &=5x^4y\cdot(-8)\cdot x^{15}y^9 \\ &=5(-8)x^{4+15}y^1+9 \\ &=-40x^{19}y^{10} \end{aligned}$$

7. $y^{-7}=\frac{1}{y^7}$

9. $t^{-4}\cdot t^{-5}=t^{-4+(-5)}=t^{-9}=\frac{1}{t^9}$

11.
$$\begin{aligned} (2a^3b^{-1})^{-4}&=2^{-4}(a^3)^{-4}(b^{-1})^{-4} \\ &=2^{-4}\cdot a^{-12}\cdot b^4 \\ &=\frac{1}{2^4}\cdot\frac{1}{a^{12}}\cdot b^4 \\ &=\frac{b^4}{16a^{12}} \end{aligned}$$

13. $3,060,000,000=3.06\times 10^9$

15.
$$\begin{aligned} \frac{5.6\times 10^6}{3.2\times 10^{-11}}&=\frac{5.6}{3.2}\times\frac{10^6}{10^{-11}} \\ &=1.75\times 10^{6-(-11)} \\ &=1.75\times 10^{17} \end{aligned}$$

17. Two monomials are added so $4x^2y-7y^3$ is a binomial.

19. $2t^3-t+7t^5+4$
The degrees of the terms are $3,1,5,0$; the leading term is $7t^5$; the leading coefficient is 7; the degree of the polynomial is 5.

21. $4a^2-6+a^2=(4+1)a^2-6=5a^2-6$

23. $3-x^2+8x+5x^2-6x-2x+4x^3$
$=4x^3+(-1+5)x^2+(8-6-2)x+3$
$=4x^3+4x^2+3$

25. $\left(x^4+\frac{2}{3}x+5\right)+\left(4x^4+5x^2+\frac{1}{3}x\right)$
$=(1+4)x^4+5x^2+\left(\frac{2}{3}x+\frac{1}{3}\right)x+5$
$=5x^4+5x^2+x+5$

27. $(t^3 - 0.3t^2 - 20) - (t^4 - 1.5t^3 + 0.3t^2 - 11)$
$= t^3 - 0.3t^2 - 20 - t^4 + 1.5t^3 - 0.3t^2 + 11$
$= -t^4 + (1 + 1.5)t^3 + (-0.3 - 0.3)t^2$
$+(-20 + 11)$
$= -t^4 + 2.5t^3 - 0.6t^2 - 9$

29. $\left(x - \frac{1}{3}\right)^2 = x^2 - 2 \cdot x \cdot \frac{1}{3} + \frac{1}{3}^2$
$= x^2 - \frac{2}{3}x + \frac{1}{9}$

31. $(3b + 5)(2b - 1) = 6b^2 - 3b + 10b - 5$
$= 6b^2 - 7b - 5$

33. $(8 - y)(6 + 5y) = 48 + 40y - 6y - 5y^2$
$= 48 + 34y - 5y^2$

35. $(8a^3 + 3)^2 = (8a^3)^2 + 2(8a^3)(3) + 3^2$
$= 64a^6 + 48a^3 + 9$

37. $2x^3y - y^3 + xy^3 + 8 - 6x^3y - x^2y^2 + 11$
$= (2 - 6)x^3y - x^2y^2 + xy^3 - y^3 + (8 + 11)$
$= -4x^3y - x^2y^2 + xy^3 - y^3 + 19$

39. $(3x^5 - y)(3x^5 + y) = (3x^5)^2 - y^2$
$= 9x^{10} - y^2$

41. $\frac{1}{5^6} = 5^{-6}$

43.
$$\begin{aligned} x^2 + (x-7)(x+4) &= 2(x-6)^2 \\ x^2 + x^2 + 4x - 7x - 28 &= 2\left[x^2 - 2 \cdot x \cdot 6 + 6^2\right] \\ x^2 + x^2 + 4x - 7x - 28 &= 2\left[x^2 - 12x + 36\right] \\ 2x^2 - 3x - 28 &= 2x^2 - 24x + 72 \\ -3x - 28 &= -24x + 72 \\ 21x - 28 &= 72 \\ 21x &= 100 \\ x &= \frac{100}{21} \end{aligned}$$

45. ***Familiarize.*** We need to multiply to determine the hours wasted. There are 60 sec in a minute and 60 min in an hr. 1 sec $= \frac{1}{60} \cdot \frac{1}{60} = \frac{1}{3600}$hr. So 4 sec $= \frac{4}{3600}$hr

Translate. We multiply the number of spam emails by the time wasted on each spam email. 12.4 billion $= 1.24 \times 10^{10}$.

Carry out
$s = 1.24 \times 10^{10} \cdot \frac{4}{3600} \approx 1.4 \times 10^7$hr

Check. We recalculate to check our solution. The answer checks.

State. About 1.4×10^7hr each day are wasted due to spam.

Chapter 5

Polynomials and Factoring

Exercise Set 5.1

1. Since $7a \cdot 5ab = 35a^2b$, choice (h) is most appropriate.

3. $5x + 10 = 5(x+2)$ and $4x + 8 = 4(x+2)$, so $x+2$ is a common factor of $5x+10$ and $4x+8$ and choice (b) is most appropriate.

5. $3x^2(3x^2 - 1) = 9x^4 - 3x^2$, so choice (c) is most appropriate.

7. $3a + 6a^2 = 3a(1+2a)$, so $1+2a$ is a factor of $3a+6a^2$ and choice (d) is most appropriate.

9. Answers may vary. $14x^3 = (14x)(x^2) = (7x^2)(2x) = (-2)(-7x^3)$

11. Answers may vary. $-15a^4 = (-15)(a^4) = (-5a)(3a^3) = (-3a^2)(5a^2)$

13. Answers may vary. $25t^5 = (5t^2)(5t^3) = (25t)(t^4) = (-5t)(-5t^4)$

15. $8x + 24 = 8 \cdot x + 8 \cdot 3$
$= 8(x+3)$

17. $6x - 30 = 6 \cdot x - 6 \cdot 5$
$= 6(x-5)$

19. $2x^2 + 2x - 8 = 2 \cdot x^2 + 2 \cdot x - 2 \cdot 4$
$= 2(x^2 + x - 4)$

21. $3t^2 + t = t \cdot 3t + t \cdot 1$
$= t(3t+1)$

23. $-5y^2 - 10y = -5y \cdot y - 5y \cdot 2$
$= -5y(y+2)$

25. $x^3 + 6x^2 = x^2 \cdot x + x^2 \cdot 6$
$= x^2(x+6)$

27. $16a^4 - 24a^2 = 8a^2 \cdot 2a^2 - 8a^2 \cdot 3$
$= 8a^2(2a^2 - 3)$

29. $-6t^6 + 9t^4 - 4t^2 = -t^2 \cdot 6t^4 - t^2(-9t^2) - t^2 \cdot 4$
$= -t^2(6t^4 - 9t^2 + 4)$

31. $6x^8 + 12x^6 - 24x^4 + 30x^2$
$= 6x^2 \cdot x^6 + 6x^2 \cdot 2x^4 - 6x^2 \cdot 4x^2 + 6x^2 \cdot 5$
$= 6x^2(x^6 + 2x^4 - 4x^2 + 5)$

33. $x^5y^5 + x^4y^3 + x^3y^3 - x^2y^2$
$= x^2y^2 \cdot x^3y^3 + x^2y^2 \cdot x^2y + x^2y^2 \cdot xy - x^2y^2 \cdot 1$
$= x^2y^2(x^3y^3 + x^2y + xy - 1)$

35. $-35a^3b^4 + 10a^2b^3 - 15a^3b^2$
$= -5a^2b^2 \cdot 7ab^2 - 5a^2b^2(-2b) - 5a^2b^2 \cdot 3a$
$= -5a^2b^2(7ab^2 - 2b + 3a)$

37. $n(n-6) + 3(n-6)$
$= (n-6)(n+3)$ Factoring out the common binomial factor $n-6$

39. $x^2(x+3) - 7(x+3)$
$= (x+3)(x^2 - 7)$ Factoring out the common binomial factor $x+3$

41. $y^2(2y-9) + (2y-9)$
$= y^2(2y-9) + 1(2y-9)$
$= (2y-9)(y^2+1)$ Factoring out the common factor $2y-9$

43. $x^3 + 2x^2 + 5x + 10$
$= (x^3 + 2x^2) + (5x + 10)$
$= x^2(x+2) + 5(x+2)$ Factoring each binomial
$= (x+2)(x^2+5)$ Factoring out the common factor $x+2$

45. $5a^3 + 15a^2 + 2a + 6$
$= (5a^3 + 15a^2) + (2a + 6)$
$= 5a^2(a+3) + 2(a+3)$ Factoring each binomial
$= (a+3)(5a^2+2)$ Factoring out the common factor $a+3$

47. $9n^3 - 6n^2 + 3n - 2$
$= 3n^2(3n-2) + 1(3n-2)$
$= (3n-2)(3n^2+1)$

49. $4t^3 - 20t^2 + 3t - 15$
$= 4t^2(t-5) + 3(t-5)$
$= (t-5)(4t^2+3)$

51. $7x^3 + 5x^2 - 21x - 15 = x^2(7x+5) - 3(7x+5)$
$= (7x+5)(x^2-3)$

53. $6a^3 + 7a^2 + 6a + 7 = a^2(6a+7) + 1(6a+7)$
$= (6a+7)(a^2+1)$

55. $2x^3 + 12x^2 - 5x - 30 = 2x^2(x+6) - 5(x+6)$
$= (x+6)(2x^2-5)$

57. We try factoring by grouping.
$p^3 + p^2 - 3p + 10 = p^2(p+1) - (3p - 10)$, or
$p^3 - 3p + p^2 + 10 = p(p^2 - 3) + p^2 + 10$
Because we cannot find a common binomial factor, this polynomial cannot be factored using factoring by grouping.

59. $y^3+8y^2-2y-16=y^2(y+8)-2(y+8)=(y+8)(y^2-2)$

61. $2x^3-8x^2-9x+36=2x^2(x-4)-9(x-4)$
$=(x-4)(2x^2-9)$

63. *Writing Exercise.* Yes; the opposite of a factor is also a factor so both can be correct.

65. $(x+2)(x+7)$
F O I L
$=x\cdot x+x\cdot 7+2\cdot x+2\cdot 7$
$=x^2+7x+2x+14$
$=x^2+9x+14$

67. $(x+2)(x-7)$
F O I L
$=x\cdot x+x\cdot(-7)+2\cdot x-2\cdot 7$
$=x^2-7x+2x+14$
$=x^2-5x-14$

69. $(a-1)(a-3)$
F O I L
$=a\cdot a-a\cdot 3-1\cdot a-1(-3)$
$=a^2-3a-a+3$
$=a^2-4a+3$

71. $(t-5)(t+10)$
F O I L
$=t\cdot t+t\cdot 10-5\cdot t-5\cdot 10$
$=t^2+10t-5t-50$
$=t^2+5t-50$

73. *Writing Exercise.* This is a good idea, because it is unlikely that Azrah will choose two replacement values that give the same value for non-equivalent expressions.

75. $4x^5+6x^2+6x^3+9=2x^2(2x^3+3)+3(2x^3+3)$
$=(2x^3+3)(2x^2+3)$

77. $2x^4+2x^3-4x^2-4x=2x(x^3+x^2-2x-2)$
$=2x\left(x^2(x+1)-2(x+1)\right)$
$=2x(x+1)(x^2-2)$

79. $5x^5-5x^4+x^3-x^2+3x-3$
$=5x^4(x-1)+x^2(x-1)+3(x-1)$
$=(x-1)(5x^4+x^2+3)$

We could also do this exercise as follows:
$5x^5-5x^4+x^3-x^2+3x-3$
$=(5x^5+x^3+3x)-(5x^4+x^2+3)$
$=x(5x^4+x^2+3)-1(5x^4+x^2+3)$
$=(5x^4+x^2+3)(x-1)$

81. Answers may vary. $8x^4y^3-24x^3y^3+16x^2y^4$

Exercise Set 5.2

1. If c is positive, then p and q have the same sign. If both are negative, then b is negative; if both are positive then c is positive. Thus we replace each blank with "positive."

3. If p is negative and q is negative, then b is negative because it is the sum of two negative numbers and c is positive because it is the product of two negative numbers.

5. Since c is negative, it is the product of a negative and a positive number. Then because c is the product of p and q and we know that p is negative, q must be positive.

7. $x^2+8x+16$

Since the constant term and the coefficient of the middle term are both positive, we look for a factorization of 16 in which both factors are positive. Their sum must be 8.

Pairs of factors	Sums of factors
1,16	17
2, 8	10
4, 4	8

The numbers we want are 4 and 4.

$x^2+8x+16=(x+4)(x+4)$

9. $x^2+11x+10$

Since the constant term and the coefficient of the middle term are both positive, we look for a factorization of 10 in which both factors are positive. Their sum must be 11.

Pairs of factors	Sums of factors
1,10	11
2, 5	10

The numbers we want are 1 and 10.

$x^2+11x+10=(x+1)(x+10)$

11. $x^2+10x+21$

Since the constant term and the coefficient of the middle term are both positive, we look for a factorization of 21 in which both factors are positive. Their sum must be 10.

Pairs of factors	Sums of factors
1, 21	22
3, 7	10

The numbers we want are 3 and 7.

$x^2+10x+21=(x+3)(x+7)$

13. $t^2-9t+14$

Since the constant term is positive and the coefficient of the middle term is negative, we look for a factorization of 14 in which both factors are negative. Their sum must be -9.

Pairs of factors	Sums of factors
−1,−14	−15
−2, −7	−9

The numbers we want are -2 and -7.

$t^2-9t+14=(t-2)(t-7)$

15. $b^2 - 5b + 4$

Since the constant term is positive and the coefficient of the middle term is negative, we look for a factorization of 4 in which both factors are negative. Their sum must be -5.

Pairs of factors	Sums of factors
$-1, -4$	-5
$-2, -2$	-4

The numbers we want are -1 and -4.

$b^2 - 5b + 4 = (b - 1)(b - 4)$.

17. $a^2 - 7a + 12$

Since the constant term is positive and the coefficient of the middle term is negative, we look for a factorization of 12 in which both factors are negative. Their sum must be -7.

Pairs of factors	Sums of factors
$-1, -12$	-13
$-2, -6$	-8
$-3, -4$	-7

The numbers we need are -3 and -4.

$a^2 - 7a + 12 = (a - 3)(a - 4)$.

19. $d^2 - 7d + 10$

Since the constant term is positive and the coefficient of the middle term is negative, we look for a factorization of 10 in which both factors are negative. Their sum must be -7.

Pairs of factors	Sums of factors
$-1, -10$	-11
$-2, -5$	-7

The numbers we want are -2 and -5.

$d^2 - 7d + 10 = (d - 2)(d - 5)$.

21. $x^2 - 2x - 15$

The constant term, -15, must be expressed as the product of a negative number and a positive number. Since the sum of those two numbers must be negative, the negative number must have the greater absolute value.

Pairs of factors	Sums of factors
$1, -15$	-14
$3, -5$	-2

The numbers we need are 3 and -5.

$x^2 - 2x - 15 = (x + 3)(x - 5)$.

23. $x^2 + 2x - 15$

The constant term, -15, must be expressed as the product of a negative number and a positive number. Since the sum of those two numbers must be positive, the positive number must have the greater absolute value.

Pairs of factors	Sums of factors
$-1, 15$	14
$-3, 5$	2

The numbers we need are -3 and 5.

$x^2 + 2x - 15 = (x - 3)(x + 5)$.

25. $2x^2 - 14x - 36 = 2(x^2 - 7x - 18)$

After factoring out the common factor, 2, we consider $x^2 - 7x - 18$. The constant term, -18, must be expressed as the product of a negative number and a positive number. Since the sum of those two numbers must be negative, the negative number must have the greater absolute value.

Pairs of factors	Sums of factors
$1, -18$	-17
$2, -9$	-7
$3, -6$	-3

The numbers we need are 2 and -9. The factorization of $x^2 - 7x - 18$ is $(x - 9)(x + 2)$. We must not forget the common factor, 2. Thus, $2x^2 - 14x - 36 = 2(x^2 - 7x - 18) = 2(x - 9)(x + 2)$.

27. $-x^3 + 6x^2 + 16x = -x(x^2 - 6x - 16)$

After factoring out the common factor, $-x$, we consider $x^2 - 6x - 16$. The constant term, -16, must be expressed as the product of a negative number and a positive number. Since the sum of those two numbers must be negative, the negative number must have the greater absolute value.

Pairs of factors	Sums of factors
$1, -16$	-15
$2, -8$	-6
$4, -4$	0

The numbers we need are 2 and -8. The factorization of $x^2 - 6x - 16$ is $(x + 2)(x - 8)$. We must not forget the common factor, $-x$. Thus, $-x^3 + 6x^2 + 16x = -x(x^2 - 6x - 16) = -x(x + 2)(x - 8)$.

29. $4y - 45 + y^2 = y^2 + 4y - 45$

The constant term, -45, must be expressed as the product of a negative number and a positive number. Since the sum of those two numbers must be positive, the positive number must have the greater absolute value.

Pairs of factors	Sums of factors
$-1, 45$	44
$-3, 15$	12
$-5, 9$	4

The numbers we need are -5 and 9.

$4y - 45 + y^2 = (y - 5)(y + 9)$

31. $x^2 - 72 + 6x = x^2 + 6x - 72$

The constant term, -72, must be expressed as the product of a negative number and a positive number. Since the sum of those two numbers must be positive, the positive number must have the greater absolute value.

Pairs of factors	Sums of factors
−1, 72	71
−2, 36	34
−3, 24	21
−4, 18	14
−6, 12	6

The numbers we need are −6 and 12.

$x^2 - 72 + 6x = (x - 6)(x + 12)$

33. $-5b^2 - 35b + 150 = -5(b^2 + 7b - 30)$

After factoring out the common factor, −5, we consider $b^2 + 7b - 30$. The constant term, −30, must be expressed as the product of a negative number and a positive number. Since the sum of those two numbers must be positive, the positive number must have the greater absolute value.

Pairs of factors	Sums of factors
−1, 30	29
−2, 15	13
−3, 10	7
−5, 6	1

The numbers we need are −3 and 10. The factorization of $b^2 + 7b - 30$ is $(b - 3)(b + 10)$. We must not forget the common factor. Thus, $-5b^2 - 35b + 150 = -5(b^2 + 7b - 30) = -5(b - 3)(b + 10)$.

35. $x^5 - x^4 - 2x^3 = x^3(x^2 - x - 2)$

After factoring out the common factor, x^3, we consider $x^2 - x - 2$. The constant term, −2, must be expressed as the product of a negative number and a positive number. Since the sum of those two numbers must be negative, the negative number must have the greater absolute value. The only possible factors that fill these requirements are 1 and −2. These are the numbers we need. The factorization of $x^2 - x - 2$ is $(x + 1)(x - 2)$. We must not forget the common factor, x^3. Thus, $x^5 - x^4 - 2x^3 = x^3(x^2 - x - 2) = x^3(x + 1)(x - 2)$.

37. $x^2 + 5x + 10$

Since the constant term and the coefficient of the middle term are both positive, we look for a factorization of 10 in which both factors are positive. Their sum must be 5. The only possible pairs of positive factors are 1 and 10, and 2 and 5 but neither sum is 5. Thus, this polynomial is not factorable into polynomials with integer coefficients. It is prime.

39. $32 + 12t + t^2 = t^2 + 12t + 32$

Since the constant term is positive and the coefficient of the middle term is positive, we look for a factorization of 32 in which both terms are positive. Their sum must be 12.

Pairs of factors	Sums of factors
1,32	33
2,16	18
4, 8	12

The numbers we want are 4 and 8.

$32 + 12t + t^2 = (t + 4)(t + 8)$.

41. $x^2 + 20x + 99$

We look for two factors, both positive, whose product is 99 and whose sum is 20.

They are 9 and 11: $9 \cdot 11 = 99$ and $9 + 11 = 20$.

$x^2 + 20 + 99 = (x + 9)(x + 11)$

43. $3x^3 - 63x^2 - 300x = 3x(x^2 - 21x - 100)$

After factoring out the common factor, $3x$, we consider $x^2 - 21x - 100$. We look for two factors, one positive, one negative, whose product is −100 and whose sum is −21.

They are −25 and 4: $-25(4) = -100$ and $-25 + 4 = -21$.

$x^2 - 21x - 100 = (x - 25)(x + 4)$, so $3x^3 - 63x^2 - 300x = 3x(x - 25)(x + 4)$.

45. $-2x^2 + 42x + 144 = -2(x^2 - 21x - 72)$

After factoring out the common factor, −2, we consider $x^2 - 21x - 72$. We look for two factors, one positive, and one negative, whose product is −72 and whose sum is −21. They are −24 and 3.

$x^2 - 21x - 72 = (x - 24)(x + 3)$, so $-2x^2 + 42x + 144 = -2(x^2 - 21x - 72) = -2(x - 24)(x + 3)$.

47. $y^2 - 20y + 96$

We look for two factors, both negative, whose product is 96 and whose sum is −20. They are −8 and −12.

$y^2 - 20y + 96 = (y - 8)(y - 12)$

49. $-a^6 - 9a^5 + 90a^4 = -a^4(a^2 + 9a - 90)$

After factoring out the common factor, $-a^4$, we consider $a^2 + 9a - 90$. We look for two factors, one positive and one negative, whose product is −90 and whose sum is 9. They are −6 and 15.

$a^2 + 9a - 90 = (a - 6)(a + 15)$, so $-a^6 - 9a^5 + 90a^4 = -a^4(a - 6)(a + 15)$.

51. $t^2 + \frac{2}{3}t + \frac{1}{9}$

We look for two factors, both positive, whose product is $\frac{1}{9}$ and whose sum is $\frac{2}{3}$. They are $\frac{1}{3}$ and $\frac{1}{3}$.

$t^2 + \frac{2}{3}t + \frac{1}{9} = \left(t + \frac{1}{3}\right)\left(t + \frac{1}{3}\right)$, or $\left(t + \frac{1}{3}\right)^2$

53. $11 + w^2 - 4w = w^2 - 4w + 11$

Since the constant term is positive and the coefficient of the middle term is negative, we look for a factorization of 11 in which both factors are negative. Their sum must be −4. The only possible pair of factors is −1 and −11, but their sum is not −4. Thus, this polynomial is not factorable into polynomials with integer coefficients. It is prime.

55. $p^2 + 7pq + 10q^2$

Think of $-7q$ as a "coefficient" of p. Then we look for factors of $10q^2$ whose sum is $-7q$. They are $-5q$ and $-2q$.

$p^2 + 7pq + 10q^2 = (p - 5q)(p - 2q)$.

57. $m^2 + 5mn + 5n^2 = m^2 + 5nm + 5n^2$

We look for factors of $5n^2$ whose sum is $5n$. The only reasonable possibilities are shown below.

Pairs of factors	Sums of factors
$5n,\ n$	$6n$
$-5n, -n$	$-6n$

There are no factors whose sum is $5n$. Thus, the polynomial is not factorable into polynomials with integer coefficients. It is prime.

59. $s^2 - 4st - 12t^2 = s^2 - 4ts - 12t^2$

We look for factors of $-12t^2$ whose sum is $-4t$. They are $-6t$ and $2t$.

$s^2 - 4st - 12t^2 = (s - 6t)(s + 2t)$

61. $6a^{10} + 30a^9 - 84a^8 = 6a^8(a^2 + 5a - 14)$

After factoring out the common factor, $6a^8$, we consider $a^2 + 5a - 14$. We look for two factors, one positive and one negative, whose product is -14 and whose sum is 5. They are -2 and 7.

$a^2 + 5a - 14 = (a - 2)(a + 7)$, so $6a^{10} + 30a^9 - 84a^8$ $= 6a^8(a - 2)(a + 7)$.

63. *Writing Exercise.* Since both constants are negative, the middle term will be negative so $(x - 17)(x - 18)$ cannot be a factorization of $x^2 + 35x + 306$.

65. $(2x + 3)(3x + 4)$

$$\begin{aligned} &\quad\ \text{F} \qquad\ \ \text{O} \qquad\quad \text{I} \qquad\ \ \text{L} \\ &= 2x \cdot 3x + 2x \cdot 4 + 3 \cdot 3x + 3 \cdot 4 \\ &= 6x^2 + 8x + 9x + 12 \\ &= 6x^2 + 17x + 12 \end{aligned}$$

67. $(2x - 3)(3x + 4)$

$$\begin{aligned} &\quad\ \text{F} \qquad\ \ \text{O} \qquad\quad \text{I} \qquad\ \ \text{L} \\ &= 2x \cdot 3x + 2x \cdot 4 - 3 \cdot 3x - 3 \cdot 4 \\ &= 6x^2 + 8x - 9x - 12 \\ &= 6x^2 - x - 12 \end{aligned}$$

69. $(5x - 1)(x - 7)$

$$\begin{aligned} &\quad\ \text{F} \qquad\ \ \text{O} \qquad\quad \text{I} \qquad\ \ \text{L} \\ &= 5x \cdot x - 5x \cdot 7 - 1 \cdot x - 1(-7) \\ &= 5x^2 - 35x - x + 7 \\ &= 5x^2 - 36x + 7 \end{aligned}$$

71. *Writing Exercise.* There is a finite number of pairs of numbers with the correct product, but there are infinitely many pairs with the correct sum.

73. $a^2 + ba - 50$

We look for all pairs of integer factors whose product is -50. The sum of each pair is represented by b.

Pairs of factors whose product is -50	Sums of factors
$-1,\ 50$	49
$1, -50$	-49
$-2,\ 25$	23
$2, -25$	-23
$-5,\ 10$	5
$5, -10$	-5

The polynomial $a^2 + ba - 50$ can be factored if b is 49, -49, 23, -23, 5, or -5.

75. $y^2 - 0.2y - 0.08$

We look for two factors, one positive and one negative, whose product is -0.08 and whose sum is -0.2. They are -0.4 and 0.2.

$y^2 - 0.2y - 0.08 = (y - 0.4)(y + 0.2)$

77. $-\frac{1}{3}a^3 + \frac{1}{3}a^2 + 2a = -\frac{1}{3}a(a^2 - a - 6)$

After factoring out the common factor, $-\frac{1}{3}a$, we consider $a^2 - a - 6$. We look for two factors, one positive and one negative, whose product is -6 and whose sum is -1. They are 2 and -3.

$a^2 - a - 6 = (a + 2)(a - 3)$, so
$-\frac{1}{3}a^3 + \frac{1}{3}a^2 + 2a = -\frac{1}{3}a(a + 2)(a - 3)$.

79. $x^{2m} + 11x^m + 28 = (x^m)^2 + 11x^m + 28$

We look for numbers p and q such that $x^{2m} + 11x^m + 28 = (x^m + p)(x^m + q)$. We find two factors, both positive, whose product is 28 and whose sum is 11. They are 4 and 7.

$x^{2m} + 11x^m + 28 = (x^m + 4)(x^m + 7)$

81. $(a + 1)x^2 + (a + 1)3x + (a + 1)2$
$= (a + 1)(x^2 + 3x + 2)$

After factoring out the common factor $a + 1$, we consider $x^2 + 3x + 2$. We look for two factors, whose product is 2 and whose sum is 3. They are 1 and 2.

$x^2 + 3x + 2 = (x + 1)(x + 2)$, so
$(a + 1)x^2 + (a + 1)3x + (a + 1)2$
$= (a + 1)(x + 1)(x + 2)$.

83. $6x^2 + 36x + 54 = 6(x^2 + 6x + 9) = 6(x + 3)(x + 3) = 6(x + 3)^2$

Since the surface area of a cube with sides is given by $6s^2$, we know that this cube has side $x + 3$. The volume of a cube with side s is given by s^3, so the volume of this cube is $(x + 3)^3$, or $x^3 + 9x^2 + 27x + 27$.

85. The shaded area consists of the area of a rectangle with sides x and $x + x$, or $2x$, and $\frac{3}{4}$ of the area of a circle with radius x. It can be expressed as follows:

$x \cdot 2x + \frac{3}{4}\pi x^2 = 2x^2 + \frac{3}{4}\pi x^2 = x^2\left(2 + \frac{3}{4}\pi\right)$, or
$\frac{1}{4}x^2(8 + 3\pi)$

87. The shaded area consists of the area of a square with side $x+x+x$, or $3x$, less the area of a semicircle with radius x. It can be expressed as follows:

$$3x \cdot 3x - \frac{1}{2}\pi x^2 = 9x^2 - \frac{1}{2}\pi x^2 = x^2\left(9 - \frac{1}{2}\pi\right)$$

89. $x^2 + 4x + 5x + 20 = x^2 + 9x + 20 = (x+4)(x+5)$

Exercise Set 5.3

1. Since $(6x-1)(2x+3) = 12x^2+16x-3$, choice (c) is correct.

3. Since $(7x+1)(2x-3) = 14x^2-19x-3$, choice (d) is correct.

5. $2x^2 + 7x - 4$

(1) There is no common factor (other than 1 or -1).

(2) Because $2x^2$ can be factored as $2x \cdot x$, we have this possibility:

$(2x + \quad)(x + \quad)$

(3) There are 3 pairs of factors of -4 and they can be listed two ways:

$-4, 1 \quad 4, -1 \quad 2, -2$

and $\quad 1, -4 \quad -1, 4 \quad -2, 2$

(4) Look for Outer and Inner products resulting from steps (2) and (3) for which the sum is $7x$. We can immediately reject all possibilities in which a factor has a common factor, such as $(2x-4)$ or $(2x+2)$, because we determined at the outset that there is no common factor other than 1 and -1. We try some possibilities:

$(2x+1)(x-4) = 2x^2 - 7x - 4$

$(2x-1)(x+4) = 2x^2 + 7x - 4$

The factorization is $(2x-1)(x+4)$.

7. $3x^2 - 17x - 6$

(1) There is no common factor (other than 1 or -1).

(2) Because $3x^2$ can be factored as $3x \cdot x$, we have this possibility:

$(3x + \quad)(x + \quad)$

(3) There are 4 pairs of factors of -6 and they can be listed two ways:

$-6, 1 \quad 6, -1 \quad -3, 2 \quad 3, -2$

and $\quad 1, -6 \quad -1, 6 \quad 2, -3 \quad -2, 3$

(4) Look for Outer and Inner products resulting from steps (2) and (3) for which the sum is $-17x$. We can immediately reject all possibilities in which either factor has a common factor, such as $(3x-6)$ or $(3x+3)$, because at the outset we determined that there is no common factor other than 1 or -1. We try some possibilities:

$(3x+2)(x-3) = 3x^2 - 7x - 6$

$(3x+1)(x-6) = 3x^2 - 17x - 6$

The factorization is $(3x+1)(x-6)$.

9. $4t^2 + 12t + 5$

(1) There is no common factor (other than 1 or -1).

(2) Because $4t^2$ can be factored as $4t \cdot t$ or $2t \cdot 2t$, we have these possibilities:

$(4t + \quad)(t + \quad)$ and $(2t + \quad)(2t + \quad)$

(3) There are 2 pairs of factors of 5 and they can be listed two ways:

$5, 1 \quad -5, -1$

and $\quad 1, 5 \quad -1, -5$

(4) Look for Outer and Inner products resulting from steps (2) and (3) for which the sum is $12t$. We try some possibilities:

$(4t+1)(t+5) = 4t^2 + 21t + 5$

$(2t+1)(2t+5) = 4t^2 + 12t + 5$

The factorization is $(2t+1)(2t+5)$.

11. $15a^2 - 14a + 3$

(1) There is no common factor (other than 1 or -1).

(2) Because $15a^2$ can be factored as $15a \cdot a$ or $5a \cdot 3a$, we have these possibilities:

$(15a + \quad)(a + \quad)$ and $(5a + \quad)(3a + \quad)$

(3) There are 2 pairs of factors of 3 and they can be listed two ways:

$3, 1 \quad -3, -1$

and $\quad 1, 3 \quad -1, -3$

(4) Look for Outer and Inner products resulting from steps (2) and (3) for which the sum is $-14a$. We can immediately reject all possibilities in which either factor has a common factor, such as $(15a+3)$ or $(3a-3)$, because at the outset we determined that there is no common factor other than 1 or -1. Since the sign of the middle term is negative and the sign of the last term is positive, the factors of 3 must both be negative. We try some possibilities:

$(15a-1)(a-3) = 15a^2 - 46a + 3$

$(5a-3)(3a-1) = 15a^2 - 14a + 3$

The factorization is $(5a-3)(3a-1)$.

13. $6x^2 + 17x + 12$

(1) There is no common factor (other than 1 or -1).

(2) Because $6x^2$ can be factored as $6x \cdot x$ and $3x \cdot 2x$, we have these possibilities:

$(6x + \quad)(x + \quad)$ and $(3x + \quad)(2x + \quad)$

(3) Since all coefficients are positive, we need consider only positive pairs of factors of 12. There are 3 pairs and they can be listed two ways:

$1, 12 \quad 2, 6 \quad 3, 4$

and $\quad 12, 1 \quad 6, 2 \quad 4, 3$

(4) We can immediately reject all possibilities in which either factor has a common factor, such as $(6x+12)$ or $(3x+3)$, because at the outset we determined that there is no common factor other than 1 or -1. We try some possibilities:

$(6x+1)(x+12) = 6x^2+73x+12$

$(3x+4)(2x+3) = 6x^2+17x+12$

The factorization is $(3x+4)(2x+3)$.

15. $6x^2 - 10x - 4$

(1) We factor out the largest common factor, 2:

$2(3x^2 - 5x - 2)$.

Then we factor the trinomial $3x^2 - 5x - 2$.

(2) Because $3x^2$ can be factored as $3x \cdot x$, we have this possibility:

$(3x+\quad)(x+\quad)$

(3) There are 2 pairs of factors of -2 and they can be listed two ways:

$-2, 1 \quad 2, -1$

and $\quad 1, -2 \quad -1, 2$

(4) Look for Outer and Inner products resulting from steps (2) and (3) for which the sum is $-5x$. We try some possibilities:

$(3x-2)(x+1) = 3x^2 + x - 2$

$(3x+2)(x-1) = 3x^2 - x - 2$

$(3x+1)(x-2) = 3x^2 - 5x - 2$

The factorization of $3x^2-5x-2$ is $(3x+1)(x-2)$. We must include the common factor in order to get a factorization of the original trinomial.

$6x^2 - 10x - 4 = 2(3x+1)(x-2)$

17. $7t^3 + 15t^2 + 2t$

(1) We factor out the common factor, t:

$t(7t^2 + 15t + 2)$.

Then we factor the trinomial $7t^2 + 15t + 2$.

(2) Because $7t^2$ can be factored as $7t \cdot t$, we have this possibility:

$(7t+\quad)(t+\quad)$

(3) Since all coefficients are positive, we need consider only positive factors of 2. There is only 1 such pair and it can be listed two ways:

$2, 1 \quad 1, 2$

(4) Look for Outer and Inner products resulting from steps (2) and (3) for which the sum is $15t$. We try some possibilities:

$(7t+2)(t+1) = 7t^2 + 9t + 2$

$(7t+1)(t+2) = 7t^2 + 15t + 2$

The factorization of $7t^2+15t+2$ is $(7t+1)(t+2)$. We must include the common factor in order to get a factorization of the original trinomial.

$7t^3 + 15t^2 + 2t = t(7t+1)(t+2)$

19. $10 - 23x + 12x^2 = 12x^2 - 23x + 10$

(1) There is no common factor (other than 1 or -1).

(2) Because $12x^2$ can be factored as $12x \cdot x$, $6x \cdot 2x$, and $4x \cdot 3x$, we have these possibilities:

$(12x+\quad)(x+\quad)$, $(6x+\quad)(2x+\quad)$, and $(4x+\quad)(3x+\quad)$

(3) Since the sign of the middle term is negative but the sign of the last term is positive, we need consider only negative factors of 10.

$-10, -1 \quad -5, -2$

and $\quad -1, -10 \quad -2, -5$

(4) We can immediately reject all possibilities in which either factor has a common factor, such as $(2x-10)$ or $(4x-2)$, because we determined at the outset that there is no common factor other than 1 or -1. We try some possibilities:

$(12x-5)(x-2) = 12x^2 - 29x + 10$

$(4x-1)(3x-10) = 12x^2 - 43x + 10$

$(4x-5)(3x-2) = 12x^2 - 23x + 10$

The factorization is $(4x-5)(3x-2)$.

21. $-35x^2 - 34x - 8$

(1) We factor out -1 in order to have a trinomial with a positive leading coefficient.

$-35x^2 - 34x - 8 = -1(35x^2 + 34x + 8)$

Now we factor $35x^2 + 34x + 8$.

(2) Because $35x^2$ can be factored as $35x \cdot x$ or $7x \cdot 5x$, we have these possibilities:

$(35x+\quad)(x+\quad)$ and $(7x+\quad)(5x+\quad)$

(3) Since all coefficients are positive, we need consider only positive pairs of factors of 8. There are 2 such pairs and they can be listed two ways:

$8, 1 \quad 4, 2$

and $\quad 1, 8 \quad 2, 4$

(4) We try some possibilities:

$(35x+8)(x+1) = 35x^2 + 43x + 8$

$(7x+8)(5x+1) = 35x^2 + 47x + 8$

$(7x+4)(5x+2) = 35x^2 + 34x + 8$

The factorization of $35x^2 + 34x + 8$ is $(7x+4)(5x+2)$.

We must include the factor of -1 in order to get a factorization of the original trinomial.

$-35x^2 - 34x - 8 = -1(7x+4)(5x+2)$, or $-(7x+4)(5x+2)$.

23. $4 + 6t^2 - 13t = 6t^2 - 13t + 4$

(1) There is no common factor (other than 1 or -1).

(2) Because $6t^2$ can be factored as $6t \cdot t$ or $3t \cdot 2t$, we have these possibilities:

$(6t+\quad)(t+\quad)$ and $(3t+\quad)(2t+\quad)$

(3) Since the sign of the middle term is negative but the sign of the last term is positive, we need to consider only negative factors of 4. There is only 1 such pair and it can be listed two ways:

$-4, -1$ and $-1, -4$

(4) We can immediately reject all possibilities in which either factor has a common factor, such as $(6t-4)$ or $(2t-4)$, because we determined at the outset that there is no common factor other than 1 or -1. We try some possibilities:

$$(6t-1)(t-4) = 6t^2 - 25t + 4$$
$$(3t-4)(2t-1) = 6t^2 - 11t + 4$$

These are the only possibilities that do not contain a common factor. Since neither is the desired factorization, we must conclude that $4 + 6t^2 - 13t$ is prime.

25. $25x^2 + 40x + 16$

(1) There is no common factor (other than 1 or -1).

(2) Because $25x^2$ can be factored as $25x \cdot x$ or $5x \cdot 5x$, we have these possibilities:

$$(25x + \quad)(x + \quad) \text{ and } (5x + \quad)(5x + \quad)$$

(3) Since all coefficients are positive, we need consider only positive pairs of factors of 16. There are 3 such pairs and two of them can be listed two ways:

16, 1   8, 2   4, 4

and   1, 16   2, 8

(4) We try some possibilities:

$$(25x+16)(x+1) = 25x^2 + 41x + 16$$
$$(5x+8)(5x+2) = 25x^2 + 50x + 16$$
$$(5x+4)(5x+4) = 25x^2 + 40x + 16$$

The factorization is $(5x+4)(5x+4)$, or $(5x+4)^2$.

27. $20y^2 + 59y - 3$

(1) There is no common factor (other than 1 or -1).

(2) Because $20y^2$ can be factored as $20y \cdot y$, $10y \cdot 2y$, or $5y \cdot 4y$, we have these possibilities:

$$(20y + \quad)(y + \quad) \text{ and } (10y + \quad)(2y + \quad) \text{ and } (5y + \quad)(4y + \quad)$$

(3) There are 2 such pairs of factors of -3, which can be listed two ways:

$-3, 1$ $3, -1$

and $1, -3$ $-1, 3$

(4) Look for Outer and Inner products resulting from steps (2) and (3) for which the sum is 59y. We try some possibilities:

$$(20y-3)(y+1) = 20y^2 + 17y - 3$$
$$(10y-3)(2y+1) = 20y^2 + 4y - 3$$
$$(5y-1)(4y+3) = 20y^2 + 11y - 3$$
$$(20y-1)(y+3) = 20y^2 + 59y - 3$$

The factorization is $(20y-1)(y+3)$.

29. $14x^2 + 73x + 45$

(1) There is no common factor (other than 1 or -1).

(2) Because $14x^2$ can be factored as $14x \cdot x$, and $7x \cdot 2x$, we have two possibilities:

$$(14x + \quad)(x + \quad), \text{ and } (7x + \quad)(2x + \quad)$$

(3) Since all coefficients are positive, we need consider only positive pairs of factors of 45. There are 3 such pairs and they can be listed two ways.

45, 1   15, 3   9, 5

and   1, 45   3, 15   5, 9

(4) Look for Outer and Inner products from steps (2) and (3) for which the sum is $73x$. We try some possibilities:

$$(14x+45)(x+1) = 14x^2 + 59x + 45$$
$$(14x+15)(x+3) = 14x^2 + 57x + 45$$
$$(7x+1)(2x+45) = 14x^2 + 317x + 45$$
$$(7x+5)(2x+9) = 14x^2 + 73x + 45$$

The factorization is $(7x+5)(2x+9)$.

31. $-2x^2 + 15 + x = -2x^2 + x + 15$

(1) We factor out -1 in order to have a trinomial with a positive leading coefficient.

$$-2x^2 + x + 15 = -1(2x^2 - x - 15)$$

Now we factor $2x^2 - x - 15$.

(2) Because $2x^2$ can be factored as $2x \cdot x$ we have this possibility:

$$(2x + \quad)(x + \quad)$$

(3) There are 4 pairs of factors of -15 and they can be listed two ways:

$-15, 1$ $15, -1$ $-5, 3$ $5, -3$

and $1, -15$ $-1, 15$ $3, -5$ $-3, 5$

(4) We try some possibilities:

$$(2x-15)(x+1) = 2x^2 - 13x - 15$$
$$(2x-5)(x+3) = 2x^2 + x - 15$$
$$(2x+5)(x-3) = 2x^2 - x - 15$$

The factorization of $2x^2 - x - 15$ is $(2x+5)(x-3)$. We must include the factor of -1 in order to get a factorization of the original trinomial.

$-2x^2 + 15 + x = -1(2x+5)(x-3)$, or $-(2x+5)(x-3)$

33. $-6x^2 - 33x - 15$

(1) Factor out -3. This not only removes the largest common factor, 3. It also produces a trinomial with a positive leading coefficient.

$$-3(2x^2 + 11x + 5)$$

Then we factor the trinomial $2x^2 + 11x + 5$.

(2) Because $2x^2$ can be factored as $2x \cdot x$ we have this possibility:

$$(2x + \quad)(x + \quad)$$

(3) Since all coefficients are positive, we need consider only positive pairs of factors of 5. There is one such pair and it can be listed two ways:

5, 1   and   1, 5

(4) We try some possibilities:

$$(2x+5)(x+1) = 2x^2 + 7x + 5$$
$$(2x+1)(x+5) = 2x^2 + 11x + 5$$

The factorization of $2x^2 + 11x + 5$ is $(2x+1)(x+5)$. We must include the common factor in order to get a factorization of the original trinomial.

$$-6x^2 - 33x - 15 = -3(2x+1)(x+5)$$

35. $10a^2 - 8a - 18$

(1) Factor out the common factor, 2:

$2(5a^2 - 4a - 9)$

Then we factor the trinomial $5a^2 - 4a - 9$.

(2) Because $5a^2$ can be factored as $5a \cdot a$, we have this possibility:

$(5a+\quad)(a+\quad)$

(3) There are 3 pairs of factors of -9, and they can be listed two ways.

$$-9,1 \quad 9,-1 \quad 3,-3$$

and $\quad 1,-9 \quad -1,9 \quad -3,3$

(4) Look for Outer and Inner products resulting from steps (2) and (3) for which the sum is $-4a$. We try some possibilities:

$$(5a-3)(a+3) = 5a^2 + 12a - 9$$
$$(5a+9)(a-1) = 5a^2 + 4a - 9$$
$$(a+1)(5a-9) = 5a^2 - 4a - 9$$

The factorization of $5a^2 - 4a - 9$ is $(a+1)(5a-9)$. We must include the common factor in order to get a factorization of the original trinomial.

$2(a+1)(5a-9)$

37. $12x^2 + 68x - 24$

(1) Factor out the common factor, 4:

$4(3x^2 + 17x - 6)$

Then we factor the trinomial $3x^2 + 17x - 6$.

(2) Because $3x^2$ can be factored as $3x \cdot x$ we have this possibility:

$(3x+\quad)(x+\quad)$

(3) There are 4 pairs of factors of -6 and they can be listed two ways:

$$6,-1 \quad -6,1 \quad 3,-2 \quad -3,2$$

and $\quad -1,6 \quad 1,-6 \quad -2,3 \quad 2,-3$

(4) We can immediately reject all possibilities in which either factor has a common factor, such as $(3x+6)$ or $(3x-3)$, because we determined at the outset that there is no common factor other than 1 or -1. We try some possibilities:

$$(3x-1)(x+6) = 3x^2 + 17x - 6$$

The factorization of $3x^2+17x-6$ is $(3x-1)(x+6)$. We must include the common factor in order to get a factorization of the original trinomial.

$12x^2 + 68x - 24 = 4(3x-1)(x+6)$

39. $4x + 1 + 3x^2 = 3x^2 + 4x + 1$

(1) There is no common factor (other than 1 or -1).

(2) Because $3x^2$ can be factored as $3x \cdot x$ we have this possibility:

$(3x+\quad)(x+\quad)$

(3) Since all coefficients are positive, we need consider only positive pairs of factors of 1. There is one such pair: 1,1.

(4) We try the possible factorization:

$$(3x+1)(x+1) = 3x^2 + 4x + 1$$

The factorization is $(3x+1)(x+1)$.

41. $x^2 + 3x - 2x - 6 = x(x+3) - 2(x+3)$
$= (x+3)(x-2)$

43. $8t^2 - 6t - 28t + 21 = 2t(4t-3) - 7(4t-3)$
$= (4t-3)(2t-7)$

45. $6x^2 + 4x + 15x + 10 = 2x(3x+2) + 5(3x+2)$
$= (3x+2)(2x+5)$

47. $2y^2 + 8y - y - 4 = 2y(y+4) - 1(y+4)$
$= (y+4)(2y-1)$

49. $6a^2 - 8a - 3a + 4 = 2a(3a-4) - 1(3a-4)$
$= (3a-4)(2a-1)$

51. $16t^2 + 23t + 7$

(1) First note that there is no common factor (other than 1 or -1).

(2) Multiply the leading coefficient, 16, and the constant, 7:

$16 \cdot 7 = 112$

(3) We look for factors of 112 that add to 23. Since all coefficients are positive, we need to consider only positive factors.

Pairs of factors	Sums of factors
1, 112	113
2, 56	58
4, 28	32
8, 14	22
16, 7	23

The numbers we need are 16 and 7.

(4) Rewrite the middle term:

$23t = 16t + 7t$

(5) Factor by grouping:

$$16t^2 + 23t + 7 = 16t^2 + 16t + 7t + 7$$
$$= 16t(t+1) + 7(t+1)$$
$$= (t+1)(16t+7)$$

53. $-9x^2 - 18x - 5$

(1) We factor out -1 in order to have a trinomial with a positive leading coefficient.

$-9x^2 - 18x - 5 = -1(9x^2 + 18x + 5)$

Now we factor $9x^2 + 18x + 5$.

(2) Multiply the leading coefficient, 9, and the constant, 5:

$9 \cdot 5 = 45$

(3) We look for factors of 45 that add to 18. Since all coefficients are positive, we need to consider only positive factors.

Pairs of factors	Sums of factors
1, 45	46
3, 15	18
5, 9	14

The numbers we need are 3 and 15.

(4) Rewrite the middle term:

$$18x = 3x + 15x$$

(5) Factor by grouping:

$$\begin{aligned} 9x^2 + 18x + 5 &= 9x^2 + 3x + 15x + 5 \\ &= 3x(3x+1) + 5(3x+1) \\ &= (3x+1)(3x+5) \end{aligned}$$

We must include the factor of -1 in order to get a factorization of the original trinomial.

$-9x^2 - 18x - 5 = -1(3x+1)(3x+5)$, or $-(3x+1)(3x+5)$

55. $10x^2 + 30x - 70$

(1) Factor out the largest common factor, 10:

$$10x^2 + 30x - 70 = 10(x^2 + 3x - 7)$$

Since x^2+3x-7 is prime, this trinomial cannot be factored further.

57. $18x^3 + 21x^2 - 9x$

(1) Factor out the largest common factor, $3x$:

$$18x^3 + 21x^2 - 9x = 3x(6x^2 + 7x - 3)$$

(2) To factor $6x^2 + 7x - 3$ by grouping we first multiply the leading coefficient, 6, and the constant, -3:

$$6(-3) = -18$$

(3) We look for factors of -18 that add to 7.

Pairs of factors	Sums of factors
−1, 18	17
1, −18	−17
−2, 9	7
2, −9	−7
−3, 6	3
3, −6	−3

The numbers we need are -2 and 9.

(4) Rewrite the middle term:

$$7x = -2x + 9x$$

(5) Factor by grouping:

$$\begin{aligned} 6x^2 + 7x - 3 &= 6x^2 - 2x + 9x - 3 \\ &= 2x(3x-1) + 3(3x-1) \\ &= (3x-1)(2x+3) \end{aligned}$$

The factorization of $6x^2+7x-3$ is $(3x-1)(2x+3)$. We must include the common factor in order to get a factorization of the original trinomial:

$$18x^3 + 21x^2 - 9x = 3x(3x-1)(2x+3)$$

59. $89x + 64 + 25x^2 = 25x^2 + 89x + 64$

(1) First note that there is no common factor (other than 1 or -1).

(2) Multiply the leading coefficient, 25, and the constant, 64:

$$25 \cdot 64 = 1600$$

(3) We look for factors of 1600 that add to 89. Since all coefficients are positive, we need to consider only positive factors. The numbers we need are 25 and 64.

(4) Rewrite the middle term:

$$89x = 25x + 64x$$

(5) Factor by grouping:

$$\begin{aligned} 25x^2 + 89x + 64 &= 25x^2 + 25x + 64x + 64 \\ &= 25x(x+1) + 64(x+1) \\ &= (x+1)(25x+64) \end{aligned}$$

61. $168x^3 + 45x^2 + 3x$

(1) Factor out the largest common factor, $3x$:

$$168x^3 + 45x^2 + 3x = 3x(56x^2 + 15x + 1)$$

(2) To factor $56x^2 + 15x + 1$ we first multiply the leading coefficient, 56, and the constant, 1:

$$56 \cdot 1 = 56$$

(3) We look for factors of 56 that add to 15. Since all coefficients are positive, we need to consider only positive factors. The numbers we need are 7 and 8.

(4) Rewrite the middle term:

$$15x = 7x + 8x$$

(5) Factor by grouping:

$$\begin{aligned} 56x^2 + 15x + 1 &= 56x^2 + 7x + 8x + 1 \\ &= 7x(8x+1) + 1(8x+1) \\ &= (8x+1)(7x+1) \end{aligned}$$

The factorization of $56x^2+15x+1$ is $(8x+1)(7x+1)$. We must include the common factor in order to get a factorization of the original trinomial:

$$168x^3 + 45x^2 + 3x = 3x(8x+1)(7x+1)$$

63. $-14t^4 + 19t^3 + 3t^2$

(1) Factor out $-t^2$. This not only removes the largest common factor, t^2. It also produces a trinomial with a positive leading coefficient.

$$-14t^4 + 19t^3 + 3t^2 = -t^2(14t^2 - 19t - 3)$$

(2) To factor $14t^2 - 19t - 3$ we first multiply the leading coefficient, 14, and the constant, -3:

$$14(-3) = -42$$

(3) We look for factors of -42 that add to -19. The numbers we need are -21 and 2.

(4) Rewrite the middle term:

$$-19t = -21t + 2t$$

(5) Factor by grouping:

$$\begin{aligned} 14t^2 - 19t - 3 &= 14t^2 - 21t + 2t - 3 \\ &= 7t(2t-3) + 1(2t-3) \\ &= (2t-3)(7t+1) \end{aligned}$$

The factorization of $14t^2-19t-3$ is $(2t-3)(7t+1)$. We must include the common factor in order to get a factorization of the original trinomial:

$-14t^4+19t^3+3t^2=-t^2(2t-3)(7t+1)$

65. $132y+32y^2-54=32y^2+132y-54$

(1) Factor out the largest common factor, 2:

$$32y^2+132y-54=2(16y^2+66y-27)$$

(2) To factor $16y^2+66y-27$ we first multiply the leading coefficient, 16, and the constant, -27:

$$16(-27)=-432$$

(3) We look for factors of -432 that add to 66. The numbers we need are 72 and -6.

(4) Rewrite the middle term:

$$66y=72y-6y$$

(5) Factor by grouping:

$$\begin{aligned}16y^2+66y-27&=16y^2+72y-6y-27\\&=8y(2y+9)-3(2y+9)\\&=(2y+9)(8y-3)\end{aligned}$$

The factorization of $16y^2+66y-27$ is $(2y+9)(8y-3)$. We must include the common factor in order to get a factorization of the original trinomial:

$132y+32y^2-54=2(2y+9)(8y-3)$

67. $2a^2-5ab+2b^2$

(1) There is no common factor (other than 1 or -1).

(2) Multiply the leading coefficient, 2, and the constant, 2:

$$2\cdot 2=4$$

(3) We look for factors of 4 that add to -5. The numbers we need are -1 and -4.

(4) Rewrite the middle term:

$$-5ab=-ab-4ab$$

(5) Factor by grouping:

$$\begin{aligned}2a^2-5ab+2b^2&=2a^2-ab-4ab+2b^2\\&=a(2a-b)-2b(2a-b)\\&=(2a-b)(a-2b)\end{aligned}$$

69. $8s^2+22st+14t^2$

(1) Factor out the largest common factor, 2:

$$8s^2+22st+14t^2=2(4s^2+11st+7t^2)$$

(2) Multiply the leading coefficient, 4, and the constant, 7:

$$4\cdot 7=28$$

(3) We look for factors of 28 that add to 11. The numbers we need are 4 and 7.

(4) Rewrite the middle term:

$$11st=4st+7st$$

(5) Factor by grouping:

$$\begin{aligned}4s^2+11st+7t^2&=4s^2+4st+7st+7t^2\\&=4s(s+t)+7t(s+t)\\&=(s+t)(4s+7t)\end{aligned}$$

The factorization of $4s^2+11st+7t^2$ is $(s+t)(4s+7t)$. We must include the common factor in order to get a factorization of the original trinomial:

$8s^2+22st+14t^2=2(s+t)(4s+7t)$

71. $27x^2-72xy+48y^2$

(1) Factor out the largest common factor, 3:

$$27x^2-72xy+48y^2=3(9x^2-24xy+16y^2)$$

(2) To factor $9x^2-24xy+16y^2$, we first multiply the leading coefficient, 9, and the constant, 16:

$$9\cdot 16=144$$

(3) We look for factors of 144 that add to -24. The numbers we need are -12 and -12.

(4) Rewrite the middle term:

$$-24xy=-12xy-12xy$$

(5) Factor by grouping:

$$\begin{aligned}9x^2-24xy+16y^2&=9x^2-12xy-12xy+16y^2\\&=3x(3x-4y)-4y(3x-4y)\\&=(3x-4y)(3x-4y)\end{aligned}$$

The factorization of $9x^2-24xy+16y^2$ is $(3x-4y)(3x-4y)$. We must include the common factor in order to get a factorization of the original trinomial:

$27x^2-72xy+48y^2=3(3x-4y)(3x-4y)$ or, $3(3x-4y)^2$

73. $-24a^2+34ab-12b^2$

(1) Factor out -2. This not only removes the largest common factor, 2. It also produces a trinomial with a positive leading coefficient.

$$-24a^2+34ab-12b^2=-2(12a^2-17ab+6b^2)$$

(2) To factor $12a^2-17ab+6b^2$, we first multiply the leading coefficient, 12, and the constant, 6:

$$12\cdot 6=72$$

(3) We look for factors of 72 that add to -17. The numbers we need are -8 and -9.

(4) Rewrite the middle term:

$$-17ab=-8ab-9ab$$

(5) Factor by grouping:

$$\begin{aligned}12a^2-17ab+6b^2&=12a^2-8ab-9ab+6b^2\\&=4a(3a-2b)-3b(3a-2b)\\&=(3a-2b)(4a-3b)\end{aligned}$$

The factorization of $12a^2-17ab+6b^2$ is $(3a-2b)(4a-3b)$. We must include the common factor in order to get a factorization of the original trinomial:

$-24a^2+34ab-12b^2=-2(3a-2b)(4a-3b)$

75. $19x^3-3x^2+14x^4=14x^4+19x^3-3x^2$

(1) Factor out the largest common factor, x^2:

$$x^2(14x^2+19x-3)$$

(2) To factor $14x^2+19x-3$ by grouping we first multiply the leading coefficient, 14, and the constant, -3:

$$14(-3)=-42$$

(3) We look for factors of -42 that add to 19. The numbers we need are 21 and -2.

(4) Rewrite the middle term:

$$19x = 21x - 2x$$

(5) Factor by grouping:

$$\begin{aligned}14x^2 + 19x - 3 &= 14x^2 + 21x - 2x - 3\\ &= 7x(2x+3) - 1(2x+3)\\ &= (2x+3)(7x-1)\end{aligned}$$

The factorization of $14x^2 + 19x - 3$ is $(2x+3)(7x-1)$. We must include the common factor in order to get a factorization of the original trinomial:

$$19x^3 - 3x^2 + 14x^4 = x^2(2x+3)(7x-1)$$

77. $18a^7 + 8a^6 + 9a^8 = 9a^8 + 18a^7 + 8a^6$

(1) Factor out the largest common factor, a^6:

$$9a^8 + 18a^7 + 8a^6 = a^6(9a^2 + 18a + 8)$$

(2) To factor $9a^2 + 18a + 8$ we first multiply the leading coefficient, 9, and the constant, 8:

$$9 \cdot 8 = 72$$

(3) Look for factors of 72 that add to 18. The numbers we need are 6 and 12.

(4) Rewrite the middle term:

$$18a = 6a + 12a$$

(5) Factor by grouping:

$$\begin{aligned}9a^2 + 18a + 8 &= 9a^2 + 6a + 12a + 8\\ &= 3a(3a+2) + 4(3a+2)\\ &= (3a+2)(3a+4)\end{aligned}$$

The factorization of $9a^2 + 18a + 8$ is $(3a+2)(3a+4)$. We must include the common factor in order to get a factorization of the original trinomial:

$$18a^7 + 8a^6 + 9a^8 = a^6(3a+2)(3a+4)$$

79. *Writing Exercise.* Kay has incorrectly changed the sign of the middle term when factoring out the largest common factor. Thus, the signs in both terms of the final factorization are wrong. The number of points that should be deducted will vary.

81. $$\begin{aligned}(x-2)^2 &= (x)^2 + 2 \cdot x(-2) + (-2)^2\\ &\qquad [(A-B)^2 = A^2 - 2AB + B^2]\\ &= x^2 - 4x + 4\end{aligned}$$

83. $$\begin{aligned}(x+2)(x-2) &= x^2 - 2^2\\ &\qquad [(A+B)(A-B) = A^2 - B^2]\\ &= x^2 - 4\end{aligned}$$

85. $$\begin{aligned}(4a+1)^2 &= (4a)^2 + 2(4a)(1) + 1^2\\ &\qquad [(A+B)^2 = A^2 + 2AB + B^2]\\ &= 16a^2 + 8a + 1\end{aligned}$$

87. $$\begin{aligned}(3c-10)^2 &= (3c)^2 - 2(3c)(10) + 10^2\\ &\qquad [(A-B)^2 = A^2 - 2AB + B^2]\\ &= 9c^2 - 60c + 100\end{aligned}$$

89. $$\begin{aligned}(8n+3)(8n-3) &= (8n)^2 - 3^2\\ &\qquad [(A+B)(A-B) = A^2 - B^2]\\ &= 64n^2 - 9\end{aligned}$$

91. For the trinomial $ax^2 + bx + c$ to be prime:

(1) Show that there is no common factor (other than 1 or -1)

(2) Multiply the leading coefficient, a, and the constant c.

$$a \cdot c = ac$$

(3) Show there are <u>no</u> factors of ac that add to b.

(4) The trinomial is prime.

93. $18x^2y^2 - 3xy - 10$

We will factor by grouping.

(1) There is no common factor (other than 1 or -1).

(2) Multiply the leading coefficient, 18, and the constant, -10:

$$18(-10) = -180$$

(3) We look for factors of -180 that add to -3. The numbers we want are -15 and 12.

(4) Rewrite the middle term:

$$-3xy = -15xy + 12xy$$

(5) Factor by grouping:

$$\begin{aligned}18x^2y^2 - 3xy - 10 &= 18x^2y^2 - 15xy + 12xy - 10\\ &= 3xy(6xy-5) + 2(6xy-5)\\ &= (6xy-5)(3xy+2)\end{aligned}$$

95. $9a^2b^3 + 25ab^2 + 16$

We cannot factor the leading term, $9a^2b^3$, in a way that will produce a middle term with variable factors ab^2, so this trinomial is prime.

97. $16t^{10} - 8t^5 + 1 = 16(t^5)^2 - 8t^5 + 1$

(1) There is no common factor (other than 1 or -1).

(2) Because $16t^{10}$ can be factored as $16t^5 \cdot t^5$ or $8t^5 \cdot 2t^5$ or $4t^5 \cdot 4t^5$, we have these possibilities:

$$(16t^5 + \quad)(t^5 + \quad) \text{ and } (8t^5 + \quad)(2t^5 + \quad)$$

and $(4t^5 + \quad)(4t^5 + \quad)$

(3) Since the last term is positive and the middle term is negative we need consider only negative factors of 1. The only negative pair of factors is -1, -1.

(4) We try some possibilities:

$$(16t^5 - 1)(t^5 - 1) = 16t^{10} - 17t^5 + 1$$
$$(8t^5 - 1)(2t^5 - 1) = 16t^{10} - 10t^5 + 1$$
$$(4t^5 - 1)(4t^5 - 1) = 16t^{10} - 8t^5 + 1$$

The factorization is $(4t^5 - 1)(4t^5 - 1)$, or $(4t^5 - 1)^2$.

99. $-15x^{2m} + 26x^m - 8 = -15(x^m)^2 + 26x^m - 8$

(1) Factor out -1 in order to have a trinomial with a positive leading coefficient.

$$-15x^{2m} + 26x^m - 8 = -1(15x^{2m} - 26x^m + 8)$$

(2) Because $15x^{2m}$ can be factored as $15x^m \cdot x^m$, or $5x^m \cdot 3x^m$, we have these possibilities:

$$(15x^m + \quad)(x^m + \quad) \text{ and } (5x^m + \quad)(3x^m + \quad)$$

(3) Since the last term is positive and the middle term is negative we need consider only negative factors of 8. There are 2 such pairs and they can be listed in two ways:

$-8, -1 \quad -4, -2$

and $\quad -1, -8 \quad -2, -4$

(4) We try some possibilities:

$(15x^m - 8)(x^m - 1) = 15x^{2m} - 9x^m + 8$

$(5x^m - 8)(3x^m - 1) = 15x^{2m} - 29x^m + 8$

$(5x^m - 2)(3x^m - 4) = 15x^{2m} - 26x^m + 8$

The factorization of $15x^{2m} - 26x^m + 8$ is $(5x^m - 2)(3x^m - 4)$. We must include the common factor to get a factorization of the original trinomial.

$-15x^{2m} + 26x^m - 8 = -1(5x^m - 2)(3x^m - 4)$, or $-(5x^m - 2)(3x^m - 4)$

101. $3a^{6n} - 2a^{3n} - 1 = 3(a^{3n})^2 - 2a^{3n} - 1$

(1) There is no common factor (other than 1 or -1).

(2) Because $3a^{6n}$ can be factored as $3a^{3n} \cdot a^{3n}$, we have this possibility:

$(3a^{3n} + \quad)(a^{3n} + \quad)$

(3) The only one pair of factors of -1: 1, -1.

(4) We try some possibilities:

$(3a^{3n} - 1)(a^{3n} + 1) = 3a^{6n} + 2a^{3n} - 1$

$(3a^{3n} + 1)(a^{3n} - 1) = 3a^{6n} - 2a^{3n} - 1$

The factorization of $3a^{6n} - 2a^{3n} - 1$ is $(3a^{3n} + 1)(a^{3n} - 1)$.

103. $7(t-3)^{2n} + 5(t-3)^n - 2 = 7[(t-3)^n]^2 + 5(t-3)^n - 2$

(1) There is no common factor (other than 1 or -1).

(2) Multiply the leading coefficient, 7, and the constant, -2:

$7(-2) = -14$

(3) Look for factors of -14 that add to 5. The numbers we want are 7 and -2.

(4) Rewrite the middle term:

$5(t-3)^n = 7(t-3)^n - 2(t-3)^n$

(5) Factor by grouping:

$$\begin{aligned} &7(t-3)^{2n} + 5(t-3)^n - 2 \\ &= 7(t-3)^{2n} + 7(t-3)^n - 2(t-3)^n - 2 \\ &= 7(t-3)^n[(t-3)^n + 1] - 2[(t-3)^n + 1] \\ &= [(t-3)^n + 1][7(t-3)^n - 2] \end{aligned}$$

The factorization of $7(t-3)^{2n} + 5(t-3)^n - 2$ is

$[(t-3)^n + 1][7(t-3)^n - 2]$

5.3 Connecting the Concepts

1. $6x^5 - 18x^2$ $\quad$ Common factor: $6x^2$

$= 6x^2 \cdot x^3 - 6x^2 \cdot 3$

$= 6x^2(x^3 - 3)$

3. $2x^2 + 13x - 7$ $\quad$ No common factor; factor with FOIL.

$= (x + 7)(2x - 1)$

5. $5x^2 + 40x - 100$ $\quad$ Common factor: 5

$= 5(x^2 + 8x - 20)$ $\quad$ Factor with FOIL

$= 5(x - 2)(x + 10)$

7. $7x^2y - 21xy - 28y$ $\quad$ Common factor: $7y$

$= 7y(x^2 - 3x - 4)$ $\quad$ Factor with FOIL

$= 7y(x - 4)(x + 1)$

9. $b^2 - 14b + 49$ $\quad$ Perfect-square trinomial

$= b^2 - 2 \cdot 7 \cdot b + 7^2$

$= (b - 7)^2$

11. $c^3 + c^2 - 4c - 4$ $\quad$ No common factor; factor with grouping.

$= c^2(c + 1) - 4(c + 1)$

$= (c + 1)(c^2 - 4)$

$= (c + 1)(c + 2)(c - 2)$

13. $t^2 + t - 10$ $\quad$ No common factor

The trinomial is prime.

15. $15p^2 + 16pq + 4q^2$ $\quad$ No common factor; factor with FOIL.

$= (3p + 2q)(5p + 2q)$

17. $x^2 + 4x - 77$ $\quad$ No common factor; factor with FOIL.

$= (x + 11)(x - 7)$

19. $5 + 3x - 2x^2$ $\quad$ Common factor: -1; write in descending order

$= -1(2x^2 - 3x - 5)$ $\quad$ Factor with FOIL

$= -1(2x - 5)(x + 1)$

Exercise Set 5.4

1. $4x^2 + 49$ is not a trinomial. It is not a difference of squares because the terms do not have different signs. There is no common factor, so $4x^2 + 49$ is a prime polynomial.

3. $t^2 - 100 = t^2 - 10^2$, so $t^2 - 100$ is a difference of squares.

5. $9x^2 + 6x + 1 = (3x)^2 + 2 \cdot 3x \cdot 1 + 1^2$, so this is a perfect-square trinomial.

7. $2t^2 + 10t + 6$ does not contain a term that is a square so it is neither a perfect-square trinomial nor a difference of squares. (We could also say that it is not a difference of squares because it is not a binomial.) There is a common factor, 2, so this is not a prime polynomial. Thus it is none of the given possibilities.

9. $16t^2 - 25 = (4t)^2 - 5^2$, so $16t^2 - 25$ is a difference of squares.

11. $x^2 + 18x + 81$

(1) Two terms, x^2 and 81, are squares.

(2) Neither x^2 nor 81 is being subtracted.

(3) Twice the product of the square roots, $2 \cdot x \cdot 9$, is $18x$, the remaining term.

Thus, $x^2 + 18x + 81$ is a perfect-square trinomial.

13. $x^2 - 10x - 25$

(1) Two terms, x^2 and 25, are squares.

(2) There is a minus sign before 25, so $x^2 - 10x - 25$ is not a perfect-square trinomial.

15. $x^2 - 3x + 9$

(1) Two terms, x^2 and 9, are squares.

(2) There is no minus sign before x^2 or 9.

(3) Twice the product of the square roots, $2 \cdot x \cdot 3$, is $6x$. This is neither the remaining term nor its opposite, so $x^2 - 3x + 9$ is not a perfect-square trinomial.

17. $9x^2 + 25 - 30x$

(1) Two terms $9x^2$, and 25, are squares.

(2) Neither $9x^2$ nor 25 is being subtracted.

(3) Twice the product of the square roots, $2 \cdot 3x \cdot 5$, is $30x$, the opposite of the remaining term, $-30x$.

Thus, $9x^2 + 25 - 30x$ is a perfect-square trinomial.

19.
$$\begin{array}{l} x^2 + 16x + 64 \\ = x^2 + 2 \cdot x \cdot 8 + 8^2 = (x+8)^2 \\ \quad \uparrow \quad \uparrow \quad \uparrow \quad \uparrow \quad \uparrow \\ = A^2 + 2 \quad A \quad B + B^2 = (A+B)^2 \end{array}$$

21.
$$\begin{array}{l} x^2 - 10x + 25 \\ = x^2 - 2 \cdot x \cdot 5 + 5^2 = (x-5)^2 \\ \quad \uparrow \quad \uparrow \quad \uparrow \quad \uparrow \quad \uparrow \\ = A^2 - 2 \quad A \quad B + B^2 = (A-B)^2 \end{array}$$

23.
$$\begin{aligned} 5p^2 + 20p + 20 &= 5(p^2 + 4p + 4) \\ &= 5(p^2 + 2 \cdot p \cdot 2 + 2^2) \\ &= 5(p+2)^2 \end{aligned}$$

25.
$$\begin{aligned} 1 - 2t + t^2 &= 1^2 - 2 \cdot 1 \cdot t + t^2 \\ &= (1-t)^2 \end{aligned}$$

We could also factor as follows:
$$\begin{aligned} 1 - 2t + t^2 &= t^2 - 2t + 1 \\ &= t^2 - 2 \cdot t \cdot 1 + 1^2 \\ &= (t-1)^2 \end{aligned}$$

27.
$$\begin{aligned} 18x^2 + 12x + 2 &= 2(9x^2 + 6x + 1) \\ &= 2[(3x)^2 + 2 \cdot 3x \cdot 1 + 1^2] \\ &= 2(3x+1)^2 \end{aligned}$$

29.
$$\begin{aligned} 49 - 56y + 16y^2 &= 16y^2 - 56y + 49 \\ &= (4y)^2 - 2 \cdot 4y \cdot 7 + 7^2 \\ &= (4y-7)^2 \end{aligned}$$

We could also factor as follows:
$$\begin{aligned} 49 - 56y + 16y^2 &= 7^2 - 2 \cdot 7 \cdot 4y + (4y)^2 \\ &= (7-4y)^2 \end{aligned}$$

31.
$$\begin{aligned} -x^5 + 18x^4 - 81x^3 &= -x^3(x^2 - 18x + 81) \\ &= -x^3(x^2 - 2 \cdot x \cdot 9 + 9^2) \\ &= -x^3(x-9)^2 \end{aligned}$$

33.
$$\begin{aligned} 2n^3 + 40n^2 + 200n &= 2n(n^2 + 20n + 100) \\ &= 2n(n^2 + 2 \cdot n \cdot 10 + 10^2) \\ &= 2n(n+10)^2 \end{aligned}$$

35.
$$\begin{aligned} 20x^2 + 100x + 125 &= 5(4x^2 + 20x + 25) \\ &= 5[(2x)^2 + 2 \cdot 2x \cdot 5 + 5^2] \\ &= 5(2x+5)^2 \end{aligned}$$

37. $49 - 42x + 9x^2 = 7^2 - 2 \cdot 7 \cdot 3x + (3x)^2 = (7-3x)^2$, or $(3x-7)^2$

39.
$$\begin{aligned} 16x^2 + 24x + 9 &= (4x)^2 + 2 \cdot 4x \cdot 3 + 3^2 \\ &= (4x+3)^2 \end{aligned}$$

41.
$$\begin{aligned} 2 + 20x + 50x^2 &= 2(1 + 10x + 25x^2) \\ &= 2[1^2 + 2 \cdot 1 \cdot 5x + (5x)^2] \\ &= 2(1+5x)^2, \text{ or } 2(5x+1)^2 \end{aligned}$$

43.
$$\begin{aligned} 9p^2 + 12pq + 4q^2 &= (3p)^2 + 2 \cdot 3p \cdot 2q + (2q)^2 \\ &= (3p+2q)^2 \end{aligned}$$

45. $a^2 - 12ab + 49b^2$

This is not a perfect square trinomial because $-2 \cdot a \cdot 7b = -14ab \neq -12ab$. Nor can it be factored using the methods of Sections 5.2 and 5.3. Thus, it is prime.

47.
$$\begin{aligned} -64m^2 - 16mn - n^2 &= -1(64m^2 + 16mn + n^2) \\ &= -1[(8m)^2 + 2 \cdot 8m \cdot n + n^2] \\ &= -1(8m+n)^2, \text{ or } -(8m+n)^2 \end{aligned}$$

49.
$$\begin{aligned} -32s^2 + 80st - 50t^2 &= -2(16s^2 - 40st + 25t^2) \\ &= -2[(4s)^2 - 2 \cdot 4s \cdot 5t + (5t)^2] \\ &= -2(4s-5t)^2 \end{aligned}$$

51. $x^2 - 100$

(1) The first expression is a square: x^2

The second expression is a square: $100 = 10^2$

(2) The terms have different signs.

Thus, $x^2 - 100$ is a difference of squares $x^2 - 10^2$.

53. $n^4 + 1$

(1) The first expression is a square: $n^4 = (n^2)^2$

The second expression is a square: $1 = 1^2$

(2) The terms do not have different signs.

Thus, $n^4 + 1$ is not a difference of squares.

55. $-1 + 64t^2$ or $64t^2 - 1$

(1) The first term is a square: $1 = 1^2$.

The second term is a square: $64t^2 = (8t)^2$.

(2) The terms have different signs.

Thus, $-1 + 64t^2$ is a difference of squares, $(8t)^2 - 1^2$.

57. $x^2 - 25 = x^2 - 5^2 = (x+5)(x-5)$

59. $p^2 - 9 = p^2 - 3^2 = (p+3)(p-3)$

61. $-49+t^2 = t^2-49 = t^2-7^2 = (t+7)(t-7)$, or $(7+t)(-7+t)$

63. $6a^2 - 24 = 6(a^2-4) = 6(a^2-2^2) = 6(a+2)(a-2)$

65. $49x^2 - 14x + 1 = (7x)^2 - 2\cdot 7x \cdot 1 + 1^2 = (7x-1)^2$

67. $200 - 2t^2 = 2(100 - t^2) = 2(10^2 - t^2)$
$= 2(10+t)(10-t)$

69. $-80a^2 + 45 = -5(16a^2 - 9) = -5[(4a^2) - 3^2]$
$= -5(4a+3)(4a-3)$

71. $5t^2 - 80 = 5(t^2 - 16) = 5(t^2 - 4^2)$
$= 5(t+4)(t-4)$

73. $8x^2 - 162 = 2(4x^2 - 81) = 2[(2x)^2 - 9^2]$
$= 2(2x+9)(2x-9)$

75. $36x - 49x^3 = x(36 - 49x^2) = x[6^2 - (7x)^2]$
$= x(6+7x)(6-7x)$

77. $49a^4 - 20$

There is no common factor (other than 1 or -1). Since 20 is not a square, this is not a difference of squares. Thus, the polynomial is prime.

79. $t^4 - 1$
$= (t^2)^2 - 1^2$
$= (t^2+1)(t^2-1)$
$= (t^2+1)(t+1)(t-1)$ Factoring further; $t^2 - 1$ is a difference of squares

81. $-3x^3 + 24x^2 - 48x = -3x(x^2 - 8x + 16)$
$= -3x(x^2 - 2\cdot x \cdot 4 + 4^2)$
$= -3x(x-4)^2$

83. $75t^3 - 27t = 3t(25t^2 - 9)$
$= 3[(5t)^2 - 3^2]$
$= 3(5t+3)(5t-3)$

85. $a^8 - 2a^7 + a^6 = a^6(a^2 - 2a + 1)$
$= a^6(a^2 - 2\cdot a \cdot 1 + 1^2)$
$= a^6(a-1)^2$

87. $10a^2 - 10b^2 = 10(a^2 - b^2)$
$= 10(a+b)(a-b)$

89. $16x^4 - y^4 = (4x^2)^2 - (y^2)^2$
$= (4x^2+y^2)(4x^2-y^2)$
$= (4x^2+y^2)(2x+y)(2x-y)$

91. $18t^2 - 8s^2 = 2(9t^2 - 4s^2)$
$= 2[(3t)^2 - (2s)^2]$
$= 2(3t+2s)(3t-2s)$

$= (a^2+9b^2)(a+3b)(a-3b)$

93. *Writing Exercise.* Two terms must be squares. There must be no minus sign before either square. The remaining term must be twice the product of the square roots of the squares or must be the opposite of that product.

95. $(2x^2y^4)^3 = 2^3(x^2)^3(y^4)^3 = 8x^6y^{12}$

97. $(x+1)(x+1)(x+1) = (x^2+x+x+1)(x+1)$ FOIL
$= (x^2+2x+1)(x+1)$
$= (x^2+2x+1)\cdot x + (x^2+2x+1)\cdot 1$
$= x^3 + 2x^2 + x + x^2 + 2x + 1$
$= x^3 + 3x^2 + 3x + 1$

99. $(p+q)^3 = (p+q)(p+q)^2$
$= (p+q)(p^2+2pq+q^2)$
$= p(p^2+2pq+q^2) + q(p^2+2pq+q^2)$
$= p^3 + 2p^2q + pq^2 + p^2q + 2pq^2 + q^3$
$= p^3 + 3p^2q + 3pq^2 + q^3$

101. *Writing Exercise.*
$(x+3)(x-3)$
$= (x-3)(x+3)$ Using a commutative law
$= x^2 - 9$

Since $x^2 - 9$ and $x^2 + 9$ are not equivalent the student's factorization of $x^2 + 9$ is incorrect. (Also it can be easily shown by multiplying that $(x+3)(x-3) \neq x^2 + 9$.) The student should recall that, if the greatest common factor has been removed, a sum of squares cannot be factored further.

103. $x^8 - 2^8 = (x^4+2^4)(x^4-2^4)$
$= (x^4+2^4)(x^2+2^2)(x^2-2^2)$
$= (x^4+2^4)(x^2+2^2)(x+2)(x-2)$, or
$(x^4+16)(x^2+4)(x+2)(x-2)$

105. $18x^3 - \frac{8}{25}x = 2x\left(9x^2 - \frac{4}{25}\right)$
$= 2x\left(3x + \frac{2}{5}\right)\left(3x - \frac{2}{5}\right)$

107. $(y-5)^4 - z^8$
$= [(y-5)^2 + z^4][(y-5)^2 - z^4]$
$= [(y-5)^2 + z^4][y-5+z^2][y-5-z^2]$
$= (y^2 - 10y + 25 + z^4)(y-5+z^2)(y-5-z^2)$

109. $-x^4 + 8x^2 + 9 = -1(x^4 - 8x^2 - 9)$
$= -1(x^2-9)(x^2+1)$
$= -1(x+3)(x-3)(x^2+1)$

111. $(y+3)^2 + 2(y+3) + 1$
$= (y+3)^2 + 2\cdot(y+3)\cdot 1 + 1^2$
$= [(y+3)+1]^2$
$= (y+4)^2$

113. $27p^3 - 45p^2 - 75p + 125$
$= 9p^2(3p - 5) - 25(3p - 5)$
$= (3p - 5)(9p^2 - 25)$
$= (3p - 5)(3p + 5)(3p - 5)$, or
$(3p - 5)^2(3p + 5)$

115. $81 - b^{4k} = 9^2 - (b^{2k})^2$
$= (9 + b^{2k})(9 - b^{2k})$
$= (9 + b^{2k})[3^2 - (b^k)^2]$
$= (9 + b^{2k})(3 + b^k)(3 - b^k)$

117. $x^2(x + 1)^2 - (x^2 + 1)^2$
$= x^2(x^2 + 2x + 1) - (x^4 + 2x^2 + 1)$
$= x^4 + 2x^3 + x^2 - x^4 - 2x^2 - 1$
$= 2x^3 + x^2 - 2x^2 - 1$
$= 2x^3 - x^2 - 1$

119. $y^2 + 6y + 9 - x^2 - 8x - 16$
$= (y^2 + 6y + 9) - (x^2 + 8x + 16)$
$= (y + 3)^2 - (x + 4)^2$
$= [(y + 3) + (x + 4)][(y + 3) - (x + 4)]$
$= (y + 3 + x + 4)(y + 3 - x - 4)$
$= (y + x + 7)(y - x - 1)$

121. For $c = a^2$, $2 \cdot a \cdot 3 = 24$. Then $a = 4$, so $c = 4^2 = 16$.

123. $(x + 1)^2 - x^2$
$= [(x + 1) + x][(x + 1) - x]$
$= 2x + 1$
$= (x + 1) + x$

Exercise Set 5.5

1. $x^3 - 1 = (x)^3 - 1^3$
This is a difference of two cubes.

3. $9x^4 - 25 = (3x^2)^2 - 5^2$
This is a difference of two squares.

5. $1000t^3 + 1 = (10t)^3 + 1^3$
This is a sum of two cubes.

7. $25x^2 + 8x$ has a common factor of x so it is not prime, but it does not fall into any of the other categories. It is classified as none of these.

9. $s^{21} - t^{15} = (s^7)^3 - (t^5)^3$
This is a difference of two cubes .

11. $x^3 - 64 = x^3 - 4^3$
$= (x - 4)(x^2 + 4x + 16)$
$A^3 - B^3 = (A - B)(A^2 + AB + B^2)$

13. $z^3 + 1 = z^3 + 1^3$
$= (z + 1)(z^2 - z + 1)$
$A^3 + B^3 = (A + B)(A^2 - AB + B^2)$

15. $t^3 - 1000 = t^3 - 10^3$
$= (t - 10)(t^2 + 10t + 100)$
$A^3 - B^3 = (A - B)(A^2 + AB + B^2)$

17. $27x^3 + 1 = (3x)^3 - 1^3$
$= (3x + 1)(9x^2 - 3x + 1)$
$A^3 + B^3 = (A + B)(A^2 - AB + B^2)$

19. $64 - 125x^3 = 4^3 - (5x)^3 = (4 - 5x)\left(16 + 20x + 25x^2\right)$

21. $x^3 - y^3 = (x - y)\left(x^2 + xy + y^2\right)$

23. $a^3 + \frac{1}{8} = a^3 + \left(\frac{1}{2}\right)^3 = \left(a + \frac{1}{2}\right)\left(a^2 - \frac{1}{2}a + \frac{1}{4}\right)$

25. $8t^3 - 8 = 8\left(t^3 - 1\right) = 8\left(t^3 - 1^3\right) = 8(t - 1)\left(t^2 + t + 1\right)$

27. $54x^3 + 2 = x\left(27x^3 + 1\right) = 2\left[(3x)^3 + 1^3\right]$
$= 2(3x + 1)\left(9x^2 - 3x + 1\right)$

29. $rs^4 + 64rs = rs\left(s^3 + 64\right)$
$= rs\left(s^3 + 4^3\right)$
$= rs(s + 4)\left(s^2 - 4s + 16\right)$

31. $5x^3 - 40z^3 = 5\left(x^3 - 8z^3\right)$
$= 5\left[x^3 - (2z)^3\right]$
$= 5(x - 2z)\left(x^2 + 2xz + 4z^2\right)$

33. $y^3 - \frac{1}{1000} = y^3 - \left(\frac{1}{10}\right)^3 = \left(y - \frac{1}{10}\right)\left(y^2 + \frac{1}{10}y + \frac{1}{100}\right)$

35. $x^3 + 0.001 = x^3 + (0.1)^3 = (x + 0.1)\left(x^2 - 0.1x + 0.01\right)$

37. $64x^6 - 8t^6 = 8\left(8x^6 - t^6\right)$
$= 8\left[\left(2x^2\right)^3 - \left(t^2\right)^3\right]$
$= 8\left(2x^2 - t^2\right)\left(4x^4 + 2x^2t^2 + t^4\right)$

39. $54y^4 - 128y = 2y\left(27y^3 - 64\right)$
$= 2y\left[(3y)^3 - 4^3\right]$
$= 2y(3y - 4)\left(9y^2 + 12y + 16\right)$

41. $z^6 - 1$
$= \left(z^3\right)^2 - 1^2$ Writing as a difference of squares
$= \left(z^3 + 1\right)\left(z^3 - 1\right)$ Factoring a difference of squares
$= (z + 1)\left(z^2 - z + 1\right)(z - 1)\left(z^2 + z + 1\right)$ Factoring a sum and a difference of cubes

43. $t^6 + 64y^6 = \left(t^2\right)^3 + \left(4y^2\right)^3 = \left(t^2 + 4y^2\right)\left(t^4 - 4t^2y^2 + 16y^4\right)$

45. $x^{12} - y^3z^{12} = (x^4)^3 - (yz^4)^3$
$= (x^4 - yz^4)(x^8 + x^4yz^4 + y^2z^8)$

47. *Writing Exercise.*
$(x + y)^3 = (x + y)(x + y)(x + y)$
$= (x^2 + 2xy + y^2)(x + y)$
$= x^3 + 3x^2y + 3xy^2 + y^3$
$\neq x^3 + y^3$

49. Two points on the line are $(-2, -5)$ and $(3, -6)$.

$$m = \frac{-6 - (-5)}{3 - (-2)} = \frac{-1}{5}$$

51. $2x - 5y = 10$

Find the y-intercept:

$$-5y = 10$$
$$y = -2$$

The y-intercept is $(0, -2)$.

Find the x-intercept:

$$2x = 10$$
$$x = 5$$

The x-intercept is $(5, 0)$.

Find find a third point, we replace y with 2 and solve for x.

$$2x - 5 \cdot 2 = 10$$
$$2x - 10 = 10$$
$$2x = 20$$
$$x = 10$$

The point is $(10, 2)$.

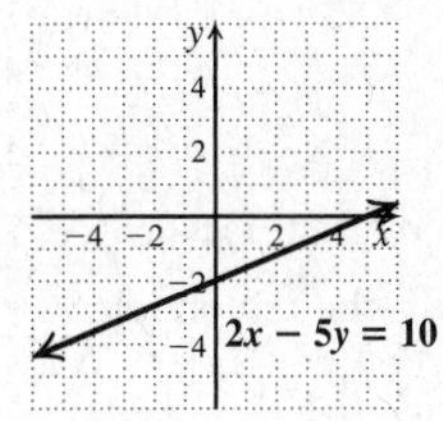

53. Graph: $y = \dfrac{2}{3}x - 1$

Because the equation is in the form $y = mx + b$, we know the y-intercept is $(0, -1)$. We find two other points on the line, substituting multiples of 3 for x to avoid fractions.

When $x = -3$, $y = \dfrac{2}{3}(-3) - 1 = -2 - 1 = -3$.

When $x = 6$, $y = \dfrac{2}{3} \cdot 6 - 1 = 4 - 1 = 3$.

x	y
0	−1
−3	−3
6	3

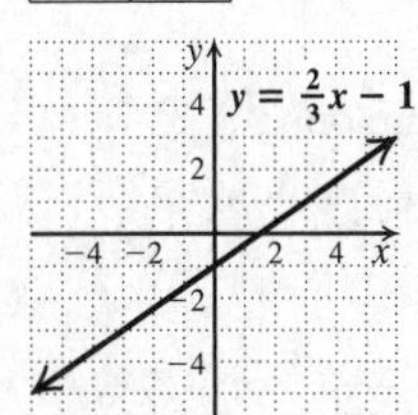

55. *Writing Exercise.* The model shows a cube with volume a^3 from which a portion whose volume is b^3 has been removed. This leaves a remaining volume which can be expressed as $a^2(a - b) + ab(a - b) + b^2(a - b)$, or $(a - b)(a^2 + ab + b^2)$. Thus, $a^3 - b^3 = (a - b)(a^2 + ab + b^2)$.

57. $x^{6a} - y^{3b} = (x^{2a})^3 - (y^b)^3$

59. $(x + 5)^3 + (x - 5)^3$ Sum of cubes

$$= [(x + 5) + (x - 5)][(x + 5)^2 - (x + 5)(x - 5) + (x - 5)^2]$$
$$= 2x[(x^2 + 10x + 25) - (x^2 - 25) + (x^2 - 10x + 25)]$$
$$= 2x(x^2 + 10x + 25 - x^2 + 25 + x^2 - 10x + 25)$$
$$= 2x(x^2 + 75)$$

61. $5x^3y^6 - \frac{5}{8}$

63. $x^{6a} - (x^{2a} + 1)^3$

$$= \left[x^{2a} - (x^{2a} + 1)\right]\left[x^{4a} + x^{2a}(x^{2a} + 1) + (x^{2a} + 1)^2\right]$$
$$= (x^{2a} - x^{2a} - 1)(x^{4a} + x^{4a} + x^{2a} + x^{4a} + 2x^{2a} + 1)$$
$$= -(3x^{4a} + 3x^{2a} + 1)$$

65.

$$t^4 - 8t^3 - t + 8$$
$$= t^3(t - 8) - (t - 8)$$
$$= (t - 8)(t^3 - 1)$$
$$= (t - 8)(t - 1)(t^2 + t + 1)$$

Exercise Set 5.6

1. common factor

3. grouping

5. $5a^2 - 125$

$= 5(a^2 - 25)$ 5 is a common factor.

$= 5(a + 5)(a - 5)$ Factoring the difference of squares

7. $y^2 + 49 - 14y$

$= y^2 - 14y + 49$ Perfect-square trinomial

$= (y - 7)^2$

9. $3t^2 + 16t + 21$

There is no common factor (other than 1 or −1). This trinomial has three terms, but it is not a perfect-square trinomial. Multiply the leading coefficient and the constant, 3 and 21: $3 \cdot 21 = 63$. Try to factor 63 so that the sum of the factors is 16. The numbers we want are 7 and 9: $7 \cdot 9 = 63$ and $7 + 9 = 16$. Split the middle term and factor by grouping.

$$3t^2 + 16t + 21 = 3t^2 + 7t + 9t + 21$$
$$= t(3t + 7) + 3(3t + 7)$$
$$= (3t + 7)(t + 3)$$

11. $x^3 + 18x^2 + 81x$

$= x(x^2 + 18x + 81)$ x is a common factor.

$= x(x^2 + 2 \cdot x \cdot 9 + 9^2)$ Perfect-square trinomial

$= x(x + 9)^2$

13. $x^3 - 5x^2 - 25x + 125$

$= x^2(x - 5) - 25(x - 5)$ Factoring by grouping

$= (x - 5)(x^2 - 25)$

$= (x - 5)(x + 5)(x - 5)$ Factoring the difference of squares

$= (x - 5)^2(x + 5)$

15. $27t^3 - 3t$
$= 3t(9t^2 - 1)$ $3t$ is a common factor.
$= 3t[(3t)^2 - 1^2]$ Difference of squares
$= 3t(3t + 1)(3t - 1)$

17. $9x^3 + 12x^2 - 45x$
$= 3x(3x^2 + 4x - 15)$ $3x$ is a common factor.
$= 3x(x + 3)(3x - 5)$ Factoring the trinomial

19. $t^2 + 25$

The polynomial has no common factor and is not a difference of squares. It is prime.

21. $6y^2 + 18y - 240$
$= 6(y^2 + 3y - 40)$ 6 is a common factor.
$= 6(y + 8)(y - 5)$ Factoring the trinomial

23. $-2a^6 + 8a^5 - 8a^4$
$= -2a^4(a^2 - 4a + 4)$ Factoring out $-2a^4$
$= -2a^4(a - 2)^2$ Factoring the perfect-square trinomial

25. $5x^5 - 80x$
$= 5x(x^4 - 16)$ $5x$ is a common factor.
$= 5x[(x^2)^2 - 4^2]$ Difference of squares
$= 5x(x^2 + 4)(x^2 - 4)$ Difference of squares
$= 5x(x^2 + 4)(x + 2)(x - 2)$

27. $t^4 - 9$ Difference of squares
$= (t^2 + 3)(t^2 - 3)$

29. $-x^6 + 2x^5 - 7x^4$
$= -x^4(x^2 - 2x + 7)$

The trinomial is prime, so this is the complete factorization.

31. $p^2 - q^2$ Difference of squares
$= (p + q)(p - q)$

33. $ax^2 + ay^2 = a(x^2 + y^2)$

35. $= 2\pi rh + 2\pi r^2$ $2\pi r$ is a common factor
$= 2\pi r(h + r)$

37. $(a + b)(5a) + (a + b)(3b)$
$= (a + b)(5a + 3b)$ $(a + b)$ is a common factor.

39. $x^2 + x + xy + y$
$= x(x + 1) + y(x + 1)$ Factoring by grouping
$= (x + 1)(x + y)$

41. $a^2 - 2a - ay + 2y$
$= a(a - 2) - y(a - 2)$ Factoring by grouping
$= (a - 2)(a - y)$

43. $3x^2 + 13xy - 10y^2 = (3x - 2y)(x + 5y)$

45. $8m^3n - 32m^2n^2 + 24mn$
$= 8mn(m^2 - 4mn + 3)$ $8mn$ is a common factor

47. $4b^2 + a^2 - 4ab$
$= 4b^2 - 4ab + a^2$
$= (2b)^2 - 2 \cdot 2b \cdot a + a^2$ Perfect-square trinomial
$= (2b - a)^2$

This result can also be expressed as $(a - 2b)^2$.

49. $16x^2 + 24xy + 9y^2$
$= (4x)^2 + 2 \cdot 4x \cdot 3y + (3y)^2$ Perfect-square trinomial
$= (4x + 3y)^2$

51. $m^2 - 5m + 8$

We cannot find a pair of factors whose product is 8 and whose sum is -5, so $m^2 - 5m + 8$ is prime.

53. $a^4b^4 - 16$
$= (a^2b^2)^2 - 4^2$ Difference of squares
$= (a^2b^2 + 4)(a^2b^2 - 4)$ Difference of squares
$= (a^2b^2 + 4)(ab + 2)(ab - 2)$

55. $80cd^2 - 36c^2d + 4c^3$
$= 4c(20d^2 - 9cd + c^2)$ $4c$ is a common factor
$= 4c(4d - c)(5d - c)$ Factoring the trinomial

57. $3b^2 + 17ab - 6a^2 = (3b - a)(b + 6a)$

59. $-12 - x^2y^2 - 8xy$
$= -x^2y^2 - 8xy - 12$
$= -1(x^2y^2 + 8xy + 12)$
$= -1(xy + 2)(xy + 6)$, or $-(xy + 2)(xy + 6)$

61. $5p^2q^2 + 25pq - 30$
$= 5(p^2q^2 + 5pq - 6)$ 5 is a common factor
$= 5(pq + 6)(pq - 1)$ Factoring the trinomial

63. $4ab^5 - 32b^4 + a^2b^6$
$= b^4(4ab - 32 + a^2b^2)$ b^4 is a common factor.
$= b^4(a^2b^2 + 4ab - 32)$
$= b^4(ab + 8)(ab - 4)$ Factoring the trinomial

65. $x^6 + x^5y - 2x^4y^2$
$= x^4(x^2 + xy - 2y^2)$ x^4 is a common factor.
$= x^4(x + 2y)(x - y)$ Factoring the trinomial

67. $36a^2 - 15a + \frac{25}{16}$
$= (6a)^2 - 2 \cdot 6a \cdot \frac{5}{4} + \left(\frac{5}{4}\right)^2$ Perfect-square trinomial
$= \left(6a - \frac{5}{4}\right)^2$

69. $\frac{1}{81}x^2 - \frac{8}{27}x + \frac{16}{9}$
$= \left(\frac{1}{9}x\right)^2 - 2 \cdot \frac{1}{9}x \cdot \frac{4}{3} + \left(\frac{4}{3}\right)^2$ Perfect-square trinomial
$= \left(\frac{1}{9}x - \frac{4}{3}\right)^2$

If we had factored out $\frac{1}{9}$ at the outset, the final result would have been $\frac{1}{9}\left(\frac{1}{3}x-4\right)^2$.

71. $1-16x^{12}y^{12}$
$= (1+4x^6y^6)(1-4x^6y^6)$ Difference of squares
$= (1+4x^6y^6)(1+2x^3y^3)(1-2x^3y^3)$ Difference of squares

73. $4a^2b^2+12ab+9$
$= (2ab)^2+2\cdot 2ab\cdot 3+3^2$ Perfect-square trinomial
$= (2ab+3)^2$

75. $z^4+6z^3-6z^2-36z$
$= z(z^3+6z^2-6z-36)$ z is a common factor
$= z\left[z^2(z+6)-6(z+6)\right]$ Factoring by grouping
$= z(z+6)(z^2-6)$

77. *Writing Exercise.* Both are correct; $(x-4)^2 = x^2-8x+16 = 16-8x+x^2 = (4-x)^2$

79. $8x-9=0$
$8x=9$ Adding 9 to both sides
$x=\frac{9}{8}$ Dividing both sides by 8

The solution is $\frac{9}{8}$.

81. $2x+7=0$
$2x=-7$ Subtracting 7 from both sides
$x=-\frac{7}{2}$ Dividing both sides by 2

The solution is $-\frac{7}{2}$.

83. $3-x=0$
$3=x$ Adding x to both sides

The solution is 3.

85. $2x-5=8x+1$
$-5=6x+1$ Subtracting $2x$ from both sides
$-6=6x$ Subtracting 1 from both sides
$-1=x$ Dividing both sides by 6

The solution is -1.

87. *Writing Exercise.* Find the product of a binomial and a trinomial. One example is found as follows:

$$(x+1)(2x^2-3x+4)$$
$$= 2x^3-3x^2+4x+2x^2-3x+4$$
$$= 2x^3-x^2+x+4$$

89. $-(x^5+7x^3-18x)$
$= -x(x^4+7x^2-18)$
$= -x(x^2+9)(x^2-2)$

91. $-x^4+7x^2+18$
$= -1(x^4-7x^2-18)$
$= -1(x^2+2)(x^2-9)$
$= -1(x^2+2)(x+3)(x-3)$, or
$-(x^2+2)(x+3)(x-3)$

93. $y^2(y+1)-4y(y+1)-21(y+1)$
$= (y+1)(y^2-4y-21)$
$= (y+1)(y-7)(y+3)$

95. $(y+4)^2+2x(y+4)+x^2$
$= (y+4)^2+2\cdot(y+4)\cdot x+x^2$ Perfect-square trinomial
$= [(y+4)+x]^2$
$= (y+4+x)^2$

97. $2(a+3)^4-(a+3)^3(b-2)-(a+3)^2(b-2)^2$
$= (a+3)^2[2(a+3)^2-(a+3)(b-2)-(b-2)^2]$
$= (a+3)^2[2(a+3)+(b-2)][(a+3)-(b-2)]$
$= (a+3)^2(2a+6+b-2)(a+3-b+2)$
$= (a+3)^2(2a+b+4)(a-b+5)$

99. $49x^4+14x^2+1-25x^6$
$= (7x^2+1)^2-25x^6$ Perfect-square trinomial
$= \left[(7x^2+1)-5x^3\right]\left[(7x^2+1+5x^3)\right]$ Difference of squares
$= (7x^2+1-5x^3)(7x^2+1+5x^3)$

Exercise Set 5.7

1. Equations of the type $ax^2+bx+c=0$, with $a\neq 0$, are quadratic, so choice (c) is correct.

3. The principle of zero products states that $A\cdot B=0$ if and only if $A=0$ or $B=0$, so choice (d) is correct.

5. $(x+2)(x+9)=0$

We use the principle of zero products.

$x+2=0$ *or* $x+9=0$

$x=-2$ *or* $x=-9$

Check:

For -2:

$(x+2)(x+9)=0$	
$(-2+2)(-2+9)$	0
$0\cdot 7$	
$0 \stackrel{?}{=} 0$ TRUE	

For -9:

$(x+2)(x+9)=0$	
$(-9+2)(-9+9)$	0
$-7\cdot 0$	
$0 \stackrel{?}{=} 0$ TRUE	

The solutions are -2 and -9.

7. $(2t-3)(t+6)=0$

$2t-3=0 \quad or \quad t+6=0$

$2t=3 \quad or \quad t=-6$

$t=\frac{3}{2} \quad or \quad t=-6$

The solutions are $\frac{3}{2}$ and -6.

9. $4(7x-1)(10x-3)=0$

$(7x-1)(10x-3)=0$ Dividing both sides by 4

$7x-1=0 \quad or \quad 10x-3=0$

$7x=1 \quad or \quad 10x=3$

$x=\frac{1}{7} \quad or \quad x=\frac{3}{10}$

The solutions are $\frac{1}{7}$ and $\frac{3}{10}$.

11. $x(x-7)=0$

$x=0 \quad or \quad x-7=0$

$x=0 \quad or \quad x=7$

The solutions are 0 and 7.

13. $\left(\frac{2}{3}x-\frac{12}{11}\right)\left(\frac{7}{4}x-\frac{1}{12}\right)=0$

$\frac{2}{3}x-\frac{12}{11}=0 \quad or \quad \frac{7}{4}x-\frac{1}{12}=0$

$\frac{2}{3}x=\frac{12}{11} \quad or \quad \frac{7}{4}x=\frac{1}{12}$

$x=\frac{3}{2}\cdot\frac{12}{11} \quad or \quad x=\frac{4}{7}\cdot\frac{1}{12}$

$x=\frac{18}{11} \quad or \quad x=\frac{1}{21}$

The solutions are $\frac{18}{11}$ and $\frac{1}{21}$.

15. $6n(3n+8)=0$

$6n=0 \quad or \quad 3n+8=0$

$n=0 \quad or \quad 3n=-8$

$n=0 \quad or \quad n=-\frac{8}{3}$

The solutions are 0 and $-\frac{8}{3}$.

17. $(20-0.4x)(7-0.1x)=0$

$20-0.4x=0 \quad or \quad 7-0.1x=0$

$-0.4x=-20 \quad or \quad -0.1x=-7$

$x=50 \quad or \quad x=70$

The solutions are 50 and 70.

19. $(3x-2)(x+5)(x-1)=0$

$3x-2=0 \quad or \quad x+5=0 \quad or \quad x-1=0$

$3x=2 \quad or \quad x=-5 \quad or \quad x=1$

$x=\frac{2}{3} \quad or \quad x=-5 \quad or \quad x=1$

The solutions are -5, $\frac{2}{3}$, and 1.

21. $x^2-7x+6=0$

$(x-6)(x-1)=0$ Factoring

$x-6=0 \quad or \quad x-1=0$ Using the principle of zero products

$x=6 \quad or \quad x=1$

The solutions are 6 and 1.

23. $x^2+4x-21=0$

$(x-3)(x+7)=0$ Factoring

$x-3=0 \quad or \quad x+7=0$ Using the principle of zero products

$x=3 \quad or \quad x=-7$

The solutions are 3 and -7.

25. $n^2+11n+18=0$

$(n+9)(n+2)=0$

$n+9=0 \quad or \quad n+2=0$

$n=-9 \quad or \quad n=-2$

The solutions are -9 and -2.

27. $x^2-10x=0$

$x(x-10)=0$

$x=0 \quad or \quad x-10=0$

$x=0 \quad or \quad x=10$

The solutions are 0 and 10.

29. $6t+t^2=0$

$t(6+t)=0$

$t=0 \quad or \quad 6+t=0$

$t=0 \quad or \quad t=-6$

The solutions are 0 and -6.

31. $x^2-36=0$

$(x+6)(x-6)=0$

$x+6=0 \quad or \quad x-6=0$

$x=-6 \quad or \quad x=6$

The solutions are -6 and 6.

33. $4t^2=49$

$4t^2-49=0$

$(2t+7)(2t-7)=0$

$2t+7=0 \quad or \quad 2t-7=0$

$2t=-7 \quad or \quad 2t=7$

$t=\frac{-7}{2} \quad or \quad t=\frac{7}{2}$

The solutions are $\frac{-7}{2}$ and $\frac{7}{2}$.

35. $0=25+x^2+10x$

$0=x^2+10x+25$ Writing in descending order

$0=(x+5)(x+5)$

$x+5=0 \quad or \quad x+5=0$

$x=-5 \quad or \quad x=-5$

The solution is -5.

37. $64 + x^2 = 16x$

$x^2 - 16x + 64 = 0$

$(x-8)(x-8) = 0$

$x - 8 = 0$ *or* $x - 8 = 0$

$x = 8$ *or* $x = 8$

The solution is 8.

39. $4t^2 = 8t$

$4t^2 - 8t = 0$

$4t(t-2) = 0$

$t = 0$ *or* $t - 2 = 0$

$t = 0$ *or* $t = 2$

The solutions are 0 and 2.

41. $4y^2 = 7y + 15$

$4y^2 - 7y - 15 = 0$

$(4y+5)(y-3) = 0$

$4y + 5 = 0$ *or* $y - 3 = 0$

$4y = -5$ *or* $y = 3$

$y = \frac{-5}{4}$ *or* $y = 3$

The solutions are $\frac{-5}{4}$ and 3.

43. $(x-7)(x+1) = -16$

$x^2 - 6x - 7 = -16$

$x^2 - 6x + 9 = 0$

$(x-3)(x-3) = 0$

$x - 3 = 0$ *or* $x - 3 = 0$

$x = 3$ *or* $x = 3$

The solution is 3.

45. $15z^2 + 7 = 20z + 7$

$15z^2 - 20z + 7 = 7$

$15z^2 - 20z = 0$

$5z(3z - 4) = 0$

$5z = 0$ *or* $3z - 4 = 0$

$z = 0$ *or* $3z = 4$

$z = 0$ *or* $z = \frac{4}{3}$

The solutions are 0 and $\frac{4}{3}$.

47. $36m^2 - 9 = 40$

$36m^2 - 49 = 0$

$(6m+7)(6m-7) = 0$

$6m + 7 = 0$ *or* $6m - 7 = 0$

$6m = -7$ *or* $6m = 7$

$m = \frac{-7}{6}$ *or* $m = \frac{7}{6}$

The solutions are $\frac{-7}{6}$ or $\frac{7}{6}$.

49. $(x+3)(3x+5) = 7$

$3x^2 + 14x + 15 = 7$

$3x^2 + 14x + 8 = 0$

$(3x+2)(x+4) = 0$

$3x + 2 = 0$ *or* $x + 4 = 0$

$3x = -2$ *or* $x = -4$

$x = \frac{-2}{3}$ *or* $x = -4$

The solutions are $-\frac{2}{3}$ and -4.

51. $3x^2 - 2x = 9 - 8x$

$3x^2 + 6x - 9 = 0$ Adding $8x$ and -9

$3(x^2 + 2x - 3) = 0$

$3(x+3)(x-1) = 0$

$x + 3 = 0$ *or* $x - 1 = 0$

$x = -3$ *or* $x = 1$

The solutions are -3 and 1.

53. $x^2(2x-1) = 3x$

$2x^3 - x^2 = 3x$

$2x^3 - x^2 - 3x = 0$

$x(2x^2 - x - 3) = 0$

$x(2x-3)(x+1) = 0$

$x = 0$ *or* $2x - 3 = 0$ *or* $x + 1 = 0$

$x = 0$ *or* $2x = 3$ *or* $x = \frac{3}{2}$

$x = 0$ *or* $x = \frac{3}{2}$ *or* $x = -1$

The solutions are -1, 0, and $\frac{3}{2}$.

55. $(2x-5)(3x^2 + 29x + 56) = 0$

$(2x-5)(3x+8)(x+7) = 0$

$2x - 5 = 0$ *or* $3x + 8 = 0$ *or* $x + 7 = 0$

$2x = 5$ *or* $3x = -8$ *or* $x = -7$

$x = \frac{5}{2}$ *or* $x = -\frac{8}{3}$ *or* $x = -7$

The solutions are -7, $-\frac{8}{3}$, and $\frac{5}{2}$.

57. The solutions of the equation are the first coordinates of the x-intercepts of the graph. From the graph we see that the x-intercepts are $(-1, 0)$ and $(4, 0)$, so the solutions of the equation are -1 and 4.

59. The solutions of the equation are the first coordinates of the x-intercepts of the graph. From the graph we see that the x-intercepts are $(-3, 0)$ and $(2, 0)$, so the solutions of the equation are -3 and 2.

61. We let $y = 0$ and solve for x.

$0 = x^2 - x - 6$

$0 = (x-3)(x+2)$

$x - 3 = 0$ *or* $x + 2 = 0$

$x = 3$ *or* $x = -2$

The x-intercepts are (3,0) and $(-2, 0)$.

63. We let $y = 0$ and solve for x.

$$0 = x^2 + 2x - 8$$
$$0 = (x+4)(x-2)$$
$$x + 4 = 0 \quad or \quad x - 2 = 0$$
$$x = -4 \quad or \quad x = 2$$

The x-intercepts are $(-4, 0)$ and $(2, 0)$.

65. We let $y = 0$ and solve for x.

$$0 = 2x^2 + 3x - 9$$
$$0 = (2x-3)(x+3)$$
$$2x - 3 = 0 \quad or \quad x + 3 = 0$$
$$2x = 3 \quad or \quad x = -3$$
$$x = \frac{3}{2} \quad or \quad x = -3$$

The x-intercepts are $\left(\frac{3}{2}, 0\right)$ and $(-3, 0)$.

67. *Writing Exercise.* The graph has no x-intercepts.

69. Let m and n represent the numbers. The sum of the two numbers is $m + n$. Thus the square of the sum of the two numbers is $(m + n)^2$.

71. Let x represent the first integer. Then $x + 1$ represents the second integer. Thus, the product is $x(x + 1)$.

73. ***Familiarize.*** We draw a picture. We let x = the measure of the second angle. Then $4x$ = the measure of the first angle, and $x - 30$ = the measure of the third angle.

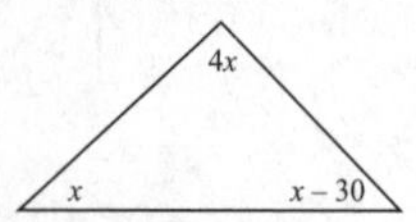

Recall that the measures of the angles of any triangle add up to 180°.

Translate.

Measure of first angle	+	Measure of 2nd angle	+	Measure of third angle	is 180°
$4x$	+	x	+	$(x-30)$	$= 180$

Carry out. we solve the equation

$$4x + x + (x - 30) = 180$$
$$6x - 30 = 180$$
$$6x = 210$$
$$x = 35$$

Possible answers for the angle measures are as follows:
First angle: $4x = 4(35°) = 140°$
Second angle: $x = 35$ deg
Third angle: $x - 30 = 35 - 30 = 5$ deg.

Check. Consider 140°, 35°and 5°. The first angle is four times the first, and the third is 30°less than the second. The sum of the angles is 180°. These numbers check.

State. The measure of the first angle is 140°, the measure of the second angle is 35°, and the measure of the third angle is 5°.

75. *Writing Exercise.* One solution of the equation is 0. Dividing both sides of the equation by x, leaving the solution $x = 3$, is equivalent to dividing by 0.

77. a)

$$x = -4 \quad or \quad x = 5$$
$$x + 4 = 0 \quad or \quad x - 5 = 0$$
$$(x+4)(x-5) = 0 \quad \text{Principle of zero products}$$
$$x^2 - x - 20 = 0 \quad \text{Multiplying}$$

b)

$$x = -1 \quad or \quad x = 7$$
$$x + 1 = 0 \quad or \quad x - 7 = 0$$
$$(x+1)(x-7) = 0$$
$$x^2 - 6x - 7 = 0$$

c)

$$x = \frac{1}{4} \quad or \quad x = 3$$
$$x - \frac{1}{4} = 0 \quad or \quad x - 3 = 0$$
$$\left(x - \frac{1}{4}\right)(x - 3) = 0$$
$$x^2 - \frac{13}{4}x + \frac{3}{4} = 0$$
$$4\left(x^2 - \frac{13}{4}x + \frac{3}{4}\right) = 4 \cdot 0 \quad \text{Multiplying both sides by 4}$$
$$4x^2 - 13x + 3 = 0$$

d)

$$x = \frac{1}{2} \quad or \quad x = \frac{1}{3}$$
$$x - \frac{1}{2} = 0 \quad or \quad x - \frac{1}{3} = 0$$
$$\left(x - \frac{1}{2}\right)\left(x - \frac{1}{3}\right) = 0$$
$$x^2 - \frac{5}{6}x + \frac{1}{6} = 0$$
$$6x^2 - 5x + 1 = 0 \quad \text{Multiplying by 6}$$

e)

$$x = \frac{2}{3} \quad or \quad x = \frac{3}{4}$$
$$x - \frac{2}{3} = 0 \quad or \quad x - \frac{3}{4} = 0$$
$$\left(x - \frac{2}{3}\right)\left(x - \frac{3}{4}\right) = 0$$
$$x^2 - \frac{17}{12}x + \frac{1}{2} = 0$$
$$12x^2 - 17x + 6 = 0 \quad \text{Multiplying by 12}$$

f)

$$x = -1 \quad or \quad x = 2 \quad or \quad x = 3$$
$$x + 1 = 0 \quad or \quad x - 2 = 0 \quad or \quad x - 3 = 0$$
$$(x+1)(x-2)(x-3) = 0$$
$$(x^2 - x - 2)(x - 3) = 0$$
$$x^3 - 4x^2 + x + 6 = 0$$

79.

$$a(9 + a) = 4(2a + 5)$$
$$9a + a^2 = 8a + 20$$
$$a^2 + a - 20 = 0 \quad \text{Subtracting } 8a \text{ and } 20$$
$$(a+5)(a-4) = 0$$

$$a+5=0 \quad or \quad a-4=0$$
$$a=-5 \quad or \quad a=4$$

The solutions are -5 and 4.

81. $$-x^2+\frac{9}{25}=0$$
$$x^2-\frac{9}{25}=0 \quad \text{Multiplying by } -1$$
$$\left(x-\frac{3}{5}\right)\left(x+\frac{3}{5}\right)=0$$
$$x-\frac{3}{5}=0 \quad or \quad x+\frac{3}{5}=0$$
$$x=\frac{3}{5} \quad or \quad x=-\frac{3}{5}$$

The solutions are $\frac{3}{5}$ and $-\frac{3}{5}$.

83. $(t+1)^2=9$

Observe that $t+1$ is a number which yields 9 when it is squared. Thus, we have

$$t+1=-3 \quad or \quad t+1=3$$
$$t=-4 \quad or \quad t=2$$

The solutions are -4 and 2.

We could also do this exercise as follows:

$$(t+1)^2=9$$
$$t^2+2t+1=9$$
$$t^2+2t-8=0$$
$$(t+4)(t-2)=0$$
$$t+4=0 \quad or \quad t-2=0$$
$$t=-4 \quad or \quad t=2$$

Again we see that the solutions are -4 and 2.

85. a) $2(x^2+10x-2)=2\cdot 0$ Multiplying (a) by 2

$2x^2+20x-4=0$

(a) and $2x^2+20x-4=0$ are equivalent.

b) $(x-6)(x+3)=x^2-3x-18$ Multiplying

(b) and $x^2-3x-18=0$ are equivalent.

c) $5x^2-5=5(x^2-1)=5(x+1)(x-1)=$
$(x+1)5(x-1)=(x+1)(5x-5)$

(c) and $(x+1)(5x-5)=0$ are equivalent.

d) $2(2x-5)(x+4)=2\cdot 0$ Multiplying (d) by 2

$2(x+4)(2x-5)=0$

$(2x+8)(2x-5)=0$

(d) and $(2x+8)(2x-5)=0$ are equivalent.

e) $4(x^2+2x+9)=4\cdot 0$ Multiplying (e) by 4

$4x^2+8x+36=0$

(e) and $4x^2+8x+36=0$ are equivalent.

f) $3(3x^2-4x+8)=3\cdot 0$ Multiplying (f) by 3

$9x^2-12x+24=0$

(f) and $9x^2-12x+24=0$ are equivalent.

87. *Writing Exercise.* Graph $y=-x^2-x+6$ and $y=4$ on the same set of axes. The first coordinates of the points of intersection of the two graphs are the solutions of $-x^2-x+6=4$.

89. 2.33, 6.77

91. $-4.59, -9.15$

93. -3.76, 0

5.7 Connecting the Concepts

1. Expression

3. Equation

5. Expression

7. $(2x^3-5x+1)+(x^2-3x-1)$
$=2x^3+x^2+(-5-3)x+(1-1)$
$=2x^3+x^2-8x$

9. $$t^2-100=0$$
$$(t+10)(t-10)=0$$
$$t+10=0 \quad or \quad t-10=0$$
$$t=-10 \quad or \quad t=10$$

The solutions are -10 and 10.

11. $n^2-10n+9=(n-1)(n-9)$ Factor with FOIL

13. $$4t^2+20t+25=0$$
$$(2t)^2+2\cdot 2t\cdot 5+5^2=0$$
$$(2t+5)(2t+5)=0$$
$$2t+5=0 \quad or \quad 2t+5=0$$
$$2t=-5 \quad or \quad 2t=-5$$
$$t=\frac{-5}{2} \quad or \quad t=\frac{-5}{2}$$

The solutions is $\frac{-5}{2}$.

15. $16x^2-81=(4x+9)(4x-9)$ Difference of squares

17. $(a^2-2)-(5a^2+a+9)$
$=a^2-2-5a^2-a-9$
$=(1-5)a^2-a+(-2-9)$
$=-4a^2-a-11$

19. $$3x^2+5x+2=0$$
$$(x+1)(3x+2)=0$$
$$x+1=0 \quad or \quad 3x+2=0$$
$$x=-1 \quad or \quad x=\frac{-2}{3}$$

The solutions are -1 and $-\frac{2}{3}$.

Exercise Set 5.8

1. ***Familiarize***. Let x = the number.

Translate. We reword the problem.

$$\underbrace{\text{The square of a number}}_{\displaystyle x^2} \quad \underset{\displaystyle -}{\text{minus}} \quad \underbrace{\text{the number}}_{\displaystyle x} \quad \underset{\displaystyle =}{\text{is}} \quad \underset{\displaystyle 6}{6.}$$

Carry out. We solve the equation.

$$x^2 - x = 6$$
$$x^2 - x - 6 = 0$$
$$(x-3)(x+2) = 0$$

$$x - 3 = 0 \quad or \quad x + 2 = 0$$
$$x = 3 \quad or \quad x = -2$$

Check. For 3: The square of 3 is 3^2, or 9, and $9 - 3 = 6$.

For -2: The square of -2, or 4 and $4 - (-2) = 4 + 2 = 6$. Both numbers check.

State. The numbers are 3 and -2.

3. ***Familiarize***. Let x = the length of the shorter leg, in m. Then $x + 2$ = the length of the longer leg.

Translate. We use the Pythagorean theorem.

$$a^2 + b^2 = c^2$$
$$x^2 + (x+2)^2 = 10^2$$

Carry out. We solve the equation.

$$x^2 + (x+2)^2 = 10^2$$
$$x^2 + x^2 + 4x + 4 = 100$$
$$2x^2 + 4x + 4 = 100$$
$$2x^2 + 4x - 96 = 0$$
$$2(x^2 + 2x - 48) = 0$$
$$2(x+8)(x-6) = 0$$

$$x + 8 = 0 \quad or \quad x - 6 = 0$$
$$x = -8 \quad or \quad x = 6$$

Check. The number -8 cannot be the length of a side because it is negative. When $x = 6$, then $x + 2 = 8$, and $6^2 + 8^2 = 36 + 64 = 100 = 10^2$, so the number 6 checks.

State.The lengths of the sides are 6 m, 8 m, and 10 m.

5. ***Familiarize***. The parking spaces are consecutive integers. Let x = the smaller integer. Then $x+1$ = the larger integer.

Translate. We reword the problem.

$$\underbrace{\text{Smaller integer}}_{\displaystyle x} \quad \underset{\displaystyle \cdot}{\text{times}} \quad \underbrace{\text{larger integer}}_{\displaystyle (x+1)} \quad \underset{\displaystyle =}{\text{is}} \quad \underset{\displaystyle 132}{132.}$$

Carry out. We solve the equation.

$$x(x+1) = 132$$
$$x^2 + x = 132$$
$$x^2 + x - 132 = 0$$
$$(x+12)(x-11) = 0$$

$$x + 12 = 0 \quad or \quad x - 11 = 0$$
$$x = -12 \quad or \quad x = 11$$

Check. The solutions of the equation are -12 and 11. Since a parking space number cannot be negative, we only need to check 11. When $x = 11$, then $x + 1 = 12$, and $11 \cdot 12 = 132$. This checks.

State. The parking space numbers are 11 and 12.

7. ***Familiarize***. Let x = the smaller even integer. Then $x + 2$ = the larger even integer.

Translate. We reword the problem.

$$\underbrace{\text{Smaller even integer}}_{\displaystyle x} \quad \underset{\displaystyle \cdot}{\text{times}} \quad \underbrace{\text{larger even integer}}_{\displaystyle (x+2)} \quad \underset{\displaystyle =}{\text{is}} \quad \underset{\displaystyle 168}{168.}$$

Carry out.

$$x(x+2) = 168$$
$$x^2 + 2x = 168$$
$$x^2 + 2x - 168 = 0$$
$$(x+14)(x-12) = 0$$

$$x + 14 = 0 \quad or \quad x - 12 = 0$$
$$x = -14 \quad or \quad x = 12$$

Check. The solutions of the equation are -14 and 12. When x is -14, then $x + 2$ is -12 and $-14(-12) = 168$. The numbers -14 and -12 are consecutive even integers which are solutions of the problem. When x is 12, then $x + 2 = 14$ and $12 \cdot 14 = 168$. The numbers 12 and 14 are also consecutive even integers which are solutions of the problem.

State. We have two solutions, each of which consists of a pair of numbers: -14 and -12 or 12 and 14.

9. ***Familiarize***. Let w = the width of the porch, in feet. Then $5w$ = the length. Recall that the area of a rectangle is Length $\cdot$ Width.

Translate.

$$\underbrace{\text{The area of the rectangle}}_{\displaystyle 5w \cdot w} \quad \underset{\displaystyle =}{\text{is}} \quad \underbrace{180\text{ ft}^2.}_{\displaystyle 180}$$

Carry out. We solve the equation.

$$5w \cdot w = 180$$
$$5w^2 = 180$$
$$5w^2 - 180 = 0$$
$$5(w^2 - 36) = 0$$
$$5(w+6)(w-6) = 0$$

$$w + 6 = 0 \quad or \quad w - 6 = 0$$
$$w = -6 \quad or \quad w = 6$$

Check. Since the width must be positive, -6 cannot be a solution. If the width is 6 ft, then the length is $5 \cdot 6$ ft, or 30 ft, and the area is $6 \text{ ft} \cdot 30 \text{ ft} = 180 \text{ ft}^2$. Thus, 6 checks.

State. The porch is 30 ft long and 6 ft wide.

11. ***Familiarize***. We make a drawing. Let $w =$ the width, in cm. Then $w + 2 =$ the length, in cm.

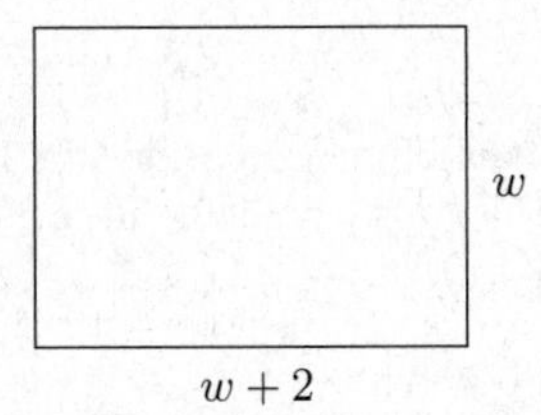

Recall that the area of a rectangle is length times width.

Translate. We reword the problem.

$$\underbrace{\text{Length}}\ \text{times}\ \text{width}\ \text{is}\ \underbrace{24 \text{ cm}^2}$$
$$(w+2) \cdot w = 24$$

Carry out. We solve the equation.

$$(w+2)w = 24$$
$$w^2 + 2w = 24$$
$$w^2 + 2w - 24 = 0$$
$$(w+6)(w-4) = 0$$
$$w + 6 = 0 \quad or \quad w - 4 = 0$$
$$w = -6 \quad or \quad w = 4$$

Check. Since the width must be positive, -6 cannot be a solution. If the width is 4 cm, then the length is $4 + 2$, or 6 cm, and the area is $6 \cdot 4$, or 24 cm^2. Thus, 4 checks.

State. The width is 4 cm, and the length is 6 cm.

13. ***Familiarize***. Using the labels shown on the drawing in the text, we let $b =$ the base, in inches, and $b - 3 =$ the height, in inches. Recall that the formula for the area of a triangle is $\frac{1}{2} \cdot (\text{base}) \cdot (\text{height})$.

Translate.

$$\frac{1}{2}\ \text{times}\ \text{base}\ \text{times}\ \text{height}\ \text{is}\ \underbrace{54 \text{ in}^2}$$
$$\frac{1}{2} \cdot (b) \cdot (b-3) = 54$$

Carry out.

$$\frac{1}{2}b(b-3) = 54$$
$$b(b-3) = 108 \quad \text{Multiplying by 2}$$
$$b^2 - 3b = 108$$
$$b^2 - 3b - 108 = 0$$
$$(b-12)(b+9) = 0$$
$$b - 12 = 0 \quad or \quad b + 9 = 0$$
$$b = 12 \quad or \quad b = -9$$

Check. Since the height must be positive, -9 cannot be a solution. If the base is 12 in, then the height is $12 - 3$, or 9 in, and the area is $\frac{1}{2} \cdot 12 \cdot 9$, or 54 in^2. Thus, 12 checks.

State. The base of the triangle is 12 in, and the height is 9 in.

15. ***Familiarize***. Using the labels show on the drawing in the text, we let $x =$ the length of the foot of the sail, in ft, and $x + 5 =$ the height of the sail, in ft. Recall that the formula for the area of a triangle is $\frac{1}{2} \cdot (\text{base}) \cdot (\text{height})$.

Translate.

$$\frac{1}{2}\ \text{times}\ \text{base}\ \text{times}\ \text{height}\ \text{is}\ \underbrace{42 \text{ ft}^2}$$
$$\frac{1}{2} \cdot x \cdot (x+5) = 42$$

Carry out.

$$\frac{1}{2}x(x+5) = 42$$
$$x(x+5) = 84 \quad \text{Multiplying by 2}$$
$$x^2 + 5x = 84$$
$$x^2 + 5x - 84 = 0$$
$$(x+12)(x-7) = 0$$
$$x + 12 = 0 \quad or \quad x - 7 = 0$$
$$x = -12 \quad or \quad x = 7$$

Check. The solutions of the equation are -12 and 7. The length of the base of a triangle cannot be negative, so -12 cannot be a solution. Suppose the length of the foot of the sail is 7 ft. Then the height is $7 + 5$, or 12 ft, and the area is $\frac{1}{2} \cdot 7 \cdot 12$, or 42 ft^2. These numbers check.

State. The length of the foot of the sail is 7 ft, and the height is 12 ft.

17. ***Familiarize and Translate***. We substitute 150 for A in the formula.

$$A = -50t^2 + 200t$$
$$150 = -50t^2 + 200t$$

Carry out. We solve the equation.

$$150 = -50t^2 + 200t$$
$$0 = -50t^2 + 200t - 150$$
$$0 = -50(t^2 - 4t + 3)$$
$$0 = -50(t-1)(t-3)$$
$$t - 1 = 0 \quad or \quad t - 3 = 0$$
$$t = 1 \quad or \quad t = 3$$

Check. Since $-50 \cdot 1^2 + 200 \cdot 1 = -50 + 200 = 150$, the number 1 checks. Since $-50 \cdot 3^2 + 200 \cdot 3 = -450 + 600 = 150$, the number 3 checks also.

State. There will be about 150 micrograms of Albuterol in the bloodstream 1 minute and 3 minutes after an inhalation.

19. ***Familiarize***. We will use the formula $N = 0.3t^2 + 0.6t$.

Translate. Substitute 36 for N.

$$36 = 0.3t^2 + 0.6t$$

Carry out.

$$36 = 0.3t^2 + 0.6t$$
$$360 = 3t^2 + 6t \quad \text{Multiplying by 10}$$
$$0 = 3t^2 + 6t - 360$$
$$0 = 3(t^2 + 2t - 120)$$
$$0 = 3(t + 12)(t - 10)$$
$$t + 12 = 0 \quad \text{or} \quad t - 10 = 0$$
$$t = -12 \quad \text{or} \quad t = 10$$

Check. The solutions of the equation are -12 and 10. Since the number of years cannot be negative, -12 cannot be a solution. However, 10 checks since $0.3(10)^2 + 0.6(10) = 30 + 6 = 36$.

State. 10 years after 1998 is the year 2008.

21. ***Familiarize***. We will use the formula $x^2 - x = N$.

Translate. Substitute 240 for N.

$$x^2 - x = 240$$

Carry out.

$$x^2 - x = 240$$
$$x^2 - x - 240 = 0$$
$$(x - 16)(x + 15) = 0$$
$$x - 16 = 0 \quad \text{or} \quad x + 15 = 0$$
$$x = 16 \quad \text{or} \quad x = -15$$

Check. The solutions of the equation are 16 and -15. Since the number of teams cannot be negative, -15 cannot be a solution. But 16 checks since $16^2 - 16 = 256 - 16 = 240$.

State. There are 16 teams in the league.

23. ***Familiarize***. We will use the formula $H = \frac{1}{2}(n^2 - n)$.

Translate. Substitute 12 for n.

$$H = \frac{1}{2}(12^2 - 12)$$

Carry out. We do the computation on the right.

$$H = \frac{1}{2}(12^2 - 12)$$
$$H = \frac{1}{2}(144 - 12)$$
$$H = \frac{1}{2}(132)$$
$$H = 66$$

Check. We can recheck the computation, or we can solve the equation $66 = \frac{1}{2}(n^2 - n)$. The answer checks.

State. 66 handshakes are possible.

25. ***Familiarize***. We will use the formula $H = \frac{1}{2}(n^2 - n)$, since "high fives" can be substituted for handshakes.

Translate. Substitute 66 for H.

$$66 = \frac{1}{2}(n^2 - n)$$

Carry out.

$$66 = \frac{1}{2}(n^2 - n)$$
$$132 = n^2 - n \quad \text{Multiplying by 2}$$
$$0 = n^2 - n - 132$$
$$0 = (n - 12)(n + 11)$$
$$n - 12 = 0 \quad or \quad n + 11 = 0$$
$$n = 12 \quad or \quad n = -11$$

Check. The solutions of the equation are 12 and -11. Since the number of players cannot be negative, -11 cannot be a solution. However, 12 checks since $\frac{1}{2}(12^2 - 12) = \frac{1}{2}(144 - 12) = \frac{1}{2}(132) = 66$.

State. 12 players were on the team.

27. ***Familiarize***. Let h = the vertical height to which each brace reaches, in feet. We have a right triangle with hypotenuse 15 ft and legs 12 ft and h.

Translate. We use the Pythagorean theorem.

$$a^2 + b^2 = c^2$$
$$12^2 + h^2 = 15^2$$

Carry out. We solve the equation.

$$12^2 + h^2 = 15^2$$
$$144 + h^2 = 225$$
$$h^2 - 81 = 0$$
$$(h + 9)(h - 9) = 0$$
$$h + 9 = 0 \quad \text{or} \quad h - 9 = 0$$
$$h = -9 \quad \text{or} \quad h = 9$$

Check. Since the vertical height must be positive, -9 cannot be a solution. If the height is 9 ft, then we have $12^2 + 9^2 = 144 + 81 = 225 = 15^2$. The number 9 checks.

State. Each brace reaches 9 ft vertically.

29. ***Familiarize***.Let w = the width of Main Street, in ft. We have a right triangle with hypotenuse 40 ft and legs of 24 ft and w.

Translate. We use the Pythagorean theorem.

$$a^2 + b^2 = c^2$$

Carry out.

$$24^2 + w^2 = 40^2$$
$$576 + w^2 = 1600$$
$$w^2 - 1024 = 0$$
$$(w - 32)(w + 32) = 0$$
$$w - 32 = 0 \quad or \quad w + 32 = 0$$
$$w = 32 \quad or \quad w = -32$$

Check. Since the width must be positive, -32 cannot be a solution. If the width is 32 ft, then we have $24^2 + 32^2 = 576 + 1024 = 1600 = 40^2$. The number 32 checks.

State. The width of Main Street is 32 ft.

31. ***Familiarize***. Let l = the length of the leg, in ft. Then $l+200$ = the length of the hypotenuse in feet.

Translate. We use the Pythagorean theorem.

$$a^2+b^2=c^2$$
$$400^2+l^2=(l+200)^2$$

Carry out.

$$400^2+l^2=l^2+400l+40,000$$
$$160,000+l^2=l^2+400l+40,000$$
$$120,000=400l$$
$$300=l$$

Check. When $l=300$, then $l+200=500$, and $400^2+300^2=160,000+90,000=250,000=500^2$, so the number 300 checks.

State. The dimensions of the garden are 300 ft by 400 ft by 500 ft.

33. ***Familiarize***. We label the drawing. Let x = the length of a side of the dining room, in ft. Then the dining room has dimensions x by x and the kitchen has dimensions x by 10. The entire rectangular space has dimension x by $x+10$. Recall that we multiply these dimensions to find the area of the rectangle.

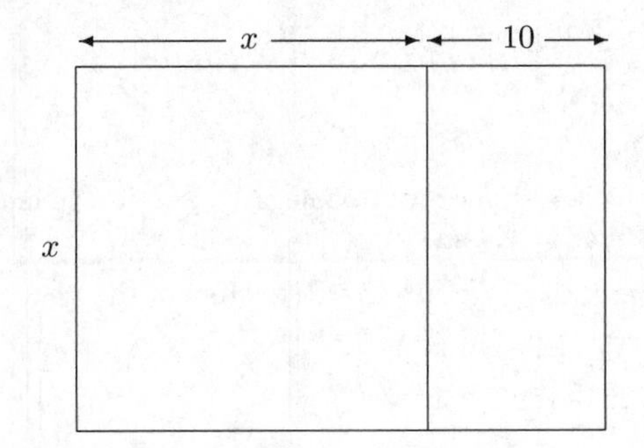

Translate.

The area of the rectangular space is 264 ft^2.

$$x(x+10) = 264$$

Carry out. We solve the equation.

$$x(x+10)=264$$
$$x^2+10x=264$$
$$x^2+10x-264=0$$
$$(x+22)(x-12)=0$$
$$x+22=0 \quad or \quad x-12=0$$
$$x=-22 \quad or \quad x=12$$

Check. Since the length of a side of the dining room must be positive, -22 cannot be a solution. If x is 12 ft, then $x+10$ is 22 ft, and the area of the space is $12\cdot 22$, or 264 ft^2. The number 12 checks.

State. The dining room is 12 ft by 12 ft, and the kitchen is 12 ft by 10 ft.

35. ***Familiarize***. We will use the formula $h=48t-16t^2$.

Translate. Substitute $\frac{1}{2}$ for t.

$$h=48\cdot\frac{1}{2}-16\left(\frac{1}{2}\right)^2$$

Carry out. We do the computation on the right.

$$h=48\cdot\frac{1}{2}-16\left(\frac{1}{2}\right)^2$$
$$h=48\cdot\frac{1}{2}-16\cdot\frac{1}{4}$$
$$h=24-4$$
$$h=20$$

Check. We can recheck the computation, or we can solve the equation $20=48t-16t^2$. The answer checks.

State. The rocket is 20 ft high $\frac{1}{2}$ sec after it is launched.

37. ***Familiarize***. We will use the formula $h=48t-16t^2$.

Translate. Substitute 32 for h.

$$32=48t-16t^2$$

Carry out. We solve the equation.

$$32=48t-16t^2$$
$$0=-16t^2+48t-32$$
$$0=-16(t^2-3t+2)$$
$$0=-16(t-1)(t-2)$$
$$t-1=0 \quad or \quad t-2=0$$
$$t=1 \quad or \quad t=2$$

Check. When $t=1$, $h=48\cdot 1-16\cdot 1^2=48-16=32$. When $t=2$, $h=48\cdot 2-16\cdot 2^2=96-64=32$. Both numbers check.

State. The rocket will be exactly 32 ft above the ground at 1 sec and at 2 sec after it is launched.

39. *Writing Exercise.* No; if we cannot factor the quadratic expression ax^2+bx+c, $a\neq 0$, then we cannot solve the quadratic equation $ax^2+bx+c=0$.

41. $\frac{-3}{5}\cdot\frac{4}{7}=-\frac{3\cdot 4}{5\cdot 7}=\frac{-12}{35}$

43. $\frac{-5}{6}-\frac{1}{6}=\frac{-5}{6}+\frac{-1}{6}=-1$

45. $\frac{-3}{8}\cdot\left(\frac{-10}{15}\right)=\frac{\cancel{3}\cdot\cancel{2}\cdot\cancel{5}}{\cancel{2}\cdot 4\cdot\cancel{3}\cdot\cancel{5}}=\frac{1}{4}$

47. $\frac{5}{24}+\frac{3}{28}=\frac{5}{24}\cdot\frac{7}{7}+\frac{3}{28}\cdot\frac{6}{6}$

$=\frac{35}{168}+\frac{18}{168}=\frac{53}{168}$

49. *Writing Exercise.* She could use the measuring sticks to draw a right angle as shown below. Then she could use the 3-ft and 4-ft sticks to extend one leg to 7 ft and the 4-ft and 5-ft sticks to extend the other leg to 9 ft.

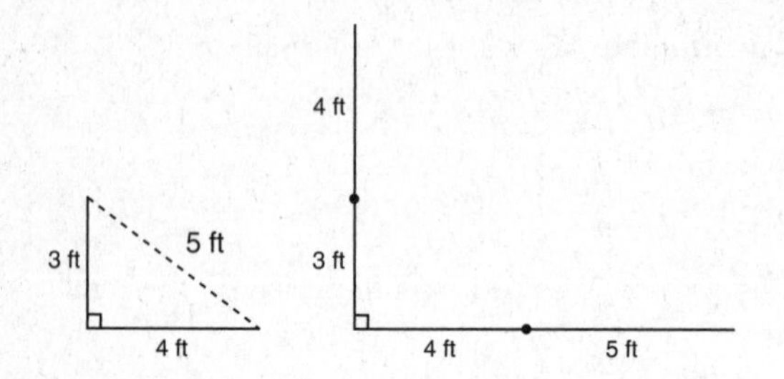

Next she could draw another right angle with either the 7-ft side or the 9-ft side as a side.

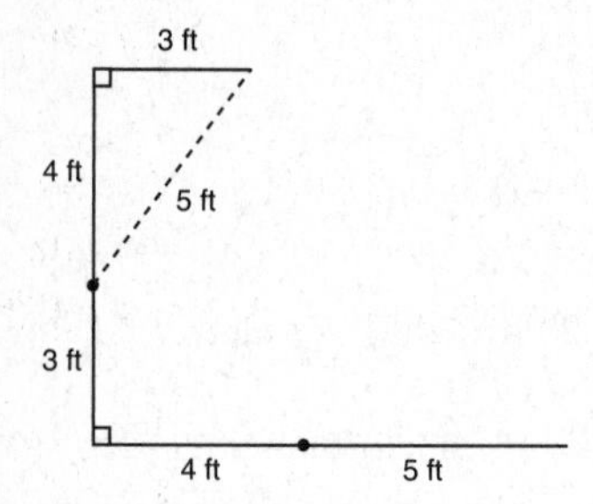

Then she could use the sticks to extend the other side to the appropriate length. Finally she would draw the remaining side of the rectangle.

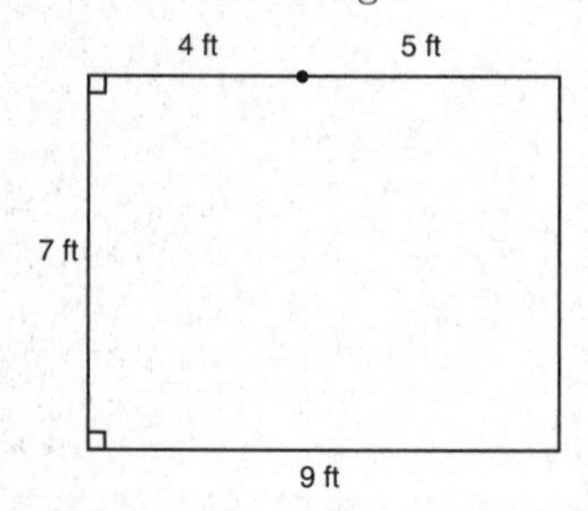

51. ***Familiarize***. First we find the length of the other leg of the right triangle. Then we find the area of the triangle, and finally we multiply by the cost per square foot of the sailcloth. Let $x =$ the length of the other leg of the right triangle, in feet.

Translate. We use the Pythagorean theorem to find x.

$$a^2 + b^2 = c^2$$

$$x^2 + 24^2 = 26^2 \quad \text{Substituting}$$

Carry out.

$$x^2 + 24^2 = 26^2$$
$$x^2 + 576 = 676$$
$$x^2 - 100 = 0$$
$$(x + 10)(x - 10) = 0$$
$$x + 10 = 0 \quad or \quad x - 10 = 0$$
$$x = -10 \quad or \quad x = 10$$

Since the length of the leg must be positive, -10 cannot be a solution. We use the number 10. Find the area of the triangle:

$$\frac{1}{2}bh = \frac{1}{2} \cdot 10 \text{ ft} \cdot 24 \text{ ft} = 120 \text{ ft}^2$$

Finally, we multiply the area, 120 ft^2, by the price per square foot of the sailcloth, \$1.50:

$$120 \cdot (1.50) = 180$$

Check. Recheck the calculations. The answer checks.

State. A new main sail costs \$180.

53. ***Familiarize***. We add labels to the drawing in the text.

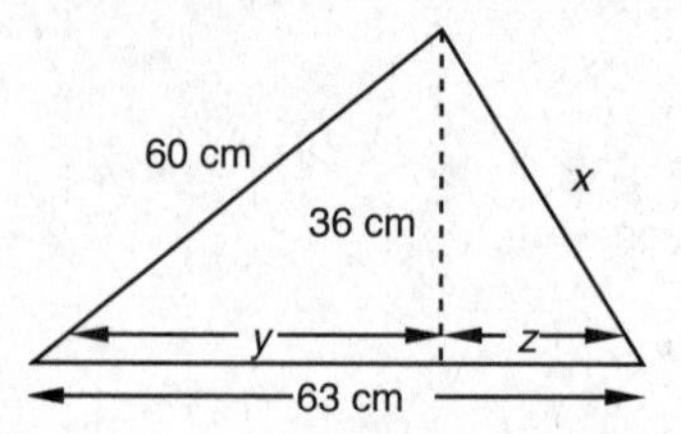

First we will use the Pythagorean theorem to find y. Then we will subtract to find z and, finally, we will use the Pythagorean theorem again to find x.

Translate. We use the Pythagorean theorem to find y.

$$a^2 + b^2 = c^2$$

$$y^2 + 36^2 = 60^2 \quad \text{Substituting}$$

Carry out.

$$y^2 + 36^2 = 60^2$$
$$y^2 + 1296 = 3600$$
$$y^2 - 2304 = 0$$
$$(y + 48)(y - 48) = 0$$
$$y + 48 = 0 \quad or \quad y - 48 = 0$$
$$y = -48 \quad or \quad y = 48$$

Since the length y cannot be negative, we use 48 cm. Then $z = 63 - 48 = 15$ cm.

Now we find x. We use the Pythagorean theorem again.

$$15^2 + 36^2 = x^2$$
$$225 + 1296 = x^2$$
$$1521 = x^2$$
$$0 = x^2 - 1521$$
$$0 = (x + 39)(x - 39)$$
$$x + 39 = 0 \quad or \quad x - 39 = 0$$
$$x = -39 \quad or \quad x = 39$$

Since the length x cannot be negative, we use 39 cm.

Check. We repeat all of the calculations. The answer checks.

State. The value of x is 39 cm.

55. ***Familiarize***. Let $w =$ the width of the side turned up. Then $20 - 2w =$ the length, in inches of the base.

Recall that we multiply these dimensions to find the area of the rectangle.

Translate.

$$\underbrace{\text{The area of the rectangular cross-section}}_{w(20-2w)} \quad \underbrace{\text{is}}_{=} \quad \underbrace{48 \text{ in}^2}_{48}.$$

$$w(20 - 2w) = 48$$

Carry out. We solve the equation.

$$w(20-2w)=48$$
$$0=2w^2-20w+48$$
$$0=2(w^2-10w+24)$$
$$0=2(w-6)(w-4)$$

$w-6=0$ *or* $w-4=0$

$w=6$ *or* $w=4$

Check. If $w=6$ in., $20-2(6)=8$ in. and the area is

6 in. $\cdot$ 8 in. = 48 in.^{2}

If $w=4$ in., $20-2(4)=12$ in. and the area is

4 in. $\cdot$ 12 in. = 48 in.^{2}

State. The possible depths of the gutter are 4 in. or 6 in.

57. ***Familiarize***. First we can use the Pythagorean theorem to find x, in ft. Then the height of the telephone pole is $x+5$.

Translate. We use the Pythagorean theorem.

$$a^2+b^2=c^2$$
$$\left(\frac{1}{2}x+1\right)^2+x^2=34^2$$

Carry out. We solve the equation.

$$\left(\frac{1}{2}x+1\right)^2+x^2=34^2$$
$$\frac{1}{4}x^2+x+1+x^2=1156$$
$$x^2+4x+4+4x^2=4624 \quad \text{Multiplying by 4}$$
$$5x^2+4x+4=4624$$
$$5x^2+4x-4620=0$$
$$(5x+154)(x-30)=0$$

$5x+154=0$ *or* $x-30=0$

$5x=-154$ *or* $x=30$

$x=-30.8$ *or* $x=30$

Check. Since the length x must be positive, -30.8 cannot be a solution. If x is 30 ft, then $\frac{1}{2}x+1$ is $\frac{1}{2}\cdot 30+1$, or 16 ft. Since $16^2+30^2=1156=34^2$, the number 30 checks. When x is 30 ft, then $x+5$ is 35 ft.

State. The height of the telephone pole is 35 ft.

59. First substitute 18 for N in the given formula.

$$18=-0.009t(t-12)^3$$

Graph $y_1=18$ and $y_2=-0.009x(x-12)^3$ in the given window and use the TRACE feature to find the first coordinates of the points of intersection of the graphs. We find $x\approx 2$ hr and $x\approx 4.2$ hr.

61. Graph $y=-0.009x(x-12)^3$ and use the TRACE feature to find the first coordinate of the highest point on the graph. We find $x=3$ hr.

Chapter 5 Review

1. False. The largest common variable factor is the smallest power of the variable in the polynomial.

3. True. see p. 329

5. False. Some quadratic equations have two different solutions.

7. True. see p 357

9. $20x^3=(4\cdot 5)\left(x\cdot x^2\right)=(-2x)\left(-10x^2\right)=(20x)\left(x^2\right)$

11. $12x^4-18x^3=6x^3\cdot 2x-6x^3\cdot 3=6x^3(2x-3)$

13. $100t^2-1=(10t+1)(10t-1)$

15. $x^2+14x+49=(x+7)(x+7)=(x+7)^2$

17. $6x^3+9x^2+2x+3$
$=3x^2\cdot 2x+3x^2\cdot 3+2x\cdot 1+3\cdot 1$
$=3x^2(2x+3)+1(2x+3)$
$=(2x+3)\left(3x^2+1\right)$

19. $25t^2+9-30t=25t^2-30t+9$
$=(5t-3)(5t-3)=(5t-3)^2$

21. $81a^4-1=\left(9a^2+1\right)\left(9a^2-1\right)$
$=\left(9a^2+1\right)(3a+1)(3a-1)$

23. $2x^3-250$
$=2(x^3-125)$
$=2(x^3-5^3)$
$=2(x-5)(x^2+5x+25)$

25. $a^2b^4-64=\left(ab^2+8\right)\left(ab^2-8\right)$

27. $75+12x^2-60x=12x^2-60x+75$
$=3\left(4x^2-20x+25\right)$
$=3(2x-5)(2x-5)=3(2x-5)^2$

29. $-t^3+t^2+42t=-t\left(t^2-t-42\right)$
$=-t(t-7)(t+6)$

31. $n^2-60-4n=n^2-4n-60$
$=(n-10)(n+6)$

33. $4t^2+13t+10=(4t+5)(t+2)$

35. $7x^3+35x^2+28x=7x\left(x^2+5x+4\right)$
$=7x(x+1)(x+4)$

37. $20x^2-20x+5=5\left(4x^2-4x+1\right)$
$=5(2x-1)(2x-1)$
$=5(2x-1)^2$

39. $15-8x+x^2=x^2-8x+15$
$=(x-3)(x-5)$

41. $x^2y^2+6xy-16=(xy+8)(xy-2)$

43. $m^2+5m+mt+5t=m(m+5)+t(m+5)$
$=(m+5)(m+t)$

45. $6m^2 + 2mn + n^2 + 3mn = 2m(3m+n) + n(n+3m)$
$= (3m+n)(2m+n)$

47. $(x-9)(x+11) = 0$

$x - 9 = 0$ *or* $x + 11 = 0$

$x = 9$ *or* $x = -11$

49.
$$16x^2 = 9$$
$$16x^2 - 9 = 0$$
$$(4x+3)(4x-3) = 0$$

$4x + 3 = 0$ *or* $4x - 3 = 0$

$x = \frac{-3}{4}$ *or* $x = \frac{3}{4}$

51.
$$2x^2 - 7x = 30$$
$$2x^2 - 7x - 30 = 0$$
$$(2x+5)(x-6) = 0$$

$2x + 5 = 0$ *or* $x - 6 = 0$

$x = \frac{-5}{2}$ *or* $x = 6$

53.
$$9t - 15t^2 = 0$$
$$3t(3-5t) = 0$$

$3t = 0$ *or* $3 - 5t = 0$

$t = 0$ *or* $t = \frac{3}{5}$

55. ***Familiarize*** Let $n =$ the number.

Translate.

The number squared is 12 more than the number

$n^2 \quad = 12 \quad + \quad n$

Carry out. We solve the equation.

$$n^2 = 12 + n$$
$$n^2 - n - 12 = 0$$
$$(n-4)(n+3) = 0$$

$n - 4 = 0$ *or* $n + 3 = 0$

$n = 4$ *or* $n = -3$

Check.

$4^2 = 12 + 4$

$16 = 16$ True

$(-3)^2 = 12 + (-3)$

$9 = 9$ True

State. The solutions are 4 and -3.

57. We let $y = 0$ and solve for x

$0 = 2x^2 - 3x - 5$

$0 = (2x-5)(x+1)$

$2x - 5 = 0$ *or* $x + 1 = 0$

$x = \frac{5}{2}$ *or* $x = -1$

The x-intercepts are $\left(\frac{5}{2}, 0\right)$ and $(-1, 0)$

59. ***Familiarize.*** Let $d =$ the diagonal of the brace.

Translate. We use the Pythagorean theorem.

$$a^2 + b^2 = c^2$$
$$8^2 + 6^2 = d^2$$

Carry out. We solve the equation.

$$8^2 + 6^2 = c^2$$
$$8^2 + 6^2 = d^2$$
$$100 = d^2$$
$$0 = d^2 - 100$$
$$0 = (d+10)(d-10)$$
$$= 0$$

$d + 10 = 0$ *or* $d - 10 = 0$

$d = -10$ *or* $d = 10$

Check. Since the diagonal must be positive, -10 cannot be a solution.

$8^2 + 6^2 = 10^2$

$100 = 100$ True

State. The diagonal is 10 holes long.

61. *Writing Exercise.* The equations solved in this chapter have an x^2-term (are quadratic), whereas those solved previously have no x^2-term (are linear). The principle of zero products is used to solve quadratic equations and is not used to solve linear equations.

63. ***Familiarize*** Let $n =$ the number.

Translate.

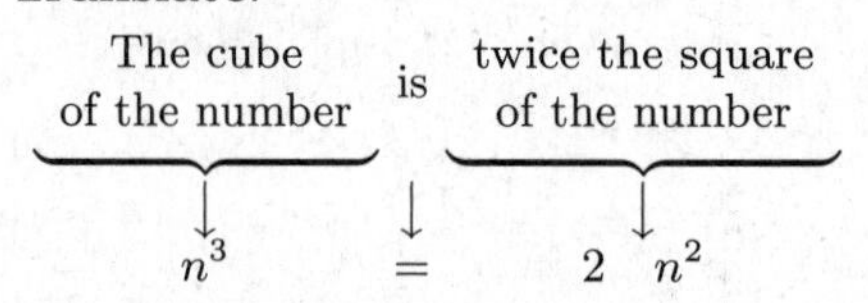

$n^3 \quad = \quad 2 \; n^2$

Carry out. We solve the equation.

$$n^3 = 2n^2$$
$$n^3 - 2n^2 = 0$$
$$n^2(n-2) = 0$$

$n^2 = 0$ *or* $n - 2 = 0$

$n = 0$ *or* $n = 2$

Check.

$n^3 = 2n^2$

$0^3 = 2(0)^2$

$0 = 0$ True

$n^3 = 2n^2$

$2^3 = 2(2)^2$

$8 = 8$ True

State. The number is 0 or 2.

65. ***Familiarize.*** Let $s =$ the side of a square, in cm, $s + 5 =$ side of the new square. Recall $A = x^2$.

Translate. The new square is $2\frac{1}{2}$ times the area of the original square

$(s+5)^2 = 2\frac{1}{4}(s)^2$

Carry out. We solve the equation.

$$\begin{aligned} 4s^2+40s+100 &= 9s^2 \\ 0 &= 5s^2-40s-100 \\ 0 &= 5(s^2-8s-20) \\ 0 &= 5(s-10)(s+2) \end{aligned}$$

$s-10=0$ *or* $s+2=0$

$s=10$ *or* $s=-2$

Check. Since the side of the square must be positive, -2 cannot be a solution.

$$\begin{aligned} (s+5)^2 &= 2\frac{1}{4}s^2 \\ (10+5)^2 &= \frac{9}{4}(10)^2 \\ 225 &= 225 \quad \text{True} \end{aligned}$$

State. The original square has side 10 cm and area of 100 cm^2. The new square has side 15 cm and area of 225 cm^2.

67. $x^2+25=0$

No real solution, the sum of two squares cannot be factored.

Chapter 5 Test

1. Answers may vary.
$(3x^2)(4x^2)$, $(-2x)(-6x^3)$, $(12x^3)(x)$

3. $$\begin{aligned} x^2+25-10x &= x^2-10x+25 \\ &= (x-5)(x-5) \\ &= (x-5)^2 \end{aligned}$$

5. $$\begin{aligned} x^3+x^2+2x+2 &= x^2(x+1)+2(x+1) \\ &= (x+1)(x^2+2) \end{aligned}$$

7. $$\begin{aligned} a^3+3a^2-4a &= a(a^2+3a-4) \\ &= a(a+4)(a-1) \end{aligned}$$

9. $4t^2-25=(2t+5)(2t-5)$

11. $$\begin{aligned} -6m^3-9m^2-3m &= -3m(2m^2+3m+1) \\ &= -3m(2m+1)(m+1) \end{aligned}$$

13. $$\begin{aligned} 45r^2+60r+20 &= 5(9r^2+12r+4) \\ &= 5(3r+2)(3r+2) \\ &= 5(3r+2)^2 \end{aligned}$$

15. $$\begin{aligned} 49t^2+36+84t &= 49t^2+84t+36 \\ &= (7t+6)(7t+6) \\ &= (7t+6)^2 \end{aligned}$$

17. x^2+3x+6 is prime.

19. $$\begin{aligned} 6t^3+9t^2-15t &= 3t(2t^2+3t-5) \\ &= 3t(2t+5)(t-1) \end{aligned}$$

21. $x^2-6x+5=0$
$(x-5)(x-1)=0$
$x-5=0$ or $x-1=0$
$x=5$ or $x=1$

23. $$\begin{aligned} 4t-10t^2 &= 0 \\ 2t(2-5t) &= 0 \end{aligned}$$
$2t=0$ or $2-5t=0$
$t=0$ or $-5t=-2$
$t=0$ or $t=\frac{2}{5}$

The solutions are 0 and $\frac{2}{5}$.

25. $$\begin{aligned} x(x-1) &= 20 \\ x^2-x &= 20 \\ x^2-x-20 &= 0 \\ (x-5)(x+4) &= 0 \end{aligned}$$
$x-5=0$ or $x+4=0$
$x=5$ or $x=-4$

The solutions are -4 and 5.

27. ***Familiarize.*** We make a drawing, Let $w=$ the width, in m, Then $w+6=$ the length in m.

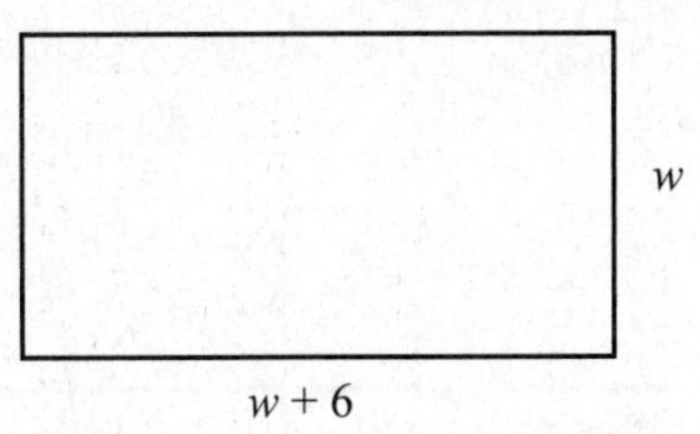

Recall the area of a rectangle is length times width.

Translate. We reword the problem

Length	times	width	is	40 m^2
$(w+6)$	$\cdot$	w	$=$	40

Carry out. We solve the equation.

$$\begin{aligned} (w+6)w &= 40 \\ w^2+6w &= 40 \\ w^2+6w-40 &= 0 \\ (w+10)(w-4) &= 0 \end{aligned}$$
$w+10=0$ or $w-4=0$
$w=-10$ or $w=4$

Check. Since the width must be positive, -10 cannot be a solution. If the width is 4m, the length is $4+6$ or 10 m, and the area is $10\cdot 4$, or 40 m^2. Thus 4 checks.
State. The width is 4 m and the length is 10 m.

29. ***Familiarize.*** From the given drawing x is the distance in feet we are looking for.
Translate. We use the Pythagorean Theorem. $3^2+4^2=x^2$
Carry out. We solve the equation:

$$\begin{aligned} 3^2+4^2 &= x^2 \\ 9+16 &= x^2 \\ 25 &= x^2 \end{aligned}$$
$x=5$ or -5

Check. The number -5 is not a solution because distance cannot be negative. If $x = 5$, $3^2 + 4^2 = 9 + 16 = 5^2$, so the answer checks.

State. The distance is 5 ft.

31. $$\begin{aligned}(a+3)^2 - 2(a+3) - 35 &= [(a+3)-7]\,[(a+3)+5] \\ &= [a+3-7][a+3+5] \\ &= (a-4)(a+8)\end{aligned}$$

Chapter 6

Rational Expressions and Equations

Exercise Set 6.1

1. $x - 2 = 0$ when $x = 2$ and $x + 3 = 0$ when $x = -3$, so choice (e) is correct.

3. $a^2 - a - 12 = (a-4)(a+3)$; $a - 4 = 0$ when $a = 4$ and $a + 3 = 0$ when $a = -3$, so choice (d) is correct.

5. $2t - 1 = 0$ when $t = \frac{1}{2}$ and $3t + 4 = 0$ when $t = -\frac{4}{3}$, so choice (c) is correct.

7. $\dfrac{18}{-11x}$

We find the real number(s) that make the denominator 0. To do so we set the denominator equal to 0 and solve for x:

$$-11x = 0$$
$$x = 0$$

The expression is undefined for $x = 0$.

9. $\dfrac{y-3}{y+5}$

Set the denominator equal to 0 and solve for y:

$$y + 5 = 0$$
$$y = -5$$

The expression is undefined for $y = -5$.

11. $\dfrac{t-5}{3t-15}$

Set the denominator equal to 0 and solve for t:

$$3t - 15 = 0$$
$$3t = 15$$
$$t = 5$$

The expression is undefined for $t = 5$.

13. $\dfrac{x^2-25}{x^2-3x-28}$

Set the denominator equal to 0 and solve for x:

$$x^2 - 3x - 28 = 0$$
$$(x-7)(x+4) = 0$$
$$x - 7 = 0 \quad or \quad x + 4 = 0$$
$$x = 7 \quad or \quad x = -4$$

The expression is undefined for $x = 7$ and $x = -4$.

15. $\dfrac{t^2+t-20}{2t^2+11t-6}$

Set the denominator equal to 0 and solve for t:

$$2t^2 + 11t - 6 = 0$$
$$(2t-1)(t+6) = 0$$
$$2t - 1 = 0 \quad or \quad t + 6 = 0$$
$$2t = 1 \quad or \quad t = -6$$
$$t = \frac{1}{2} \quad or \quad t = -6$$

The expression is undefined for $t = \frac{1}{2}$ and $t = -6$.

17. $\dfrac{50a^2b}{40ab^3}$

$= \dfrac{5a \cdot 10ab}{4b^2 \cdot 10ab}$ — Factoring the numerator and denominator. Note the common factor of $10ab$.

$= \dfrac{5a}{4b^2} \cdot \dfrac{10ab}{10ab}$ — Rewriting as a product of two rational expressions

$= \dfrac{5a}{4b^2} \cdot 1$ — $\dfrac{10ab}{10ab} = 1$

$= \dfrac{5a}{4b^2}$ — Removing the factor 1

19. $\dfrac{6t+12}{6t-18} = \dfrac{\not{6}(t+2)}{\not{6}(t-3)} = \dfrac{(t+2)}{(t-3)}$

21. $\dfrac{21t-7}{24t-8} = \dfrac{7\cancel{(3t-1)}}{8\cancel{(3t-1)}} = \dfrac{7}{8}$

23.
$$\frac{a^2-9}{a^2+4a+3} = \frac{(a+3)(a-3)}{(a+3)(a+1)}$$
$$= \frac{a+3}{a+3} \cdot \frac{a-3}{a+1}$$
$$= 1 \cdot \frac{a-3}{a+1}$$
$$= \frac{a-3}{a+1}$$

25.
$$\frac{-36x^8}{54x^5} = \frac{-2x^3 \cdot 18x^5}{3 \cdot 18x^5}$$
$$= \frac{-2x^3}{3} \cdot \frac{18x^5}{18x^5}$$
$$= \frac{-2x^3}{3}$$

Check: Let $x = 1$.

$$\frac{-36x^8}{54x^5} = \frac{-36 \cdot 1^8}{54 \cdot 1^5} = \frac{-36}{54} = \frac{-2}{3}$$

$$\frac{-2x^3}{3} = \frac{-2 \cdot 1^3}{3} = \frac{-2}{3}$$

The answer is probably correct.

27. $\dfrac{-2y+6}{-8y} = \dfrac{-2(y-3)}{-2\cdot 4y}$

$= \dfrac{-2}{-2}\cdot\dfrac{y-3}{4y}$

$= 1\cdot\dfrac{y-3}{4y}$

$= \dfrac{y-3}{4y}$

Check: Let $x=2$.

$\dfrac{-2y+6}{-8y} = \dfrac{-2\cdot 2+6}{-8\cdot 2} = \dfrac{2}{-16} = -\dfrac{1}{8}$

$\dfrac{y-3}{4y} = \dfrac{2-3}{4\cdot 2} = \dfrac{-1}{8} = -\dfrac{1}{8}$

The answer is probably correct.

29. $\dfrac{t^2-16}{t^2-t-20} = \dfrac{(t-4)(t+4)}{(t-5)(t+4)}$

$= \dfrac{t-4}{t-5}\cdot\dfrac{t+4}{t+4}$

$= \dfrac{t-4}{t-5}\cdot 1$

$= \dfrac{t-4}{t-5}$

Check: Let $t=1$.

$\dfrac{t^2-16}{t^2-t-20} = \dfrac{1^2-16}{1^2-1-20} = \dfrac{-15}{-20} = \dfrac{3}{4}$

$\dfrac{t-4}{t-5} = \dfrac{1-4}{1-5} = \dfrac{3}{4}$

The answer is probably correct.

31. $\dfrac{3a^2+9a-12}{6a^2-30a+24} = \dfrac{3(a^2+3a-4)}{6(a^2-5a+4)}$

$= \dfrac{3(a+4)(a-1)}{3\cdot 2(a-4)(a-1)}$

$= \dfrac{3(a-1)}{3(a-1)}\cdot\dfrac{a+4}{2(a-4)}$

$= 1\cdot\dfrac{a+4}{2(a-4)}$

$= \dfrac{a+4}{2(a-4)}$

Check: Let $a=2$.

$\dfrac{3a^2+9a-12}{6a^2-30a+24} = \dfrac{3\cdot 2^2+9\cdot 2-12}{6\cdot 2^2-30\cdot 2+24} = \dfrac{18}{-12} = -\dfrac{3}{2}$

$\dfrac{a+4}{2(a-4)} = \dfrac{2+4}{2(2-4)} = \dfrac{6}{-4} = -\dfrac{3}{2}$

The answer is probably correct.

33. $\dfrac{x^2-8x+16}{x^2-16} = \dfrac{(x-4)(x-4)}{(x+4)(x-4)}$

$= \dfrac{x-4}{x+4}\cdot\dfrac{x-4}{x-4}$

$= \dfrac{x-4}{x+4}\cdot 1$

$= \dfrac{x-4}{x+4}$

Check: Let $x=1$.

$\dfrac{x^2-8x+16}{x^2-16} = \dfrac{1^2-8\cdot 1+16}{1^2-16} = \dfrac{1-8+16}{1-16} = \dfrac{9}{-15} = -\dfrac{3}{5}$

$\dfrac{x-4}{x+4} = \dfrac{1-4}{1+4} = -\dfrac{3}{5}$

The answer is probably correct.

35. $\dfrac{n-2}{n^3-8} = \dfrac{n-2}{(n-2)(n^2+2n+4)}$

$= \dfrac{n-2}{n-2}\cdot\dfrac{1}{n^2+2n+4}$

$= \dfrac{1}{n^2+2n+4}$

37. $\dfrac{t^2-1}{t+1} = \dfrac{(t+1)(t-1)}{t+1}$

$= \dfrac{t+1}{t+1}\cdot\dfrac{t-1}{1}$

$= 1\cdot\dfrac{t-1}{1}$

$= t-1$

Check: Let $t=2$.

$\dfrac{t^2-1}{t+1} = \dfrac{2^2-1}{2+1} = \dfrac{3}{3} = 1$

$t-1 = 2-1 = 1$

The answer is probably correct.

39. $\dfrac{y^2+4}{y+2}$ cannot be simplified.

Neither the numerator nor the denominator can be factored.

41. $\dfrac{5x^2+20}{10x^2+40} = \dfrac{5(x^2+4)}{10(x^2+4)}$

$= \dfrac{1\cdot \cancel{5}\cdot \cancel{(x^2+4)}}{2\cdot \cancel{5}\cdot \cancel{(x^2+4)}}$

$= \dfrac{1}{2}$

Check: Let $x=1$.

$\dfrac{5x^2+20}{10x^2+40} = \dfrac{5\cdot 1^2+20}{10\cdot 1^2+40} = \dfrac{25}{50} = \dfrac{1}{2}$

$\dfrac{1}{2} = \dfrac{1}{2}$

The answer is probably correct.

43. $\dfrac{y^2+6y}{2y^2+13y+6} = \dfrac{y(y+6)}{(2y+1)(y+6)}$
$= \dfrac{y}{2y+1} \cdot \dfrac{y+6}{y+6}$
$= \dfrac{y}{2y+1} \cdot 1$
$= \dfrac{y}{2y+1}$

Check: Let $y = 1$.
$\dfrac{y^2+6y}{2y^2+13y+6} = \dfrac{1^2+6\cdot 1}{2\cdot 1^2+13\cdot 1+6} = \dfrac{7}{21} = \dfrac{1}{3}$
$\dfrac{y}{2y+1} = \dfrac{1}{2\cdot 1+1} = \dfrac{1}{3}$
The answer is probably correct.

45. $\dfrac{4x^2-12x+9}{10x^2-11x-6} = \dfrac{(2x-3)(2x-3)}{(2x-3)(5x+2)}$
$= \dfrac{2x-3}{2x-3} \cdot \dfrac{2x-3}{5x+2}$
$= 1 \cdot \dfrac{2x-3}{5x+2}$
$= \dfrac{2x-3}{5x+2}$

Check: Let $t = 1$.
$\dfrac{4x^2-12x+9}{10x^2-11x-6} = \dfrac{4\cdot 1^2-12\cdot 1+9}{10\cdot 1^2-11\cdot 1-6} = \dfrac{1}{-7} = -\dfrac{1}{7}$
$\dfrac{2x-3}{5x+2} = \dfrac{2\cdot 1-3}{5\cdot 1+2} = \dfrac{-1}{7} = -\dfrac{1}{7}$
The answer is probably correct.

47. $\dfrac{10-x}{x-10} = \dfrac{-(x-10)}{x-10}$
$= \dfrac{-1}{1} \cdot \dfrac{x-10}{x-10} = -1 \cdot 1 = -1$

Check: Let $x = 1$.
$\dfrac{10-x}{x-10} = \dfrac{10-1}{1-10} = \dfrac{9}{-9} = -1$

The answer is probably correct.

49. $\dfrac{7t-14}{2-t} = \dfrac{7(t-2)}{-(t-2)}$
$= \dfrac{7}{-1} \cdot \dfrac{t-2}{t-2}$
$= \dfrac{7}{-1} \cdot 1$
$= -7$

Check: Let $t = 1$.
$\dfrac{7t-14}{2-t} = \dfrac{7\cdot 1-14}{2-1} = \dfrac{-7}{1} = -7$
The answer is probably correct.

51. $\dfrac{a-b}{4b-4a} = \dfrac{a-b}{-4(a-b)}$
$= \dfrac{1}{-4} \cdot \dfrac{a-b}{a-b}$
$= -\dfrac{1}{4} \cdot 1$
$= -\dfrac{1}{4}$

Check: Let $a = 2$ and $b = 1$.
$\dfrac{a-b}{4b-4a} = \dfrac{2-1}{4\cdot 1-4\cdot 2} = \dfrac{1}{4-8} = -\dfrac{1}{-4}$
The answer is probably correct.

53. $\dfrac{3x^2-3y^2}{2y^2-2x^2} = \dfrac{3(x^2-y^2)}{2(y^2-x^2)}$
$= \dfrac{3(x^2-y^2)}{2(-1)(x^2-y^2)}$
$= \dfrac{3}{2(-1)} \cdot \dfrac{x^2-y^2}{x^2-y^2}$
$= \dfrac{3}{2(-1)} \cdot 1$
$= -\dfrac{3}{2}$

Check: Let $x = 1$ and $y = 2$.
$\dfrac{3x^2-3y^2}{2y^2-2x^2} = \dfrac{3\cdot 1^2-3\cdot 2^2}{2\cdot 2^2-2\cdot 1^2} = \dfrac{-9}{6} = -\dfrac{3}{2}$

The answer is probably correct.

55. $\dfrac{7s^2-28t^2}{28t^2-7s^2}$
Note that the numerator and denominator are opposites. Thus, we have an expression divided by its opposite, so the result is -1.

57. *Writing Exercise.* Simplifying removes a factor equal to 1, allowing us to rewrite an expression $a \cdot 1$ as a.

59. $-\dfrac{2}{15} \cdot \dfrac{10}{7} = -\dfrac{2\cdot 10}{15\cdot 7}$
$= -\dfrac{2\cdot 2\cdot \cancel{5}}{3\cdot \cancel{5}\cdot 7}$
$= -\dfrac{4}{21}$

61. $\dfrac{5}{8} \div \left(-\dfrac{1}{6}\right) = \dfrac{5}{8} \cdot (-6)$
$= -\dfrac{5\cdot 6}{8}$
$= -\dfrac{5\cdot \cancel{2}\cdot 3}{\cancel{2}\cdot 4}$
$= -\dfrac{15}{4}$

63. $\dfrac{7}{9} - \dfrac{2}{3} \cdot \dfrac{6}{7} = \dfrac{7}{9} - \dfrac{4}{7} = \dfrac{7}{9} \cdot \dfrac{7}{7} - \dfrac{4}{7} \cdot \dfrac{9}{9}$
$= \dfrac{49}{63} - \dfrac{36}{63} = \dfrac{13}{63}$

65. *Writing Exercise.* Although a rational expression has been simplified incorrectly, it is possible that there are one or more values of the variable(s) for which the two expressions are the same. For example, $\dfrac{x^2+x-2}{x^2+3x+2}$ could be simplified incorrectly as $\dfrac{x-1}{x+2}$, but evaluating the expressions for $x = 1$ gives 0 in each case. $\left(\text{The correct simplification is } \dfrac{x-1}{x+1}.\right)$

67. $\dfrac{16y^4 - x^4}{(x^2+4y^2)(x-2y)}$

$= \dfrac{(4y^2+x^2)(4y^2-x^2)}{(x^2+4y^2)(x-2y)}$

$= \dfrac{(4y^2+x^2)(2y+x)(2y-x)}{(x^2+4y^2)(x-2y)}$

$= \dfrac{(x^2+4y^2)(2y+x)(-1)(x-2y)}{(x^2+4y^2)(x-2y)}$

$= \dfrac{(x^2+4y^2)(x-2y)}{(x^2+4y^2)(x-2y)} \cdot \dfrac{(2y+x)(-1)}{1}$

$= -2y - x$, or $-x - 2y$, or $-(2y+x)$

69. $\dfrac{x^5 - 2x^3 + 4x^2 - 8}{x^7 + 2x^4 - 4x^3 - 8}$

$= \dfrac{x^3(x^2-2)+4(x^2-2)}{x^4(x^3+2)-4(x^3+2)}$

$= \dfrac{(x^2-2)(x^3+4)}{(x^3+2)(x^4-4)}$

$= \dfrac{(x^2-2)(x^3+4)}{(x^3+2)(x^2+2)(x^2-2)}$

$= \dfrac{\cancel{(x^2-2)}(x^3+4)}{(x^3+2)(x^2+2)\cancel{(x^2-2)}}$

$= \dfrac{x^3+4}{(x^3+2)(x^2+2)}$

71. $\dfrac{(t^4-1)(t^2-9)(t-9)^2}{(t^4-81)(t^2+1)(t+1)^2}$

$= \dfrac{(t^2+1)(t+1)(t-1)(t+3)(t-3)(t-9)(t-9)}{(t^2+9)(t+3)(t-3)(t^2+1)(t+1)(t+1)}$

$= \dfrac{\cancel{(t^2+1)}\cancel{(t+1)}(t-1)\cancel{(t+3)}\cancel{(t-3)}(t-9)(t-9)}{(t^2+9)\cancel{(t+3)}\cancel{(t-3)}\cancel{(t^2+1)}\cancel{(t+1)}(t+1)}$

$= \dfrac{(t-1)(t-9)(t-9)}{(t^2+9)(t+1)}$, or $\dfrac{(t-1)(t-9)^2}{(t^2+9)(t+1)}$

73. $\dfrac{(x^2-y^2)(x^2-2xy+y^2)}{(x+y)^2(x^2-4xy-5y^2)}$

$= \dfrac{(x+y)(x-y)(x-y)(x-y)}{(x+y)(x+y)(x-5y)(x+y)}$

$= \dfrac{\cancel{(x+y)}(x-y)(x-y)(x-y)}{\cancel{(x+y)}(x+y)(x-5y)(x+y)}$

$= \dfrac{(x-y)^3}{(x+y)^2(x-5y)}$

75. *Writing Exercise.*

$\dfrac{5(2x+5)-25}{10} = \dfrac{10x+25-25}{10}$

$= \dfrac{10x}{10}$

$= x$

You get the same number you selected.

A person asked to select a number and then perform these operations would probably be surprised that the result is the original number.

Exercise Set 6.2

---

1. $\dfrac{3x}{8} \cdot \dfrac{x+2}{5x-1} = \dfrac{3x(x+2)}{8(5x-1)}$

3. $\dfrac{a-4}{a+6} \cdot \dfrac{a+2}{a+6} = \dfrac{(a-4)(a+2)}{(a+6)(a+6)}$, or $\dfrac{(a-4)(a+2)}{(a+6)^2}$

5. $\dfrac{2x+3}{4} \cdot \dfrac{x+1}{x-5} = \dfrac{(2x+3)(x+1)}{4(x-5)}$

7. $\dfrac{n-4}{n^2+4} \cdot \dfrac{n+4}{n^2-4} = \dfrac{(n-4)(n+4)}{(n^2+4)(n^2-4)}$

9. $\dfrac{y+6}{1+y} \cdot \dfrac{y-3}{y+3} = \dfrac{(y+6)(y-3)}{(1+y)(y+3)}$

11. $\dfrac{8t^3}{5t} \cdot \dfrac{3}{4t}$

$= \dfrac{8t^3 \cdot 3}{5t \cdot 4t}$ Multiplying the numerators and the denominators

$= \dfrac{2 \cdot 4 \cdot t \cdot t \cdot t \cdot 3}{5 \cdot t \cdot 4 \cdot t}$ Factoring the numerator and the denominator

$= \dfrac{2 \cdot \cancel{4} \cdot t \cdot \cancel{t} \cdot \cancel{t} \cdot 3}{5 \cdot \cancel{t} \cdot \cancel{4} \cdot \cancel{t}}$ Removing a factor equal to 1

$= \dfrac{6t}{5}$ Simplifying

13. $\dfrac{3c}{d^2} \cdot \dfrac{8d}{6c^3}$

$= \dfrac{3c \cdot 8d}{d^2 \cdot 6c^3}$ Multiplying the numerators and the denominators

$= \dfrac{3 \cdot c \cdot 2 \cdot 4 \cdot d}{d \cdot d \cdot 3 \cdot 2 \cdot c \cdot c \cdot c}$ Factoring the numerator and the denominator

$= \dfrac{\cancel{3} \cdot \cancel{c} \cdot \cancel{2} \cdot 4 \cdot \cancel{d}}{\cancel{d} \cdot d \cdot \cancel{3} \cdot \cancel{2} \cdot \cancel{c} \cdot c \cdot c}$

$= \dfrac{4}{dc^2}$

15. $\dfrac{x^2-3x-10}{(x-2)^2} \cdot (x-2) = \dfrac{(x^2-3x-10)(x-2)}{(x-2)^2}$

$= \dfrac{(x-5)(x+2)(x-2)}{(x-2)(x-2)}$

$= \dfrac{(x-5)(x+2)\cancel{(x-2)}}{(x-2)\cancel{(x-2)}}$

$= \dfrac{(x-5)(x+2)}{x-2}$

17. $\dfrac{n^2-6n+5}{n+6} \cdot \dfrac{n-6}{n^2+36} = \dfrac{(n^2-6n+5)(n-6)}{(n+6)(n^2+36)}$

$= \dfrac{(n-5)(n-1)(n-6)}{(n+6)(n^2+36)}$

(No simplification is possible.)

19. $\dfrac{a^2-9}{a^2} \cdot \dfrac{7a}{a^2+a-12} = \dfrac{(a+3)(a-3) \cdot 7 \cdot a}{a \cdot a(a+4)(a-3)}$

$= \dfrac{(a+3)\cancel{(a-3)} \cdot 7 \cdot \cancel{a}}{\cancel{a} \cdot a(a+4)\cancel{(a-3)}}$

$= \dfrac{7(a+3)}{a(a+4)}$

21. $\dfrac{4v-8}{5v}\cdot\dfrac{15v^2}{4v^2-16v+16}$

$=\dfrac{(4v-8)15v^2}{5v(4v^2-16v+16)}$

$=\dfrac{\cancel{4(v-2)}\cdot\cancel{5}\cdot 3\cdot v\cdot\cancel{v}}{\cancel{5v}\cdot\cancel{4(v-2)}(v-2)}$

$=\dfrac{3v}{v-2}$

23. $\dfrac{t^2+2t-3}{t^2+4t-5}\cdot\dfrac{t^2-3t-10}{t^2+5t+6}$

$=\dfrac{(t^2+2t-3)(t^2-3t-10)}{(t^2+4t-5)(t^2+5t+6)}$

$=\dfrac{(t+3)(t-1)(t-5)(t+2)}{(t+5)(t-1)(t+3)(t+2)}$

$=\dfrac{\cancel{(t+3)}\cancel{(t-1)}(t-5)\cancel{(t+2)}}{(t+5)\cancel{(t-1)}\cancel{(t+3)}\cancel{(t+2)}}$

$=\dfrac{t-5}{t+5}$

25. $\dfrac{12y+12}{5y+25}\cdot\dfrac{3y^2-75}{8y^2-8}$

$=\dfrac{(12y+12)(3y^2-75)}{(5y+25)(8y^2-8)}$

$=\dfrac{3\cdot\cancel{4(y+1)}\,3\cancel{(y+5)}(y-5)}{5\cancel{(y+5)}\,2\cdot\cancel{4(y+1)}(y-1)}$

$=\dfrac{9(y-5)}{10(y-1)}$

27. $\dfrac{x^2+4x+4}{(x-1)^2}\cdot\dfrac{x^2-2x+1}{(x+2)^2}=\dfrac{(x+2)^2(x-1)^2}{(x-1)^2(x+2)^2}=1$

29. $\dfrac{t^2-4t+4}{t^2-7t+6}\cdot\dfrac{2t^2+7t-15}{t^2-10t+25}$

$=\dfrac{(t^2-4t+4)(2t^2+7t-15)}{(2t^2-7t+6)(t^2-10t+25)}$

$=\dfrac{(t-2)(t-2)(2t-3)(t+5)}{(2t-3)(t-2)(t-5)(t-5)}$

$=\dfrac{\cancel{(t-2)}(t-2)\cancel{(2t-3)}(t+5)}{\cancel{(2t-3)}\cancel{(t-2)}(t-5)(t-5)}$

$=\dfrac{(t-2)(t+5)}{(t-5)^2}$

31. $(10x^2-x-2)\cdot\dfrac{4x^2-8x+3}{10x^2-11x-6}$

$=\dfrac{(10x^2-x-2)(4x^2-8x+3)}{(10x^2-11x-6)}$

$=\dfrac{\cancel{(5x+2)}(2x-1)(2x-1)\cancel{(2x-3)}}{\cancel{(5x+2)}\cancel{(2x-3)}}$

$=(2x-1)^2$

33. $\dfrac{c^3+8}{c^5-4c^3}\cdot\dfrac{c^6-4c^5+4c^4}{c^2-2c+4}$

$=\dfrac{(c+2)(c^2-2c+4)}{c^3(c^2-4)}\cdot\dfrac{c^4(c^2-4c+4)}{c^2-2c+4}$

$=\dfrac{\cancel{(c+2)}\cancel{(c^2-2c+4)}(\cancel{c^3}\cdot c)\cancel{(c-2)}(c-2)}{\cancel{c^3}\cancel{(c+2)}\cancel{(c-2)}\cancel{(c^2-2c+4)}}$

$=c(c-2)$

35. The reciprocal of $\dfrac{2x}{9}$ is $\dfrac{9}{2x}$ because $\dfrac{2x}{9}\cdot\dfrac{9}{2x}=1$.

37. The reciprocal of a^4+3a is $\dfrac{1}{a^4+3a}$ because $\dfrac{a^4+3a}{1}\cdot\dfrac{1}{a^4+3a}=1$.

39. $\dfrac{5}{9}\div\dfrac{3}{4}$

$=\dfrac{5}{9}\cdot\dfrac{4}{3}$ Multiplying by the reciprocal of the divisor

$=\dfrac{5\cdot 4}{9\cdot 3}$

$=\dfrac{20}{27}$

No simplification is possible.

41. $\dfrac{x}{4}\div\dfrac{5}{x}$

$=\dfrac{x}{4}\cdot\dfrac{x}{5}$ Multiplying by the reciprocal of the divisor

$=\dfrac{x\cdot x}{4\cdot 5}$

$=\dfrac{x^2}{20}$

43. $\dfrac{a^5}{b^4}\div\dfrac{a^2}{b}=\dfrac{a^5}{b^4}\cdot\dfrac{b}{a^2}$

$=\dfrac{a^5\cdot b}{b^4\cdot a^2}$

$=\dfrac{a^2\cdot a^3\cdot b}{b\cdot b^3\cdot a^2}$

$=\dfrac{a^2b}{a^2b}\cdot\dfrac{a^3}{b^3}$

$=\dfrac{a^3}{b^3}$

45. $\dfrac{t-3}{6}\div\dfrac{t+1}{8}=\dfrac{t-3}{6}\cdot\dfrac{8}{t+1}$

$=\dfrac{(t-3)(8)}{6\cdot(t+1)}$

$=\dfrac{(t-3)\cdot 4\cdot\cancel{2}}{\cancel{2}\cdot 3(t+1)}$

$=\dfrac{4(t-3)}{3(t+1)}$

47.
$$\begin{aligned}\frac{4y-8}{y+2}\div\frac{y-2}{y^2-4}&=\frac{4y-8}{y+2}\cdot\frac{y^2-4}{y-2}\\&=\frac{(4y-8)(y^2-4)}{(y+2)(y-2)}\\&=\frac{4\cancel{(y-2)}\cancel{(y+2)}(y-2)}{\cancel{(y+2)}\cancel{(y-2)}(1)}\\&=4(y-2)\end{aligned}$$

49.
$$\begin{aligned}\frac{a}{a-b}\div\frac{b}{b-a}&=\frac{a}{a-b}\cdot\frac{b-a}{b}\\&=\frac{a(b-a)}{(a-b)(b)}\\&=\frac{a(-1)\cancel{(a-b)}}{\cancel{(a-b)}(b)}\\&=\frac{-a}{b}=-\frac{a}{b}\end{aligned}$$

51.
$$\begin{aligned}(n^2+5n+6)\div\frac{n^2-4}{n+3}&=\frac{(n^2+5n+6)}{1}\cdot\frac{(n+3)}{n^2-4}\\&=\frac{(n^2+5n+6)\,(n+3)}{n^2-4}\\&=\frac{(n+3)\,\cancel{(n+2)}\,(n+3)}{\cancel{(n+2)}\,(n-2)}\\&=\frac{(n+3)^2}{n-2}\end{aligned}$$

53.
$$\begin{aligned}\frac{-3+3x}{16}\div\frac{x-1}{5}&=\frac{3x-3}{16}\cdot\frac{5}{x-1}\\&=\frac{(3x-3)\cdot 5}{16(x-1)}\\&=\frac{3(x-1)\cdot 5}{16(x-1)}\\&=\frac{3\cancel{(x-1)}\cdot 5}{16\cancel{(x-1)}}\\&=\frac{15}{16}\end{aligned}$$

55.
$$\begin{aligned}\frac{x-1}{x+2}\div\frac{1-x}{4+x^2}&=\frac{x-1}{x+2}\cdot\frac{4+x^2}{1-x}\\&=\frac{(x-1)\,(4+x^2)}{(x+2)\,(1-x)}\\&=\frac{\cancel{(x-1)}\,(x^2+4)}{-1\,(x+2)\,\cancel{(x-1)}}\\&=-\frac{x^2+4}{x+2}\text{ or }\frac{-x^2-4}{x+2}\end{aligned}$$

57.
$$\begin{aligned}\frac{a+2}{a-1}\div\frac{3a+6}{a-5}&=\frac{a+2}{a-1}\cdot\frac{a-5}{3a+6}\\&=\frac{(a+2)(a-5)}{(a-1)(3a+6)}\\&=\frac{(a+2)(a-5)}{(a-1)\cdot 3\cdot(a+2)}\\&=\frac{\cancel{(a+2)}(a-5)}{(a-1)\cdot 3\cdot\cancel{(a+2)}}\\&=\frac{a-5}{3(a-1)}\end{aligned}$$

59.
$$\begin{aligned}&(2x-1)\div\frac{2x^2-11x+5}{4x^2-1}\\&=\frac{2x-1}{1}\cdot\frac{4x^2-1}{2x^2-11x+5}\\&=\frac{(2x-1)(4x^2-1)}{1\cdot(2x^2-11x+5)}\\&=\frac{(2x-1)(2x+1)(2x-1)}{1\cdot(2x-1)(x-5)}\\&=\frac{\cancel{(2x-1)}(2x+1)(2x-1)}{1\cdot\cancel{(2x-1)}(x-5)}\\&=\frac{(2x-1)(2x+1)}{x-5}\end{aligned}$$

61.
$$\begin{aligned}&\frac{w^2-14w+49}{2w^2-3w-14}\div\frac{3w^2-20w-7}{w^2-6w-16}\\&=\frac{w^2-14w+49}{2w^2-3w-14}\cdot\frac{w^2-6w-16}{3w^2-20w-7}\\&=\frac{(w^2-14w+49)\,(w^2-6w-16)}{(2w^2-3w-14)\,(3w^2-20w-7)}\\&=\frac{(w-7)\,\cancel{(w-7)}\,(w-8)\,\cancel{(w+2)}}{(2w-7)\,\cancel{(w+2)}\,(3w+1)\,\cancel{(w-7)}}\\&=\frac{(w-7)\,(w-8)}{(2w-7)\,(3w+1)}\end{aligned}$$

63.
$$\begin{aligned}&\frac{c^2+10c+21}{c^2-2c-15}\div(5c^2+32c-21)\\&=\frac{c^2+10c+21}{c^2-2c-15}\cdot\frac{1}{5c^2+32c-21}\\&=\frac{(c^2+10c+21)\cdot 1}{(c^2-2c-15)(5c^2+32c-21)}\\&=\frac{(c+7)(c+3)}{(c-5)(c+3)(5c-3)(c+7)}\\&=\frac{(c+7)(c+3)}{(c+7)(c+3)}\cdot\frac{1}{(c-5)(5c-3)}\\&=\frac{1}{(c-5)(5c-3)}\end{aligned}$$

65.
$$\begin{aligned}&\frac{x-y}{x^2+2xy+y^2}\div\frac{x^2-y^2}{x^2-5xy+4y^2}\\&=\frac{x-y}{x^2+2xy+y^2}\cdot\frac{x^2-5xy+4y^2}{x^2-y^2}\\&=\frac{\cancel{(x-y)}(x-y)(x-4y)}{(x+y)(x+y)(x+y)\cancel{(x-y)}}\\&=\frac{(x-y)(x-4y)}{(x+y)^3}\end{aligned}$$

67.
$$\begin{aligned}&\frac{x^3-64}{x^3+64}\div\frac{x^2-16}{x^2-4x+16}\\&=\frac{x^3-64}{x^3+64}\cdot\frac{x^2-4x+16}{x^2-16}\\&=\frac{(x-4)\,(x^2+4x+16)\,(x^2-4x+16)}{(x+4)\,(x^2-4x+16)\,(x+4)\,(x-4)}\\&=\frac{\cancel{(x-4)}\,(x^2+4x+16)\,(x^2-\cancel{4x}+16)}{(x+4)\,(x^2-\cancel{4x}+16)\,(x+4)\,\cancel{(x-4)}}\\&=\frac{(x^2+4x+16)}{(x+4)^2}\end{aligned}$$

69. $\dfrac{8a^3+b^3}{2a^2+3ab+b^2} \div \dfrac{8a^2-4ab+2b^2}{4a^2+4ab+b^2}$

$$= \frac{(2a+b)\left(4a^2-2ab+b^2\right)}{(2a+b)(a+b)} \cdot \frac{4a^2+4ab+b^2}{8a^2-4ab+2b^2}$$

$$= \frac{(2a+b)\left(4a^2-2ab+b^2\right)(2a+b)(2a+b)}{(2a+b)(a+b)(2)\left(4a^2-2ab+b^2\right)}$$

$$= \frac{(2a+b)^2}{2(a+b)}$$

71. *Writing Exercise.* Parentheses are required to ensure that numerators and denominators are multiplied correctly. That is, the product of $(x+2)$ and $(3x-1)$ and the product of $(5x-7)$ and $(x+4)$ in the denominator.

73. $\dfrac{3}{4}+\dfrac{5}{6} = \dfrac{3}{4}\cdot\dfrac{3}{3}+\dfrac{5}{6}\cdot\dfrac{2}{2}$

$$= \frac{9}{12}+\frac{10}{12}$$

$$= \frac{19}{12}$$

75. $\dfrac{2}{9}-\dfrac{1}{6} = \dfrac{2}{9}\cdot\dfrac{2}{2}-\dfrac{1}{6}\cdot\dfrac{3}{3}$

$$= \frac{4}{18}-\frac{3}{18}$$

$$= \frac{1}{18}$$

77. $2x^2-x+1-(x^2-x-2) = 2x^2-x+1-x^2+x+2 = x^2+3$

79. *Writing Exercise.* Yes; consider the product $\dfrac{a}{b}\cdot\dfrac{c}{d}=\dfrac{ac}{bd}$. The reciprocal of the product is $\dfrac{bd}{ac}$. This is equal to the product of the reciprocals of the two original factors: $\dfrac{b}{a}\cdot\dfrac{d}{c}=\dfrac{bd}{ac}$.

81. The reciprocal of $2\frac{1}{3}x$ is $\dfrac{1}{2\frac{1}{3}x} = \dfrac{1}{\frac{7x}{3}} = 1\div\dfrac{7x}{3} = 1\cdot\dfrac{3}{7x} = \dfrac{3}{7x}$

83. $(x-2a)\div\dfrac{a^2x^2-4a^4}{a^2x+2a^3} = \dfrac{x-2a}{1}\cdot\dfrac{a^2x+2a^3}{a^2x^2-4a^4}$

$$= \frac{(x-2a)(a^2x^2+2a^3)}{(a^2x^2-4a^4)}$$

$$= \frac{(x-2a)a^2(x+2a)}{a^2(x-2a)(x+2a)} = 1$$

85. $\dfrac{3x^2-2xy-y^2}{x^2-y^2} \div (3x^2+4xy+y^2)^2$

$$= \frac{3x^2-2xy-y^2}{x^2-y^2}\cdot\frac{1}{(3x^2+4xy+y^2)^2}$$

$$= \frac{(3x+y)(x-y)\cdot 1}{(x+y)(x-y)(3x+y)(3x+y)(x+y)(x+y)}$$

$$= \frac{1}{(x+y)^3(3x+y)}$$

87. $\dfrac{a^2-3b}{a^2+2b}\cdot\dfrac{a^2-2b}{a^2+3b}\cdot\dfrac{a^2+2b}{a^2-3b}$

Note that $\dfrac{a^2-3b}{a^2+2b}\cdot\dfrac{a^2+2b}{a^2-3b}$ is the product of reciprocals and thus is equal to 1. Then the product in the original exercise is the remaining factor, $\dfrac{a^2-2b}{a^2+3b}$.

89. $\dfrac{z^2-8z+16}{z^2+8z+16} \div \dfrac{(z-4)^5}{(z+4)^5} \div \dfrac{3z+12}{z^2-16}$

$$= \frac{(z-4)^2}{(z+4)^2}\cdot\frac{(z+4)^5}{(z-4)^5}\cdot\frac{(z+4)(z-4)}{3(z+4)}$$

$$= \frac{(z-4)^2(z+4)^2(z+4)^3(z+4)(z-4)}{(z+4)^2(z-4)^2(z-4)(z-4)^2(3)(z+4)}$$

$$= \frac{(z+4)^3}{3(z-4)^2}$$

91. $\dfrac{a^4-81b^4}{a^2c-6abc+9b^2c}\cdot\dfrac{a+3b}{a^2+9b^2}\div\dfrac{a^2+6ab+9b^2}{(a-3b)^2}$

$$= \frac{(a^2+9b^2)(a+3b)(a-3b)}{c(a-3b)^2}\cdot\frac{a+3b}{a^2+9b^2}\cdot\frac{(a-3b)^2}{(a+3b)^2}$$

$$= \frac{a-3b}{c}$$

93. Enter $y_1 = \dfrac{x-1}{x^2+2x+1} \div \dfrac{x^2-1}{x^2-5x+4}$ and $y_2 = \dfrac{x^2-5x+4}{(x+1)^3}$, display the values of y_1 and y_2 in a table, and compare the values. (See the Technology Connection on page 385 in the text.)

Exercise Set 6.3

---

1. To add two rational expressions when the denominators are the same, add numerators and keep the common denominator. (See page 389 in the text.)

3. The least common multiple of two denominators is usually referred to as the least common denominator and is abbreviated LCD. (See page 390 in the text.)

5. $\dfrac{3}{t}+\dfrac{5}{t}=\dfrac{8}{t}$ Adding numerators

7. $\dfrac{x}{12}+\dfrac{2x+5}{12}=\dfrac{3x+5}{12}$ Adding numerators

9. $\dfrac{4}{a+3}+\dfrac{5}{a+3}=\dfrac{9}{a+3}$

11. $\dfrac{11}{4x-7}-\dfrac{3}{4x-7}=\dfrac{8}{4x-7}$ Subtracting numerators

13. $\dfrac{3y+8}{2y}-\dfrac{y+1}{2y}$

$$= \frac{3y+8-(y+1)}{2y}$$

$$= \frac{3y+8-y-1}{2y} \quad \text{Removing parentheses}$$

$$= \frac{2y+7}{2y}$$

15. $\dfrac{5x+7}{x+3}+\dfrac{x+11}{x+3}$

$$= \frac{6x+18}{x+3} \quad \text{Adding numerators}$$

$$= \frac{6(x+3)}{x+3} \quad \text{Factoring}$$

$$= \frac{6(x+3)}{x+3} \quad \text{Removing a factor equal to 1}$$

$$= 6$$

17. $\frac{5x+7}{x+3} - \frac{x+11}{x+3} = \frac{5x+7-(x+11)}{x+3}$
$= \frac{5x+7-x-11}{x+3}$
$= \frac{4x-4}{x+3}$
$= \frac{4(x-1)}{x+3}$

19. $\frac{a^2}{a-4} + \frac{a-20}{a-4} = \frac{a^2+a-20}{a-4}$
$= \frac{(a+5)(a-4)}{a-4}$
$= \frac{(a+5)\cancel{(a-4)}}{\cancel{a-4}}$
$= a+5$

21. $\frac{y^2}{y+2} - \frac{5y+14}{y+2} = \frac{y^2-(5y+14)}{y+2}$
$= \frac{y^2-5y-14}{y+2}$
$= \frac{(y-7)(y+2)}{y+2}$
$= \frac{(y-7)\cancel{(y+2)}}{\cancel{y+2}}$
$= y-7$

23. $\frac{t^2-5t}{t-1} + \frac{5t-t^2}{t-1}$
Note that the numerators are opposites, so their sum is 0. Then we have $\frac{0}{t-1}$, or 0.

25. $\frac{x-6}{x^2+5x+6} + \frac{9}{x^2+5x+6} = \frac{x+3}{x^2+5x+6}$
$= \frac{x+3}{(x+3)(x+2)}$
$= \frac{\cancel{x+3}}{\cancel{(x+3)}(x+2)}$
$= \frac{1}{x+2}$

27. $\frac{t^2-5t}{t^2+6t+9} + \frac{4t-12}{t^2+6t+9} = \frac{t^2-t-12}{t^2+6t+9}$
$= \frac{(t-4)(t+3)}{(t+3)^2}$
$= \frac{(t-4)\cancel{(t+3)}}{(t+3)\cancel{(t+3)}}$
$= \frac{t-4}{t+3}$

29. $\frac{2y^2+3y}{y^2-7y+12} - \frac{y^2+4y+6}{y^2-7y+12}$
$= \frac{2y^2+3y-(y^2+4y+6)}{y^2-7y+12}$
$= \frac{2y^2+3y-y^2-4y-6}{y^2-7y+12}$
$= \frac{y^2-y-6}{y^2-7y+12}$
$= \frac{(y-3)(y+2)}{(y-3)(y-4)}$
$= \frac{\cancel{(y-3)}(y+2)}{\cancel{(y-3)}(y-4)}$
$= \frac{y+2}{y-4}$

31. $\frac{3-2x}{x^2-6x+8} + \frac{7-3x}{x^2-6x+8}$
$= \frac{10-5x}{x^2-6x+8}$
$= \frac{5(2-x)}{(x-4)(x-2)}$
$= \frac{5(-1)(x-2)}{(x-4)(x-2)}$
$= \frac{5(-1)\cancel{(x-2)}}{(x-4)\cancel{(x-2)}}$
$= \frac{-5}{x-4}$, or $-\frac{5}{x-4}$, or $\frac{5}{4-x}$

33. $\frac{x-9}{x^2+3x-4} - \frac{2x-5}{x^2+3x-4}$
$= \frac{x-9-(2x-5)}{x^2+3x-4}$
$= \frac{x-9-2x+5}{x^2+3x-4}$
$= \frac{-x-4}{x^2+3x-4}$
$= \frac{-(x+4)}{(x+4)(x-1)}$
$= \frac{-1\cancel{(x+4)}}{\cancel{(x+4)}(x-1)}$
$= \frac{-1}{x-1}$, or $-\frac{1}{x-1}$, or $\frac{1}{1-x}$

35. $15 = 3 \cdot 5$
$36 = 2 \cdot 2 \cdot 3 \cdot 3$
$\text{LCM} = 2 \cdot 2 \cdot 3 \cdot 3 \cdot 5 = 180$

37. $8 = 2 \cdot 2 \cdot 2$
$9 = 3 \cdot 3$
$\text{LCM} = 2 \cdot 2 \cdot 2 \cdot 3 \cdot 3$, or 72

39. $6 = 2 \cdot 3$

$12 = 2 \cdot 2 \cdot 3$

$15 = 3 \cdot 5$

$\text{LCM} = 2 \cdot 2 \cdot 3 \cdot 5 = 60$

41. $18t^2 = 2 \cdot 3 \cdot 3 \cdot t \cdot t$

$6t^5 = 2 \cdot 3 \cdot t \cdot t \cdot t \cdot t \cdot t$

$\text{LCM} = 2 \cdot 3 \cdot 3 \cdot t \cdot t \cdot t \cdot t \cdot t = 18t^5$

43. $15a^4b^7 = 3 \cdot 5 \cdot a \cdot a \cdot a \cdot a \cdot b \cdot b \cdot b \cdot b \cdot b \cdot b \cdot b$

$10a^2b^8 = 2 \cdot 5 \cdot a \cdot a \cdot b \cdot b \cdot b \cdot b \cdot b \cdot b \cdot b \cdot b$

$\text{LCM} = 2 \cdot 3 \cdot 5 \cdot a \cdot a \cdot a \cdot a \cdot b \cdot b \cdot b \cdot b \cdot b \cdot b \cdot b \cdot b$, or $30a^4b^8$

45. $2(y-3) = 2 \cdot (y-3)$

$6(y-3) = 2 \cdot 3 \cdot (y-3)$

$\text{LCM} = 2 \cdot 3 \cdot (y-3)$, or $6(y-3)$

47. $x^2 - 2x - 15 = (x-5)(x+3)$

$x^2 - 9 = (x-3)(x+3)$

$\text{LCM} = (x-5)(x-3)(x+3)$

49. $t^3 + 4t^2 + 4t = t(t^2 + 4t + 4) = t(t+2)(t+2)$

$t^2 - 4t = t(t-4)$

$\text{LCM} = t(t+2)(t+2)(t-4) = t(t+2)^2(t-4)$

51. $6xz^2 = 2 \cdot 3 \cdot x \cdot z \cdot z$

$8x^2y = 2 \cdot 2 \cdot 2 \cdot x \cdot x \cdot y$

$15y^3z = 3 \cdot 5 \cdot y \cdot y \cdot y \cdot z$

$\text{LCM} = 2 \cdot 2 \cdot 2 \cdot 3 \cdot 5 \cdot x \cdot x \cdot y \cdot y \cdot y \cdot z \cdot z = 120x^2y^3z^2$

53. $a + 1 = a + 1$

$(a-1)^2 = (a-1)(a-1)$

$a^2 - 1 = (a+1)(a-1)$

$\text{LCM} = (a+1)(a-1)(a-1) = (a+1)(a-1)^2$

55. $2n^2 + n - 1 = (2n-1)(n+1)$

$2n^2 + 3n - 2 = (2n-1)(n+2)$

$\text{LCM} = (2n-1)(n+1)(n+2)$

57. $6x^3 - 24x^2 + 18x = 6x(x^2 - 4x + 3)$

$= 2 \cdot 3 \cdot x(x-1)(x-3)$

$4x^5 - 24x^4 + 20x^3 = 4x^3(x^2 - 6x + 5)$

$= 2 \cdot 2 \cdot x \cdot x \cdot x(x-1)(x-5)$

$\text{LCM} = 2 \cdot 2 \cdot 3 \cdot x \cdot x \cdot x(x-1)(x-3)(x-5)$

$= 12x^3(x-1)(x-3)(x-5)$

59. $2x^3 - 2 = 2(x^3 - 1) = 2(x-1)(x^2 + x + 1)$

$x^2 - 1 = (x+1)(x-1)$

$\text{LCM} = 2(x+1)(x-1)(x^2 + x + 1)$

61. $6t^4 = 2 \cdot 3 \cdot t \cdot t \cdot t \cdot t$

$18t^2 = 2 \cdot 3 \cdot 3 \cdot t \cdot t$

The LCD is $2 \cdot 3 \cdot 3 \cdot t \cdot t \cdot t \cdot t$, or $18t^4$.

$$\frac{5}{6t^4} \cdot \frac{3}{3} = \frac{15}{18t^4} \text{ and}$$

$$\frac{s}{18t^2} \cdot \frac{t^2}{t^2} = \frac{st^2}{18t^4}$$

63. $3x^4y^2 = 3 \cdot x \cdot x \cdot x \cdot x \cdot y \cdot y$

$9xy^3 = 3 \cdot 3 \cdot x \cdot y \cdot y \cdot y$

The LCD is $3 \cdot 3 \cdot x \cdot x \cdot x \cdot x \cdot y \cdot y \cdot y$, or $9x^4y^3$.

$$\frac{7}{3x^4y^2} \cdot \frac{3y}{3y} = \frac{21y}{9x^4y^3} \text{ and}$$

$$\frac{4}{9xy^3} \cdot \frac{x^3}{x^3} = \frac{4x^3}{9x^4y^3}$$

65. The LCD is $(x+2)(x-2)(x+3)$. (See Exercise 47.)

$$\frac{2x}{x^2-4} = \frac{2x}{(x+2)(x-2)} \cdot \frac{x+3}{x+3} = \frac{2x(x+3)}{(x+2)(x-2)(x+3)}$$

$$\frac{4x}{x^2+5x+6} = \frac{4x}{(x+3)(x+2)} \cdot \frac{x-2}{x-2} = \frac{4x(x-2)}{(x+3)(x+2)(x-2)}$$

67. *Writing Exercise.* If the numbers have a common factor, their product contains that factor more than the greatest number of times it occurs in any one factorization. In this case, their product is not their least common multiple.

69. $-\frac{5}{8} = \frac{-5}{8} = \frac{5}{-8}$

71. $-(x-y) = -x + y$, or $y - x$

73. $-1(2x-7) = -2x + 7$ or $7 - 2x$

75. *Writing Exercise.* The polynomials contain no common factors other than constants.

77.
$$\frac{6x-1}{x-1} + \frac{3(2x+5)}{x-1} + \frac{3(2x-3)}{x-1} = \frac{6x-1+6x+15+6x-9}{x-1} = \frac{18x+5}{x-1}$$

79.
$$\frac{x^2}{3x^2-5x-2} - \frac{2x}{3x+1} \cdot \frac{1}{x-2}$$

$$= \frac{x^2}{(3x+1)(x-2)} - \frac{2x}{(3x+1)(x-2)}$$

$$= \frac{x^2-2x}{(3x+1)(x-2)}$$

$$= \frac{x(\cancel{x-2})}{(3x+1)(\cancel{x-2})}$$

$$= \frac{x}{3x+1}$$

81. The smallest number of strands that can be used is the LCM of 10 and 3.

$10 = 2 \cdot 5$

$3 = 3$

$\text{LCM} = 2 \cdot 5 \cdot 3 = 30$

The smallest number of strands that can be used is 30.

83. If the number of strands must also be a multiple of 4, we find the smallest multiple of 30 that is also a multiple of 4.

$1 \cdot 30 = 30$, not a multiple of 4

$2 \cdot 30 = 60 = 15 \cdot 4$, a multiple of 4

The smallest number of strands that can be used is 60.

85. $4x^2 - 25 = (2x+5)(2x-5)$

$6x^2 - 7x - 20 = (3x+4)(2x-5)$

$(9x^2 + 24x + 16)^2 = [(3x+4)(3x+4)]^2$
$= (3x+4)(3x+4)(3x+4)(3x+4)$

$\text{LCM} = (2x+5)(2x-5)(3x+4)^4$

87. The first copier prints 20 pages per minute, which is $\frac{20}{60}$ or $\frac{1}{3}$ copy per second. The second copier prints 18 pages per minutes, which is $\frac{18}{60}$ or $\frac{3}{10}$ copy per second. The time it takes until the machines begin copying a page at exactly the same time again is the LCM of their copying rates.

$3 = 3$

$10 = 2 \cdot 5$

$\text{LCM} = 2 \cdot 3 \cdot 5 = 30$

It takes 30 seconds

89. The number of minutes after 5:00 A.M. when the shuttles will first leave at the same time again is the LCM of their departure intervals, 25 minutes and 35 minutes.

$25 = 5 \cdot 5$

$35 = 5 \cdot 7$

$\text{LCM} = 5 \cdot 5 \cdot 7$, or 175

Thus, the shuttles will leave at the same time 175 minutes after 5:00 A.M., or at 7:55 A.M.

91. *Writing Exercise.* Evaluate both expressions for some value of the variable for which both are defined. If the results are the same, we can conclude that the answer is probably correct.

Exercise Set 6.4

1. To add or subtract when denominators are different, first find the LCD.

3. Add or subtract the numerators, as indicated. Write the sum or difference over the LCD.

5. $\frac{3}{x^2} + \frac{5}{x} = \frac{3}{x \cdot x} + \frac{5}{x}$ LCD $= x \cdot x$, or x^2

$$= \frac{3}{x \cdot x} + \frac{5}{x} \cdot \frac{x}{x}$$

$$= \frac{3+5x}{x^2}$$

7. $\left.\begin{array}{l} 6r = 2 \cdot 3 \cdot r \\ 8r = 2 \cdot 2 \cdot 2 \cdot r \end{array}\right\}$ LCD $= 2 \cdot 2 \cdot 2 \cdot 3 \cdot r$, or $24r$

$$\frac{1}{6r} - \frac{3}{8r} = \frac{1}{6r} \cdot \frac{4}{4} - \frac{3}{8r} \cdot \frac{3}{3}$$

$$= \frac{4-9}{24r}$$

$$= \frac{-5}{24r}, \text{ or } -\frac{5}{24r}$$

9. $\left.\begin{array}{l} uv^2 = u \cdot v \cdot v \\ u^3v = u \cdot u \cdot u \cdot v \end{array}\right\}$ LCD $= u \cdot u \cdot u \cdot v \cdot v$, or u^3v^2

$$\frac{3}{uv^2} + \frac{4}{u^3v} = \frac{3}{uv^2} \cdot \frac{u^2}{u^2} + \frac{4}{u^3v} \cdot \frac{v}{v} = \frac{3u^2+4v}{u^3v^2}$$

11. $\left.\begin{array}{l} 3xy^2 = 3 \cdot x \cdot y \cdot y \\ x^2y^3 = x \cdot x \cdot y \cdot y \cdot y \end{array}\right\}$ LCD $= 3 \cdot x \cdot x \cdot y \cdot y \cdot y$, or $3x^2y^3$

$$\frac{-2}{3xy^2} - \frac{6}{x^2y^3} = \frac{-2}{3xy^2} \cdot \frac{xy}{xy} - \frac{6}{x^2y^3} \cdot \frac{3}{3} = \frac{-2xy-18}{3x^2y^3}$$

13. $\left.\begin{array}{l} 8 = 2 \cdot 2 \cdot 2 \\ 6 = 2 \cdot 3 \end{array}\right\}$ LCD $= 2 \cdot 2 \cdot 2 \cdot 3$, or 24

$$\frac{x+3}{8} + \frac{x-2}{6} = \frac{x+3}{8} \cdot \frac{3}{3} + \frac{x-2}{6} \cdot \frac{4}{4}$$

$$= \frac{3(x+3)+4(x-2)}{24}$$

$$= \frac{3x+9+4x-8}{24}$$

$$= \frac{7x+1}{24}$$

15. $\left.\begin{array}{l} 6 = 2 \cdot 3 \\ 3 = 3 \end{array}\right\}$ LCD $= 2 \cdot 3$, or 6

$$\frac{x-2}{6} - \frac{x+1}{3} = \frac{x-2}{6} - \frac{x+1}{3} \cdot \frac{2}{2}$$

$$= \frac{x-2}{6} - \frac{2x+2}{6}$$

$$= \frac{x-2-(2x+2)}{6}$$

$$= \frac{x-2-2x-2}{6}$$

$$= \frac{-x-4}{6}, \text{ or } \frac{-(x+4)}{6}$$

17. $\left.\begin{array}{l} 15a = 3 \cdot 5 \cdot a \\ 3a^2 = 3 \cdot a \cdot a \end{array}\right\}$ LCD $= 5 \cdot 3a \cdot a$, or $15a^2$

$$\frac{a+3}{15a} + \frac{2a-1}{3a^2} = \frac{a+3}{15a} \cdot \frac{a}{a} + \frac{2a-1}{3a^2} \cdot \frac{5}{5}$$

$$= \frac{a^2+3a+10a-5}{15a^2}$$

$$= \frac{a^2+13a-5}{15a^2}$$

19. $\left.\begin{array}{l}3z = 3\cdot z\\ 4z = 2\cdot 2\cdot z\end{array}\right\}$ LCD $= 2\cdot 2\cdot 3\cdot z$, or $12z$

$$\frac{4z-9}{3z}-\frac{3z-8}{4z}=\frac{4z-9}{3z}\cdot\frac{4}{4}-\frac{3z-8}{4z}\cdot\frac{3}{3}$$
$$=\frac{16z-36}{12z}-\frac{9z-24}{12z}$$
$$=\frac{16z-36-(9z-24)}{12z}$$
$$=\frac{16z-36-9z+24}{12z}$$
$$=\frac{7z-12}{12z}$$

21. $\left.\begin{array}{l}cd^2 = c\cdot d\cdot d\\ c^2d = c\cdot c\cdot d\end{array}\right\}$ LCD $= c\cdot c\cdot d\cdot d$, or c^2d^2

$$\frac{3c+d}{cd^2}+\frac{c-d}{c^2d}=\frac{3c+d}{cd^2}\cdot\frac{c}{c}+\frac{c-d}{c^2d}\cdot\frac{d}{d}$$
$$=\frac{c(3c+d)+d(c-d)}{c^2d^2}$$
$$=\frac{3c^2+cd+cd-d^2}{c^2d^2}$$
$$=\frac{3c^2+2cd-d^2}{c^2d^2}$$

23. $\left.\begin{array}{l}3xt^2 = 3\cdot x\cdot t\cdot t\\ x^2t = x\cdot x\cdot t\end{array}\right\}$ LCD $= 3\cdot x\cdot x\cdot t\cdot t$, or $3x^2t^2$

$$\frac{4x+2t}{3xt^2}-\frac{5x-3t}{x^2t}$$
$$=\frac{4x+2t}{3xt^2}\cdot\frac{x}{x}-\frac{5x-3t}{x^2t}\cdot\frac{3t}{3t}$$
$$=\frac{4x^2+2tx}{3x^2t^2}-\frac{15xt-9t^2}{3x^2t^2}$$
$$=\frac{4x^2+2tx-(15xt-9t^2)}{3x^2t^2}$$
$$=\frac{4x^2+2tx-15xt+9t^2}{3x^2t^2}$$
$$=\frac{4x^2-13xt+9t^2}{3x^2t^2}$$

(Although $4x^2-13xt+9t^2$ can be factored, doing so will not enable us to simplify the result further.)

25. The denominators cannot be factored, so the LCD is their product, $(x-2)(x+2)$.

$$\frac{3}{x-2}+\frac{3}{x+2}=\frac{3}{x-2}\cdot\frac{x+2}{x+2}+\frac{3}{x+2}\cdot\frac{x-2}{x-2}$$
$$=\frac{3(x+2)+3(x-2)}{(x-2)(x+2)}$$
$$=\frac{3x+6+3x-6}{(x-2)(x+2)}$$
$$=\frac{6x}{(x-2)(x+2)}$$

27. $\dfrac{t}{t+3}-\dfrac{1}{t-1}$ LCD $= (t+3)(t-1)$

$$=\frac{t}{t+3}\cdot\frac{t-1}{t-1}-\frac{1}{t-1}\cdot\frac{t+3}{t+3}$$
$$=\frac{t^2-t}{(t+3)(t-1)}-\frac{t+3}{(t+3)(t-1)}$$
$$=\frac{t^2-t-(t+3)}{(t+3)(t-1)}$$
$$=\frac{t^2-t-t-3}{(t+3)(t-1)}$$
$$=\frac{t^2-2t-3}{(t+3)(t-1)}$$

(Although t^2-2t-3 can be factored, doing so will not enable us to simplify the result further.)

29. $\left.\begin{array}{l}3x = 3\cdot x\\ x+1 = x+1\end{array}\right\}$ LCD $= 3x(x+1)$

$$\frac{3}{x+1}+\frac{2}{3x}=\frac{3}{x+1}\cdot\frac{3x}{3x}+\frac{2}{3x}\cdot\frac{x+1}{x+1}$$
$$=\frac{9x+2(x+1)}{3x(x+1)}$$
$$=\frac{9x+2x+2}{3x(x+1)}$$
$$=\frac{11x+2}{3x(x+1)}$$

31. $\dfrac{3}{2t^2-2t}-\dfrac{5}{2t-2}$

$$=\frac{3}{2t(t-1)}-\frac{5}{2(t-1)}$$
$$=\frac{3}{2t(t-1)}-\frac{5}{2(t-1)}\cdot\frac{t}{t}$$
$$=\frac{3-5t}{2t(t-1)}$$

33. $\dfrac{3a}{a^2-9}+\dfrac{a}{a+3}$

$$=\frac{3a}{(a+3)(a-3)}+\frac{a}{a+3}$$

LCD $= (a+3)(a-3)$

$$=\frac{3a}{(a+3)(a-3)}+\frac{a}{a+3}\cdot\frac{a-3}{a-3}$$
$$=\frac{3a+a(a-3)}{(a+3)(a-3)}$$
$$=\frac{3a+a^2-3a}{(a+3)(a-3)}$$
$$=\frac{a^2}{(a+3)(a-3)}$$

35. $\frac{6}{z+4} - \frac{2}{3z+12} = \frac{6}{z+4} - \frac{2}{3(z+4)}$

LCD $= 3(z+4)$

$= \frac{6}{z+4}\cdot\frac{3}{3} - \frac{2}{3(z+4)}$

$= \frac{18}{3(z+4)} - \frac{2}{3(z+4)}$

$= \frac{16}{3(z+4)}$

37. $\frac{5}{q-1} + \frac{2}{(q-1)^2}$

$= \frac{5}{q-1}\cdot\frac{q-1}{q-1} + \frac{2}{(q-1)^2}$

$= \frac{5(q-1)+2}{(q-1)^2}$

$= \frac{5q-5+2}{(q-1)^2}$

$= \frac{5q-3}{(q-1)^2}$

39. $\frac{t-3}{t^3-1} - \frac{2}{1-t^3} = \frac{t-3}{t^3-1} + \frac{2}{t^3-1}$

$= \frac{t-3+2}{(t-1)(t^2+t+1)} = \frac{t-1}{(t-1)(t^2+t+1)}$

$= \frac{1}{t^2+t+1}$

41. $\frac{3a}{4a-20} + \frac{9a}{6a-30}$

$= \frac{3a}{2\cdot 2(a-5)} + \frac{9a}{2\cdot 3(a-5)}$

LCD $= 2\cdot 2\cdot 3(a-5)$

$= \frac{3a}{2\cdot 2(a-5)}\cdot\frac{3}{3} + \frac{9a}{2\cdot 3(a-5)}\cdot\frac{2}{2}$

$= \frac{9a+18a}{2\cdot 2\cdot 3(a-5)}$

$= \frac{27a}{2\cdot 2\cdot 3(a-5)}$

$= \frac{\cancel{3}\cdot 9\cdot a}{2\cdot 2\cdot \cancel{3}(a-5)}$

$= \frac{9a}{4(a-5)}$

43. $\frac{x}{x-5} + \frac{x}{5-x} = \frac{x}{x-5} + \frac{x}{5-x}\cdot\frac{-1}{-1}$

$= \frac{x}{x-5} + \frac{-x}{x-5}$

$= 0$

45. $\frac{6}{a^2+a-2} + \frac{4}{a^2-4a+3}$

$= \frac{6}{(a+2)(a-1)} + \frac{4}{(a-3)(a-1)}$

LCD $= (a+2)(a-1)(a-3)$

$= \frac{6}{(a+2)(a-1)}\cdot\frac{a-3}{a-3} + \frac{4}{(a-3)(a-1)}\cdot\frac{a+2}{a+2}$

$= \frac{6(a-3)+4(a+2)}{(a+2)(a-1)(a-3)}$

$= \frac{6a-18+4a+8}{(a+2)(a-1)(a-3)}$

$= \frac{10a-10}{(a+2)(a-1)(a-3)}$

$= \frac{10\cancel{(a-1)}}{(a+2)\cancel{(a-1)}(a-3)}$

$= \frac{10}{(a+2)(a-3)}$

47. $\frac{x}{x^2+9x+20} - \frac{4}{x^2+7x+12}$

$= \frac{x}{(x+4)(x+5)} - \frac{4}{(x+3)(x+4)}$

LCD $= (x+3)(x+4)(x+5)$

$= \frac{x}{(x+4)(x+5)}\cdot\frac{x+3}{x+3} - \frac{4}{(x+3)(x+4)}\cdot\frac{x+5}{x+5}$

$= \frac{x(x+3)-4(x+5)}{(x+3)(x+4)(x+5)}$

$= \frac{x^2+3x-4x-20}{(x+3)(x+4)(x+5)}$

$= \frac{x^2-x-20}{(x+3)(x+4)(x+5)}$

$= \frac{\cancel{(x+4)}(x-5)}{(x+3)\cancel{(x+4)}(x+5)}$

$= \frac{x-5}{(x+3)(x+5)}$

49. $\frac{3z}{z^2-4z+4} + \frac{10}{z^2+z-6}$

$= \frac{3z}{(z-2)^2} + \frac{10}{(z-2)(z+3)}$,

LCD $= (z-2)^2(z+3)$

$= \frac{3z}{(z-2)^2}\cdot\frac{z+3}{z+3} + \frac{10}{(z-2)(z+3)}\cdot\frac{z-2}{z-2}$

$= \frac{3z(z+3)+10(z-2)}{(z-2)^2(z+3)}$

$= \frac{3z^2+9z+10z-20}{(z-2)^2(z+3)}$

$= \frac{3z^2+19z-20}{(z-2)^2(z+3)}$

51. $\frac{-7}{x^2+25x+24} - \frac{0}{x^2+11x+10}$

Note that $\dfrac{0}{x^2+11x+10} = 0$, so the difference is $\dfrac{-7}{x^2+25x+24}$.

53. $$\frac{5x}{4} - \frac{x-2}{-4} = \frac{5x}{4} - \frac{x-2}{-4}\cdot\frac{-1}{-1}$$
$$= \frac{5x}{4} - \frac{2-x}{4}$$
$$= \frac{5x-(2-x)}{4}$$
$$= \frac{5x-2+x}{4}$$
$$= \frac{6x-2}{4}$$
$$= \frac{2(3x-1)}{2\cdot 2}$$
$$= \frac{\cancel{2}(3x-1)}{\cancel{2}\cdot 2}$$
$$= \frac{3x-1}{2}$$

55. $$\frac{y^2}{y-3} + \frac{9}{3-y} = \frac{y^2}{y-3} + \frac{9}{3-y}\cdot\frac{-1}{-1}$$
$$= \frac{y^2}{y-3} + \frac{-9}{-3+y}$$
$$= \frac{y^2-9}{y-3}$$
$$= \frac{(y+3)\cancel{(y-3)}}{\cancel{y-3}}$$
$$= y+3$$

57. $$\frac{c-5}{c^2-64} + \frac{c-5}{64-c^2} = \frac{c-5}{c^2-64} + \frac{c-5}{64-c^2}\cdot\frac{-1}{-1}$$
$$= \frac{c-5}{c^2-64} + \frac{5-c}{c^2-64}$$
$$= \frac{c-5+5-c}{c^2-64}$$
$$= \frac{0}{c^2-64}$$
$$= 0$$

59. $$\frac{4-p}{25-p^2} + \frac{p+1}{p-5}$$
$$= \frac{4-p}{(5+p)(5-p)} + \frac{p+1}{p-5}$$
$$= \frac{4-p}{(5+p)(5-p)}\cdot\frac{-1}{-1} + \frac{p+1}{p-5}$$
$$= \frac{p-4}{(p+5)(p-5)} + \frac{p+1}{p-5} \quad \text{LCD} = (p+5)(p-5)$$
$$= \frac{p-4}{(p+5)(p-5)} + \frac{p+1}{p-5}\cdot\frac{p+5}{p+5}$$
$$= \frac{p-4+p^2+6p+5}{(p+5)(p-5)}$$
$$= \frac{p^2+7p+1}{(p+5)(p-5)}$$

61. $$\frac{x}{x-4} - \frac{3}{16-x^2}$$
$$= \frac{x}{x-4} - \frac{3}{(4+x)(4-x)}$$
$$= \frac{x}{x-4}\cdot\frac{-1}{-1} - \frac{3}{(4+x)(4-x)}$$
$$= \frac{-x}{4-x} - \frac{3}{(4+x)(4-x)} \quad \text{LCD} = (4-x)(4+x)$$
$$= \frac{-x}{4-x}\cdot\frac{4+x}{4+x} - \frac{3}{(4+x)(4-x)}$$
$$= \frac{-x(4+x)-3}{(4-x)(4+x)}$$
$$= \frac{-4x-x^2-3}{(4-x)(4+x)}$$
$$= \frac{-x^2-4x-3}{(4-x)(4+x)}, \text{ or } \frac{x^2+4x+3}{(x+4)(x-4)}$$

(Although x^2+4x+3 can be factored, doing so will not enable us to simplify the result further.)

63. $$\frac{a}{a^2-1} + \frac{2a}{a-a^2} = \frac{a}{a^2-1} + \frac{2\cdot\cancel{a}}{\cancel{a}(1-a)}$$
$$= \frac{a}{(a+1)(a-1)} + \frac{2}{1-a}$$
$$= \frac{a}{(a+1)(a-1)} + \frac{2}{1-a}\cdot\frac{-1}{-1}$$
$$= \frac{a}{(a+1)(a-1)} + \frac{-2}{a-1}$$
$$\text{LCD} = (a+1)(a-1)$$
$$= \frac{a}{(a+1)(a-1)} + \frac{-2}{a-1}\cdot\frac{a+1}{a+1}$$
$$= \frac{a-2a-2}{(a+1)(a-1)}$$
$$= \frac{-a-2}{(a+1)(a-1)}, \text{ or}$$
$$= \frac{a+2}{(1+a)(1-a)}$$

65. $$\frac{4x}{x^2-y^2} - \frac{6}{y-x}$$
$$= \frac{4x}{(x+y)(x-y)} - \frac{6}{y-x}$$
$$= \frac{4x}{(x+y)(x-y)} - \frac{6}{y-x}\cdot\frac{-1}{-1}$$
$$= \frac{4x}{(x+y)(x-y)} - \frac{-6}{x-y} \quad \text{LCD} = (x+y)(x-y)$$
$$= \frac{4x}{(x+y)(x-y)} - \frac{-6}{x-y}\cdot\frac{x+y}{x+y}$$
$$= \frac{4x-(-6)(x+y)}{(x+y)(x-y)}$$
$$= \frac{4x+6x+6y}{(x+y)(x-y)}$$
$$= \frac{10x+6y}{(x+y)(x-y)}$$

(Although $10x+6y$ can be factored, doing so will not enable us to simplify the result further.)

67. $\dfrac{x-3}{2-x}-\dfrac{x+3}{x+2}+\dfrac{x+6}{4-x^2}$

$=\dfrac{x-3}{2-x}-\dfrac{x+3}{x+2}+\dfrac{x+6}{(2+x)(2-x)}$

LCD $=(2+x)(2-x)$

$=\dfrac{x-3}{2-x}\cdot\dfrac{2+x}{2+x}-\dfrac{x+3}{x+2}\cdot\dfrac{2-x}{2-x}+\dfrac{x+6}{(2+x)(2-x)}$

$=\dfrac{(x-3)(2+x)-(x+3)(2-x)+(x+6)}{(2+x)(2-x)}$

$=\dfrac{x^2-x-6-(-x^2-x+6)+x+6}{(2+x)(2-x)}$

$=\dfrac{x^2-x-6+x^2+x-6+x+6}{(2+x)(2-x)}$

$=\dfrac{2x^2+x-6}{(2+x)(2-x)}$

$=\dfrac{(2x-3)\cancel{(x+2)}}{\cancel{(2+x)}(2-x)}$

$=\dfrac{2x-3}{2-x}$

69. $\dfrac{2x+5}{x+1}+\dfrac{x+7}{x+5}-\dfrac{5x+17}{(x+1)(x+5)}$

LCD $=(x+1)(x+5)$

$=\dfrac{(2x+5)(x+5)+(x+7)(x+1)-(5x+17)}{(x+1)(x+5)}$

$=\dfrac{2x^2+15x+25+x^2+8x+7-5x-17}{(x+1)(x+5)}$

$=\dfrac{3x^2+18x+15}{(x+1)(x+5)}$

$=\dfrac{3\cancel{(x+1)}\cancel{(x+5)}}{\cancel{(x+1)}\cancel{(x+5)}}$

$=3$

71. $\dfrac{1}{x+y}+\dfrac{1}{x-y}-\dfrac{2x}{x^2-y^2}$

LCD $=(x+y)(x-y)$

$=\dfrac{1}{x+y}\cdot\dfrac{x-y}{x-y}+\dfrac{1}{x-y}\cdot\dfrac{x+y}{x+y}-\dfrac{2x}{(x+y)(x-y)}$

$=\dfrac{(x-y)+(x+y)-2x}{(x+y)(x-y)}$

$=0$

73. *Writing Exercise.* Using the least common denominator usually reduces the complexity of computations and requires less simplification of the sum or difference.

75. $\dfrac{-3}{8}\div\dfrac{11}{4}=\dfrac{-3}{8}\cdot\dfrac{4}{11}=-\dfrac{3\cdot\cancel{4}}{2\cdot\cancel{4}\cdot 11}=\dfrac{-3}{22}$

77. $\dfrac{\frac{3}{4}}{\frac{5}{6}}=\dfrac{3}{4}\cdot\dfrac{6}{5}=\dfrac{3\cdot 3\cdot\cancel{2}}{\cancel{2}\cdot 2\cdot 5}=\dfrac{9}{10}$

79. $\dfrac{2x+6}{x-1}\div\dfrac{3x+9}{x-1}=\dfrac{2x+6}{x-1}\cdot\dfrac{x-1}{3x+9}=\dfrac{2\cancel{(x+3)}\cancel{(x-1)}}{\cancel{(x-1)}3\cancel{(x+3)}}=\dfrac{2}{3}$

81. *Writing Exercise.* Their sum is zero. Another explanation is that $-\left(\dfrac{1}{3-x}\right)=\dfrac{1}{-(3-x)}=\dfrac{1}{x-3}$.

83. $P=2\left(\dfrac{3}{x+4}\right)+2\left(\dfrac{2}{x-5}\right)$

$=\dfrac{6}{x+4}+\dfrac{4}{x-5}$ LCD $=(x+4)(x-5)$

$=\dfrac{6}{x+4}\cdot\dfrac{x-5}{x-5}+\dfrac{4}{x-5}\cdot\dfrac{x+4}{x+4}$

$=\dfrac{6x-30+4x+16}{(x+4)(x-5)}$

$=\dfrac{10x-14}{(x+4)(x-5)}$, or $\dfrac{10x-14}{x^2-x-20}$

$A=\left(\dfrac{3}{x+4}\right)\left(\dfrac{2}{x-5}\right)=\dfrac{6}{(x+4)(x-5)}$

85. $\dfrac{x^2}{3x^2-5x-2}-\dfrac{2x}{3x+1}\cdot\dfrac{1}{x-2}$

$=\dfrac{x^2}{(3x+1)(x-2)}-\dfrac{2x}{(3x+1)(x-2)}$

$=\dfrac{x^2-2x}{(3x+1)(x-2)}$

$=\dfrac{x(x-2)}{(3x+1)(x-2)}$

$=\dfrac{x}{3x+1}\cdot\dfrac{x-2}{x-2}$

$=\dfrac{x}{3x+1}$

87. We recognize that this is the product of the sum and difference of two terms: $(A+B)(A-B)=A^2-B^2$.

$\left(\dfrac{x}{x+7}-\dfrac{3}{x+2}\right)\left(\dfrac{x}{x+7}+\dfrac{3}{x+2}\right)$

$=\dfrac{x^2}{(x+7)^2}-\dfrac{9}{(x+2)^2}$ LCD $=(x+7)^2(x+2)^2$

$=\dfrac{x^2}{(x+7)^2}\cdot\dfrac{(x+2)^2}{(x+2)^2}-\dfrac{9}{(x+2)^2}\cdot\dfrac{(x+7)^2}{(x+7)^2}$

$=\dfrac{x^2(x+2)^2-9(x+7)^2}{(x+7)^2(x+2)^2}$

$=\dfrac{x^2(x^2+4x+4)-9(x^2+14x+49)}{(x+7)^2(x+2)^2}$

$=\dfrac{x^4+4x^3+4x^2-9x^2-126x-441}{(x+7)^2(x+2)^2}$

$=\dfrac{x^4+4x^3-5x^2-126x-441}{(x+7)^2(x+2)^2}$

89. $\left(\frac{a}{a-b}+\frac{b}{a+b}\right)\left(\frac{1}{3a+b}+\frac{2a+6b}{9a^2-b^2}\right)$

$=\frac{a}{(a-b)(3a+b)}+\frac{a(2a+6b)}{(a-b)(9a^2-b^2)}+$

$\frac{b}{(a+b)(3a+b)}+\frac{b(2a+6b)}{(a+b)(9a^2-b^2)}$

$=\frac{a}{(a-b)(3a+b)}+\frac{2a^2+6ab}{(a-b)(3a+b)(3a-b)}+$

$\frac{b}{(a+b)(3a+b)}+\frac{2ab+6b^2}{(a+b)(3a+b)(3a-b)}$

LCD $=(a-b)(a+b)(3a+b)(3a-b)$

$=[a(a+b)(3a-b)+(2a^2+6ab)(a+b)+$

$b(a-b)(3a-b)+(2ab+6b^2)(a-b)]/$

$[(a-b)(a+b)(3a+b)(3a-b)]$

$=(3a^3+2a^2b-ab^2+2a^3+8a^2b+6ab^2+b^3-$

$4ab^2+3a^2b+4ab^2-6b^3+2a^2b)/$

$[(a-b)(a+b)(3a+b)(3a-b)]$

$=\frac{5a^3+15a^2b+5ab^2-5b^3}{(a-b)(a+b)(3a+b)(3a-b)}$

$=\frac{5\cancel{(a+b)}(a^2+2ab-b^2)}{(a-b)\cancel{(a+b)}(3a+b)(3a-b)}$

$=\frac{5(a^2+2ab-b^2)}{(a-b)(3a+b)(3a-b)}$

91. Answers may vary. $\frac{a}{a-b}+\frac{3b}{b-a}$

93. *Writing Exercise.* Both y_1 and y_2 are undefined when $x=5$.

6.4 Connecting the Concepts

1. $=\frac{3}{5x}+\frac{2}{x^2}$ Addition

$=\frac{3}{5x}\cdot\frac{x}{x}+\frac{2}{x^2}\cdot\frac{5}{5}$ LCD $=5x^2$

$=\frac{3x+10}{5x^2}$

3. $=\frac{3}{5x}\div\frac{2}{x^2}$ Division

$=\frac{3}{5x}\cdot\frac{x^2}{2}$

$=\frac{3x}{10}\cdot\frac{x}{x}$

$=\frac{3x}{10}$

5. $=\frac{2x-6}{5x+10}\cdot\frac{x+2}{6x-12}$ Multiplication

$=\frac{\cancel{2}(x-3)\cancel{(x+2)}}{5\cancel{(x+2)}\cancel{2}\cdot 3(x-2)}$

$=\frac{(x-3)}{15(x-2)}$

7. $=\frac{2}{x-5}\div\frac{6}{x-5}$ Division

$=\frac{2}{x-5}\cdot\frac{x-5}{6}$

$=\frac{1}{3}\cdot\frac{2(x-5)}{2(x-5)}$

$=\frac{1}{3}$

9. $=\frac{2}{x+3}+\frac{3}{x+4}$ Addition

$=\frac{2}{x+3}\cdot\frac{x+4}{x+4}+\frac{3}{x+4}\cdot\frac{x+3}{x+3}$ LCD $=(x+3)(x+4)$

$=\frac{2(x+4)+3(x+3)}{(x+3)(x+4)}$

$=\frac{2x+8+3x+9}{(x+3)(x+4)}$

$=\frac{5x+17}{(x+3)(x+4)}$

11. $=\frac{3}{x-4}-\frac{2}{4-x}$ Subtraction

$=\frac{3}{x-4}+\frac{2}{x-4}$

$=\frac{5}{x-4}$

13. $=\frac{a}{6a-9b}-\frac{b}{4a-6b}$ Subtraction

$=\frac{a}{3(2a-3b)}\cdot\frac{2}{2}-\frac{b}{2(2a-3b)}\cdot\frac{3}{3}$ LCD $=6(2a-3b)$

$=\frac{2a-3b}{6(2a-3b)}$

$=\frac{1}{6}$

15. $=\frac{x+1}{x^2-7x+10}+\frac{3}{x^2-x-2}$ Addition

$=\frac{x+1}{(x-5)(x-2)}\cdot\frac{x+1}{x+1}$

$+\frac{3}{(x-2)(x+1)}\cdot\frac{x-5}{x-5}$ LCD$=(x+1)(x-2)(x-5)$

$=\frac{x^2+2x+1+3x-15}{(x+1)(x-2)(x-5)}$

$=\frac{x^2+5x-14}{(x+1)(x-2)(x-5)}$

$=\frac{(x+7)\cancel{(x-2)}}{(x+1)\cancel{(x-2)}(x-5)}$

$=\frac{x+7}{(x+1)(x-5)}$

17. $=\frac{t+2}{10}+\frac{2t+1}{15}$ Addition

$=\frac{3(t+2)}{3(10)}+\frac{2(2t+1)}{2(15)}$ LCD $=30$

$=\frac{3t+6+4t+2}{30}=\frac{7t+8}{30}$

19. $=\frac{a^2-2a+1}{a^2-4}\div(a^2-3a+2)$ Division

$=\frac{(a-1)\cancel{(a-1)}}{(a+2)(a-2)}\cdot\frac{1}{(a-2)\cancel{(a-1)}}$

$=\frac{a-1}{(a+2)(a-2)^2}$

Exercise Set 6.5

1. The LCD is the LCM of x^2, x, 2, and $4x$. It is $4x^2$.

$$\frac{\frac{5}{x^2}+\frac{1}{x}}{\frac{7}{2}-\frac{3}{4x}}\cdot\frac{4x^2}{4x^2}=\frac{\frac{5}{x^2}\cdot 4x^2+\frac{1}{x}\cdot 4x^2}{\frac{7}{2}\cdot 4x^2-\frac{3}{4x}\cdot 4x^2}$$

Choice (d) is correct.

3. We subtract to get a single rational expression in the numerator and add to get a single rational expression in the denominator.

$$\frac{\frac{4}{5x}-\frac{1}{10}}{\frac{8}{x^2}+\frac{7}{2}}=\frac{\frac{4}{5x}\cdot\frac{2}{2}-\frac{1}{10}\cdot\frac{x}{x}}{\frac{8}{x^2}\cdot\frac{2}{2}+\frac{7}{2}\cdot\frac{x^2}{x^2}}$$

$$=\frac{\frac{8}{10x}-\frac{x}{10x}}{\frac{16}{2x^2}+\frac{7x^2}{2x^2}}$$

$$=\frac{\frac{8-x}{10x}}{\frac{16+7x^2}{2x^2}}$$

Choice (b) is correct.

5. $\dfrac{1+\frac{1}{4}}{2+\frac{3}{4}}$ LCD is 4

$=\dfrac{1+\frac{1}{4}}{2+\frac{3}{4}}\cdot\dfrac{4}{4}$ Multiplying by $\frac{4}{4}$

$=\dfrac{\left(1+\frac{1}{4}\right)4}{\left(2+\frac{3}{4}\right)4}$ Multiplying numerator and denominator by 4

$=\dfrac{1\cdot 4+\frac{1}{4}\cdot 4}{2\cdot 4+\frac{3}{4}\cdot 4}$

$=\dfrac{4+1}{8+3}$

$=\dfrac{5}{11}$

7. $\dfrac{\frac{1}{2}+\frac{1}{3}}{\frac{1}{4}-\frac{1}{6}}$

$=\dfrac{\frac{1}{2}\cdot\frac{3}{3}+\frac{1}{3}\cdot\frac{2}{2}}{\frac{1}{4}\cdot\frac{3}{3}-\frac{1}{6}\cdot\frac{2}{2}}$ Getting a common denominator in numerator and in denominator

$=\dfrac{\frac{3}{6}+\frac{2}{6}}{\frac{3}{12}-\frac{2}{12}}$

$=\dfrac{\frac{5}{6}}{\frac{1}{12}}$ Adding in the numerator; subtracting in the denominator

$=\dfrac{5}{6}\cdot\dfrac{12}{1}$ Multiplying by the reciprocal of the divisor

$=\dfrac{5\cdot 6\cdot 2}{6}$

$=\dfrac{5\cdot \not{6}\cdot 2}{\not{6}}$

$=10$

9. $\dfrac{\frac{x}{4}+x}{\frac{4}{x}+x}$ LCD is $4x$

$=\dfrac{\frac{x}{4}+x}{\frac{4}{x}+x}\cdot\dfrac{4x}{4x}$

$=\dfrac{\left(\frac{x}{4}+x\right)(4x)}{\left(\frac{4}{x}+x\right)(4x)}$

$=\dfrac{x^2+4x^2}{16+4x^2}$

$\dfrac{5x^2}{16+4x^2}$

11. $\dfrac{\frac{10}{t}}{\frac{2}{t^2}-\frac{5}{t}}$

$=\dfrac{\frac{10}{t}}{\frac{2}{t^2}-\frac{5}{t}}\cdot\dfrac{t^2}{t^2}$

$=\dfrac{\frac{10}{t}\cdot t^2}{\left(\frac{2}{t^2}-\frac{5}{t}\right)t^2}$

$=\dfrac{10t}{\frac{2}{t^2}\cdot t^2-\frac{5}{t}\cdot t^2}$

$=\dfrac{10t}{2-5t}$, or $\dfrac{-10t}{5t-2}$

13. $\dfrac{\dfrac{2a-5}{3a}}{\dfrac{a-7}{6a}}$

$= \dfrac{2a-5}{3a} \cdot \dfrac{6a}{a-7}$ Multiplying by the reciprocal of the divisor

$= \dfrac{(2a-5)\cdot 2\cdot 3a}{3a\cdot(a-7)}$

$= \dfrac{(2a-5)\cdot 2\cdot \cancel{3a}}{\cancel{3a}\cdot(a-7)}$

$= \dfrac{2(2a-5)}{a-7}$

$= \dfrac{4a-10}{a-7}$

15. $\dfrac{\dfrac{x}{6}-\dfrac{3}{x}}{\dfrac{1}{3}+\dfrac{1}{x}}$ LCD is $6x$

$= \dfrac{\dfrac{x}{6}-\dfrac{3}{x}}{\dfrac{1}{3}+\dfrac{1}{x}} \cdot \dfrac{6x}{6x}$

$= \dfrac{\dfrac{x}{6}\cdot 6x - \dfrac{3}{x}\cdot 6x}{\dfrac{1}{3}\cdot 6x + \dfrac{1}{x}\cdot 6x}$

$= \dfrac{x^2-18}{2x+6}$

17. $\dfrac{\dfrac{1}{s}-\dfrac{1}{5}}{\dfrac{s-5}{s}}$ LCD is $5s$

$= \dfrac{\dfrac{1}{s}-\dfrac{1}{5}}{\dfrac{s-5}{s}} \cdot \dfrac{5s}{5s}$

$= \dfrac{\dfrac{1}{s}\cdot 5s - \dfrac{1}{5}\cdot 5s}{\left(\dfrac{s-5}{s}\right)(5s)}$

$= \dfrac{5-s}{(s-5)(5)}$

$= \dfrac{-\cancel{(s-5)}}{\cancel{(s-5)}(5)}$

$= -\dfrac{1}{5}$

19. $\dfrac{\dfrac{1}{t^2}+1}{\dfrac{1}{t}-1}$ LCD is t^2

$= \dfrac{\dfrac{1}{t^2}+1}{\dfrac{1}{t}-1} \cdot \dfrac{t^2}{t^2}$

$= \dfrac{\dfrac{1}{t^2}\cdot t^2 + 1\cdot t^2}{\dfrac{1}{t}\cdot t^2 - 1\cdot t^2}$

$= \dfrac{1+t^2}{t-t^2}$

(Although the denominator can be factored, doing so will not enable us to simplify further.)

21. $\dfrac{\dfrac{x^2}{x^2-y^2}}{\dfrac{x}{x+y}}$

$= \dfrac{x^2}{x^2-y^2} \cdot \dfrac{x+y}{x}$ Multiplying by the reciprocal of the divisor

$= \dfrac{x^2(x+y)}{(x^2-y^2)(x)}$

$= \dfrac{x\cdot x\cdot(x+y)}{(x+y)(x-y)(x)}$

$= \dfrac{\cancel{x}\cdot x\cdot\cancel{(x+y)}}{\cancel{(x+y)}(x-y)(\cancel{x})}$

$= \dfrac{x}{x-y}$

23. $\dfrac{\dfrac{7}{c^2}+\dfrac{4}{c}}{\dfrac{6}{c}-\dfrac{3}{c^3}}$ LCD is c^3

$= \dfrac{\dfrac{7}{c^2}+\dfrac{4}{c}}{\dfrac{6}{c}-\dfrac{3}{c^3}} \cdot \dfrac{c^3}{c^3}$

$= \dfrac{\dfrac{7}{c^2}\cdot c^3 + \dfrac{4}{c}\cdot c^3}{\dfrac{6}{c}\cdot c^3 - \dfrac{3}{c^3}\cdot c^3}$

$= \dfrac{7c+4c^2}{6c^2-3}$

(Although the numerator and the denominator can be factored, doing so will not enable us to simplify further.)

25. $\dfrac{\dfrac{2}{7a^4}-\dfrac{1}{14a}}{\dfrac{3}{5a^2}+\dfrac{2}{15a}} = \dfrac{\dfrac{2}{7a^4}\cdot\dfrac{2}{2}-\dfrac{1}{14a}\cdot\dfrac{a^3}{a^3}}{\dfrac{3}{5a^2}\cdot\dfrac{3}{3}+\dfrac{2}{15a}\cdot\dfrac{a}{a}}$

$= \dfrac{\dfrac{4-a^3}{14a^4}}{\dfrac{9+2a}{15a^2}}$

$= \dfrac{4-a^3}{14a^4}\cdot\dfrac{15a^2}{9+2a}$

$= \dfrac{15\cdot\cancel{a^2}(4-a^3)}{14a^2\cdot\cancel{a^2}(9+2a)}$

$= \dfrac{15(4-a^3)}{14a^2(9+2a)}$, or $\dfrac{60-15a^3}{126a^2+28a^3}$

27. $\dfrac{\dfrac{x}{5y^3}+\dfrac{3}{10y}}{\dfrac{3}{10y}+\dfrac{x}{5y^3}}$

Observe that, by the commutative law of addition, the numerator and denominator are equivalent, so the result is 1.

29. $$\frac{\frac{3}{ab^4}+\frac{4}{a^3b}}{\frac{5}{a^3b}-\frac{3}{ab}} = \frac{\frac{3}{ab^4}\cdot\frac{a^2}{a^2}+\frac{4}{a^3b}\cdot\frac{b^3}{b^3}}{\frac{5}{a^3b}-\frac{3}{ab}\cdot\frac{a^2}{a^2}}$$

$$= \frac{\frac{3a^2+4b^3}{a^3b^4}}{\frac{5-3a^2}{a^3b}}$$

$$= \frac{3a^2+4b^3}{a^3b^4}\cdot\frac{a^3b}{5-3a^2}$$

$$= \frac{\cancel{a^3b}(3a^2+4b^3)}{\cancel{a^3b}\cdot b^3(5-3a^2)}$$

$$= \frac{3a^2+4b^3}{b^3(5-3a^2)},\text{ or }\frac{3a^2+4b^3}{5b^3-3a^2b^3}$$

31. $$\frac{t-\frac{9}{t}}{t+\frac{4}{t}} = \frac{t\cdot\frac{t}{t}-\frac{9}{t}}{t\cdot\frac{t}{t}+\frac{4}{t}}$$

$$= \frac{\frac{t^2-9}{t}}{\frac{t^2+4}{t}}$$

$$= \frac{t^2-9}{t}\cdot\frac{t}{t^2+4}$$

$$= \frac{\cancel{t}(t^2-9)}{\cancel{t}(t^2+4)}$$

$$= \frac{t^2-9}{t^2+4}$$

33. $$\frac{\frac{1}{a}+\frac{1}{b}}{\frac{1}{a^3}+\frac{1}{b^3}} = \frac{\frac{1}{a}+\frac{1}{b}}{\frac{1}{a^3}+\frac{1}{b^3}}\cdot\frac{a^3b^3}{a^3b^3}$$

$$= \frac{\frac{1}{a}\cdot a^3b^3+\frac{1}{b}\cdot a^3b^3}{\frac{1}{a^3}\cdot a^3b^3+\frac{1}{b^3}\cdot a^3b^3}$$

$$= \frac{a^2b^3+a^3b^2}{b^3+a^3}$$

$$= \frac{a^2b^2\,(b+a)}{(b+a)\,(b^2-ab+a^2)}$$

$$= \frac{a^2b^2}{b^2-ab+a^2}$$

35. $$\frac{3+\frac{4}{ab^3}}{\frac{3+a}{a^2b}} = \frac{3+\frac{4}{ab^3}}{\frac{3+a}{a^2b}}\cdot\frac{a^2b^3}{a^2b^3}$$

$$= \frac{3\cdot a^2b^3+\frac{4}{ab^3}\cdot a^2b^3}{\frac{3+a}{a^2b}\cdot a^2b^3}$$

$$= \frac{3a^2b^3+4a}{b^2(3+a)},\text{ or }\frac{3a^2b^3+4a}{3b^2+ab^2}$$

37. $$\frac{t+5+\frac{3}{t}}{t+2+\frac{1}{t}}\qquad\text{LCD is } t$$

$$= \frac{t+5+\frac{3}{t}}{t+2+\frac{1}{t}}\cdot\frac{t}{t}$$

$$= \frac{t\cdot t+5\cdot t+\frac{3}{t}\cdot t}{t\cdot t+2\cdot t+\frac{1}{t}\cdot t}$$

$$= \frac{t^2+5t+3}{t^2+2t+1}$$

$$= \frac{t^2+5t+3}{(t+1)^2}$$

39. $$\frac{x-2-\frac{1}{x}}{x-5-\frac{4}{x}} = \frac{x-2-\frac{1}{x}}{x-5-\frac{4}{x}}\cdot\frac{x}{x}$$

$$= \frac{x\cdot x-2\cdot x-\frac{1}{x}\cdot x}{x\cdot x-5\cdot x-\frac{4}{x}\cdot x}$$

$$= \frac{x^2-2x-1}{x^2-5x-4}$$

41. *Writing Exercise.* Yes; Method 2, multiplying by the LCD, does not require division of rational expressions.

43.
$$3x-5+2(4x-1) = 12x-3$$
$$3x-5+8x-2 = 12x-3$$
$$11x-7 = 12x-3$$
$$-7 = x-3$$
$$-4 = x$$

The solution is -4.

45. $$\frac{3}{4}x-\frac{5}{8} = \frac{3}{8}x+\frac{7}{4}\qquad\text{LCD is 8}$$
$$8\left(\frac{3}{4}x-\frac{5}{8}\right) = 8\left(\frac{3}{8}x+\frac{7}{4}\right)$$
$$8\cdot\frac{3}{4}x-8\cdot\frac{5}{8} = 8\cdot\frac{3}{8}x+8\cdot\frac{7}{4}$$
$$6x-5 = 3x+14$$
$$3x-5 = 14$$
$$3x = 19$$
$$x = \frac{19}{3}$$

The solution is $\frac{19}{3}$.

47.
$$x^2-7x+12 = 0$$
$$(x-3)\,(x-4) = 0$$
$$x-3=0,\text{ or } x-4=0$$
$$x=3\text{ or }x=4$$

49. *Writing Exercise.* Although either method could be used, Method 2 requires fewer steps.

51. $\dfrac{\dfrac{x-5}{x-6}}{\dfrac{x-7}{x-8}}$

This expression is undefined for any value of x that makes a denominator 0. We see that $x-6=0$ when $x=6$, $x-7=0$ when $x=7$, and $x-8=0$ when $x=8$, so the expression is undefined for the x-values 6, 7, and 8.

53. $\dfrac{\dfrac{2x+3}{5x+4}}{\dfrac{3}{7}-\dfrac{x^2}{21}}$

This expression is undefined for any value of x that makes a denominator 0. First we find the value of x for which $5x+4=0$.

$$5x+4=0$$
$$5x=-4$$
$$x=-\frac{4}{5}$$

Then we find the value of x for which $\dfrac{3}{7}-\dfrac{x^2}{21}=0$:

$$\frac{3}{7}-\frac{x^2}{21}=0$$
$$21\left(\frac{3}{7}-\frac{x^2}{21}\right)=21\cdot 0$$
$$21\cdot\frac{3}{7}-21\cdot\frac{x^2}{21}=0$$
$$9-x^2=0$$
$$9=x^2$$
$$\pm 3=x$$

The expression is undefined for the x-values $-\dfrac{4}{5}$, -3 and 3.

55. For the complex rational expression $\dfrac{\dfrac{A}{B}}{\dfrac{C}{D}}$ the LCD is BD.

$$=\frac{\dfrac{A}{B}\cdot BD}{\dfrac{C}{D}\cdot BD}$$
$$=\frac{\dfrac{ABD}{B}}{\dfrac{CBD}{D}}=\frac{\dfrac{A\cancel{B}D}{\cancel{B}}}{\dfrac{BC\cancel{D}}{\cancel{D}}}$$
$$=\frac{AD}{BC}$$
$$=\frac{A}{B}\cdot\frac{D}{C}$$

57.
$$\frac{\dfrac{x}{x+5}+\dfrac{3}{x+2}}{\dfrac{2}{x+2}-\dfrac{x}{x+5}}=\frac{\dfrac{x}{x+5}+\dfrac{3}{x+2}}{\dfrac{2}{x+2}-\dfrac{x}{x+5}}\cdot\frac{(x+5)(x+2)}{(x+5)(x+2)}$$
$$=\frac{x(x+2)+3(x+5)}{2(x+5)-x(x+2)}$$
$$=\frac{x^2+2x+3x+15}{2x+10-x^2-2x}$$
$$=\frac{x^2+5x+15}{-x^2+10}$$

59. $\left[\dfrac{\dfrac{x-1}{x-1}-1}{\dfrac{x+1}{x-1}+1}\right]^5$

Consider the numerator of the complex rational expression:

$$\frac{x-1}{x-1}-1=1-1=0$$

Since the denominator, $\dfrac{x+1}{x-1}+1$ is not equal to 0, the simplified form of the original expression is 0.

61.
$$\frac{\dfrac{z}{1-\dfrac{z}{2+2z}}-2z}{\dfrac{2z}{5z-2}-3}=\frac{\dfrac{z}{\dfrac{2+2z-z}{2+2z}}-2z}{\dfrac{2z-15z+6}{5z-2}}$$
$$=\frac{\dfrac{z}{\dfrac{2+z}{2+2z}}-2z}{\dfrac{-13z+6}{5z-2}}$$
$$=\frac{z\cdot\dfrac{2+2z}{2+z}-2z}{\dfrac{-13z+6}{5z-2}}$$
$$=\frac{\dfrac{z(2+2z)-2z(2+z)}{2+z}}{\dfrac{-13z+6}{5z-2}}$$
$$=\frac{\dfrac{2z+2z^2-4z-2z^2}{2+z}}{\dfrac{-13z+6}{5z-2}}$$
$$=\frac{\dfrac{-2z}{2+z}}{\dfrac{-13z+6}{5z-2}}$$
$$=\frac{-2z}{2+z}\cdot\frac{5z-2}{-13z+6}$$
$$=\frac{-2z(5z-2)}{(2+z)(-13z+6)},\text{ or}$$
$$\frac{2z(5z-2)}{(2+z)(13z-6)}$$

63. *Writing Exercise.* When a variable appears only in the numerator(s) of the rational expression(s) that are in the numerator of the complex rational expression, there will be no restrictions on the variables.

Exercise Set 6.6

1. The statement is false. See Example 2(c).

3. The statement is true. See page 416 in the text.

5. Because no variable appears in a denominator, no restrictions exist.

$$\frac{3}{5}-\frac{2}{3}=\frac{x}{6}, \text{ LCD} = 30$$

$$30\left(\frac{3}{5}-\frac{2}{3}\right)=30\cdot\frac{x}{6}$$

$$30\cdot\frac{3}{5}-30\cdot\frac{2}{3}=30\cdot\frac{x}{6}$$

$$18-20=5x$$

$$-2=5x$$

$$\frac{-2}{5}=x$$

Check:

$\frac{3}{5}-\frac{2}{3}=\frac{x}{6}$	
$\frac{3}{5}-\frac{2}{3}$	$\frac{\frac{-2}{5}}{6}$
$\frac{18}{30}-\frac{20}{30}$	$\frac{-2}{5}\cdot\frac{1}{6}$

$$-\frac{2}{30}\overset{?}{=}\frac{-2}{30} \quad \text{TRUE}$$

This checks, so the solution is $-\frac{2}{5}$.

7. Note that x cannot be 0.

$$\frac{1}{3}+\frac{5}{6}=\frac{1}{x}, \text{ LCD} = 6x$$

$$6x\left(\frac{1}{3}+\frac{5}{6}\right)=6x\cdot\frac{1}{x}$$

$$6x\cdot\frac{1}{3}+6x\cdot\frac{5}{6}=6x\cdot\frac{1}{x}$$

$$2x+5x=6$$

$$7x=6$$

$$x=\frac{6}{7}$$

Check:

$\frac{1}{3}+\frac{5}{6}=\frac{1}{x}$	
$\frac{1}{3}+\frac{5}{6}$	$\frac{1}{\frac{6}{7}}$
$\frac{2}{6}+\frac{5}{6}$	$1\cdot\frac{7}{6}$

$$\frac{7}{6}\overset{?}{=}\frac{7}{6} \quad \text{TRUE}$$

This checks, so the solution is $\frac{6}{7}$.

9. Note that t cannot be 0.

$$\frac{1}{8}+\frac{1}{12}=\frac{1}{t}, \text{ LCD } = 48t$$

$$48t\left(\frac{1}{8}+\frac{1}{12}\right)=48t\cdot\frac{1}{t}$$

$$48t\cdot\frac{1}{8}+48t\cdot\frac{1}{12}=48t\cdot\frac{1}{t}$$

$$6t+4t=48$$

$$10t=48$$

$$t=\frac{24}{5}$$

Check:

$\frac{1}{8}+\frac{1}{12}=\frac{1}{t}$	
$\frac{1}{8}+\frac{1}{12}$	$\frac{1}{\frac{24}{5}}$
$\frac{3}{24}+\frac{2}{24}$	$1\cdot\frac{5}{24}$

$$\frac{5}{24}\overset{?}{=}\frac{5}{24} \quad \text{TRUE}$$

This checks, so the solution is $\frac{24}{5}$.

11. Note that y cannot be 0.

$$y+\frac{4}{y}=-5 \text{ LCD } = y$$

$$y\left(y+\frac{4}{y}\right)=-5\cdot y$$

$$y\cdot y+y\cdot\frac{4}{y}=-5\cdot y$$

$$y^2+4=-5y$$

$$y^2+5y+4=0$$

$$(y+4)(y+1)=0$$

$$y+4=0 \text{ or } y+1=0$$

$$y=-4 \text{ or } y=-1$$

Check:

$y+\frac{4}{y}=-5$	
$-4+\frac{4}{-4}$	-5
$-4-1$	

$$-5\overset{?}{=}-5 \quad \text{TRUE}$$

$y+\frac{4}{y}=-5$	
$-1+\frac{4}{-1}$	-5
$-1-4$	

$$-5\overset{?}{=}-5 \quad \text{TRUE}$$

Both of these check, so the solutions are -4 and -1.

13. Note that x cannot be 0.

$$\frac{x}{6} - \frac{6}{x} = 0, \text{ LCD} = 6x$$
$$6x\left(\frac{x}{6} - \frac{6}{x}\right) = 6x \cdot 0$$
$$6x \cdot \frac{x}{6} - 6x \cdot \frac{6}{x} = 6x \cdot 0$$
$$x^2 - 36 = 0$$
$$(x+6)(x-6) = 0$$
$$x+6 = 0 \quad or \quad x-6 = 0$$
$$x = -6 \quad or \quad x = 6$$

Check:

$$\frac{x}{6} - \frac{6}{x} = 0$$
$$\frac{-6}{6} - \frac{6}{-6} \;\Big|\; 0$$
$$-1+1$$
$$0 \stackrel{?}{=} 0 \quad \text{TRUE}$$

$$\frac{x}{6} - \frac{6}{x} = 0$$
$$\frac{6}{6} - \frac{6}{6} \;\Big|\; 0$$
$$1-1$$
$$0 \stackrel{?}{=} 0 \quad \text{TRUE}$$

Both of these check, so the two solutions are -6 and 6.

15. Note that x cannot be 0.

$$\frac{2}{x} = \frac{5}{x} - \frac{1}{4}, \text{ LCD } = 4x$$
$$4x \cdot \frac{2}{x} = 4x\left(\frac{5}{x} - \frac{1}{4}\right)$$
$$4x \cdot \frac{2}{x} = 4x \cdot \frac{5}{x} - 4x \cdot \frac{1}{4}$$
$$8 = 20 - x$$
$$-12 = -x$$
$$12 = x$$

Check:

$$\frac{2}{x} = \frac{5}{x} - \frac{1}{4}$$
$$\frac{12}{12} \;\Big|\; \frac{5}{12} - \frac{1}{4}$$
$$\frac{5}{12} - \frac{3}{12}$$
$$\frac{2}{12}$$
$$\frac{2}{12} \stackrel{?}{=} \frac{2}{12} \quad \text{TRUE}$$

This checks, so the solution is 12.

17. Note that t cannot be 0.

$$\frac{5}{3t} + \frac{3}{t} = 1, \text{ LCD} = 3t$$
$$3t\left(\frac{5}{3t} + \frac{3}{t}\right) = 3t \cdot 1$$
$$3t \cdot \frac{5}{3t} + 3t \cdot \frac{3}{t} = 3t \cdot 1$$
$$5 + 9 = 3t$$
$$14 = 3t$$
$$\frac{14}{3} = t$$

Check:

$$\frac{5}{3t} + \frac{3}{t} = 1$$
$$\frac{5}{3 \cdot \frac{14}{3}} + \frac{3}{\frac{14}{3}} \;\Big|\; 1$$
$$\frac{5}{14} + \frac{9}{14}$$
$$\frac{14}{14}$$
$$1 \stackrel{?}{=} 1 \quad \text{TRUE}$$

This checks, so the solution is $\frac{14}{3}$.

19. To avoid the division by 0, we must have $n - 6 \neq 0$, or $n \neq 6$.

$$\frac{n+2}{n-6} = \frac{1}{2}, \text{ LCD } = 2(n-6)$$
$$2(n-6) \cdot \frac{n+2}{n-6} = 2(n-6) \cdot \frac{1}{2}$$
$$2(n+2) = n - 6$$
$$2n + 4 = n - 6$$
$$n = -10$$

Check:

$$\frac{n+2}{n-6} = \frac{1}{2}$$
$$\frac{-10+2}{-10-6} \;\Big|\; \frac{1}{2}$$
$$\frac{-8}{-16}$$
$$\frac{1}{2} \stackrel{?}{=} \frac{1}{2} \quad \text{TRUE}$$

This checks, so the solution is -10.

21. Note that x cannot be 0.

$$x + \frac{12}{x} = -7, \text{ LCD is } x$$
$$x\left(x + \frac{12}{x}\right) = x \cdot (-7)$$
$$x \cdot x + x \cdot \frac{12}{x} = -7x$$
$$x^2 + 12 = -7x$$
$$x^2 + 7x + 12 = 0$$
$$(x+3)(x+4) = 0$$
$$x + 3 = 0 \quad or \quad x + 4 = 0$$
$$x = -3 \quad or \quad x = -4$$

Both numbers check, so the solutions are -3 and -4.

23. To avoid division by 0, we must have $x-4 \neq 0$ and $x+1 \neq 0$, or $x \neq 4$ and $x \neq -1$.

$$\frac{3}{x-4} = \frac{5}{x+1}, \text{ LCD } = (x-4)(x+1)$$
$$(x-4)(x+1) \cdot \frac{3}{x-4} = (x-4)(x+1) \cdot \frac{5}{x+1}$$
$$3(x+1) = 5(x-4)$$
$$3x + 3 = 5x - 20$$
$$23 = 2x$$
$$\frac{23}{2} = x$$

This checks, so the solution is $\frac{23}{2}$.

25. Because no variable appears in a denominator, no restrictions exist.

$$\frac{a}{6} - \frac{a}{10} = \frac{1}{6}, \text{ LCD} = 30$$
$$30\left(\frac{a}{6} - \frac{a}{10}\right) = 30 \cdot \frac{1}{6}$$
$$30 \cdot \frac{a}{6} - 30 \cdot \frac{a}{10} = 30 \cdot \frac{1}{6}$$
$$5a - 3a = 5$$
$$2a = 5$$
$$a = \frac{5}{2}$$

This checks, so the solution is $\frac{5}{2}$.

27. Because no variable appears in a denominator, no restrictions exist.

$$\frac{x+1}{3} - 1 = \frac{x-1}{2}, \text{ LCD} = 6$$
$$6\left(\frac{x+1}{3} - 1\right) = 6 \cdot \frac{x-1}{2}$$
$$6 \cdot \frac{x+1}{3} - 6 \cdot 1 = 6 \cdot \frac{x-1}{2}$$
$$2(x+1) - 6 = 3(x-1)$$
$$2x + 2 - 6 = 3x - 3$$
$$2x - 4 = 3x - 3$$
$$-1 = x$$

This checks, so the solution is -1.

29. To avoid division by 0, we must have $y - 3 \neq 0$, or $y \neq 3$.

$$\frac{y+3}{y-3} = \frac{6}{y-3}, \text{ LCD } = y - 3$$
$$(y-3) \cdot \frac{y+3}{y-3} = (y-3) \cdot \frac{6}{y-3}$$
$$y + 3 = 6$$
$$y = 3$$

Because of the restriction $y \neq 3$, the number 3 must be rejected as a solution. The equation has no solution.

31. To avoid division by 0, we must have $x+4 \neq 0$ and $x \neq 0$, or $x \neq -4$ and $x \neq 0$.

$$\frac{3}{x+4} = \frac{5}{x}, \text{ LCD} = x(x+4)$$
$$x(x+4) \cdot \frac{3}{x+4} = x(x+4) \cdot \frac{5}{x}$$
$$3x = 5(x+4)$$
$$3x = 5x + 20$$
$$-2x = 20$$
$$x = -10$$

This checks, so the solution is -10.

33. To avoid division by 0, we must have $n+2 \neq 0$ and $n+1 \neq 0$, or $n \neq -2$ and $n \neq -1$.

$$\frac{n+1}{n+2} = \frac{n-3}{n+1}, \text{ LCD } = (n+2)(n+1)$$
$$(n+2)(n+1) \cdot \frac{n+1}{n+2} = (n+2)(n+1) \cdot \frac{n-3}{n+1}$$
$$(n+1)(n+1) = (n+2)(n-3)$$
$$n^2 + 2n + 1 = n^2 - n - 6$$
$$3n = -7$$
$$n = \frac{-7}{3}$$

This checks, so the solution is $\frac{-7}{3}$.

35. To avoid division by 0, we must have $t - 2 \neq 0$, or $t \neq 2$.

$$\frac{5}{t-2} + \frac{3t}{t-2} = \frac{4}{t^2-4t+4}, \text{ LCD is } (t-2)^2$$
$$(t-2)^2\left(\frac{5}{t-2} + \frac{3t}{t-2}\right) = (t-2)^2 \cdot \frac{4}{(t-2)^2}$$
$$5(t-2) + 3t(t-2) = 4$$
$$5t - 10 + 3t^2 - 6t = 4$$
$$3t^2 - t - 10 = 4$$
$$3t^2 - t - 14 = 0$$
$$(3t-7)(t+2) = 0$$
$$3t - 7 = 0 \quad or \quad t + 2 = 0$$
$$3t = 7 \quad or \quad t = -2$$
$$t = \frac{7}{3} \quad or \quad t = -2$$

Both numbers check. The solutions are $\frac{7}{3}$ and -2.

37. To avoid division by 0, we must have $x+5 \neq 0$ and $x-5 \neq 0$, or $x \neq -5$ and $x \neq 5$.

$$\frac{x}{x+5}-\frac{5}{x-5}=\frac{14}{x^2-25}.$$
$\text{LCD}=(x+5)(x-5)$

$$(x+5)(x-5)\cdot\left(\frac{x}{x+5}-\frac{5}{x-5}\right)=(x+5)(x-5)\cdot\frac{14}{(x+5)(x-5)}$$
$$x(x-5)-5(x+5)=14$$
$$x^2-5x-5x-25=14$$
$$x^2-10x-39=0$$
$$(x-13)(x+3)=0$$
$$x-13=0 \text{ or } x+3=0$$
$$x=13 \text{ or } x=-3$$

Both numbers check. The solutions are $-3, 13$.

39. To avoid division by 0, we must have $t-3\neq 0$ and $t+3\neq 0$, or $t\neq 3$ and $t\neq -3$.

$$\frac{5}{t-3}-\frac{30}{t^2-9}=1,$$
$\text{LCD}=(t-3)(t+3)$

$$(t-3)(t+3)\cdot\left(\frac{5}{t-3}-\frac{30}{t^2-9}\right)=(t-3)(t+3)\cdot 1$$
$$5(t+3)-30=(t-3)(t+3)$$
$$5t+15-30=t^2-9$$
$$0=t^2-5t+6$$
$$0=(t-3)(t-2)$$
$$t-3=0 \text{ or } t-2=0$$
$$t=3 \text{ or } t=2$$

Because of the restriction $t\neq 3$, we must reject the number 3 as a solution. The number 2 checks, so it is the solution.

41. To avoid division by 0, we must have $6-a\neq 0$ (or equivalently $a-6\neq 0$) or $a\neq 6$.

$$\frac{7}{6-a}=\frac{a+1}{a-6}$$
$$\frac{-1}{-1}\cdot\frac{7}{6-a}=\frac{a+1}{a-6}$$
$$\frac{-7}{a-6}=\frac{a+1}{a-6},\quad \text{LCD}=a-6$$
$$(a-6)\cdot\frac{-7}{a-6}=(a-6)\cdot\frac{a+1}{a-6}$$
$$-7=a+1$$
$$-8=a$$

This checks.

43. $\dfrac{-2}{x+2}=\dfrac{x}{x+2}$

To avoid division by 0, we must have $x+2\neq 0$, or $x\neq -2$. Now observe that the denominators are the same, so the numerators must be the same. Thus, we have $-2=x$, but because of the restriction $x\neq -2$ this cannot be a solution. The equation has no solution.

45. Note that x cannot be 0.

$$\frac{12}{x}=\frac{x}{3},\quad \text{LCD}=3x$$
$$3x\cdot\frac{12}{x}=3x\cdot\frac{x}{3}$$
$$36=x^2$$
$$0=x^2-36$$
$$0=(x+6)(x-6)$$
$$x+6=0 \text{ or } x-6=0$$
$$x=-6 \text{ or } x=6$$

This checks.

47. *Writing Exercise.* When solving rational equations, we multiply each side by the LCM of the denominators in order to clear fractions.

49. ***Familiarize.*** Let $x=$ the first odd integer. Then $x+2=$ the next odd integer.

Translate.

$$\underbrace{\text{The sum of two consecutive odd integers}}_{x+(x+2)}\ \underbrace{\text{is}}_{=}\ \underbrace{276.}_{276}$$

Carry out. We solve the equation.

$$x+(x+2)=276$$
$$2x+2=276$$
$$2x=274$$
$$x=137$$

When $x=137$, then $x+2=137+2=139$.

Check. The numbers 137 and 139 are consecutive odd integers and $137+139=276$. These numbers check.

State. The integers are 137 and 139.

51. ***Familiarize.*** Let $b=$ the base of the triangle, in cm. Then $b+3=$ the height. Recall that the area of a triangle is given by $\frac{1}{2}\times$ base $\times$ height.

Translate.

$$\underbrace{\text{The area of the triangle}}_{\frac{1}{2}\cdot b\cdot(b+3)}\ \underbrace{\text{is}}_{=}\ \underbrace{54\text{ cm}^2.}_{54}$$

Carry out. We solve the equation.

$$\frac{1}{2}b(b+3)=54$$
$$2\cdot\frac{1}{2}b(b+3)=2\cdot 54$$
$$b(b+3)=108$$
$$b^2+3b=108$$
$$b^2+3b-108=0$$
$$(b-9)(b+12)=0$$
$$b-9=0 \text{ or } b+12=0$$
$$b=9 \text{ or } b=-12$$

Check. The length of the base cannot be negative so we need to check only 9. If the base is 9 cm, then the height is $9+3$, or 12 cm, and the area is $\frac{1}{2}\cdot 9\cdot 12$, or 54 cm^2. The answer checks.

State. The base measures 9 cm, and the height measures 12 cm.

53. To find the rate, in centimeters per day, we divide the amount of growth by the number of days. From June 9 to June 24 is $24-9=15$ days.

$$\begin{aligned}\text{Rate, in cm per day} &= \frac{0.9\text{ cm}}{15\text{ days}}\\ &= 0.06\text{ cm/day}\\ &= 0.06\text{ cm per day}\end{aligned}$$

55. *Writing Exercise.* Begin with an equation. Then divide on both sides of the equation by an expression whose value is zero for at least one solution of the equation. See Exercises 43 and 44 for examples.

57. To avoid division by 0, we must have $x-3\neq 0$, or $x\neq 3$.

$$\begin{aligned}1+\frac{x-1}{x-3} &= \frac{2}{x-3}-x,\ \text{LCD}=x-3\\ (x-3)\left(1+\frac{x-1}{x-3}\right) &= (x-3)\left(\frac{2}{x-3}-x\right)\\ (x-3)\cdot 1+(x-3)\cdot\frac{x-1}{x-3} &= (x-3)\cdot\frac{2}{x-3}-(x-3)x\\ x-3+x-1 &= 2-x^2+3x\\ 2x-4 &= 2-x^2+3x\\ x^2-x-6 &= 0\\ (x-3)(x+2) &= 0\end{aligned}$$

$$x-3=0 \quad or \quad x+2=0$$
$$x=3 \quad or \quad x=-2$$

Because of the restriction $x\neq 3$, we must reject the number 3 as a solution. The number -2 checks, so it is the solution.

59. To avoid division by 0, we must have $x+2\neq 0$ and $x-2\neq 0$, or $x\neq -2$ and $x\neq 2$.

$$\begin{aligned}\frac{12-6x}{x^2-4} &= \frac{3x}{x+2}-\frac{3-2x}{2-x},\\ \frac{12-6x}{(x+2)(x-2)} &= \frac{3x}{x+2}-\frac{3-2x}{2-x}\cdot\frac{-1}{-1}\\ \frac{12-6x}{(x+2)(x-2)} &= \frac{3x}{x+2}-\frac{2x-3}{x-2},\\ &\text{LCD}=(x+2)(x-2)\\ (x+2)(x-2)\cdot\frac{12-6x}{(x+2)(x-2)} &=\\ &(x+2)(x-2)\left(\frac{3x}{x+2}-\frac{2x-3}{x-2}\right)\\ 12-6x &=\\ &3x(x-2)-(x+2)(2x-3)\\ 12-6x &=\\ &3x^2-6x-2x^2-x+6\\ 0 &= x^2-x-6\\ 0 &= (x-3)(x+2)\end{aligned}$$

$$x-3=0 \quad or \quad x+2=0$$
$$x=3 \quad or \quad x=-2$$

Because of the restriction $x\neq -2$, we must reject the number -2 as a solution. The number 3 checks, so it is the solution.

61. To avoid division by 0, we must have $a+3\neq 0$, or $a\neq -3$.

$$\begin{aligned}7-\frac{a-2}{a+3} &= \frac{a^2-4}{a+3}+5,\ \text{LCD}=a+3\\ (a+3)\left(7-\frac{a-2}{a+3}\right) &= (a+3)\left(\frac{a^2-4}{a+3}+5\right)\\ 7(a+3)-(a-2) &= a^2-4+5(a+3)\\ 7a+21-a+2 &= a^2-4+5a+15\\ 6a+23 &= a^2+5a+11\\ 0 &= a^2-a-12\\ 0 &= (a-4)(a+3)\end{aligned}$$

$$a-4=0 \quad or \quad a+3=0$$
$$a=4 \quad or \quad a=-3$$

Because of the restriction $a\neq -3$, we must reject the number -3 as a solution. The number 4 checks, so it is the solution.

63. To avoid division by 0, we must have $x-1\neq 0$, or $x\neq 1$.

$$\begin{aligned}\frac{1}{x-1}+x-5 &= \frac{5x-4}{x-1}-6,\ \text{LCD}=x-1\\ (x-1)\left(\frac{1}{x-1}+x-5\right) &= (x-1)\left(\frac{5x-4}{x-1}-6\right)\\ 1+x(x-1)-5(x-1) &= 5x-4-6(x-1)\\ 1+x^2-x-5x+5 &= 5x-4-6x+6\\ x^2-6x+6 &= -x+2\\ x^2-5x+4 &= 0\\ (x-1)(x-4) &= 0\end{aligned}$$

$$x-1=0 \quad or \quad x-4=0$$
$$x=1 \quad or \quad x=4$$

Because of the restriction $x\neq 1$, we must reject the number 1 as a solution. The number 4 checks, so it is the solution.

65. Note that x cannot be 0.

$$\begin{aligned}\frac{\frac{1}{x}+1}{x} &= \frac{\frac{1}{x}}{2}\\ \left(\frac{1}{x}+1\right)\cdot\frac{1}{x} &= \frac{1}{x}\cdot\frac{1}{2}\\ \frac{1}{x^2}+\frac{1}{x} &= \frac{1}{2x},\ \text{LCD}=2x^2\\ 2+2x &= x\\ 2 &= -x\\ -2 &= x\end{aligned}$$

This checks.

67. [writing exercise icon]

6.6 Connecting the Concepts

1. Expression; $\dfrac{4x^2-8x}{4x^2+4x}=\dfrac{4x\,(x-2)}{4x\,(x+1)}=\dfrac{x-2}{x+1}$

3. Equation; $\frac{3}{y} - \frac{1}{4} = \frac{1}{y}$ Note: $y \neq 0$

$$4y\left(\frac{3}{y} - \frac{1}{4}\right) = 4y\left(\frac{1}{y}\right) \qquad \text{LCD } = 4y$$
$$12 - y = 4$$
$$-y = -8$$
$$y = 8$$

The solution is 8.

5. Equation; $\frac{5}{x+3} = \frac{3}{x+2}$ Note $x \neq -2, -3$

$$(x+2)(x+3) \cdot \frac{5}{x+3} = (x+2)(x+3) \cdot \frac{3}{x+2}$$
$$\text{LCD } = (x+2)(x+3)$$
$$5(x+2) = 3(x+3)$$
$$5x + 10 = 3x + 9$$
$$5x = 3x - 1$$
$$2x = -1$$
$$x = \frac{-1}{2}$$

7. Expression; $\frac{2a}{a+1} - \frac{4a}{1-a^2}$

$$= \frac{2a}{a+1} - \frac{4a}{a^2-1}$$
$$= \frac{2a(a-1) + 4a}{(a+1)(a-1)} \qquad \text{LCD } = (a+1)(a-1)$$
$$= \frac{2a^2 - 2a + 4}{(a+1)(a-1)}$$
$$= \frac{2a^2 - 2a}{(a+1)(a-1)}$$
$$= \frac{2a(a-1)}{(a+1)(a-1)}$$
$$= \frac{2a}{a+1}$$

9. Expression; $\frac{18x^2}{25} \div \frac{12x}{5} = \frac{18x^2}{25} \cdot \frac{5}{12x}$

$$= \frac{3x}{10} \cdot \frac{5 \cdot 6x}{5 \cdot 6x}$$
$$= \frac{3x}{10}$$

Exercise Set 6.7

1. 1 cake in 2 hours $= \frac{1 \text{ cake}}{2 \text{ hr}} = \frac{1}{2}$ cake per hour

3. Sandy: $\frac{1 \text{ cake}}{2 \text{ hr}} = \frac{1}{2}$ cake per hour

   Eric: $\frac{1 \text{ cake}}{3 \text{ hr}} = \frac{1}{3}$ cake per hour

   Together: $\frac{1}{2} + \frac{1}{3} = \frac{3}{6} + \frac{2}{6} = \frac{5}{6}$ cake per hour

5. 1 lawn in 3 hours $= \frac{1 \text{ lawn}}{3 \text{ hr}} = \frac{1}{3}$ lawn per hour

7. ***Familiarize***. Let t represent the number of hours it takes Bryan and Caroline to refinish the floors working together.

   Translate. Bryan takes 8 hr and Caroline takes 6 hr to complete the job, so we have $\frac{t}{8} + \frac{t}{6} = 1$

   Carry out. We solve the equation. Multiply on both sides by the LCD, 24.

   Check. If Bryan does the job alone in 8 hr, then in $3\frac{3}{7}$ hr he does $\frac{24/7}{8}$, or $\frac{3}{7}$ of the job. If Caroline does the job alone in 6 hr, then in $3\frac{3}{7}$ hr she does $\frac{24/7}{6}$, or $\frac{4}{7}$ of the job. Together, they do $\frac{3}{7} + \frac{4}{7}$, or 1 entire job. The result checks.

   State. It would take Bryan and Caroline $3\frac{3}{7}$ hr to finish the job working together.

9. ***Familiarize***. The job takes Aficio 7 minutes working alone and MX 6 minutes working alone. Then in 1 minute Aficio does $\frac{1}{7}$ of the job and MX does $\frac{1}{6}$ of the job. Working together, they can do $\frac{1}{7} + \frac{1}{6}$, or $\frac{13}{42}$ of the job in 1 minute. In 2 minutes, Aficio does $2\left(\frac{1}{7}\right)$ of the job and MX does $2\left(\frac{1}{6}\right)$ of the job. Working together they can do $2\left(\frac{1}{7}\right) + 2\left(\frac{1}{6}\right)$, or $\frac{13}{21}$ of the job in 2 minutes. In 4 minutes Aficio does $4\left(\frac{1}{7}\right)$ of the job and MX does $4\left(\frac{1}{6}\right)$ of the job. Working together they can do $4\left(\frac{1}{7}\right) + 4\left(\frac{1}{6}\right)$, or $\frac{26}{21}$ of the job which is more of the job then needs to be done. The answer is somewhere between 2 minutes and 4 minutes.

   Translate. If they work together t minutes, then Aficio does $t\left(\frac{1}{7}\right)$ of the job and MX does $t\left(\frac{1}{6}\right)$ of the job. We want some number t such that

$$\left(\frac{1}{7} + \frac{1}{6}\right)t = 1, \text{ or } \frac{13}{42}t = 1.$$

   Carry out. We solve the equation.

$$\frac{13}{42}t = 1$$
$$\frac{42}{13} \cdot \frac{13}{42}t = \frac{42}{13} \cdot 1$$
$$t = \frac{42}{13}$$

   Check. We repeat computations. The answer checks. We also expected the result to be between 2 minutes and 4 minutes.

   State. Working together, it takes Aficio and MX 42/13 mins, or 3 3/13 mins.

11. ***Familiarize***. In 1 minute the Wayne pump does $\frac{1}{42}$ of the job and the Craftsman pump does $\frac{1}{35}$ of the job. Working together, they do $\frac{1}{42} + \frac{1}{35}$ of the job in 1 minute. Suppose it takes t minutes to do the job working together.

    Translate. We find t such that $t\left(\frac{1}{42}\right) + t\left(\frac{1}{35}\right) = 1$, or $\frac{t}{42} + \frac{t}{35} = 1$.

    Carry out. We solve the equation. We multiply both sides by the LCD, 210.

$$210\left(\frac{t}{42} + \frac{t}{35}\right) = 210 \cdot 1$$
$$5t + 6t = 210$$
$$11t = 210$$
$$t = \frac{210}{11}$$

Check. In $\frac{210}{11}$ min the Wayne pump does $\frac{210}{11} \cdot \frac{1}{42}$, or $\frac{5}{11}$ of the job and the Craftsman pump does $\frac{210}{11} \cdot \frac{1}{35}$, or $\frac{6}{11}$ of the job. Together they do $\frac{5}{11} + \frac{6}{11}$, or 1 entire job. The answer checks.

State. The two pumps can pump out the basement in $\frac{210}{11}$ min, or $19\frac{1}{11}$ min, working together.

13. ***Familiarize***. Let t represent the time, in minutes, that it takes the K5400 to print the brochures working alone. Then $2t$ represents the time it takes the H470 to do the job, working alone. In 1 minute the K5400 does $\frac{1}{t}$ of the job and the H470 does $\frac{1}{2t}$ of the job.

Translate. Working together, they can do the entire job in 45 min, so we want to find t such that $45\left(\frac{1}{t}\right)+45\left(\frac{1}{2t}\right) = 1$, or $\frac{45}{t} + \frac{45}{2t} = 1$.

Carry out. We solve the equation. We multiply both sides by the LCD, $2t$.

$$2t\left(\frac{45}{t} + \frac{45}{2t}\right) = 2t \cdot 1$$
$$90 + 45 = 2t$$
$$135 = 2t$$
$$\frac{135}{2} = t, \text{ or } 67\tfrac{1}{2}$$

Check. If the K5400 can do the job in $\frac{135}{2}$ min, then in 45 min it does $45 \cdot \frac{1}{135/2}$, or $\frac{2}{3}$ of the job. If it takes the H470 $2 \cdot \frac{135}{2}$, or 135 min, to do the job, then in 45 min it does $45 \cdot \frac{1}{135}$, or $\frac{1}{3}$ of the job. Working together, the two machines do $\frac{2}{3} + \frac{1}{3}$, or 1 entire job, in 45 min.

State. Working alone, it takes the K5400 $67\frac{1}{2}$ min and the H470 135 min to print the brochure.

15. ***Familiarize***. Let t represent the number of minutes it takes the Airgle machine to purify the air working alone. Then $t-15$ represents the time it takes the Austin machine to purify the air, working alone. In 1 minute the Airgle does $\frac{1}{t}$ of the job and the Austin does $\frac{1}{t-15}$ of the job.

Translate. Working together, the two machines can purify the air in 10 min to find t such that $10\left(\frac{1}{t}\right) + 10\left(\frac{1}{t-15}\right) = 1$, or $\frac{10}{t} + \frac{10}{t-15} = 1$.

Carry out. We solve the equation. First we multiply both sides by the LCD, $t(t-15)$.

$$t(t-15)\left(\frac{10}{t} + \frac{10}{t-15}\right) = t(t-15) \cdot 1$$
$$10(t-15) + 10t = t(t-15)$$
$$10t - 150 + 10t = t^2 - 15t$$
$$0 = t^2 - 35t + 150$$
$$0 = (t-5)(t-30)$$
$$t = 5 \text{ or } t = 30$$

Check. If $t = 5$, then $t - 15 = 5 - 15 = -10$. Since negative time has no meaning in this application, 5 cannot be a solution. If $t = 30$, then $t - 15 = 30 - 15 = 15$. In 10 min the Airgle machine does $10 \cdot \frac{1}{30}$, or $\frac{1}{3}$ of the job. In 10 min the Austin does $10 \cdot \frac{1}{15}$, or $\frac{2}{3}$ of the job. Together they do $\frac{1}{3} + \frac{2}{3}$, or 1 entire job. The answer checks.

State. Working alone, the Airgle machine can purify the air in 30 min and the Austin machine can purify the air in 15 min.

17. ***Familiarize***.Let t represent the number of minutes it takes Chris to do the job working alone. Then $t + 120$ represents the time it takes Kim to do the job working alone. We will convert hours to minutes:

Translate. In 175 min Chris and Kim will do one entire job, so we have $175\left(\frac{1}{t}\right)+175\left(\frac{1}{t+120}\right) = 1$, or $\frac{175}{t}+\frac{175}{t+120} = 1$

Carry out.We solve the equation. Multiply on both sides by the LCD, $t(t + 120)$.

$$t(t+120)\left(\frac{175}{t} + \frac{175}{t+120}\right) = t(t+120) \cdot 1$$
$$175(t+120) + 175t = t(t+120)$$
$$175t + 21,000 + 175t = t^2 + 120t$$
$$0 = t^2 - 230t - 21,000$$
$$0 = (t-300)(t+70)$$
$$t = 300 \text{ or } t = -70$$

Check.Since negative time has no meaning in this problem -70 is not a solution of the original problem. If Chris does the job alone in 300 min, then in 175 min he does $\frac{175}{300} = \frac{7}{12}$ of the job. If Kim does the job alone in 300 + 120, or 420 min, then in 175 min she does $\frac{175}{420} = \frac{5}{12}$ of the job. Together, they do $\frac{7}{12} + \frac{5}{12}$, or 1 entire job, in 175 min. The result checks.

State. It would take Chris 300 min, or 5 hours to do the job alone.

19. ***Familiarize***. We complete the table shown in the text.

$d = r \cdot t$

	Distance	Speed	Time
B & M	330	$r - 14$	$\frac{330}{r-14}$
AMTRAK	400	r	$\frac{400}{r}$

Translate. Since the time must be the same for both trains, we have the equation

$$\frac{330}{r-14} = \frac{400}{r}.$$

Carry out. We first multiply by the LCD, $r(r-14)$.

$$r(r-14) \cdot \frac{330}{r-14} = r(r-14) \cdot \frac{400}{r}$$
$$330r = 400(r-14)$$
$$330r = 400r - 5600$$
$$-70r = -5600$$
$$r = 80$$

If the speed of the AMTRAK train is 80 km/h, then the speed of the B & M train is 80 – 14, or 66 km/h.

Check. The speed of the B&M train is 14 km/h slower than the speed of the AMTRAK train. At 66 km/h the B&M train travels 330 km in 330/66, or 5 hr. At 80 km/h the AMTRAK train travels 400 km in 400/80, or 5 hr. The times are the same, so the answer checks.

State. The speed of the AMTRAK train is 80 km/h, and the speed of the B&M freight train is 66 km/h.

21. ***Familiarize.*** We first make a drawing. Let r = the kayak's speed in still water in mph. Then $r - 3$ = the speed upstream and $r + 3$ = the speed downstream.

We organize the information in a table. The time is the same both upstream and downstream so we use t for each time.

	Distance	Speed	Time
Upstream	4	$r - 3$	t
Downstream	10	$r + 3$	t

Translate. Using the formula Time = Distance/Rate in each row of the table and the fact that the times are the same, we can write an equation. $\frac{4}{r-3} = \frac{10}{r+3}$

Carry out. We solve the equation.

$$\frac{4}{r-3} = \frac{10}{r+3}, \text{ LCD is } (r-3)(r+3)$$
$$(r-3)(r+3)\cdot\frac{4}{r-3} = (r-3)(r+3)\cdot\frac{10}{r+3}$$
$$4(r+3) = 10(r-3)$$
$$4r + 12 = 10r - 30$$
$$42 = 6r$$
$$7 = r$$

Check. If $r = 7$ mph, then $r - 3$ is 4 mph and $r + 3$ is 10 mph. The time upstream is $\frac{4}{4}$, or 1 hour. The time downstream is $\frac{10}{10}$, or 1 hour. Since the times are the same, the answer checks.

State. The speed of the kayak in still water is 7 mph.

23. ***Familiarize.*** We first make a drawing. Let r = Roslyn's speed on a nonmoving sidewalk in ft/sec. Then her speed moving forward on the moving sidewalk is r + 1.8, and her speed in the opposite direction is $r - 1.8$.

We organize the information in a table. The time is the same both forward and in the opposite direction, so we use t for each time.

	Distance	Speed	Time
Forward	105	$r + 1.8$	t
Opposite direction	51	$r - 1.8$	t

Translate. Using the formula Time = Distance/Rate in each row of the table and the fact that the times are the same, we can write an equation. $\frac{105}{r+1.8} = \frac{51}{r-1.8}$

Carry out. We solve the equation.

$$\frac{105}{r+1.8} = \frac{51}{r-1.8}$$
$$\text{LCD is } (r+1.8)(r-1.8)$$
$$(r+1.8)(r-1.8)\frac{105}{r+1.8} = (r+1.8)(r-1.8)\frac{51}{r-1.8}$$
$$105(r-1.8) = 51(r+1.8)$$
$$105r - 189 = 51r + 91.8$$
$$54r = 280.8$$
$$r = 5.2$$

Check. If Roslyn's speed on a nonmoving sidewalk is 5.2 ft/sec, then her speed moving forward on the moving sidewalk is 5.2 + 1.8, or 7 ft/sec, and her speed moving in the opposite direction on the sidewalk is 5.2 - 1.8, or 3.4 ft/sec. Moving 105 ft at 7 ft/sec takes $\frac{105}{7} = 15$ sec. Moving 51 ft at 3.4 ft/sec takes $\frac{51}{3.4} = 15$ sec. Since the times are the same, the answer checks.

State. Roslyn's would be walking 5.2 ft/sec on a nonmoving sidewalk.

25. ***Familiarize.*** Let t = the time it takes Caledonia to drive to town and organize the given information in a table.

	Distance	Speed	Time
Caledonia	15	r	t
Manley	20	r	$t + 1$

Translate. We can replace the r's in the table above using the formula $r = d/t$.

	Distance	Speed	Time
Caledonia	15	$\frac{15}{t}$	t
Manley	20	$\frac{20}{t+1}$	$t + 1$

Since the speeds are the same for both riders, we have the equation

$$\frac{15}{t} = \frac{20}{t+1}.$$

Carry out. We multiply by the LCD, $t(t+1)$.

$$t(t+1)\cdot\frac{15}{t} = t(t+1)\cdot\frac{20}{t+1}$$
$$15(t+1) = 20t$$
$$15t + 15 = 20t$$
$$15 = 5t$$
$$3 = t$$

If $t = 3$, then $t + 1 = 3 + 1$, or 4.

Check. If Caledonia's time is 3 hr and Manley's time is 4 hr, then Manley's time is 1 hr more than Caledonia's. Caledonia's speed is 15/3, or 5 mph. Manley's speed is 20/4, or 5 mph. Since the speeds are the same, the answer checks.

State. It takes Caledonia 3 hr to drive to town.

27. ***Familiarize.*** We let r = the speed of the river. Then $15+r$ = LeBron's speed downstream in km/h and $15-r$ = his speed upstream in km/h. The times are the same. Let t represent the time. We organize the information in a table.

	Distance	Speed	Time
Upstream	140	$15 + r$	t
Downstream	35	$15 - r$	t

Translate. Using the formula Time = Distance/Rate in each row of the table and the fact that the times are the same, we can write an equation. $\frac{140}{15+r} = \frac{35}{15-r}$

Carry out. We solve the equation.

$$\frac{140}{15+r} = \frac{35}{15-r}, \text{ LCD is } (15+r)(15-r)$$
$$(15+r)(15-r)\cdot\frac{140}{15+r} = (15+r)(15-r)\cdot\frac{35}{15-r}$$
$$140(15-r) = 35(15+r)$$
$$2100 - 140r = 525 + 35r$$
$$1575 = 175r$$
$$9 = r$$

Check. If $r = 9$, then the speed downstream is 15 + 9, or 24 km/h and the speed upstream is 15 - 9, or 6 km/h. The time for the trip downstream is $\frac{140}{24}$, or $5\frac{5}{6}$ hours. The time for the trip upstream is $\frac{35}{6}$, or $5\frac{5}{6}$ hours. The times are the same. The values check.

State. The speed of the river is 9 km/h.

29. ***Familiarize***. Let $c =$ the speed of the current, in km/h. Then $7+c =$ the speed downriver and $7-c =$ the speed upriver. We organize the information in a table.

	Distance	Speed	Time
Downriver	45	$7+c$	t_1
Upriver	45	$7-c$	t_2

Translate. Using the formula Time = Distance/Rate we see that $t_1 = \frac{45}{7+c}$ and $t_2 = \frac{45}{7-c}$. The total time upriver and back is 14 hr, so $t_1 + t_2 = 14$, or $\frac{45}{7+c} + \frac{45}{7-c} = 14$.

Carry out. We solve the equation. Multiply both sides by the LCD, $(7+c)(7-c)$.

$$(7+c)(7-c)\left(\frac{45}{7+c}+\frac{45}{7-c}\right) = (7+c)(7-c)14$$
$$45(7-c) + 45(7+c) = 14(49-c^2)$$
$$315 - 45c + 315 + 45c = 686 - 14c^2$$
$$14c^2 - 56 = 0$$
$$14(c+2)(c-2) = 0$$
$$c+2 = 0 \quad or \quad c-2 = 0$$
$$c = -2 \quad or \quad c = 2$$

Check. Since speed cannot be negative in this problem, -2 cannot be a solution of the original problem. If the speed of the current is 2 km/h, the barge travels upriver at 7 - 2, or 5 km/h. At this rate it takes $\frac{45}{5}$, or 9 hr, to travel 45 km. The barge travels downriver at 7 + 2, or 9 km/h. At this rate it takes $\frac{45}{9}$, or 5 hr, to travel 45 km. The total travel time is 9 + 5, or 14 hr. The answer checks.

State. The speed of the current is 2 km/h.

31. ***Familiarize***. Let $r =$ the speed at which the train actually traveled in mph, and let $t =$ the actual travel time in hours. We organize the information in a table.

	Distance	Speed	Time
Actual speed	120	r	t
Faster speed	120	$r+10$	$t-2$

Translate. From the first row of the table we have $120 = rt$, and from the second row we have $120 = (r+10)(t-2)$. Solving the first equation for t, we have $t = \frac{120}{r}$. Substituting for t in the second equation, we have $120 = (r+10)\left(\frac{120}{r} - 2\right)$.

Carry out. We solve the equation.

$$120 = (r+10)\left(\frac{120}{r}-2\right)$$
$$120 = 120 - 2r + \frac{1200}{r} - 20$$
$$20 = -2r + \frac{1200}{r}$$
$$r\cdot 20 = r\left(-2r + \frac{1200}{r}\right)$$
$$20r = -2r^2 + 1200$$
$$2r^2 + 20r - 1200 = 0$$
$$2(r^2 + 10r - 600) = 0$$
$$2(r+30)(r-20) = 0$$
$$r = -30 \ or \ r = 20$$

Check. Since speed cannot be negative in this problem, cannot be a solution of the original problem. If the speed is 20 mph, it takes $\frac{120}{20}$, or 6 hr, to travel 120 mi. If the speed is 10 mph faster, or 30 mph, it takes $\frac{120}{30}$, or 4 hr, to travel 120 mi. Since 4 hr is 2 hr less time than 6 hr, the answer checks.

State. The speed was 20 mph.

33. We write a proportion and then solve it.

$$\frac{b}{6} = \frac{7}{4}$$
$$b = \frac{7}{4}\cdot 6$$
$$b = \frac{42}{4}, \text{ or } 10.5$$

(Note that the proportions $\frac{6}{b} = \frac{4}{7}$, $\frac{b}{7} = \frac{6}{4}$, or $\frac{7}{b} = \frac{4}{6}$ could also be used.)

35. We write a proportion and then solve it.

$$\frac{4}{f} = \frac{6}{4}$$
$$4f\cdot\frac{4}{f} = 4f\cdot\frac{6}{4}$$
$$16 = 6f$$
$$\frac{8}{3} = f \qquad \text{Simplifying}$$

(One of the following proportions could also be used: $\frac{f}{4} = \frac{4}{6}$, $\frac{4}{f} = \frac{9}{6}$, $\frac{f}{4} = \frac{6}{9}$, $\frac{4}{9} = \frac{f}{6}$, $\frac{9}{4} = \frac{6}{f}$)

37. From the blueprint we see that 9 in. represents 36 ft and that p in. represent 15 ft. We use a proportion to find p.

$$\frac{9}{36} = \frac{p}{15}$$
$$180\cdot\frac{9}{36} = 180\cdot\frac{p}{15}$$
$$45 = 12p$$
$$\frac{15}{4} = p, \text{ or}$$
$$3\frac{3}{4} = p$$

The length of p is $3\frac{3}{4}$ in.

39. From the blueprint we see that 9 in. represents 36 ft and that 5 in. represents r ft. We use a proportion to find r.

$$\frac{9}{36} = \frac{5}{r}$$

$$36r \cdot \frac{9}{36} = 36r \cdot \frac{5}{r}$$

$$9r = 180$$

$$r = 20$$

The length of r is 20 ft.

41. Consider the two similar right triangles in the drawing. One has legs 4 ft and 6 ft. The other has legs 10ft and l ft. We use a proportion to find l.

$$\frac{4}{6} = \frac{10}{l}$$

$$6l \cdot \frac{4}{6} = 6l \cdot \frac{10}{l}$$

$$4l = 60$$

$$l = 15\text{ft}$$

43. Consider the two similar right triangles in the drawing. One has legs 5 and 7. The other has legs 9 and r. We use a proportion to find r.

$$\frac{5}{7} = \frac{9}{r}$$

$$7r \cdot \frac{5}{7} = 7r \cdot \frac{9}{r}$$

$$5r = 63$$

$$r = \frac{63}{5}, \text{ or } 12.6$$

45. ***Familiarize***. Brett had 384 text messages in 8 days. Let $n =$ the number of text messages in 30 days.

Translate.

$$\begin{array}{r} \text{Messages} \rightarrow \\ \text{Days} \rightarrow \end{array} \frac{384}{8} = \frac{n}{30} \begin{array}{l} \leftarrow \text{Messages} \\ \leftarrow \text{Days} \end{array}$$

Carry out. We solve the proportion.

$$120 \cdot \frac{384}{8} = 120 \cdot \frac{n}{30}$$

$$5760 = 4n$$

$$1440 = n$$

Check. $\frac{384}{8} = 48$, $\frac{1440}{30} = 48$

The ratios are the same so the answer checks.

State. He will send or receive 1440 messages in 30 days.

47. ***Familiarize***. Persons caught on 295 mi stretch is 12,334. Let $n =$ the number caught on 5525 mi border.

Translate.

$$\begin{array}{r} \text{Persons} \rightarrow \\ \text{distance} \rightarrow \end{array} \frac{12334}{295} = \frac{n}{5525} \begin{array}{l} \leftarrow \text{Persons} \\ \leftarrow \text{distance} \end{array}$$

Carry out. We solve the proportion.

$$325,975 \cdot \frac{12334}{295} = 325,975 \cdot \frac{n}{5525}$$

$$13,629,070 = 59n$$

$$231,001 \approx n$$

Check. $\frac{12334}{295} \approx 41.81$, $\frac{231001}{5525} \approx 41.81$

The ratios are the same, so the answer checks.

State. Over the entire border about 231,001 people may be caught.

49. ***Familiarize***. Let $g =$ the number if gal of gas for a 810 mi trip. We can use a proportion to solve for g.

Translate.

$$\begin{array}{r} \text{gal} \rightarrow \\ \text{miles} \rightarrow \end{array} \frac{4}{180} = \frac{g}{810} \begin{array}{l} \leftarrow \text{gal} \\ \leftarrow \text{miles} \end{array}$$

Carry out. We solve the proportion.

We multiply by 1620 to get g alone.

$$1620 \cdot \frac{4}{180} = 1620 \cdot \frac{g}{810}$$

$$36 = 2g$$

$$18 = g$$

Check.

$\frac{4}{180} \approx 0.02$, $\frac{18}{810} \approx 0.02$

The ratios are the same, so the answer checks.

State. For a trip of 810 mi, 18 gal of gas are needed.

51. ***Familiarize***. Let $w =$ the wing width of a stork, in cm. We can use a proportion.

Translate. We translate to a proportion.

$$\begin{array}{r} \text{wing span} \rightarrow \\ \text{wing width} \rightarrow \end{array} \frac{180}{24} = \frac{200}{w} \begin{array}{l} \leftarrow \text{wing span} \\ \leftarrow \text{wing width} \end{array}$$

Carry out. We solve the proportion.

$$24w \cdot \frac{180}{24} = 24w \cdot \frac{200}{w}$$

$$180w = 4800$$

$$w = \frac{80}{3} = 26\frac{2}{3}\text{cm}$$

Check.

$\frac{180}{24} = 7.5$, $\frac{200}{26\frac{2}{3}} = 7.5$

The ratios are the same, so the answer checks.

State. The wing width of a stork is $26\frac{2}{3}$cm.

53. ***Familiarize***. Let $D =$ the number of defective bulbs in a batch of 1430 bulbs. We can use a proportion to find D.

Translate.

$$\begin{array}{r} \text{defective bulbs} \rightarrow \\ \text{batch size} \rightarrow \end{array} \frac{8}{220} = \frac{D}{1430} \begin{array}{l} \leftarrow \text{defective bulbs} \\ \leftarrow \text{batch size} \end{array}$$

Carry out. We solve the proportion.

$$2860 \cdot \frac{8}{220} = 2860 \cdot \frac{D}{1430}$$
$$104 = 2D$$
$$52 = D$$

Check. $\frac{8}{220} = 0.0\overline{36}, \ \frac{52}{1430} = 0.0\overline{36}$

The ratios are the same, so the answer checks.

State. In a batch of 1430 bulbs, 52 defective bulbs can be expected.

55. ***Familiarize.*** Let z = the number of ounces of water needed by a Bolognese. We can use a proportion to solve for z.

Translate. We translate to a proportion.

$$\begin{array}{l}\text{dog weight} \longrightarrow \\ \text{water} \longrightarrow\end{array} \frac{8}{12} = \frac{5}{z} \begin{array}{l}\longleftarrow \text{dog weight} \\ \longleftarrow \text{water}\end{array}$$

Carry out. We solve the proportion.

$$12z \cdot \frac{8}{12} = 12z \cdot \frac{5}{z}$$
$$8z = 60$$
$$z = \frac{60}{8} = \frac{15}{2} = 7\frac{1}{2}\text{oz}$$

Check.

$\frac{8}{12} = 0.\overline{6}, \ \frac{5}{7\frac{1}{2}} = 0.\overline{6}$

The ratios are the same, so the answer checks.

State. For a 5-lb Bolognese, approximately $7\frac{1}{2}$oz of water is required per day.

57. ***Familiarize.*** Let p = the number of Whale in the pod. We use a proportion to solve for p.

Translate.

$$\begin{array}{l}\text{sighted} \rightarrow \\ \text{pod} \rightarrow\end{array} \frac{12}{27} = \frac{40}{p} \begin{array}{l}\leftarrow \text{sighted} \\ \leftarrow \text{pod}\end{array}$$

Carry out. We solve the proportion.

$$27p \cdot \frac{12}{27} = 27p \cdot \frac{40}{p}$$
$$12p = 1080$$
$$p = 90$$

Check. $\frac{12}{27} = \frac{4}{9}, \ \frac{40}{90} = \frac{4}{p}$

The ratios are the same, so the answer checks.

State. There are 90 whales in the pod, when 40 whales are sighted.

59. ***Familiarize.*** The ratio of the weight of an object on the moon to the weight of an object on Earth is 0.16 to 1.

a) We wish to find how much a 12-ton rocket would weigh on the moon.

b) We wish to find how much a 180-lb astronaut would weigh on the moon.

Translate. We translate to proportions.

a) $$\begin{array}{r}\text{Weight} \\ \text{on the moon} \rightarrow \\ \text{Weight} \rightarrow \\ \text{on Earth}\end{array} \frac{0.16}{1} = \frac{T}{12} \begin{array}{l}\text{Weight} \\ \leftarrow \text{on the moon} \\ \leftarrow \text{Weight} \\ \text{on Earth}\end{array}$$

b) $$\begin{array}{r}\text{Weight} \\ \text{on the moon} \rightarrow \\ \text{Weight} \rightarrow \\ \text{on Earth}\end{array} \frac{0.16}{1} = \frac{P}{180} \begin{array}{l}\text{Weight} \\ \leftarrow \text{on the moon} \\ \leftarrow \text{Weight} \\ \text{on Earth}\end{array}$$

Carry out. We solve each proportion.

a) $$\frac{0.16}{1} = \frac{T}{12}$$
$$12(0.16) = T$$
$$1.92 = T$$

b) $$\frac{0.16}{1} = \frac{P}{180}$$
$$180(0.16) = P$$
$$28.8 = P$$

Check. $\frac{0.16}{1} = 0.16, \ \frac{1.92}{12} = 0.16, \ \frac{28.8}{180} = 0.16$

The ratios are the same, so the answer checks.

State.

a) A 12-ton rocket would weigh 1.92 tons on the moon.

b) A 180-lb astronaut would weigh 28.8 lb on the moon.

61. *Writing Exercise.* No. If the workers work at different rates, two workers will complete a task in more than half the time of the faster person working alone but in less than half the slower person's time. This is illustrated in Example 1.

63. Graph: $y = 2x - 6$.

We select some x-values and compute y-values.

If $x = 1$, then $y = 2 \cdot 1 - 6 = -4$.

If $x = 3$, then $y = 2 \cdot 3 - 6 = 0$.

If $x = 5$, then $y = 2 \cdot 5 - 6 = 4$.

x	y	(x, y)
1	-4	$(1, -4)$
3	0	$(3, 0)$
5	4	$(5, 4)$

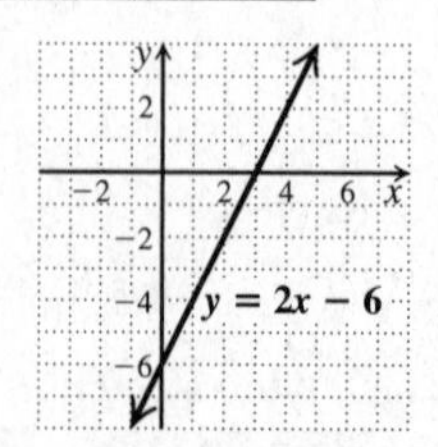

65. Graph: $3x + 2y = 12$.

We can replace either variable with a number and then calculate the other coordinate. We will find the intercepts and one other point.

If $y = 0$, we have:

$$3x + 2 \cdot 0 = 12$$
$$3x = 12$$
$$x = 4$$

The x-intercept is $(4, 0)$.

If $x = 0$, we have:

$$3 \cdot 0 + 2y = 12$$
$$2y = 12$$
$$y = 6$$

The y-intercept is $(0, 6)$.

If $y = -3$, we have:

$$3x + 2(-3) = 12$$
$$3x - 6 = 12$$
$$3x = 18$$
$$x = 6$$

The point $(6, -3)$ is on the graph.

We plot these points and draw a line through them.

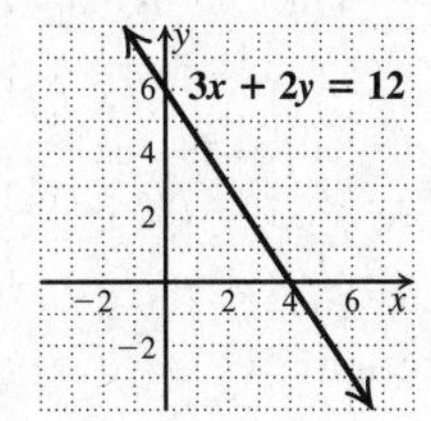

67. Graph: $y = -\frac{3}{4}x + 2$

We select some x-values and compute y-values. We use multiples of 4 to avoid fractions.

If $x = -4$, then $y = -\frac{3}{4}(-4) + 2 = 5$.

If $x = 0$, then $y = -\frac{3}{4} \cdot 0 + 2 = 2$.

If $x = 4$, then $y = -\frac{3}{4} \cdot 4 + 2 = -1$.

x	y	(x, y)
-4	5	$(-4, 5)$
0	2	$(0, 2)$
4	-1	$(4, -1)$

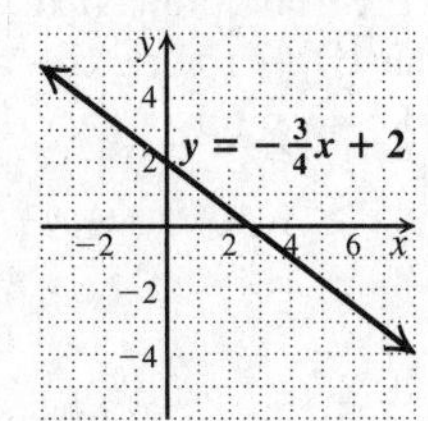

69. *Writing Exercise.* Yes; if the steamrollers working together take more than half as long as the slower steam-roller would working alone, then they do more than one entire job. That is, in half the time it takes the slower steamroller to do the job alone, the faster steamroller can do more than half of the job alone, and together they do more than $\frac{1}{2} + \frac{1}{2}$, or 1 entire job in that time.

71. ***Familiarize.*** If the drainage gate is closed, $\frac{1}{9}$ of the bog is filled in 1 hr. If the bog is not being filled, $\frac{1}{11}$ of the bog is drained in 1 hr. If the bog is being filled with the drainage gate left open, $\frac{1}{9} - \frac{1}{11}$ of the bog is filled in 1 hr. Let t = the time it takes to fill the bog with the drainage gate left open.

Translate. We want to find t such that $t\left(\frac{1}{9} - \frac{1}{11}\right) = 1$, or $\frac{t}{9} - \frac{t}{11} = 1$.

Carry out. We solve the equation. First we multiply by the LCD, 99.

Check. In $\frac{99}{2}$ hr, we have

$$\frac{99}{2}\left(\frac{1}{9} - \frac{1}{11}\right) = \frac{11}{2} - \frac{9}{2} = \frac{2}{2} = 1$$

full bog.

State. It will take $\frac{99}{2}$, or $49\frac{1}{2}$ hr, to fill the bog.

73. ***Familiarize.*** Let p = the number of people per hour moved by the 60 cm-wide escalator. Then $2p$ = the number of people per hour moved by the 100 cm-wide escalator. We convert 1575 people per 14 minutes to people per hour: $\frac{1575 \text{ people}}{14 \text{ min}} \cdot \frac{60 \text{ min}}{1 \text{ hr}} = 6750$ people/hr

Translate. We use the information that together the escalators move 6750 people per hour to write an equation. $p + 2p = 6750$

Carry out. We solve the equation.

$$p + 2p = 6750$$
$$3p = 6750$$
$$p = 2250$$

Check. If the 60 cm-wide escalator moves 2250 people per hour, then the 100 cm-wide escalator moves , or 4500 people per hour. Together, they move , or 6750 people per hour. The answer checks.

State. The 60 cm-wide escalator moves 2250 people per hour.

75. ***Familiarize.*** Let d = the distance, in miles, the paddle-boat can cruise upriver before it is time to turn around. The boat's speed upriver is 12 - 5, or 7 mph, and its speed downriver is 12 + 5, or 17 mph. We organize the information in a table.

	Distance	Speed	Time
Upriver	d	7	t_1
Downriver	d	17	t_2

Translate. Using the formula Time = Distance/Rate we see that $t_1 = \frac{d}{7}$ and $t_2 = \frac{d}{17}$. The time upriver and back is 3 hr, so $t_1 + t_2 = 3$, or $\frac{d}{7} + \frac{d}{17} = 3$.

Carry out. We solve the equation.

$$7 \cdot 17\left(\frac{d}{7} + \frac{d}{17}\right) = 7 \cdot 17 \cdot 3$$
$$17d + 7d = 357$$
$$24d = 357$$
$$d = \frac{119}{8}$$

Check. Traveling $\frac{119}{8}$ mi upriver at a speed of 7 mph takes $\frac{119/8}{7} = \frac{17}{8}$ hr. Traveling $\frac{119}{8}$ mi downriver at a speed of 17 mph takes $\frac{119/8}{17} = \frac{7}{8}$ hr. The total time is $\frac{17}{8} + \frac{7}{8} = \frac{24}{8} = 3$ hr. The answer checks.

State. The pilot can go $\frac{119}{8}$, or $14\frac{7}{8}$ mi upriver before it is time to turn around.

77. We will begin by finding how long it will take Alma and Kevin to grade a batch of exams, working together. Then we will find what percentage of the job was done by Alma.

Solve: $\left(\frac{1}{3} + \frac{1}{4}\right)t = 1$, or $\frac{7}{12} \cdot t = 1$

$t = \frac{12}{7}$ hr

Now, since Alma can do the job alone in 3 hr, she does $\frac{1}{3}$ of the job in 1 hr and in $\frac{12}{7}$ hr she does $\frac{12}{7} \cdot \frac{1}{3} \approx 0.57 \approx 57\%$ of the job.

79. ***Familiarize***. Let t represent the time it takes the printers to print 500 pages working together.

Translate. The faster machine can print 500 pages in 40 min, and it takes the slower printer 50 min to do the same job. Then we have $\frac{t}{40} + \frac{t}{50} = 1$.

Carry out. We solve the equation.

$$\frac{t}{40} + \frac{t}{50} = 1, \text{ LCD is } 200$$
$$200\left(\frac{t}{40} + \frac{t}{50}\right) = 200 \cdot 1$$
$$5t + 4t = 200 \cdot 1$$
$$9t = 200$$
$$t = \frac{200}{9}$$

In $\frac{200}{9}$ min, the faster printer does $\frac{200/9}{40}$, or $\frac{200}{9} \cdot \frac{1}{40}$, or $\frac{5}{9}$ of the job. Then starting at page 1, it would print $\frac{5}{9} \cdot 500$, or $277\frac{7}{9}$ pages. Thus, in $\frac{200}{9}$ min, the two machines will meet on page 278.

Check. We can check to see that the slower machine is also printing page 278 after $\frac{200}{9}$ min. In $\frac{200}{9}$ min, the slower machine does $\frac{200/9}{50}$, or $\frac{200}{9} \cdot \frac{1}{50}$, or $\frac{4}{9}$ of the job. Then it would print $\frac{4}{9} \cdot 500$, or $222\frac{2}{9}$ pages. Working backward from page 500, this machine would be on page $500 - 222\frac{2}{9}$, or $277\frac{2}{9}$. Thus, both machines are printing page 278 after $\frac{200}{9}$ min. The answer checks.

State. The two machines will meet on page 278.

81. Find a second proportion:

$\frac{A}{B} = \frac{C}{D}$ Given

$\frac{D}{A} \cdot \frac{A}{B} = \frac{D}{A} \cdot \frac{C}{D}$ Multiplying by $\frac{D}{A}$

$\frac{D}{B} = \frac{C}{A}$

Find a third proportion:

$\frac{A}{B} = \frac{C}{D}$ Given

$\frac{B}{C} \cdot \frac{A}{B} = \frac{B}{C} \cdot \frac{C}{D}$ Multiplying by $\frac{B}{C}$

$\frac{A}{C} = \frac{B}{D}$

Find a fourth proportion:

$\frac{A}{B} = \frac{C}{D}$ Given

$\frac{DB}{AC} \cdot \frac{A}{B} = \frac{DB}{AC} \cdot \frac{C}{D}$ Multiplying by $\frac{DB}{AC}$

$\frac{D}{C} = \frac{B}{A}$

83. ***Familiarize***. Let r = the speed in mph Garry would have to travel for the last half of the trip in order to average a speed of 45 mph for the entire trip. We organize the information in a table.

	Distance	Speed	Time
First half	50	40	t_1
Last half	50	r	t_2

The total distance is 50 + 50, or 100 mi. The total time is $t_1 + t_2$, or $\frac{50}{40} + \frac{50}{r}$, or $\frac{5}{4} + \frac{50}{r}$. The average speed is 45 mph.

Translate.

Average speed $= \frac{\text{Totaldistance}}{\text{Totaltime}}$

$45 = \frac{100}{\frac{5}{4} + \frac{50}{r}}$

Carry out. We solve the equation.

$$45 = \frac{100}{\frac{5}{4} + \frac{50}{r}}$$
$$45 = \frac{100}{\frac{5r+200}{4r}}$$
$$45 = 100 \cdot \frac{4r}{5r + 200}$$
$$45 = \frac{400r}{5r + 200}$$
$$(5r + 200)(45) = (5r + 200) \cdot \frac{400r}{5r + 200}$$
$$225r + 9000 = 400r$$
$$9000 = 175r$$
$$\frac{360}{7} = r$$

Check. Traveling 50 mi at 40 mph takes $\frac{50}{40}$, or $\frac{5}{4}$ hr. Traveling 50 mi at $\frac{360}{7}$ mph takes $\frac{50}{360/7}$, or $\frac{35}{36}$ hr. Then the total time is $\frac{5}{4} + \frac{35}{36} = \frac{80}{36} = \frac{20}{9}$ hr. The average speed when traveling 100 mi for $\frac{20}{9}$ hr is $\frac{100}{20/9} = 45$ mph. The answer checks.

State. Garry would have to travel at a speed of $\frac{360}{7}$, or $51\frac{3}{7}$ mph for the last half of the trip so that the average speed for the entire trip would be 45 mph.

85. *Writing Exercise.*

$$\frac{A+B}{B} = \frac{C+D}{D}$$
$$\frac{A}{B} + \frac{B}{B} = \frac{C}{D} + \frac{D}{D}$$
$$\frac{A}{B} + 1 = \frac{C}{D} + 1$$
$$\frac{A}{B} = \frac{C}{D}$$

The equations are equivalent.

Chapter 6 Review

1. False; some rational expressions like $\dfrac{y^2+4}{y+2}$ cannot be simplified.

3. False; when $t=3$, then $\dfrac{t-3}{t^2-4}=\dfrac{3-3}{3^2-4}=\dfrac{0}{5}=0$.

5. True; see page 384 in the text.

7. False; see page 390 in the text.

9. $\dfrac{17}{-x^2}$

Set the denominator equal to 0 and solve for x.
$-x^2=0$
$x=0$
The expression is undefined for $x=0$.

11. $\dfrac{x-5}{x^2-36}$

Set the denominator equal to 0 and solve for x.

$$\begin{aligned} x^2-36&=0\\ (x+6)(x-6)&=0\\ x+6=0 &\text{ or } x-6=0\\ x=-6 &\text{ or } x=6 \end{aligned}$$

The expression is undefined for $x=-6$ and $x=6$.

13. $\dfrac{-6}{(t+2)^2}$

Set the denominator equal to 0 and solve for t.

$$\begin{aligned} (t+2)^2&=0\\ t+2&=0\\ t&=-2 \end{aligned}$$

The expression is undefined for $t=-2$.

15. $\dfrac{14x^2-x-3}{2x^2-7x+3}=\dfrac{(2x-1)(7x+3)}{(2x-1)(x-3)}=\dfrac{7x+3}{x-3}$

17. $\dfrac{5x^2-20y^2}{2y-x}=\dfrac{-5(4y^2-x^2)}{(2y-x)}$
$=\dfrac{-5(2y+x)(2y-x)}{(2y-x)}=-5(2y+x)$

19.
$$\begin{aligned} \frac{6y-12}{2y^2+3y-2}\cdot\frac{y^2-4}{8y-8}&=\frac{6(y-2)(y-2)(y+2)}{(2y-1)(y+2)(8)(y-1)}\\ &=\frac{3(y-2)^2}{4(2y-1)(y-1)}\cdot\frac{2(y+2)}{2(y+2)}\\ &=\frac{3(y-2)^2}{4(2y-1)(y-1)} \end{aligned}$$

21.
$$\begin{aligned} \frac{4x^4}{x^2-1}\div\frac{2x^3}{x^2-2x+1}&=\frac{4x^4}{(x+1)(x-1)}\cdot\frac{(x-1)(x-1)}{2x^3}\\ &=\frac{2x(x-1)}{x+1}\cdot\frac{2x^3(x-1)}{2x^3(x-1)}\\ &=\frac{2x(x-1)}{x+1} \end{aligned}$$

23.
$$\begin{aligned} (t^2+3t-4)\div\frac{t^2-1}{t+4}&=(t+4)(t-1)\cdot\frac{(t+4)}{(t+1)(t-1)}\\ &=\frac{(t+4)^2}{t+1}\cdot\frac{t-1}{t-1}\\ &=\frac{(t+4)^2}{t+1} \end{aligned}$$

25. $x^2-x=x(x-1)$
$x^5-x^3=x^3(x^2-1)=x\cdot x\cdot x\cdot(x+1)(x-1)$
$x^4=x\cdot x\cdot x\cdot x$
$\text{LCM}=x\cdot x\cdot x\cdot x\cdot(x+1)(x-1)=x^4(x+1)(x-1)$

27. $\dfrac{x+6}{x+3}+\dfrac{9-4x}{x+3}=\dfrac{x+6+9-4x}{x+3}=\dfrac{-3x+15}{x+3}$

29.
$$\begin{aligned} \frac{3x-1}{2x}-\frac{x-3}{x}&=\frac{3x-1}{2x}-\frac{x-3}{x}\cdot\frac{2}{2}\\ \text{LCD}&=2x\\ &=\frac{3x-1-2(x-3)}{2x}\\ &=\frac{3x-1-2x+6}{2x}\\ &=\frac{x+5}{2x} \end{aligned}$$

31.
$$\begin{aligned} \frac{y^2}{y-2}+\frac{6y-8}{2-y}&=\frac{y^2}{y-2}+\frac{8-6y}{y-2}\\ &=\frac{y^2-6y+8}{y-2}\\ &=\frac{(y-4)(y-2)}{y-2}\\ &=y-4 \end{aligned}$$

33.
$$\begin{aligned} \frac{d^2}{d-2}+\frac{4}{2-d}&=\frac{d^2}{d-2}-\frac{4}{d-2}\\ &=\frac{d^2-4}{d-2}\\ &=\frac{(d+2)(d-2)}{d-2}\\ &=d+2 \end{aligned}$$

35.
$$\begin{aligned} &\frac{3x}{x+2}-\frac{x}{x-2}+\frac{8}{x^2-4}\\ &=\frac{3x}{x+2}\cdot\frac{x-2}{x-2}-\frac{x}{x-2}\cdot\frac{x+2}{x+2}+\frac{8}{(x+2)(x-2)}\\ &\text{LCD}=(x+2)(x-2)\\ &=\frac{3x^2-6x-x^2-2x+8}{(x+2)(x-2)}\\ &=\frac{2x^2-8x+8}{(x+2)(x-2)}=\frac{2(x^2-4x+4)}{(x+2)(x-2)}\\ &=\frac{2(x-2)(x-2)}{(x+2)(x-2)}\\ &=\frac{2(x-2)}{x+2} \end{aligned}$$

37.
$$\begin{aligned} \frac{\frac{1}{z}+1}{\frac{1}{z^2}-1}&=\frac{\frac{1}{z}+1}{\frac{1}{z^2}-1}\cdot\frac{z^2}{z^2}\\ \text{LCD}&=z^2\\ &=\frac{z+z^2}{1-z^2}\\ &=\frac{z(1+z)}{(1-z)(1+z)}\\ &=\frac{z}{1-z} \end{aligned}$$

39. $\dfrac{\frac{c}{d}-\frac{d}{c}}{\frac{1}{c}+\frac{1}{d}} = \dfrac{\frac{c}{d}-\frac{d}{c}}{\frac{1}{c}+\frac{1}{d}} \cdot \dfrac{cd}{cd}$

LCD $= cd$

$$= \frac{c^2-d^2}{d+c}$$
$$= \frac{(c-d)(c+d)}{c+d}$$
$$= c-d$$

41. $\dfrac{3}{x+4} = \dfrac{1}{x-1}$ Note $x \neq 1, -4$

$$(x-1)(x+4)\frac{3}{x+4} = (x-1)(x+4)\frac{1}{x-1}$$

LCD $= (x-1)(x+4)$

$$3(x-1) = x+4$$
$$3x-3 = x+4$$
$$3x = x+7$$
$$2x = 7$$
$$x = \frac{7}{2}$$

43. ***Familiarize.*** The job takes Jackson 12 hours working alone and Charis 9 hours working alone. Then in 1 hour Jackson does $\frac{1}{12}$ of the job and Charis does $\frac{1}{9}$ of the job. Working together, they can do $\frac{1}{9}+\frac{1}{12}$, or $\frac{7}{36}$ of the job in 1 hour.

Translate. If they work together t hours, then Jackson does $t\left(\frac{1}{9}\right)$ of the job and Charis does $t\left(\frac{1}{12}\right)$ of the job. We want some number t such that

$$\left(\frac{1}{9}+\frac{1}{12}\right)t = 1, \text{ or } \frac{7}{36}t = 1.$$

Carry out. We solve the equation.

$$\frac{7}{36}t = 1$$
$$\frac{36}{7}\cdot\frac{7}{36}t = \frac{36}{7}\cdot 1$$
$$t = \frac{36}{7} \text{ or } 5\frac{1}{7}$$

Check. The check can be done by repeating the computation.

State. Working together, it takes them $5\frac{1}{7}$ hrs to complete the job.

45. ***Familiarize.*** Let r = the rate by rail, in km/hr. The car's speed is $r+15$. Also set t = the time, in hours that the train and car travel. We organize the information in a table.

	Distance	Speed	Time
train	60	r	t
car	70	$r+15$	t

Translate. We can replace the t's in the table above using the formula $r = d/t$.

	Distance	Speed	Time
train	60	r	$\frac{60}{r}$
car	70	$r+15$	$\frac{70}{r+15}$

Since the times are the same for both vehicles, we have the equation

$$\frac{60}{r} = \frac{70}{r+15}.$$

Carry out. We multiply by the LCD, $r(r+15)$.

$$r(r+15)\cdot\frac{60}{r} = r(r+15)\cdot\frac{70}{r+15}$$
$$60(r+15) = 70r$$
$$60r+900 = 70r$$
$$900 = 10r$$
$$90 = r$$

If $r = 90$, then $r+15 = 105$.

Check. If the train's speed is 90 km/hr and the car's speed is 105 km/hr, then the car's speed is 15 km/hr faster than the train. The car's time is 105/70 or 1.5hr. The train's time is 90/60, or 1.5hr. Since the speeds are the same, the answer checks.

State. The car travels at 105 km/hr and the train travels at 90 km/hr.

47. ***Familiarize.*** The ratio of seal tagged to the total number of seal in the harbor, T, is $\frac{33}{T}$. Of the 40 seals caught later, there were 24 are tagged. The ratio of tagged seals to seals caught is $\frac{24}{40}$.

Translate. We translate to a proportion.

Seals originally tagged → 33, Seals in harbor → T; Tagged seals caught later ← 24, Seals caught later ← 40

$$\frac{33}{T} = \frac{24}{40}$$

Carry out. We solve the proportion.

$$40T\cdot\frac{33}{T} = 40T\cdot\frac{24}{40}$$
$$1320 = 24T$$
$$55 = T$$

Check.

$$\frac{33}{55} = 0.6, \quad \frac{24}{40} = 0.6$$

The ratios are the same, so the answer checks.

State. We estimate that there are 55 seals in the harbor.

49. *Writing Exercise.* The LCM of denominators is used to clear fractions when simplifying complex rational expressions using the method of multiplying by the LCD, and when solving rational equations.

51. $\dfrac{2a^2+5a-3}{a^2}\cdot\dfrac{5a^3+30a^2}{2a^2+7a-4}\div\dfrac{a^2+6a}{a^2+7a+12}$

$=\dfrac{\cancel{(2a-1)}(a+3)}{\cancel{a^2}}\cdot\dfrac{\cancel{5a^2}\cancel{(a+6)}}{\cancel{(2a-1)}\cancel{(a+4)}}\cdot\dfrac{(a+3)\cancel{(a+4)}}{a\cancel{(a+6)}}$

$=\dfrac{5(a+3)^2}{a}$

53. $\dfrac{5\cancel{(x-y)}}{\cancel{(x-y)}(x+2y)}-\dfrac{5\cancel{(x-3y)}}{(x+2y)\cancel{(x-3y)}}$

$=\dfrac{5}{x+2y}-\dfrac{5}{x+2y}=0$

Chapter 6 Test

1. $\dfrac{2-x}{5x}$

We find the number which makes the denominator 0.

$5x=0$

$x=0$

The expression is undefined for $x=0$.

3. $\dfrac{x-7}{x^2-1}$

We find the number which makes the denominator 0.

$x^2-1=0$

$(x+1)(x-1)=0$

$x+1=0$ or $x-1=0$

$x=-1$ or $x=1$

The expression is undefined for $x=-1$ and $x=1$.

5. $\dfrac{6x^2+17x+7}{2x^2+7x+3}=\dfrac{(3x+7)(2x+1)}{(x+3)(2x+1)}=\dfrac{(3x+7)\cancel{(2x+1)}}{(x+3)\cancel{(2x+1)}}$

$=\dfrac{3x+7}{x+3}$

7. $\dfrac{25y^2-1}{9y^2-6y}\div\dfrac{5y^2+9y-2}{3y^2+y-2}$

$=\dfrac{25y^2-1}{9y^2-6y}\cdot\dfrac{3y^2+y-2}{5y^2+9y-2}$

$=\dfrac{(5y+1)(5y-1)}{3y(3y-2)}\cdot\dfrac{(3y-2)(y+1)}{(5y-1)(y+2)}$

$=\dfrac{(5y+1)\cancel{(5y-1)}\cancel{(3y-2)}(y+1)}{3y\cancel{(3y-2)}\cancel{(5y-1)}(y+2)}$

$=\dfrac{(5y+1)(y+1)}{3y(y+2)}$

9. $(x^2+6x+9)\cdot\dfrac{(x-3)^2}{x^2-9}$

$=\dfrac{(x+3)(x+3)}{1}\cdot\dfrac{(x-3)(x-3)}{(x+3)(x-3)}$

$=\dfrac{\cancel{(x+3)}(x+3)\cancel{(x-3)}(x-3)}{\cancel{(x+3)}\cancel{(x-3)}}$

$=(x+3)(x-3)$

11. $\dfrac{2+x}{x^3}+\dfrac{7-4x}{x^3}=\dfrac{2+x+7-4x}{x^3}=\dfrac{-3x+9}{x^3}$

13. $\dfrac{2x-4}{x-3}+\dfrac{x-1}{3-x}=\dfrac{2x-4}{-1(3-x)}+\dfrac{x-1}{-1(3-x)}$

$=\dfrac{-(2x-4)}{3-x}+\dfrac{x-1}{3-x}$

$=\dfrac{-2x+4+x-1}{3-x}$

$=\dfrac{3-x}{3-x}$

$=1$

15. $\dfrac{7}{t-2}+\dfrac{4}{t}$ LCD is $t(t-2)$

$=\dfrac{7}{t-2}\cdot\dfrac{t}{t}+\dfrac{4}{t}\cdot\dfrac{t-2}{t-2}$

$=\dfrac{7t}{t(t-2)}+\dfrac{4(t-2)}{t(t-2)}$

$=\dfrac{7t+4t-8}{t(t-2)}$

$=\dfrac{11t-8}{t(t-2)}$

17. $\dfrac{1}{x-1}+\dfrac{4}{x^2-1}-\dfrac{2}{x^2-2x+1}=$

$\dfrac{1}{x-1}+\dfrac{4}{(x+1)(x-1)}-\dfrac{2}{(x-1)(x-1)}$

LCD is $(x-1)(x-1)(x+1)$

$=\dfrac{1}{x-1}\cdot\dfrac{(x-1)(x+1)}{(x-1)(x+1)}+$

$\dfrac{4}{(x-1)(x+1)}\cdot\dfrac{x-1}{x-1}-\dfrac{2(x+1)}{(x+1)(x-1)^2}$

$=\dfrac{(x-1)(x+1)}{(x+1)(x-1)^2}+\dfrac{4(x-1)}{(x+1)(x-1)^2}-\dfrac{2(x+1)}{(x+1)(x-1)^2}$

$=\dfrac{(x-1)(x+1)+4(x-1)-2(x+1)}{(x+1)(x-1)^2}$

$=\dfrac{x^2-1+4x-4-2x-2}{(x+1)(x-1)^2}$

$=\dfrac{x^2+2x-7}{(x+1)(x-1)^2}$

19. $\dfrac{\frac{x}{8}-\frac{8}{x}}{\frac{1}{8}+\frac{1}{x}}$ LCD is $8x$

$=\dfrac{8x}{8x}\cdot\dfrac{\frac{x}{8}-\frac{8}{x}}{\frac{1}{8}+\frac{1}{x}}=\dfrac{\frac{8x^2}{8}-\frac{64x}{x}}{\frac{8x}{8}+\frac{8x}{x}}$

$=\dfrac{x^2-64}{x+8}=\dfrac{(x+8)(x-8)}{x+8}=\dfrac{\cancel{(x+8)}(x-8)}{\cancel{x+8}}$

$=x-8$

21. To avoid division by 0, we must have $x\neq 0$ and $x-2\neq 0$, or $x\neq 0$ and $x\neq 2$.

$\dfrac{15}{x}-\dfrac{15}{x-2}=-2$ LCD $=x(x-2)$

$$x(x-2)\left(\frac{15}{x}-\frac{15}{x-2}\right)=x(x-2)(-2)$$
$$15(x-2)-15x=-2x(x-2)$$
$$15x-30-15x=-2x^2+4x$$
$$2x^2-4x-30=0$$
$$2(x^2-2x-15)=0$$
$$2(x-5)(x+3)=0$$

$x-5=0$ or $x+3=0$

$x=5$ or $x=-3$

The solutions are -3 and 5.

23. ***Familiarize.*** Burning 320 calories corresponds to walking 4 mi, and we wish to find the number of miles m that correspond to burning 100 calories. We can use a proportion.

Translate.

calories burned → $\frac{320}{4}=\frac{100}{m}$ ← calories burned
miles walked → ← miles walked

Carry out. We solve the proportion.

$$4m\cdot\frac{320}{4}=4m\cdot\frac{100}{m}$$
$$320m=400$$
$$m=\frac{5}{4}=1\frac{1}{4}$$

Check. $\frac{320}{4}=80$, $\frac{100}{5/4}=80$

The ratios are the same sot he answer checks.

State. Walking $1\frac{1}{4}$ mi corresponds to burning 100 calories.

25. ***Familiarize.*** Let $t=$ the number of hours it would take Rema to mulch the flower beds, working alone. Then $t+6$ = the number of hours it would take Perez, working alone. Note that $2\frac{6}{7}=\frac{20}{7}$ In $\frac{20}{7}$ hr Rema does $\frac{20}{7}\cdot\frac{1}{t}$ of the job and Perez does $\frac{20}{7}\cdot\frac{1}{t+6}$ of the job, and together they do 1 complete job.

Translate. We use the information above to write an equation. $\frac{20}{7}\cdot\frac{1}{t}+\frac{20}{7}\cdot\frac{1}{t+6}=1$

Carry out. We solve the equation.

$$\frac{20}{7}\cdot\frac{1}{t}+\frac{20}{7}\cdot\frac{1}{t+6}=1,\ \text{LCD}=7t(t+6)$$
$$7t(t+6)\cdot\left(\frac{20}{7}\cdot\frac{1}{t}+\frac{20}{7}\cdot\frac{1}{t+6}\right)=7t(t+6)\cdot 1$$
$$20(t+6)+20t=7t(t+6)$$
$$20t+120+20t=7t^2+42t$$
$$0=7t^2+2x-120$$
$$0=(7t+30)(t-4)$$

$7t+30=0$ or $t-4=0$

$t=-\frac{30}{7}$ or $t=4$

Check. Time cannot be negative in this application, so we check only 4. If Rema can mulch the flower beds working alone in 4 hr, then it would take Perez 4 + 6, or 10 hr, working alone. In $\frac{20}{7}$ hr they would do $\frac{20}{7}\cdot\frac{1}{4}+\frac{20}{7}\cdot\frac{1}{10}=\frac{5}{7}+\frac{2}{7}=1$ complete job. The answer checks.

State. It would take Rema 4 hr to mulch the flower beds working alone, and it would take Perez 10 hr working alone.

27. ***Familiarize.*** Let $x=$ the number. Then $-\frac{1}{x}$ is the opposite of the numbers reciprocal.

Translate. The square of the number, x^2 is equivalent to $-\frac{1}{x}$, so we write a proportion. $x^2=-\frac{1}{x}$

Carry out. We solve the equation.

$$x\cdot x^2=x\cdot\left(-\frac{1}{x}\right)$$
$$x^3=-1$$
$$x=\sqrt[3]{-1}=-1$$

Check. $(-1)^2=1$, $-\frac{1}{-1}=1$, so the ratios are equivalent.

State. The number is -1.

Chapter 7

Functions and Graphs

Exercise Set 7.1

1. A function is a special kind of correspondence between two sets.

3. For any function, the set of all inputs, or first values, is called the domain.

5. When a function is graphed, members of the domain are located on the horizontal axis.

7. The notation $f(3)$ is read "f of 3," "f at 3," or "the value of f at 3."

9. The correspondence is a function, because each member of the domain corresponds to exactly one member of the range.

11. The correspondence is a function, because each member of the domain corresponds to exactly one member of the range.

13. The correspondence is not a function because a member of the domain (June 9 or October 5) corresponds to more than one member of the range.

15. The correspondence is a function, because each member of the domain corresponds to exactly one member of the range.

17. The correspondence is a function, because each flash drive has only one storage capacity.

19. This correspondence is a function, because each player has only one uniform number.

21. a) Locate 1 on the horizontal axis and then find the point on the graph for which 1 is the first coordinate. From that point, look to the vertical axis to find the corresponding y-coordinate, –1. Thus, $f(1)=-1$.

    b) To determine which member(s) of the domain are paired with 2, locate 2 on the vertical axis. From there look left and right to the graph to find any points for which 2 is the second coordinate. One such point exists. Its first coordinate is –3. Thus, the x-value for which $f(x)=2$ is –3.

23. a) Locate 1 on the horizontal axis and then find the point on the graph for which 1 is the first coordinate. From that point, look to the vertical axis to find the corresponding y-coordinate, 3. Thus, $f(1)=3$.

    b) To determine which member(s) of the domain are paired with 2, locate 2 on the vertical axis. From there look left and right to the graph to find any points for which 2 is the second coordinate. One such point exists. Its first coordinate is 3. Thus, the x-value for which $f(x)=2$ is 3.

25. a) Locate 1 on the horizontal axis and then find the point on the graph for which 1 is the first coordinate. From that point, look to the vertical axis to find the corresponding y-coordinate, 3. Thus, $f(1)=3$.

    b) To determine which member(s) of the domain are paired with 2, locate 2 on the vertical axis. From there look left and right to the graph to find any points for which 2 is the second coordinate. One such point exists. Its first coordinate is 0. Thus, the x-value for which $f(x)=2$ is 0.

27. a) Locate 1 on the horizontal axis and then find the point on the graph for which 1 is the first coordinate. From that point, look to the vertical axis to find the corresponding y-coordinate, 3. Thus, $f(1)=3$.

    b) To determine which member(s) of the domain are paired with 2, locate 2 on the vertical axis. From there look left and right to the graph to find any points for which 2 is the second coordinate. One such point exists. Its first coordinate is –3. Thus, the x-value for which $f(x)=2$ is –3.

29. a) Locate 1 on the horizontal axis and then find the point on the graph for which 1 is the first coordinate. From that point, look to the vertical axis to find the corresponding y-coordinate, 1. Thus, $f(1)=1$.

    b) To determine which member(s) of the domain are

paired with 2, locate 2 on the vertical axis. From there look left and right to the graph to find any points for which 2 is the second coordinate. One such point exists. Its first coordinate is 3. Thus, the x-value for which $f(x)=2$ is 3.

31. a) Locate 1 on the horizontal axis and then find the point on the graph for which 1 is the first coordinate. From that point, look to the vertical axis to find the corresponding y-coordinate, 4. Thus, $f(1)=4$.

b) To determine which member(s) of the domain are paired with 2, locate 2 on the vertical axis. From there look left and right to the graph to find any points for which 2 is the second coordinate. There are two such points, (–1, 2) and (3, 2). Thus, the x-values for which $f(x)=2$ are –1 and 3.

33. a) Locate 1 on the horizontal axis and then find the point on the graph for which 1 is the first coordinate. From that point, look to the vertical axis to find the corresponding y-coordinate, 2. Thus, $f(1)=2$.

b) To determine which member(s) of the domain are paired with 2, locate 2 on the vertical axis. From there look left and right to the graph to find any points for which 2 is the second coordinate. All points in the set $\{x|0<x\leq 2\}$ satisfy this condition. These are the x-values for which $f(x)=2$.

35. We can use the vertical-line test:

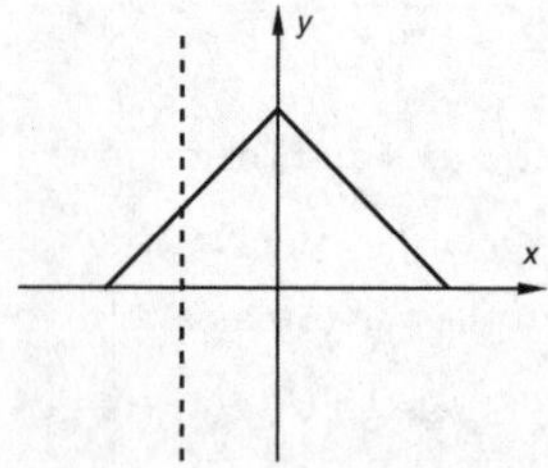

Visualize moving this vertical line across the graph. No vertical line will intersect the graph more than once. Thus, the graph is a graph of a function.

37. We can use the vertical-line test:

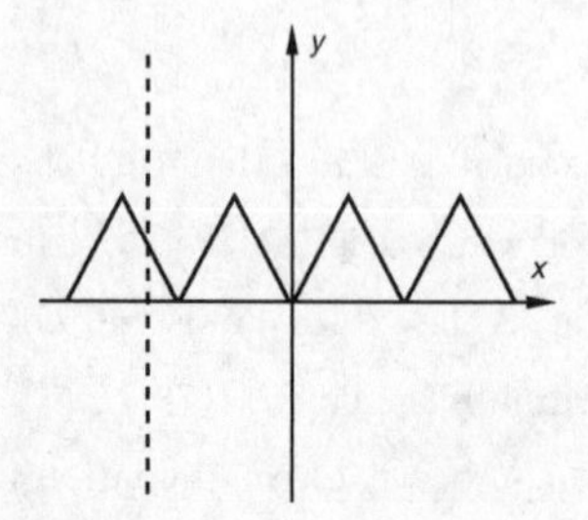

Visualize moving this vertical line across the graph. No vertical line will intersect the graph more than once. Thus, the graph is a graph of a function.

39. We can use the vertical line test.

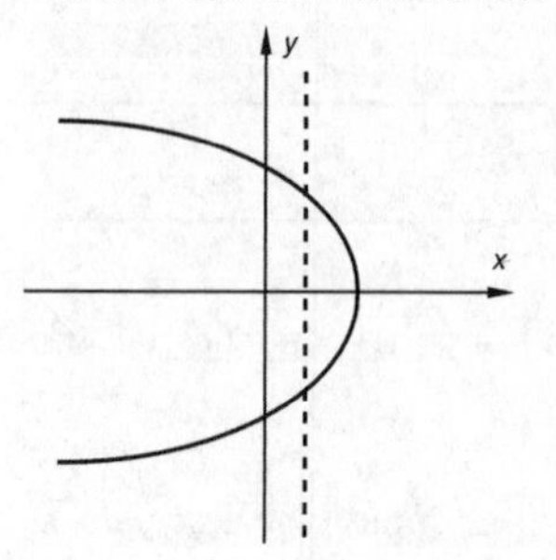

It is possible for a vertical line to intersect the graph more than once. Thus this is not the graph of a function.

41. $g(x)=2x+5$

a) $g(0)=2(0)+5=0+5=5$

b) $g(-4)=2(-4)+5=-8+5=-3$

c) $g(-7)=2(-7)+5=-14+5=-9$

d) $g(8)=2(8)+5=16+5=21$

e) $g(a+2)=2(a+2)+5=2a+4+5=2a+9$

f) $g(a)+2=2(a)+5+2=2a+7$

43. $f(n)=5n^2+4n$

a) $f(0)=5\cdot 0^2+4\cdot 0=0+0=0$

b) $f(-1)=5(-1)^2+4(-1)=5-4=1$

c) $f(3)=5\cdot 3^2+4\cdot 3=45+12=57$

d) $f(t)=5t^2+4t$

e) $f(2a)=5(2a)^2+4\cdot 2a=5\cdot 4a^2+8a=20a^2+8a$

f) $f(3)-9=5\cdot 3^2+4\cdot 3-9=5\cdot 9+4\cdot 3-9$
$=45+12-9=48$

45. $f(x)=\dfrac{x-3}{2x-5}$

a) $f(0)=\dfrac{0-3}{2\cdot 0-5}=\dfrac{-3}{0-5}=\dfrac{-3}{-5}=\dfrac{3}{5}$

b) $f(4)=\dfrac{4-3}{2\cdot 4-5}=\dfrac{1}{8-5}=\dfrac{1}{3}$

c) $f(-1)=\dfrac{-1-3}{2(-1)-5}=\dfrac{-4}{-2-5}=\dfrac{-4}{-7}=\dfrac{4}{7}$

d) $f(3)=\dfrac{3-3}{2\cdot 3-5}=\dfrac{0}{6-5}=\dfrac{0}{1}=0$

e) $f(x+2)=\dfrac{x+2-3}{2(x+2)-5}=\dfrac{x-1}{2x+4-5}=\dfrac{x-1}{2x-1}$

f) $f(a+h)=\dfrac{a+h-3}{2(a+h)-5}=\dfrac{a+h-3}{2a+2h-5}$

47. $A(s)=s^2\frac{\sqrt{3}}{4}$

$A(4)=4^2\frac{\sqrt{3}}{4}=4\sqrt{3}\approx 6.93$

The area is $4\sqrt{3}\text{ cm}^2\approx 6.93\text{ cm}^2$.

49. $V(r)=4\pi r^2$

$V(3)=4\pi(3)^2=36\pi$

The area is $36\pi\text{ in}^2\approx 113.10\text{ in}^2$.

51. $H(x)=2.75x+71.48$

$H(34)=2.75(34)+71.48=164.98$

The predicted height is 164.98 cm.

53. $F(C)=\frac{9}{5}C+32$

$F(-5)=\frac{9}{5}(-5)+32=-9+32=23$

The equivalent temperature is 23°F.

55. Locate the point that is directly above 225. Then estimate its second coordinate by moving horizontally from the point to the vertical axis. The rate is about 75 heart attacks per 10,000 men.

57. Locate the point on the graph that is directly above '00. Then estimate its second coordinate by moving horizontally from the point to the vertical axis. In 2000, about 500 movies were released. That is $F(2000)\approx 500$.

59. Plot and connect the points, using the wattage of the incandescent as the first coordinate and the wattage of the CFL as the second coordinate.

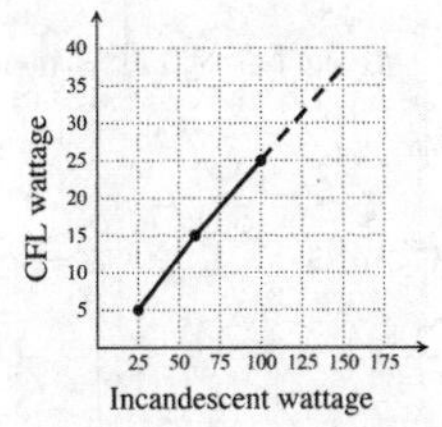

To estimate the wattage of the CFL bulb that creates light equivalent to a 75-watt incandescent bulb, first locate the point directly above 75. Then estimate the second coordinate by moving horizontally from the point to the vertical axis. Read the approximate function value there. The wattage is about 19 watts.

To predict the wattage of the CFL bulb that creates light equivalent to a 120-watt incandescent bulb, extend the graph and extrapolate. The wattage is about 30 watts.

61. Plot and connect the points, using body weight as the first coordinate and the corresponding number of drinks as the second coordinate.

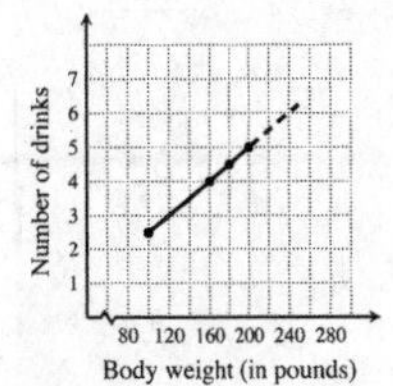

To estimate the number of drinks that a 140-lb person would have to drink to be considered intoxicated, first locate the point that is directly above 140. Then estimate its second coordinate by moving horizontally from the point to the vertical axis. Read the approximate function value there. The estimated number of drinks is 3.5.

To predict the number of drinks it would take for a 230-lb person to be considered intoxicated, extend the graph and extrapolate. It appears that it would take about 6 drinks.

63. Plot and connect the points, using the year as the first coordinate and the total sales as the second coordinate.

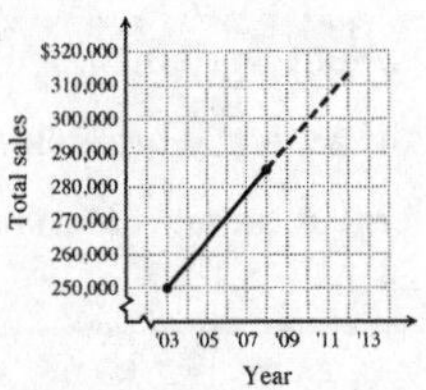

To estimate the total sales for 2004, first locate the point directly above 2004. Then estimate its second coordinate by moving horizontally to the vertical axis. Read the approximate function value there. The estimated 2004 sales total is about $257,000.

To estimate the sales for 2011, extend the graph and extrapolate. We estimate the sales for 2011 to be about $306,000.

65. *Writing Exercise.* You could measure time in years, because it seems reasonable that each graph represents a ten-year span. Answers may vary.

67. $2(x-5)-3=4-(x-1)$

$2x-10-3=4-x+1$

$2x-13=5-x$

$3x-13=5$

$3x=18$

$x=6$

The solution is 6.

69. $\frac{1}{x} = -2$

$\frac{1}{x} \cdot x = -2 \cdot x$

$1 = -2x$

$-\frac{1}{2} = x$

The solution is $-\frac{1}{2}$.

71. $(x-2)(x+3) = 6$

$x^2 + x - 6 = 6$

$x^2 + x - 12 = 0$

$(x+4)(x-3) = 0$

$x = -4$ *or* $x = 3$

The solutions are –4 and 3.

73. $\frac{x+1}{x} = 8$

$\frac{x+1}{x} \cdot x = 8 \cdot x$

$x + 1 = 8x$

$1 = 7x$

$\frac{1}{7} = x$

The solution is $\frac{1}{7}$.

75. *Writing Exercise.* The independent variable should be chosen as the number of fish in an aquarium, since the survival of the fish is dependent upon an adequate food supply.

77. To find $f(g(-4))$, we first find $g(-4)$:

$g(-4) = 2(-4) + 5 = -8 + 5 = -3$.

Then

$f(g(-4)) = f(-3) = 3(-3)^2 - 1 = 3 \cdot 9 - 1 = 27 - 1 = 26$.

To find $g(f(-4))$, we first find $f(-4)$:

$f(-4) = 3(-4)^2 - 1 = 3 \cdot 16 - 1 = 48 - 1 = 47$.

Then $g(f(-4)) = g(47) = 2 \cdot 47 + 5 = 94 + 5 = 99$.

79. $f(\text{tiger}) = \text{dog}$

$f(\text{dog}) = f(f(\text{tiger})) = \text{cat}$

$f(\text{cat}) = f(f(f(\text{tiger}))) = \text{fish}$

$f(\text{fish}) = f(f(f(f(\text{tiger})))) = \text{worm}$

81. Locate the highest point on the graph. Then move vertically to the horizontal axis and read the corresponding time. It is about 2 min, 50 sec.

83. The two largest contractions occurred at about 2 minutes, 50 seconds and 5 minutes, 40 seconds. The difference in these times, is 2 minutes, 50 seconds, so the frequency is about 1 every 3 minutes.

85.

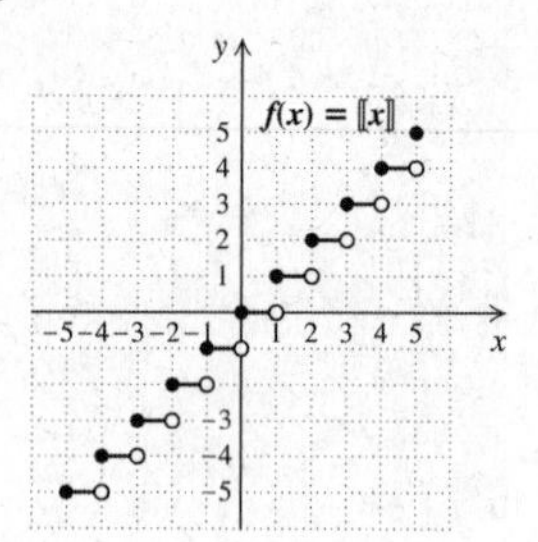

Exercise Set 7.2

1. Since $-1 \le 0 < 10$, function (c) would be used.

3. Since $10 \ge 10$, function (d) would be used.

5. Since $-1 \le -1 < 10$, function (c) would be used.

7. The domain is the set of all the first coordinates, $\{2, 9, -2, -4\}$.
The range is the set of all second coordinates, $\{8, 3, 10, 4\}$.

9. The domain is the set of all the first coordinates, $\{0, 4, -5, -1\}$.
The range is the set of all second coordinates, $\{0, -2\}$.

11. The function of f can be written as $\{(-4,-2), (-2,-1), (0, 0), (2, 1), (4, 2)\}$.
The domain is the set of all first coordinates, $\{-4, -2, 0, 2, 4\}$ and the range is the set of all second coordinates, $\{-2, -1, 0, 1, 2\}$.

13. The function of f can be written as $\{(-5, -1), (-3, -1), (-1, -1), (0, 1), (2, 1), (4, 1)\}$.
The domain is the set of all first coordinates, $\{-5, -3, -1, 0, 2, 4\}$ and the range is the set of all second coordinates, $\{-1, 1\}$.

15. The domain of the function is the set of all x-values that are in the graph, $\{x \mid -4 \le x \le 3\}$, or $[-4, 3]$.
The range is the set of all y-values that are in the graph, $\{y \mid -3 \le y \le 4\}$, or $[-3, 4]$.

17. The domain of the function is the set of all x-values that are in the graph, $\{x \mid -4 \le x \le 5\}$, or $[-4, 5]$.
The range is the set of all y-values that are in the graph, $\{y \mid -2 \le y \le 4\}$, or $[-2, 4]$.

19. The domain of the function is the set of all x-values that are in the graph, $\{x|-4 \le x \le 4\}$, or $[-4, 4]$.

The range is the set of all y-values that are in the graph, $\{-3, -1, 1\}$.

21. For any x-value and for any y-value there is a point on the graph. Thus,

Domain of $f = \{x|x \text{ is a real number}\}$, or $\mathbb{R}$ and

Range of $f = \{y|y \text{ is a real number}\}$, or $\mathbb{R}$.

23. For any x-value there is a point on the graph. Thus,

Domain of $f = \{x|x \text{ is a real number}\}$, or $\mathbb{R}$.

The only y-value on the graph is 4. Thus,

Range of $f = \{4\}$.

25. For any x-value there is a point on the graph. Thus,

Domain of $f = \{x|x \text{ is a real number}\}$, or $\mathbb{R}$.

The function has no y-values less than 1 and every y-value greater than or equal to 1 corresponds to a member of the domain. Thus,

Range of $f = \{y|y \ge 1\}$, or $[1, \infty)$.

27. The hole in the graph at $(-2, -4)$ indicates that the function is not defined for $x = -2$. For any other x-value there is a point on the graph. Thus,

Domain of $f = \{x|x \text{ is a real number } and\ x \ne -2\}$.

There is no function value at $(-2, -4)$, so -4 is not in the range of the function. For any other y-value there is a point on the graph. Thus,

Range of $f = \{y|y \text{ is a real number } and\ y \ne -4\}$.

29. The function has no x-values less than 0 and every x-value greater than or equal to 0 corresponds to a member of the domain. Thus,

Domain of $f = \{x|x \ge 0\}$, or $[0, \infty)$.

The function has no y-values less than 0 and every y-value greater than or equal to 0 corresponds to a member of the range. Thus,

Range of $f = \{y|y \ge 0\}$, or $[0, \infty)$.

31. $f(x) = \dfrac{5}{x-3}$

Since $\dfrac{5}{x-3}$ cannot be computed when the denominator is 0, we find the x-value that causes $x - 3$ to be 0:

$x - 3 = 0$

$x = 3$ Adding 3 to both sides

Thus, 3 is not in the domain of f, while all other real numbers are. The domain of f is

$\{x|x \text{ is a real number } and\ x \ne 3\}$.

33. $f(x) = \dfrac{x}{2x-1}$

Since $\dfrac{x}{2x-1}$ cannot be computed when the denominator is 0, we find the x-value that causes $2x - 1$ to be 0:

$2x - 1 = 0$

$2x = 1$

$x = \dfrac{1}{2}$

Thus, $\dfrac{1}{2}$ is not in the domain of f, while all other real numbers are. The domain of f is

$\left\{x \middle| x \text{ is a real number } and\ x \ne \dfrac{1}{2}\right\}$.

35. $f(x) = 2x + 1$

Since we can compute $2x + 1$ for any real number x, the domain is the set of all real numbers.

37. $g(x) = |5 - x|$

Since we can compute $|5 - x|$ for any real number x, the domain is the set of all real numbers.

39. $f(x) = \dfrac{5}{x^2 - 9}$

The expression $\dfrac{5}{x^2 - 9}$ is undefined when $x^2 - 9 = 0$.

$x^2 - 9 = 0$

$(x+3)(x-3) = 0$

$x + 3 = 0$ *or* $x - 3 = 0$

$x = -3$ *or* $x = 3$

Thus, Domain of $f = \{x|x \text{ is a real number } and\ x \ne -3\ and\ x \ne 3\}$.

41. $f(x) = x^2 - 9$

Since we can compute $x^2 - 9$ for any real number x, the domain is the set of all real numbers.

43. $f(x) = \dfrac{2x-7}{x^2 + 8x + 7}$

The expression $\dfrac{2x-7}{x^2 + 8x + 7}$ is undefined when $x^2 + 8x + 7 = 0$.

$x^2 + 8x + 7 = 0$

$(x+1)(x+7) = 0$

$x + 1 = 0$ *or* $x + 7 = 0$

$x = -1$ *or* $x = -7$

Thus, Domain of $f = \{x|x \text{ is a real number } and\ x \ne -1\ and\ x \ne -7\}$.

45. $R(t) = 46.8 - 0.075t$

If we assume the function is not valid for years before 1930, we must have $t \geq 0$. In addition, $R(t)$ must be positive, so we have:

$$46.8 - 0.075t > 0$$
$$-0.075t > -46.8$$
$$t < 624$$

Then the domain of the function is $\{t | 0 \leq t < 624\}$ or $[0, 624)$.

47. $A(p) = -2.5p + 26.5$

The price must be positive, so we have $p > \$0$. In addition $A(p)$ must be nonnegative, so we have:

$$-2.5p + 26.5 \geq 0$$
$$26.5 \geq 2.5p$$
$$10.6 \geq p$$

Then the domain of the function is $\{p | 0 \leq p \leq \$10.60\}$ or $[0, 10.60]$.

49. $P(d) = 0.03d + 1$

The depth must be positive, so we have $d \geq 0$. In addition $P(d)$ must be nonnegative, so we have:

$$0.03d + 1 \geq 0$$
$$0.03d \geq -1$$
$$d \geq -33.\overline{3}$$

Then we have $d \geq 0$ *and* $d \geq -33.\overline{3}$, so the domain of the function is $\{d | d \geq 0\}$ or $[0, \infty)$.

51. $h(t) = -16t^2 + 64t + 80$

The time cannot be negative, so we have $t \geq 0$. The height cannot be negative either, so an upper limit for t will be the positive value of t for which $h(t) = 0$.

$$-16t^2 + 64t + 80 = 0$$
$$-16(t^2 - 4t - 5) = 0$$
$$-16(t-5)(t+1) = 0$$
$$t - 5 = 0 \quad or \quad t + 1 = 0$$
$$t = 5 \quad or \quad t = -1$$

We know that –1 is not in the domain of the function. We also see that 5 is an upper limit for t. Then the domain of the function is $\{t | 0 \leq t \leq 5\}$ or $[0, 5]$.

53. $f(x) = \begin{cases} x, & \text{if } x < 0 \\ 2x+1, & \text{if } x \geq 0 \end{cases}$

a) Since $-5 < 0$, we use the equation $f(x) = x$. Thus, $f(-5) = -5$.

b) Since $0 \geq 0$, we use the equation $f(x) = 2x + 1$. $f(0) = 2 \cdot 0 + 1 = 0 + 1 = 1$

c) Since $10 \geq 0$, we use the equation $f(x) = 2x + 1$. $f(10) = 2 \cdot 10 + 1 = 20 + 1 = 21$

55. $G(x) = \begin{cases} x - 5, & \text{if } x < -1 \\ x, & \text{if } -1 \leq x \leq 2 \\ x + 2, & \text{if } x > 2 \end{cases}$

a) Since $-1 \leq 0 \leq 2$, we use the equation $G(x) = x$. $G(0) = 0$

b) Since $-1 \leq 2 \leq 2$, we use the equation $G(x) = x$. $G(2) = 2$

c) Since $5 > 2$, we use the equation $G(x) = x + 2$. $G(5) = 5 + 2 = 7$

57. $f(x) = \begin{cases} x^2 - 10, & \text{if } x < -10 \\ x^2, & \text{if } -10 \leq x \leq 10 \\ x^2 + 10, & \text{if } x > 10 \end{cases}$

a) Since $-10 \leq -10 \leq 10$, we use the equation $f(x) = x^2$. $f(-10) = (-10)^2 = 100$

b) Since $-10 \leq 10 \leq 10$, we use the equation $f(x) = x^2$. $f(10) = 10^2 = 100$

c) Since $11 > 10$, we use the equation $f(x) = x^2 + 10$. $f(11) = 11^2 + 10 = 121 + 10 = 131$

59. *Writing Exercise.* The expression $\frac{x+3}{2}$ is defined for all real numbers, but the expression $\frac{2}{x+3}$ is not defined for $x = -3$.

61. $y = 2x - 3$

First we plot the y-intercept, (0, –3). We can think of the slope as $\frac{2}{1}$. Starting at (0, –3), find a second point by moving up 2 units and to the right 1 unit to the point $(1, -1)$. In a similar manner we can move from $(1, -1)$ to (2, 1). Then we connect the points.

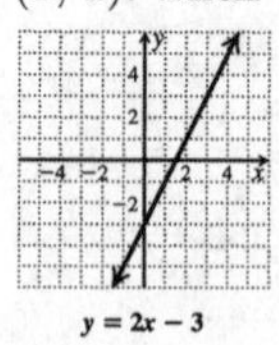

$y = 2x - 3$

63. $\frac{2}{3}x - 4$

The slope is $\frac{2}{3}$, and the y-intercept is (0, –4).

65. $y = \frac{4}{3}x$, or $y = \frac{4}{3}x + 0$

The slope is $\frac{4}{3}$ and the y-intercept is (0, 0).

67. *Writing Exercise.* No; the domain of $f(x)$ is $\{x|x \text{ is a real number } and \ x \neq 0\}$, but the domain of $g(x)$ is $\{x|x \text{ is a real number}\}$.

69. Answers may vary.

Domain: $\mathbb{R}$; range: $\mathbb{R}$

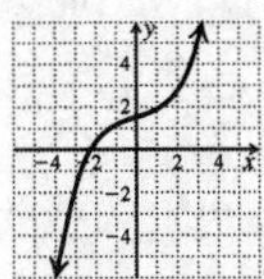

71. Answers may vary.

Domain: $\{x|1 \leq x \leq 5\}$; range: $\{y|0 \leq y \leq 2\}$

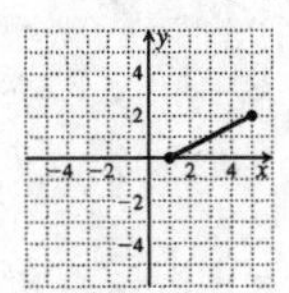

73. The graph indicates that the function is not defined for $x = 0$. For any other x-value there is a point on the graph. Thus,

Domain of $f = \{x|x \text{ is a real number } and \ x \neq 0\}$.

The graph also indicates that the function is not defined for $y = 0$. For any other y-value there is a point on the graph. Thus,

Range of $f = \{y|y \text{ is a real number } and \ y \neq 0\}$

75. The function has no x-values for $-2 \leq x \leq 0$. For any other x-value there is a point on the graph. Thus, the domain of the function is $\{x|x < -2 \ or \ x > 0\}$.

The function has no y-values for $-2 \leq y \leq 3$. Every other y-value corresponds to a member of the range. Then the range is $\{y|y < -2 \ or \ y > 3\}$.

77. From the graph below, we see that the domain of f is $\{x|x \text{ is a real number}\}$ and the range is $\{y|y \geq 0\}$.

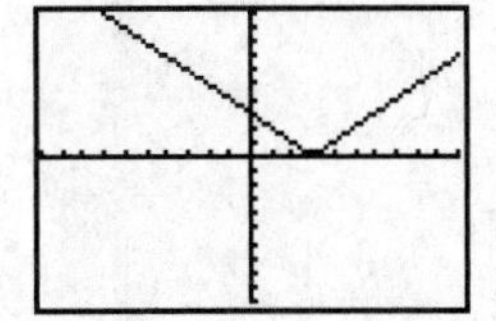

79. From the graph below, we see that the domain of f is $\{x|x \text{ is a real number } and\ x \neq 2\}$ and the range is $\{y|y \text{ is a real number } and\ y \neq 0\}$.

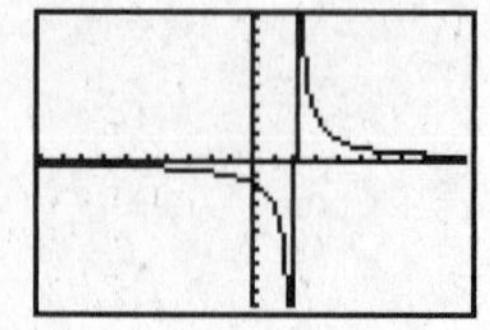

81. We graph the function $h(t) = -16t^2 + 64t + 80$ in the window [0, 5,–5, 150] with Xscl = 1 and Yscl = 15.

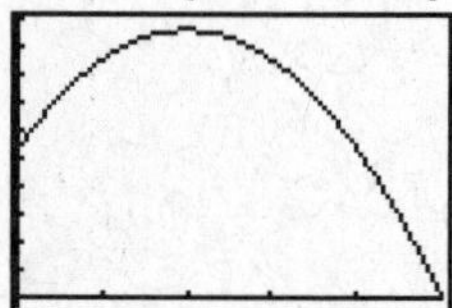

From the graph we estimate that the range of the function is $\{h|0 \leq h \leq 144\}$.

83. From Exercise 53: $f(x) = \begin{cases} x, & \text{if } x < 0 \\ 2x+1, & \text{if } x \geq 0 \end{cases}$

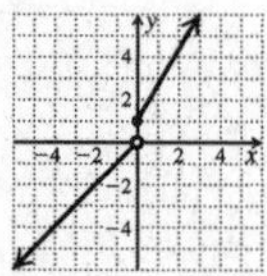

85. From Exercise 55: $G(x) = \begin{cases} x-5, & \text{if } x < -1 \\ x, & \text{if } -1 \leq x \leq 2 \\ x+2, & \text{if } x > 2 \end{cases}$

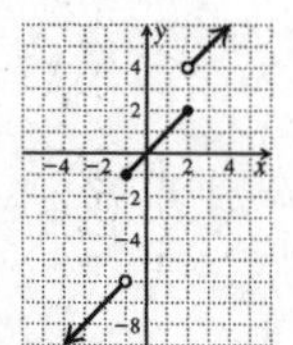

87. From Exercise 57: $f(x) = \begin{cases} x^2-10, & \text{if } x < -10 \\ x^2, & \text{if } -10 \leq x \leq 10 \\ x^2+10, & \text{if } x > 10 \end{cases}$

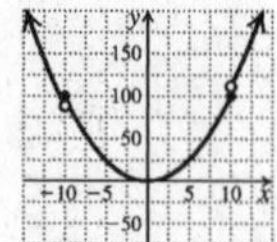

89. *Graphing Calculator Exercise*

Exercise Set 7.3

1. False; the vertical-line test states that if it is possible for a vertical line to cross a graph more than once, then the graph is not the graph of a function.

3. False; unless restricted, the domain of a constant function is the set of all real numbers.

5. True

7. Graph $y = 2x - 1$.

The slope is 2 and the y-intercept is (0, –1). We plot (0,–1). Then, thinking of the slope as $\frac{2}{1}$, we start at (0, –1) and move up 2 units and right 1 unit to the point (1, 1). Alternatively, we can think of the slope as $\frac{-2}{-1}$ and, starting at (0,–1), move down 2 units and left 1 unit to (–1,–3). Using the points found we draw the graph.

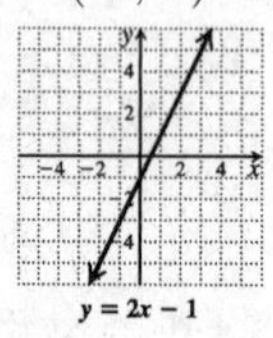

$y = 2x - 1$

9. Graph $y = -\frac{2}{3}x + 3$.

The slope is $-\frac{2}{3}$ and the y-intercept is (0, 3). We plot (0, 3). Then, thinking of the slope as $\frac{-2}{3}$, we start at (0, 3) and move down 2 units and right 3 units to the point (3, 1). Alternatively, we can think of the slope as $\frac{2}{-3}$ and, starting at (0, 3), move up 2 units and left 3 units to (–3, 5). Using the points found we draw the graph.

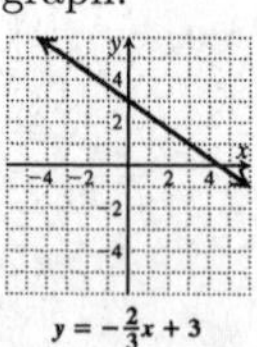

$y = -\frac{2}{3}x + 3$

11. Graph $3y = 6 - 4x$.

First we write the equation in slope-intercept form.

$$3y = 6 - 4x$$
$$y = \frac{1}{3}(6 - 4x)$$
$$y = 2 - \frac{4}{3}x$$
$$y = -\frac{4}{3}x + 2$$

The slope is $-\frac{4}{3}$ and the y-intercept is (0, 2). We plot (0, 2). Then, thinking of the slope as $\frac{-4}{3}$, we start at (0, 2) and move down 4 units and right 3 units to the point (3, –2). Alternatively, we can think of the slope as $\frac{4}{-3}$ and, starting at (0, 2), move up 4 units and left 3 units to (–3, 6). Using the points found we draw the graph.

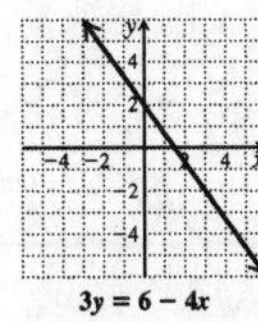

13. Graph $x-y=4$.

First we determine the intercepts.

Let $y = 0$ to find the x-intercept.

$$\begin{aligned} x-0&=4\\ x&=4 \end{aligned}$$

The x-intercept is (4, 0).

Let $x = 0$ to find the y-intercept.

$$\begin{aligned} 0-y&=4\\ -y&=4\\ y&=-4 \end{aligned}$$

The y-intercept is (0, –4).

We plot (4, 0) and (0,–4) and draw the line.

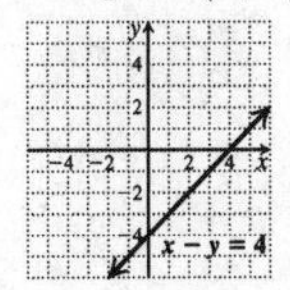

15. Graph $y=-2$.

For every input x, the value of y is –2. The graph is a horizontal line.

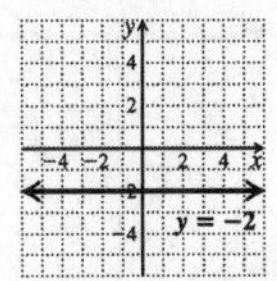

17. Graph $x=4$.

The value of x is 4 for any value of y. The graph is a vertical line.

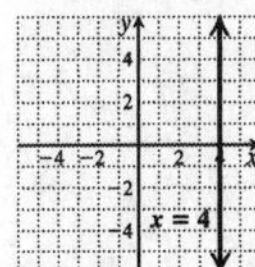

19. Graph $f(x)=x+3$.

The slope is 1, or $\frac{1}{1}$ and the y-intercept is (0, 3). We plot (0, 3) and move up 1 unit and right 1 unit to the point (1, 4). After we have sketched the line, a third point can be calculated as a check.

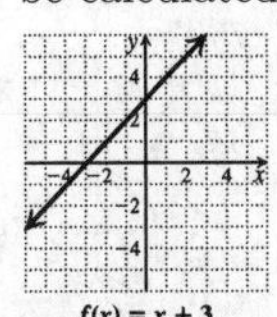

21. Graph $f(x)=\frac{3}{4}x+1$.

The slope is $\frac{3}{4}$ and the y-intercept is (0, 1). We plot (0, 1) and move up 3 units and right 4 unit to the point (4, 4). After we have sketched the line, a third point can be calculated as a check.

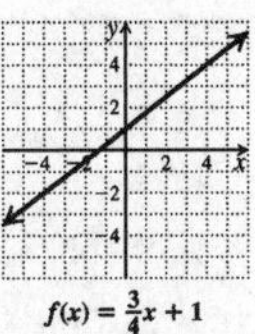

23. Graph $g(x)=4$.

For every input x, the output is 4. The graph is a horizontal line.

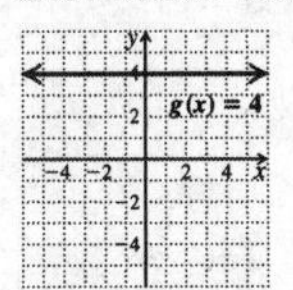

25. ***Familiarize.*** After an initial fee of \$30, an additional fee of \$0.75 is charged each mile. After one mile, the total cost is \$30 + \$0.75 = \$30.75. After two miles, the total cost is $30+2\cdot 0.75=\$31.50$. We can generalize this with a model, letting $C(d)$ represent the total cost, in dollars, for d miles of driving.

Translate. We reword the problem and translate.

Total Cost	is	initial cost	plus	\$0.75 per mile.
↓	↓	↓	↓	↓
$C(d)$	$=$	30	$+$	$0.75d$

where $d \geq 0$ since we cannot have negative miles driven. So the charge for a one-day rental of a truck driven d miles is $C(d)=0.75d+30$.

Carry out. To determine the number of miles required to reach a total cost of \$75, we substitute 75 for $C(d)$ and solve for d.

$$\begin{aligned} C(d)&=0.75d+30 & &\\ 75&=0.75d+30 & &\text{Substituting}\\ 45&=0.75d & &\text{Subtracting 30 from both sides}\\ 60&=d & &\text{Dividing both sides by 0.75} \end{aligned}$$

Check. We evaluate

$$C(60)=0.75(60)+30=45+30=75.$$

State. The function is $C(d)=0.75d+30$ and 60 miles were driven for a total cost of \$75.

27. ***Familiarize.*** Lauren's hair starts at a length of 5 in. After one month, the total length is $5+\frac{1}{2}=5\frac{1}{2}$. After two months, the total length is $5+2\cdot\frac{1}{2}=6$. We can generalize this with a model, letting $L(t)$ represent the total length, in inches, after t months.

Translate. We reword the problem and translate.

$$\underbrace{\text{Total length}}_{\downarrow\; L(t)} \quad \underbrace{\text{is}}_{\downarrow\; =} \quad \underbrace{\text{initial length}}_{\downarrow\; 5} \quad \underbrace{\text{plus}}_{\downarrow\; +} \quad \underbrace{\tfrac{1}{2}\text{ in. per month}}_{\downarrow\; \frac{1}{2}t}.$$

where $t \geq 0$ since we cannot have negative months.

So the length of Lauren's hair t months after the haircut is $L(t)=\frac{1}{2}t+5$.

Carry out. To determine the number of months required to reach a total length of 15 in., we substitute 15 for $L(t)$ and solve for t.

$$\begin{aligned} L(t)&=\tfrac{1}{2}t+5 \\ 15&=\tfrac{1}{2}t+5 && \text{Substituting} \\ 10&=\tfrac{1}{2}t && \text{Subtracting 5 from both sides} \\ 20&=t && \text{Multiplying both sides by 2} \end{aligned}$$

Check. We evaluate

$$L(20)=\tfrac{1}{2}(20)+5=10+5=15.$$

State. The function is $L(t)=\frac{1}{2}t+5$ and it will take 20 months after Lauren's haircut for her hair to be 15 in. long.

29. ***Familiarize.*** In 2006, there are 5960 acres of organic cotton. After one year, the total acres is 5960 + 849 = 6809. After two years, the total acres is $5960+2\cdot 849=7658$. We can generalize this with a model, letting $A(t)$ represent the total acres t years after 2006.

Translate. We reword the problem and translate.

$$\underbrace{\text{Total acres}}_{\downarrow\; A(t)} \quad \underbrace{\text{is}}_{\downarrow\; =} \quad \underbrace{\text{initial acres}}_{\downarrow\; 5960} \quad \underbrace{\text{plus}}_{\downarrow\; +} \quad \underbrace{\text{849 per year}}_{\downarrow\; 849t}.$$

where $t \geq 0$ since we cannot have negative years.

So the linear function for the number of acres after t years after 2006 is $A(t)=849t+5960$.

Carry out. To determine the number of years required to reach a total acreage of 10,205, we substitute 10,205 for $A(t)$ and solve for t.

$$\begin{aligned} A(t)&=849t+5960 \\ 10{,}205&=849t+5960 \\ 4245&=849t \\ 5&=t \end{aligned}$$

Check. We evaluate

$$A(5)=849(5)+5960=4245+5960=10{,}205.$$

State. The function is $A(t)=849t+5960$ and 5 years after 2006, or in 2011 are required to grow 10,205 acres of organic cotton.

31. a) Let t represent the number of years after 2000, and form the pairs (2, 14.5) and (7, 19). First we find the slope of the function that fits the data:

$$m=\frac{19-14.5}{7-2}=\frac{4.5}{5}=0.9$$

Use the point-slope equation with $m = 0.9$ and the point (7, 19).

$$\begin{aligned} a-19&=0.9(t-7) \\ a-19&=0.9t-6.3 \\ a&=0.9t+12.7 \\ a(t)&=0.9t+12.7 \end{aligned}$$

b) In 2013, $t = 2013 - 2000 = 13$.

$$a(13)=0.9(13)+12.7=24.4$$

In 2013, there will be 24.4 million cars.

c) Substitute 25 for $a(t)$ and solve for t.

$$\begin{aligned} a(t)&=0.9t+12.7 \\ 25&=0.9t+12.7 \\ 12.3&=0.9t \\ 14&\approx t \end{aligned}$$

2000 + 14 = 2014

In about 2014 there will be 25 million cars.

33. a) Let t represent the number of years after 1990, and form the pairs (4, 79.0) and (14, 80.4). First we find the slope of the function that fits the data:

$$m=\frac{80.4-79.0}{14-4}=\frac{1.4}{10}=0.14$$

Use the point-slope equation with $m = 0.14$ and the point (4, 79.0).

$$\begin{aligned} E-79.0&=0.14(t-4) \\ E-79.0&=0.14t-0.56 \\ E&=0.14t+78.44 \\ E(t)&=0.14t+78.44 \end{aligned}$$

b) In 2012, $t = 2012 - 1990 = 22$.

$$E(22)=0.14(22)+78.44=3.08+78.44=81.52$$

In 2012, life expectancy will be 81.52 years.

35. a) Let t represent the number of years after 2000, and form the pairs (2, 282) and (6, 372.1). First we find the slope of the function that fits the data:

$m=\frac{372.1-282}{6-2}=\frac{90.1}{4}=22.525$

Use the point-slope equation with m = 22.525 and the point (2, 282).

$A-282=22.525(t-2)$
$A-282=22.525t-45.05$
$A=22.525t+236.95$
$A(t)=22.525t+236.95$

b) In 2010, t = 2010 – 2000 = 10.

$A(10)=22.525(10)+236.95=462.2$

In 2010, there will be $462.2 in PAC contributions.

37. a) Let t represent the number of years after 2000, and form the pairs (0, 16) and (5, 63). First we find the slope of the function that fits the data:

$m=\frac{63-16}{5-0}=\frac{47}{5}=9.4$

Use the point-slope equation with m = 9.4 and the point (0, 16).

$N-16=9.4(t-0)$
$N-16=9.4t$
$N=9.4t+16$
$N(t)=9.4t+16$

b) In 2010, t = 2010 – 2000 = 10.

$N(10)=9.4(10)+16=110$

In 2010, there will be 110 million Americans using online banking.

c) Substitute 157 for $N(t)$ and solve for t.

$N(t)=9.4t+16$
$157=9.4t+16$
$141=9.4t$
$15=t$

2000 + 15 = 2015

In 2015 there will be 157 million Americans using online banking.

39. a) Let t represent the number of years after 1990, and form the pairs (4, 74.9) and (15, 79). First we find the slope of the function that fits the data:

$m=\frac{79-74.9}{15-4}=\frac{4.1}{11}=\frac{41}{110}$

Use the point-slope equation with $m=\frac{41}{110}$ and the point (15, 79).

$A-79=\frac{41}{110}(t-15)$
$A-79=\frac{41}{110}t-\frac{123}{22}$
$A=\frac{41}{110}t+\frac{1615}{22}$
$A(t)=\frac{41}{110}t+\frac{1615}{22}$

b) In 2010, t = 2010 – 1990 = 20.

$A(20)=\frac{41}{110}(20)+\frac{1615}{22}\approx 80.9$

In 2010, there will be about 80.9 million acres in the National Park system.

41. $f(x)=\frac{1}{3}x-7$

The function is in the form $f(x)=mx+b$, so it is a linear function. We can compute $\frac{1}{3}x-7$ for any value of x, so the domain is the set of all real numbers.

43. $p(x)=x^2+x+1$

The function is in the form $f(x)=ax^2+bx+c$, $a\neq 0$, so it is a quadratic function. We can compute x^2+x+1 for any value of x, so the domain is the set of all real numbers.

45. $f(t)=\frac{12}{3t+4}$

The function is described by a rational equation so it is a rational function. The expression $\frac{12}{3t+4}$ is undefined when $t=-\frac{4}{3}$, so the domain is

$\left\{t\middle| t \text{ is a real number } and\ t\neq -\frac{4}{3}\right\}$.

47. $f(x)=0.02x^4-0.1x+1.7$

The function is described by a polynomial equation that is neither linear nor quadratic, so it is a polynomial function. We can compute $0.02x^4-0.1x+1.7$ for any value of x, so the domain is the set of all real numbers.

49. $f(x)=\frac{x}{2x-5}$

The function is described by a rational equation, so it is a rational function. The expression $\frac{x}{2x-5}$ is undefined when $x=\frac{5}{2}$, so the domain is

$\left\{x\middle| x \text{ is a real number } and\ x\neq \frac{5}{2}\right\}$.

51. $f(n)=\frac{4n-7}{n^2+3n+2}$

The function is described by a rational equation so it is a rational function. The expression $\frac{4n-7}{n^2+3n+2}$ is undefined for values of n that make the denominator 0. We find those values:

$$n^2+3n+2=0$$
$$(n+1)(n+2)=0$$
$$n+1=0 \quad or \quad n+2=0$$
$$n=-1 \quad or \quad n=-2$$

Then the domain is $\{n|n$ is a real number *and* $n\neq-1$ *and* $n\neq-2\}$.

53. $f(n)=200-0.1n$

The function can be written in the form $f(n)=mn+b$, so it is a linear function. We can compute 200 – $0.1n$ for any value of n, so the domain is the set of all real numbers.

55. The function has no y-values less than 0 and every y-value greater than or equal to 0 corresponds to a member of the domain. Thus, the range is $\{y|y\geq 0\}$.

57. Every y-value corresponds to a member of the domain, so the range is the set of all real numbers.

59. The function has no y-values greater than 0 and every y-value less than or equal to 0 corresponds to a member of the domain. Thus, the range is $\{y|y\leq 0\}$.

61. Graph $f(x)=x+3$.

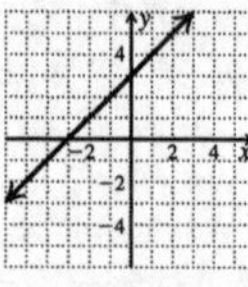

$f(x) = x + 3$

For any x-value and for any y-value there is a point on the graph. Thus,
Domain of $f=\{x|x$ is a real number$\}$, or $\mathbb{R}$ and
Range of $f=\{y|y$ is a real number$\}$, or $\mathbb{R}$.

63. Graph $f(x)=-1$.

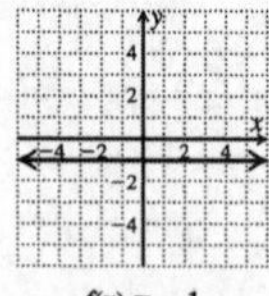

$f(x) = -1$

For any x-value there is a point on the graph, so
Domain of $f=\{x|x$ is a real number$\}$, or $\mathbb{R}$.
The only y-value on the graph is –1, so
Range of $f=\{-1\}$.

65. Graph $f(x)=|x|+1$.

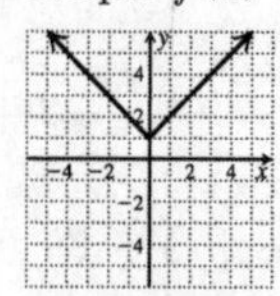

$f(x) = |x| + 1$

For any x-value there is a point on the graph, so
Domain of $f=\{x|x$ is a real number$\}$, or $\mathbb{R}$.
There is no y-value less than 1 and every y-value greater than or equal to 1 corresponds to a member of the domain. Thus,
Range of $f=\{y|y\geq 1\}$.

67. Graph $g(x)=x^2$.

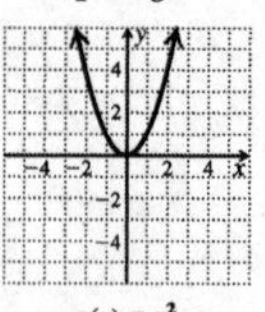

$g(x) = x^2$

For any x-value there is a point on the graph, so
Domain of $g=\{x|x$ is a real number$\}$, or $\mathbb{R}$.
There is no y-value less than 0 and every y-value greater than or equal to 0 corresponds to a member of the domain. Thus, Range of $g=\{y|y\geq 0\}$.

69. *Writing Exercise.* Using the counter example, $g(x)=x^2$, as in Exercise 67, the graph indicates that the range is restricted to $g=\{y|y\geq 0\}$, not $\mathbb{R}$.

71.
$$(x^2+2x+7)+(3x^2-8)$$
$$=(x^2+3x^2)+2x+(7-8)$$
$$=4x^2+2x-1$$

73.
$$(2x+1)(x-7)=2x^2-14x+x-7$$
$$=2x^2-13x-7$$

75.
$$(x^3+x^2-4x+7)-(3x^2-x+2)$$
$$=x^3+x^2-4x+7-3x^2+x-2$$
$$=x^3+(x^2-3x^2)+(-4x+x)+(7-2)$$
$$=x^3-2x^2-3x+5$$

77. *Writing Exercise.* In the 2 years from 2002 to 2004 PAC contributions increased \$28.5 million. In Exercise 35(b) we predicted that contributions would increase \$90.1 million in the 4 years from 2006 to 2010. This information indicates that the answer to the exercise might be high.

79. Let $c = 1$ and $d = 2$. Then

$f(c+d) = f(1+2) = f(3) = 3m+b$, but

$f(c)+f(d) = (m+b)+(2m+b) = 3m+2b$.

The given statement is false.

81. Let $k = 2$. Then $f(kx) = f(2x) = 2mx+b$, but

$kf(x) = 2(mx+b) = 2mx+2b$.

The given statement is false.

83. ***Familiarize.*** Celsius temperature C corresponding to a Fahrenheit temperature F can be modeled by a line that contains the points (32, 0) and (212, 100).

Translate. We find an equation relating C and F.

$$m = \frac{100-0}{212-32} = \frac{100}{180} = \frac{5}{9}$$

$$C - 0 = \frac{5}{9}(F-32)$$

$$C = \frac{5}{9}(F-32)$$

Carry out. Using function notation we have

$C(F) = \frac{5}{9}(F-32)$. Now we find $C(70)$:

$$C(70) = \frac{5}{9}(70-32) = \frac{5}{9}(38) \approx 21.1.$$

Check. We can repeat the calculations. We could also graph the function and determine that (70, 21.1) is on the graph.

State. A temperature of about 21.1°C corresponds to a temperature of 70°F.

85. ***Familiarize.*** The total cost C of the phone, in dollars, after t months, can be modeled by a line that contains the points (5, 410) and (9, 690).

Translate. We find an equation relating C and t.

$$m = \frac{690-410}{9-5} = \frac{280}{4} = 70$$

$$C - 410 = 70(t-5)$$

$$C - 410 = 70t - 350$$

$$C = 70t + 60$$

Carry out. Using function notation we have

$C(t) = 70t+60$. To find the costs already incurred when the service began we find $C(0)$:

$$C(0) = 70 \cdot 0 + 60 = 60$$

Check. We can repeat the calculations. We could also graph the function and determine that (0, 60) is on the graph.

State. Tam had already incurred $60 in costs when the service just began.

87. a) We have two pairs, (3, –5) and (7, –1). Use the point-slope form:

$$m = \frac{-1-(-5)}{7-3} = \frac{-1+5}{4} = \frac{4}{4} = 1$$

$$y-(-5) = 1(x-3)$$

$$y+5 = x-3$$

$$y = x-8$$

$$g(x) = x-8 \quad \text{Using function notation}$$

b) $g(-2) = -2-8 = -10$

c) $g(a) = a-8$

If $g(a) = 75$, we have

$a - 8 = 75$

$a = 83.$

Connecting the Concepts

1. Domain of $f: \{-1, 0, 3, 4\}$

3. The input of –1 corresponds to the output –2.

$f(-1) = -2$

5. For any x-value there is a point on the graph, so

Domain of $g = \{x | x \text{ is a real number}\}$, or $\mathbb{R}$.

7. $h(x) = \frac{2}{x}$

$h(10) = \frac{2}{10} = \frac{1}{5}$

9.
$$(g \cdot h)(-2) = g(-2)h(-2)$$
$$= (-2-1)\left(\frac{2}{-2}\right)$$
$$= (-3)(-1)$$
$$= 3$$

11. The function is in the form $f(x) = mx+b$, so it is a linear function.

13. Graph $g(x) = x-1$.

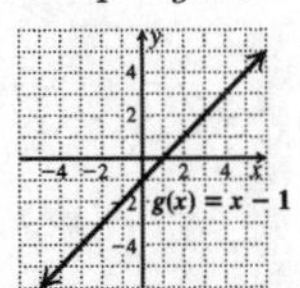

For any y-value there is a point on the graph, so

Range of $g = \{y | y \text{ is a real number}\}$, or $\mathbb{R}$.

15. The input of 2 corresponds to the output –4.

$F(2) = -4$

17. The function has no y-values less than –4 and every y-value greater than or equal to –4 corresponds to a member of the domain. Thus, the range is $\{y|y \geq -4\}$, or $[-4, \infty)$.

19. Since 1 = 1, $G(1) = 10$

Exercise Set 7.4

1. sum; see page 475 in the text.

3. evaluate; see page 475 in the text.

5. excluding; see page 479 in the text.

7. Since $f(3) = -2 \cdot 3 + 3 = -3$ and $g(3) = 3^2 - 5 = 4$, we have $f(3) + g(3) = -3 + 4 = 1$.

9. Since $f(1) = -2 \cdot 1 + 3 = 1$ and $g(1) = 1^2 - 5 = -4$, we have $f(1) - g(1) = 1 - (-4) = 5$.

11. Since $f(-2) = -2 \cdot (-2) + 3 = 7$ and $g(-2) = (-2)^2 - 5 = -1$ we have $f(-2) \cdot g(-2) = 7(-1) = -7$.

13. Since $f(-4) = -2 \cdot (-4) + 3 = 11$ and $g(-4) = (-4)^2 - 5 = 11$, we have $\dfrac{f(-4)}{g(-4)} = \dfrac{11}{11} = 1$.

15. Since $g(1) = 1^2 - 5 = -4$ and $f(1) = -2 \cdot 1 + 3 = 1$, we have $g(1) - f(1) = -4 - 1 = -5$.

17.
$$\begin{aligned}(f+g)(x) &= f(x) + g(x) = (-2x+3) + (x^2 - 5) \\ &= x^2 - 2x - 2\end{aligned}$$

19.
$$\begin{aligned}(F+G)(x) &= F(x) + G(x) \\ &= x^2 - 2 + 5 - x \\ &= x^2 - x + 3\end{aligned}$$

21.
$$\begin{aligned}(F-G)(x) &= F(x) - G(x) \\ &= x^2 - 2 - (5 - x) \\ &= x^2 - 2 - 5 + x \\ &= x^2 + x - 7\end{aligned}$$

Then we have
$$\begin{aligned}(F-G)(3) &= 3^2 + 3 - 7 \\ &= 9 + 3 - 7 \\ &= 5.\end{aligned}$$

23.
$$\begin{aligned}(F \cdot G)(x) &= F(x) \cdot G(x) \\ &= (x^2 - 2)(5 - x) \\ &= 5x^2 - x^3 - 10 + 2x\end{aligned}$$

Then we have
$$\begin{aligned}(F \cdot G)(-3) &= 5(-3)^2 - (-3)^3 - 10 + 2(-3) \\ &= 5 \cdot 9 - (-27) - 10 - 6 \\ &= 45 + 27 - 10 - 6 \\ &= 56.\end{aligned}$$

25.
$$\begin{aligned}(F/G)(x) &= F(x)/G(x) \\ &= \frac{x^2 - 2}{5 - x},\ x \neq 5\end{aligned}$$

27. Using our work in Exercise 25, we have
$$(G/F)(-2) = \frac{5-(-2)}{(-2)^2 - 2} = \frac{5+2}{4-2} = \frac{7}{2}.$$

29. $(F+F)(x) = F(x) + F(x) = (x^2 - 2) + (x^2 - 2) = 2x^2 - 4$
$(F+F)(x) = 2(1)^2 - 4 = -2$

31.
$$\begin{aligned}N(2004) &= (C+B)(2004) = C(2004) + B(2004) \\ &\approx 1.2 + 2.9 = 4.1 \text{ million}\end{aligned}$$

We estimate the number of births in 2004 to be 4.1 million.

33. $(P-L)(2) = P(2) - L(2) \approx 26.5\% - 22.5\% \approx 4\%$

35.
$$\begin{aligned}(p+r)('05) &= p('05) + r('05) \\ &\approx 25 + 70 = 95 \text{ million}\end{aligned}$$

This represents the number of tons of municipal solid waste that was composted or recycled in 2005.

37. $F('96) \approx 215$ million

This represents the number of tons of municipal solid waste in 1996.

39.
$$\begin{aligned}(F-p)('04) &= F('04) - p('04) \\ &\approx 260 - 30 = 230 \text{ million}\end{aligned}$$

This represents the number of tons of municipal solid waste that was not composted in 2004.

41. The domain of f and of g is all real numbers. Thus,
$$\begin{aligned}\text{Domain of } f+g = \text{Domain of } f-g &= \text{Domain of } f \cdot g \\ &= \{x \mid x \text{ is a real number}\}.\end{aligned}$$

43. Because division by 0 is undefined, we have
Domain of $f = \{x \mid x \text{ is a real number } \textit{and } x \neq -5\}$,
and Domain of $g = \{x \mid x \text{ is a real number}\}$.
Thus,
$$\begin{aligned}\text{Domain of } f+g = \text{Domain of } f-g &= \text{Domain of } f \cdot g \\ &= \{x \mid x \text{ is a real number } \textit{and } x \neq -5\}.\end{aligned}$$

45. Because division by 0 is undefined, we have

Domain of $f = \{x \mid x \text{ is a real number } and \ x \neq 0\}$,

and Domain of $g = \{x \mid x \text{ is a real number}\}$.

Thus, Domain of

$f + g =$ Domain of $f - g =$ Domain of $f \cdot g$
$= \{x \mid x \text{ is a real number } and \ x \neq 0\}$.

47. Because division by 0 is undefined, we have

Domain of $f = \{x \mid x \text{ is a real number } and \ x \neq 1\}$,

and Domain of $g = \{x \mid x \text{ is a real number}\}$.

Thus,

Domain of $f + g =$ Domain of $f - g =$ Domain of $f \cdot g$
$= \{x \mid x \text{ is a real number } and \ x \neq 1\}$.

49. Because division by 0 is undefined, we have

Domain of $f = \left\{x \middle| x \text{ is a real number } and \ x \neq -\frac{9}{2}\right\}$,

and Domain of $g = \{x \mid x \text{ is a real number } and \ x \neq 1\}$.

Thus, Domain of $f + g =$ Domain of $f - g =$ Domain of $f \cdot g$

$= \left\{x \middle| x \text{ is a real number } and \ x \neq -\frac{9}{2} \ and \ x \neq 1\right\}$.

51. Domain of $f =$ Domain of $g = \{x \mid x \text{ is a real number}\}$.

Since $g(x) = 0$ when $x - 3 = 0$, we have $g(x) = 0$ when $x = 3$. We conclude that

Domain of $f / g = \{x \mid x \text{ is a real number } and \ x \neq 3\}$.

53. Domain of $f =$ Domain of $g = \{x \mid x \text{ is a real number}\}$.

Since $g(x) = 0$ when $2x + 8 = 0$, we have $g(x) = 0$ when $x = -4$. We conclude that

Domain of $f / g = \{x \mid x \text{ is a real number } and \ x \neq -4\}$.

55. Domain of $f = \{x \mid x \text{ is a real number and } x \neq 4\}$.

Domain of $g = \{x \mid x \text{ is a real number}\}$.

Since $g(x) = 0$ when $5 - x = 0$, we have $g(x) = 0$ when $x = 5$. We conclude that Domain of

$f / g = \{x \mid x \text{ is a real number and } x \neq 4 \text{ and } x \neq 5\}$.

57. Domain of $f = \{x \mid x \text{ is a real number and } x \neq -1\}$.

Domain of $g = \{x \mid x \text{ is a real number}\}$.

Since $g(x) = 0$ when $2x + 5 = 0$, we have $g(x) = 0$ when $x = -\frac{5}{2}$. We conclude that Domain of f / g

$= \left\{x \middle| x \text{ is a real number and } x \neq -1 \text{ and } x \neq -\frac{5}{2}\right\}$.

59. $(F + G)(5) = F(5) + G(5) = 1 + 3 = 4$
$(F + G)(7) = F(7) + G(7) = -1 + 4 = 3$

61. $(G - F)(7) = G(7) - F(7) = 4 - (-1) = 4 + 1 = 5$
$(G - F)(3) = G(3) - F(3) = 1 - 2 = -1$

63. From the graph we see that Domain of $F = \{x \mid 0 \leq x \leq 9\}$ and Domain of $G = \{x \mid 3 \leq x \leq 10\}$. Then Domain of $F + G = \{x \mid 3 \leq x \leq 9\}$. Since $G(x)$ is never 0, Domain of $F / G = \{x \mid 3 \leq x \leq 9\}$.

65. We use $(F + G)(x) = F(x) + G(x)$.

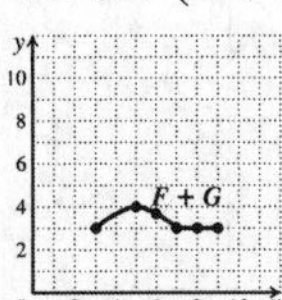

67. *Writing Exercise.* For the years from 1985 through 2000, Americans consumed more soft drinks than juice, bottled water and milk combined. We are using the approximation of $S(t) - [M(t) + J(t) + W(t)]$.

69. $ac = b$
$c = \frac{b}{a}$

71. $pq - rq = st$
$q(p - r) = st$
$q = \frac{st}{p - r}$

73. $ab - cd = 3b + d$
$ab - 3b - cd = d$
$ab - 3b = cd + d$
$b(a - 3) = cd + d$
$b = \frac{cd + d}{a - 3}$

75. *Writing Exercise.* First draw four graphs with the number of hours after the first dose is taken on the horizontal axis and the amount of Advil absorbed, in mg, on the vertical axis. Each graph would show the absorption of one dose of Advil. Then superimpose the four graphs and, finally, add the amount of Advil absorbed to create the final graph.

77. Domain of $F = \{x \mid x \text{ is a real number and } x \neq 4\}$.

Domain of $G = \{x \mid x \text{ is a real number and } x \neq 3\}$.

$G(x) = 0$ when $x^2 - 4 = 0$, or when $x = 2$ or $x = -2$. Then Domain of $F / G = \{x | x$ is a real number and $x \neq 4$ and $x \neq 3$ and $x \neq 2$ and $x \neq -2\}$.

79. Answers may vary.

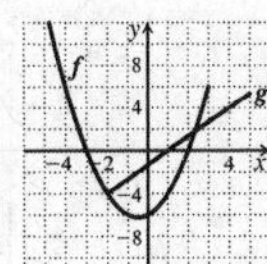

81. The problem states that Domain of $m = \{x | -1 < x < 5\}$. Since $n(x) = 0$ when $2x - 3 = 0$, we have $n(x) = 0$ when $x = \frac{3}{2}$. We conclude that Domain of m/n $= \left\{x \middle| x \text{ is a real number and } -1 < x < 5 \text{ and } x \neq \frac{3}{2}\right\}$.

83. Answers may vary. $f(x) = \frac{1}{x+2}$, $g(x) = \frac{1}{x-5}$

85. *Graphing Calculator Exercise*

Exercise Set 7.5

1. (d) LCD

3. (e) Product

5. (a) Directly

7. As the number of painters increases, the time required to scrape the house decreases, so we have inverse variation.

9. As the number of laps increases, the time required to swim them increases, so we have direct variation.

11. As the number of volunteers increases, the time required to wrap the toys decreases, so we have inverse variation.

13.
$$f = \frac{L}{d}$$
$$df = L \quad \text{Multiplying by } d$$
$$d = \frac{L}{f} \quad \text{Dividing by } f$$

15.
$$s = \frac{(v_1 + v_2)t}{2}$$
$$2s = (v_1 + v_2)t \quad \text{Multiplying by 2}$$
$$\frac{2s}{t} = v_1 + v_2 \quad \text{Dividing by } t$$
$$\frac{2s}{t} - v_2 = v_1$$

This result can also be expressed as $v_1 = \frac{2s - tv_2}{t}$.

17.
$$\frac{t}{a} + \frac{t}{b} = 1$$
$$ab\left(\frac{t}{a} + \frac{t}{b}\right) = ab \cdot 1 \quad \text{Multiplying by the LCD}$$
$$ab \cdot \frac{t}{a} + ab \cdot \frac{t}{b} = ab$$
$$bt + at = ab$$
$$at = ab - bt$$
$$at = b(a - t) \quad \text{Factoring}$$
$$\frac{at}{a - t} = b$$

19.
$$R = \frac{gs}{g+s}$$
$$(g+s) \cdot R = (g+s) \cdot \frac{gs}{g+s} \quad \text{Multiplying by the LCD}$$
$$Rg + Rs = gs$$
$$Rs = gs - Rg$$
$$Rs = g(s - R) \quad \text{Factoring out } g$$
$$\frac{Rs}{s - R} = g \quad \text{Multiplying by } \frac{1}{s-R}$$

21.
$$I = \frac{nE}{R + nr}$$
$$I(R + nr) = \frac{nE}{R + nr} \cdot (R + nr) \quad \text{Multiplying by the LCD}$$
$$IR + Inr = nE$$
$$IR = nE - Inr$$
$$IR = n(E - Ir)$$
$$\frac{IR}{E - Ir} = n$$

23.
$$\frac{1}{p} + \frac{1}{q} = \frac{1}{f}$$
$$pqf\left(\frac{1}{p} + \frac{1}{q}\right) = pqf \cdot \frac{1}{f} \quad \text{Multiplying by the LCD}$$
$$qf + pf = pq$$
$$pf = pq - qf$$
$$pf = q(p - f)$$
$$\frac{pf}{p - f} = q$$

25.
$$S = \frac{H}{m(t_1 - t_2)}$$
$$(t_1 - t_2)S = \frac{H}{m} \quad \text{Multiplying by } t_1 - t_2$$
$$t_1 - t_2 = \frac{H}{Sm} \quad \text{Dividing by } S$$
$$t_1 = \frac{H}{Sm} + t_2, \text{ or } \frac{H + Smt_2}{Sm}$$

27.
$$\frac{E}{e} = \frac{R + r}{r}$$
$$er \cdot \frac{E}{e} = er \cdot \frac{R + r}{r} \quad \text{Multiplying by the LCD}$$
$$Er = e(R + r)$$
$$Er = eR + er$$
$$Er - er = eR$$
$$r(E - e) = eR$$
$$r = \frac{eR}{E - e}$$

29.
$$S = \frac{a}{1 - r}$$
$$(1 - r)S = a \quad \text{Multiplying by the LCD, } 1 - r$$
$$1 - r = \frac{a}{S} \quad \text{Dividing by } S$$
$$1 - \frac{a}{S} = r \quad \text{Adding } r \text{ and } -\frac{a}{S}$$

This result can also be expressed as $r = \frac{S - a}{S}$.

31.
$$c = \frac{f}{(a+b)c}$$
$$\frac{a+b}{c} \cdot c = \frac{a+b}{c} \cdot \frac{f}{(a+b)c}$$
$$a+b = \frac{f}{c^2}$$

33.
$$P = \frac{A}{1+r}$$
$$P(1+r) = \frac{A}{1+r} \cdot (1+r)$$
$$P(1+r) = A$$
$$1+r = \frac{A}{P}$$
$$r = \frac{A}{P} - 1, \text{ or } \frac{A-P}{P}$$

35.
$$v = \frac{d_2 - d_1}{t_2 - t_1}$$
$$(t_2 - t_1)v = (t_2 - t_1) \cdot \frac{d_2 - d_1}{t_2 - t_1}$$
$$(t_2 - t_1)v = d_2 - d_1$$
$$t_2 - t_1 = \frac{d_2 - d_1}{v}$$
$$-t_1 = -t_2 + \frac{d_2 - d_1}{v}$$
$$t_1 = t_2 - \frac{d_2 - d_1}{v}, \text{ or } \frac{t_2 v - d_2 + d_1}{v}$$

37.
$$\frac{1}{t} = \frac{1}{a} + \frac{1}{b}$$
$$tab \cdot \frac{1}{t} = tab\left(\frac{1}{a} + \frac{1}{b}\right)$$
$$ab = tb + ta$$
$$ab = t(b+a)$$
$$\frac{ab}{b+a} = t$$

39.
$$A = \frac{2Tt + Qq}{2T + Q}$$
$$(2T+Q) \cdot A = (2T+Q) \cdot \frac{2Tt + Qq}{2T + Q}$$
$$2AT + AQ = 2Tt + Qq$$
$$AQ - Qq = 2Tt - 2AT \quad \text{Adding } -2AT \text{ and } -Qq$$
$$Q(A-q) = 2Tt - 2AT$$
$$Q = \frac{2Tt - 2AT}{A-q}$$

41.
$$p = \frac{-98.42 + 4.15c - 0.082w}{w}$$
$$pw = -98.42 + 4.15c - 0.082w$$
$$pw + 0.082w = -98.42 + 4.15c$$
$$w(p + 0.082) = -98.42 + 4.15c$$
$$w = \frac{-98.42 + 4.15c}{p + 0.082}$$

43. $y = kx$

$30 = k \cdot 5$ Substituting

$6 = k$

The variation constant is 6.

The equation of variation is $y = 6x$.

45. $y = kx$

$3.4 = k \cdot 2$ Substituting

$1.7 = k$

The variation constant is 1.7.

The equation of variation is $y = 1.7x$.

47. $y = kx$

$2 = k \cdot \frac{1}{5}$ Substituting

$10 = k$ Multiplying by 5

The variation constant is 10.

The equation of variation is $y = 10x$.

49. $y = \frac{k}{x}$

$5 = \frac{k}{20}$ Substituting

$100 = k$

The variation constant is 100.

The equation of variation is $y = \frac{100}{x}$.

51. $y = \frac{k}{x}$

$11 = \frac{k}{4}$ Substituting

$44 = k$

The variation constant is 44.

The equation of variation is $y = \frac{44}{x}$.

53. $y = \frac{k}{x}$

$27 = \frac{k}{\frac{1}{3}}$ Substituting

$9 = k$

The variation constant is 9.

The equation of variation is $y = \frac{9}{x}$.

55. ***Familiarize.*** Because of the phrase "d ... varies directly as ... m," we express the distance as a function of the mass. Thus we have $d(m) = km$. We know that $d(3) = 20$.

Translate. We find the variation constant and then find the equation of variation.

$$d(m) = km$$
$$d(3) = k \cdot 3 \quad \text{Replacing } m \text{ with } 3$$
$$20 = k \cdot 3 \quad \text{Replacing } d(3) \text{ with } 20$$
$$\frac{20}{3} = k \quad \text{Variation constant}$$

The equation of variation is $d(m) = \frac{20}{3}m$.

Carry out. We compute $d(5)$.

$$d(m) = \frac{20}{3}m$$
$$d(5) = \frac{20}{3} \cdot 5 \quad \text{Replacing } m \text{ with } 5$$
$$= \frac{100}{3}, \text{ or } 33\frac{1}{3}$$

Check. Reexamine the calculations.

State. The distance is $33\frac{1}{3}$ cm.

57. ***Familiarize.*** Because T varies inversely as P, we write $T(P) = k/P$. We know that $T(7) = 5$.

Translate. We find the variation constant and the equation of variation.

$$T(P) = \frac{k}{P}$$
$$T(7) = \frac{k}{7} \quad \text{Replacing } P \text{ with } 7$$
$$5 = \frac{k}{7} \quad \text{Replacing } T(P) \text{ with } 5$$
$$35 = k \quad \text{Variation constant}$$
$$T(P) = \frac{35}{P} \quad \text{Equation of variation}$$

Carry out. We find $T(10)$.

$$T(10) = \frac{35}{10} = 3.5$$

Check. Reexamine the calculations.

State. It would take 3.5 hr for 10 volunteers to complete the job.

59. ***Familiarize.*** Because W varies directly as S, we write $W(S) = kS$. We know that $W(150) = 16.8$.

Translate. We find the variation constant and the equation of variation.

$$W(S) = kS$$
$$W(150) = k \cdot 150 \quad \text{Replacing } S \text{ with } 150$$
$$16.8 = k \cdot 150 \quad \text{Replacing } W(150) \text{ with } 16.8$$
$$0.112 = k \quad \text{Variation constant}$$
$$W(S) = 0.112S \quad \text{Equation of variation}$$

Carry out. We find $W(500)$.

$$W(500) = 0.112 \cdot 500 = 56$$

Check. Reexamine the calculations.

State. 56 in. of water will replace 500 in. of snow.

61. Since we have direct variation and $48 = \frac{1}{2} \cdot 96$, then the result is $\frac{1}{2} \cdot 64$ kg, or 32 kg. We could also do this problem as follows.

Familiarize. Because W varies directly as the total mass, we write $W(m) = km$. We know that $W(96) = 64$.

Translate.

$$W(m) = km$$
$$W(96) = k \cdot 96 \quad \text{Replacing } m \text{ with } 96$$
$$64 = k \cdot 96 \quad \text{Replacing } W(96) \text{ with } 64$$
$$\frac{2}{3} = k \quad \text{Variation constant}$$
$$W(m) = \frac{2}{3}m \quad \text{Equation of variation}$$

Carry out. Find $W(48)$.

$$W(m) = \frac{2}{3}m$$
$$W(48) = \frac{2}{3} \cdot 48 = 32$$

Check. Reexamine the calculations.

State. There are 32 kg of water in a 64 kg person.

63. ***Familiarize.*** Because the frequency, f varies inversely as length L, we write $f(L) = k/L$. We know that $f(33) = 260$.

Translate. We find the variation constant and the equation of variation.

$$f(L) = \frac{k}{L}$$
$$f(33) = \frac{k}{33} \quad \text{Replacing } L \text{ with } 33$$
$$260 = \frac{k}{33} \quad \text{Replacing } f(33) \text{ with } 260$$
$$8580 = k \quad \text{Variation constant}$$
$$f(L) = \frac{8580}{L} \quad \text{Equation of variation}$$

Carry out. We find $f(30)$.

$$f(30) = \frac{8580}{30} = 286$$

Check. Reexamine the calculations.

State. If the string was shortened to 30 cm the new frequency would be 286 Hz.

65. ***Familiarize.*** Because of the phrase "t varies inversely as ...u," we write $t(u) = k/u$. We know that $t(4) = 75$.

Translate. We find the variation constant and then we find the equation of variation.

$$t(u) = \frac{k}{u}$$
$$t(4) = \frac{k}{4} \quad \text{Replacing } u \text{ with } 4$$
$$75 = \frac{k}{4} \quad \text{Replacing } t(4) \text{ with} 75$$
$$300 = k \quad \text{Variation constant}$$
$$t(u) = \frac{300}{u} \quad \text{Equation of variation}$$

Carry out. We find $t(14)$.

$$t(14) = \frac{300}{14} \approx 21$$

Check. Reexamine the calculations. Note that, as expected, as the UV rating increases, the time it takes to burn goes down.

State. It will take about 21 min to burn when the UV rating is 14.

67. ***Familiarize***. The amount A of carbon monoxide released, in tons, varies directly as the population P. We write A as a function of P: $A(P) = kP$. We know that $A(2.6) = 0.94$.

Translate.

$$\begin{aligned} A(P) &= kP \\ A(2.6) &= k \cdot 2.6 && \text{Replacing } P \text{ with } 2.6 \\ 0.94 &= k \cdot 2.6 && \text{Replacing } A(2.6) \text{ with } 0.94 \\ \frac{94}{260} &= k && \text{Variation constant} \\ A(P) &= \frac{94}{260}P && \text{Equation of variation} \end{aligned}$$

Carry out. Find $A(305{,}000{,}000)$.

$$\begin{aligned} A(P) &= \frac{94}{260}P \\ A(305{,}000{,}000) &= \frac{94}{260}(305{,}000{,}000) \\ &\approx 110{,}000{,}000 \end{aligned}$$

Check. Reexamine the calculations. Answers may vary slightly due to rounding differences.

State. About 110,000,000 tons of carbon monoxide were released nationally.

69.
$$\begin{aligned} y &= kx^2 \\ 50 &= k(10)^2 && \text{Substituting} \\ 50 &= k \cdot 100 \\ \frac{1}{2} &= k && \text{Variation constant} \end{aligned}$$

The equation of variation is $y = \frac{1}{2}x^2$.

71.
$$\begin{aligned} y &= \frac{k}{x^2} \\ 50 &= \frac{k}{(10)^2} && \text{Substituting} \\ 50 &= \frac{k}{100} \\ 5000 &= k && \text{Variation constant} \end{aligned}$$

The equation of variation is $y = \frac{5000}{x^2}$.

73.
$$\begin{aligned} y &= kxz \\ 105 &= k \cdot 14 \cdot 5 && \text{Substituting} \\ 105 &= k \cdot 70 \\ 1.5 &= k && \text{Variation constant} \end{aligned}$$

The equation of variation is $y = 1.5xz$.

75.
$$\begin{aligned} y &= k \cdot \frac{wx^2}{z} \\ 49 &= k \cdot \frac{3 \cdot 7^2}{12} && \text{Substituting} \\ 4 &= k && \text{Variation constant} \end{aligned}$$

The equation of variation is $y = \frac{4wx^2}{z}$.

77. ***Familiarize***. Because the stopping distance d, in feet varies directly as the square of the speed r, in mph, we write $d = kr^2$. We know that $d = 138$ when $r = 60$.

Translate. Find k and the equation of variation.

$$\begin{aligned} d &= kr^2 \\ 138 &= k(60)^2 \\ \frac{23}{600} &= k \\ d &= \frac{23}{600}r^2 && \text{Equation of variation} \end{aligned}$$

Carry out. We find the value of d when r is 40.

$$d = \frac{23}{600}(40)^2 \approx 61.3 \text{ ft}$$

Check. Reexamine the calculations.

State. It would take a car going 40 mph about 61.3 ft to stop.

79. ***Familiarize***. Because V varies directly as T and inversely as P, we write $V = kT/P$. We know that $V = 231$ when $T = 300$ and $P = 20$.

Translate. Find k and the equation of variation.

$$\begin{aligned} V &= \frac{kT}{P} \\ 231 &= \frac{k \cdot 300}{20} \\ \frac{20}{300} \cdot 231 &= k \\ 15.4 &= k \\ V &= \frac{15.4T}{P} && \text{Equation of variation} \end{aligned}$$

Carry out. Substitute 320 for T, and 16 for P and find V.

$$V = \frac{15.4(320)}{16} = 308$$

Check. Reexamine the calculations.

State. The volume is 308 cm^3 when $T = 320$ K and $P = 16 \text{ lb/cm}^2$.

81. ***Familiarize***. The drag W varies jointly as the surface area A and velocity v, so we write $W = kAv$. We know that $W = 222$ when $A = 37.8$ and $v = 40$.

Translate. Find k.

$$\begin{aligned} W &= kAv \\ 222 &= k(37.8)(40) \\ \frac{222}{37.8(40)} &= k \\ \frac{37}{252} &= k \\ W &= \frac{37}{252}Av && \text{Equation of variation} \end{aligned}$$

Carry out. Substitute 51 for A, 430 for W, solve for v.

$$430 = \frac{37}{252} \cdot 51 \cdot v$$
$$57.42 \text{ mph} \approx v$$

(If we had used the rounded value 0.1468 for k, the resulting speed would have been approximately 57.43 mph.)

Check. Reexamine the calculations.

State. The car must travel about 57.42 mph.

83. *Writing Exercise.* Yes; let $y(x) = kx$. Then $y(2x) = k(2x) = k \cdot 2x = 2 \cdot kx = 2 \cdot y(x)$. Thus, doubling x causes y to be doubled.

85. $x - 6y = 3$
$$-6y = -x + 3$$
$$y = \frac{1}{6}x - \frac{1}{2}$$

87. $5x + 2y = -3$
$$2y = -5x - 3 \quad \text{Subtracting } 5x$$
$$\frac{1}{2} \cdot 2y = \frac{1}{2}(-5x - 3) \quad \text{Multiplying by } \frac{1}{2}$$
$$y = -\frac{5}{2}x - \frac{3}{2}$$

89. Let n represent the number; $2n + 5 = 49$.

91. Let n represent the first integer; $x + (x + 1) = 145$.

93. *Writing Exercise.* Let C represent the number of complaints, and let E represent the number of employees. We then have
$$C(E) = \frac{k}{E},$$
where k is a positive constant. This can be graphed as shown.

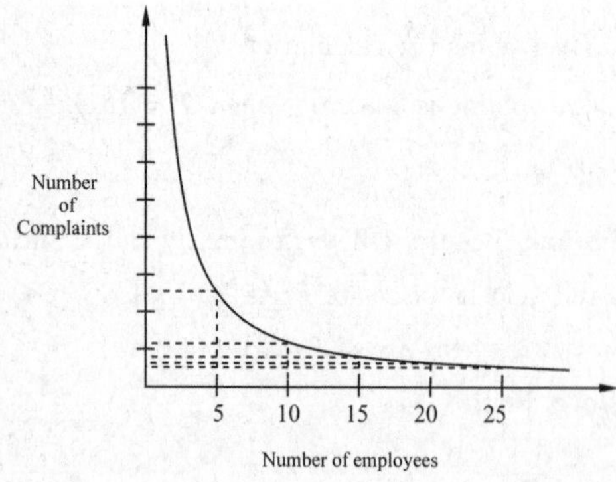

Note that regardless of the scale used on the vertical axis $C(10) - C(5) > C(25) - C(20)$. Thus, a greater reduction in the number of complaints occurs when the firm expands from 5 to 10 employees.

95. Use the result of Example 2.
$$h = \frac{2R^2 g}{V^2} - R$$

We have $V = 6.5$ mi/sec, $R = 3960$ mi, and $g = 32.2$ ft/sec^2. We must convert 32.2 ft/sec^2 to mi/sec^2 so all units of length are the same.
$$32.2 \frac{\text{ft}}{\text{sec}^2} \cdot \frac{1 \text{ mi}}{5280 \text{ ft}} \approx 0.0060984 \frac{\text{mi}}{\text{sec}^2}$$

Now we substitute and compute.
$$h = \frac{2(3960)^2(00060984)}{(65)^2} - 3960$$
$$h \approx 567$$

The satellite is about 567 mi from the surface of Earth.

97. $c = \frac{a}{a+12} \cdot d$
$$c = \frac{2a}{2a+12} \cdot d \quad \text{Doubling } a$$
$$= \frac{2a}{2(a+6)} \cdot d$$
$$= \frac{a}{a+6} \cdot d \quad \text{Simplifying}$$

The ratio of the larger dose to the smaller dose is
$$\frac{\frac{a}{a+6} \cdot d}{\frac{a}{a+12} \cdot d} = \frac{\frac{ad}{a+6}}{\frac{ad}{a+12}}$$
$$= \frac{ad}{a+6} \cdot \frac{a+12}{ad}$$
$$= \frac{\cancel{ad}(a+12)}{(a+6)\cancel{ad}}$$
$$= \frac{a+12}{a+6}$$

The amount by which the dosage increases is
$$\frac{a}{a+6} \cdot d - \frac{a}{a+12} \cdot d$$
$$= \frac{ad}{a+6} - \frac{ad}{a+12}$$
$$= \frac{ad}{a+6} \cdot \frac{a+12}{a+12} - \frac{ad}{a+12} \cdot \frac{a+6}{a+6}$$
$$= \frac{ad(a+12) - ad(a+6)}{(a+6)(a+12)}$$
$$= \frac{a^2d + 12ad - a^2d - 6ad}{(a+6)(a+12)}$$
$$= \frac{6ad}{(a+6)(a+12)}$$

Then the percent by which the dosage increases is
$$\frac{\frac{6ad}{(a+6)(a+12)}}{\frac{a}{a+12} \cdot d} = \frac{\frac{6ad}{(a+6)(a+12)}}{\frac{ad}{a+12}}$$
$$= \frac{6ad}{(a+6)(a+12)} \cdot \frac{a+12}{ad}$$
$$= \frac{6 \cdot \cancel{ad} \cdot \cancel{(a+12)}}{(a+6)\cancel{(a+12)} \cdot \cancel{ad}}$$
$$= \frac{6}{a+6}$$

This is a decimal representation for the percent of increase. To give the result in percent notation we multiply by 100 and use a percent symbol. We have

$\frac{6}{a+6}\cdot 100\%$, or $\frac{600}{a+6}\%$.

99. $$a=\frac{\frac{d_4-d_3}{t_4-t_3}-\frac{d_2-d_1}{t_2-t_1}}{t_4-t_2}$$

$$a(t_4-t_2)=\frac{d_4-d_3}{t_4-t_3}-\frac{d_2-d_1}{t_2-t_1} \quad \text{Multiplying by } t_4-t_2$$

$$a(t_4-t_2)(t_4-t_3)(t_2-t_1)=(d_4-d_3)(t_2-t_1)-(d_2-d_1)(t_4-t_3)$$

Multiplying by $(t_4-t_3)(t_2-t_1)$

$$a(t_4-t_2)(t_4-t_3)(t_2-t_1)-(d_4-d_3)(t_2-t_1) = -(d_2-d_1)(t_4-t_3)$$

$$(t_2-t_1)[a(t_4-t_2)(t_4-t_3)-(d_4-d_3)] = -(d_2-d_1)(t_4-t_3)$$

$$t_2-t_1=\frac{-(d_2-d_1)(t_4-t_3)}{a(t_4-t_2)(t_4-t_3)-(d_4-d_3)}$$

$$t_2+\frac{(d_2-d_1)(t_4-t_3)}{a(t_4-t_2)(t_4-t_3)+d_3-d_4}=t_1$$

101. Let w = the wattage of the bulb. Then we have $I=\frac{kw}{d^2}$.

Now substitute $2w$ for w and $2d$ for d.

$$I=\frac{k(2w)}{(2d)^2}=\frac{2kw}{4d^2}=\frac{kw}{2d^2}=\frac{1}{2}\cdot\frac{kw}{d^2}$$

We see that the intensity is halved.

103. ***Familiarize.*** We write $T=kml^2f^2$. We know that $T=100$ when $m=5$, $l=2$, and $f=80$.

Translate. Find k.

$$T=kml^2f^2$$
$$100=k(5)(2)^2(80)^2$$
$$0.00078125=k$$
$$T=0.00078125ml^2f^2$$

Carry out. Substitute 72 for T, 5 for m, and 80 for f and solve for l.

$$72=000078125(5)(l^2)(80)^2$$
$$2.88=l^2$$
$$1.697\approx l$$

Check. Recheck the calculations.

State. The string should be about 1.697 m long.

105. ***Familiarize.*** Because d varies inversely as s, we write $d(s)=k/s$. We know that $d(0.56)=50$.

Translate.

$$d(s)=\frac{k}{s}$$
$$d(0.56)=\frac{k}{0.56} \quad \text{Replacing } s \text{ with } 0.56$$
$$50=\frac{k}{0.56} \quad \text{Replacing } d(0.56) \text{ with } 50$$
$$28=k$$
$$d(s)=\frac{28}{s} \quad \text{Equation of variation}$$

Carry out. Find $d(0.40)$.

$$d(0.40)=\frac{28}{0.40}=70$$

Check. Reexamine the calculations. Also observe that, as expected, when d decreases, then s increases.

State. The equation of variation is $d(s)=\frac{28}{s}$. The distance is 70 yd.

Chapter 7 Review

1. True

3. False; (9, 5) and (7, 5) can represent a function and pass the vertical-line test.

5. False; the *vertical*-line test is a quick way to determine whether a graph represents a function.

7. True; $(f+g)(x)=f(x)+g(x)$ represents the addition of functions f and g.

9. True

11. a) Locate 2 on the horizontal axis and find the point on the graph for which 2 is the first coordinate. From this point, look to the vertical axis to find the corresponding y-coordinate, 3. Thus $f(2)=3$.

b) The set of all x-values in the graph extends from –2 to 4, so the domain is $\{x|-2\le x\le 4\}$, or [–2, 4].

c) To determine which member(s) of the domain are with 2, locate 2 on the vertical axis. From there, look left and right to the graph to find any points for which 2 is the second coordinate. One such point exists. Its first coordinate is –1. Thus $f(-1)=2$.

d) The set of all y-values in the graph extends from 1 to 5, so the range is $\{y|1\le y\le 5\}$, or [1, 5].

13. $$f(x)=x^2+2x-3$$
$$f(2a)=(2a)^2+2(2a)-3$$
$$=4a^2+4a-3$$

15. Plot and connect the points, using the year as the first coordinate and the corresponding minimum hourly wage as the second coordinate.

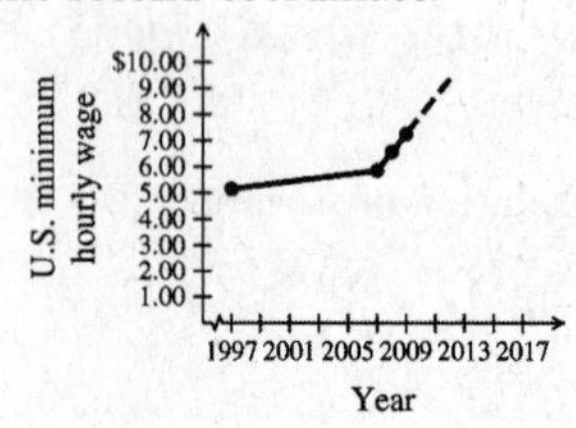

To predict the minimum hourly wage in 2012, extend the graph and extrapolate. It appears that it will be about \$9.30.

17. a) No, the graph is not a function since it fails the vertical-line test.

19. a) Yes, it is a function.

b) For any x-value there is a point on the graph. Thus, Domain of $f = \{x|x \text{ is a real number}\}$, or $\mathbb{R}$.
The only y-value on the graph is –2. Thus, Range of $f = \{-2\}$.

21. $g(x) = \dfrac{x^2}{x-1}$

Since $x - 1$ cannot be computer when the denominator is 0, we find x-values that cause $x - 1$ to be 0:

$$x - 1 = 0$$
$$x = 1$$

Thus, 1 is not in the domain of g, while all other real numbers are. The domain of g is $\{x|x \text{ is a real number } and\ x \neq 1\}$.

23. $r(t) = 900 - 15t$

The time must be nonnegative, so we have $t \geq 0$. In addition, $r(t)$ must be nonnegative so we have:

$$900 - 15t \geq 0$$
$$-15t \geq -900$$
$$t \leq 60$$

Then the domain of the function is $\{t|0 \leq t \leq 60\}$, or $[0, 60]$.

25. ***Familiarize.*** After an initial fee of \$90, an additional fee of \$30 is charged each month. After one month, the total cost is \$90 + \$30 = \$120. After two months, the total cost is $90 + 2\cdot 30 = \$150$. We can generalize this with a model, letting $C(t)$ represent the total cost, in dollars, for t months.

Translate. We reword the problem and translate.

Total Cost is initial cost plus \$30 per month.

$$C(t) \quad = \quad 90 \quad + \quad 30t$$

where $t \geq 0$ since we cannot have negative months.

Carry out. To determine the number of months required to reach a total cost of \$300, we substitute 300 for $C(t)$ and solve for t.

$$C(t) = 30t + 90$$
$$300 = 30t + 90 \quad \text{Substituting}$$
$$210 = 30t \quad \text{Subtracting 90 from both sides}$$
$$7 = t \quad \text{Dividing both sides by 30}$$

Check. We evaluate

$$C(7) = 30(7) + 90 = 210 + 30 = 300.$$

State. The function is $C(t) = 30t + 90$ and it will take 7 months to reach a total cost of \$300.

27. $f(x) = |3x - 7|$

The function is described by an absolute-value equation, so it is an absolute-value function.

29. $p(x) = x^2 + x - 10$

The function is in the form $f(x) = ax^2 + bx + c$, $a \neq 0$, so it is a quadratic function.

31. $s(t) = \dfrac{t+1}{t+2}$

The function is described by a rational equation so it is a rational function.

33. Graph $f(x) = 2x + 1$.

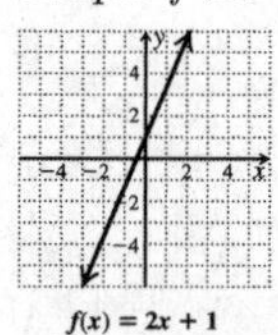

For any x-value and for any y-value there is a point on the graph. Thus,
Domain of $f = \{x|x \text{ is a real number}\}$, or $\mathbb{R}$ and
Range of $f = \{y|y \text{ is a real number}\}$, or $\mathbb{R}$.

35. $(g \cdot h)(4) = g(4)\cdot h(4) = [3(4) - 6][(4)^2 + 1]$
$= [12 - 6][16 + 1]$
$= 6 \cdot 17 = 102$

37. $\left(\dfrac{g}{h}\right)(-1) = \dfrac{g(-1)}{h(-1)} = \dfrac{3(-1) - 6}{(-1)^2 + 1} = \dfrac{-3-6}{1+1} = -\dfrac{9}{2}$

39. Domain of g = Domain of h = $\{x|x \text{ is a real number}\}$.

Since $g(x)=0$ when $3x-6=0$, we have $g(x)=0$ when $x=2$. We conclude that

Domain of $h/g = \{x|x \text{ is a real number and } x \neq 2\}$.

41.

$$S = \frac{H}{m(t_1-t_2)}$$
$$m(t_1-t_2)\cdot S = m(t_1-t_2)\cdot\frac{H}{m(t_1-t_2)}$$
$$mS(t_1-t_2) = H$$
$$m = \frac{H}{S(t_1-t_2)}$$

43.

$$T = \frac{A}{v(t_2-t_1)}$$
$$v(t_2-t_1)\cdot T = v(t_2-t_1)\cdot\frac{A}{v(t_2-t_1)}$$
$$vTt_2 - vTt_1 = A$$
$$vTt_2 - A = vTt_1$$
$$\frac{vTt_2 - A}{vT} = t_1, \text{ or } t_1 = t_2 - \frac{A}{vT}$$

45. $y = \frac{k}{x}$

$3 = \frac{k}{\frac{1}{4}}$ Substituting

$\frac{3}{4} = k$

The variation constant is $\frac{3}{4}$.

The equation of variation is $y = \frac{\frac{3}{4}}{x}$, or $\frac{3}{4x}$.

47. ***Familiarize.*** Because of the phrase "t varies inversely as ...u," we write $t(u) = k/u$. We know that $t(6) = 10$.

Translate. We find the variation constant and then we find the equation of variation.

$t(u) = \frac{k}{u}$

$t(6) = \frac{k}{6}$ Replacing u with 6

$10 = \frac{k}{6}$ Replacing $t(6)$ with 10

$60 = k$ Variation constant

$t(u) = \frac{60}{u}$ Equation of variation

Carry out. We find $t(4)$.

$$t(4) = \frac{60}{4} = 15$$

Check. Reexamine the calculations. Note that, as expected, as the UV rating increases, the time it takes to burn goes down.

State. It will take 15 min to burn when the UV rating is 4.

49. ***Familiarize.*** Because the time t, in seconds varies inversely as the current I, in amperes, we write $t = \frac{k}{I^2}$.

We know that $t = 3.4$ when $I = 0.089$.

Translate. Find k and the equation of variation.

$t = \frac{k}{I^2}$

$3.4 = \frac{k}{(0.089)^2}$

$0.0269314 = k$

$t = \frac{0.0269314}{I^2}$ Equation of variation

Carry out. We find the value of t when I is 0.096.

$$t = \frac{0.0269314}{(0.096)^2} \approx 2.9 \text{ sec}$$

Check. Reexamine the calculations.

State. A 0.096-amp current would be deadly after about 2.9 sec.

51. *Writing Exercise.* Jenna is not correct. Any value of the variable that makes a denominator 0 is not in the domain; 0 itself may or may not make a denominator 0.

53. The function has no x-values less than –4. The hold in the graph at (2, 3) indicates that the function is not defined for $x = 2$. So for any other x-values greater than or equal to –4 there is a point on the graph. Thus,

Domain of $f = \{x|x \geq -4 \text{ and } x \neq 2\}$.

The function has no y-values less than 0. There is no function value at (2, 3), so 3 is not in the range of the function. For any other y-values greater than or equal to 0, there is a point on the graph. Thus,

Range of $f = \{y|y \geq 0 \text{ and } y \neq 3\}$.

Chapter 7 Test

1. a) $f(-2) = 1$

b) Domain is $\{x|-3 \leq x \leq 4\}$, or $[-3, 4]$.

c) If $f(x) = \frac{1}{2}$, then $x = 3$.

d) Range is $\{y|-1 \leq y \leq 2\}$, or $[-1, 2]$.

3. a) Yes; it passes the vertical-line test.

b) For any x-value and for any y-value there is a point on the graph. Thus,

Domain of $f = \{x|x \text{ is a real number}\}$, or $\mathbb{R}$ and

Range of $f = \{y|y \text{ is a real number}\}$, or $\mathbb{R}$.

5. a) No; it fails the vertical-line test.

7. $f(x)=\begin{cases} x^2, \text{ if } x<-2 \\ 3x-5, \text{ if } 0\le x\le 2 \\ x+7, \text{ if } x>2 \end{cases}$

a) Since $0\le 0\le 2$, we use the equation $f(x)=3x-5$.

$f(0)=3(0)-5=-5$

b) Since $3>2$, we use the equation $f(x)=x+7$.

$f(3)=3+7=10$

9. a) Let m = the number of miles, $C(m)$ represents the cost. We form the pairs (250, 100) and (300, 115).

$$\text{Slope}=\frac{115-100}{300-250}=\frac{15}{50}=\frac{3}{10}=0.3$$

$$\begin{aligned} y-y_1&=m(x-x_1) \\ C-100&=0.3(m-250) \\ C&=0.3m+25 \\ C(m)&=0.3m+25 \end{aligned}$$

b) $C(500)=0.3(500)+25=\$175$

11. $g(x)=\dfrac{3}{x^2-16}$

The function is described by a rational equation so it is a rational function. The expression $\dfrac{3}{x^2-16}$ is undefined for values of x that make the denominator 0. We find those values:

$$\begin{aligned} x^2-16&=0 \\ (x+4)(x-4)&=0 \\ x+4=0 \quad &or \quad x-4=0 \\ x=-4 \quad &or \quad x=4 \end{aligned}$$

Then the domain is $\{x|x \text{ is a real number } and\ x\ne -4\ and\ x\ne 4\}$.

13. Graph $f(x)=\frac{1}{3}x-2$.

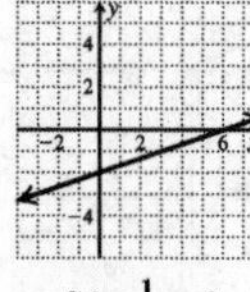

For any x-value and for any y-value there is a point on the graph. Thus,

Domain of $f=\{x|x \text{ is a real number}\}$, or $\mathbb{R}$ and

Range of $f=\{y|y \text{ is a real number}\}$, or $\mathbb{R}$.

15. Graph $h(x)=-\frac{1}{2}$.

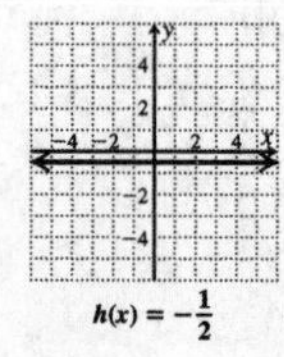

For any x-value there is a point on the graph, so Domain of $h=\{x|x \text{ is a real number}\}$, or $\mathbb{R}$.

The only y-value on the graph is $-\frac{1}{2}$, so

Range of $h=\left\{-\frac{1}{2}\right\}$.

17. $h(x)=2x+1$

$h(5a)=2(5a)+1=10a+1$

19. Domain of g: $\{x|x \text{ is a real number and } x\ne 0\}$

21. $h(x)=2x+1=0$, if $x=-\frac{1}{2}$

Domain of g/h:

$\left\{x\middle|x \text{ is a real number and } x\ne 0 \text{ and } x\ne -\frac{1}{2}\right\}$

23.
$$\begin{aligned} y&=kx \\ 10&=k\cdot 20 \quad \text{Substituting} \\ \frac{1}{2}&=k \end{aligned}$$

The variation constant is $\frac{1}{2}$.

The equation of variation is $y=\frac{1}{2}x$.

25. ***Familiarize.*** Because the surface area A, in square inches varies directly as the square of the radius r, in inches, we write $A=kr^2$. We know that A = 325 when r = 5.

Translate. Find k and the equation of variation.

$$\begin{aligned} A&=kr^2 \\ 325&=k(5)^2 \\ 13&=k \\ d&=13r^2 \quad \text{Equation of variation} \end{aligned}$$

Carry out. We find the value of d when r is 7.

$$A=13(7)^2=637 \text{ in}^2$$

Check. Reexamine the calculations.

State. The area would be 637 in^2 when the radius is 7 in.

27. Answers may vary. In order to have the restriction on the domain of $f/g/h$ that $x\ne\frac{3}{4}$ and $x\ne\frac{2}{7}$, we need to find some function $h(x)$ such that $h(x)=0$ when $x=\frac{2}{7}$.

$$\begin{aligned} x&=\frac{2}{7} \\ 7x&=2 \\ 7x-2&=0 \end{aligned}$$

One possible answer is $h(x)=7x-2$.

Chapter 8

Systems of Linear Equations and Problem Solving

Exercise Set 8.1

1. False; see Example 4(b).

3. True; see page 508 in the text.

5. True; see Example 4(b).

7. False; see page 511 in the text.

9. We use alphabetical order for the variables. We replace x by 2 and y by 3.

$$\begin{array}{c|c} \multicolumn{2}{c}{2x-y=1} \\ \hline 2\cdot 2-3 & 1 \\ 4-3 & \\ \multicolumn{2}{c}{1 \stackrel{?}{=} 1 \quad \text{TRUE}} \end{array} \qquad \begin{array}{c|c} \multicolumn{2}{c}{5x-3y=1} \\ \hline 5\cdot 2-3\cdot 3 & 1 \\ 10-9 & \\ \multicolumn{2}{c}{1 \stackrel{?}{=} 1 \quad \text{TRUE}} \end{array}$$

The pair (2, 3) makes both equations true, so is it a solution of the system.

11. We use alphabetical order for the variables. We replace x by –5 and y by 1.

$$\begin{array}{c|c} \multicolumn{2}{c}{x+5y=0} \\ \hline -5+5\cdot 1 & 0 \\ -5+5 & \\ \multicolumn{2}{c}{0 \stackrel{?}{=} 0 \quad \text{TRUE}} \end{array} \qquad \begin{array}{c|c} \multicolumn{2}{c}{y=2x+9} \\ \hline 1 & 2(-5)+9 \\ & -10+9 \\ \multicolumn{2}{c}{1 \stackrel{?}{=} -1 \quad \text{FALSE}} \end{array}$$

The pair (–5, 1) is not a solution of $y = 2x + 9$. Therefore, it is not a solution of the system of equations.

13. We replace x by 0 and y by –5.

$$\begin{array}{c|c} \multicolumn{2}{c}{x-y=5} \\ \hline 0-(-5) & 5 \\ 0+5 & \\ \multicolumn{2}{c}{5 \stackrel{?}{=} 5 \quad \text{TRUE}} \end{array} \qquad \begin{array}{c|c} \multicolumn{2}{c}{y=3x-5} \\ \hline -5 & 3\cdot 0-5 \\ & 0-5 \\ \multicolumn{2}{c}{-5 \stackrel{?}{=} -5 \quad \text{TRUE}} \end{array}$$

The pair (0, –5) makes both equations true, so is it a solution of the system.

15. Observe that if we multiply both sides of the first equation by 2, we get the second equation. Thus, if we find that the given points makes the one equation true, we will also know that it makes the other equation true. We replace x by 3 and y by –1 in the first equation.

$$\begin{array}{c|c} \multicolumn{2}{c}{3x-4y=13} \\ \hline 3\cdot 3-4(-1) & 13 \\ 9+4 & \\ \multicolumn{2}{c}{13 \stackrel{?}{=} 13 \quad \text{TRUE}} \end{array}$$

The pair (3, –1) makes both equations true, so is it a solution of the system.

17. Graph both equations.

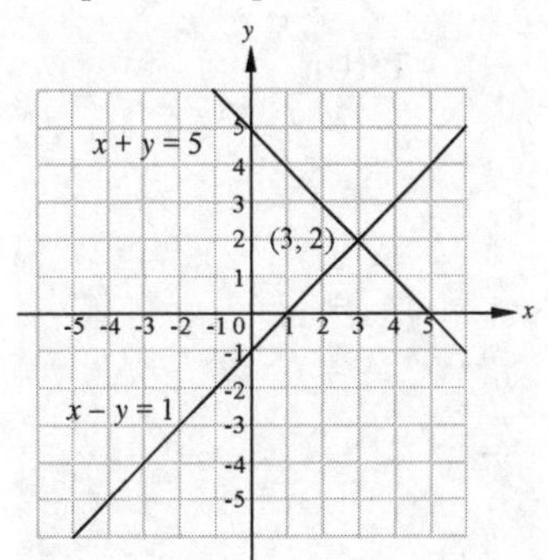

The solution (point of intersection) is apparently (3, 2).

Check:

$$\begin{array}{c|c} \multicolumn{2}{c}{x-y=1} \\ \hline 3-2 & 1 \\ \multicolumn{2}{c}{1 \stackrel{?}{=} 1 \quad \text{TRUE}} \end{array} \qquad \begin{array}{c|c} \multicolumn{2}{c}{x+y=5} \\ \hline 3+2 & 5 \\ \multicolumn{2}{c}{5 \stackrel{?}{=} 5 \quad \text{TRUE}} \end{array}$$

The solution is (3, 2).

19. Graph the equations.

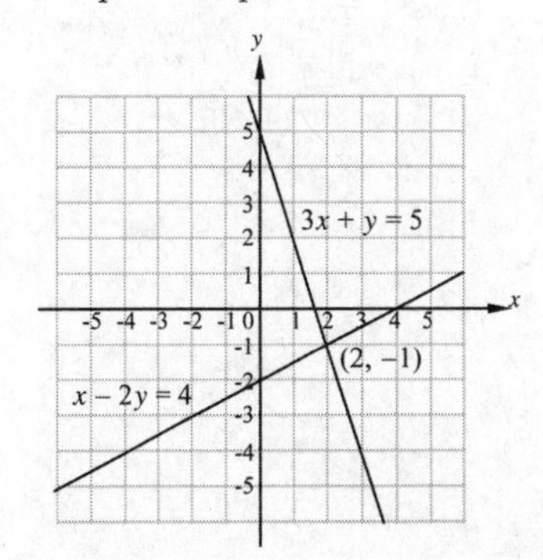

The solution (point of intersection) is apparently (2, –1).

Check:

$$\begin{array}{c|c} \multicolumn{2}{c}{3x+y=5} \\ \hline 3\cdot 2+(-1) & 5 \\ 6-1 & \\ \multicolumn{2}{c}{5 \stackrel{?}{=} 5 \quad \text{TRUE}} \end{array} \qquad \begin{array}{c|c} \multicolumn{2}{c}{x-2y=4} \\ \hline 2-2(-1) & 4 \\ 2+2 & \\ \multicolumn{2}{c}{4 \stackrel{?}{=} 4 \quad \text{TRUE}} \end{array}$$

The solution is (2, –1).

21. Graph both equations.

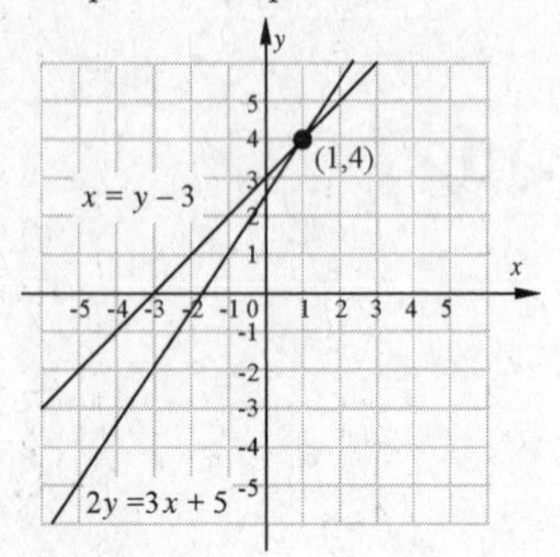

The solution (point of intersection) is apparently (1, 4).

Check:

$$\begin{array}{c|c} 2y = 3x+5 & \\ \hline 2\cdot 4 & 3\cdot 1+5 \\ 8 & 3+5 \end{array}$$

$8 \stackrel{?}{=} 8$ TRUE

$$\begin{array}{c|c} x = y-3 & \\ \hline 1 & 4-3 \end{array}$$

$1 \stackrel{?}{=} 1$ TRUE

The solution is (1, 4).

23. Graph both equations.

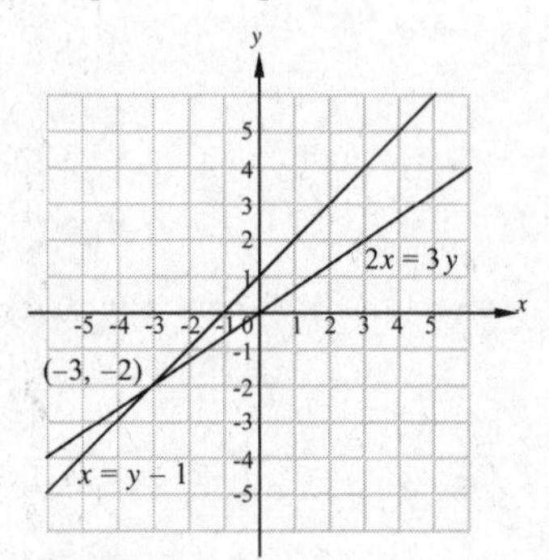

The solution (point of intersection) is apparently (–3, –2).

Check:

$$\begin{array}{c|c} x = y-1 & \\ \hline -3 & -2-1 \end{array}$$

$-3 \stackrel{?}{=} -3$ TRUE

$$\begin{array}{c|c} 2x = 3y & \\ \hline 2(-3) & 3(-2) \end{array}$$

$-6 \stackrel{?}{=} -6$ TRUE

The solution is (–3, –2).

25. Graph both equations.

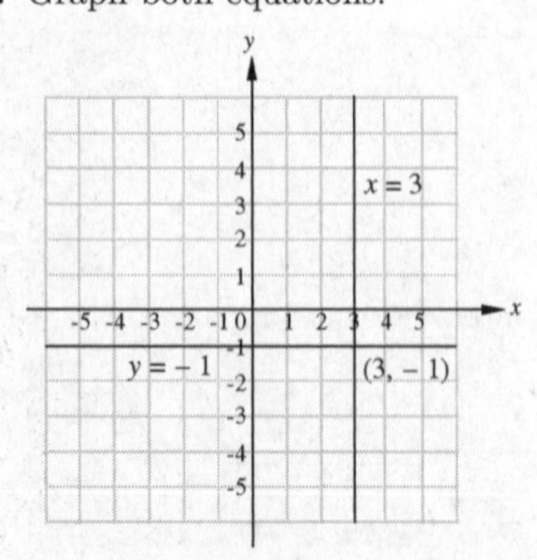

The ordered pairs (3, –1) checks in both equations. It is the solution.

27. Graph both equations.

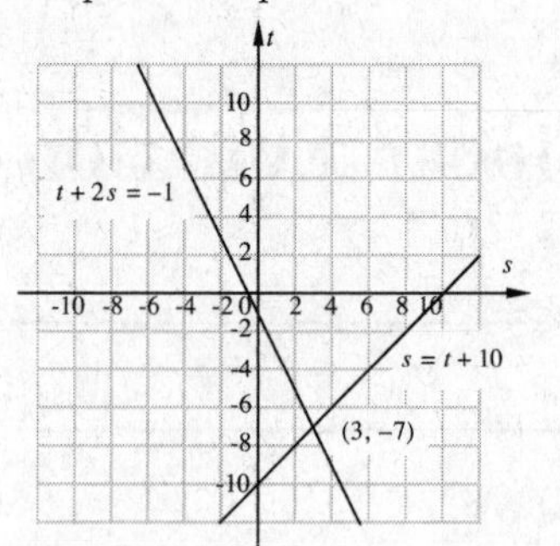

The solution (point of intersection) is apparently (3, –7).

Check:

$$\begin{array}{c|c} t+2s = -1 & \\ \hline -7+2\cdot 3 & -1 \\ -7+6 & \end{array}$$

$-1 \stackrel{?}{=} -1$ TRUE

$$\begin{array}{c|c} s = t+10 & \\ \hline 3 & -7+10 \end{array}$$

$3 \stackrel{?}{=} 3$ TRUE

The solution is (3, –7).

29. Graph both equations.

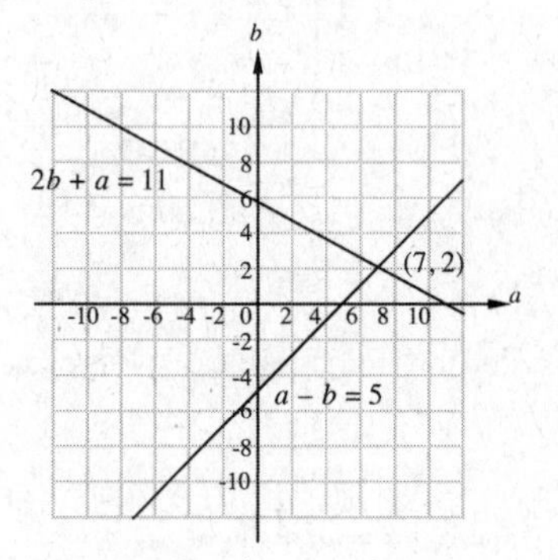

The solution (point of intersection) is apparently (7, 2).

Check:

$$\begin{array}{c|c} 2b+a = 11 & \\ \hline 2\cdot 2+7 & 11 \\ 4+7 & \end{array}$$

$11 \stackrel{?}{=} 11$ TRUE

$$\begin{array}{c|c} a-b = 5 & \\ \hline 7-2 & 5 \end{array}$$

$5 \stackrel{?}{=} 5$ TRUE

The solution is (7, 2).

31. Graph both equations.

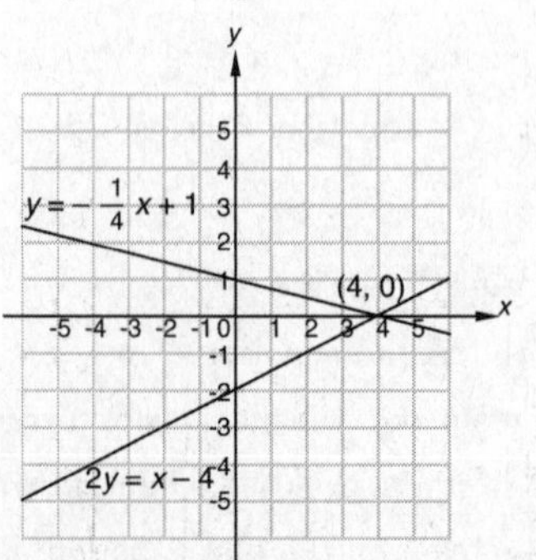

The solution (point of intersection) is apparently (4, 0).

Check:

$$\begin{array}{c|c} y=-\frac{1}{4}x+1 & \\ \hline 0 & -\frac{1}{4}\cdot 4+1 \\ & -1+1 \\ \end{array}$$
$0 \overset{?}{=} 0$ TRUE

$$\begin{array}{c|c} 2y=x-4 & \\ \hline 2\cdot 0 & 4-4 \\ \end{array}$$
$0 \overset{?}{=} 0$ TRUE

The solution is (4, 0).

33. Graph both equations.

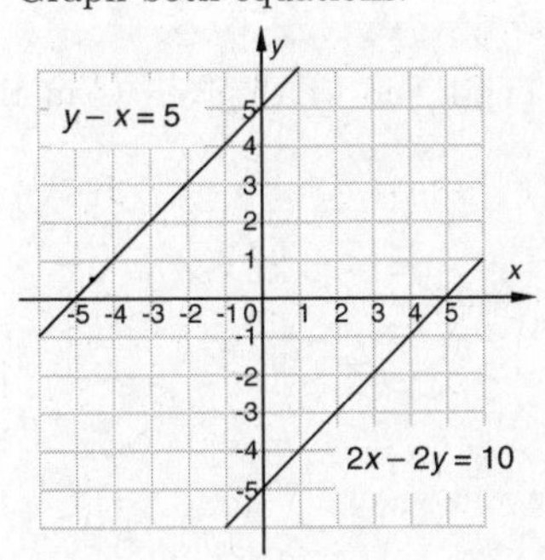

The lines are parallel. The system has no solution.

35. Graph both equations.

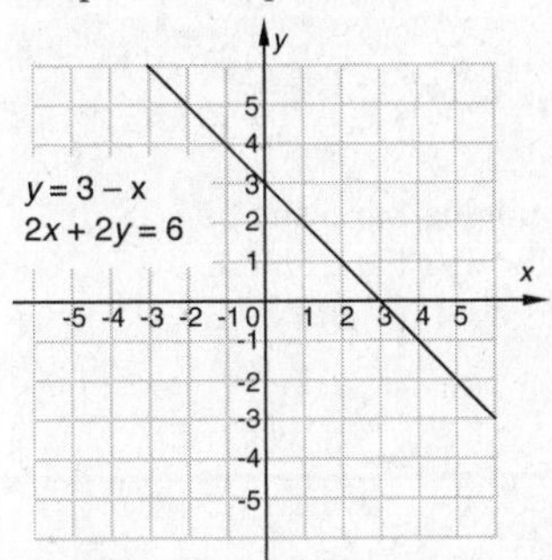

The graphs are the same. Any solution of one equation is a solution of the other. Each equation has infinitely many solutions. The solution set is the set of all pairs $(x,\ y)$ for which $y=3-x$, or $\{(x,\ y)|y=3-x\}$. (In place of $y=3-x$ we could have used $2x+2y=6$ since the two equations are equivalent.)

37. A system of equations is consistent if it has at least one solution. Of the systems under consideration, only the one in Exercise 33 has no solution. Therefore, all except the system in Exercise 33 are consistent.

39. A system of two equations in two variables is dependent if it has infinitely many solutions. Only the system in Exercise 35 is dependent.

41. ***Familiarize.*** Let $x =$ the first number and $y =$ the second number.

Translate.

The sum of the numbers is 10.

$$x+y = 10$$

The first number is $\frac{2}{3}$ of the second number.

$$x = \frac{2}{3} \times y$$

We have a system of equations:

$$x+y=10,$$
$$x=\frac{2}{3}y.$$

43. ***Familiarize.*** Let $p =$ the number or personal e-mails and $b =$ the number if business e-mails.

Translate.

Personal e-mails and Business e-mails is 578.

$$p + b = 578$$

Business e-mails was 30 more than personal e-mails.

$$b = 30 + p$$

We have a system of equations:

$$p+b=578,$$
$$b=p+30.$$

45. ***Familiarize.*** Let $x =$ the measure of one angle and $y =$ the measure of the other angle.

Translate.

Two angles are supplementary.

Rewording: The sum of the measures is 180°.

$$x+y = 180$$

One angle is 3 less than twice the other.

Rewording: One angle is twice the other angle minus 3°.

$$x = 2y - 3$$

We have a system of equations:

$$x+y=180,$$
$$x=2y-3$$

47. Familiarize. Let $g =$ the number of field goals and $t =$ the number of foul shots made.

Translate. We organize the information in a table.

Kind of shot	Field goal	Foul shot	Total
Number scored	g	t	64
Points per score	2	1	
Points scored	$2g$	t	100

From the "Number scored" row of the table we get one equation:

$g+t=64$

The "Points scored" row gives us another equation:

$2g+t=100$

We have a system of equations:

$$g+t=64,$$
$$2g+t=100$$

49. ***Familiarize.*** Let $x =$ the number of hats sold and $y =$ the number of tee shirts sold.

Translate. We organize the information in a table.

Souvenir	Hat	Tee shirt	Total
Number sold	x	y	45
Price	\$14.50	\$19.50	
Amount taken in	$14.50x$	$19.50y$	697.50

The "Number sold" row of the table gives us one equation:

$x+y=45.$

The "Amount taken in" row gives us a second equation:

$14.50x+19.50y=697.50.$

We can multiply both sides of the second equation by 10 to clear the decimals.

$$x+y=45,$$
$$145x+195y=6975.$$

51. ***Familiarize.*** Let $h =$ the number of vials of Humalog sold and $n =$ the number of vials of Lantus sold.

Translate. We organize the information in a table.

Brand	Humalog	Lantus	Total
Number sold	h	n	50
Price	\$83.29	\$76.76	
Amount taken in	$83.29h$	$76.76n$	3981.66

The "Number sold" row of the table gives us one equation:

$h+n=50.$

The "Amount taken in" row gives us a second equation:

$83.29h+76.76n=3981.66.$

We can multiply both sides of the second equation by 100 to clear the decimals.

$$h+n=50,$$
$$8329h+7676n=398{,}166.$$

53. ***Familiarize.*** The lacrosse field is a rectangular with perimeter 340 yd. Let $l =$ the length, in yards, and $w =$ the width, in yards. Recall that for a rectangle with length l and width w, the perimeter P is given by

$P=2l+2w.$

Translate. The formula for perimeter gives us one equation:

$2l+2w=340.$

The statement relating length and width gives us another equation:

$l=w+50.$

We have a system of equations:

$$2l+2w=340,$$
$$l=w+50.$$

55. *Writing Exercise.* Answers may vary.

The Fever made 37 field goals in a basketball game, some 3-pointers and some 2-pointers. Altogether the 37 baskets counted for 80 points. How many of each type of field goal was made?

57.
$$\begin{aligned} 3x+2(5x-1)&=6 & \\ 3x+10x-2&=6 & \text{Removing parentheses} \\ 13x-2&=6 & \text{Collecting like terms} \\ 13x&=8 & \text{Adding 2 to both sides} \\ x&=\frac{8}{13} & \end{aligned}$$

The solution is $\frac{8}{13}$.

59.
$$\begin{aligned} 9y&=5-(y+6) & \\ 9y&=5-y-6 & \text{Removing parentheses} \\ 9y&=-y-1 & \text{Collecting like terms} \\ 10y&=-1 & \text{Adding } y \text{ to both sides} \\ y&=-\frac{1}{10} & \end{aligned}$$

The solution is $-\frac{1}{10}$.

61.
$$\begin{aligned} 3x-y&=4 & \\ -y&=-3x+4 & \text{Adding } -3x \text{ to both sides} \\ y&=3x-4 & \text{Multiplying both sides by } -1 \end{aligned}$$

63. *Writing Exercise.* At the point where the lines representing TV and Radio intersect, there was no clear leader because those media had the same market share. Overall, the graph clearly indicates that Newspapers have maintained a higher advertising market share.

65. a) There are many correct answers. One can be found by expressing the sum and difference of the two numbers:

$$x+y=6,$$
$$x-y=4$$

b) There are many correct answers. For example, write an equation in two variables. Then write a second equation by multiplying the left side of the first equation by one nonzero constant and multiplying the right side by another nonzero constant.

$$x+y=1,$$
$$2x+2y=3$$

c) There are many correct answers. One can be found by writing an equation in two variables and then writing a nonzero constant multiple of that equation:

$$x+y=1,$$
$$2x+2y=2$$

67. Substitute 4 for x and –5 for y in the first equation:

$$A(4)-6(-5)=13$$
$$4A+30=13$$
$$4A=-17$$
$$A=-\frac{17}{4}$$

Substitute 4 for x and –5 for y in the second equation:

$$4-B(-5)=-8$$
$$4+5B=-8$$
$$5B=-12$$
$$B=-\frac{12}{5}$$

We have $A=-\frac{17}{4}$, $B=-\frac{12}{5}$.

69. ***Familiarize.*** Let x = the number of years Dell has taught and y = the number of years Juanita has taught. Two years ago, Dell and Juanita had taught $x-2$ and $y-2$ years, respectively.

Translate.

Together, the number of years of service is 46.

$$x+y=46$$

Two years ago Dell had taught 2.5 times as many years as Juanita.

$$x-2=2.5(y-2)$$

We have a system of equations:

$$x+y=46,$$
$$x-2=2.5(y-2)$$

71. ***Familiarize.*** Let b = the number of ounces of baking soda and v = the number of ounces of vinegar to be used. The amount of baking soda in the mixture will be four times the amount of vinegar.

Translate.

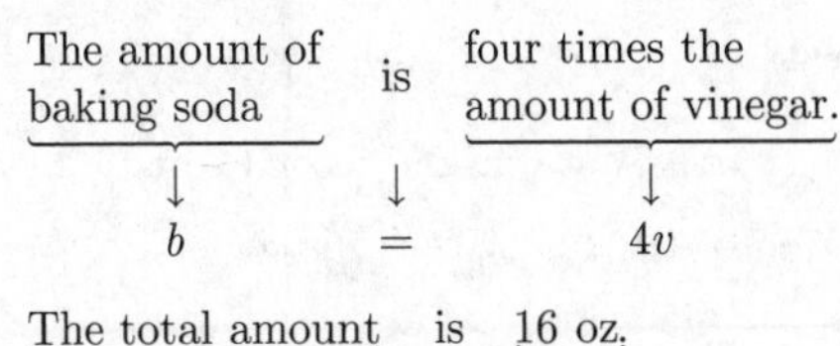

The total amount is 16 oz.

$$b+v=16$$

We have a system of equations.

$$b=4v,$$
$$b+v=16$$

73. From Exercise 44, graph both equations:

$$v+m=16$$
$$m=2v+4$$

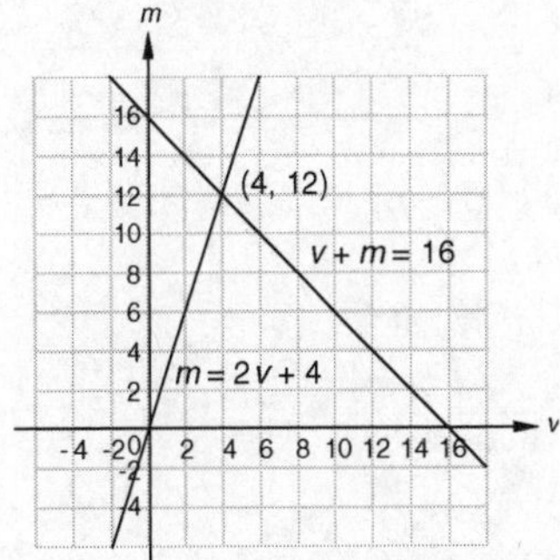

The lines intersect at $v=4$, $m=12$. The solution is 4 oz of vinegar and 12 oz of mineral oil.

75. Graph both equations.

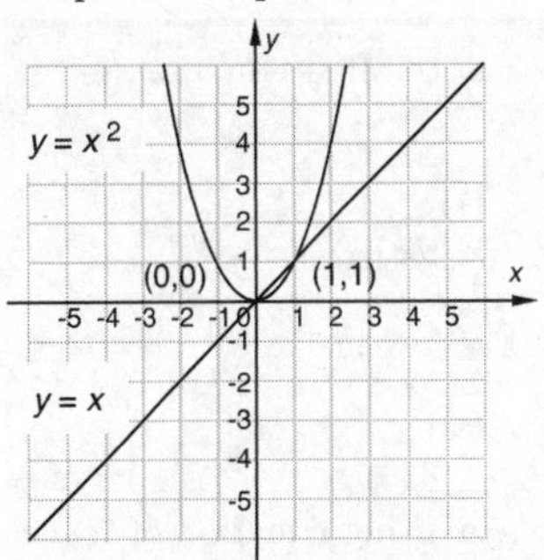

The solutions are apparently (0, 0) and (1, 1). Both pairs check.

77. (0.07, –7.95)

79. (0.00, 1.25)

Exercise Set 8.2

1. Adding the equations, we get $8x=7$, so choice (d) is correct.

3. Multiplying the first equation by –5 gives us the system of equations in (a), so choice (a) is correct.

5. Substituting $4x - 7$ for y in the second equation gives us $6x+3(4x-7)=19$, so choice (c) is correct.

7. $y=3-2x$, (1)
$3x+y=5$ (2)

We substitute $3 - 2x$ for y into the second equation and solve for x.

$$\begin{aligned} 3x+y&=5 && (2)\\ 3x+(3-2x)&=5 && \text{Substituting}\\ x+3&=5\\ x&=2 \end{aligned}$$

Next substitute 2 for x in either equation of the original system and solve for y.

$$\begin{aligned} y&=3-2x && (1)\\ y&=3-2\cdot 2 && \text{Substitute}\\ y&=3-4\\ y&=-1 \end{aligned}$$

We check the ordered pair (2, –1).

$y=3-2x$	
-1	$3-2\cdot 2$
	$3-4$

$-1\overset{?}{=}-1$ TRUE

$3x+y=5$	
$3\cdot 2+(-1)$	5
$6-1$	

$5\overset{?}{=}5$ TRUE

Since (2, –1) checks, it is the solution.

9. $3x+5y=3$, (1)
$x=8-4y$ (2)

We substitute $8 - 4y$ for x in the first equation and solve for y.

$$\begin{aligned} 3x+5y&=3 && (1)\\ 3(8-4y)+5y&=3 && \text{Substituting}\\ 24-12y+5y&=3\\ 24-7y&=3\\ -7y&=-21\\ y&=3 \end{aligned}$$

Next we substitute 3 for y in either equation of the original system and solve for x.

$x=8-4y$ (2)
$x=8-4\cdot 3=8-12=-4$

We check the ordered pair (–4, 3).

$x=8-4y$	
-4	$8-4\cdot 3$
	$8-12$

$-4\overset{?}{=}-4$ TRUE

$3x+5y=3$	
$3(-4)+5\cdot 3$	3
$-12+15$	

$3\overset{?}{=}3$ TRUE

Since (–4, 3) checks, it is the solution.

11. $3s-4t=14$, (1)
$5s+t=8$ (2)

We solve the second equation for t.

$$\begin{aligned} 5s+t&=8 && (2)\\ t&=8-5s && (3) \end{aligned}$$

We substitute $8 - 5s$ for t in the first equation and solve for s.

$$\begin{aligned} 3s-4t&=14 && (1)\\ 3s-4(8-5s)&=14 && \text{Substituting}\\ 3s-32+20s&=14\\ 23s-32&=14\\ 23s&=46\\ s&=2 \end{aligned}$$

Next we substitute 2 for s in Equation (1), (2), or (3). It is easiest to use Equation (3) since it is already solved for t.

$$t=8-5\cdot 2=8-10=-2$$

We check the ordered pair (2, –2).

$3s-4t=14$	
$3\cdot 2-4(-2)$	14
$6+8$	

$14\overset{?}{=}14$ TRUE

$5s+t=8$	
$5\cdot 2+(-2)$	8
$10-2$	

$8\overset{?}{=}8$ TRUE

Since (2, –2) checks, it is the solution.

13. $4x-2y=6$, (1)
$2x-3=y$ (2)

We substitute $2x - 3$ for y in the first equation and solve for x.

$$\begin{aligned} 4x-2y&=6 && (1)\\ 4x-2(2x-3)&=6\\ 4x-4x+6&=6\\ 6&=6 \end{aligned}$$

We have an identity, or an equation that is always true. The equations are dependent and the solution set is infinite: $\{(x,\ y)\,|\,2x-3=y\}$.

15. $-5s+t=11$, (1)
$4s+12t=4$ (2)

We solve the first equation for t.

$$\begin{aligned} -5s+t&=11 && (1)\\ t&=5s+11 && (3) \end{aligned}$$

We substitute $5s + 11$ for t in the second equation and solve for s.

$$\begin{aligned} 4s+12t&=4 && (2)\\ 4s+12(5s+11)&=4\\ 4s+60s+132&=4\\ 64s+132&=4\\ 64s&=-128\\ s&=-2 \end{aligned}$$

Next we substitute –2 for s in Equation (3).

$$t=5s+11=5(-2)+11=-10+11=1$$

We check the ordered pair (–2, 1).

$$\begin{array}{c|c} -5s+t=11 & \\ \hline -5(-2)+1 & 11 \\ 10+1 & \\ 11 \overset{?}{=} 11 & \text{TRUE} \end{array} \qquad \begin{array}{c|c} 4s+12t=4 & \\ \hline 4(-2)+12\cdot 1 & 4 \\ -8+12 & \\ 4 \overset{?}{=} 4 & \text{TRUE} \end{array}$$

Since (–2, 1) checks, it is the solution.

17. $2x+2y=2,$ (1)
$3x-y=1$ (2)

We solve the second equation for y.

$$\begin{aligned} 3x-y&=1 && (2) \\ -y&=-3x+1 \\ y&=3x-1 && (3) \end{aligned}$$

We substitute $3x-1$ for y in the first equation and solve for x.

$$\begin{aligned} 2x+2y&=2 && (1) \\ 2x+2(3x-1)&=2 \\ 2x+6x-2&=2 \\ 8x-2&=2 \\ 8x&=4 \\ x&=\tfrac{1}{2} \end{aligned}$$

Next we substitute $\frac{1}{2}$ for x in Equation (3).

$$y=3x-1=3\cdot\frac{1}{2}-1=\frac{3}{2}-1=\frac{1}{2}$$

The ordered pair $\left(\frac{1}{2}, \frac{1}{2}\right)$ checks in both equations. It is the solution.

19. $2a+6b=4,$ (1)
$3a-b=6$ (2)

We solve the second equation for b.

$$\begin{aligned} 3a-b&=6 && (2) \\ -b&=-3a+6 \\ b&=3a-6 && (3) \end{aligned}$$

We substitute $3a-6$ for b in the first equation and solve for a.

$$\begin{aligned} 2a+6b&=4 && (1) \\ 2a+6(3a-6)&=4 \\ 2a+18a-36&=4 \\ 20a-36&=4 \\ 20a&=40 \\ a&=2 \end{aligned}$$

We substitute 2 for a in Equation 3 and solve for y.

$$\begin{aligned} b&=3a-6 && (3) \\ b&=3\cdot 2-6 \\ b&=6-6 \\ b&=0 \end{aligned}$$

The ordered pair (2, 0) checks in both equations. It is the solution.

21. $2x-3=y$ (1)
$y-2x=1,$ (2)

We substitute $2x-3$ for y in the second equation and solve for x.

$$\begin{aligned} y-2x&=1 && (2) \\ 2x-3-2x&=1 && \text{Substituting} \\ -3&=1 && \text{Collecting like terms} \end{aligned}$$

We have a contradiction, or an equation that is always false. Therefore, there is no solution.

23.
$$\begin{aligned} x+3y&=7 && (1) \\ -x+4y&=7 && (2) \\ \hline 0+7y&=14 && \text{Adding} \\ 7y&=14 \\ y&=2 \end{aligned}$$

Substitute 2 for y in one of the original equations and solve for x.

$$\begin{aligned} x+3y&=7 && (1) \\ x+3\cdot 2&=7 && \text{Substituting} \\ x+6&=7 \\ x&=1 \end{aligned}$$

Check:

$$\begin{array}{c|c} x+3y=7 & \\ \hline 1+3\cdot 2 & 7 \\ 1+6 & \\ 7 \overset{?}{=} 7 & \text{TRUE} \end{array} \qquad \begin{array}{c|c} -x+4y=7 & \\ \hline -1+4\cdot 2 & 7 \\ -1+8 & \\ 7 \overset{?}{=} 7 & \text{TRUE} \end{array}$$

Since (1, 2) checks, it is the solution.

25.
$$\begin{aligned} x-2y&=11 && (1) \\ 3x+2y&=17 && (2) \\ \hline 4x\quad&=28 && \text{Adding} \\ x&=7 \end{aligned}$$

Substitute 7 for x in Equation (1) and solve for y.

$$\begin{aligned} x-2y&=11 && (1) \\ 7-2y&=11 && \text{Substituting} \\ -2y&=4 \\ y&=-2 \end{aligned}$$

We obtain (7, –2). This checks, so it is the solution.

27.
$$\begin{aligned} 9x+3y&=-3 && (1) \\ 2x-3y&=-8 && (2) \\ \hline 11x+0&=-11 && \text{Adding} \\ 11x&=-11 \\ x&=-1 \end{aligned}$$

Substitute –1 for x in Equation (1) and solve for y.

$$\begin{aligned} 9x+3y&=-3 \\ 9(-1)+3y&=-3 && \text{Substituting} \\ -9+3y&=-3 \\ 3y&=6 \\ y&=2 \end{aligned}$$

We obtain (–1, 2). This checks, so it is the solution.

29. $5x+3y=19,$ (1)
$x-6y=11$ (2)

We multiply Equation (1) by 2 to make two terms become opposites.

$$\begin{aligned} 10x+6y&=38 \quad \text{Multiplying (1) by 2} \\ x-6y&=11 \\ \hline 11x \quad &=49 \\ x&=\frac{49}{11} \end{aligned}$$

Substitute $\frac{49}{11}$ for x in Equation (1) and solve for y.

$$\begin{aligned} 5x+3y&=19 \\ 5\cdot\frac{49}{11}+3y&=19 \\ 3y&=-\frac{36}{11} \\ y&=-\frac{12}{11} \end{aligned}$$

We obtain $\left(\frac{49}{11}, -\frac{12}{11}\right)$. This checks, so it is the solution.

31. $5r-3s=24$, (1)
$3r+5s=28$ (2)

We multiply twice to make two terms become additive inverses.

$$\begin{aligned} \text{From (1): } 25r-15s&=120 \quad \text{Multiplying by 5} \\ \text{From (2): } 9r+15s&=84 \quad \text{Multiplying by 3} \\ \hline 34r+\ 0\ &=204 \quad \text{Adding} \\ r&=6 \end{aligned}$$

Substitute 6 for r in Equation (2) and solve for s.

$$\begin{aligned} 3r+5s&=28 \\ 3\cdot6+5s&=28 \quad \text{Substituting} \\ 18+5s&=28 \\ 5s&=10 \\ s&=2 \end{aligned}$$

We obtain (6, 2). This checks, so it is the solution.

33. $6s+9t=12$, (1)
$4s+6t=5$ (2)

We multiply twice to make two terms become opposites.

$$\begin{aligned} \text{From (1): } 12s+18t&=24 \quad \text{Multiplying by 2} \\ \text{From (2): } -12s-18t&=-15 \quad \text{Multiplying by } -3 \\ \hline 0&=9 \end{aligned}$$

We get a contradiction, or an equation that is always false. The system has no solution.

35. $\frac{1}{2}x-\frac{1}{6}y=10$, (1)
$\frac{2}{5}x+\frac{1}{2}y=8$ (2)

We first multiply each equation by the LCM of the denominators to clear fractions.

$3x-y=60$, (3) Multiplying (1) by 6
$4x+5y=80$ (4) Multiplying (2) by 10

We multiply Equation (1) by 5 and then add.

$$\begin{aligned} 15x-5y&=300 \quad \text{Multiplying (3) by 5} \\ 4x+5y&=80 \quad (4) \\ \hline 19x \quad &=380 \\ x&=20 \end{aligned}$$

Substitute 20 for x in one of the equations in which the fractions were cleared and solve for y.

$$\begin{aligned} 3x-y&=60 \quad (3) \\ 3\cdot20-y&=60 \\ 60-y&=60 \\ -y&=0 \\ y&=0 \end{aligned}$$

We obtain (20, 0). This checks, so it is the solution.

37. $\frac{x}{2}+\frac{y}{3}=\frac{7}{6}$, (1)
$\frac{2x}{3}+\frac{3y}{4}=\frac{5}{4}$ (2)

We first multiply each equation by the LCM of the denominators to clear fractions.

$3x+2y=7$ (3) Multiplying (1) by 6
$8x+9y=15$ (4) Multiplying (2) by 12

We multiply twice to make two terms become opposites.

$$\begin{aligned} \text{From (3): } 27x+18y&=63 \quad \text{Multiplying by 9} \\ \text{From (4): } -16x-18y&=-30 \quad \text{Multiplying by } -2 \\ \hline 11x \quad &=33 \quad \text{Adding} \\ x&=3 \end{aligned}$$

Substitute 3 for x in one of the equations in which the fractions were cleared and solve for y.

$$\begin{aligned} 3x+2y&=7 \quad (3) \\ 3\cdot3+2y&=7 \quad \text{Substituting} \\ 9+2y&=7 \\ 2y&=-2 \\ y&=-1 \end{aligned}$$

We obtain (3, –1). This checks, so it is the solution.

39. $12x-6y=-15$, (1)
$-4x+2y=5$ (2)

Observe that, if we multiply Equation (1) by $-\frac{1}{3}$, we obtain Equation (2). Thus, any pair that is a solution of Equation (1) is also a solution of Equation (2). The equations are dependent and the solution set is infinite: $\{(x, y)|-4x+2y=5\}$.

41. $0.3x+0.2y=0.3$,
$0.5x+0.4y=0.4$

We first multiply each equation by 10 to clear decimals.

$3x+2y=3$, (1)
$5x+4y=4$ (2)

We multiply Equation (1) by –2.

$$\begin{aligned} -6x-4y&=-6 \quad \text{Multiplying (1) by } -2 \\ 5x+4y&=4 \quad (2) \\ \hline -x \quad &=-2 \\ x&=2 \end{aligned}$$

Substitute 2 for x in Equation (1) and solve for y.

$$\begin{aligned} 3x+2y&=3 \quad (1)\\ 3\cdot 2+2y&=3\\ 6+2y&=3\\ 2y&=-3\\ y&=-\frac{3}{2} \end{aligned}$$

We obtain $\left(2,-\frac{3}{2}\right)$. This checks, so it is the solution.

43. $a-2b=16,\quad (1)$
$b+3=3a\quad (2)$

We will use the substitution method. First solve Equation (1) for a.

$$\begin{aligned} a-2b&=16\\ a&=2b+16 \quad (3) \end{aligned}$$

Now substitute $2b + 16$ for a in Equation (2) and solve for b.

$$\begin{aligned} b+3&=3a && (2)\\ b+3&=3(2b+16) && \text{Substituting}\\ b+3&=6b+48\\ -45&=5b\\ -9&=b \end{aligned}$$

Substitute –9 for b in Equation (3).

$$a=2(-9)+16=-2$$

We obtain (–2, –9). This checks, so it is the solution.

45. $10x+y=306,\quad (1)$
$10y+x=90\quad (2)$

We will use the substitution method. First solve Equation (1) for y.

$$\begin{aligned} 10x+y&=306\\ y&=-10x+306 \quad (3) \end{aligned}$$

Now substitute $-10x + 306$ for y in Equation (2) and solve for y.

$$\begin{aligned} 10y+x&=90 && (2)\\ 10(-10x+306)+x&=90 && \text{Substituting}\\ -100x+3060+x&=90\\ -99x+3060&=90\\ -99x&=-2970\\ x&=30 \end{aligned}$$

Substitute 30 for x in Equation (3).

$$y=-10\cdot 30+306=6$$

We obtain (30, 6). This checks, so it is the solution.

47. $6x-3y=3,\quad (1)$
$4x-2y=2\quad (2)$

Observe that, if we multiply Equation (1) by $\frac{3}{2}$, we obtain Equation (2). Thus, any pair that is a solution of Equation (1) is also a solution of Equation (2). The equations are dependent and the solution set infinite: $\{(x,\ y)|4x-2y=2\}$.

49. $3s-7t=5,$
$7t-3s=8$

First we rewrite the second equation with the variables in a different order. Then we use the elimination method.

$$\begin{aligned} 3s-7t&=5 \quad (1)\\ -3s+7t&=8 \quad (2)\\ \hline 0&=13 \end{aligned}$$

We get a contradiction, so the system has no solution.

51. $0.05x+0.25y=22,$
$0.15x+0.05y=24$

We first multiply each equation by 100 to clear decimals.

$$\begin{aligned} 5x+25y&=2200, \quad (1)\\ 15x+5y&=2400 \quad (2) \end{aligned}$$

We multiply by –5 on both sides of the second equation and add.

$$\begin{aligned} 5x+25y&=2200 && (1)\\ -75x-25y&=-12{,}000 && \text{Multiplying (2) by } -5\\ \hline -70x\quad\ &=-9800 && \text{Adding}\\ x&=\frac{-9800}{-70}\\ x&=140 \end{aligned}$$

Substitute 140 for x in one of the equations in which the decimals were cleared and solve for y.

$$\begin{aligned} 5x+25y&=2200 && (1)\\ 5\cdot 140+25y&=2200 && \text{Substituting}\\ 700+25y&=2200\\ 25y&=1500\\ y&=60 \end{aligned}$$

We obtain (140, 60). This checks, so it is the solution.

53. $13a-7b=9,\quad (1)$
$2a-8b=6\quad (2)$

We will use the elimination method. First we multiply the equations so that the b-terms can be eliminated.

$$\begin{aligned} \text{From (1): } 104a-56b&=72 && \text{Multiplying by 8}\\ \text{From (2): } -14a+56b&=-42 && \text{Multiplying by } -7\\ \hline 90a\quad\ &=30 && \text{Adding}\\ a&=\frac{1}{3} \end{aligned}$$

Substitute $\frac{1}{3}$ for a in one of the equations and solve for b.

$$\begin{aligned} 2a-8b&=6 && (1)\\ 2\cdot\frac{1}{3}-8b&=6 && \text{Substituting}\\ \frac{2}{3}-8b&=6\\ -8b&=\frac{16}{3}\\ b&=-\frac{2}{3} \end{aligned}$$

We obtain $\left(\frac{1}{3},-\frac{2}{3}\right)$. This checks, so it is the solution.

55. *Writing Exercise.* Answers may vary. Form a linear expression in two variables and set it equal to two different constants.

57. ***Familiarize.*** Let $w=$ the number of kilowatt hours of electricity used each month by the toaster oven. Then $4w=$ the number of kilowatt hours used by the convection oven. The sum of these two numbers is 15 kilowatt hours.

Translate.

kilowatt hours for toaster oven	plus	kilowatt hours for convection oven	is	15.
↓	↓	↓	↓	↓
w	$+$	$4w$	$=$	15

Carry out. We solve the equation.

$$w+4w=15$$
$$5w=15$$
$$w=3$$

If $w=3$, then $4w=12$.

Check. We have 3 kWh + 4(3) kWh = 15 kWh. The answer checks.

State. For the month, the toaster oven uses 3 kWh and the convection oven uses 12 kWh.

59. ***Familiarize.*** Let $p=$ the original price of the house. The reduced price is $\frac{9}{10}p$ or \$94,500.

Translate.

reduced price	is	$\frac{9}{10}$	of	original price.
↓	↓	↓	↓	↓
$94,500$	$=$	$\frac{9}{10}$	$\cdot$	p

Carry out. We solve the equation.

$$94{,}500=\frac{9}{10}p$$
$$945{,}000=9p$$
$$105{,}000=p$$

Check. The reduced price is found by multiplying 105,000 by $\frac{9}{10}$ obtaining \$94,500. The answer checks.

State. The original price of the home was \$105,000.

61. ***Familiarize.*** Let $l=$ the length of the first piece in inches. Then $2l=$ the length of the second piece and $\frac{1}{10}(2l)$ or $\frac{1}{5}l=$ the length of the third piece. The sum of these three lengths is 96 in.

Translate.

First length	plus	second length	plus	third length	is	96.
↓	↓	↓	↓	↓	↓	↓
l	$+$	$2l$	$+$	$\frac{1}{5}l$	$=$	96

Carry out. We solve the equation.

$$2l+l+\frac{1}{5}l=96$$
$$5l+10l+l=480 \quad \text{Clearing fractions}$$
$$16l=480$$
$$l=30$$

If $l=30$, then $2l=60$, and $\frac{1}{5}l=6$. The pieces are 30 in., 60 in. and 6 in.

Check. We have $96=30+2(30)+\frac{1}{5}(30)$. The answer checks.

State. The pieces are 30 in., 60 in., and 6 in.

63. *Writing Exercise.* Answers may vary. See the box on page 521 of the text.

65. First write $f(x)=mx+b$ as $y=mx+b$. Then substitute 1 for x and 2 for y to get one equation and also substitute –3 for x and 4 for y to get a second equation:

$$2=m\cdot 1+b$$
$$4=m(-3)+b$$

Solve the resulting system of equations.

$$2=m+b$$
$$4=-3m+b$$

Multiply the second equation by –1 and add.

$$2=m+b$$
$$\underline{-4=3m-b}$$
$$-2=4m$$
$$-\frac{1}{2}=m$$

Substitute $-\frac{1}{2}$ for m in the first equation and solve for b.

$$2=-\frac{1}{2}+b$$
$$\frac{5}{2}=b$$

Thus, $m=-\frac{1}{2}$ and $b=\frac{5}{2}$.

67. Substitute –4 for x and –3 for y in both equations and solve for a and b.

$$-4a-3b=-26, \quad (1)$$
$$-4b+3a=7 \quad (2)$$

$$-12a-9b=-78 \quad \text{Multiplying (1) by 3}$$
$$\underline{12a-16b=28} \quad \text{Multiplying (2) by 4}$$
$$-25b=-50$$
$$b=2$$

Substitute 2 for b in Equation (2).

$$\begin{aligned} -4\cdot 2+3a&=7\\ 3a&=15\\ a&=5 \end{aligned}$$

Thus, $a = 5$ and $b = 2$.

69. $\frac{x+y}{2}-\frac{x-y}{5}=1,$

$\frac{x-y}{2}+\frac{x+y}{6}=-2$

After clearing fractions we have:

$3x+7y=10,$ (1)
$4x-2y=-12$ (2)

$$\begin{array}{ll} 6x+14y=20 & \text{Multiplying (1) by 2}\\ \underline{28x-14y=-84} & \text{Multiplying (2) by 7}\\ 34x\qquad=-64 & \\ x=-\frac{32}{17} & \end{array}$$

Substitute $-\frac{32}{17}$ for x in Equation (1).

$$\begin{aligned} 3\left(-\frac{32}{17}\right)+7y&=10\\ 7y&=\frac{266}{17}\\ y&=\frac{38}{17} \end{aligned}$$

The solution is $\left(-\frac{32}{17}, \frac{38}{17}\right)$.

71. $\frac{2}{x}+\frac{1}{y}=0,$ $\qquad 2\cdot\frac{1}{x}+\frac{1}{y}=0,$

or

$\frac{5}{x}+\frac{2}{y}=-5$ $\qquad 5\cdot\frac{1}{x}+2\cdot\frac{1}{y}=-5$

Substitute u for $\frac{1}{x}$ and v for $\frac{1}{y}$.

$2u+v=0,$ (1)
$5u+2v=-5$ (2)

$$\begin{array}{ll} -4u-2v=0 & \text{Multiplying (1) by } -2\\ \underline{5u+2v=-5} & (2)\\ u=-5 & \end{array}$$

Substitute –5 for u in Equation (1).

$$\begin{aligned} 2(-5)+v&=0\\ -10+v&=0\\ v&=10 \end{aligned}$$

If $u = -5$, then $\frac{1}{x}=-5$. Thus $x=-\frac{1}{5}$.

If $v = 10$, then $\frac{1}{y}=10$. Thus $y=\frac{1}{10}$.

The solution is $\left(-\frac{1}{5}, \frac{1}{10}\right)$.

73. *Writing Exercise.* One way is to show the student algebraically that the equations have different intercepts. Another is to solve the system using the elimination method and observe that a contradiction results.

Connecting the Concepts

1. $x=y,$ (1)
$x+y=2$ (2)

Substitute y for x in Equation (2) and solve for y.

$$\begin{array}{ll} x+y=2 & (2)\\ y+y=2 & \text{Substituting}\\ 2y=2 & \\ y=1 & \end{array}$$

Substitute 1 for y in Equation (1) and solve for x.

$$\begin{array}{ll} x=y & (1)\\ x=1 & \end{array}$$

We obtain $(1, 1)$ as the solution.

3. $y=\frac{1}{2}x+1,$ (1)
$y=2x-5$ (2)

Substitute $2x-5$ for y in Equation (1) and solve for x.

$$\begin{array}{ll} y=\frac{1}{2}x+1 & (1)\\ 2x-5=\frac{1}{2}x+1 & \text{Substituting}\\ 4x-10=x+2 & \text{Clearing fractions}\\ 3x=12 & \\ x=4 & \end{array}$$

Substitute 4 for x in Equation (2) and solve for y.

$$\begin{array}{ll} y=2x-5 & (2)\\ y=2(4)-5=3 & \end{array}$$

We obtain $(4, 3)$ as the solution.

5. $x=5,$ (1)
$y=10$ (2)

We obtain $(5, 10)$ as the solution.

7. $2x-y=1,$ (1)
$2y-4x=3$ (2)

We multiply Equation (1) by 2.

$$\begin{array}{ll} 4x-2y=2 & \text{Multiplying (1) by 2}\\ \underline{-4x+2y=3} & (2)\\ 0=5 & \text{FALSE} \end{array}$$

There is no solution.

9. $x+2y=3,$ (1)
$3x=4-y$ (2)

We solve Equation (2) for y.

$$\begin{array}{ll} 3x=4-y & (2)\\ y=4-3x & \end{array}$$

Substitute $4-3x$ for y in Equation (1) and solve for x.

$$\begin{array}{ll} x+2y=3 & (1)\\ x+2(4-3x)=3 & \text{Substituting}\\ x+8-6x=3 & \\ -5x=-5 & \\ x=1 & \end{array}$$

Substitute 1 for x in Equation (2) and solve for y.

$$y = 4 - 3x \quad (1)$$
$$y = 4 - 3(1) = 1$$

We obtain $(1,\ 1)$ as the solution.

11. $10x + 20y = 40, \quad (1)$
$x - y = 7 \quad (2)$

We multiply Equation (2) by 20.

$$\begin{array}{rl} 10x + 20y = 40 & (1) \\ 20x - 20y = 140 & \text{Multiplying (2) by 20} \\ \hline 30x \quad\quad = 180 & \\ x = 6 & \end{array}$$

Substitute 6 for x in Equation (2) and solve for y.

$$\begin{aligned} x - y &= 7 \quad (2) \\ 6 - y &= 7 \\ -y &= 1 \\ y &= -1 \end{aligned}$$

We obtain $(6, -1)$ as the solution.

13. $2x - 5y = 1, \quad (1)$
$3x + 2y = 11 \quad (2)$

We multiply Equation (1) by 2 and Equation (2) by 5.

$$\begin{array}{rl} 4x - 10y = 2 & \text{Multiplying (1) by 2} \\ 15x + 10y = 55 & \text{Multiplying (2) by 5} \\ \hline 19x \quad\quad = 57 & \\ x = 3 & \end{array}$$

Substitute 3 for x in Equation (2) and solve for y.

$$\begin{aligned} 3x + 2y &= 11 \quad (2) \\ 3(3) + 2y &= 11 \\ 9 + 2y &= 11 \\ 2y &= 2 \\ y &= 1 \end{aligned}$$

We obtain $(3,\ 1)$ as the solution.

15.
$$\begin{array}{rl} 1.1x - 0.3y = 0.8 & (1) \\ 2.3x + 0.3y = 2.6 & (2) \\ \hline 3.4x \quad\quad = 3.4 & \text{Adding} \\ x = 1 & \end{array}$$

Substitute 1 for x in Equation (1) and solve for y.

$$\begin{aligned} 1.1x - 0.3y &= 0.8 \quad (1) \\ 1.1(1) - 0.3y &= 0.8 \quad \text{Substituting} \\ -0.3y &= -0.3 \\ y &= 1 \end{aligned}$$

We obtain $(1,\ 1)$ as the solution.

17. $x - 2y = 5, \quad (1)$
$3x - 15 = 6y \quad (2)$

We solve Equation (1) for x.

$$\begin{aligned} x - 2y &= 5 \quad (1) \\ x &= 2y + 5 \end{aligned}$$

Substitute $2y + 5$ for x in Equation (2) and solve for y.

$$\begin{aligned} 3x - 15 &= 6y \quad (2) \\ 3(2y + 5) - 15 &= 6y \quad \text{Substituting} \\ 6y + 15 - 15 &= 6y \\ 0 &= 0 \end{aligned}$$

There are many solutions.

The solution set is $\{(x,\ y) | x - 2y = 5\}$.

19. $0.2x + 0.7y = 1.2,$
$0.3x - 0.1y = 2.7$

We first multiply each equation by 10 to clear decimals.

$2x + 7y = 12, \quad (1)$
$3x - y = 27 \quad (2)$

We multiply Equation (2) by 7.

$$\begin{array}{rl} 2x + 7y = 12 & (1) \\ 21x - 7y = 189 & \text{Multiplying (2) by 7} \\ \hline 23x \quad\quad = 201 & \\ x = \frac{201}{23} & \end{array}$$

Substitute $\frac{201}{23}$ for x in Equation (2) and solve for y.

$$\begin{aligned} 3x - y &= 27 \quad (2) \\ 3\left(\frac{201}{23}\right) - y &= 27 \\ \frac{603}{23} - y &= 27 \\ -y &= \frac{18}{23} \\ y &= -\frac{18}{23} \end{aligned}$$

We obtain $\left(\frac{201}{23}, -\frac{18}{23}\right)$ as the solution.

Exercise Set 8.3

1. The Familiarize and Translate steps were done in Exercise 41 of Exercise Set 8.1.

Carry out. We solve the system of equations.

$$\begin{aligned} x + y &= 10, \quad (1) \\ x &= \frac{2}{3}y \quad (2) \end{aligned}$$

where x is the first number and y is the second number.

We use substitution.

$$\begin{aligned} \frac{2}{3}y + y &= 10 \\ \frac{5}{3}y &= 10 \\ y &= 6 \end{aligned}$$

Now substitute 6 for y in Equation (2).

$$x = \frac{2}{3}(6) = 4$$

Check. The sum of the numbers is 4 + 6, or 10 and $\frac{2}{3}$ times the second number, 6, is the first number 4. The answer checks.

State. The first number is 4, and the second number is 6.

3. The Familiarize and Translate steps were done in Exercise 43 of Exercise Set 8.1.

Carry out. We solve the system of equations.

$$p+b=578, \quad (1)$$
$$b=p+30 \quad (2)$$

where p and b represent the number of personal e-mails and business e-mails, respectively. We use substitution. Substitute p + 30 for b in Equation (1) and solve for p.

$$p+(p+30)=578$$
$$2p+30=578$$
$$2p=548$$
$$p=274$$

Now substitute 274 for p in Equation (2).

$$b=274+30=304$$

Check. The sum of the personal e-mails and the business e-mails is 274 + 304, or 578. The answer checks.

State. There are 274 personal e-mails and 304 business e-mails.

5. The Familiarize and Translate steps were done in Exercise 45 of Exercise Set 8.1.

Carry out. We solve the system of equations

$$x+y=180, \quad (1)$$
$$x=2y-3 \quad (2)$$

where x = the measure of one angle and y = the measure of the other angle. We use substitution.

Substitute $2y$ – 3 for x in (1) and solve for y.

$$2y-3+y=180$$
$$3y-3=180$$
$$3y=183$$
$$y=61$$

Now substitute 61 for y in (2).

$$x=2\cdot 61-3=122-3=119$$

Check. The sum of the angle measures is 119°+61°, or 180°, so the angles are supplementary. Also $2\cdot 61°-3°=122°-3°=119°$. The answer checks.

State. The measures of the angles are 119° and 61°.

7. The Familiarize and Translate steps were done in Exercise 47 of Exercise Set 8.1.

Carry out. We solve the system of equations

$$g+t=64, \quad (1)$$
$$2g+t=100 \quad (2)$$

where g = the number of field goals and t = the number of foul shots Chamberlain made. We use elimination.

$$-g-t=-64 \quad \text{Multiplying (1) by} -1$$
$$2g+t=100$$
$$g=36$$

Substitute 36 for g in (1) and solve for t.

$$36+t=64$$
$$t=28$$

Check. The total number of scores was 36 + 28, or 64. The total number of points was $2\cdot 36+28=72+28=100$. The answer checks.

State. Chamberlain made 36 field goals and 28 foul shots.

9. The Familiarize and Translate steps were done in Exercise 49 of Exercise Set 8.1.

Carry out. We solve the system of equations.

$$x+y=45, \quad (1)$$
$$145x+195y=6975 \quad (2)$$

where x is the number of hats and y is the number of tee shirts sold. We use elimination. Begin by multiplying Equation (1) by –145.

$$-145x-145y=-6525 \quad \text{Multiplying (1) by } -145$$
$$145x+195y=6975 \quad (2)$$
$$50y=450$$
$$y=9$$

Substitute 9 for y in Equation (1) and solve for x.

$$x+9=45$$
$$x=36$$

Check. The number of hats and tee shirts sold is 9 + 36, or 45. The amount taken in was $\$14.50(36)+\$19.50(9)=\$522+\$175.50=\$697.50$. The answer checks.

State. 36 hats and 9 tee shirts were sold.

11. The Familiarize and Translate steps were done in Exercise 51 of Exercise Set 8.1.

Carry out. We solve the system of equations.

$$h+n=50, \quad (1)$$
$$8329h+7676n=398{,}166 \quad (2)$$

where h = the number or vials of Humalog sold and n = the number of vials of Lantus sold. We use elimination.

$$-7676h-7676n=-383{,}800 \quad \text{Multiplying (1) by } -7676$$
$$8329h+7676n=398{,}166 \quad (2)$$
$$653h=14{,}366$$
$$h=22$$

Substitute 22 for h in Equation (1) and solve for n.

$$22+n=50$$
$$n=28$$

Check. A total of 22 + 28, or 50 vials, were sold. The amount collected was

$\$83.29(22)+\$76.76(28)=\$1832.38+\$2149.28=\$3981.66$

The answer checks.

State. 22 vials of Humalog and 28 vials of Lantus were sold.

13. The Familiarize and Translate steps were done in Exercise 53 of Exercise Set 8.1.

Carry out. We solve the system of equations.

$$2l+2w=340, \quad (1)$$
$$l=w+50 \quad (2)$$

where $l=$ the length, in yards, and $w=$ the width, in yards of the lacrosse field. We use substitution. We substitute $w+50$ for l in Equation (1) and solve for w.

$$2(w+50)+2w=340$$
$$2w+100+2w=340$$
$$4w+100=340$$
$$4w=240$$
$$w=60$$

Now substitute 60 for w in Equation (2).

$$l=60+50=110$$

Check. The perimeter is $2\cdot110+2\cdot60=220+120=340$. The length, 110 yards, is 50 yards more than the width, 60 yards. The answer checks.

State. The length of the lacrosse field is 110 yards, and the width is 60 yards.

15. ***Familiarize.*** Let $x=$ the number of reams of regular paper used and $y=$ the number of reams of recycled paper.

Translate. We organize the information in a table.

	Regular	Recycled paper	Total
Number used	x	y	116
Price	\$3.79	\$5.49	
Total cost	$3.79x$	$5.49y$	582.44

We get one equation from the "Number used" row of the table: $x+y=116$

The "Total cost" row yields a second equation. All costs are expressed in dollars.

$$3.79x+5.49y=582.44.$$

We have the problem translated to a system of equations.

$$x+y=116 \quad (1)$$
$$3.79x+5.49y=582.44 \quad (2)$$

Carry out. We use the elimination method to solve the system of equations.

$$-379x-379y=-43{,}964 \quad \text{Multiplying (1) by } -379$$
$$379x+549y=58{,}244 \quad \text{Multiplying (2) by 100}$$
$$170y=14{,}280$$
$$y=84$$

Substitute 84 for y in Equation (1) and solve for x.

$$x+84=116$$
$$x=32$$

Check. A total of 32 + 84, or 116 reams of paper were used. The total cost was \$3.79(32) + \$5.49(84) = \$121.28 + \$461.16 = \$582.44. The answer checks.

State. 32 reams of regular paper and 84 reams of recycled paper were used.

17. ***Familiarize.*** Let $x=$ the number of 13-watt bulbs and $y=$ the number of 18-watt bulbs purchased.

Translate. We organize the information in a table.

	13-watt bulbs	18-watt bulbs	Total
Number purchased	x	y	200
Price	\$5	\$6	
Total cost	$5x$	$6y$	1140

We get our equation from the "Number purchased" row of the table:

$$x+y=200$$

The "Total cost" row yields a second equation:

$$5x+6y=1140$$

We have translated to a system of equations:

$$x+y=200 \quad (1)$$
$$5x+6y=1140 \quad (2)$$

Carry out. We use the elimination method to solve the system of equations.

$$-5x-5y=-1000 \quad \text{Multiplying (1) by } -5$$
$$5x+6y=1140 \quad (2)$$
$$y=140$$

Substitute 140 for y in Equation (1) and solve for x.

$$x+140=200$$
$$x=60$$

Check. A total of 60 + 140, or 200 bulbs, were purchased. The total cost was $\$5(60)+\$6(140)=\$300+\$840=\$1140$. The answer checks.

State. 60 13-watt bulbs and 140 18-watt bulbs were purchased.

19. ***Familiarize.*** Let $a=$ the number of Apple cartridges purchased and $h=$ the number of HP cartridges.

Translate.

	Apple	HP	Total
Number purchased	a	h	450
Price	\$58.99	\$64.99	
Total cost	$58.99a$	$64.99h$	27,625.50

We get one equation from the "Number purchased" row of the table:

$$a+h=450$$

The "Total cost" row yields a second equation:

$$58.99a+64.99h=27{,}625.50$$

We have translated to a system of equations:

$$a+h=450 \quad (1)$$
$$58.99a+64.99h=27{,}625.50 \quad (2)$$

Carry out. We use the elimination method to solve the system of equations.

$$\begin{aligned} -5899a-5899h &= -2{,}654{,}550 && \text{Multiplying (1) by } -5899 \\ 5899a+6499h &= 2{,}762{,}550 && \text{Multiplying (2) by } 100 \\ \hline 600h &= 108{,}000 \\ h &= 180 \end{aligned}$$

Substitute 180 for h in Equation (1) and solve for a.

$$a+180=450$$
$$a=270$$

Check. A total of 270 + 180, or 450 cartridges, were sold. The total cost was $\$58.99(270)+\$64.99(180)$ $=\$15{,}927.30+\$11{,}698.20=\$27{,}625.50$. The answer checks.

State. 270 Apple cartridges and 180 HP cartridges were purchased.

21. An immediate solution can be determined from the fact that \$12 is the average of equal parts of \$11 and \$13. So there is 14 lb of Mexican and 14 lb of Peruvian.

Or we can solve using the usual method.

Familiarize. Let m = the number of pounds of Mexican coffee and p = the number of pounds of Peruvian coffee to be used in the mixture. The value of the mixture will be \$12(28), or \$336.

Translate. We organize the information in a table.

	Mexican	Peruvian	Mixture
Number of pounds	m	p	28
Price per pound	\$13	\$11	\$12
Value of coffee	$13m$	$11p$	336

The "Number of pounds" row of the table gives us one equation:

$$m+p=28$$

The "Value of coffee" row yields a second equation:

$$13m+11p=336$$

We have translated to a system of equations:

$$m+p=28 \quad (1)$$
$$13m+11p=336 \quad (2)$$

Carry out. We use the elimination method to solve the system of equations.

$$\begin{aligned} -11m-11p &= -308 && \text{Multiplying (1) by } -11 \\ 13m+11p &= 336 && (2) \\ \hline 2m &= 28 \\ m &= 14 \end{aligned}$$

Substitute 14 for m in Equation (1) and solve for p.

$$14+p=28$$
$$p=14$$

Check. The total mixture contains 14 lb + 14 lb, or 28 lb. Its value is $\$13\cdot 14+\$11\cdot 14$ $=\$182+\$154=\$336$. The answer checks.

State. 14 lb of Mexican coffee and 14 lb of Peruvian coffee should be used.

23. ***Familiarize.*** Let x = the number of ounces of custom printed M&Ms and y = the number of ounces of bulk M&Ms. The value of the mixture will be \$0.32(20), or \$6.40. Converting lb to oz: 20 lb = 20(16) = 320 oz.

Translate. We organize the information in a table.

	Custom printed	Bulk	Mixture
Number of ounces	x	y	320
Price per ounce	\$0.60	\$0.25	\$0.32
Value of M&Ms	$0.60x$	$0.25y$	102.40

The "Number of ounces" row of the table gives us one equation:

$$x+y=320$$

The "Value of M&Ms" row yields a second equation:

$$0.60x+0.25y=102.40$$

After clearing decimals, we have the problem translated to a system of equations:

$$x+y=320 \quad (1)$$
$$60x+25y=10{,}240 \quad (2)$$

Carry out. We use the elimination method to solve the system of equations.

$$\begin{aligned} -25x-25y &= -8000 && \text{Multiplying (1) by } -25 \\ 60x+25y &= 10{,}240 && (2) \\ \hline 35x \quad &= 2240 \\ x &= 64 \end{aligned}$$

Substitute 64 for x in Equation (1) and solve for y.

$$64+y=320$$
$$y=256$$

Check. The total mixture contains 64 oz + 256 oz, or 320 oz. Its value is $\$0.60\cdot 64+\$0.25\cdot 256=\$102.40$. The answer checks.

State. 64 oz of custom-printed M&Ms and 256 oz of bulk M&Ms should be used.

25. ***Familiarize.*** Let x = the number of pounds of 50% chocolate candy and y = the number of pounds of 10% chocolate candy. The amount of chocolate in the mixture is 25%(20 lb), or 0.25(20 lb) =5 lb.

Translate. We organize the information in a table.

	50% chocolate	10% chocolate	Mixture
Number of pounds	x	y	20
Percent of chocolate	50%	10%	25%
Amount of chocolate	$0.50x$	$0.10y$	5

The "Number of pounds" row of the table gives us one equation:

$$x+y=20$$

The last row of the table yields a second equation:

$$0.50x+0.10y=5$$

After clearing decimals, we have the problem translated to a system of equations:

$$\begin{aligned} x+y&=20 \quad (1)\\ 5x+y&=50 \quad (2) \end{aligned}$$

Carry out. We use the elimination method to solve the system of equations.

$$\begin{aligned} -x-y&=-20 \quad \text{Multiplying (1) by } -1\\ 5x+y&=50 \quad (2)\\ \hline 4x\quad&=30\\ x&=7.5 \end{aligned}$$

Substitute 7.5 for x in Equation (1) and solve for y.

$$\begin{aligned} 7.5+y&=20\\ y&=12.5 \end{aligned}$$

Check. The amount of the mixture is 7.5 lb + 12.5 lb, or 20 lb. The amount of chocolate is $0.50(7.5)+0.10(12.5)=3.75\text{ lb}+1.25\text{ lb}=5\text{ lb}$. The answer checks.

State. 7.5 lb of 50% chocolate and 12.5 lb of 10% chocolate should be used.

27. ***Familiarize.*** Let $x=$ the number of pounds of Deep Thought Granola and $y=$ the number of pounds of Oat Dream Granola to be used in the mixture. The amount of nuts and dried fruit in the mixture is 19%(20 lb), or $0.19(20\text{ lb})=3.8\text{ lb}$.

Translate. We organize the information in a table.

	Deep Thought	Oat Dream	Mixture
Number of pounds	x	y	20
Percent of nuts and dried fruit	25%	10%	19%
Amount of nuts and dried fruit	$0.25x$	$0.10y$	3.8 lb

We get one equation from the "Number of pounds" row of the table:

$$x+y=20$$

The last row of the table yields a second equation:

$$0.25x+0.1y=3.8$$

After clearing decimals, we have the problem translated to a system of equations:

$$\begin{aligned} x+y&=20, \quad (1)\\ 25x+10y&=380 \quad (2) \end{aligned}$$

Carry out. We use the elimination method to solve the system of equations.

$$\begin{aligned} -10x-10y&=-200 \quad \text{Multiplying (1) by } -10\\ 25x+10y&=380\\ \hline 15x\quad&=180\\ x&=12 \end{aligned}$$

Substitute 12 for x in (1) and solve for y.

$$\begin{aligned} 12+y&=20\\ y&=8 \end{aligned}$$

Check. The amount of the mixture is 12 lb + 8 lb, or 20 lb. The amount of nuts and dried fruit in the mixture is $0.25(12\text{ lb})+0.1(8\text{ lb})=3\text{ lb}+0.8\text{ lb}=3.8\text{ lb}$. The answer checks.

State. 12 lb of Deep Thought Granola and 8 lb of Oat Dream Granola should be mixed.

29. ***Familiarize.*** Let $x=$ the amount of the 6.5% loan and $y=$ the amount of the 7.2% loan. Recall that the formula for simple interest is

Interest = Principal × Rate × Time

Translate. We organize the information in a table.

	6.5% loan	7.2% loan	Total
Principal	x	y	12,000
Interest rate	6.5%	7.2%	
Time	1 yr	1 yr	
Interest	$0.065x$	$0.072y$	811.50

The "Principal" row of the table gives us one equation:

$$x+y=12{,}000$$

The last row of the table yields a second equation:

$$0.065x+0.072y=811.50$$

After clearing decimals, we have the problem translated to a system of equations:

$$\begin{aligned} x+y&=12{,}000 \quad (1)\\ 65x+72y&=811{,}500 \quad (2) \end{aligned}$$

Carry out. We use the elimination method to solve the system of equations.

$$\begin{aligned} -65x-65y&=-780{,}000 \quad \text{Multiplying (1) by } -65\\ 65x+72y&=811{,}500 \quad (2)\\ \hline 7y&=31{,}500\\ y&=4500 \end{aligned}$$

Substitute 4500 for y in Equation (1) and solve for x.

$$x + 4500 = 12{,}000$$
$$x = 7500$$

Check. The loans total \$7500 + \$4500, or \$12,000. The total interest is $0.065(\$7500) + 0.072(\$4500)$ $= \$487.50 + \$324 = \$811.50$. The answer checks.

State. The 6.5% loan was for \$7500 and the 7.2% loan was for \$4500.

31. ***Familiarize.*** Let $x =$ the number of liters of Steady State and $y =$ the number of liters of Even Flow in the mixture. The amount of alcohol in the mixture is $0.15(20\text{ L}) = 3\text{ L}$.

Translate. We organize the information in a table.

	18% solution	10% solution	Mixture
Number of liters	x	y	20
Percent of alcohol	18%	10%	15%
Amount of alcohol	$0.18x$	$0.10y$	3

We get one equation from the "Number of liters" row of the table:

$$x + y = 20\,.$$

The last row of the table yields a second equation:

$$0.18x + 0.1y = 3$$

After clearing decimals we have the problem translated to a system of equations:

$$x + y = 20, \quad (1)$$
$$18x + 10y = 300 \quad (2)$$

Carry out. We use the elimination method to solve the system of equations.

$$-10x - 10y = -200 \quad \text{Multiplying (1) by } -10$$
$$18x + 10y = 300 \quad (2)$$
$$8x = 100$$
$$x = 12.5$$

Substitute 12.5 for x in (1) and solve for y.

$$12.5 + y = 20$$
$$y = 7.5$$

Check. The total amount of the mixture is $12.5\text{ L} + 7.5\text{ L}$ or 20 L. The amount of alcohol in the mixture is $0.18(12.5\text{ L}) + 0.1(7.5\text{ L}) = 2.25\text{ L} + 0.75\text{ L} = 3\text{ L}$. The answer checks.

State. 12.5 L of Steady State and 7.5 L of Even Flow should be used.

33. ***Familiarize.*** Let $x =$ the number of gallons of 87-octane gas and $y =$ the number of gallons of 95-octane gas in the mixture. The amount of octane in the mixture can be expressed as $93(10)$, or 930.

Translate. We organize the information in a table.

	87-octane	95-octane	Mixture
Number of gallons	x	y	10
Octane rating	87	95	93
Total octane	$87x$	$95y$	930

We get one equation from the "Number of gallons" row of the table :

$$x + y = 10$$

The last row of the table yields a second equation:

$$87x + 95y = 930$$

We have a system of equations:

$$x + y = 10 \quad (1)$$
$$87x + 95y = 930 \quad (2)$$

Carry out. We use the elimination method to solve the system of equations.

$$-87x - 87y = -870 \quad \text{Multiplying (1) by } -87$$
$$87x + 95y = 930 \quad (2)$$
$$8y = 60$$
$$y = 7.5$$

Substitute 7.5 for y in Equation (1) and solve for x.

$$x + 7.5 = 10$$
$$x = 2.5$$

Check. The total amount of the mixture is 2.5 gal + 7.5 gal, or 10 gal. The amount of octane can be expressed as $87(2.5) + 95(7.5) = 217.5 + 712.5 = 930$. The answer checks.

State. 2.5 gal of 87-octane gas and 7.5 gal of 95-octane gas should be used.

35. ***Familiarize.*** From the bar graph we see that whole milk is 4% milk fat, milk for cream cheese is 8% milk fat, and cream is 30% milk fat. Let $x =$ the number of pounds of whole milk and $y =$ the number of pounds of cream to be used. The mixture contains 8%(200 lb), or $0.08(200\text{ lb}) = 16\text{ lb}$ of milk fat.

Translate. We organize the information in a table.

	Whole milk	Cream	Mixture
Number of pounds	x	y	200
Percent of milk fat	4%	30%	8%
Amount of milk fat	$0.04x$	$0.30y$	16 lb

We get one equation from the "Number of pounds" row of the table:

$$x + y = 200$$

The last row of the table yields a second equation:

$0.04x + 0.3y = 16$

After clearing decimals, we have the problem translated to a system of equations:

$$x + y = 200, \quad (1)$$
$$4x + 30y = 1600 \quad (2)$$

Carry out. We use the elimination method to solve the system of equations.

$$-4x - 4y = -800 \quad \text{Multiplying (1) by } -4$$
$$4x + 30y = 1600 \quad (2)$$
$$26y = 800$$
$$y = \frac{400}{13}, \quad \text{or } 30\frac{10}{13}$$

Substitute $\frac{400}{13}$ for y in (1) and solve for x.

$$x + \frac{400}{13} = 200$$
$$x = \frac{2200}{13}, \text{ or } 169\frac{3}{13}$$

Check. The total amount of the mixture is $\frac{2200}{13}\text{ lb} + \frac{400}{13}\text{ lb} = \frac{2600}{13}\text{ lb} = 200\text{ lb}$. The amount of milk fat in the mixture is $0.04\left(\frac{2200}{13}\text{ lb}\right) + 0.3\left(\frac{400}{13}\text{ lb}\right) = \frac{88}{13}\text{ lb} + \frac{120}{13}\text{ lb} = \frac{208}{13}\text{ lb} = 16\text{ lb}$. The answer checks.

State. $169\frac{3}{13}$ lb of whole milk and $30\frac{10}{13}$ lb of cream should be mixed.

37. ***Familiarize.*** We first make a drawing.

Slow train,
d kilometers, 75 km/h $(t+2)$ hr

Fast train,
d kilometers, 125 km/h t hr

From the drawing we see that the distances are the same. Now complete the chart.

$d = r \cdot t$

	Distance	Rate	Time	
Slow train	d	75	$t+2$	$\to d = 75(t+2)$
Fast train	d	125	t	$\to d = 125t$

Translate. Using $d = rt$ in each row of the table, we get a system of equations:

$$d = 75(t+2),$$
$$d = 125t$$

Carry out. We solve the system of equations.

$$125t = 75(t+2) \quad \text{Using substitution}$$
$$125t = 75t + 150$$
$$50t = 150$$
$$t = 3$$

Then $d = 125t = 125 \cdot 3 = 375$

Check. At 125 km/h, in 3 hr the fast train will travel $125 \cdot 3 = 375$ km. At 75 km/h, in 3 + 2, or 5 hr the slow train will travel $75 \cdot 5 = 375$ km. The numbers check.

State. The trains will meet 375 km from the station.

39. ***Familiarize.*** We first make a drawing. Let d = the distance and r = the speed of the canoe in still water. Then when the canoe travels downstream, its speed is $r + 6$, and its speed upstream is $r - 6$. From the drawing we see that the distances are the same.

Downstream, 6 mph current
d mi, $r + 6$, 4 hr

Upstream, 6 mph current
d mi, $r - 6$, 10 hr

Organize the information in a table.

	Distance	Rate	Time
Downstream	d	$r+6$	4
Upstream	d	$r-6$	10

Translate. Using $d = rt$ in each row of the table, we get a system of equations:

$$d = 4(r+6), \qquad d = 4r + 24,$$
$$\text{or}$$
$$d = 10(r-6) \qquad d = 10r - 60$$

Carry out. Solve the system of equations.

$$4r + 24 = 10r - 60$$
$$24 = 6r - 60$$
$$84 = 6r$$
$$14 = r$$

Check. When $r = 14$, then $r + 6 = 14 + 6 = 20$, and the distance traveled in 4 hours is $4 \cdot 20 = 80$ km. Also $r - 6 = 14 - 6 = 8$, and the distance traveled in 10 hours is $10 \cdot 8 = 80$ km. The answer checks.

State. The speed of the canoe in still water is 14 km/h.

41. ***Familiarize.*** We make a drawing. Note that the plane's speed traveling toward London is $360 + 50$, or 410 mph, and the speed traveling toward New York City is $360 - 50$, or 310 mph. Also, when the plane is d mi from New York City, it is $3458 - d$ mi from London.

New York City 310 mph t hours — t hours London 410 mph

3458 mi

d — 3458 mi $- d$

Organize the information in a table.

	Distance	Rate	Time
Toward NYC	d	310	t
Toward London	$3458 - d$	410	t

Translate. Using $d = rt$ in each row of the table, we get a system of equations:

$$d = 310t, \quad (1)$$
$$3458 - d = 410t \quad (2)$$

Carry out. We solve the system of equations.

$$3458 - 310t = 410t \quad \text{Using substitution}$$
$$3458 = 720t$$
$$4.8028 \approx t$$

Substitute 4.8028 for t in (1).

$$d \approx 310(4.8028) \approx 1489$$

Check. If the plane is 1489 mi from New York City, it can return to New York City, flying at 310 mph, in $1489/310 \approx 4.8$ hr . If the plane is $3458 - 1489$, or 1969 mi from London, it can fly to London, traveling at 410 mph, in $1969/410 \approx 4.8$ hr . Since the times are the same, the answer checks.

State. The point of no return is about 1489 mi from New York City.

43. ***Familiarize.*** Let $l =$ the length, in feet, and $w =$ the width, in feet. Recall that the formula for the perimeter P of a rectangle with length l and width w is $P = 2l + 2w$.

Translate.

The perimeter is 860 ft.

$$2l + 2w = 860$$

The length is 100 ft. more than the width.

$$l = 100 + w$$

We have translated to a system of equations:

$$2l + 2w = 860, \quad (1)$$
$$l = 100 + w \quad (2)$$

Carry out. We use the substitution method to solve the system of equations.

Substitute 100 + w for l in (1) and solve for w.

$$2(100 + w) + 2w = 860$$
$$200 + 2w + 2w = 860$$
$$200 + 4w = 860$$
$$4w = 660$$
$$w = 165$$

Now substitute 165 for w in (2).

$$l = 100 + 165 = 265$$

Check. The perimeter is $2 \cdot 265 \text{ ft} + 2 \cdot 165 \text{ ft} = 530 \text{ ft} + 330 \text{ ft} = 860 \text{ ft}$. The length, 265 ft, is 100 ft more than the width, 165 ft. The answer checks.

State. The length is 265 ft, and the width is 165 ft.

45. ***Familiarize.*** Let $x =$ the number of Wii game machines and $y =$ the number of PlayStation consoles.

Translate.

Number of Wiis is three times number of PlayStations.

$$x = 3 \times y$$

Total number sold is 4.84 million.

$$x + y = 4.84$$

We have a system of equations:

$$x = 3y \quad (1)$$
$$x + y = 4.84 \quad (2)$$

Carry out. We use the substitution method to solve the system of equations.

Substitute $3y$ for x in Equation (2) and solve for y.

$$3y + y = 4.84$$
$$4y = 4.84$$
$$y = 1.21$$

Now substitute 1.21 for y in Equation (1).

$$x = 3(1.21) = 3.63$$

Check. The total number of machines sold is 3.63 million + 1.21 million = 4.84 million. The number of Wii game machines sold, 3.63 million is 3 times 1.21 million, the number of PlayStation consoles sold. The answer checks.

State. 3.63 million Wii game machines and 1.21 million PlayStation consoles were sold.

47. ***Familiarize.*** Let $x =$ the number of unlimited 1 DVD plans for \$8.99 and $y =$ the number of limit of 2 DVD per month plans for \$4.99.

Translate.

Total number of plans is 250.

$$x + y = 250$$

Total value of plans is 1975.50.

$$8.99x + 4.99y = 1975.50$$

After clearing decimals, we have a system of equations:

$$x + y = 250 \quad (1)$$
$$899x + 499y = 197,550 \quad (2)$$

Carry out. We use the elimination method to solve the system of equations.

$$-499x - 499y = -124,750 \quad \text{Multiplying (1) by } -499$$
$$899x + 499y = 197,550 \quad (2)$$
$$400x = 72,800$$
$$x = 182$$

Substitute 182 for x in Equation (1) and solve for y.

$$182+y=250$$
$$y=68$$

Check. The total number of plans is 182 + 68, or 250. The total value of the plans is $8.99(182)+4.99(68)$ $=1636.18+339.32=1975.50$. The answer checks.

State. 182 of the \$8.99 plans and 68 of the \$4.99 plans were purchased.

49. ***Familiarize.*** The change from the \$9.25 purchase is \$20 – \$9.25, or \$10.75. Let $x=$ the number of quarters and $y=$ the number of fifty-cent pieces. The total value of the quarters, in dollars, is $0.25x$ and the total value of the fifty-cent pieces, in dollars, is $0.50y$.

Translate.

The total number of coins is 30.

$$x+y = 30$$

The total value of the coins is \$10.75.

$$0.25x+0.50y = 10.75$$

After clearing decimals we have the following system of equations:

$$x+y=30, \quad (1)$$
$$25x+50y=1075 \quad (2)$$

Carry out. We use the elimination method to solve the system of equations.

$$-25x-25y=-750 \quad \text{Multiplying (1) by } -25$$
$$25x+50y=1075$$
$$25y=325$$
$$y=13$$

Substitute 13 for y in (1) and solve for x.

$$x+13=30$$
$$x=17$$

Check. The total number of coins is 17 + 13, or 30. The total value of the coins is $\$0.25(17)+\$0.50(13)$ $=\$4.25+\$6.50=\$10.75$. The answer checks.

State. There were 17 quarters and 13 fifty-cent pieces.

51. *Writing Exercise.* Both can be considered mixture problems or total value problems. In each system the first equation pertains to a total amount and the second pertains to a total value.

53. $2x-3y-z=2(5)-3(2)-3$
$=10-6-3=1$

55. $x+y+2z=1+(-4)+2(-5)$
$=1-4-10=-13$

57. $a-2b-3c=-2-2(3)-3(-5)$
$=-2-6+15=7$

59. *Writing Exercise.* We might not have found l after first finding a. However, in order to check, it is necessary to find l, so the method of solving need not change.

61. ***Familiarize.*** Let $x=$ the number of reams of 0% post-consumer fiber paper purchased and $y=$ the number of reams of 30% post-consumer fiber paper.

Translate. We organize the information in a table.

	0% post-consumer	30% post-consumer	Total
Reams purchased	x	y	60
Percent of post-consumer fiber	0%	30%	20%
Total post-consumer fiber	$0\cdot x$, or 0	$0.3y$	$0.2(60)$, or 12

We get one equation from the "Reams purchased" row of the table:

$$x+y=60$$

The last row of the table yields a second equation:

$$0x+0.3y=12\text{, or } 0.3y=12$$

After clearing the decimal we have the problem translated to a system of equations.

$$x+y=60, \quad (1)$$
$$3y=120 \quad (2)$$

Carry out. First we solve (2) for y.

$$3y=120$$
$$y=40$$

Now substitute 40 for y in (1) and solve for x.

$$x+40=60$$
$$x=20$$

Check. The total purchase is 20 + 40, or 60 reams. The post-consumer fiber can be expressed as $0\cdot 20+0.3(40)=12$. The answer checks.

State. 20 reams of 0% post-consumer fiber paper and 40 reams of 30% post-consumer fiber paper would have to be purchased.

63. ***Familiarize.*** Let $x=$ the number of ounces of pure silver.

Translate. We organize the information in a table.

	Coin	Pure silver	New Mixture
Number of ounces	32	x	$32+x$
Percent silver	90%	100%	92.5%
Amount of silver	$0.9(32)$	$1\cdot x$	$0.925(32+x)$

The last row gives the equation

$0.9(32)+x=0.925(32+x).$

Carry out. We solve the equation.

$$28.8+x=29.6+0.925x$$
$$0.075x=0.8$$
$$x=\frac{32}{3}=10\frac{2}{3}$$

Check. 90% of 32, or 28.8 + 100% of $10\frac{2}{3}$, or $10\frac{2}{3}$ is $39\frac{7}{15}$ and 92.5% of $\left(32+10\frac{2}{3}\right)$ is $0.925\left(42\frac{2}{3}\right)$, or $39\frac{7}{15}$.

The answer checks.

State. $10\frac{2}{3}$ ounces of pure silver should be added.

65. ***Familiarize.*** Let $x=$ the number of 3-volume sets and $y=$ the number of single volumes. Note that $3x$ is the number of volumes in the set.

Translate.

Total number of volumes is 51.

$$3x+y=51$$

Total value sales is 1641.

$$88x+39y=1641$$

We have a system of equations:

$$3x+y=51 \quad (1)$$
$$88x+39y=1641 \quad (2)$$

Carry out. We use the elimination method to solve the system of equations.

$$-117x-39y=-1989 \quad \text{Multiplying (1) by } -39$$
$$88x+39y=1641 \quad (2)$$
$$-29x=-348$$
$$x=12$$

Although the problem asks for the number of 3-volume sets, we will also find y in order to check. Substitute 12 for x in Equation (1) and solve for y.

$$3(12)+y=51$$
$$y=15$$

Check. The total number of books is $3(12)+15=36+15=51$. The total sales is $\$88(12)+\$39(15)=\$1056+\$585=\$1641$. The answer checks.

State. 12 3-volume sets were ordered.

67. ***Familiarize.*** Let $x=$ the number of gallons of pure brown and $y=$ the number of gallons of neutral stain that should be added to the original 0.5 gal. Note that a total of 1 gal of stain needs to be added to bring the amount of stain up to 1.5 gal. The original 0.5 gal of stain contains 20%(0.5 gal), or $0.2(0.5\text{ gal})=0.1$ gal of brown stain. The final solution contains 60%(1.5 gal), or $0.6(1.5\text{ gal})=0.9$ gal of brown stain. This is composed of the original 0.1 gal and the x gal that are added.

Translate.

The amount of stain added was 1 gal.

$$x+y=1$$

The amount of brown stain in the final solution is 0.9 gal.

$$0.1+x=0.9$$

We have a system of equations.

$$x+y=1, \quad (1)$$
$$0.1+x=0.9 \quad (2)$$

Carry out. First we solve (2) for x.

$$0.1+x=0.9$$
$$x=0.8$$

Then substitute 0.8 for x in (1) and solve for y.

$$0.8+y=1$$
$$y=0.2$$

Check. Total amount of stain: 0.5 + 0.8 + 0.2 = 1.5 gal

Total amount of brown stain: 0.1 + 0.8 = 0.9 gal

Total amount of neutral stain:

$0.8(0.5)+0.2=0.4+0.2=0.6\text{ gal}=0.4(1.5\text{ gal})$

The answer checks.

State. 0.8 gal of pure brown and 0.2 gal of neutral stain should be added.

69. ***Familiarize.*** Let x and y represent the number of city miles and highway miles that were driven, respectively. Then in city driving, $\frac{x}{18}$ gallons of gasoline are used; in highway driving, $\frac{y}{24}$ gallons are used.

Translate. We organize the information in a table.

Type of driving	City	Highway	Total
Number of miles	x	y	465
Gallons of gasoline used	$\frac{x}{18}$	$\frac{y}{24}$	23

The first row of the table gives us one equation:

$$x+y=465$$

The second row gives us another equation:

$$\frac{x}{18}+\frac{y}{24}=23$$

After clearing fractions, we have the following system of equations:

$x+y=465,\quad (1)$
$24x+18y=9936\quad (2)$

Carry out. We solve the system of equations using the elimination method.

$$\begin{array}{rl} -18x-18y=-8370 & \text{Multiplying (1) by } -18 \\ 24x+18y=9936 & (2) \\ \hline 6x\qquad\ =1566 & \\ x=261 & \end{array}$$

Now substitute 261 for x in Equation (1) and solve for y.

$$261+y=465$$
$$y=204$$

Check. The total mileage is 261 + 204, or 465. In 216 city miles, 261 / 18 , or 14.5 gal of gasoline are used; in 204 highway miles, 204 / 24 , or 8.5 gal are used. Then a total of 14.5 + 8.5, or 23 gal of gasoline are used. The answer checks.

State. 261 miles were driven in the city, and 204 miles were driven on the highway.

71. The 1.5 gal mixture contains 0.1 + x gal of pure brown stain. (See Exercise 67.). Thus, the function $P(x)=\dfrac{0.1+x}{1.5}$ gives the percentage of brown in the mixture as a decimal quantity. Using the Intersect feature, we confirm that when x = 0.8, then $P(x)=0.6$ or 60%.

Exercise Set 8.4

1. The equation is equivalent to one in the form $Ax + By + Cz = D$, so the statement is true.

3. False; see page 541.

5. True; see Example 6.

7. Substitute (2, –1, –2) into the three equations, using alphabetical order.

$$\begin{array}{r|l} \multicolumn{2}{c}{x+y-2z=5} \\ \hline 2+(-1)-2(-2) & 5 \\ 2-1+4 & \\ \multicolumn{2}{c}{5\overset{?}{=}5 \quad \text{TRUE}} \end{array}$$

$$\begin{array}{r|l} \multicolumn{2}{c}{2x-y-z=7} \\ \hline 2\cdot 2-(-1)-(-2) & 7 \\ 4+1+2 & \\ \multicolumn{2}{c}{7\overset{?}{=}7 \quad \text{TRUE}} \end{array}$$

$$\begin{array}{r|l} \multicolumn{2}{c}{-x-2y-3z=6} \\ \hline -2-2(-1)-3(-2) & 6 \\ -2+2+6 & \\ \multicolumn{2}{c}{6\overset{?}{=}6 \quad \text{TRUE}} \end{array}$$

Since the triple (2, –1, –2) is true in all three equations, it is a solution of the system.

9.
$$\begin{array}{rl} x-y-z=0, & (1) \\ 2x-3y+2z=7, & (2) \\ -x+2y+z=1 & (3) \end{array}$$

1., 2. The equations are already in standard form with no fractions or decimals.

3. Use Equations (1) and (2) to eliminate x.

$$\begin{array}{rl} -2x+2y+2z=0 & \text{Multiplying (1) by } -2 \\ 2x-3y+2z=7 & (2) \\ \hline -y+4z=7 & \text{(4) Adding} \end{array}$$

4. Use a different pair of equations and eliminate x.

$$\begin{array}{rl} x-y-z=0 & (1) \\ -x+2y+z=1 & (3) \\ \hline y\qquad\ =1 & \text{Adding} \end{array}$$

5. When we used Equation (1) and (3) to eliminate x, we also eliminated z and found y = 1. Substitute 1 for y in Equation (4) to find z.

$$\begin{array}{rl} -1+4z=7 & \text{Substituting 1 for} y \text{ in (4)} \\ 4z=8 & \\ z=2 & \end{array}$$

6. Substitute in one of the original equations to find x.

$$x-1-2=0$$
$$x=3$$

We obtain (3, 1, 2). This checks, so it is the solution.

11.
$$\begin{array}{rl} x-y-z=1, & (1) \\ 2x+y+2z=4, & (2) \\ x+y+3z=5 & (3) \end{array}$$

1., 2. The equations are already in standard form with no fractions or decimals.

3. Use Equations (1) and (2) to eliminate y.

$$\begin{array}{rl} x-y\ -z=1 & (1) \\ 2x+y+2z=4 & (2) \\ \hline 3x\qquad +z=5 & \text{(4) Adding} \end{array}$$

4. Use a different pair of equations and eliminate y.

$$\begin{array}{rl} x-y\ -z=1 & (1) \\ x+y+3z=5 & (3) \\ \hline 2x\qquad +2z=6 & \text{(5) Adding} \end{array}$$

5. Now solve the system of Equations (4) and (5).

$$\begin{array}{rl} 3x+z=5 & (4) \\ 2x+2z=6 & (5) \end{array}$$

$$\begin{array}{rl} -6x-2z=-10 & \text{Multiplying (4) by } -2 \\ 2x+2z=6 & (5) \\ \hline -4x\qquad =-4 & \text{Adding} \\ x=1 & \end{array}$$

$3 \cdot 1+z=5$ Substituting 1 for x in (4)
$3+z=5$
$z=2$

6. Substitute in one of the original equations to find y.

$1+y+3 \cdot 2=5$ Substituting 1 for x and 2 for z in (3)
$1+y+6=5$
$y+7=5$
$y=-2$

We obtain (1, –2, 2). This checks, so it is the solution.

13. $3x+4y-3z=4,$ (1)
$5x-y+2z=3,$ (2)
$x+2y-z=-2$ (3)

1., 2. The equations are already in standard form with no fractions or decimals.

3., 4. We eliminate y from two different pairs of equations.

$3x+4y-3z=4$ (1)
$20x-4y+8z=12$ Multiplying (2) by 4
$23x+5z=16$ (4)

$10x-2y+4z=6$ Multiplying (2) by 2
$x+2y-z=-2$ (3)
$11x+3z=4$ (5)

5. Now solve the system of Equations (4) and (5).

$23x+5z=16$ (4)
$11x+3z=4$ (5)

$-69x-15z=-48$ Multiplying (4) by -3
$55x+15z=20$ Multiplying (5) by 5
$-14x=-28$ Adding
$x=2$

$11 \cdot 2+3z=4$ Substituting 2 for x in (5)
$3z=-18$
$z=-6$

6. Substitute in one of the original equations to find y.

$3 \cdot 2+4y-3(-6)=4$ Substituting 2 for x and -6 for z in (1)
$6+4y+18=4$
$4y=-20$
$y=-5$

We obtain (2, –5, –6). This checks, so it is the solution.

15. $x+y+z=0,$ (1)
$2x+3y+2z=-3,$ (2)
$-x-2y-z=1$ (3)

1., 2. The equations are already in standard form with no fractions or decimals.

3., 4. We eliminate x from two different pairs of equations.

$-2x-2y-2z=0$ Multiplying (1) by -2
$2x+3y+2z=-3$ (2)
$y=-3$

We eliminated not only x but also z and found that $y=-3$.

5., 6. Substitute –3 for y in two of the original equations to produce a system of two equations in two variables. Then solve this system.

$x-3+z=0$ Substituting in (1)
$-x-2(-3)-z=1$ Substituting in (3)

Simplifying we have

$x+z=3$
$-x-z=-5$
$0=-2$

We get a false equation, so there is no solution.

17. $2x-3y-z=-9,$ (1)
$2x+5y+z=1,$ (2)
$x-y+z=3$ (3)

1., 2. The equations are already in standard form with no fractions or decimals.

3., 4. We eliminate z from two different pairs of equations.

$2x-3y-z=-9$ (1)
$2x+5y+z=1$ (2)
$4x+2y=-8$ (4)

$2x-3y-z=-9$ (1)
$x-y+z=3$ (2)
$3x-4y=-6$ (5)

5. Now solve the system of Equations (4) and (5).

$4x+2y=-8$ (4)
$3x-4y=-6$ (5)

$8x+4y=-16$ Multiplying (4) by 2
$3x-4y=-6$ (5)
$11x=-22$ Adding
$x=-2$

$4(-2)+2y=-8$ Substituting -2 for x in (4)
$-8+2y=-8$
$y=0$

6. Substitute in one of the original equations to find z.

$2(-2)+5 \cdot 0+z=1$ Substituting -2 for x and 0 for y in (1)
$-4+z=1$
$z=5$

We obtain (–2, 0, 5). This checks, so it is the solution.

19. $a+b+c=5,$ (1)
$2a+3b-c=2,$ (2)
$2a+3b-2c=4$ (3)

1., 2. The equations are already in standard form with no fractions or decimals.

3., 4. We eliminate a from two different pairs of equations.

$-2a-2b-2c=-10$ Multiplying (1) by -2
$2a+3b-c=2$ (2)
$b-3c=-8$ (4)

$$\begin{array}{rl} -2a-3b+c=-2 & \text{Multiplying (1) by } -1 \\ 2a+3b-2c=4 & (3) \\ \hline -c=2 & \\ c=-2 & \end{array}$$

We eliminate not only a, but also b and found $c=-2$.

5. Substitute –2 for c in Equation (4) to find b.

$$\begin{array}{rl} b-3(-2)=-8 & \text{Substituting } -2 \text{ for } c \text{ in (4)} \\ b+6=-8 & \\ b=-14 & \end{array}$$

6. Substitute in one of the original equations to find a.

$$\begin{array}{rl} a-14-2=5 & \text{Substituting } -2 \text{ for } c \text{ and} \\ & -14 \text{ for } b \text{ in (1)} \\ a-16=5 & \\ a=21 & \end{array}$$

We obtain (21, –14, –2). This checks, so it is the solution.

21. $$\begin{array}{rl} -2x+8y+2z=4, & (1) \\ x+6y+3z=4, & (2) \\ 3x-2y+z=0 & (3) \end{array}$$

1. , 2. The equations are already in standard form with no fractions or decimals.

3., 4. We eliminate z from two different pairs of equations.

$$\begin{array}{rl} -2x+8y+2z=4 & (1) \\ -6x+4y-2z=0 & \text{Multiplying (3) by } -2 \\ \hline -8x+12y\quad=4 & (4) \end{array}$$

$$\begin{array}{rl} x+6y+3z=4 & (2) \\ -9x+6y-3z=0 & \text{Multiplying (3) by } -3 \\ \hline -8x+12y\quad=4 & (5) \end{array}$$

5. Now solve the system of Equations (4) and (5).

$$\begin{array}{rl} -8x+12y=4 & (4) \\ -8x+12y=4 & (5) \end{array}$$

$$\begin{array}{rl} -8x+12y=4 & (4) \\ 8x-12y=-4 & \text{Multiplying (5) by } -1 \\ \hline 0=0 & (6) \end{array}$$

Equation (6) indicates that Equations (1), (2), and (3) are dependent. (Note that if Equation (1) is subtracted from Equation (2), the result is Equation (3).) We could also have concluded that the equations are dependent by observing that Equations (4) and (5) are identical.

23. $$\begin{array}{rl} 2u-4v-w=8, & (1) \\ 3u+2v+w=6, & (2) \\ 5u-2v+3w=2 & (3) \end{array}$$

1., 2. The equations are already in standard form with no fractions or decimals.

3., 4. We eliminate w from two different pairs of equations.

$$\begin{array}{rl} 2u-4v-w=8 & (1) \\ 3u+2v+w=6 & (2) \\ \hline 5u-2v\quad=14 & (4) \end{array}$$

$$\begin{array}{rl} 6u-12v-3w=24 & \text{Multiplying (1) by 3} \\ 5u-2v+3w=2 & (3) \\ \hline 11u-14v\quad=26 & (5) \end{array}$$

5. Now solve the system of Equations (4) and (5).

$$\begin{array}{rl} 5u-2v=14 & (4) \\ 11u-14v=26 & (5) \end{array}$$

$$\begin{array}{rl} -35u+14v=-98 & \text{Multiplying (4) by } -7 \\ 11u-14v=26 & (5) \\ \hline -24u\quad=-72 & \\ u=3 & \end{array}$$

$$\begin{array}{rl} 5\cdot 3-2v=14 & \text{Substituting 3 for } u \text{ in (4)} \\ 15-2v=14 & \\ -2v=-1 & \\ v=\frac{1}{2} & \end{array}$$

6. Substitute in one of the original equations to find v.

$$\begin{array}{rl} 2\cdot 3-4\left(\frac{1}{2}\right)-2=8 & \text{Substituting 3 for } u \text{ and } \frac{1}{2} \\ & \text{for } v \text{ in (1)} \\ 6-2-w=8 & \\ w=-4 & \end{array}$$

We obtain $\left(3,\ \frac{1}{2},\ -4\right)$. This checks, so it is the solution.

25. $$\begin{array}{l} r+\frac{3}{2}s+6t=2, \\ 2r-3s+3t=0.5, \\ r+s+t=1 \end{array}$$

1. All equations are already in standard form.

2. Multiply the first equation by 2 to clear the fraction. Also, multiply the second equation by 10 to clear the decimal.

$$\begin{array}{rl} 2r+3s+12t=4, & (1) \\ 20r-30s+30t=5, & (2) \\ r+s+t=1 & (3) \end{array}$$

3., 4. We eliminate s from two different pairs of equations.

$$\begin{array}{rl} 20r+30s+120t=40 & \text{Multiplying (1) by 10} \\ 20r-30s+30t=5 & (2) \\ \hline 40r\quad+150t=45 & \text{(4) Adding} \end{array}$$

$$\begin{array}{rl} 20r+30s+120t=40 & \text{Multiplying (1) by 10} \\ -30r-30s-30t=-30 & \text{Multiplying (3) by } -30 \\ \hline -10r\quad+90t=10 & \text{(5) Adding} \end{array}$$

5. Solve the system of Equations (4) and (5).

$$\begin{array}{rl} 40r+150t=45, & (4) \\ -10r+90t=10, & (5) \end{array}$$

$$\begin{array}{rl} 40r+150t=45 & (4) \\ -40r+360t=40 & \text{Multiplying (5) by4} \\ \hline 510t=85 & \\ t=\frac{85}{510} & \\ t=\frac{1}{6} & \end{array}$$

$40r + 150\left(\frac{1}{6}\right) = 45$ Substituting $\frac{1}{6}$ for t in (4)

$40r + 25 = 45$

$40r = 20$

$r = \frac{1}{2}$

6. Substitute in one of the original equations to find s.

$\frac{1}{2} + s + \frac{1}{6} = 1$ Substituting $\frac{1}{2}$ for r and $\frac{1}{6}$ for t in (3)

$s + \frac{2}{3} = 1$

$s = \frac{1}{3}$

We obtain $\left(\frac{1}{2}, \frac{1}{3}, \frac{1}{6}\right)$. This checks, so it is the solution.

27. $4a + 9b = 8$, (1)

$8a + 6c = -1$, (2)

$6b + 6c = -1$ (3)

1., 2. The equations are already in standard form with no fractions or decimals.

3., 4. Note that there is no c in Equation (1). We will use Equations (2) and (3) to obtain another equation with no c-term.

$8a + 6c = -1$ (2)

$-6b - 6c = 1$ Multiplying (3) by -1

$8a - 6b = 0$ (4) Adding

5. Now solve the system of Equations (1) and (4).

$-8a - 18b = -16$ Multiplying (1) by -2

$8a - 6b = 0$

$-24b = -16$

$b = \frac{2}{3}$

$8a - 6\left(\frac{2}{3}\right) = 0$ Substituting $\frac{2}{3}$ for b in (4)

$8a - 4 = 0$

$8a = 4$

$a = \frac{1}{2}$

6. Substitute in Equation (2) or (3) to find c.

$8\left(\frac{1}{2}\right) + 6c = -1$ Substituting $\frac{1}{2}$ for a in (2)

$4 + 6c = -1$

$6c = -5$

$c = -\frac{5}{6}$

We obtain $\left(\frac{1}{2}, \frac{2}{3}, -\frac{5}{6}\right)$. This checks, so it is the solution.

29. $x + y + z = 57$, (1)

$-2x + y = 3$, (2)

$x - z = 6$ (3)

1., 2. The equations are already in standard form with no fractions or decimals.

3., 4. Note that there is no z in Equation (2). We will use Equations (1) and (3) to obtain another equation with no z-term.

$x + y + z = 57$ (1)

$x - z = 6$ (3)

$2x + y = 63$ (4)

5. Now solve the system of Equations (2) and (4).

$-2x + y = 3$ (2)

$2x + y = 63$ (4)

$2y = 66$

$y = 33$

$2x + 33 = 63$ Substituting 33 for y in (4)

$2x = 30$

$x = 15$

6. Substitute in Equation (1) or (3) to find z.

$15 - z = 6$ Substituting 15 for x in (3)

$9 = z$

We obtain (15, 33, 9). This checks, so it is the solution.

31. $a - 3c = 6$, (1)

$b + 2c = 2$, (2)

$7a - 3b - 5c = 14$ (3)

1., 2. The equations are already in standard form with no fractions or decimals.

3., 4. Note that there is no b in Equation (1). We will use Equations (2) and (3) to obtain another equation with no b-term.

$3b + 6c = 6$ Multiplying (2) by 3

$7a - 3b - 5c = 14$ (3)

$7a + c = 20$ (4)

5. Now solve the system of Equations (1) and (4).

$a - 3c = 6$ (1)

$7a + c = 20$ (4)

$a - 3c = 6$ (1)

$21a + 3c = 60$ Multiplying (4) by 3

$22a = 66$

$a = 3$

$3 - 3c = 6$ Substituting in (1)

$-3c = 3$

$c = -1$

6. Substitute in Equation (2) or (3) to find b.

$b + 2(-1) = 2$ Substituting in (2)

$b - 2 = 2$

$b = 4$

We obtain (3, 4, −1). This checks, so it is the solution.

33. $x + y + z = 83$, (1)

$y = 2x + 3$, (2)

$z = 40 + x$ (3)

Observe, from Equations (2) and (3), that we can substitute $2 + 3x$ for y and $40 + x$ for z in Equation (1) and solve for x.

$$\begin{aligned} x+y+z&=83 \\ x+(2x+3)+(40+x)&=83 \\ 4x+43&=83 \\ 4x&=40 \\ x&=10 \end{aligned}$$

Now substitute 10 for x in Equation (2).

$$y=2x+3=2\cdot10+3=20+3=23$$

Finally, substitute 10 for x in Equation (3).

$$z=40+x=40+10=50\,.$$

We obtain (10, 23, 50). This checks, so it is the solution.

35.
$$\begin{aligned} x\qquad +z&=0, \quad (1) \\ x+y+2z&=3, \quad (2) \\ y+z&=2 \quad\ (3) \end{aligned}$$

1., 2. The equations are already in standard form with no fractions or decimals.

3., 4. Note that there is no y in Equation (1). We use Equations (2) and (3) to obtain another equation with no y-term.

$$\begin{aligned} x+y+2z&=3 \qquad (2) \\ -y-z&=-2 \quad \text{Multiplying (3) by } -1 \\ \hline x\qquad +z&=1 \qquad (4)\ \text{Adding} \end{aligned}$$

5. Now solve the system of Equations (1) and (4).

$$\begin{aligned} x+z&=0 \qquad (1) \\ -x-z&=-1 \quad \text{Multiplying (4) by } -1 \\ \hline 0&=-1 \quad \text{Adding} \end{aligned}$$

We get a false equation, or contradiction. There is no solution.

37.
$$\begin{aligned} x+y+z&=1, \quad (1) \\ -x+2y+z&=2, \quad (2) \\ 2x-y\qquad &=-1 \quad (3) \end{aligned}$$

1., 2. The equations are already in standard form with no fractions or decimals.

3. Note that there is no z in Equation (3). We will use Equations (1) and (2) to eliminate z:

$$\begin{aligned} x+y+z&=1 \qquad (1) \\ x-2y-z&=-2 \quad \text{Multiplying (2) by } -1 \\ \hline 2x-y\qquad &=-1 \quad \text{Adding} \end{aligned}$$

Equations (3) and (4) are identical, so Equations (1), (2), and (3) are dependent. (We have seen that if Equation (2) is multiplied by –1 and added to Equation (1), the result is Equation (3).)

39. *Writing Exercise.* This approach will work, since any of the variables may be selected to be eliminated. Sometimes the coefficients of x may be larger, which will make the calculations a bit more difficult.

41. Let x and y represent the numbers; $x=\frac{1}{2}y$

43. Let x represent the first number,

Let $x + 1$ represent the second number,

Let $x + 2$ represent the third number.

$x + (x + 1) + (x + 2) = 100$

45. Let x, y, and z represent the numbers; $xy = 5z$

47. *Writing Exercise.* No; the graph of each equation is a plane and all three planes can only have no points in common, exactly one point in common, or infinitely many points in common.

49.
$$\begin{aligned} \frac{x+2}{3}-\frac{y+4}{2}+\frac{z+1}{6}&=0, \\ \frac{x-4}{3}+\frac{y+1}{4}-\frac{z-2}{2}&=-1, \\ \frac{x+1}{2}+\frac{y}{2}+\frac{z-1}{4}&=\frac{3}{4} \end{aligned}$$

1., 2. We clear fractions and write each equation in standard form.

To clear fractions, we multiply both sides of each equation by the LCM of its denominators. The LCM's are 6, 12, and 4, respectively.

$$\begin{aligned} 6\left(\frac{x+2}{3}-\frac{y+4}{2}+\frac{z+1}{6}\right)&=6\cdot0 \\ 2(x+2)-3(y+4)+(z+1)&=0 \\ 2x+4-3y-12+z+1&=0 \\ 2x-3y+z&=7 \end{aligned}$$

$$\begin{aligned} 12\left(\frac{x-4}{3}+\frac{y+1}{4}-\frac{z-2}{2}\right)&=12\cdot(-1) \\ 4(x-4)+3(y+1)-6(z-2)&=-12 \\ 4x-16+3y+3-6z+12&=-12 \\ 4x+3y-6z&=-11 \end{aligned}$$

$$\begin{aligned} 4\left(\frac{x+1}{2}+\frac{y}{2}+\frac{z-1}{4}\right)&=4\cdot\frac{3}{4} \\ 2(x+1)+2(y)+(z-1)&=3 \\ 2x+2+2y+z-1&=3 \\ 2x+2y+z&=2 \end{aligned}$$

The resulting system is

$$\begin{aligned} 2x-3y+z&=7, \quad (1) \\ 4x+3y-6z&=-11, \quad (2) \\ 2x+2y+z&=2 \quad (3) \end{aligned}$$

3., 4. We eliminate z from two different pairs of equations.

$$\begin{aligned} 12x-18y+6z&=42 \quad \text{Multiplying (1) by 6} \\ 4x+3y-6z&=-11 \quad (2) \\ \hline 16x-15y\qquad &=31 \quad (4)\ \text{Adding} \end{aligned}$$

$$\begin{aligned} 2x-3y+z&=7 \quad (1) \\ -2x-2y-z&=-2 \quad \text{Multiplying (3) by } -1 \\ \hline -5y\qquad &=5 \quad (5)\ \text{Adding} \end{aligned}$$

5. Solve (5) for y: $-5y=5$

$$y=-1$$

Substitute –1 for y in (4):

$$\begin{aligned} 16x-15(-1)&=31\\ 16x+15&=31\\ 16x&=16\\ x&=1 \end{aligned}$$

6. Substitute 1 for x and –1 for y in (1):

$$\begin{aligned} 2\cdot1-3(-1)+z&=7\\ 5+z&=7\\ z&=2 \end{aligned}$$

We obtain (1, –1, 2). This checks, so it is the solution.

51.
$$\begin{aligned} w+x+y+z&=2, &(1)\\ w+2x+2y+4z&=1, &(2)\\ w-x+y+z&=6, &(3)\\ w-3x-y+z&=2 &(4) \end{aligned}$$

The equations are already in standard form with no fractions or decimals.

Start by eliminating w from three different pairs of equations.

$$\begin{array}{ll} w+x+y+z=2 & (1)\\ \underline{-w-2x-2y-4z=-1} & \text{Multiplying (2) by } -1\\ -x-y-3z=1 & \text{(5) Adding} \end{array}$$

$$\begin{array}{ll} w+x+y+z=2 & (1)\\ \underline{-w+x-y-z=-6} & \text{Multiplying (3) by } -1\\ 2x=-4 & \text{(6) Adding} \end{array}$$

$$\begin{array}{ll} w+x+y+z=2 & (1)\\ \underline{-w+3x+y-z=-2} & \text{Multiplying (4) by } -1\\ 4x+2y=0 & \text{(7) Adding} \end{array}$$

We can solve (6) for x:

$$\begin{aligned} 2x&=-4\\ x&=-2 \end{aligned}$$

Substitute –2 for x in (7):

$$\begin{aligned} 4(-2)+2y&=0\\ -8+2y&=0\\ 2y&=8\\ y&=4 \end{aligned}$$

Substitute –2 for x and 4 for y in (5):

$$\begin{aligned} -(-2)-4-3z&=1\\ -2-3z&=1\\ -3z&=3\\ z&=-1 \end{aligned}$$

Substitute –2 for x, 4 for y, and –1 for z in (1):

$$\begin{aligned} w-2+4-1&=2\\ w+1&=2\\ w&=1 \end{aligned}$$

We obtain (1, –2, 4, –1). This checks, so it is the solution.

53.
$$\begin{aligned} \frac{2}{x}-\frac{1}{y}-\frac{3}{z}&=-1,\\ \frac{2}{x}-\frac{1}{y}+\frac{1}{z}&=-9,\\ \frac{1}{x}+\frac{2}{y}-\frac{4}{z}&=17 \end{aligned}$$

Let u represent $\frac{1}{x}$, v represent $\frac{1}{y}$, and w represent $\frac{1}{z}$.

Substituting, we have

$$\begin{aligned} 2u-v-3w&=-1, &(1)\\ 2u-v+w&=-9, &(2)\\ u+2v-4w&=17 &(3) \end{aligned}$$

1., 2. The equations in u, v, and w are in standard form with no fractions or decimals.

3., 4. We eliminate v from two different pairs of equations.

$$\begin{array}{ll} 2u-v-3w=-1 & (1)\\ \underline{-2u+v-w=9} & \text{Multiplying (2) by } -1\\ -4w=8 & \text{(4) Adding} \end{array}$$

$$\begin{array}{ll} 4u-2v-6w=-2 & \text{Multiplying (1) by 2}\\ \underline{u+2v-4w=17} & (3)\\ 5u-10w=15 & \text{(5) Adding} \end{array}$$

5. We can solve (4) for w:

$$\begin{aligned} -4w&=8\\ w&=-2 \end{aligned}$$

Substitute –2 for w in (5):

$$\begin{aligned} 5u-10(-2)&=15\\ 5u+20&=15\\ 5u&=-5\\ u&=-1 \end{aligned}$$

6. Substitute –1 for u and –2 for w in (1):

$$\begin{aligned} 2(-1)-v-3(-2)&=-1\\ -v+4&=-1\\ -v&=-5\\ v&=5 \end{aligned}$$

Solve for x, y, and z. We substitute –1 for u, 5 for v and –2 for w.

$$\begin{array}{lll} u=\frac{1}{x} & v=\frac{1}{y} & w=\frac{1}{z}\\ -1=\frac{1}{x} & 5=\frac{1}{y} & -2=\frac{1}{z}\\ x=-1 & y=\frac{1}{5} & z=-\frac{1}{2} \end{array}$$

We obtain $\left(-1, \frac{1}{5}, -\frac{1}{2}\right)$. This checks, so it is the solution.

55.
$$\begin{aligned} 5x-6y+kz&=-5, &(1)\\ x+3y-2z&=2, &(2)\\ 2x-y+4z&=-1 &(3) \end{aligned}$$

Eliminate y from two different pairs of equations.

$$\begin{array}{ll} 5x-6y+kz=-5 & (1)\\ \underline{2x+6y-4z=4} & \text{Multiplying (2) by 2}\\ 7x+(k-4)z=-1 & (4) \end{array}$$

$$\begin{array}{ll} x+3y-2z=2 & (2) \\ 6x-3y+12z=-3 & \text{Multiplying (3) by 3} \\ \hline 7x\qquad +10z=-1 & (5) \end{array}$$

Solve the system of Equations (4) and (5).

$$\begin{array}{ll} 7x+(k-4)z=-1 & (4) \\ 7x+10z=-1 & (5) \end{array}$$

$$\begin{array}{ll} -7x-(k-4)z=1 & \text{Multiplying (4) by} -1 \\ 7x\qquad +10z=-1 & (5) \\ \hline (-k+14)z=0 & (6) \end{array}$$

The system is dependent for the value of k that makes Equation (6) true. This occurs when $-k+14$ is 0. We solve for k:

$$-k+14=0$$
$$14=k$$

57. $z=b-mx-ny$

Three solutions are (1, 1, 2), (3, 2, –6), and $\left(\frac{3}{2}, 1, 1\right)$.

We substitute for x, y, and z and then solve for b, m, and n.

$$2=b-m-n,$$
$$-6=b-3m-2n,$$
$$1=b-\tfrac{3}{2}m-n$$

1., 2. Write the equations in standard form. Also, clear the fraction in the last equation.

$$\begin{array}{ll} b-m-n=2, & (1) \\ b-3m-2n=-6, & (2) \\ 2b-3m-2n=2 & (3) \end{array}$$

3., 4. Eliminate b from two different pairs of equations.

$$\begin{array}{ll} b-m-n=2 & (1) \\ -b+3m+2n=6 & \text{Multiplying (2) by} -1 \\ \hline 2m+n=8 & \text{(4) Adding} \end{array}$$

$$\begin{array}{ll} -2b+2m+2n=-4 & \text{Multiplying (1) by} -2 \\ 2b-3m-2n=2 & (3) \\ \hline -m\qquad =-2 & \text{(5) Adding} \end{array}$$

5. We solve Equation (5) for m:

$$-m=-2$$
$$m=2$$

Substitute in Equation (4) and solve for n.

$$2\cdot 2+n=8$$
$$4+n=8$$
$$n=4$$

6. Substitute in one of the original equations to find b.

$$b-2-4=2 \quad \text{Substituting 2 for } m \text{ and 4 for } n \text{ in (1)}$$
$$b-6=2$$
$$b=8$$

The solution is (8, 2, 4), so the equation is $z=8-2x-4y$.

59. *Writing Exercise.*

$$\begin{array}{ll} x+2y-z=1, & (1) \\ -x-2y+z=3, & (2) \\ 2x+4y-2z=2 & (3) \end{array}$$

Kadi determined that Equations (1) and (3) are dependent, since multiplying Equation (1) by 2 gives Equation (3). Ahmed found there is no solution by adding Equations (1) and (2) resulted in a contradiction.

Exercise Set 8.5

1. ***Familiarize.*** Let x = the first number, y = the second number, and z = the third number.

Translate.

The sum of the three numbers is 85.
↓ ↓ ↓
$x+y+z = 85$

The second is seven more than the first.
↓ ↓ ↓ ↓ ↓
$y = 7 + x$

The third is two more than four times the second.
↓ ↓ ↓ ↓ ↓ ↓ ↓
$z = 2 + 4 \times y$

We have a system of equations:

$$\begin{array}{lcl} x+y+z=85, & \text{or} & x+y+z=85 \\ y=7+x & & -x+y\quad =7 \\ z=2+4y & & -4y+z=2 \end{array}$$

Carry out. Solving the system we get (8, 15, 62).

Check. The sum of the three numbers is 8 + 15 + 62, or 85. The second number 15, is 7 more than the first number 8. The third number, 62 is 2 more than four times the first number, 8. The numbers check.

State. The numbers are 8, 15, and 62.

3. ***Familiarize.*** Let x = the first number, y = the second number, and z = the third number.

Translate.

The sum of three numbers is 26.
↓ ↓ ↓
$x+y+z = 26$

Twice the first minus the second is the third less 2
↓ ↓ ↓ ↓ ↓ ↓ ↓
$2x - y = z - 2$

The third is the second minus 3 times the first.
↓ ↓ ↓ ↓ ↓
$z = y - 3x$

We now have a system of equations.

$$x+y+z=26, \quad \text{or} \quad x+y+z=26,$$
$$2x-y=z-2, \quad\quad 2x-y-z=-2,$$
$$z=y-3x \quad\quad 3x-y+z=0$$

Carry out. Solving the system we get (8, 21, –3).

Check. The sum of the numbers is $8+21-3$, or 26. Twice the first minus the second is $2\cdot8-21$, or –5, which is 2 less than the third. The second minus three times the first is $21-3\cdot8$, or –3, which is the third. The numbers check.

State. The numbers are 8, 21, and –3.

5. ***Familiarize.*** We first make a drawing.

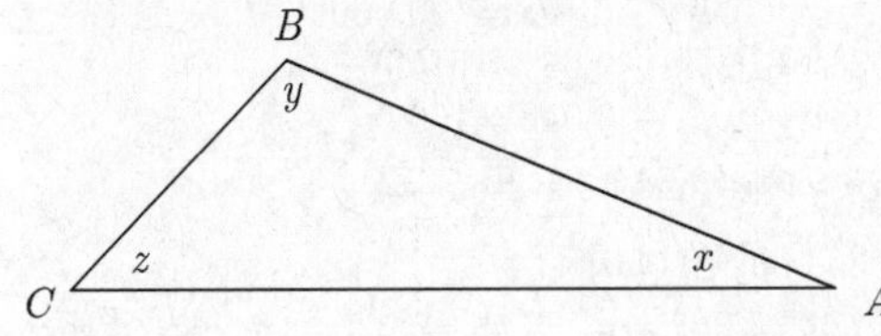

We let x, y, and z represent the measures of angles A, B, and C, respectively. The measures of the angles of a triangle add up to 180°.

Translate.

The sum of the measures is 180.
$$x+y+z = 180$$

The measure of angle B is three times the measure of angle A.
$$y = 3x$$

The measure of angle C is 20 more than the measure of angle A.
$$z = x+20$$

We now have a system of equations.

$$x+y+z=180,$$
$$y=3x,$$
$$z=x+20$$

Carry out. Solving the system we get (32, 96, 52).

Check. The sum of the measures is $32°+96°+52°$, or 180°. Three times the measure of angle A is $3\cdot32°$, or 96°, the measure of angle B. 20 more than the measure of angle A is $32°+20°$, or 52°, the measure of angle C. The numbers check.

State. The measures of angles A, B, and C are 32°, 96°, and 52°, respectively.

7. ***Familiarize.*** Let x, y and z represent the SAT critical reading score, mathematics score, and writing score, respectively.

Translate.

The average total score is 1511.
$$x+y+z = 1511$$

average math score is 13 more than reading score.
$$y = 13 + x$$

average writing score is reading score less 8.
$$z = x - 8$$

We have a system of equations:

$$x+y+z=1511, \quad \text{or} \quad x+y+z=1511$$
$$y=13+x \quad\quad -x+y=13$$
$$z=x-8 \quad\quad -x+z=-8$$

Carry out. Solving the system we get (502, 515, 494).

Check. The sum of the scores is 502 + 515 + 494, or 1511. The mathematics score, 515, is 13 more than the reading score, 502. The writing score, 494 is 8 less than the reading score, 502. The answer checks.

State. The average score for reading was 502, for mathematics 515, and for writing 494.

9. ***Familiarize.*** Let x, y, and z represent the number of grams of fiber in 1 bran muffin, 1 banana, and a 1-cup serving of Wheaties, respectively.

Translate.

Two bran muffins, 1 banana, and a 1-cup serving of Wheaties contain 9 g of fiber, so we have

$$2x+y+z=9.$$

One bran muffin, 2 bananas, and a 1-cup serving of Wheaties contain 10.5 g of fiber, so we have

$$x+2y+z=10.5$$

Two bran muffins and a 1-cup serving of Wheaties contain 6 g of fiber, so we have

$$2x+z=6.$$

We now have a system of equations.

$$2x+y+z=9,$$
$$x+2y+z=10.5,$$
$$2x+z=6$$

Carry out. Solving the system, we get (1.5, 3, 3).

Check. Two bran muffins, 1 banana, and a 1-cup serving of Wheaties contain $2(1.5)+3+3$, or 9 g of fiber. One bran muffin, 2 bananas, and a 1-cup serving of Wheaties contain $1.5+2\cdot3+3$, or 10.5 g of fiber. Two bran muffins and a 1-cup serving of Wheaties contain $2(1.5)+3$, or 6 g of fiber. The answer checks.

State. A bran muffin has 1.5 g of fiber, a banana has 3 g, and a 1-cup serving of Wheaties has 3 g.

11. Observe that the basic model plus tow package costs \$30,815 and when a rear backup camera is added the price rises to \$31,565. This tells us that the price of a rear backup camera is \$31,565 – \$30,815, or \$750. Now observe that the basic model and rear backup camera costs \$31,360 so the basic model costs \$31,360 – \$750, or \$30,610. Finally we know the tow package is \$30,815 – \$30,610, or \$205.

13. ***Familiarize.*** Let $x =$ the number of 12-oz cups, $y =$ the number of 16-oz cups, and $z =$ the number of 20-oz cups that Reba filled. Note that six 144-oz brewers contain $6 \cdot 144$, or 864 oz of coffee. Also, x 12-oz cups contain a total of $12x$ oz of coffee and bring in \$1.65x, y 16-oz cups contain $16y$ oz and bring in \$1.85y, and z 20-oz cups contain $20z$ oz and bring in \$1.95z.

Translate.

The total number of coffees served was 55.

$$x+y+z = 55$$

The total amount of coffee served was 864 oz.

$$12x+16y+20z=864$$

The total amount collected was \$99.65.

$$1.65x+1.85y+1.95z = 99.65$$

Now we have a system of equations.

$$\begin{aligned} x+y+z&=55,\\ 12x+16y+20z&=864,\\ 1.65x+1.85y+1.95z&=99.65 \end{aligned}$$

Carry out. Solving the system we get (17, 25, 13).

Check. The total number of coffees served was $17+25+13$, or 55. The total amount of coffee served was $12\cdot17+16\cdot25+20\cdot13=204+400+260=864$ oz. The total amount collected was \$1.65(17)+\$1.85(25)+\$1.95(13)=\$28.05+\$46.25+\$25.35 = \$85.90. The numbers check.

State. Reba filled 17 12-oz cups, 25 16-oz cups, and 13 20-oz cups.

15. ***Familiarize.*** Let $x =$ the amount of the loan at 8%, $y =$ the amount of the loan at 5%, and $z =$ the amount of the loan at 4%.

Translate.

Total of the three loans is \$120,000.

$$x+y+z = 120,000$$

Total interest due is \$5750.

$$0.08x+0.05y+0.04z = 5750$$

Mortgage interest is \$1600 more than bank loan interest.

$$0.04z = 1600 + 0.08x$$

We have a system of equations:

$$\begin{aligned} x \quad +y \quad +z&=120,000\\ 0.08x+0.05y+0.04z&=5750\\ -0.08x \qquad +0.04z&=1600 \end{aligned}$$

Carry out. Solving the system we get (15,000, 35,000, 70,000).

Check. The total of the three loans is \$15,000 + \$35,000 + \$70,000 = \$120,000. The total interest due is 0.08(\$15,000)+0.05(\$35,000)+0.04(\$70,000) =\$1200+\$1750+\$2800=\$5750. The interest from the mortgage, 0.04(\$70,000), or \$2800 is \$1600 more than the bank loan interest 0.08(\$15,000), or \$1200. The numbers check.

State. The bank loan was \$15,000, the small-business loan was \$35,000, and the mortgage loan was \$70,000.

17. ***Familiarize.*** Let $x =$ the price of 1 g of gold, $y =$ the the price of 1 g of silver, and $z =$ the price of 1 g of copper.

Translate.

Cost of 100 g of red gold is \$2265.40.

$$100(0.75x+0.05y+0.20z) = 2265.40$$

Cost of 100 g of yellow gold is \$2287.75.

$$100(0.75x+0.125y+0.125z) = 2287.75$$

Cost of 100 g of white gold is \$1312.50.

$$100(0.375x+0.625y) = 1312.50$$

We have a system of equations:

$$\begin{aligned} 75x \quad +5y \quad +20z&=2265.40\\ 750x+125y+125z&=22,877.50\\ 375x+625y \qquad &=13,125.00 \end{aligned}$$

Carry out. Solving the system we get (30, 3, 0.02).

Check. The cost of 100 g of red gold is 100[0.75(\$30)+0.05(\$3)+0.20(\$0.02)]

$= \$2250 + \$15 + \$0.40 = \$2265.40$. The cost of 100 g of yellow gold is $100[0.75(\$30) + 0.125(\$3) + 0.125(\$0.02)]$ $= \$2250 + \$37.50 + \$0.25 = \$2287.75$. The cost of 100 g of white gold is $100[0.375(\$30) + 0.625(\$3)]$ $= \$1125 + \$187.50 = \$1312.50$. The numbers check.

State. The price of 1 g of gold is \$30, of silver is \$3 and of copper is \$0.02.

19. ***Familiarize.*** Let $r =$ the number of servings of roast beef, $p =$ the number of baked potatoes, and $b =$ the number of servings of broccoli. Then r servings of roast beef contain $300r$ Calories, $20r$ g of protein, and no vitamin C. In p baked potatoes there are $100p$ Calories, $5p$ g of protein, and $20p$ mg of vitamin C. And b servings of broccoli contain $50b$ Calories, $5b$ g of protein, and $100b$ mg of vitamin C. The patient requires 800 Calories, 55 g of protein, and 220 mg of vitamin C.

Translate. Write equations for the total number of calories, the total amount of protein, and the total amount of vitamin C.

$$300r + 100p + 50b = 800 \quad \text{(Calories)}$$
$$20r + 5p + 5b = 55 \quad \text{(protein)}$$
$$20p + 100b = 220 \quad \text{(vitamin C)}$$

We now have a system of equations.

Carry out. Solving the system we get (2, 1, 2).

Check. Two servings of roast beef provide 600 Calories, 40 g of protein, and no vitamin C. One baked potato provides 100 Calories, 5 g of protein, and 20 mg of vitamin C. And 2 servings of broccoli provide 100 Calories, 10 g of protein, and 200 mg of vitamin C. Together, then, they provide 800 Calories, 55 g of protein, and 220 mg of vitamin C. The values check.

State. The dietician should prepare 2 servings of roast beef, 1 baked potato, and 2 servings of broccoli.

21. ***Familiarize.*** Let x, y, and z be the number of tickets sold for the first mezzanine, main floor and second mezzanine, respectively.

Translate.

Total number of tickets	is	40.
↓	↓	↓
$x + y + z$	$=$	40

Number of tickets for first mezzanine and main floor	is	number of tickets for second mezzanine.
↓	↓	↓
$x + y$	$=$	z

Total cost of tickets	is	\$1432.
↓	↓	↓
$52x + 38y + 28z$	$=$	1432

We have a system of equations:

$$x + y + z = 40$$
$$x + y = z$$
$$52x + 38y + 28z = 1432$$

Carry out. Solving the system we get (8, 12, 20).

Check. The total number of tickets is 8 + 12 + 20, or 40. The sum of first mezzanine and main floor is 8 + 12, is the number of second mezzanine, 20. The total cost is $\$52(8) + \$38(12) + \$28(20) = \$416 + \$456 + \$560 = \$1432$. The numbers check.

State. There were 8 first mezzanine tickets, 12 main floor tickets and 20 second mezzanine tickets sold.

23. ***Familiarize.*** Let x, y, and z represent the populations of Asia, Africa, and the rest of the world, respectively, in billions, in 2050.

Translate.

The total world population	will be	9.4 billion.
↓	↓	↓
$x + y + z$	$=$	9.4

Population of Asia	will be	3.5 billion	more than	Population of Africa
↓	↓	↓	↓	↓
x	$=$	3.5	$+$	y

Population of the rest of the world	will be	two-fifths	of	population of Asia	less	0.3 billion.
↓	↓	↓	↓	↓	↓	↓
z	$=$	$\frac{2}{5}$	$\times$	x	$-$	0.3

We have a system of equations.

$$x + y + z = 9.4$$
$$x = 3.5 + y$$
$$z = \frac{2}{5}x - 0.3$$

Carry out. Solving the system we get (5.5, 2.0, 1.9).

Check. The total population will be 5.5 + 2.0 + 1.9, or 9.4 billion. The population of Asia, 5.5 billion, is 3.5 billion more than the population of Africa, 2.0 billion. Also, the rest of the world population, 1.9 billion is 0.3 billion less than $\frac{2}{5}$ of 5.5, or 2.2 billion. The numbers check.

State. In 2050, the population of Asia will be 5.5 billion, the population of Africa will be 2.0 billion and the population of the rest of the world will be 1.9 billion.

25. *Writing Exercise.* Problems like Exercises 13 and 14 have three unknowns whereas those in Section 3.3 have two unknowns.

27. $-2(2x-3y)=-4x+6y$

29. $-6(x-2y)+(6x-5y)=-6x+12y+6x-5y=7y$

31. $-(2a-b-6c)=-2a+b+6c$

33. $-2(3x-y+z)+3(-2x+y-2z)$
$=-6x+2y-2z-6x+3y-6z$
$=-12x+5y-8z$

35. *Writing Exercise.* If the Knicks made 19 more 2-pointers than foul shots, and no foul shots were made, then there would have been 19 2-pointers. If there were 50 baskets and 19 were 2-pointers and 31 3-pointers yield a total of $2\cdot 19+3\cdot 31$, or 131 points, rather than 92 points. Thus, if no foul shots were made, there is no solution.

37. ***Familiarize.*** Let w, x, y, and z represent the monthly rates for an applicant, a spouse, the first child and the second child, respectively.

Translate.

Rate for applicant and spouse is \$174.

$$w+x = 174$$

Rate for applicant, spouse and one child is \$221.

$$w+x+y = 221$$

Rate for applicant, spouse and two children is \$263.

$$w+x+y+z = 263$$

Rate for applicant, and one child is \$134.

$$w+y = 134$$

We have a system of equations:

$$\begin{aligned} w+x&=174 \quad (1)\\ w+x+y&=221 \quad (2)\\ w+x+y+z&=263 \quad (3)\\ w+y&=134 \quad (4)\end{aligned}$$

Carry out. We solve the system of equations. First substitute 221 for $w + x + y$ in (3) and solve for z.

$$\begin{aligned} 221+z&=263\\ z&=42\end{aligned}$$

Next substitute 174 for $w + x$ in (2) and solve for y.

$$\begin{aligned} 174+y&=221\\ y&=47\end{aligned}$$

Substitute 47 for y in (4) and solve for w.

$$\begin{aligned} w+47&=134\\ w&=87\end{aligned}$$

Finally, substitute 87 for w in (1) and solve for x.

$$\begin{aligned} 87+x&=174\\ x&=87\end{aligned}$$

The solution is (87, 87, 47, 42).

Check. The check is left to the student.

State. The separate monthly rates for an applicant, a spouse, the first child, and the second child are \$87, \$87, \$47, and \$42, respectively.

39. ***Familiarize.*** Let w, x, y, and z represent the ages of Tammy, Carmen, Dennis, and Mark respectively.

Translate.

Tammy's age is the sum of the ages of Carmen and Dennis, so we have

$$w=x+y\,.$$

Carmen's age is 2 more than the sum of the ages of Dennis and Mark, so we have

$$x=2+y+z\,.$$

Dennis's age is four times Mark's age, so we have

$$y=4z\,.$$

The sum of all four ages is 42, so we have

$$w+x+y+z=42\,.$$

Now we have a system of equations.

$$\begin{aligned} w&=x+y, && (1)\\ x&=2+y+z, && (2)\\ y&=4z, && (3)\\ w+x+y+z&=42 && (4)\end{aligned}$$

Carry out. We solve the system of equations. First we will express w, x, and y in terms of z and then solve for z. From (3) we know that $y=4z$. Substitute $4z$ for y in (2):

$$x=2+4z+z=2+5z\,.$$

Substitute $2+5z$ for x and $4z$ for y in (1):

$$w=2+5z+4z=2+9z\,.$$

Now substitute $2+9z$ for w, $2+5z$ for x, and $4z$ for y in (4) and solve for z.

$$\begin{aligned} 2+9z+2+5z+4z+z&=42\\ 19z+4&=42\\ 19z&=38\\ z&=2\end{aligned}$$

Then we have:

$$\begin{aligned} w&=2+9z=2+9\cdot 2=20,\\ x&=2+5z=2+5\cdot 2=12, \text{ and}\\ y&=4z=4\cdot 2=8\end{aligned}$$

Although we were asked to find only Tammy's age, we found all of the ages so that we can check the result.

Check. The check is left to the student.

State. Tammy is 20 years old.

41. Let T, G, and H represent the number of tickets Tom, Gary, and Hal begin with, respectively. After Hal gives tickets to Tom and Gary, each has the following number of tickets:

$$\begin{aligned} \text{Tom}: &\quad T+T, \text{ or } 2T, \\ \text{Gary}: &\quad G+G, \text{ or } 2G, \\ \text{Hal}: &\quad H-T-G. \end{aligned}$$

After Tom gives tickets to Gary and Hal, each has the following number of tickets:

$$\begin{aligned} \text{Gary}: &\quad 2G+2G, \text{ or } 4G, \\ \text{Hal}: &\quad (H-T-G)+(H-T-G), \text{ or } \\ &\quad 2(H-T-G), \\ \text{Tom}: &\quad 2T-2G-(H-T-G), \text{ or } \\ &\quad 3T-H-G \end{aligned}$$

After Gary gives tickets to Hal and Tom, each has the following number of tickets:

$$\begin{aligned} \text{Hal}: &\quad 2(H-T-G)+2(H-T-G), \text{ or } \\ &\quad 4(H-T-G) \\ \text{Tom}: &\quad (3T-H-G)+(3T-H-G), \text{ or } \\ &\quad 2(3T-H-G), \\ \text{Gary}: &\quad 4G-2(H-T-G)-(3T-H-G), \text{ or } \\ &\quad 7G-H-T. \end{aligned}$$

Since Hal, Tom, and Gary each finish with 40 tickets, we write the following system of equations:

$$\begin{aligned} 4(H-T-G) &= 40, \\ 2(3T-H-G) &= 40, \\ 7G-H-T &= 40 \end{aligned}$$

Solving the system we find that $T = 35$, so Tom started with 35 tickets.

Exercise Set 8.6

1. matrix; see page 552 in the text.

3. entry; see page 554 in the text.

5. rows; see page 554 in the text.

7. $x+2y=11,$
$3x-y=5$

Write a matrix using only the constants.

$$\left[\begin{array}{cc|c} 1 & 2 & 11 \\ 3 & -1 & 5 \end{array}\right]$$

Multiply the first row by –3 and add it to the second row.

$$\left[\begin{array}{cc|c} 1 & 2 & 11 \\ 0 & -7 & -28 \end{array}\right] \quad \text{New Row 2} = -3(\text{Row 1})+\text{Row 2}$$

Reinserting the variables, we have

$$\begin{aligned} x+2y &= 11, \quad (1) \\ -7y &= -28 \quad (2) \end{aligned}$$

Solve Equation (2) for y.

$$\begin{aligned} -7y &= -28 \\ y &= 4 \end{aligned}$$

Substitute 4 for y in Equation (1) and solve for x.

$$\begin{aligned} x+2(4) &= 11 \\ x+8 &= 11 \\ x &= 3 \end{aligned}$$

The solution is (3, 4).

9. $3x+y=-1,$
$6x+5y=13$

We first write a matrix using only the constants.

$$\left[\begin{array}{cc|c} 3 & 1 & -1 \\ 6 & 5 & 13 \end{array}\right]$$

Multiply the first row by –2 and add it to the second row.

$$\left[\begin{array}{cc|c} 3 & 1 & -1 \\ 0 & 3 & 15 \end{array}\right] \quad \text{New Row 2} = -2(\text{Row 1})+\text{Row 2}$$

Reinserting the variables, we have

$$\begin{aligned} 3x+y &= -1, \quad (1) \\ 3y &= 15 \quad (2) \end{aligned}$$

Solve Equation (2) for y.

$$\begin{aligned} 3y &= 15 \\ y &= 5 \end{aligned}$$

Substitute 5 for y in Equation (1) and solve for x.

$$\begin{aligned} 3x+5 &= -1 \\ 3x &= -6 \\ x &= -2 \end{aligned}$$

The solution is (–2, 5).

11. $6x-2y=4,$
$7x+y=13$

Write a matrix using only the constants.

$$\left[\begin{array}{cc|c} 6 & -2 & 4 \\ 7 & 1 & 13 \end{array}\right]$$

Multiply the second row by 6 to make the first number in row 2 a multiple of 6.

$$\left[\begin{array}{cc|c} 6 & -2 & 4 \\ 42 & 6 & 78 \end{array}\right] \quad \text{New Row 2} = 6(\text{Row 2})$$

Now multiply the first row by -7 and add it to the second row.

$$\left[\begin{array}{cc|c} 6 & -2 & 4 \\ 0 & 20 & 50 \end{array}\right] \quad \text{New Row 2} = -7(\text{Row 1})+\text{Row 2}$$

Reinserting the variables, we have

$$\begin{aligned} 6x-2y &= 4, \quad (1) \\ 20y &= 50. \quad (2) \end{aligned}$$

Solve Equation (2) for y.

$$20y = 50$$
$$y = \frac{5}{2}$$

Substitute $\frac{5}{2}$ for y in Equation (1) and solve for x.

$$6x - 2y = 4$$
$$6x - 2\left(\frac{5}{2}\right) = 4$$
$$6x - 5 = 4$$
$$6x = 9$$
$$x = \frac{3}{2}$$

The solution is $\left(\frac{3}{2}, \frac{5}{2}\right)$.

13. $3x + 2y + 2z = 3,$
$x + 2y - z = 5,$
$2x - 4y + z = 0$

We first write a matrix using only the constants.

$$\left[\begin{array}{ccc|c} 3 & 2 & 2 & 3 \\ 1 & 2 & -1 & 5 \\ 2 & -4 & 1 & 0 \end{array}\right]$$

First interchange rows 1 and 2 so that each number below the first number in the first row is a multiple of that number.

$$\left[\begin{array}{ccc|c} 1 & 2 & -1 & 5 \\ 3 & 2 & 2 & 3 \\ 2 & -4 & 1 & 0 \end{array}\right]$$

Multiply row 1 by –3 and add it to row 2.

Multiply row 1 by –2 and add it to row 3.

$$\left[\begin{array}{ccc|c} 1 & 2 & -1 & 5 \\ 0 & -4 & 5 & -12 \\ 0 & -8 & 3 & -10 \end{array}\right]$$

Multiply row 2 by –2 and add it to row 3.

$$\left[\begin{array}{ccc|c} 1 & 2 & -1 & 5 \\ 0 & -4 & 5 & -12 \\ 0 & 0 & -7 & 14 \end{array}\right]$$

Reinserting the variables, we have

$$x + 2y - z = 5, \quad (1)$$
$$-4y + 5z = -12, \quad (2)$$
$$-7z = 14. \quad (3)$$

Solve (3) for z.

$$-7z = 14$$
$$z = -2$$

Substitute –2 for z in (2) and solve for y.

$$-4y + 5(-2) = -12$$
$$-4y - 10 = -12$$
$$-4y = -2$$
$$y = \frac{1}{2}$$

Substitute $\frac{1}{2}$ for y and –2 for z in (1) and solve for x.

$$x + 2 \cdot \frac{1}{2} - (-2) = 5$$
$$x + 1 + 2 = 5$$
$$x + 3 = 5$$
$$x = 2$$

The solution is $\left(2, \frac{1}{2}, -2\right)$.

15. $p - 2q - 3r = 3,$
$2p - q - 2r = 4,$
$4p + 5q + 6r = 4$

We first write a matrix using only the constants.

$$\left[\begin{array}{ccc|c} 1 & -2 & -3 & 3 \\ 2 & -1 & -2 & 4 \\ 4 & 5 & 6 & 4 \end{array}\right]$$

$$\left[\begin{array}{ccc|c} 1 & -2 & -3 & 3 \\ 0 & 3 & 4 & -2 \\ 0 & 13 & 18 & -8 \end{array}\right] \quad \begin{array}{l} \\ \text{New Row 2} = -2(\text{Row 1}) + \text{Row 2} \\ \text{New Row 3} = -4(\text{Row 1}) + \text{Row 3} \end{array}$$

$$\left[\begin{array}{ccc|c} 1 & -2 & -3 & 3 \\ 0 & 3 & 4 & -2 \\ 0 & 39 & 54 & -24 \end{array}\right] \quad \text{New Row 3} = 3(\text{Row 3})$$

$$\left[\begin{array}{ccc|c} 1 & -2 & -3 & 3 \\ 0 & 3 & 4 & -2 \\ 0 & 0 & 2 & 2 \end{array}\right] \quad \text{New Row 3} = -13(\text{Row 2}) + \text{Row 3}$$

Reinserting the variables, we have

$$p - 2q - 3r = 3, \quad (1)$$
$$3q + 4r = -2, \quad (2)$$
$$2r = 2 \quad (3)$$

Solve (3) for r.

$$2r = 2$$
$$r = 1$$

Substitute 1 for r in (2) and solve for q.

$$3q + 4 \cdot 1 = -2$$
$$3q + 4 = -2$$
$$3q = -6$$
$$q = -2$$

Substitute –2 for q and 1 for r in (1) and solve for p.

$$p - 2(-2) - 3 \cdot 1 = 3$$
$$p + 4 - 3 = 3$$
$$p + 1 = 3$$
$$p = 2$$

The solution is (2, –2, 1).

17. $3p + 2r = 11,$
$q - 7r = 4,$
$p - 6q = 1$

We first write a matrix using only the constants.

$$\left[\begin{array}{ccc|c} 3 & 0 & 2 & 11 \\ 0 & 1 & -7 & 4 \\ 1 & -6 & 0 & 1 \end{array}\right]$$

$$\left[\begin{array}{ccc|c} 1 & -6 & 0 & 1 \\ 0 & 1 & -7 & 4 \\ 3 & 0 & 2 & 11 \end{array}\right] \quad \begin{array}{l} \text{Interchange} \\ \text{Row 1 and Row 3} \end{array}$$

$$\left[\begin{array}{ccc|c} 1 & -6 & 0 & 1 \\ 0 & 1 & -7 & 4 \\ 0 & 18 & 2 & 8 \end{array}\right] \quad \text{New Row 3} = -3(\text{Row 1}) + \text{Row 3}$$

$$\left[\begin{array}{ccc|c} 1 & -6 & 0 & 1 \\ 0 & 1 & -7 & 4 \\ 0 & 0 & 128 & -64 \end{array}\right] \quad \text{New Row 3} = -18(\text{Row 2}) + \text{Row 3}$$

Reinserting the variables, we have

$$\begin{aligned} p - 6q &= 1, \quad (1) \\ q - 7r &= 4, \quad (2) \\ 128r &= -64. \quad (3) \end{aligned}$$

Solve (3) for r.

$$\begin{aligned} 128r &= -64 \\ r &= -\frac{1}{2} \end{aligned}$$

Substitute $-\frac{1}{2}$ for r in (2) and solve for q.

$$\begin{aligned} q - 7r &= 4 \\ q - 7\left(-\frac{1}{2}\right) &= 4 \\ q + \frac{7}{2} &= 4 \\ q &= \frac{1}{2} \end{aligned}$$

Substitute $\frac{1}{2}$ for q in (1) and solve for p.

$$\begin{aligned} p - 6 \cdot \frac{1}{2} &= 1 \\ p - 3 &= 1 \\ p &= 4 \end{aligned}$$

The solution is $\left(4, \frac{1}{2}, -\frac{1}{2}\right)$.

19. We will rewrite the equations with the variables in alphabetical order:

$$\begin{aligned} -2w + 2x + 2y - 2z &= -10, \\ w + x + y + z &= -5, \\ 3w + x - y + 4z &= -2, \\ w + 3x - 2y + 2z &= -6 \end{aligned}$$

Write a matrix using only the constants.

$$\left[\begin{array}{cccc|c} -2 & 2 & 2 & -2 & -10 \\ 1 & 1 & 1 & 1 & -5 \\ 3 & 1 & -1 & 4 & -2 \\ 1 & 3 & -2 & 2 & -6 \end{array}\right]$$

$$\left[\begin{array}{cccc|c} -1 & 1 & 1 & -1 & -5 \\ 1 & 1 & 1 & 1 & -5 \\ 3 & 1 & -1 & 4 & -2 \\ 1 & 3 & -2 & 2 & -6 \end{array}\right] \quad \text{New Row 1} = \frac{1}{2}(\text{Row 1})$$

$$\left[\begin{array}{cccc|c} -1 & 1 & 1 & -1 & -5 \\ 0 & 2 & 2 & 0 & -10 \\ 0 & 4 & 2 & 1 & -17 \\ 0 & 4 & -1 & 1 & -11 \end{array}\right] \quad \begin{array}{l} \text{New Row 2} = \text{Row 1} + \text{Row 2} \\ \text{New Row 3} = 3(\text{Row 1}) + \text{Row 3} \\ \text{New Row 4} = \text{Row 1} + \text{Row 4} \end{array}$$

$$\left[\begin{array}{cccc|c} -1 & 1 & 1 & -1 & -5 \\ 0 & 2 & 2 & 0 & -10 \\ 0 & 0 & -2 & 1 & 3 \\ 0 & 0 & -5 & 1 & 9 \end{array}\right] \quad \begin{array}{l} \text{New Row 3} = -2(\text{Row 2}) + \text{Row 3} \\ \text{New Row 4} = -2(\text{Row 2}) + \text{Row 4} \end{array}$$

$$\left[\begin{array}{cccc|c} -1 & 1 & 1 & -1 & -5 \\ 0 & 2 & 2 & 0 & -10 \\ 0 & 0 & -2 & 1 & 3 \\ 0 & 0 & -10 & 2 & 18 \end{array}\right] \quad \text{New Row 4} = 2(\text{Row 4})$$

$$\left[\begin{array}{cccc|c} -1 & 1 & 1 & -1 & -5 \\ 0 & 2 & 2 & 0 & -10 \\ 0 & 0 & -2 & 1 & 3 \\ 0 & 0 & 0 & -3 & 3 \end{array}\right] \quad \text{New Row 4} = -5(\text{Row 3}) + \text{Row 4}$$

Reinserting the variables, we have

$$\begin{aligned} -w + x + y - z &= -5, \quad (1) \\ 2x + 2y &= -10, \quad (2) \\ -2y + z &= 3, \quad (3) \\ -3z &= 3. \quad (4) \end{aligned}$$

Solve (4) for z.

$$\begin{aligned} -3z &= 3 \\ z &= -1 \end{aligned}$$

Substitute –1 for z in (3) and solve for y.

$$\begin{aligned} -2y + (-1) &= 3 \\ -2y &= 4 \\ y &= -2 \end{aligned}$$

Substitute –2 for y in (2) and solve for x.

$$\begin{aligned} 2x + 2(-2) &= -10 \\ 2x - 4 &= -10 \\ 2x &= -6 \\ x &= -3 \end{aligned}$$

Substitute –3 for x, –2 for y, and –1 for z in (1) and solve for w.

$$\begin{aligned} -w + (-3) + (-2) - (-1) &= -5 \\ -w - 3 - 2 + 1 &= -5 \\ -w - 4 &= -5 \\ -w &= -1 \\ w &= 1 \end{aligned}$$

The solution is (1, –3, –2, –1).

21. ***Familiarize.*** Let d = the number of dimes and n = the number of nickels. The value of d dimes is \$0.10d, and the value of n nickels is \$0.05n.

Translate.

<u>Total number of coins</u> is 42.

$$\begin{array}{ccc} \downarrow & \downarrow & \downarrow \\ d + n & = & 42 \end{array}$$

<u>Total value of coins</u> is \$3.

$$\begin{array}{ccc} \downarrow & \downarrow & \downarrow \\ 0.10d + 0.05n & = & 3 \end{array}$$

After clearing decimals, we have this system.

$$\begin{aligned} d + n &= 42, \\ 10d + 5n &= 300 \end{aligned}$$

Carry out. Solve using matrices.

$$\left[\begin{array}{cc|c} 1 & 1 & 42 \\ 10 & 5 & 300 \end{array}\right]$$

$$\left[\begin{array}{cc|c} 1 & 1 & 42 \\ 0 & -5 & -120 \end{array}\right] \quad \text{New Row 2} = -10(\text{Row 1}) + \text{Row 2}$$

Reinserting the variables, we have

$d + n = 42,\quad (1)$
$-5n = -120\quad (2)$

Solve (2) for n.

$-5n = -120$
$n = 24$
$d + 24 = 42\quad$ Back – substituting
$d = 18$

Check. The sum of the two numbers is 42. The total value is $\$0.10(18) + \$0.05(24) = \$1.80 + \$1.20 = \$3$. The numbers check.

State. There are 18 dimes and 24 nickels.

23. ***Familiarize.*** Let x = the number of pounds of dried fruit and y = the number of pounds of macadamia nuts. We organize the information in a table.

	Dried fruit	Macadamia nuts	Mixture
Number of pounds	x	y	15
Price per pound	\$5.80	\$14.75	\$9.38
Value of coffee	$5.80x$	$14.75y$	140.70

Translate.

The total number of pounds is 15.

$x + y = 15$

The total value of the mixture is \$140.70.

$5.80x + 14.75y = 140.70$

After clearing decimals, we have this system:

$x + y = 15,$
$580x + 1475y = 14{,}070$

Carry out. Solve using matrices.

$$\left[\begin{array}{cc|c} 1 & 1 & 15 \\ 580 & 1475 & 14{,}070 \end{array}\right]$$

Multiply the first row by –580 and add it to the second row.

$$\left[\begin{array}{cc|c} 1 & 1 & 15 \\ 0 & 895 & 5370 \end{array}\right] \quad \text{New Row 2} = -580(\text{Row 1}) + \text{Row 2}$$

Reinserting the variables, we have

$x + y = 15$
$895y = 5370$
$y = 6$

Back-substitute 6 for y in Equation (1) and solve for x.

$x + 6 = 15$
$x = 9$

Check. The sum of the numbers, 6 + 9 = 15. The total value is $\$5.80(9) + \$14.75(6)$, or $\$52.20 + \$88.50$, or \$140.70. The numbers check.

State. 9 pounds of dried fruit and 6 pounds of macadamia nuts should be used.

25. ***Familiarize.*** We let x, y, and z represent the amounts invested at 7%, 8%, and 9%, respectively. Recall the formula for simple interest:

Interest = Principal × Rate × Time

Translate. We organize the information in a table.

	First Investment	Second Investment	Third Investment	Total
P	x	y	z	\$2500
R	7%	8%	9%	
T	1 yr	1 yr	1 yr	
I	$0.07x$	$0.08y$	$0.09z$	\$212

The first row gives us one equation:

$x + y + z = 2500$

The last row gives a second equation:

$0.07x + 0.08y + 0.09z = 212$

Amount invested at 9%	is	\$1100	more than	amount invested at 8%.
↓	↓	↓	↓	↓
z	=	\$1100	+	y

After clearing decimals, we have this system:

$x + y + z = 2500,$
$7x + 8y + 9z = 21{,}200,$
$-y + z = 1100$

Carry out. Solve using matrices.

$$\left[\begin{array}{ccc|c} 1 & 1 & 1 & 2500 \\ 7 & 8 & 9 & 21{,}200 \\ 0 & -1 & 1 & 1100 \end{array}\right]$$

$$\left[\begin{array}{ccc|c} 1 & 1 & 1 & 2500 \\ 0 & 1 & 2 & 3700 \\ 0 & -1 & 1 & 1100 \end{array}\right] \quad \text{New Row 2} = -7(\text{Row 1}) + \text{Row 2}$$

$$\left[\begin{array}{ccc|c} 1 & 1 & 1 & 2500 \\ 0 & 1 & 2 & 3700 \\ 0 & 0 & 3 & 4800 \end{array}\right] \quad \text{New Row 3} = \text{Row 2} + \text{Row 3}$$

Reinserting the variables, we have

$x + y + z = 2500,\quad (1)$
$y + 2z = 3700,\quad (2)$
$3z = 4800\quad (3)$

Solve (3) for z.

$3z = 4800$
$z = 1600$

Back-substitute 1600 for z in (2) and solve for y.

$y + 2 \cdot 1600 = 3700$
$y + 3200 = 3700$
$y = 500$

Back-substitute 500 for y and 1600 for z in (1) and solve for x.

$x + 500 + 1600 = 2500$
$x + 2100 = 2500$
$x = 400$

Check. The total investment is \$400 + \$500 + \$1600, or

\$2500. The total interest is

$0.07(\$400)+0.08(\$500)+0.09(\$1600)$

$=\$28+\$40+\$144=\$212$. The amount invested at 9%, \$1600, is \$1100 more than the amount invested at 8%, \$500. The numbers check.

State. \$400 is invested at 7%, \$500 is invested at 8%, and \$1600 is invested at 9%.

27. *Writing Exercise.* Row-equivalent operations yield a row containing all 0's. (This corresponds to an equation that is true for all values of the variables.)

29. $3(-1)-(-4)(5)=-3-(-20)=-3+20=17$

31. $-2(5\cdot3-4\cdot6)-3(2\cdot7-15)+4(3\cdot8-5\cdot4)$
$=-2(15-24)-3(14-15)+4(24-20)$
$=-2(-9)-3(-1)+4(4)$
$=18+3+16$
$=21+16$
$=37$

33. *Writing Exercise.* No; two systems of equations with different coefficients and constants can have the same solution. Because the coefficients and constants are different, corresponding entries in the matrices used to solve the systems are not all equal to each other.

35. ***Familiarize.*** Let w, x, y, and z represent the thousand's, hundred's, ten's, and one's digits, respectively.

Translate.

The sum of the digits	is	10.
↓	↓	↓
$w+x+y+z$	$=$	10

Twice the sum of the thousand's and ten's digits	is	the sum of the hundred's and one's digits	less one.
↓	↓	↓	↓
$2(w+y)$	$=$	$x+z$	-1

The ten's digit	is	twice	the thousand's digit.
↓	↓	↓	↓
y	$=$	$2\cdot$	w

The one's digit	equals	the sum of the thousand's and hundred's digits.
↓	↓	↓
z	$=$	$w+x$

We have a system of equations which can be written as

$$\begin{aligned} w+x+y+z&=10,\\ 2w-x+2y-z&=-1,\\ -2w+y&=0,\\ w+x-z&=0. \end{aligned}$$

Carry out. We can use matrices to solve the system. We get (1, 3, 2, 4).

Check. The sum of the digits is 10. Twice the sum of 1 and 2 is 6. This is one less than the sum of 3 and 4. The ten's digit, 2, is twice the thousand's digit, 1. The one's digit, 4, equals 1 + 3. The numbers check.

State. The number is 1324.

Exercise Set 8.7

1. True; see page 556 in the text.

3. True; see page 556 in the text.

5. False; see page 557 in the text.

7. $\begin{vmatrix} 3 & 5 \\ 4 & 8 \end{vmatrix}=3\cdot8-4\cdot5=24-20=4$

9. $\begin{vmatrix} 10 & 8 \\ -5 & -9 \end{vmatrix}=10(-9)-8(-5)=-90+40=-50$

11. $\begin{vmatrix} 1 & 4 & 0 \\ 0 & -1 & 2 \\ 3 & -2 & 1 \end{vmatrix}$

$=1\begin{vmatrix} -1 & 2 \\ -2 & 1 \end{vmatrix}-0\begin{vmatrix} 4 & 0 \\ -2 & 1 \end{vmatrix}+3\begin{vmatrix} 4 & 0 \\ -1 & 2 \end{vmatrix}$

$=1[-1\cdot1-(-2)\cdot2]-0+3[4\cdot2-(-1)\cdot0]$
$=1\cdot3-0+3\cdot8$
$=3-0+24$
$=27$

13. $\begin{vmatrix} -1 & -2 & -3 \\ 3 & 4 & 2 \\ 0 & 1 & 2 \end{vmatrix}$

$=-1\begin{vmatrix} 4 & 2 \\ 1 & 2 \end{vmatrix}-3\begin{vmatrix} -2 & -3 \\ 1 & 2 \end{vmatrix}+0\begin{vmatrix} -2 & -3 \\ 4 & 2 \end{vmatrix}$

$=-1[4\cdot2-1\cdot2]-3[-2\cdot2-1(-3)]+0$
$=-1\cdot6-3\cdot(-1)+0$
$=-6+3+0$
$=-3$

15. $\begin{vmatrix} -4 & -2 & 3 \\ -3 & 1 & 2 \\ 3 & 4 & -2 \end{vmatrix}$

$=-4\begin{vmatrix} 1 & 2 \\ 4 & -2 \end{vmatrix}-(-3)\begin{vmatrix} -2 & 3 \\ 4 & -2 \end{vmatrix}+3\begin{vmatrix} -2 & 3 \\ 1 & 2 \end{vmatrix}$

$=-4[1(-2)-4\cdot2]+3[-2(-2)-4\cdot3]$
$\quad+3[-2\cdot2-1\cdot3]$
$=-4(-10)+3(-8)+3(-7)$
$=40-24-21=-5$

17. $5x+8y=1,$
$3x+7y=5$

We compute D, D_x, and D_y.

$$D=\begin{vmatrix}5 & 8\\3 & 7\end{vmatrix}=35-24=11$$

$$D_x=\begin{vmatrix}1 & 8\\5 & 7\end{vmatrix}=7-40=-33$$

$$D_y=\begin{vmatrix}5 & 1\\3 & 5\end{vmatrix}=25-3=22$$

Then,

$$x=\frac{D_x}{D}=\frac{-33}{11}=-3$$

and

$$y=\frac{D_y}{D}=\frac{22}{11}=2.$$

The solution is (–3, 2).

19. $5x-4y=-3,$
$7x+2y=6$

We compute D, D_x, and D_y.

$$D=\begin{vmatrix}5 & -4\\7 & 2\end{vmatrix}=10-(-28)=38$$

$$D_x=\begin{vmatrix}-3 & -4\\6 & 2\end{vmatrix}=-6-(-24)=18$$

$$D_y=\begin{vmatrix}5 & -3\\7 & 6\end{vmatrix}=30-(-21)=51$$

Then,

$$x=\frac{D_x}{D}=\frac{18}{38}=\frac{9}{19}$$

and

$$y=\frac{D_y}{D}=\frac{51}{38}.$$

The solution is $\left(\frac{9}{19}, \frac{51}{38}\right)$.

21. $3x - y + 2z = 1,$
$x - y + 2z = 3,$
$-2x + 3y + z = 1$

We compute D, D_x, and D_y.

$$D=\begin{vmatrix}3 & -1 & 2\\1 & -1 & 2\\-2 & 3 & 1\end{vmatrix}$$

$$=3\begin{vmatrix}-1 & 2\\3 & 1\end{vmatrix}-1\begin{vmatrix}-1 & 2\\3 & 1\end{vmatrix}-2\begin{vmatrix}-1 & 2\\-1 & 2\end{vmatrix}$$

$$=3(-7)-1(-7)-2(0)$$
$$=-21+7-0$$
$$=-14$$

$$D_x=\begin{vmatrix}1 & -1 & 2\\3 & -1 & 2\\1 & 3 & 1\end{vmatrix}$$

$$=1\begin{vmatrix}-1 & 2\\3 & 1\end{vmatrix}-3\begin{vmatrix}-1 & 2\\3 & 1\end{vmatrix}+1\begin{vmatrix}-1 & 2\\-1 & 2\end{vmatrix}$$

$$=1(-7)-3(-7)+1(0)$$
$$=-7+21+0$$
$$=14$$

$$D_y=\begin{vmatrix}3 & 1 & 2\\1 & 3 & 2\\-2 & 1 & 1\end{vmatrix}$$

$$=3\begin{vmatrix}3 & 2\\1 & 1\end{vmatrix}-1\begin{vmatrix}1 & 2\\1 & 1\end{vmatrix}-2\begin{vmatrix}1 & 2\\3 & 2\end{vmatrix}$$

$$=3\cdot1-1(-1)-2(-4)$$
$$=3+1+8$$
$$=12$$

Then,

$$x=\frac{D_x}{D}=\frac{14}{-14}=-1$$

and

$$y=\frac{D_y}{D}=\frac{12}{-14}=-\frac{6}{7}.$$

Substitute in the third equation to find z.

$$-2(-1)+3\left(-\frac{6}{7}\right)+z=1$$
$$2-\frac{18}{7}+z=1$$
$$-\frac{4}{7}+z=1$$
$$z=\frac{11}{7}$$

The solution is $\left(-1, -\frac{6}{7}, \frac{11}{7}\right)$.

23. $2x - 3y + 5z = 27,$
$x + 2y - z = -4,$
$5x - y + 4z = 27$

We compute D, D_x, and D_y.

$$D=\begin{vmatrix}2 & -3 & 5\\1 & 2 & -1\\5 & -1 & 4\end{vmatrix}$$

$$=2\begin{vmatrix}2 & -1\\-1 & 4\end{vmatrix}-1\begin{vmatrix}-3 & 5\\-1 & 4\end{vmatrix}+5\begin{vmatrix}-3 & 5\\2 & -1\end{vmatrix}$$

$$=2(7)-1(-7)+5(-7)$$
$$=14+7-35=-14$$

$$D_x=\begin{vmatrix}27 & -3 & 5\\-4 & 2 & -1\\27 & -1 & 4\end{vmatrix}$$

$$=27\begin{vmatrix}2 & -1\\-1 & 4\end{vmatrix}-(-4)\begin{vmatrix}-3 & 5\\-1 & 4\end{vmatrix}+27\begin{vmatrix}-3 & 5\\2 & -1\end{vmatrix}$$

$$=27(7)+4(-7)+27(-7)$$
$$=189-28-189$$
$$=-28$$

$$D_y=\begin{vmatrix}2 & 27 & 5\\1 & -4 & -1\\5 & 27 & 4\end{vmatrix}$$

$$=2\begin{vmatrix}-4 & -1\\27 & 4\end{vmatrix}-1\begin{vmatrix}27 & 5\\27 & 4\end{vmatrix}+5\begin{vmatrix}27 & 5\\-4 & -1\end{vmatrix}$$

$$=2(11)-1(-27)+5(-7)$$
$$=22+27-35$$
$$=14$$

Then,

$$x = \frac{D_x}{D} = \frac{-28}{-14} = 2,$$

and

$$y = \frac{D_y}{D} = \frac{14}{-14} = -1.$$

We substitute in the second equation to find z.

$$\begin{aligned} 2+2(-1)-z &= -4 \\ 2-2-z &= -4 \\ -z &= -4 \\ z &= 4 \end{aligned}$$

The solution is (2, –1, 4).

25. $r-2s+3t=6,$
$2r-s-t=-3,$
$r+s+t=6$

We compute D, D_r, and D_s.

$$D = \begin{vmatrix} 1 & -2 & 3 \\ 2 & -1 & -1 \\ 1 & 1 & 1 \end{vmatrix}$$

$$\begin{aligned} &= 1\begin{vmatrix} -1 & -1 \\ 1 & 1 \end{vmatrix} - 2\begin{vmatrix} -2 & 3 \\ 1 & 1 \end{vmatrix} + 1\begin{vmatrix} -2 & 3 \\ -1 & -1 \end{vmatrix} \\ &= 1(0)-2(-5)+1(5) \\ &= 0+10+5 \\ &= 15 \end{aligned}$$

$$D_r = \begin{vmatrix} 6 & -2 & 3 \\ -3 & -1 & -1 \\ 6 & 1 & 1 \end{vmatrix}$$

$$\begin{aligned} &= 6\begin{vmatrix} -1 & -1 \\ 1 & 1 \end{vmatrix} - (-3)\begin{vmatrix} -2 & 3 \\ 1 & 1 \end{vmatrix} + 6\begin{vmatrix} -2 & 3 \\ -1 & -1 \end{vmatrix} \\ &= 6(0)+3(-5)+6(5) \\ &= 0-15+30 \\ &= 15 \end{aligned}$$

$$D_s = \begin{vmatrix} 1 & 6 & 3 \\ 2 & -3 & -1 \\ 1 & 6 & 1 \end{vmatrix}$$

$$\begin{aligned} &= 1\begin{vmatrix} -3 & -1 \\ 6 & 1 \end{vmatrix} - 2\begin{vmatrix} 6 & 3 \\ 6 & 1 \end{vmatrix} + 1\begin{vmatrix} 6 & 3 \\ -3 & -1 \end{vmatrix} \\ &= 1(3)-2(-12)+1(3) \\ &= 3+24+3 \\ &= 30 \end{aligned}$$

Then,

$$r = \frac{D_r}{D} = \frac{15}{15} = 1,$$

and

$$s = \frac{D_s}{D} = \frac{30}{15} = 2.$$

Substitute in the third equation to find t.

$$\begin{aligned} 1+2+t &= 6 \\ 3+t &= 6 \\ t &= 3 \end{aligned}$$

The solution is (1, 2, 3).

27. *Writing Exercise.* Answers may vary. One pattern for Cramer's rule is that the system is dependent if the denominator is 0 and all the numerators are also 0.

29. $f(90) = 80(90) + 2500 = 7200 + 2500 = 9700$

31. $$\begin{aligned} (g-f)(10) &= g(10) - f(10) \\ &= 150(10) - [80(10) + 2500] \\ &= 1500 - [800 + 2500] \\ &= 1500 - 3300 \\ &= -1800 \end{aligned}$$

33. $$\begin{aligned} f(x) &= g(x) \\ 80x + 2500 &= 150x \\ 2500 &= 70x \\ \frac{250}{7} &= x \end{aligned}$$

35. *Writing Exercise.* If $a_1x + b_1y = c_1$ and $a_2x + b_2y = c_2$ are dependent, then one equation is a multiple of the other. That is, $a_1 = ka_2$ and $b_1 = kb_2$ for some constant k. Then

$$\begin{aligned} \begin{vmatrix} a_1 & b_1 \\ a_2 & b_2 \end{vmatrix} &= \begin{vmatrix} ka_2 & kb_2 \\ a_2 & b_2 \end{vmatrix} \\ &= ka_2(b_2) - a_2(kb_2) \\ &= 0 \end{aligned}$$

37. $\begin{vmatrix} y & -2 \\ 4 & 3 \end{vmatrix} = 44$

$$\begin{aligned} y \cdot 3 - 4(-2) &= 44 \quad \text{Evaluating the determinant} \\ 3y + 8 &= 44 \\ 3y &= 36 \\ y &= 12 \end{aligned}$$

39. $\begin{vmatrix} m+1 & -2 \\ m-2 & 1 \end{vmatrix} = 27$

$$\begin{aligned} (m+1)(1) - (m-2)(-2) &= 27 \quad \text{Evaluating the determinant} \\ m+1+2m-4 &= 27 \\ 3m &= 30 \\ m &= 10 \end{aligned}$$

Exercise Set 8.8

1. b; see page 561 in the text.

3. h; see page 562 in the text.

5. e; see page 561 in the text.

7. c; see page 563 in the text.

9. $C(x)=35x+200{,}000 \quad R(x)=55x$

a) $P(x)=R(x)-C(x)$
$=55x-(35x+200{,}000)$
$=55x-35x-200{,}000$
$=20x-200{,}000$

b) Solve the system
$R(x)=55x,$
$C(x)=35x+200{,}000.$

Since both $R(x)$ and $C(x)$ are in dollars and they are equal at the break-even point, we can rewrite the system:

$d=55x, \quad (1)$
$d=35x+200{,}000 \quad (2)$

We solve using substitution.

$55x=35x+200{,}000$ Substituting $55x$ for d in (2)
$20x=200{,}000$
$x=10{,}000$

Thus, 10,000 units must be produced and sold in order to break even.

The revenue will be $R(10{,}000)=55\cdot 10{,}000=550{,}000$.

The break-even point is (10,000 units, $550,000).

11. $C(x)=15x+3100 \quad R(x)=40x$

a) $P(x)=R(x)-C(x)$
$=40x-(15x+3100)$
$=40x-15x-3100$
$=25x-3100$

b) Solve the system
$R(x)=40x,$
$C(x)=15x+3100.$

Since both $R(x)$ and $C(x)$ are in dollars and they are equal at the break-even point, we can rewrite the system:

$d=40x, \quad (1)$
$d=15x+3100 \quad (2)$

We solve using substitution.

$40x=15x+3100$ Substituting $40x$ for d in (2)
$25x=3100$
$x=124$

Thus, 124 units must be produced and sold in order to break even.

The revenue will be $R(124)=40\cdot 124=4960$.

The break-even point is (124 units, $4960).

13. $C(x)=40x+22{,}500 \quad R(x)=85x$

a) $P(x)=R(x)-C(x)$
$=85x-(40x+22{,}500)$
$=85x-40x-22{,}500$
$=45x-22{,}500$

b) Solve the system
$R(x)=85x,$
$C(x)=40x+22{,}500.$

Since both $R(x)$ and $C(x)$ are in dollars and they are equal at the break-even point, we can rewrite the system:

$d=85x, \quad (1)$
$d=40x+22{,}500 \quad (2)$

We solve using substitution.

$85x=40x+22{,}500$ Substituting $85x$ for d in (2)
$45x=22{,}500$
$x=500$

Thus, 500 units must be produced and sold in order to break even.

The revenue will be $R(500)=85\cdot 500=42{,}500$.

The break-even point is (500 units, $42,500).

15. $C(x)=24x+50{,}000 \quad R(x)=40x$

a) $P(x)=R(x)-C(x)$
$=40x-(24x+50{,}000)$
$=40x-24x-50{,}000$
$=16x-50{,}000$

b) Solve the system
$R(x)=40x,$
$C(x)=24x+50{,}000.$

Since both $R(x)$ and $C(x)$ are in dollars and they are equal at the break-even point, we can rewrite the system:

$d=40x, \quad (1)$
$d=24x+50{,}000 \quad (2)$

We solve using substitution.

$40x=24x+50{,}000$ Substituting $40x$ for d in (2)
$16x=50{,}000$
$x=3125$

Thus, 3125 units must be produced and sold in order to break even.

The revenue will be $R(3125)=40\cdot 3125=\$125{,}000$.

The break-even point is (3125 units, $125,000).

17. $C(x)=75x+100{,}000 \quad R(x)=125x$

a) $P(x)=R(x)-C(x)$
$=125x-(75x+100{,}000)$
$=125x-75x-100{,}000$
$=50x-100{,}000$

b) Solve the system
$R(x)=125x,$
$C(x)=75x+100{,}000.$

Since $R(x)=C(x)$ at the break-even point, we can rewrite the system:

$R(x)=125x, \quad (1)$
$C(x)=75x+100{,}000 \quad (2)$

We solve using substitution.

$$\begin{aligned} 125x &= 75x + 100{,}000 \quad \text{Substituting } 125x \text{ for } R(x) \text{ in (2)} \\ 50x &= 100{,}000 \\ x &= 2000 \end{aligned}$$

To break even 2000 units must be produced and sold.

The revenue will be $R(2000) = 125 \cdot 2000 = 250{,}000$. The break-even point is (2000 units, $250,000).

19. $D(p) = 2000 - 15p$,
$S(p) = 740 + 6p$

Rewrite the system:

$$\begin{aligned} q &= 2000 - 15p, \quad (1) \\ q &= 740 + 6p \quad (2) \end{aligned}$$

Substitute $2000 - 15p$ for q in (2) and solve.

$$\begin{aligned} 2000 - 15p &= 740 + 6p \\ 1260 &= 21p \\ 60 &= p \end{aligned}$$

The equilibrium price is $60 per unit.

To find the equilibrium quantity we substitute $60 into either $D(p)$ or $S(p)$.

$$D(60) = 2000 - 15(60) = 2000 - 900 = 1100$$

The equilibrium quantity is 1100 units.

The equilibrium point is ($60, 1100).

21. $D(p) = 760 - 13p$,
$S(p) = 430 + 2p$

Rewrite the system:

$$\begin{aligned} q &= 760 - 13p, \quad (1) \\ q &= 430 + 2p \quad (2) \end{aligned}$$

Substitute $760 - 13p$ for q in (2) and solve.

$$\begin{aligned} 760 - 13p &= 430 + 2p \\ 330 &= 15p \\ 22 &= p \end{aligned}$$

The equilibrium price is $22 per unit.

To find the equilibrium quantity we substitute $22 into either $D(p)$ or $S(p)$.

$$S(22) = 430 + 2(22) = 430 + 44 = 474$$

The equilibrium quantity is 474 units.

The equilibrium point is ($22, 474).

23. $D(p) = 7500 - 25p$,
$S(p) = 6000 + 5p$

Rewrite the system:

$$\begin{aligned} q &= 7500 - 25p, \quad (1) \\ q &= 6000 + 5p \quad (2) \end{aligned}$$

Substitute $7500 - 25p$ for q in (2) and solve.

$$\begin{aligned} 7500 - 25p &= 6000 + 5p \\ 1500 &= 30p \\ 50 &= p \end{aligned}$$

The equilibrium price is $50 per unit.

To find the equilibrium quantity we substitute $50 into either $D(p)$ or $S(p)$.

$$D(50) = 7500 - 25(50) = 7500 - 1250 = 6250$$

The equilibrium quantity is 6250 units.

The equilibrium point is ($50, 6250).

25. $D(p) = 1600 - 53p$,
$S(p) = 320 + 75p$

Rewrite the system:

$$\begin{aligned} q &= 1600 - 53p, \quad (1) \\ q &= 320 + 75p \quad (2) \end{aligned}$$

Substitute $1600 - 53p$ for q in (2) and solve.

$$\begin{aligned} 1600 - 53p &= 320 + 75p \\ 1280 &= 128p \\ 10 &= p \end{aligned}$$

The equilibrium price is $10 per unit.

To find the equilibrium quantity we substitute $10 into either $D(p)$ or $S(p)$.

$$S(10) = 320 + 75(10) = 320 + 750 = 1070$$

The equilibrium quantity is 1070 units.

The equilibrium point is ($10, 1070).

27. a) $C(x) = \text{Fixed costs} + \text{Variable costs}$
$C(x) = 45{,}000 + 40x$,

where x is the number of MP3/cell phones.

b) Each MP3/cell phone sells for $130. The total revenue is 130 times the number of MP3/cell phone sold. We assume that all MP3/cell phones produced are sold.

$$R(x) = 130x$$

c)
$$\begin{aligned} P(x) &= R(x) - C(x) \\ P(x) &= 130x - (45{,}000 + 40x) \\ &= 130x - 45{,}000 - 40x \\ &= 90x - 45{,}000 \end{aligned}$$

d)
$$\begin{aligned} P(3000) &= 90(3000) - 45{,}000 \\ &= 270{,}000 - 45{,}000 \\ &= \$225{,}000 \end{aligned}$$

The company will realize a profit of $225,000 when 3000 MP3/cell phones are produced and sold.

$$\begin{aligned} P(400) &= 90(400) - 45{,}000 \\ &= 36{,}000 - 45{,}000 \\ &= -\$9000 \end{aligned}$$

The company will realize a $9000 loss when 400 MP3/cell phones are produced and sold.

e) Solve the system

$$\begin{aligned} R(x) &= 130x, \\ C(x) &= 45{,}000 + 40x. \end{aligned}$$

Since both $R(x)$ and $C(x)$ are in dollars and they are equal at the break-even point, we can rewrite the system:

$$\begin{aligned} d &= 130x, \quad (1) \\ d &= 45{,}000 + 40x \quad (2) \end{aligned}$$

We solve using substitution.

$130x = 45{,}000 + 40x$ Substituting $130x$ for d in (2)

$90x = 45{,}000$

$x = 500$

The firm will break even if it produces and sells 500 MP3/cell phones and takes in a total of $R(500) = 130 \cdot 500 = \$65{,}000$ in revenue. Thus, the break-even point is (500 MP3/cell phones, \$65,000).

29. a) $C(x) = \text{Fixed costs} + \text{Variable costs}$

$C(x) = 10{,}000 + 30x,$

where x is the number of pet car seats produced.

b) Each pet car seat sells for \$80. The total revenue is 80 times the number of seats sold. We assume that all seats produced are sold.

$R(x) = 80x$

c) $P(x) = R(x) - C(x)$

$$\begin{aligned} P(x) &= 80x - (10{,}000 + 30x) \\ &= 80x - 10{,}000 - 30x \\ &= 50x - 10{,}000 \end{aligned}$$

d) $$\begin{aligned} P(2000) &= 50(2000) - 10{,}000 \\ &= 100{,}000 - 10{,}000 \\ &= 90{,}000 \end{aligned}$$

The company will realize a profit of \$90,000 when 2000 seats are produced and sold.

$$\begin{aligned} P(50) &= 50(50) - 10{,}000 \\ &= 2500 - 10{,}000 \\ &= -7500 \end{aligned}$$

The company will realize a loss of \$7500 when 50 seats are produced and sold.

e) Solve the system

$R(x) = 80x,$

$C(x) = 10{,}000 + 30x.$

Since both $R(x)$ and $C(x)$ are in dollars and they are equal at the break-even point, we can rewrite the system:

$d = 80x,$ (1)

$d = 10{,}000 + 30x$ (2)

We solve using substitution.

$80x = 10{,}000 + 30x$ Substituting $80x$ for d in (2)

$50x = 10{,}000$

$x = 200$

The firm will break even if it produces and sells 200 seats and takes in a total of $R(200) = 80 \cdot 200 = \$16{,}000$ in revenue. Thus, the break-even point is (200 seats, \$16,000).

31. *Writing Exercise.* Since $P(x) = R(x) - C(x)$, we can also say $R(x) = P(x) + C(x)$. Thus, the slope of the line representing revenue is the sum of the slopes of the other two lines, which represent cost and profit.

33. $$\begin{aligned} 4x - 3 &= 21 \\ 4x &= 24 \\ x &= 6 \end{aligned}$$

35. $$\begin{aligned} 3x - 5 &= 12x + 6 \\ -11 &= 9x \\ -\frac{11}{9} &= x \end{aligned}$$

37. $$\begin{aligned} 3 - (x + 2) &= 7 \\ 3 - x - 2 &= 7 \\ 1 - x &= 7 \\ -x &= 6 \\ x &= -6 \end{aligned}$$

39. *Writing Exercise.* No; there will be variable costs in the production of the birdbaths, so Rosie will need more than $\$300 \cdot 10$, or \$3000, in revenue in order to break even.

41. The supply function contains the points (\$2, 100) and (\$8, 500). We find its equation:

$$m = \frac{500 - 100}{8 - 2} = \frac{400}{6} = \frac{200}{3}$$

$y - y_1 = m(x - x_1)$ Point-slope form

$$\begin{aligned} y - 100 &= \frac{200}{3}(x - 2) \\ y - 100 &= \frac{200}{3}x - \frac{400}{3} \\ y &= \frac{200}{3}x - \frac{100}{3} \end{aligned}$$

We can equivalently express supply S as a function of price p:

$$S(p) = \frac{200}{3}p - \frac{100}{3}$$

The demand function contains the points (\$1, 500) and (\$9,100). We find its equation:

$$m = \frac{100 - 500}{9 - 1} = \frac{-400}{8} = -50$$

$$\begin{aligned} y - y_1 &= m(x - x_1) \\ y - 500 &= -50(x - 1) \\ y - 500 &= -50x + 50 \\ y &= -50x + 550 \end{aligned}$$

We can equivalently express demand D as a function of price p:

$$D(p) = -50p + 550$$

We have a system of equations

$$S(p) = \frac{200}{3}p - \frac{100}{3},$$
$$D(p) = -50p + 550.$$

Rewrite the system:

$$q = \frac{200}{3}p - \frac{100}{3}, \quad (1)$$
$$q = -50p + 550 \quad (2)$$

Substitute $\frac{200}{3}p - \frac{100}{3}$ for q in (2) and solve.

$$\begin{aligned}\frac{200}{3}p-\frac{100}{3}&=-50p+550\\ 200p-100&=-150p+1650 \quad \text{Multiplying by 3 to clear fractions}\\ 350p-100&=1650\\ 350p&=1750\\ p&=5\end{aligned}$$

The equilibrium price is \$5 per unit.

To find the equilibrium quantity, we substitute \$5 into either $S(p)$ or $D(p)$.

$$D(5)=-50(5)+550=-250+550=300$$

The equilibrium quantity is 300 yo-yo's.

The equilibrium point is (\$5, 300 yo-yo's).

43. a) Use a graphing calculator to find the first coordinate of the point of intersection of $y_1=-14.97x+987.35$ and $y_2=98.55x-5.13$, to the nearest hundredth. It is 8.74, so the price per unit that should be charged is \$8.74.

b) Use a graphing calculator to find the first coordinate of the point of intersection of $y_1=87{,}985+5.15x$ and $y_2=8.74x$. It is about 24,508.4, so 24,509 units must be sold in order to break even.

Chapter 8 Review

1. substitution; see Section 8.2.

3. graphical; see Sections 8.1 and 8.2.

5. inconsistent; see Section 8.1.

7. parallel; see Section 8.1.

9. determinant; see Section 8.7.

11. Graph the equations.

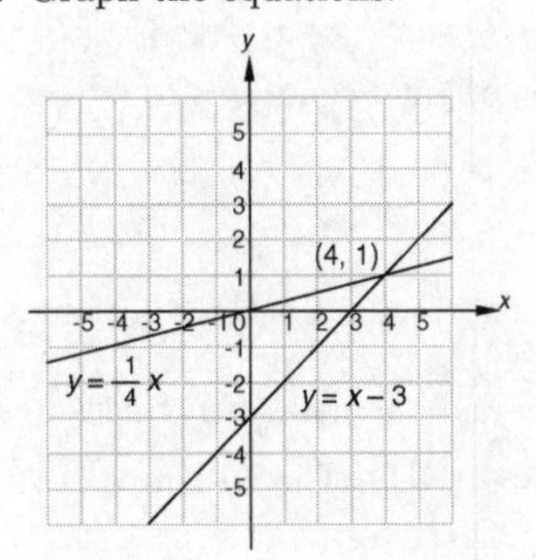

The solution (point of intersection) is (4, 1).

13. $5x-2y=4$, (1)
$x=y-2$ (2)

We substitute $y-2$ for x in Equation (1) and solve for y.

$$\begin{aligned}5x-2y&=4 \quad (1)\\ 5(y-2)-2y&=4 \quad \text{Substituting}\\ 5y-10-2y&=4\\ 3y-10&=4\\ 3y&=14\\ y&=\frac{14}{3}\end{aligned}$$

Next we substitute $\frac{14}{3}$ for y in either equation of the original system and solve for x.

$$\begin{aligned}x&=y-2 \quad (1)\\ x&=\frac{14}{3}-2=\frac{8}{3}\end{aligned}$$

Since $\left(\frac{8}{3}, \frac{14}{3}\right)$ checks, it is the solution.

15. $x-3y=-2$, (1)
$7y-4x=6$ (2)

We solve Equation (1) for x.

$$\begin{aligned}x-3y&=-2\\ x&=3y-2 \quad (3)\end{aligned}$$

We substitute $3y-2$ for x in the second equation and solve for y.

$$\begin{aligned}7y-4x&=6 \quad (2)\\ 7y-4(3y-2)&=6 \quad \text{Substituting}\\ 7y-12y+8&=6\\ -5y+8&=6\\ -5y&=-2\\ y&=\frac{2}{5}\end{aligned}$$

We substitute $\frac{2}{5}$ for y in Equation (3) and solve for x.

$$\begin{aligned}x&=3y-2 \quad (3)\\ x&=3\cdot\frac{2}{5}-2=\frac{6}{5}-2=-\frac{4}{5}\end{aligned}$$

Since $\left(-\frac{4}{5}, \frac{2}{5}\right)$ checks, it is the solution.

17. $4x-7y=18$, (1)
$9x+14y=40$ (2)

We multiply Equation (1) by 2.

$$\begin{aligned}8x-14y&=36 \quad \text{Multiplying (1) by 2}\\ 9x+14y&=40 \quad (2)\\ \hline 17x\qquad&=76 \quad \text{Adding}\\ x&=\frac{76}{17}\end{aligned}$$

Substitute $\frac{76}{17}$ for x in Equation (1) and solve for y.

$$\begin{aligned}4x-7y&=18\\ 4\left(\frac{76}{17}\right)-7y&=18 \quad \text{Substituting}\\ \frac{304}{17}-7y&=18\\ -7y&=\frac{2}{17}\\ y&=-\frac{2}{119}\end{aligned}$$

We obtain $\left(\frac{76}{17},-\frac{2}{119}\right)$. This checks, so it is the solution.

19. $1.5x-3=-2y,$ (1)
$3x+4y=6$ (2)

Rewriting the equations in standard form.

$1.5x+2y=3,$ (1)
$3x+4y=6$ (2)

Observe that, if we multiply Equation (1) by 2, we obtain Equation (2). Thus, any pair that is a solution of Equation (1) is also a solution of Equation (2). The equations are dependent and the solution set is infinite: $\{(x,\ y)\,|\,3x+4y=6\}$.

21. ***Familiarize.*** Let t = the number of hours for the passenger train, and $t+1$ = the number of hours for the freight train.

Now complete the chart.

$d = r \cdot t$

	Distance	Rate	Time	
Passenger train	d	55	t	$\rightarrow d=55t$
Freight train	d	44	$t+1$	$\rightarrow d=44(t+1)$

Translate. Using $d = rt$ in each row of the table, we get a system of equations:

$d=55t,$
$d=44(t+1)$

Carry out. We solve the system of equations.

$55t=44(t+1)$ Using substitution
$55t=44t+44$
$11t=44$
$t=4$

Check. At 55 mph, in 4 hr the passenger train will travel $55\cdot4=220$ mi. At 44 mph, in 4 + 1, or 5 hr the freight train will travel $44\cdot5=220$ mi. The numbers check.

State. The passenger train overtakes the freight train after 4 hours.

23. $x+4y+3z=2,$ (1)
$2x+y+z=10,$ (2)
$-x+y+2z=8$ (3)

1., 2. The equations are already in standard form with no fractions or decimals.

3. Use Equations (1) and (3) to eliminate x.

$x+4y+3z=2$ (1)
$-x+y+2z=8$ (3)
$5y+5z=10$ (4) Adding

4. Use a different pair of equations and eliminate x.

$2x+y+z=10$ (2)
$-2x+2y+4z=16$ Multiplying (3) by 2
$3y+5z=26$ (5) Adding

5. Now solve the system of Equations (4) and (5).

$5y+5z=10$ (4)
$3y+5z=26$ (5)

$5y+5z=10$ (4)
$-3y-5z=-26$ Multiplying (5) by -1
$2y=-16$ Adding
$y=-8$
$5(-8)+5z=10$ Substituting -8 for y in (4)
$-40+5z=10$
$5z=50$
$z=10$

6. Substitute in one of the original equations to find x.

$x+4(-8)+3(10)=2$ Substituting -8 for y and 10 for z in (1)
$x-2=2$
$x=4$

We obtain (4, –8, 10). This checks, so it is the solution.

25. $2x-5y-2z=-4,$ (1)
$7x+2y-5z=-6,$ (2)
$-2x+3y+2z=4$ (3)

1., 2. The equations are already in standard form with no fractions or decimals.

3. Use Equations (1) and (2) to eliminate x.

$14x-35y-14z=-28$ Multiplying (1) by 7
$-14x-4y+10z=12$ Multiplying (2) by -2
$-39y-4z=-16$ (4) Adding

4. Use Equations (1) and (3) to eliminate x.

$2x-5y-2z=-4$ (1)
$-2x+3y+2z=4$ (3)
$-2y=0$ (5) Adding
$y=0$

5. When we used Equation (1) and (3) to eliminate x, we also eliminated z and found $y=0$. Substitute 0 for y in Equation (4) to find z.

$-39\cdot0-4z=-16$ Substituting 0 for y in (4)
$-4z=-16$
$z=4$

6. Substitute in one of the original equations to find x.

$2x-5\cdot0-2\cdot4=-4$
$2x=4$
$x=2$

We obtain (2, 0, 4). This checks, so it is the solution.

27. ***Familiarize.*** We let x, y, and z represent the measures of angles A, B, and C, respectively. The measures of the angles of a triangle add up to 180°.

Translate.

The sum of the measures is 180.
$x+y+z = 180$

The measure of angle A	is	four times the measure of angle C.
$\downarrow$	$\downarrow$	$\downarrow$
x	$=$	$4z$

The measure of angle B	is	45° more than the measure of angle C.
$\downarrow$	$\downarrow$	$\downarrow$
y	$=$	$z+45$

We now have a system of equations.

$$x+y+z=180,$$
$$x=4z,$$
$$y=z+45$$

Carry out. Solving the system we get (90, 67.5, 22.5).

Check. The sum of the measures is $90°+67.5°+22.5°$, or 180°. Four times the measure of angle C is $4\cdot 22.5°$, or 90°, the measure of angle A. 45 more than the measure of angle C is $45°+22.5°$, or 67.5°, the measure of angle B. The numbers check.

State. The measures of angles A, B, and C are 90°, 67.5°, and 22.5°, respectively.

29. ***Familiarize.*** Let x, y, and z represent the average number of cries for a man, woman and a one-year-old child, respectively.

Translate.

The sum for each month is 56.7, so we have

$$x+y+z=56.7.$$

The number of cries for a woman is 3.9 more than the man, so we have

$$y=3.9+x.$$

The number of cries for the child is 43.3 more than the sum for the man and woman, so we have

$$z=43.3+x+y.$$

We now have a system of equations.

$$x+y+z=56.7,$$
$$y=3.9+x,$$
$$z=43.3+x+y$$

Carry out. Solving the system, we get (1.4, 5.3, 50).

Check. The sum of the average number of cry times for a man, a woman, and a one-year-old child is $1.4+5.3+50$, or 56.7. A woman cries 3.9 + 1.4, or 5.3 times a month. A one-year-old child cries 43.3 + 1.4 + 5.3, or 50 times a month. The answer checks.

State. The monthy average number of cries for a man is 1.4, for a woman is 5.3, and a one-year-old child is 50.

31.
$$3x-y+z=-1,$$
$$2x+3y+z=4,$$
$$5x+4y+2z=5$$

We first write a matrix using only the constants.

$$\left[\begin{array}{ccc|c} 3 & -1 & 1 & -1 \\ 2 & 3 & 1 & 4 \\ 5 & 4 & 2 & 5 \end{array}\right]$$

$$\left[\begin{array}{ccc|c} 3 & -1 & 1 & -1 \\ 0 & 11 & 1 & 14 \\ 0 & 17 & 1 & 20 \end{array}\right] \begin{array}{l} \\ \text{New Row 2} = -2(\text{Row 1}) + 3(\text{Row 2}) \\ \text{New Row 3} = -5(\text{Row 1}) + 3(\text{Row 3}) \end{array}$$

$$\left[\begin{array}{ccc|c} 3 & -1 & 1 & -1 \\ 0 & 11 & 1 & 14 \\ 0 & 187 & 11 & 220 \end{array}\right] \quad \text{New Row 3} = 11(\text{Row 3})$$

$$\left[\begin{array}{ccc|c} 3 & -1 & 1 & -1 \\ 0 & 11 & 1 & 14 \\ 0 & 0 & -6 & -18 \end{array}\right] \quad \text{New Row 3} = -17(\text{Row 2}) + \text{Row 3}$$

Reinserting the variables, we have

$$3x-y+z=-1, \quad (1)$$
$$11y+z=14, \quad (2)$$
$$-6z=-18 \quad (3)$$

Solve (3) for z.

$$-6z=-18$$
$$z=3$$

Substitute 3 for z in (2) and solve for y.

$$11y+3=14$$
$$11y=11$$
$$y=1$$

Substitute 1 for y and 3 for z in (1) and solve for x.

$$3x-1+3=-1$$
$$3x+2=-1$$
$$3x=-3$$
$$x=-1$$

The solution is (−1, 1, 3).

33.
$$\begin{vmatrix} 2 & 3 & 0 \\ 1 & 4 & -2 \\ 2 & -1 & 5 \end{vmatrix}$$

$$=2\begin{vmatrix} 4 & -2 \\ -1 & 5 \end{vmatrix} - 1\begin{vmatrix} 3 & 0 \\ -1 & 5 \end{vmatrix} + 2\begin{vmatrix} 3 & 0 \\ 4 & -2 \end{vmatrix}$$

$$=2[4\cdot 5-(-2)(-1)]-1[3\cdot 5-0(-1)]$$
$$+2[3(-2)-0\cdot 4]$$
$$=2(18)-1(15)+2(-6)$$
$$=36-15-12=9$$

35.
$$2x + y + z = -2,$$
$$2x - y + 3z = 6,$$
$$3x - 5y + 4z = 7$$

We compute D, D_x, and D_y.

$$D=\begin{vmatrix} 2 & 1 & 1 \\ 2 & -1 & 3 \\ 3 & -5 & 4 \end{vmatrix}$$

$$=2\begin{vmatrix} -1 & 3 \\ -5 & 4 \end{vmatrix} - 2\begin{vmatrix} 1 & 1 \\ -5 & 4 \end{vmatrix} + 3\begin{vmatrix} 1 & 1 \\ -1 & 3 \end{vmatrix}$$

$$=2(11)-2(9)+3(4)=22-18+12=16$$

$$D_x = \begin{vmatrix} -2 & 1 & 1 \\ 6 & -1 & 3 \\ 7 & -5 & 4 \end{vmatrix}$$
$$= -2\begin{vmatrix} -1 & 3 \\ -5 & 4 \end{vmatrix} - 6\begin{vmatrix} 1 & 1 \\ -5 & 4 \end{vmatrix} + 7\begin{vmatrix} 1 & 1 \\ -1 & 3 \end{vmatrix}$$
$$= -2(11) - 6(9) + 7(4)$$
$$= -22 - 54 + 28 = -48$$
$$D_y = \begin{vmatrix} 2 & -2 & 1 \\ 2 & 6 & 3 \\ 3 & 7 & 4 \end{vmatrix}$$
$$= 2\begin{vmatrix} 6 & 3 \\ 7 & 4 \end{vmatrix} - 2\begin{vmatrix} -2 & 1 \\ 7 & 4 \end{vmatrix} + 3\begin{vmatrix} -2 & 1 \\ 6 & 3 \end{vmatrix}$$
$$= 2(3) - 2(-15) + 3(-12)$$
$$= 6 + 30 - 36 = 0$$

Then,
$$x = \frac{D_x}{D} = \frac{-48}{16} = -3,$$
and
$$y = \frac{D_y}{D} = \frac{0}{16} = 0.$$

We substitute in the second equation to find z.
$$2(-3) + 0 + z = -2$$
$$-6 + z = -2$$
$$z = 4$$

The solution is (–3, 0, 4).

37. a) $C(x) = 4.75x + 54{,}000$, where x is the number of pints of honey.

b) Each pint of honey sells for \$9.25. The total revenue is 9.25 times the number of pints of honey. We assume all the honey is sold. $R(x) = 9.25x$

c) $P(x) = R(x) - C(x)$
$$= 9.25x - (4.75x + 54{,}000)$$
$$= 9.25x - 4.75x - 54{,}000$$
$$= 4.5x - 54{,}000$$

d) $P(5000) = 4.5(5000) - 54{,}000$
$$= 22{,}500 - 54{,}000$$
$$= -31{,}500$$

Danae will realize a \$31,500 loss when 5000 pints of honey are produced and sold.
$$P(15{,}000) = 4.5(15{,}000) - 54{,}000$$
$$= 67{,}500 - 54{,}000$$
$$= 13{,}500$$

Danae will realize a profit of \$13,500 when 15,000 pints of honey are produced and sold.

e) Solve the system
$$R(x) = 9.25x,$$
$$C(x) = 4.75x + 54{,}000.$$

Since both $R(x)$ and $C(x)$ are in dollars and they are equal at the break-even point, we can rewrite the system:
$$d = 9.25x, \quad (1)$$
$$d = 4.75x + 54{,}000 \quad (2)$$

We solve using substitution.
$$9.25x = 4.75x + 54{,}000$$ Substituting $9.25x$ for d in (2)
$$4.5x = 54{,}000$$
$$x = 12{,}000$$

Thus, 12,000 units must be produced and sold in order to break even.

The revenue will be $R(12{,}000) = 9.25 \cdot 12{,}000 = \$111{,}000$.

The break-even point is (12,000 pints, \$111,000).

39. *Writing Exercise.* A system of equations can be both dependent and inconsistent if it is equivalent to a system with fewer equations that has no solution. An example is a system of three equations in three unknowns in which two of the equations represent the same plane, and the third represents a parallel plane.

41. ***Familiarize.*** Let $x =$ the number of 2-count packs of pencils and $y =$ the number of 12-count packs of pencils.

Translate.

The total number of pencils is 138, so we have
$$2x + 12y = 138.$$

The purchase price is \$157.26, so we have
$$5.99x + 7.49y = 157.26.$$

After clearing decimals, we have the problem translated to a system of equations:
$$2x + 12y = 138 \quad (1)$$
$$599x + 749y = 15{,}726 \quad (2)$$

Carry out. We use the elimination method to solve the system of equations.
$$-1198x - 7188y = -82{,}662$$ Multiplying (1) by -599
$$1198x + 1498y = 31{,}452$$ Multiplying (2) by 2
$$-5690y = -51{,}210$$
$$y = 9$$

Substitute 9 for y in Equation (1) and solve for x.
$$2x + 12 \cdot 9 = 138$$
$$2x + 108 = 138$$
$$2x = 30$$
$$x = 15$$

Check. The total number of pencils is $12(15) + 12(9) = 30 + 108$, or 138 pencils. The purchase price is $\$5.99(15) + \$7.49(9) = \$89.85 + \$67.41 = \$157.26$. The answer checks.

State. Wiese Accounting purchased 15 packs of Round Stic Grip and 9 packs of Matic Grip pencils.

43. $f(x)=ax^2+bx+c$

The three points are (–2, 3), (1, 1) and (0, 3).

We substitute for x and $f(x)$ and solve for a, b, and c.

$$3=4a-2b+c$$
$$1=a+b+c$$
$$3=c$$

1., 2. We put the equations in standard form. There are no fractions or decimals.

$$4a-2b+c=3,\quad (1)$$
$$a+b+c=1,\quad (2)$$
$$c=3\quad (3)$$

3., 4. Note that there is no a or b in Equation (3). We will use Equations (1) and (2) to obtain another equation with no b-term.

$$\begin{aligned} 4a-2b+c&=3 && (1)\\ 2a+2b+2c&=2 && \text{Multiplying (2) by 2}\\ \hline 6a\quad+3c&=5 && (4) \end{aligned}$$

5. Now substitute 3 for c in (4) to solve for a.

$$\begin{aligned} 6a+3\cdot 3&=5 && \text{Substituting 3 for } c \text{ in (4)}\\ 6a&=-4\\ a&=-\frac{2}{3} \end{aligned}$$

6. Substitute in Equation (2) to find b.

$$\begin{aligned} -\frac{2}{3}+b+3&=1 && \text{Substituting}\\ b&=-\frac{4}{3} \end{aligned}$$

We obtain $\left(-\frac{2}{3},-\frac{4}{3},\ 3\right)$. So, the function is

$f(x)=-\frac{2}{3}x^2-\frac{4}{3}x+3$.

Chapter 8 Test

1. Graph the equations.

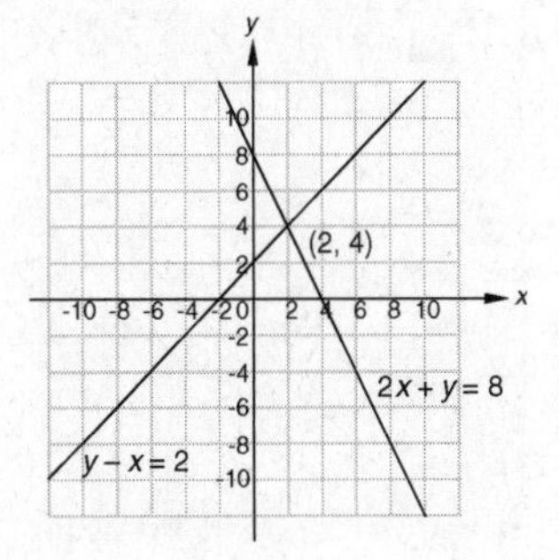

3. $2x-4y=-6\quad (1)$
$x=2y-3\quad (2)$

We substitute $2y-3$ for x in the first equation and solve for y.

$$\begin{aligned} 2x-4y&=-6\\ 2(2y-3)-4y&=-6\\ 4y-6-4y&=-6\\ -6&=-6 \end{aligned}$$

We have an identity, or an equation that is always true. The equations are dependent and the solution set is infinite: $\{(x,\ y)|2y-3=x\}$.

5. $4y+2x=18,$
$3x+6y=26$

Rewrite the equations in standard form.

$$2x+4y=18,\quad (1)$$
$$3x+6y=26\quad (2)$$

We multiply Equation (1) by –3 and Equation (2) by 2.

$$\begin{aligned} -6x-12y&=-54 && \text{Multiplying (1) by } -3\\ 6x+12y&=52 && \text{Multiplying (2) by 2}\\ \hline 0&=-2 && \text{Adding} \end{aligned}$$

We have a contradiction, or an equation that is always false. Therefore, there is no solution.

7. ***Familiarize.*** Let w = the width of the basketball court and let l = the length of the basketball court. Recall that the perimeter of a rectangle is given by the formula $P=2w+2l$.

Translate.

The perimeter of the court is 288 ft, so we have

$$2w+2l=288.$$

The length if 44 longer than the width, so we have

$$l=44+w.$$

We now have a system of equations.

$$2w+2l=288.\quad (1)$$
$$l=44+w\quad (2)$$

Carry out. Substitute 44 + w for l in the first equation and solve for w.

$$\begin{aligned} 2w+2l&=288\\ 2w+2(44+w)&=288\\ 2w+88+2w&=288\\ 4w+88&=288\\ 4w&=200\\ w&=50 \end{aligned}$$

Finally substitute 50 for w in Equation (2) and solve for l.

$$l=44+w$$
$$l=44+50=94$$

Check. The perimeter of the court is $2(94\text{ ft})+2(50\text{ ft})=188$ ft + 100 ft = 288 ft. The length, 94 ft, is 44 ft more than the width, or 44 + 50 = 94.

State. The basketball court is 94 ft long and 50 ft wide.

9. ***Familiarize.*** Let d = the distance. Let t = the time it takes the car to travel at 65 mph. Then t + 1 is the time it takes the car to travel at 55 mph. The distances are the same.

Organize the information in a table.

	Distance	Rate	Time
Truck	d	55	$t+1$
Car	d	65	t

Translate. Using $d = rt$ in each row of the table, we get a system of equations:

$$d = 55(t+1)$$
$$d = 65t$$

Carry out. Solve the system of equations.

$$55(t+1) = 65t \quad \text{Substituting}$$
$$55t + 55 = 65t$$
$$55 = 10t$$
$$5.5 = t$$

Check. At 55 mph in 5.5 + 1, or 6.5 the truck travels 55(6.5), or 357.5 miles. At 65 mph, in 5.5 hr, the car travels 65(5.5), or 357.5 miles. The answer checks.

State. The car will catch up to the truck in 5.5 hr.

11. $6x + 2y - 4z = 15,$ (1)
$-3x - 4y + 2z = -6,$ (2)
$4x - 6y + 3z = 8$ (3)

1., 2. The equations are already in standard form with no fractions or decimals.

3., 4. We eliminate x from two different pairs of equations.

$$6x + 2y - 4z = 15 \quad (1)$$
$$-6x - 8y + 4z = -12 \quad \text{Multiplying (2) by 2}$$
$$-6y = 3 \quad (4)$$
$$y = -\frac{1}{2}$$

$$-12x - 16y + 8z = -24 \quad \text{Multiplying (2) by 4}$$
$$12x - 18y + 9z = 24 \quad \text{Multiplying (3) by 3}$$
$$-34y + 17z = 0 \quad (5)$$

5. Now substitute $-\frac{1}{2}$ for y in (5) to solve for z.

$$-34\left(-\frac{1}{2}\right) + 17z = 0 \quad \text{Substituting } -\frac{1}{2} \text{ for } y \text{ in (5)}$$
$$17 + 17z = 0$$
$$z = -1$$

6. Substitute in Equation (1) to find x.

$$6x + 2\left(-\frac{1}{2}\right) - 4(-1) = 15 \quad \text{Substituting}$$
$$6x - 1 + 4 = 15$$
$$6x = 12$$
$$x = 2$$

We obtain $\left(2, -\frac{1}{2}, -1\right)$. This checks, so it is the solution.

13. $3x + 3z = 0,$ (1)
$2x + 2y = 2,$ (2)
$3y + 3z = 3$ (3)

1., 2. The equations are already in standard form with no fractions or decimals.

3. , 4. Note that there is no z in Equation (2). We will use Equations (1) and (3) to obtain another equation with no z-term.

$$3x + 3z = 0 \quad (1)$$
$$-3y - 3z = -3 \quad \text{Multiplying (3) by } -1$$
$$3x - 3y = -3 \quad (4)$$

5. Now solve the system of Equations (2) and (4).

$$2x + 2y = 2 \quad (2)$$
$$3x - 3y = -3 \quad (4)$$

$$6x + 6y = 6 \quad \text{Multiplying (2) by 3}$$
$$-6x + 6y = 6 \quad \text{Multiplying (4) by } -2$$
$$12y = 12$$
$$y = 1$$
$$2x + 2\cdot 1 = 2 \quad \text{Substituting in (2)}$$
$$2x = 0$$
$$x = 0$$

6. Substitute in Equation (1) or (3) to find z.

$$3\cdot 0 + 3z = 0 \quad \text{Substituting in (1)}$$
$$3z = 0$$
$$z = 0$$

We obtain (0, 1, 0). This checks, so it is the solution.

15. $x + 3y - 3z = 12,$
$3x - y + 4z = 0,$
$-x + 2y - z = 1$

We first write a matrix using only the constants.

$$\left[\begin{array}{ccc|c} 1 & 3 & -3 & 12 \\ 3 & -1 & 4 & 0 \\ -1 & 2 & -1 & 1 \end{array}\right]$$

$$\left[\begin{array}{ccc|c} 1 & 3 & -3 & 12 \\ 0 & -10 & 13 & -36 \\ 0 & 5 & -4 & 13 \end{array}\right] \quad \begin{array}{l} \\ \text{New Row 2} = -3(\text{Row 1}) + \text{Row 2} \\ \text{New Row 3} = (\text{Row 1}) + \text{Row 3} \end{array}$$

$$\left[\begin{array}{ccc|c} 1 & 3 & -3 & 12 \\ 0 & 5 & -4 & 13 \\ 0 & -10 & 13 & -36 \end{array}\right] \quad \begin{array}{l} \text{Interchange} \\ \text{Row 2 and Row 3} \end{array}$$

$$\left[\begin{array}{ccc|c} 1 & 3 & -3 & 12 \\ 0 & 5 & -4 & 13 \\ 0 & 0 & 5 & -10 \end{array}\right] \quad \text{New Row 3} = 2(\text{Row 2}) + \text{Row 3}$$

Reinserting the variables, we have

$$x + 3y - 3z = 12, \quad (1)$$
$$5y - 4z = 13, \quad (2)$$
$$5z = -10. \quad (3)$$

Solve (3) for z.

$$5z = -10$$
$$z = -2$$

Substitute –2 for z in (2) and solve for y.

$$5y-4z=13$$
$$5y-4(-2)=13$$
$$5y=5$$
$$y=1$$

Substitute 1 for y in (1) and solve for x.

$$x+3\cdot1-3(-2)=12$$
$$x+3+6=12$$
$$x=3$$

The solution is (3, 1, –2).

17. $\begin{vmatrix} 3 & 4 & 2 \\ -2 & -5 & 4 \\ 0 & 5 & -3 \end{vmatrix}$

$$=3\begin{vmatrix} -5 & 4 \\ 5 & -3 \end{vmatrix}-(-2)\begin{vmatrix} 4 & 2 \\ 5 & -3 \end{vmatrix}+0\begin{vmatrix} 4 & 2 \\ -5 & 4 \end{vmatrix}$$
$$=3[(-5)(-3)-5\cdot4]+2[4(-3)-5\cdot2]$$
$$+0[4\cdot4-2(-5)]$$
$$=3(-5)+2(-22)+0$$
$$=-15-44=-59$$

19. ***Familiarize.*** Let x, y, and z represent the number of hours for the electrician, carpenter and plumber, respectively.

Translate.

The total number of hours worked is 21.5, so we have

$$x+y+z=21.5.$$

The total earnings is \$673, so we have

$$30x+28.50y+34z=673.$$

The plumber worked 2 more hours than the carpenter, so we have

$$z=2+y.$$

We have a system of equations:

$$x+y+z=21.5,$$
$$30x+28.50y+34z=673,$$
$$z=2+y$$

Carry out. Solving the system we get (3.5, 8, 10).

Check. The total number of hours is 3.5 + 8 + 10, or 21.5 h. The total amount earned is $\$30(3.5)+\$28.50(8)+\$34(10)=\$105+\$228+\$340$, or \$673. The plumber worked 10 h, which is 2 h more than the 8 h worked by the carpenter. The numbers check.

State. The electrician worked 3.5 h, the carpenter worked 8 h and the plumber worked 10 h.

21. a) $C(x)=25x+44{,}000$, where x is the number of hammocks produced.

b) $R(x)=80x$ We assume all hammocks are sold.

c) $P(x)=R(x)-C(x)$
$$=80x-(25x+44{,}000)$$
$$=80x-25x-44{,}000$$
$$=55x-44{,}000$$

d) $P(300)=55(300)-44{,}000=16{,}500-44{,}000$
$$=-27{,}500$$

The company will realize a \$27,500 loss when 300 hammocks are produced and sold.

$$P(900)=55(900)-44{,}000=49{,}500-44{,}000$$
$$=5500$$

The company will realize a profit of \$5500 when 900 hammocks are produced and sold.

e) Solve the system

$$R(x)=80x,$$
$$C(x)=25x+44{,}000.$$

Since both $R(x)$ and $C(x)$ are in dollars and they are equal at the break-even point, we can rewrite the system:

$$d=80x, \quad (1)$$
$$d=25x+44{,}000 \quad (2)$$

We solve using substitution.

$$80x=25x+44{,}000$$ Substituting $80x$ for d in (2)
$$55x=44{,}000$$
$$x=800$$

Thus, 800 hammocks must be produced and sold in order to break even.

The revenue will be $R(800)=80\cdot800=\$64{,}000$.

The break-even point is (800 hammocks, \$64,000).

23. ***Familiarize.*** Let x = the number of pounds of Kona coffee

Translate. We organize the information in a table.

	Kona	Mexican	Mixture
Number of pounds	x	40	$x+40$
Percent Kona	100%	0%	30%
Amount of Kona	$1.00x$	0	$0.30(x+40)$

The last row of the table gives us one equation:

$$1.00x+0=0.3(x+40).$$

Carry out. After clearing decimals, we solve the equation.

$$100x+0=30(x+40)$$
$$100x=30x+1200$$
$$70x=1200$$
$$x=\frac{120}{7}$$

Check. 30% of $\left(\frac{120}{7}+40\right)$ lb is $0.3\left(\frac{400}{7}\right)$, or $\frac{120}{7}$ lb. of Kona coffee. The answer checks.

State. At least $\frac{120}{7}$ lb of Kona coffee is added to the 40 lb of Mexican coffee to market the mixture as Kona Blend.

Chapter 9

Inequalities and Problem Solving

Exercise Set 9.1

1. If we add $3x$ to both sides of the equation $5x + 7 = 6 - 3x$, we get the equation $8x + 7 = 6$, so these are equivalent equations.

3. If we add 7 to both sides of the inequality $x - 7 > -2$, we get the inequality $x > 5$ so these are equivalent inequalities.

5. The solution set of $-4t \leq 12$ is $\{t \mid t \geq -3\}$ and the solution set of $t \leq -3$ is $\{t \mid t \leq -3\}$. The solution sets are not the same, so the inequalities are not equivalent.

7. The expressions are equivalent by the distributive law.

9. The solution set of $-\frac{1}{2}x < 7$ is $\{x \mid x > -14\}$ and the solution set of $x > 14$ is $\{x \mid x > 14\}$. The solution sets are not the same, so the inequalities are not equivalent.

11.
$$\begin{aligned} 3x+1 &< 7 \\ 3x &< 6 \quad \text{Adding } -1 \\ x &< 2 \quad \text{Dividing by 3} \end{aligned}$$
The solution set is $\{x \mid x < 2\}$, or $(-\infty, 2)$.

13.
$$\begin{aligned} 3-x &\geq 12 \\ -x &\geq 9 \quad \text{Adding } -3 \\ x &\leq -9 \quad \text{Dividing by } -1 \text{ and reversing the inequality symbol} \end{aligned}$$
The solution set is $\{x \mid x \leq -9\}$, or $(-\infty, -9]$.

15.
$$\begin{aligned} \frac{2x+7}{5} &< -9 \\ 5 \cdot \frac{2x+7}{5} &< 5(-9) \quad \text{Multiplying by 5} \\ 2x+7 &< -45 \\ 2x &< -52 \quad \text{Adding } -7 \\ x &< -26 \quad \text{Dividing by 2} \end{aligned}$$
The solution set is $\{x \mid x < -26\}$, or $(-\infty, -26)$.

17.
$$\begin{aligned} \frac{3t-7}{-4} &\leq 5 \\ -4 \cdot \frac{3t-7}{-4} &\geq -4 \cdot 5 \quad \text{Multiplying by } -4 \text{ and reversing the inequality symbol} \\ 3t-7 &\geq -20 \\ 3t &\geq -13 \quad \text{Adding 7} \\ t &\geq -\frac{13}{3} \quad \text{Dividing by 3} \end{aligned}$$
The solution set is $\left\{t \middle| t \geq -\frac{13}{3}\right\}$, or $\left[-\frac{13}{3}, \infty\right)$.

19.
$$\begin{aligned} 3-8y &\geq 9-4y \\ -4y+3 &\geq 9 \\ -4y &\geq 6 \\ y &\leq -\frac{3}{2} \end{aligned}$$
The solution set is $\left\{y \middle| y \leq -\frac{3}{2}\right\}$, or $\left(-\infty, -\frac{3}{2}\right]$.

21.
$$\begin{aligned} 5(t-3)+4t &< 2(7+2t) \\ 5t-15+4t &< 14+4t \\ 9t-15 &< 14+4t \\ 5t-15 &< 14 \\ 5t &< 29 \\ t &< \frac{29}{5} \end{aligned}$$
The solution set is $\left\{t \middle| t < \frac{29}{5}\right\}$, or $\left(-\infty, \frac{29}{5}\right)$.

23.
$$\begin{aligned} 5[3m-(m+4)] &> -2(m-4) \\ 5(3m-m-4) &> -2(m-4) \\ 5(2m-4) &> -2(m-4) \\ 10m-20 &> -2m+8 \\ 12m-20 &> 8 \\ 12m &> 28 \\ m &> \frac{28}{12} \\ m &> \frac{7}{3} \end{aligned}$$
The solution set is $\left\{m \middle| m > \frac{7}{3}\right\}$, or $\left(\frac{7}{3}, \infty\right)$.

25. The graphs intersect at $(2, 3)$. Since the inequality symbol is $\geq$, we include the point in the solution. From the graph, we see that for values to the right of the point $(2, 3)$, $f(x) \geq g(x)$.

The solution set is $\{x \mid x \geq 2\}$, or $[2, \infty)$.

27. The graphs intersect at (0, 3). Since the inequality symbol is <, we do not include the point in the solution. From the graph, we see that for values to the right of the point (0, 3), $f(x) < g(x)$.
The solution set is $\{x|x > 0\}$, or $(0, \infty)$.

29. Solve graphically: $x - 3 < 4$.
We graph and solve the system of equations
$y = x - 3,$
$y = 4$

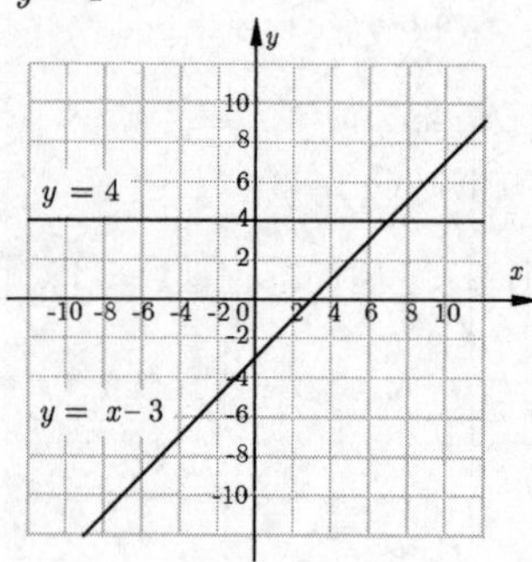

By substitution, we have
$x - 3 = 4$
$x = 7$
Then $y = 7 - 3 = 4$.
Thus, the point of intersection is (7, 4). Since the inequality symbol is <, we do not include this point in the solution.
We see from the graph, $x - 3 < 4$ is to the left of the point of intersection.
Thus, the solution set is $\{x|x < 7\}$, or $(-\infty, 7)$.

31. Solve graphically: $2x - 3 \geq 1$.
We graph and solve the system of equations
$y = 2x - 3,$
$y = 1$

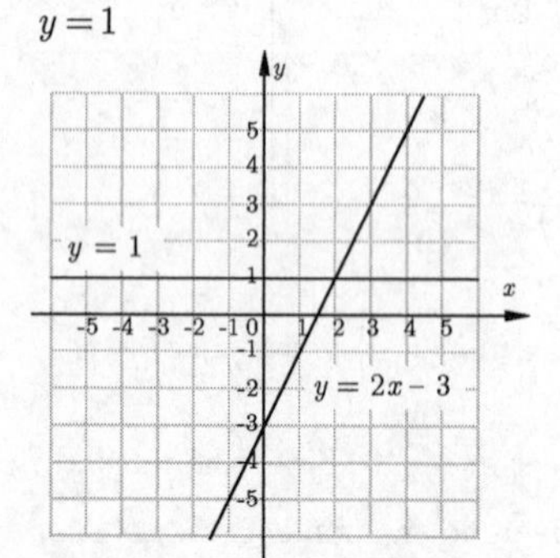

By substitution, we have
$2x - 3 = 1$
$2x = 4$
$x = 2$
Then $y = 2(2) - 3 = 1$.
Thus, the point of intersection is (2, 1). Since the inequality symbol is $\geq$, we include this point in the solution.
We see from the graph, $2x - 3 \geq 1$ is to the right of the point of intersection.
Thus, the solution set is $\{x|x \geq 2\}$, or $[2, \infty)$.

32. Solve graphically: $3x + 1 < 1$.
We graph and solve the system of equations
$y = 3x + 1,$
$y = 1$

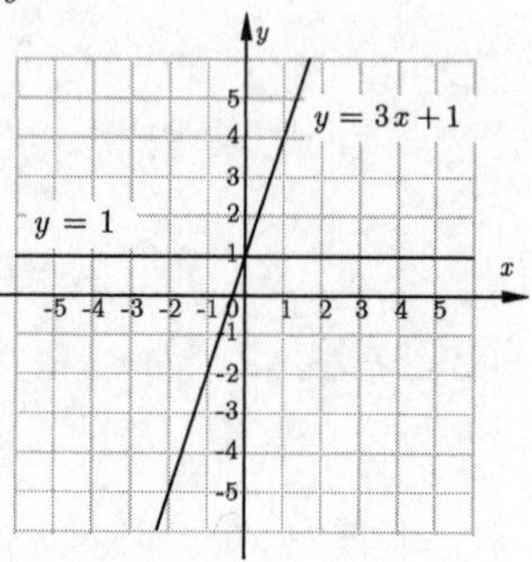

By substitution, we have
$3x + 1 = 1$
$3x = 0$
$x = 0$
Then $y = 3(0) + 1 = 1$.
Thus, the point of intersection is (0, 1). Since the inequality symbol is <, we do not include this point in the solution.
We see from the graph, $3x + 1 < 1$ is to the left of the point of intersection.
Thus, the solution set is $\{x|x < 0\}$, or $(-\infty, 0)$.

33. Solve graphically: $x + 3 > 2x - 5$.
We graph and solve the system of equations
$y = x + 3,$
$y = 2x - 5$

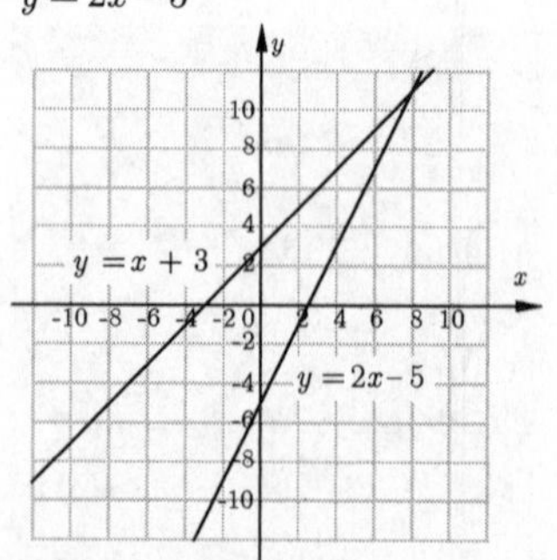

By substitution, we have
$x + 3 = 2x - 5$
$3 = x - 5$
$8 = x$
Then $y = 8 + 3 = 11$.
Thus, the point of intersection is (8, 11). Since the

inequality symbol is $>$, we do not include this point in the solution.

We see from the graph, $x+3>2x-5$ is to the left of the point of intersection.

Thus, the solution set is $\{x|x<8\}$, or $(-\infty, 8)$.

35. Solve graphically: $\frac{1}{2}x-2\leq 1-x$.

We graph and solve the system of equations

$$y=\frac{1}{2}x-2,$$
$$y=1-x$$

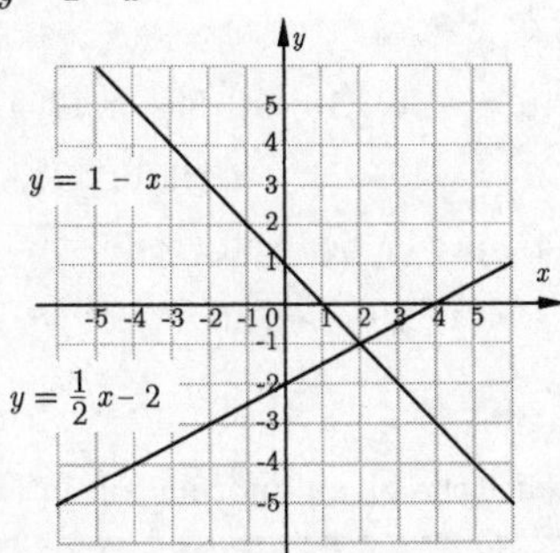

By substitution, we have

$$\frac{1}{2}x-2=1-x$$
$$\frac{3}{2}x-2=1$$
$$\frac{3}{2}x=3$$
$$x=2$$

Then $y = 1 - 2 = -1$.

Thus, the point of intersection is $(2, -1)$. Since the inequality symbol is $\leq$, we include this point in the solution.

We see from the graph, $\frac{1}{2}x-2\leq 1-x$ is to the left of the point of intersection.

Thus, the solution set is $\{x|x\leq 2\}$, or $(-\infty, 2]$.

37. $f(x)=7-3x$, $g(x)=2x-3$

$f(x)\leq g(x)$
$7-3x\leq 2x-3$
$7-5x\leq -3$ Adding $-2x$
$-5x\leq -10$ Adding -7
$x\geq 2$ Multiplying by $-\frac{1}{5}$ and reversing the inequality symbol

The solution set is $\{x|x\geq 2\}$, or $[2, \infty)$.

−2 0 2

39. $y_1=2x-7$, $y_2=5x-9$

$y_1<y_2$
$2x-7<5x-9$
$-3x-7<-9$ Adding $-5x$
$-3x<-2$ Adding 7
$x>\frac{2}{3}$ Dividing by -3

The solution set is $\left\{x\,|\,x>\frac{2}{3}\right\}$, or $\left(\frac{2}{3}, \infty\right)$.

0 $\frac{2}{3}$ 1

41. $f(x)=\sqrt{x-10}$

$x-10\geq 0$ $x-10$ must be nonnegative.
$x\geq 10$ Adding 10

When $x\geq 10$, the expression $x-10$ is nonnegative. Thus the domain of f is $\{x\mid x\geq 10\}$, or $[10, \infty)$.

43. $f(x)=\sqrt{3-x}$

$3-x\geq 0$ $3-x$ must be nonnegative.
$-x\geq -3$ Adding -3
$x\leq 3$ Multiplying by -1 and reversing the inequality symbol

When $x\leq 3$, the expression $3-x$ is nonnegative. Thus the domain of f is $\{x\mid x\leq 3\}$, or $(-\infty, 3]$.

45. $f(x)=\sqrt{2x+7}$

$2x+7\geq 0$ $2x+7$ must be nonnegative.
$2x\geq -7$ Adding -7
$x\geq -\frac{7}{2}$ Dividing by 2

When $x\geq -\frac{7}{2}$, the expression $2x+7$ is nonnegative.

Thus the domain of f is $\left\{x\middle|x\geq -\frac{7}{2}\right\}$, or $\left[-\frac{7}{2}, \infty\right)$.

47. $f(x)=\sqrt{8-2x}$

$8-2x\geq 0$ $8-2x$ must be nonnegative.
$-2x\geq -8$ Adding -8
$x\leq 4$ Dividing by -2 and reversing the inequality symbol

When $x\leq 4$, the expression $8-2x$ is nonnegative. Thus the domain of f is $\{x\mid x\leq 4\}$, or $(-\infty, 4]$.

49. $f(x)=\sqrt{\frac{1}{3}x+5}$

$\frac{1}{3}x+5\geq 0$ $\frac{1}{3}x+5$ must be nonnegative.
$\frac{1}{3}x\geq -5$
$x\geq -15$

Thus the domain of f is $\{x|x\geq -15\}$, or $[-15, \infty)$.

51. $f(x)=\sqrt{\dfrac{x-5}{4}}$

$x-5\geq 0$ $\quad$ $x-5$ must be nonnegative.

$x\geq 5$

Thus the domain of f is $\{x|x\geq 5\}$, or $[5,\infty)$.

53. ***Familiarize.*** Let $n=$ the number of hours. Then the total fee using the hourly plan is $120n$.

Translate. We write an inequality stating that the hourly plan costs less than the flat fee.

$$120n < 900$$

Carry out.

$$120n<900$$
$$n<\frac{15}{2},\text{ or } 7\frac{1}{2}$$

Check. We can do a partial check by substituting a value for n less than $\frac{15}{2}$. When $n=7$, the hourly plan costs 120(7), or \$840, so the hourly plan is less than the flat fee of \$900. When $n=8$, the hourly plan costs 120(8), or \$960, so the hourly plan is more expensive than the flat fee of \$900.

State. The hourly rate is less expensive for lengths of time less than $7\frac{1}{2}$ hours.

55. ***Familiarize.*** Let $n=$ the number of correct answers. Then the points earned are $2n$, and the points deducted are $\frac{1}{2}$ of the rest of the questions, $80-n$, or $\frac{1}{2}(80-n)$.

Translate. We write an inequality stating the score is at least 100.

$$2n-\frac{1}{2}(80-n)\geq 100.$$

Carry out.

$$2n-\frac{1}{2}(80-n)\geq 100$$
$$2n-40+\frac{1}{2}n\geq 100$$
$$\frac{5}{2}n-40\geq 100$$
$$\frac{5}{2}n\geq 140$$
$$n\geq 56$$

Check. When $n=56$, the score earned is $2(56)-\frac{1}{2}(80-56)$, or $112-\frac{1}{2}(24)$, or $112-12$, or 100.

When $n=58$, the score earned is $2(58)-\frac{1}{2}(80-58)$, or $116-\frac{1}{2}(22)$, or $116-11$, or 105. Since the score is exactly 100 when 56 questions are answered correctly and more than 100 when 58 questions are correct, we have performed a partial check.

State. At least 56 questions are correct for a score of at least 100.

57. ***Familiarize.*** Find the values of t for which $C(t)<1750$.

Translate. $-40.5t+2159<1750$

Carry out.

$$-40.5t+2159<1750$$
$$-40.5t<-409$$
$$t>\frac{818}{81},\text{ or } 10\frac{8}{81}$$

Check. $C\left(\frac{818}{81}\right)=1750.$

When $t=10$, $C(10)=-40.5(10)+2159$, or 1754.

When $t=11$, $C(11)=-40.5(11)+2159$, or 1713.5.

State. The domestic production will drop below 1750 million barrels later than 10 years after 2000, or the years after 2010.

59. ***Familiarize.*** We list the given information in a table.

Plan A: Monthly Income	Plan B: Monthly Income
\$400 salary 8% of sales Total: 400+8% of sales	\$610 5% of sales Total: 610+5% of sales

Suppose Toni had gross sales of \$5000 one month. Then under plan A she would earn

$$\$400+0.08(\$5000),\text{ or }\$800.$$

Under plan B she would earn

$$\$610+0.05(\$5000),\text{ or }\$860.$$

This shows that, for gross sales of \$5000, plan B is better. If Toni had gross sales of \$10,000 one month, then under plan A she would earn

$$\$400+0.08(\$10{,}000),\text{ or }\$1200.$$

Under plan B she would earn

$$\$610+0.05(\$10{,}000),\text{ or }\$1110.$$

This shows that, for gross sales of \$10,000, plan A is better. To determine all values for which plan A is better we solve an inequality.

Translate.

Income from plan A	is greater than	Income from plan B.
↓	↓	↓
$400+0.08s$	$>$	$610+0.05s$

Carry out.

$$400+0.08s>610+0.05s$$
$$400+0.03s>610$$
$$0.03s>210$$
$$s>7000$$

Check. For $s = \$7000$, the income from plan A is

$\$400 + 0.08(\$7000)$, or \$960

and the income from plan B is

$\$610 + 0.05(\$7000)$, or \$960.

This shows that for sales of \$7000 Toni's income is the same from each plan. In the Familiarize step we show that, for a value less than \$7000, plan B is better and, for a value greater than \$7000, plan A is better. Since we cannot check all possible values, we stop here.

State. Toni should select plan A for gross sales greater than \$7000.

61. ***Familiarize.*** Let m = the medical bill. Then the "Green Badge" medical insurance plan will cost \$2000 + 0.30(\$$m$ - \$2000) and the "Blue Seal" plan will cost \$2500 + 0.20(\$$m$ - \$2500).

Translate. We write an inequality stating that the "Blue Seal" plan costs less than the "Green Badge" plan.

$$2000 + 0.30(m - 2000) > 2500 + 0.20(m - 2500)$$

Carry out.

$$\begin{aligned} 2000 + 0.30(m-2000) &> 2500 + 0.20(m-2500) \\ 2000 + 0.3m - 600 &> 2500 + 0.2m - 500 \\ 1400 + 0.3m &> 2000 + 0.2m \\ 0.1m &> 600 \\ m &> 6000 \end{aligned}$$

Check. When m = 5000, the "Green Badge" plan cost is \$2000 + 0.30(\$5000 - \$2000), or \$2000 + 0.30(\$3000), or \$2900 and the "Blue Seal" plan cost is \$2500 + 0.20(\$5000 - \$2500), or \$2500 + 0.20(\$2500), or \$3000. So the "Green Badge" plan is less expensive. When m = 7000, the "Green Badge" plan cost is \$2000 + 0.30(\$7000 - \$2000), or \$2000 + 0.30(\$5000), or \$3500 and the "Blue Seal" plan is \$2500 + 0.20(\$7000 - \$2500), or \$2500 + 0.20(\$4500), or \$3400. So the "Blue Seal" plan is less expensive.

State. For medical bills of more than \$6000, the "Blue Seal" plan is less expensive.

63. a) ***Familiarize.*** Find the values of d for which $F(d) > 25$.

Translate.

$$\left(\frac{4.95}{d} - 4.50\right) \times 100 > 25$$

Carry out.

$$\begin{aligned} \left(\frac{4.95}{d} - 4.50\right) \times 100 &> 25 \\ \frac{495}{d} - 450 &> 25 \\ \frac{495}{d} &> 475 \\ 495 &> 475d \\ \frac{495}{475} &> d \\ \frac{99}{95} &> d, \text{ or } d < 1.04 \end{aligned}$$

Check. When $d = 1$,

$$F(1) = \left(\frac{4.95}{1} - 4.50\right) \times 100, \text{ or 45 percent.}$$

When $d = 1.05$,

$$F(1.05) = \left(\frac{4.95}{1.05} - 4.50\right) \times 100, \text{ or 21 percent.}$$

State. A man is considered obese for body density less than $\frac{99}{95}$ kg/L, or about 1.04 kg/L.

b) ***Familiarize.*** Find the values of d for which $F(d) > 32$.

Translate.

$$\left(\frac{4.95}{d} - 4.50\right) \times 100 > 32$$

Carry out.

$$\begin{aligned} \left(\frac{4.95}{d} - 4.50\right) \times 100 &> 32 \\ \frac{495}{d} - 450 &> 32 \\ \frac{495}{d} &> 482 \\ 495 &> 482d \\ \frac{495}{482} &> d, \text{ or } d < 1.03 \end{aligned}$$

Check. Our check from part (a) leads to the result that $F(d) > 32$ when $d < 1.03$.

State. A woman is considered obese for body density less than $\frac{495}{482}$ kg/L, or about 1.03 kg/L.

65. a) ***Familiarize.*** Find the values of x for which $R(x) < C(x)$.

Translate.

$$48x < 90{,}000 + 25x$$

Carry out.

$$\begin{aligned} 23x &< 90{,}000 \\ x &< 3913\tfrac{1}{23} \end{aligned}$$

Check. $R\left(3913\frac{1}{23}\right) = \$187{,}826.09 = C\left(3913\frac{1}{23}\right)$.

Calculate $R(x)$ and $C(x)$ for some x greater than $3913\frac{1}{23}$ and for some x less than $3913\frac{1}{23}$.

Suppose $x=4000$:
$R(x)=48(4000)=192{,}000$ and
$C(x)=90{,}000+25(4000)=190{,}000$.

In this case $R(x)>C(x)$.

Suppose $x=3900$:
$R(x)=48(3900)=187{,}200$ and
$C(x)=90{,}000+25(3900)=187{,}500$.

In this case $R(x)<C(x)$.

Then for $x<3913\frac{1}{23}$, $R(x)<C(x)$.

State. We will state the result in terms of integers, since the company cannot sell a fraction of a lamp. For 3913 or fewer lamps the company loses money.

b) Our check in part a) shows that for $x>3913\frac{1}{23}$, $R(x)>C(x)$ and the company makes a profit. Again, we will state the result in terms of an integer. For more than 3913 lamps the company makes money.

67. *Writing Exercise.* In solving the equation our solution is one value of x that makes the statement true. In solving either inequality, our solution is an infinite set of numbers bounded by one value of x. Solving the equation, we have $x = 5$. Solving the inequality $x + 3 > 8$, we have $x > 5$, so 5 is a lower bound of the solution set. Solving the inequality $x + 3 < 8$, we have $x < 5$, so 5 is an upper bound of the solution set.

69. $f(x)=\frac{5}{x}$, the domain is

$\{x \mid x \text{ is a real number and } x \neq 0\}$.

71. $f(x)=\frac{x-2}{2x+1}$, $2x+1\neq 0$, so $x\neq -\frac{1}{2}$. the domain is

$\left\{x \middle| x \text{ is a real number and } x\neq -\frac{1}{2}\right\}$.

73. $f(x)=\frac{x+10}{8}$ The domain is $\{x \mid x \text{ is a real number}\}$.

75. *Writing Exercise.* Since the percentage cannot exceed 100%, the function $P(t)$ will make sense only for the years where $P(t)<100\%$, or about 11 years.

77.
$$\begin{aligned}
3ax+2x &\geq 5ax-4\\
2x-2ax &\geq -4\\
2x(1-a) &\geq -4\\
x(1-a) &\geq -2\\
x &\leq -\frac{2}{1-a}, \text{ or } \frac{2}{a-1}
\end{aligned}$$

We reversed the inequality symbol when we divided because when $a>1$, then $1-a<0$.

The solution set is $\left\{x \middle| x\leq \frac{2}{a-1}\right\}$.

79.
$$\begin{aligned}
a(by-2) &\geq b(2y+5)\\
aby-2a &\geq 2by+5b\\
aby-2by &\geq 2a+5b\\
y(ab-2b) &\geq 2a+5b\\
y &\geq \frac{2a+5b}{ab-2b}, \text{ or } \frac{2a+5b}{b(a-2)}
\end{aligned}$$

The inequality symbol remained unchanged when we divided because when $a>2$ and $b>0$, then $ab-2b>0$.

The solution set is $\left\{y \middle| y\geq \frac{2a+5b}{b(a-2)}\right\}$.

81.
$$\begin{aligned}
c(2-5x)+dx &> m(4+2x)\\
2c-5cx+dx &> 4m+2mx\\
-5cx+dx-2mx &> 4m-2c\\
x(-5c+d-2m) &> 4m-2c\\
x[d-(5c+2m)] &> 4m-2c\\
x &> \frac{4m-2c}{d-(5c+2m)}
\end{aligned}$$

The inequality symbol remained unchanged when we divided because when $5c+2m<d$, then $d-(5c+2m)>0$.

The solution set is $\left\{x \middle| x> \frac{4m-2c}{d-(5c+2m)}\right\}$.

83. False. If $a=2$, $b=3$, $c=4$, and $d=5$, then $2<3$ and $4<5$ but $2-4=3-5$.

85. *Writing Exercise.* Not equivalent

0 is a solution of $x<3$ but not of $x+\frac{1}{x}<3+\frac{1}{x}$.

87. $x+5\leq 5+x$

$5\leq 5$ Subtracting x

We get an inequality that is true for all real numbers x. Thus the solution set is all real numbers.

0

89. $0^2=0$, $x^2>0$ for $x\neq 0$

The solution is $\{x \mid x \text{ is a real number } \textit{and } x\neq 0\}$.

0

91. From the graph, we see that $y_1\geq y_2$ to the left of the point of intersection, (6, –1), which is included in the solution set.

The solution is $\{x \mid x\leq 6\}$ or $(-\infty, 6]$.

93. a) The graph of y_1 lies above the graph of y_2 for x-values to the left of the point of intersection, or in the interval $(-\infty, 4)$.

b) The graph of y_2 lies on or below the graph of y_3 for x-values at and to the right of the point of intersection, or in the interval $[2, \infty)$.

c) The graph of y_3 lies on or above the graph of y_1 at and to the right of the point of intersection, or in the interval $[3.2, \infty)$.

95. *Graphing Calculator Exercise*

Exercise Set 9.2

1. h

3. f

5. e

7. b

9. c

11. $\{2, 4, 16\} \cap \{4, 16, 256\}$

The numbers 4 and 16 are common to both sets, so the intersection is $\{4, 16\}$.

13. $\{0, 5, 10, 15\} \cup \{5, 15, 20\}$

The numbers in either or both sets are 0, 5, 10, 15, and 20, so the union is $\{0, 5, 10, 15, 20\}$.

15. $\{a, b, c, d, e, f\} \cap \{b, d, f\}$

The letters b, d, and f are common to both sets, so the intersection is $\{b, d, f\}$.

17. $\{x, y, z\} \cup \{u, v, x, y, z\}$

The letters in either or both sets are u, v, x, y, and z, so the union is $\{u, v, x, y, z\}$.

19. $\{3, 6, 9, 12\} \cap \{5, 10, 15\}$

There are no numbers common to both sets, so the solution set has no members. It is $\varnothing$.

21. $\{1, 3, 5\} \cup \varnothing$

The numbers in either or both sets are 1, 3, and 5, so the union is $\{1, 3, 5\}$.

23. $1 < x < 3$

This inequality is an abbreviation for the conjunction $1 < x$ *and* $x < 3$. The graph is the intersection of two separate solution sets:

$\{x|1 < x\} \cap \{x|x < 3\} = \{x|1 < x < 3\}$.

Interval notation: $(1, 3)$

25. $-6 \le y \le 0$

This inequality is an abbreviation for the conjunction $-6 \le y$ and $y \le 0$.

Interval notation: $[-6, 0]$

27. $x \le -1$ *or* $x > 4$

The graph of this disjunction is the union of the graphs of the individual solution sets $\{x \mid x \le -1\}$ and $\{x \mid x > 4\}$.

Interval notation: $(-\infty, -1] \cup (4, \infty)$

29. $x \le -2$ *or* $x > 1$

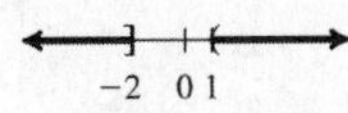

Interval notation: $(-\infty, -2] \cup (1, \infty)$

31.
$-4 \le -x < 2$
$4 \ge x > -2$ Multiplying by -1 and reversing the inequality symbols
$-2 < x \le 4$ Rewriting

Interval notation: $(-2, 4]$

33. $x > -2$ *and* $x < 4$

This conjunction can be abbreviated as $-2 < x < 4$.

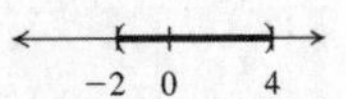

Interval notation: $(-2, 4)$

35. $5 > a$ *or* $a > 7$

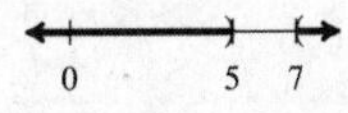

Interval notation: $(-\infty, 5) \cup (7, \infty)$

37. $x \ge 5$ *or* $-x \ge 4$

Multiplying the second inequality by -1 and reversing the inequality symbols, we get $x \ge 5$ or $x \le -4$.

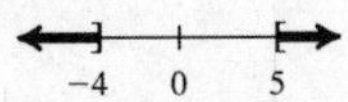

Interval notation: $(-\infty, -4] \cup [5, \infty)$

39. $7 > y$ *and* $y \ge -3$

This conjunction can be abbreviated as $-3 \le y < 7$.

Interval notation: $[-3, 7)$

41. $-x < 7$ *and* $-x \ge 0$

Multiplying the inequalities by -1 and reversing the

inequality symbols, we get $x > -7$ *and* $x \le 0$.

Interval notation: $(-7, 0]$

43. $t < 2$ *or* $t < 5$

Observe that every number that is less than 2 is also less than 5. Then $t < 2$ *or* $t < 5$ is equivalent to $t < 5$ and the graph of this disjunction is the set $\{t | t < 5\}$.

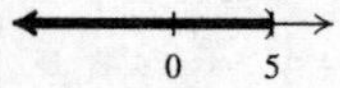

Interval notation: $(-\infty, 5)$

45.
$$-3 \le x+2 < 9$$
$$-3-2 \le x < 9-2$$
$$-5 \le x < 7$$
The solution set is $\{x \mid -5 \le x < 7\}$ or $[-5, 7)$.

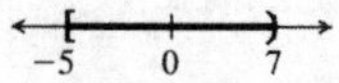

47.
$$0 < t-4 \quad and \quad t-1 \le 7$$
$$4 < t \quad and \quad t \le 8$$
We can abbreviate the answer as $4 < t \le 8$.

The solution set is $\{t \mid 4 < t \le 8\}$, or $(4, 8]$.

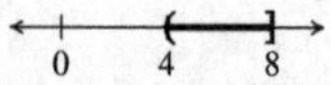

49.
$$-7 \le 2a-3 \quad and \quad 3a+1 < 7$$
$$-4 \le 2a \quad and \quad 3a < 6$$
$$-2 \le a \quad and \quad a < 2$$
We can abbreviate the answer as $-2 \le a < 2$. The solution set is $\{a | -2 \le a < 2\}$, or $[-2, 2)$.

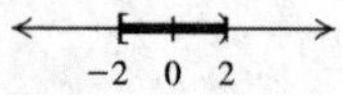

51. $x + 3 \le -1$ *or* $x + 3 > -2$

Observe that any real number is either less than or equal to -1 or greater than or equal to -2. Then the solution set is $\{x \mid x$ is a real number$\}$, or $(-\infty, \infty)$.

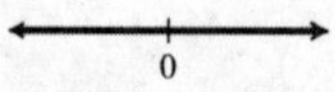

53.
$$-10 \le 3x-1 \le 5$$
$$-9 \le 3x \le 6$$
$$-3 \le x \le 2$$
The solution set is $\{x \mid -3 \le x \le 2\}$, or $[-3, 2]$.

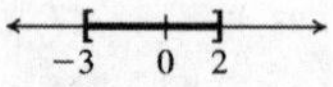

55.
$$5 > \frac{x-3}{4} > 1$$
$$20 > x-3 > 4 \quad \text{Multiplying by 4}$$
$$23 > x > 7, \text{ or}$$
$$7 < x < 23$$
The solution set is $\{x | 7 < x < 23\}$, or $(7, 23)$.

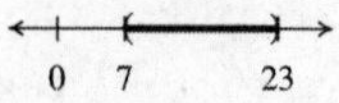

57.
$$-2 \le \frac{x+2}{-5} \le 6$$
$$10 \ge x+2 \ge -30 \quad \text{Multiplying by } -5$$
$$8 \ge x \ge -32, \text{ or}$$
$$-32 \le x \le 8$$
The solution set is $\{x \mid -32 \le x \le 8\}$, or $[-32, 8]$.

59.
$$2 \le 3x-1 \le 8$$
$$3 \le 3x \le 9$$
$$1 \le x \le 3$$
The solution set is $\{x | 1 \le x \le 3\}$, or $[1, 3]$.

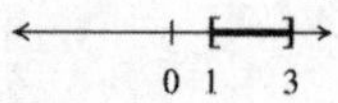

61.
$$-21 \le -2x-7 < 0$$
$$-14 \le -2x < 7$$
$$7 \ge x > -\frac{7}{2}, \text{ or}$$
$$-\frac{7}{2} < x \le 7$$
The solution set is $\left\{x \middle| -\frac{7}{2} < x \le 7\right\}$, or $\left(-\frac{7}{2}, 7\right]$.

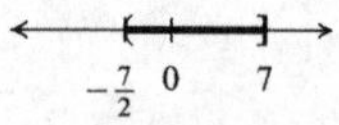

63.
$$5t+3 < 3 \quad or \quad 5t+3 > 8$$
$$5t < 0 \quad or \quad 5t > 5$$
$$t < 0 \quad or \quad t > 1$$
The solution set is $\{t \mid t < 0 \text{ or } t > 1\}$, or $(-\infty, 0) \cup (1, \infty)$.

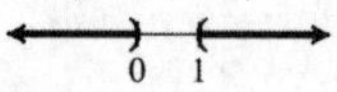

65.
$$6 > 2a-1 \quad or \quad -4 \le -3a+2$$
$$7 > 2a \quad or \quad -6 \le -3a$$
$$\frac{7}{2} > a \quad or \quad 2 \ge a$$
The solution set is $\left\{a \middle| \frac{7}{2} > a\right\} \cup \{a | 2 \ge a\} = \left\{a \middle| \frac{7}{2} > a\right\}$, or $\left\{a \middle| a < \frac{7}{2}\right\}$, or $\left(-\infty, \frac{7}{2}\right)$.

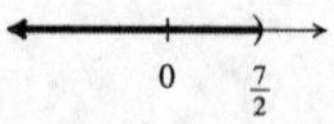

67.
$$a+3 < -2 \quad and \quad 3a-4 < 8$$
$$a < -5 \quad and \quad 3a < 12$$
$$a < -5 \quad and \quad a < 4$$
The solution set is $\{a | a < -5\} \cap \{a | a < 4\} = \{a | a < -5\}$, or $(-\infty, -5)$.

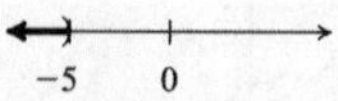

69.
$$3x+2 < 2 \quad and \quad 3-x < 1$$
$$3x < 0 \quad and \quad -x < -2$$
$$x < 0 \quad and \quad x > 2$$
The solution set is $\{x | x < 0\} \cap \{x | x > 2\} = \varnothing$.

71. $2t-7\le 5$ *or* $5-2t>3$
$2t\le 12$ *or* $-2t>-2$
$t\le 6$ *or* $t<1$

The solution set is $\{t|t\le 6\}\cup\{t|t<1\}=\{t|t\le 6\}$, or $(-\infty, 6]$.

0 6

73. $f(x)=\dfrac{9}{x+6}$

$f(x)$ cannot be computed when the denominator is 0.

Since $x + 6 = 0$ is equivalent to $x = -6$, we have

Domain of $f=\{x|x$ is a real number *and* $x\ne -6\}$
$=(-\infty,-6)\cup(-6, \infty)$.

75. $f(x)=\dfrac{1}{x}$

$f(x)$ cannot be computed when the denominator is 0.

We have Domain of f
$=\{x|x$ is a real number and $x\ne 0\}$ $=(-\infty, 0)\cup(0, \infty)$.

77. $f(x)=\dfrac{x+3}{2x-8}$

$f(x)$ cannot be computed when the denominator is 0.

Since $2x-8=0$ is equivalent to $x=4$, we have

Domain of $f=\{x|x$ is a real number *and* $x\ne 4\}$, or

$(-\infty, 4)\cup(4, \infty)$.

79. *Writing Exercise.* By definition, the notation $2 < x < 5$ indicates that $2 < x$ *and* $x < 5$. A solution of the disjunction $2 < x$ *or* $x < 5$ must be in at least one of these sets but not necessarily in both, so the disjunction cannot be written as $2 < x < 5$.

81. $g(x)=2x$

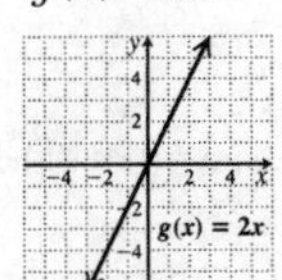

83. $g(x)=-3$

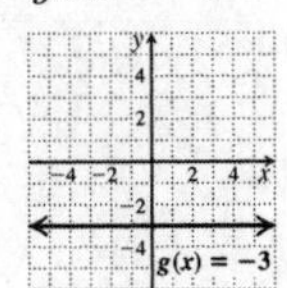

85. Graph both equations.

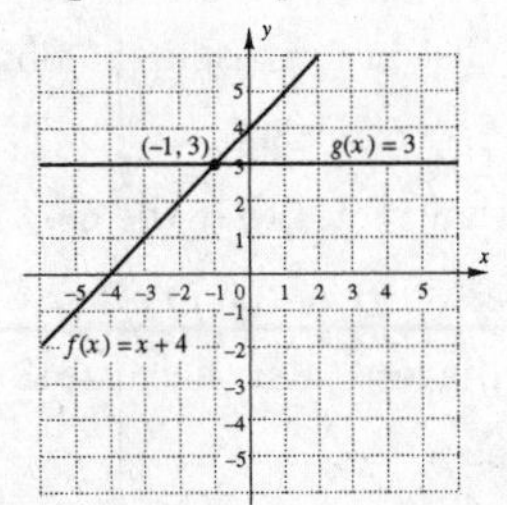

The point of intersection is (–1, 3).

$y=x+4$		$y=3$	
3	$-1+4$	3	3
$3\overset{?}{=}3$	TRUE	$3\overset{?}{=}3$	TRUE

The solution is –1.

87. *Writing Exercise.* When $[a, b]\cup[c, d]=[a, d]$, we know that $a < b$, $c < d$, and $b \ge c$.

89. From the graph we observe that the values of x for which $2x - 5 > -7$ *and* $2x - 5 < 7$ are $\{x \mid -1 < x < 6\}$, or $(-1, 6)$.

91. Solve $18{,}000 < S(t) < 21{,}000$, or

$$18{,}000<500t+16{,}500<21{,}000$$
$$1500<500t<4500$$
$$3<t<9$$

Thus, from 3 through 9 years after 2000, or between 2003 and 2009.

93. Solve $32<f(x)<46$, or $32<2(x+10)<46$.

$$32<2(x+10)<46$$
$$32<2x+20<46$$
$$12<2x<26$$
$$6<x<13$$

For U.S. dress sizes between 6 and 13, dress sizes in Italy will be between 32 and 46.

95. Solve $25\le F(d)\le 31$.

$$25\le(4.95/d-4.50)\times 100\le 31$$
$$25\le\frac{495}{d}-450\le 31$$
$$475\le\frac{495}{d}\le 481$$
$$\frac{1}{475}\ge\frac{d}{495}\ge\frac{1}{481}$$
$$\frac{495}{475}\ge d\ge\frac{495}{481},$$
$$\text{or } 1.03\le d\le 1.04$$

Acceptable body densities are between 1.03 kg/L and 1.04 kg/L.

97. Let $c =$ the number of crossings in six months. Then at the \$6 per crossing rate, the total cost of c crossings is

$6c. A six-month pass costs $50 and additional $2 per crossing toll brings the total cost of c crossings to $50 + $2c. An unlimited crossing pass costs $300.

We write an inequality that states that the cost of c crossings using six-month passes is less than the cost using $6 per crossing toll and is less than the cost of using the unlimited-trip pass. Solve:

$$50+2c<6c \quad and \quad 50+2c<300.$$

We get $12.5 < c$ *and* $c < 125$.

For more than 12 crossings but less than 125 crossings in six months, the reduced-fare pass is more economical.

99. $4m-8>6m+5$ *or* $5m-8<-2$

$-13>2m$ *or* $5m<6$

$-\frac{13}{2}>m$ *or* $m<\frac{6}{5}$

$\left\{m\middle|m<\frac{6}{5}\right\}$, or $\left(-\infty, \frac{6}{5}\right)$

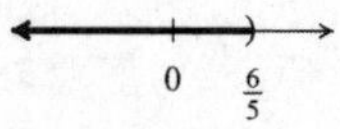

101. $3x<4-5x<5+3x$

$0<4-8x<5$

$-4<-8x<1$

$\frac{1}{2}>x>-\frac{1}{8}$

The solution set is $\left\{x\middle|-\frac{1}{8}<x<\frac{1}{2}\right\}$, or $\left(-\frac{1}{8}, \frac{1}{2}\right)$.

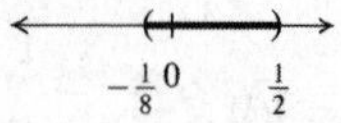

103. Let $a=b=c=2$. Then $a\le c$ and $c\le b$, but $b\not> a$. The given statement is false.

105. If $-a<c$, then $-1(-a)>-1\cdot c$, or $a>-c$. Then if $a>-c$ and $-c>b$, we have $a>-c>b$, so $a>b$ and the given statement is true.

107. $f(x)=\dfrac{\sqrt{3-4x}}{x+7}$

$3-4x\ge 0$ is equivalent to $x\le\frac{3}{4}$ and $x+7=0$ is equivalent to $x=-7$. Then we have Domain of $f=\left\{x\middle|x\le\frac{3}{4} \text{ and } x\ne-7\right\}$, or $(-\infty,-7)\cup\left(-7, \frac{3}{4}\right]$.

109. Observe that the graph of y_2 lies on or above the graph of y_1 and below the graph of y_3 for x in the interval $[-3, 4)$.

111. *Graphing Calculator Exercise*

Exercise Set 9.3

1. $|\,x\,|\ge 0$, so the statement is true.

3. True; see page 597 in the text.

5. True; see page 599 in the text.

7. False; see page 600 in the text.

9. g

11. d

13. a

15. $|\,x\,| = 10$

$x = -10$ *or* $x = 10$ Using the absolute value principle

The solution set is $\{-10, 10\}$.

17. $|\,x\,| = -1$

The absolute value of a number is always nonnegative. Therefore, the solution set is $\varnothing$.

19. $|\,p\,| = 0$

The only number whose absolute value is 0 is 0. The solution set is $\{0\}$.

21. $|2x-3|=4$

$2x-3=-4$ *or* $2x-3=4$ Absolute-value principle

$2x=-1$ *or* $2x=7$

$x=-\frac{1}{2}$ *or* $x=\frac{7}{2}$

The solution set is $\left\{-\frac{1}{2}, \frac{7}{2}\right\}$.

23. $|\,3x+5\,| = -8$

Absolute value is always nonnegative, so the equation has no solution. The solution set is $\varnothing$.

25. $|x-2|=6$

$x-2=-6$ *or* $x-2=6$ Absolute-value principle

$x=-4$ *or* $x=8$

The solution set is $\{-4, 8\}$.

27. $|x-7|=1$

$x-7=-1$ *or* $x-7=1$

$x=6$ *or* $x=8$

The solution set is $\{6, 8\}$.

29. $|t|+1.1=6.6$

$|t|=5.5$ Adding -1.1

$t=-5.5$ *or* $t=5.5$

The solution set is $\{-5.5, 5.5\}$.

31. $|5x|-3=37$

$|5x|=40$ Adding 3

$5x=-40$ *or* $5x=40$

$x=-8$ *or* $x=8$

The solution set is $\{-8, 8\}$.

33. $7|q|+2=9$

$7|q|=7$ Adding -2

$|q|=1$ Multiplying by $\frac{1}{7}$

$q=-1$ or $q=1$

The solution set is $\{-1, 1\}$.

35. $\left|\frac{2x-1}{3}\right|=4$

$\frac{2x-1}{3}=-4$ *or* $\frac{2x-1}{3}=4$

$2x-1=-12$ *or* $2x-1=12$

$2x=-11$ *or* $2x=13$

$x=-\frac{11}{2}$ *or* $x=\frac{13}{2}$

The solution set is $\left\{-\frac{11}{2}, \frac{13}{2}\right\}$.

37. $|5-m|+9=16$

$|5-m|=7$ Adding -9

$5-m=-7$ *or* $5-m=7$

$-m=-12$ *or* $-m=2$

$m=12$ *or* $m=-2$

The solution set is $\{-2, 12\}$.

39. $5-2|3x-4|=-5$

$-2|3x-4|=-10$

$|3x-4|=5$

$3x-4=-5$ *or* $3x-4=5$

$3x=-1$ *or* $3x=9$

$x=-\frac{1}{3}$ *or* $x=3$

The solution set is $\left\{-\frac{1}{3}, 3\right\}$.

41. $|2x+6|=8$

$2x+6=-8$ *or* $2x+6=8$

$2x=-14$ *or* $2x=2$

$x=-7$ *or* $x=1$

The solution set is $\{-7, 1\}$.

43. $|x|-3=5.7$

$|x|=8.7$

$x=-8.7$ *or* $x=8.7$

The solution set is $\{-8.7, 8.7\}$.

45. $\left|\frac{1-2x}{5}\right|=2$

$\frac{1-2x}{5}=-2$ *or* $\frac{1-2x}{5}=2$

$1-2x=-10$ *or* $1-2x=10$

$-2x=-11$ *or* $-2x=9$

$x=\frac{11}{2}$ *or* $x=-\frac{9}{2}$

The solution set is $\left\{-\frac{9}{2}, \frac{11}{2}\right\}$.

47. $|x-7|=|2x+1|$

$x-7=2x+1$ *or* $x-7=-(2x+1)$

$-x=8$ *or* $x-7=-2x-1$

$x=-8$ *or* $3x=6$

$x=2$

The solution set is $\{-8, 2\}$.

49. $|x+4|=|x-3|$

$x+4=x-3$ *or* $x+4=-(x-3)$

$4=-3$ *or* $x+4=-x+3$

False $\quad 2x=-1$

$x=-\frac{1}{2}$

The solution set is $\left\{-\frac{1}{2}\right\}$.

51. $|3a-1|=|2a+4|$

$3a-1=2a+4$ *or* $3a-1=-(2a+4)$

$a-1=4$ *or* $3a-1=-2a-4$

$a=5$ *or* $5a-1=-4$

$5a=-3$

$a=-\frac{3}{5}$

The solution set is $\left\{-\frac{3}{5}, 5\right\}$.

53. Since $|n-3|$ and $|3-n|$ are equivalent expressions, the solution set of $|n-3|=|3-n|$ is the set of all real numbers.

55. $|7-4a|=|4a+5|$

$7-4a=4a+5$ *or* $7-4a=-(4a+5)$

$-8a=-2$ *or* $7-4a=-4a-5$

$a=\frac{1}{4}$ *or* $7=-5$

False

The solution set is $\left\{\frac{1}{4}\right\}$.

57. $|a|\le 3$

$-3\le a\le 3$ Part (b)

The solution set is $\{a|-3\le a\le 3\}$, or $[-3, 3]$.

−3 0 3

59. $|t|>0$

$t<0$ or $0<t$ Part (c)

The solution set is $\{t|t<0 \text{ or } t>0\}$, or $\{t|t\neq 0\}$, or $(-\infty, 0)\cup(0, \infty)$.

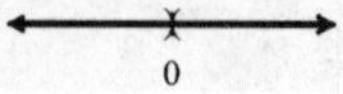

61. $|x-1|<4$

$-4<x-1<4$ Part (b)

$-3<x<5$

The solution set is $\{x|-3<x<5\}$, or $(-3, 5)$.

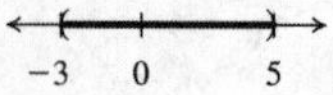

63. $|n+2|\leq 6$

$-6\leq n+2\leq 6$ Part (b)

$-8\leq n\leq 4$

The solution set is $\{n|-8\leq n\leq 4\}$, or $[-8, 4]$.

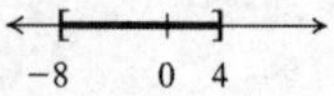

65. $|x-3|+2>7$

$|x-3|>5$ Adding -2

$x-3<-5$ *or* $5<x-3$ Part(c)

$x<-2$ *or* $8<x$

The solution set is $\{x|x<-2 \text{ or } x>8\}$, or $(-\infty,-2)\cup(8, \infty)$.

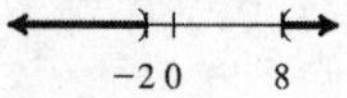

67. $|2y-9|>-5$

Since absolute value is never negative, any value of $2y-9$, and hence any value of y, will satisfy the inequality. The solution set is the set of all real numbers, or $(-\infty, \infty)$.

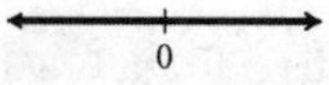

69. $|3a+4|+2\geq 8$

$|3a+4|\geq 6$ Adding -2

$3a+4\leq -6$ *or* $6\leq 3a+4$ Part (c)

$3a\leq -10$ *or* $2\leq 3a$

$a\leq -\frac{10}{3}$ *or* $\frac{2}{3}\leq a$

The solution set is $\left\{a\middle|a\leq -\frac{10}{3} \text{ or } a\geq \frac{2}{3}\right\}$, or $\left(-\infty,-\frac{10}{3}\right]\cup\left[\frac{2}{3}, \infty\right)$.

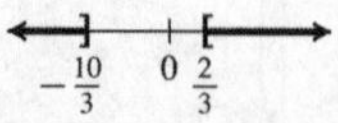

71. $|y-3|<12$

$-12<y-3<12$ Part (b)

$-9<y<15$ Adding 3

The solution set is $\{y|-9<y<15\}$, or $(-9, 15)$.

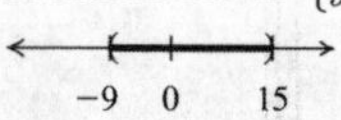

73. $9-|x+4|\leq 5$

$-|x+4|\leq -4$

$|x+4|\geq 4$

$x+4\leq -4$ *or* $4\leq x+4$ Part (c)

$x\leq -8$ *or* $0\leq x$

The solution set is $\{x|x\leq -8 \text{ or } x\geq 0\}$, or $(-\infty,-8]\cup[0, \infty)$.

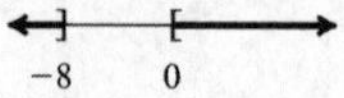

75. $6+|3-2x|>10$

$|3-2x|>4$

$3-2x<-4$ *or* $4<3-2x$

$-2x<-7$ *or* $1<-2x$

$x>\frac{7}{2}$ *or* $-\frac{1}{2}>x$

The solution set is $\left\{x\middle|x<-\frac{1}{2} \text{ or } x>\frac{7}{2}\right\}$, or $\left(-\infty,-\frac{1}{2}\right)\cup\left(\frac{7}{2}, \infty\right)$.

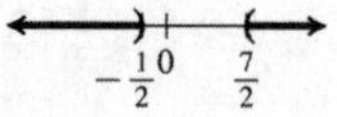

77. $|5-4x|<-6$

Absolute value is always nonnegative, so the inequality has no solution. The solution set is $\varnothing$.

79. $\left|\frac{1+3x}{5}\right|>\frac{7}{8}$

$\frac{1+3x}{5}<-\frac{7}{8}$ *or* $\frac{7}{8}<\frac{1+3x}{5}$

$1+3x<-\frac{35}{8}$ *or* $\frac{35}{8}<1+3x$

$3x<-\frac{43}{8}$ *or* $\frac{27}{8}<3x$

$x<-\frac{43}{24}$ *or* $\frac{9}{8}<x$

The solution set is $\left\{x\middle|x<-\frac{43}{24} \text{ or } x>\frac{9}{8}\right\}$, or $\left(-\infty,-\frac{43}{24}\right)\cup\left(\frac{9}{8}, \infty\right)$.

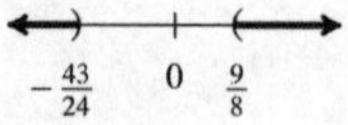

81. $|m+3|+8\leq 14$

$|m+3|\leq 6$ Adding -8

$-6\leq m+3\leq 6$

$-9\leq m\leq 3$

The solution set is $\{m|-9\leq m\leq 3\}$, or $[-9, 3]$.

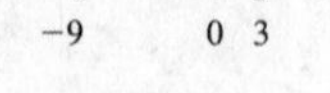

83. $25-2|a+3|>19$

$-2|a+3|>-6$

$|a+3|<3$ Multiplying by $-\frac{1}{2}$

$-3<a+3<3$ Part(b)

$-6<a<0$

The solution set is $\{a|-6<a<0\}$, or $(-6, 0)$.

85. $|2x-3|\leq 4$

$-4\leq 2x-3\leq 4$ Part (b)
$-1\leq 2x\leq 7$ Adding 3
$-\frac{1}{2}\leq x\leq \frac{7}{2}$ Multiplying by $\frac{1}{2}$

The solution set is $\left\{x\middle|-\frac{1}{2}\leq x\leq \frac{7}{2}\right\}$, or $\left[-\frac{1}{2}, \frac{7}{2}\right]$.

87. $5+|3x-4|\geq 16$
$|3x-4|\geq 11$

$3x-4\leq -11$ *or* $11\leq 3x-4$ Part (c)
$3x\leq -7$ *or* $15\leq 3x$
$x\leq -\frac{7}{3}$ *or* $5\leq x$

The solution set is $\left\{x\middle|x\leq -\frac{7}{3} \text{ or } x\geq 5\right\}$, or $\left(-\infty,-\frac{7}{3}\right]\cup[5, \infty)$.

89. $7+|2x-1|<16$
$|2x-1|<9$

$-9<2x-1<9$ Part (b)
$-8<2x<10$
$-4<x<5$

The solution set is $\{x|-4<x<5\}$, or $(-4, 5)$.

91. *Writing Exercise.* The solutions of $|x|<5$ are those numbers whose distance from zero is less than 5. Since the distance of –7 from 0 is not less than 5, then –7 is not a solution of the inequality.

93. $3x-y=6$

x	y
0	–6
2	0

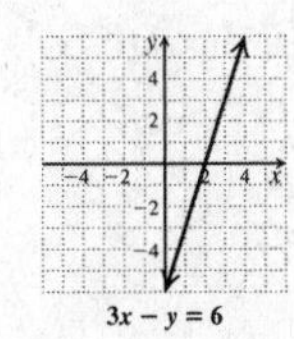

95. $x=-2$

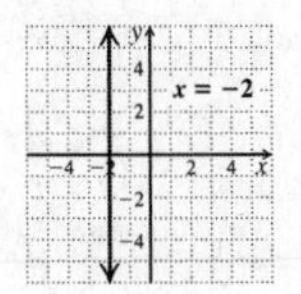

97. Solve using elimination.

$$\begin{aligned} x-3y&=8 \quad (1)\\ 2x+3y&=4 \quad (2)\\ \hline 3x\quad\ &=12\\ x&=4 \end{aligned}$$

Substituting 4 for x in Equation (1).

$4-3y=8$
$-3y=4$
$y=-\frac{4}{3}$

The solution is $\left(4,-\frac{4}{3}\right)$.

99. Solve using substitution.

$y=1-5x$ (1)
$2x-y=4$ (2)

Substitute $1-5x$ for y into Equation (2).

$2x-(1-5x)=4$
$2x-1+5x=4$
$7x=5$
$x=\frac{5}{7}$

When $x=\frac{5}{7}$, then $y=1-5x=1-5\left(\frac{5}{7}\right)=-\frac{18}{7}$.

The solution is $\left(\frac{5}{7},-\frac{18}{7}\right)$.

101. *Writing Exercise.* Graph $y=g(x)$ and $y=c$ on the same axes. The solution set consists of x-values for which $(x, g(x))$ is below the horizontal line $y=c$.

103. From the definition of absolute value, $|3t-5|=3t-5$ only when $3t-5\geq 0$. Solve $3t-5\geq 0$.

$3t-5\geq 0$
$3t\geq 5$
$t\geq \frac{5}{3}$

The solution set is $\left\{t\middle|t\geq \frac{5}{3}\right\}$, or $\left[\frac{5}{3}, \infty\right)$.

105. $|x+2|>x$

The inequality is true for all $x<0$ (because absolute value must be nonnegative). The solution set in this case is $\{x|x<0\}$. If $x=0$, we have $|0+2|>0$, which is true. The solution set in this case is $\{0\}$. If $x>0$, we have the following:

$x+2<-x$ *or* $x<x+2$
$2x<-2$ *or* $0<2$
$x<-1$

Although $x>0$ *and* $x<-1$ yields no solution, $x>0$ *and* $2>0$ (true for all x) yields the solution set $\{x|x>0\}$ in this case. The solution set for the inequality is $\{x|x<0\} \cup \{0\} \cup \{x|x>0\}$, or $\{x|x \text{ is a real number}\}$, or $(-\infty, \infty)$.

107. $|5t-3| = 2t+4$

From the definition of absolute value, we know that $2t+4 \geq 0$, or $t \geq -2$. So we have

$t \geq -2$ and

$$\begin{array}{rcr} 5t-3=-(2t+4) & or & 5t-3=2t+4 \\ 5t-3=-2t-4 & or & 3t=7 \\ 7t=-1 & or & t=\frac{7}{3} \\ t=-\frac{1}{7} & or & t=\frac{7}{3} \end{array}$$

Since $-\frac{1}{7} \geq -2$ *and* $\frac{7}{3} \geq -2$, the solution set is $\left\{-\frac{1}{7}, \frac{7}{3}\right\}$.

109. Using part (b), we find that $-3<x<3$ is equivalent to $|x|<3$.

111. $x \leq -6$ or $6 \leq x$

$|x| \geq 6$ Using part (c)

113.

$$\begin{array}{rcll} x<-8 & or & 2<x & \\ x+3<-5 & or & 5<x+3 & \text{Adding 3} \\ |x+3|>5 & & & \text{Using part (c)} \end{array}$$

115. The distance from x to 7 is $|x-7|$ or $|7-x|$, so we have $|x-7|<2$, or $|7-x|<2$.

117. The length of the segment from –1 to 7 is $|-1-7|=|-8|=8$ units. The midpoint of the segment is $\frac{-1+7}{2}=\frac{6}{2}=3$. Thus, the interval extends 8/2, or 4, units on each side of 3. An inequality for which the closed interval is the solution set is then $|x-3| \leq 4$.

119. The length of the segment from –7 to –1 is $|-7-(-1)|=|-6|=6$ units. The midpoint of the segment is $\frac{-7+(-1)}{2}=\frac{-8}{2}=-4$. Thus, the interval extends 6/2, or 3, units on each side of –4. An inequality for which the open interval is the solution set is $|x-(-4)|<3$, or $|x+4|<3$.

121. $|d-60\text{ft}| \leq 10\text{ft}$

$$\begin{array}{c} -10\text{ft} \leq d-60\text{ft} \leq 10\text{ft} \\ 50\text{ft} \leq d \leq 70\text{ft} \end{array}$$

When the bungee jumper is 50 ft above the river, she is $150-50$, or 100 ft, from the bridge. When she is 70 ft above the river, she is $150-70$, or 80 ft, from the bridge. Thus, at any given time, the bungee jumper is between 80 ft and 100 ft from the bridge.

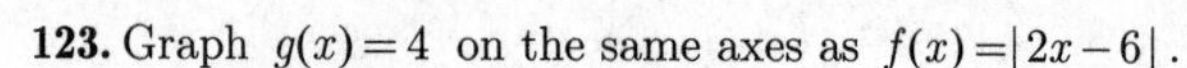

123. Graph $g(x)=4$ on the same axes as $f(x)=|2x-6|$.

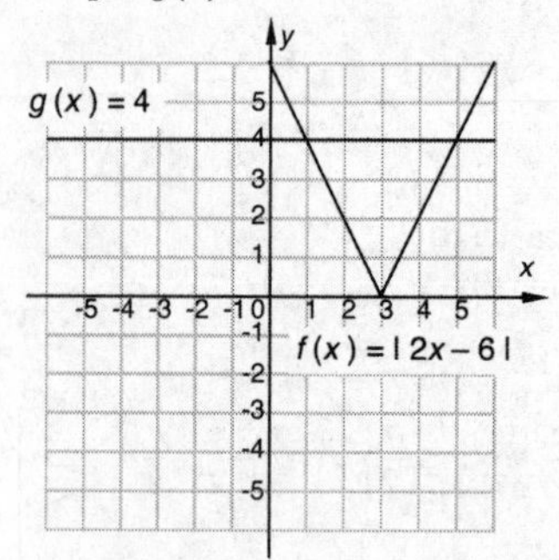

The solution set consists of the x-values for which $(x, f(x))$ is on or below the horizontal line $g(x)=4$. These x-values comprise the interval $[1, 5]$.

125. *Writing Exercise.* No portion of the graph of the function $y = \text{abs}(x-3)$, the V-shaped graph, can lie below the x-axis, because absolute value is always nonnegative.

Connecting the Concepts

1.
$$\begin{aligned} 2x+3&=7 \\ 2x&=4 \\ x&=2 \end{aligned}$$

The solution is 2.

3.
$$\begin{aligned} 3(t-5)&=4-(t+1) \\ 3t-15&=4-t-1 \\ 3t-15&=-t+3 \\ 4t-15&=3 \\ 4t&=18 \\ t&=\frac{18}{4}, \text{ or } \frac{9}{2} \end{aligned}$$

The solution is $\frac{9}{2}$.

5. $-x \leq 6$

$x \geq -6$ Reversing the inequality symbol

The solution is $\{x|x \geq -6\}$, or $[-6, \infty)$.

7.
$$\begin{aligned} 2(3n+6)-n&=4-3(n+1) \\ 6n+12-n&=4-3n-3 \\ 5n+12&=-3n+1 \\ 8n&=-11 \\ n&=-\frac{11}{8} \end{aligned}$$

The solution is $-\frac{11}{8}$.

9. $2+|3x|=10$

$|3x|=8$

$3x=-8 \quad or \quad 3x=8$

$x=-\frac{8}{3} \quad or \quad x=\frac{8}{3}$

The solution is $\left\{-\frac{8}{3}, \frac{8}{3}\right\}$.

11. $\frac{1}{2}x-7=\frac{3}{4}+\frac{1}{4}x$

$\frac{1}{4}x=\frac{31}{4}$

$x=31$

The solution is 31.

13. $|2x+5|+1\geq 13$

$|2x+5|\geq 12$

$2x+5\leq -12 \quad or \quad 12\leq 2x+5$

$2x\leq -17 \quad or \quad 7\leq 2x$

$x\leq -\frac{17}{2} \quad or \quad \frac{7}{2}\leq x$

The solution is $\left\{x\middle|x\leq -\frac{17}{2} \text{ or } x\geq \frac{7}{2}\right\}$, or $\left(-\infty,-\frac{17}{2}\right]\cup\left[\frac{7}{2}, \infty\right)$.

15. $|m+6|-8<10$

$|m+6|<18$

$-18<m+6<18$

$-24<m<12$

The solution is $\{m|-24<m<12\}$, or $(-24, 12)$.

17. $4-|7-t|\leq 1$

$-|7-t|\leq -3$

$|7-t|\geq 3$

$7-t\leq -3 \quad or \quad 3\leq 7-t$

$-t\leq -10 \quad or \quad -4\leq -t$

$t\geq 10 \quad or \quad 4\geq t$

The solution is $\{t|t\leq 4 \text{ or } t\geq 10\}$, or $(-\infty, 4]\cup[10, \infty)$.

19. $8-5|a+6|>3$

$-5|a+6|>-5$

$|a+6|<1$

$-1<a+6<1$

$-7<a<-5$

The solution is $\{a|-7<a<-5\}$, or $(-7, -5)$.

Exercise Set 9.4

1. e; see pages 606 and 611 in the text.

3. d; see pages 610-612 in the text.

5. b; see page 612 in the text.

7. We replace x with -2 and y with 3.

$$\begin{array}{c|c} 2x-3y & -4 \\ \hline 2(-2)-3 & -4 \end{array}$$

$-7 \overset{?}{<} -4$ FALSE

Since $-7 < -4$ is false, $(-2, 3)$ is not a solution.

9. We replace x with 5 and y with 8.

$$\begin{array}{c|c} 3y-5x & 0 \\ \hline 3\cdot 8-5\cdot 5 & 0 \\ 24-25 & \end{array}$$

$-1 \overset{?}{\leq} 0$ TRUE

Since $-1 \leq 0$ is true, $(5, 8)$ is a solution.

11. Graph: $y\geq \frac{1}{2}x$

We first graph the line $y=\frac{1}{2}x$. We draw the line solid since the inequality symbol is $\geq$. To determine which half-plane to shade, test a point not on the line, $(0, 1)$:

$$\begin{array}{c|c} y & \frac{1}{2}x \\ \hline 1 & \frac{1}{2}\cdot 0 \end{array}$$

$1 \overset{?}{\geq} 0$ TRUE

Since $1 \geq 0$ is true, $(0, 1)$ is a solution as are all of the points in the half-plane containing $(0, 1)$. We shade that half-plane and obtain the graph.

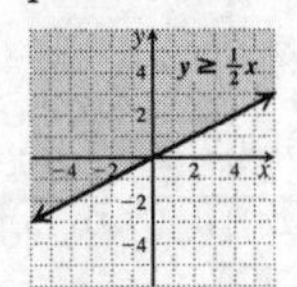

13. Graph: $y > x - 3$.

First graph the line $y = x - 3$. Draw it dashed since the inequality symbol is $>$. Test the point $(0, 0)$ to determine if it is a solution.

$$\begin{array}{c|c} y & x-3 \\ \hline 0 & 0-3 \end{array}$$

$0 \overset{?}{>} -3$ TRUE

Since $0 > -3$ is true, we shade the half-plane that contains $(0, 0)$ and obtain the graph.

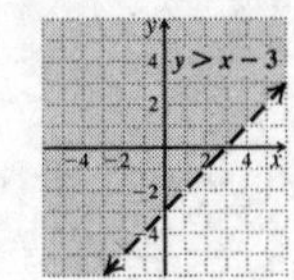

15. Graph: $y \leq x + 2$.

First graph the line $y = x + 2$. Draw it solid since the inequality symbol is $\leq$. Test the point $(0, 0)$ to determine if it is a solution.

$$\begin{array}{c|c} \multicolumn{2}{c}{y \leq x+2} \\ \hline 0 & 0+2 \end{array}$$

$0 \overset{?}{\leq} 2$ TRUE

Since $0 \leq 2$ is true, we shade the half-plane that contains (0, 0) and obtain the graph.

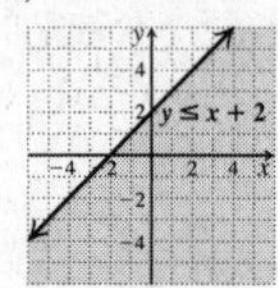

17. Graph: $x - y \leq 4$

First graph the line $x - y = 4$. Draw a solid line since the inequality symbol is $\leq$. Test the point (0, 0) to determine if it is a solution.

$$\begin{array}{c|c} \multicolumn{2}{c}{x - y \leq 4} \\ \hline 0-0 & 4 \end{array}$$

$0 \overset{?}{\leq} 4$ TRUE

Since $0 \leq 4$ is true, we shade the half-plane that contains (0, 0) and obtain the graph.

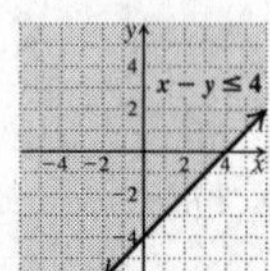

19. Graph: $2x + 3y < 6$

First graph $2x + 3y = 6$. Draw the line dashed since the inequality symbol is <. Test the point (0, 0) to determine if it is a solution.

$$\begin{array}{c|c} \multicolumn{2}{c}{2x + 3y < 6} \\ \hline 2\cdot 0 + 3 \cdot 0 & 6 \end{array}$$

$0 \overset{?}{<} 6$ TRUE

Since $0 < 6$ is true, we shade the half-plane containing (0, 0) and obtain the graph.

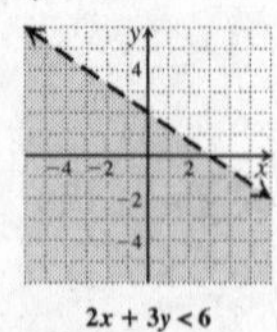
2x + 3y < 6

21. Graph: $2y - x \leq 4$

We first graph $2y - x = 4$. Draw the line solid since the inequality symbol is $\leq$. Test the point (0, 0) to determine if it is a solution.

$$\begin{array}{c|c} \multicolumn{2}{c}{2y - x \leq 4} \\ \hline 2\cdot 0 - 0 & 4 \end{array}$$

$0 \overset{?}{\leq} 4$ TRUE

Since $0 \leq 4$ is true, we shade the half-plane containing (0, 0) and obtain the graph.

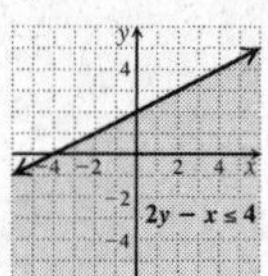

23. Graph: $2x - 2y \geq 8 + 2y$
$2x - 4y \geq 8$

First graph $2x - 4y = 8$. Draw the line solid since the inequality symbol is $\geq$. Test the point (0, 0) to determine if it is a solution.

$$\begin{array}{c|c} \multicolumn{2}{c}{2x - 4y \geq 8} \\ \hline 2\cdot 0 - 4 \cdot 0 & 8 \end{array}$$

$0 \overset{?}{\geq} 8$ FALSE

Since $0 \geq 8$ is false, we shade the half-plane that does not contain (0, 0) and obtain the graph.

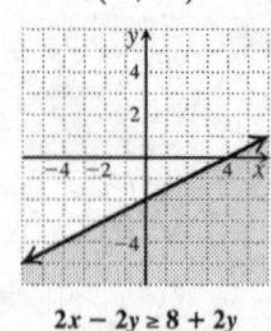
2x − 2y ≥ 8 + 2y

25. Graph: $x > -2$.

We first graph $x = -2$. We draw the line dashed since the inequality symbol is >. Test the point (0, 0) to determine if it is a solution.

$$\begin{array}{c|c} \multicolumn{2}{c}{x > -2} \\ \hline 0 & -2 \end{array}$$

$0 \overset{?}{>} -2$ TRUE

Since $0 > -2$ is true, we shade the half-plane that contains (0, 0) and obtain the graph.

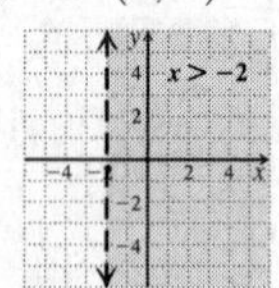

27. Graph: $y \leq 6$.

We first graph $y = 6$. We draw the line solid since the inequality symbol is $\leq$. Test the point (0, 0) to determine if it is a solution.

$$\begin{array}{c|c} \multicolumn{2}{c}{y \leq 6} \\ \hline 0 & 6 \end{array}$$

$0 \overset{?}{\leq} 6$ TRUE

Since $0 \leq 6$ is true, we shade the half-plane that contains (0, 0) and obtain the graph.

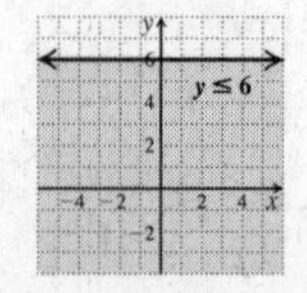

29. Graph: $-2 < y < 7$

This is a system of inequalities:

$-2 < y,$
$y < 7$

The graph of $-2 < y$ is the half-plane above the line $-2 = y$; the graph of $y < 7$ is the half-plane below the line $y = 7$. We shade the intersection of these graphs.

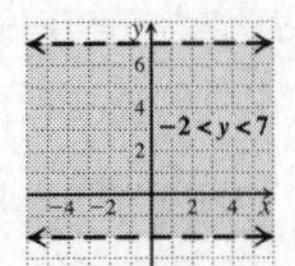

31. Graph: $-5 \leq x < 4$.

This is a system of inequalities:

$-5 \leq x$
$x < 4$

Graph $-5 \leq x$ and $x < 4$. Then shade the intersection of these graphs.

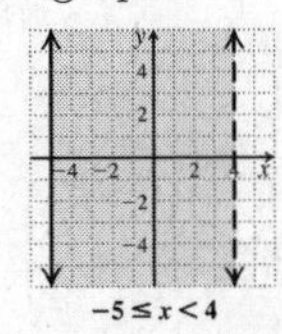

33. Graph: $0 \leq y \leq 3$

This is a system of inequalities:

$0 \leq y,$
$y \leq 3$

Graph $0 \leq y$ and $y \leq 3$.

Then we shade the intersection of these graphs.

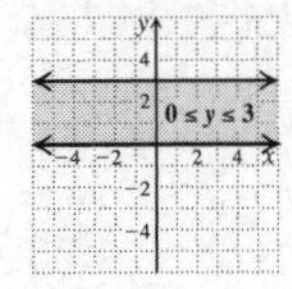

35. Graph: $y > x,$
$y < -x + 3.$

We graph the lines $y = x$ and $y = -x + 3$, using dashed lines. We indicate the region for each inequality by the arrows at the ends of the lines. Note where the regions overlap and shade the region of solutions.

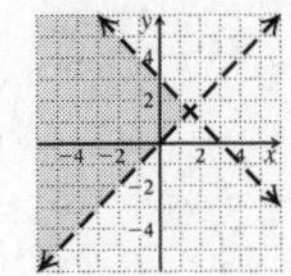

37. Graph: $y \leq x,$
$y \leq 2x - 5.$

We graph the lines $y = x$ and $y = 2x - 5$, using solid lines. Indicate the region for each inequality by arrows and shade the region where they overlap

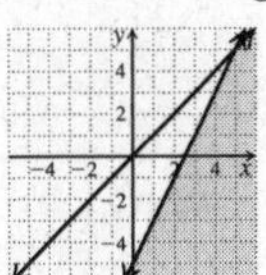

39. Graph: $y \leq -3,$
$x \geq -1$

Graph $y = -3$ and $x = -1$ using solid lines. Indicate the region for each inequality by arrows, and shade the region where they overlap.

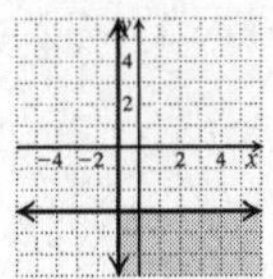

41. Graph: $x > -4,$
$y < -2x + 3$

Graph the lines $x = -4$ and $y = -2x + 3$, using dashed lines. Indicate the region for each inequality by arrows, and shade the region where they overlap.

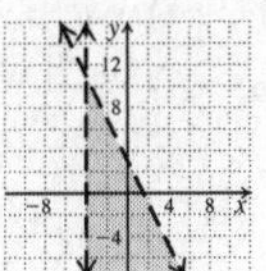

43. Graph: $y \leq 5,$
$y \geq -x + 4$

Graph the lines $y = 5$ and $y = -x + 4$, using solid lines. Indicate the region for each inequality by arrows, and shade the region where they overlap.

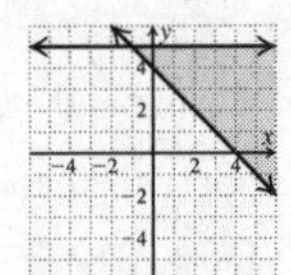

45. Graph: $x + y \leq 6,$
$x - y \leq 4$

Graph the lines $x + y = 6$ and $x - y = 4$, using solid lines. Indicate the region for each inequality by arrows, and shade the region where they overlap.

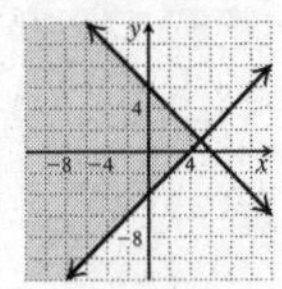

47. Graph: $y + 3x > 0,$
$y + 3x < 2$

Graph the lines $y + 3x = 0$ and $y + 3x = 2$, using dashed lines. Indicate the region for each inequality by arrows, and shade the region where they overlap.

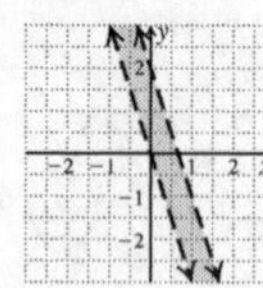

49. Graph: $y \le 2x-3,$ (1)
$y \ge -2x+1,$ (2)
$x \le 5$ (3)

Graph the lines $y=2x-3$, $y=-2x+1$, and $x=5$ using solid lines. Indicate the region for each inequality by arrows, and shade the region where they overlap.

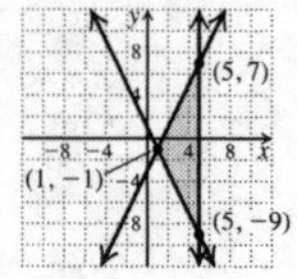

To find the vertex we solve three different systems of related equations.

From (1) and (2) we have $y=2x-3,$
$y=-2x+1.$

Solving, we obtain the vertex $(1,-1)$.

From (1) and (3) we have $y=2x-3,$
$x=5.$

Solving, we obtain the vertex $(5,\ 7)$.

From (2) and (3) we have $y=-2x+1,$
$x=5.$

Solving, we obtain the vertex $(5,-9)$.

51. Graph: $x+2y \le 12,$ (1)
$2x+y \le 12$ (2)
$x \ge 0,$ (3)
$y \ge 0$ (4)

Graph the lines $x+2y=12$, $2x+y=12$, $x=0$, and $y=0$ using solid lines. Indicate the region for each inequality by arrows, and shade the region where they overlap.

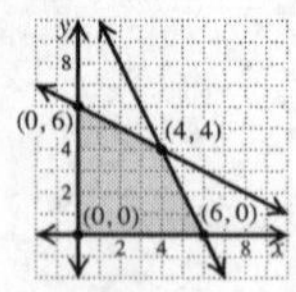

To find the vertices we solve four different systems of equations.

From (1) and (2) we have $x+2y=12,$
$2x+y=12.$

Solving, we obtain the vertex $(4,\ 4)$.

From (1) and (3) we have $x+2y=12,$
$x=0.$

Solving, we obtain the vertex $(0,\ 6)$.

From (2) and (4) we have $2x+y=12,$
$y=0.$

Solving, we obtain the vertex $(6,\ 0)$.

From (3) and (4) we have $x=0,$
$y=0.$

Solving, we obtain the vertex $(0,\ 0)$.

53. Graph: $8x+5y \le 40,$ (1)
$x+2y \le 8$ (2)
$x \ge 0,$ (3)
$y \ge 0$ (4)

Graph the lines $8x+5y=40$, $x+2y=8$, $x=0$, and $y=0$ using solid lines. Indicate the region for each inequality by arrows, and shade the region where they overlap.

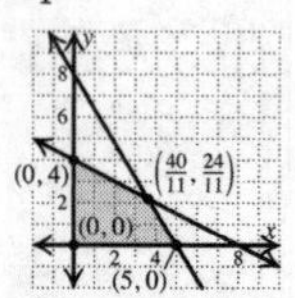

To find the vertices we solve four different systems of equations.

From (1) and (2) we have $8x+5y=40,$
$x+2y=8.$

Solving, we obtain the vertex $(\frac{40}{11},\frac{24}{11})$.

From (1) and (4) we have $8x+5y=40,$
$y=0.$

Solving, we obtain the vertex $(5,0)$.

From (2) and (3) we have $x+2y=8,$
$x=0.$

Solving, we obtain the vertex $(0,4)$.

From (3) and (4) we have $x=0,$
$y=0.$

Solving, we obtain the vertex $(0,0)$.

55. Graph: $y-x \ge 2,$ (1)
$y-x \le 4,$ (2)
$2 \le x \le 5$ (3)

Think of (3) as two inequalities:
$2 \le x,$ (4)
$x \le 5$ (5)

Graph the lines $y-x=2$, $y-x=4$, $x=2$, and $x=5$, using solid lines. Indicate the region for each inequality by arrows, and shade the region where they overlap.

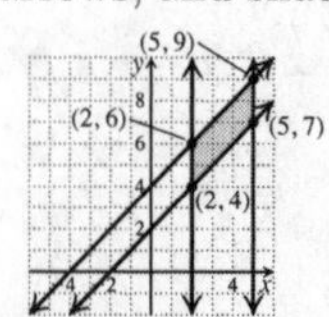

To find the vertices we solve four different systems of equations.

From (1) and (4) we have $y-x=2,$
$x=2.$

Solving, we obtain the vertex $(2,\ 4)$.

From (1) and (5) we have $y-x=2,$
$x=5.$

Solving, we obtain the vertex (5, 7).

From (2) and (4) we have $y-x=4,$
$x=2.$

Solving, we obtain the vertex (2, 6).

From (2) and (5) we have $y-x=4,$
$x=5.$

Solving, we obtain the vertex (5, 9).

57. *Writing Exercise.* The boundary line is drawn dashed for the symbols $<$ and $>$ to show that the line is a border to the solution, but not part of the solution. The boundary line is drawn solid for the symbols $\leq$ and $\geq$ to show that both the line and the half-plane are the solution.

59. ***Familiarize.*** The formula for interest I, with principal P, rate r, and time t is $I = Prt$.

Translate. Substitute \$25.35 for I, \$1560 for P and $\frac{1}{2}$ for t in the formula.

$$I = Prt$$
$$\$25.35 = \$1560 \cdot r \cdot \frac{1}{2}$$

Carry out. We solve the equation.

$$25.35 = 1560 \cdot r \cdot \frac{1}{2}$$
$$25.35 = 780r \quad \text{Multiplying}$$
$$0.0325 = r \quad \text{Dividing by 780 on both sides}$$

Check. Interest is $\$15.60(0.0325)\left(\frac{1}{2}\right)$ or \$25.35. These numbers check.

State. The rate of interest is 0.0325, or 3.25%.

61. ***Familiarize.*** Let x = the amount invested at 5% and y = the amount invested at 3%.

Translate. The interest from the first investment is $0.05x$ and the interest from the second investment is $0.03y$.

The sum of the investments is

$$x + y = 10{,}000.$$

The sum of the interests is

$$0.05x + 0.03y = 428.$$

Carry out. We solve the system of equations.

$$x + y = 10{,}000, \quad (1)$$
$$0.05x + 0.03y = 428. \quad (2)$$

We solve equation (1) for y, and substitute into equation (2), and solve for x.

$$0.05x + 0.03(10{,}000 - x) = 428$$
$$0.05x + 300 - 0.03x = 428$$
$$0.02x + 300 = 428$$
$$0.02x = 128$$
$$x = 6400$$

$$y = 10{,}000 - x = 10{,}000 - 6400 = 3600$$

Check. The interest of the first investment is 0.05(\$6400), or \$320 and the interest of the second investment is 0.03(\$3600), or \$108. The total interest is \$320 + \$108, or \$428. These numbers check.

State. \$6400 is invested at 5% and \$3600 is invested at 3%.

63. ***Familiarize.*** Let x = the number of student tickets sold and y = the number of adult tickets sold. We arrange the information in a table.

	Student	Adult	Total
Price	\$1	\$3	
Number sold	x	y	170
Money taken in	$1x$	$3y$	386

Translate. The last two rows of the table give us two equations. The total number of tickets sold was 170, so we have

$$x + y = 170.$$

The total amount of money collected was \$386, so we have

$$1x + 3y = 386.$$

Carry out. Solve the system using the elimination method.

$$x + y = 170, \quad (1)$$
$$x + 3y = 386. \quad (2)$$

$$-x - y = -170 \quad \text{Multiplying (1) by } -1$$
$$x + 3y = 386$$
$$2y = 216$$
$$y = 108$$

We go back to Equation (1) and substitute 108 for y.

$$x + y = 170$$
$$x + 108 = 170$$
$$x = 62$$

Check. The number of tickets sold was 108+62, or 170. The money collected was \$1(62) + \$3(108), or \$62+\$324, or \$386. These numbers check.

State. 62 student tickets and 108 adult tickets were sold.

65. *Writing Exercise.* The intersection of the individual graphs is a single point. An example is the following system:

$y \le -x,$
$x \ge 0,$
$y \ge 0$

The solution set is (0, 0).

67. Graph: $x + y > 8,$
$x + y \le -2$

Graph the line $x + y = 8$ using a dashed line and graph $x + y = -2$, using a solid line. Indicate the region for each inequality by arrows. The regions do not overlap (the solution set is $\varnothing$), so we do not shade any portion of the graph.

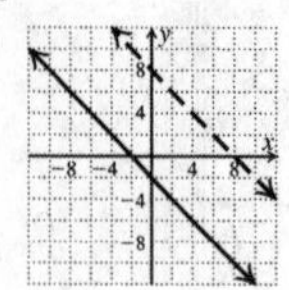

69. Graph: $x - 2y \le 0,$
$-2x + y \le 2,$
$x \le 2,$
$y \le 2,$
$x + y \le 4$

Graph the five inequalities above, and shade the region where they overlap.

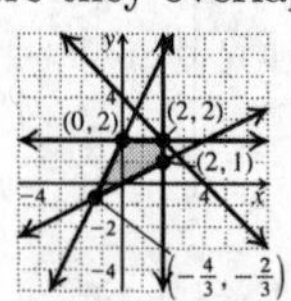

71. Both the width and the height must be positive, so we have

$w > 0,$
$h > 0.$

To be checked as luggage, the sum of the width, height, and length cannot exceed 62 in., so we have

$w + h + 30 \le 62,$ or
$w + h \le 32.$

The girth is represented by $2w + 2h$ and the length is 30 in. In order to meet postal regulations the sum of the girth and the length cannot exceed 130 in., so we have:

$2w + 2h + 30 < 130,$ or
$2w + 2h \le 100,$ or
$w + h \le 50$

Thus, have a system of inequalities:

$w > 0,$
$h > 0$
$w + h \le 32,$
$w + h \le 50$

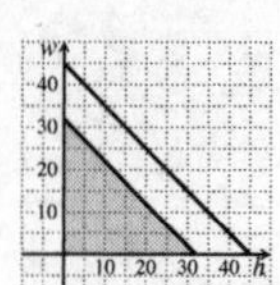

73. We graph the following inequalities:

$q \ge 700$
$v \ge 400$
$q + v \ge 1150$
$q \le 800$
$v \le 800$

75. Graph: $35c + 75a > 1000,$
$c \ge 0,$
$a \ge 0$

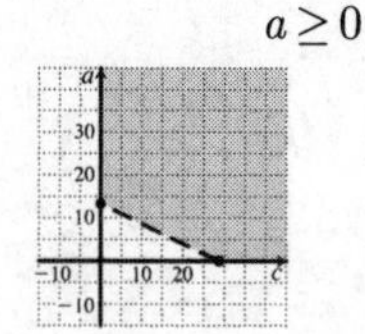

77. $h < 2w$
$w \le 1.5h$
$h \le 3200$
$h \ge 0$
$w \ge 0$

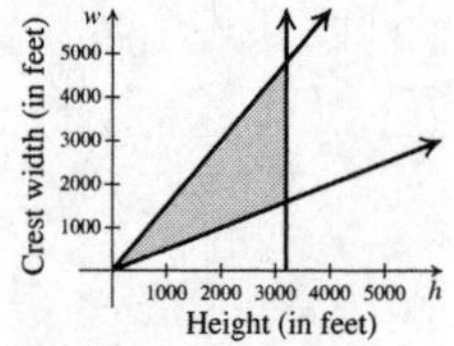

79. a) $3x + 6y > 2$

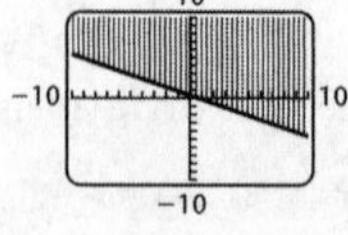

b) $x - 5y \le 10$

c) $13x - 25y + 10 \le 0$

d) $2x + 5y > 0$

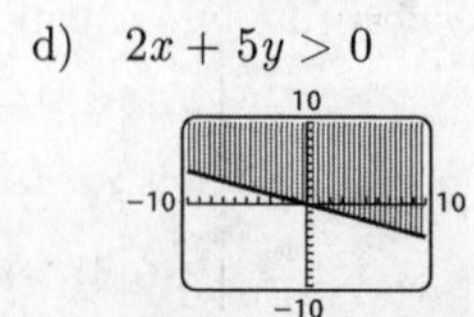

Connecting the Concepts

1. $x+2=7$

$x=5$

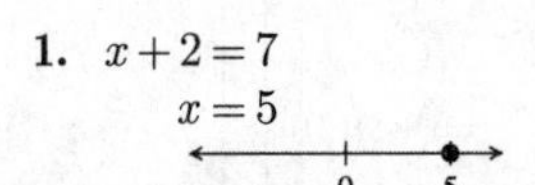

3. $x+2\leq 7$

$x\leq 5$

5. $6-2x\geq 8$

$-2x\geq 2$

$x\leq -1$

7. $x+y=2$

$y=-x+2$

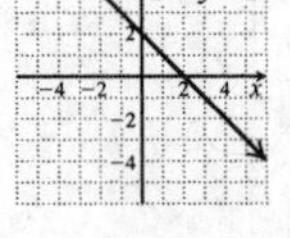

9. $x+y\geq 2$

$y\geq -x+2$

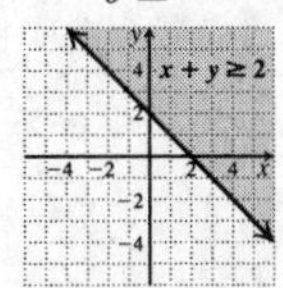

11. $x=4$

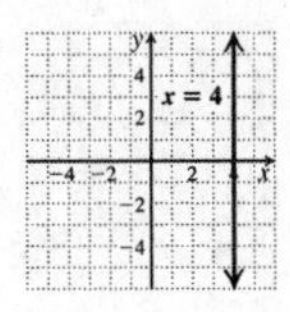

13. $x+y=1, \quad \Rightarrow \quad y=-x+1,$

$x-y=1 \quad \Rightarrow \quad y=x-1$

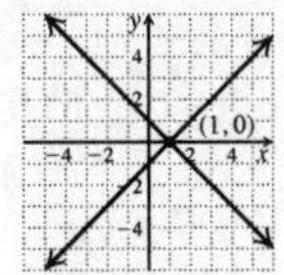

15. $2x+y<6$

$y<-2x+6$

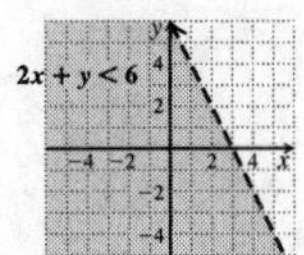

17. $4x=3y$

$y=\frac{4}{3}x$

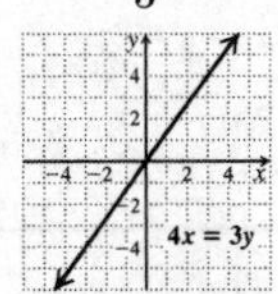

19. $x-y\leq 3, \quad \Rightarrow \quad y\geq x-3,$

$y\geq 2x, \quad \Rightarrow \quad y\geq 2x,$

$2y-x\leq 2 \quad \Rightarrow \quad y\leq \frac{1}{2}x+1$

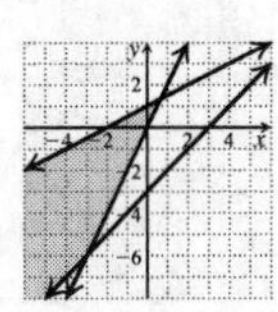

Exercise Set 9.5

1. Objective; see page 618 in the text.

3. Corner; see page 619 in the text.

5. Vertices; see page 619 in the text.

7. Find the maximum and minimum values of

$$F=2x+14y\ ,$$

subject to

$$\begin{aligned} 5x+3y&\leq 34, \quad (1)\\ 3x+5y&\leq 30, \quad (2)\\ x&\geq 0, \quad (3)\\ y&\geq 0. \quad (4)\end{aligned}$$

Graph the system of inequalities and find the coordinates of the vertices.

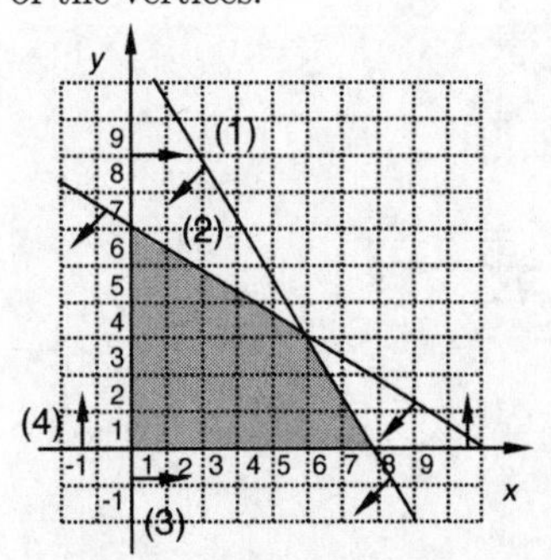

To find one vertex we solve the system

$x=0,$

$y=0.$

This vertex is $(0,0)$.

To find a second vertex we solve the system

$5x+3y=34,$

$y=0.$

This vertex is $\left(\frac{34}{5}, 0\right)$.

To find a third vertex we solve the system

$$5x+3y=34,$$
$$3x+5y=30.$$

This vertex is $(5, 3)$.

To find the fourth vertex we solve the system

$$3x+5y=30,$$
$$x=0.$$

This vertex is $(0, 6)$.

Now find the value of F at each of these points.

Vertex (x, y)	$F=2x+14y$	
$(0, 0)$	$2\cdot0+14\cdot0=0+0=0$	⟵Minimum
$\left(\frac{34}{5}, 0\right)$	$2\cdot\frac{34}{5}+14\cdot0=\frac{68}{5}+0=13\frac{3}{5}$	
$(5, 3)$	$2\cdot5+14\cdot3=10+42=52$	
$(0, 6)$	$2\cdot0+14\cdot6=0+84=84$	⟵Maximum

The maximum value of F is 84 when $x=0$ and $y=6$.

The minimum value of F is 0 when $x=0$ and $y=0$.

9. Find the maximum and minimum values of

$$P=8x-y+20,$$

subject to

$$6x+8y\le48, \quad (1)$$
$$0\le y\le4, \quad (2)$$
$$0\le x\le7. \quad (3)$$

Think of (2) as $0\le y$, (4)
$y\le4$. (5)

Think of (3) as $0\le x$, (6)
$x\le7$. (7)

Graph the system of inequalities.

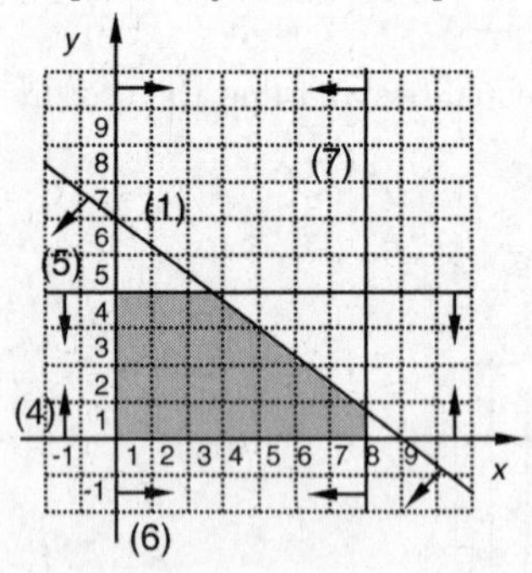

To determine the coordinates of the vertices, we solve the following systems:

$x=0,\ y=0;$ $x=7,\ y=0;$ $6x+8y=48,\ x=7;$

$6x+8y=48,\ y=4;$ $x=0,\ y=4$

The vertices are $(0, 0)$, $(7, 0)$, $\left(7, \frac{3}{4}\right)$, $\left(\frac{8}{3}, 4\right)$, and $(0, 4)$, respectively. Compute the value of P at each of these points.

Vertex (x, y)	$P=8x-y+20$	
$(0, 0)$	$8\cdot0-0+20$ $=0-0+20=20$	
$(7, 0)$	$8\cdot7-0+20$ $=56-0+20=76$	← Maximum
$\left(7, \frac{3}{4}\right)$	$8\cdot7-\frac{3}{4}+20$ $=56-\frac{3}{4}+20=75\frac{1}{4}$	
$\left(\frac{8}{3}, 4\right)$	$8\cdot\frac{8}{3}-4+20$ $=\frac{64}{3}-4+20=37\frac{1}{3}$	
$(0,4)$	$8\cdot0-4+20$ $=0-4+20=16$	← Minimum

The maximum is 76 when $x=7$ and $y=0$. The minimum is 16 when $x=0$ and $y=4$.

11. Find the maximum and minimum values of

$$F=2y-3x,$$

subject to

$$y\le2x+1, \quad (1)$$
$$y\ge-2x+3, \quad (2)$$
$$x\le3 \quad (3)$$

Graph the system of inequalities and find the coordinates of the vertices.

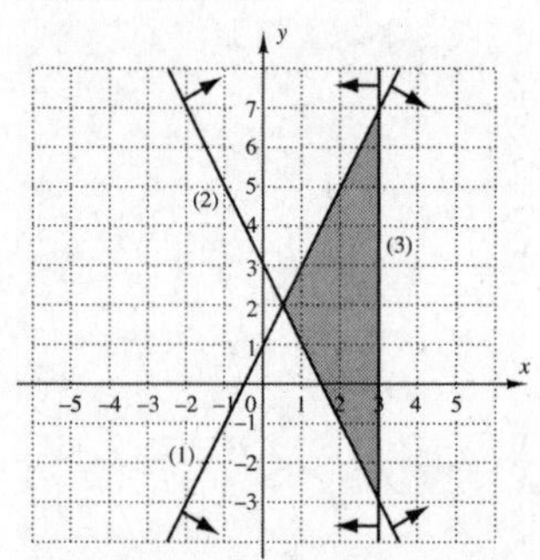

To determine the coordinates of the vertices, we solve the following systems:

$y=2x+1,\ y=-2x+3;$ $y=2x+1,\ x=3;$ $y=-2x+3,\ x=3$

The solutions of the systems are $\left(\frac{1}{2}, 2\right)$, $(3, 7)$, and $(3,-3)$, respectively. Now find the value of F at each of these points.

Vertex (x, y)	$F=2y-3x$	
$\left(\frac{1}{2}, 2\right)$	$2\cdot2-3\cdot\frac{1}{2}=\frac{5}{2}$	
$(3, 7)$	$2\cdot7-3\cdot3=5$	← Maximum
$(3,-3)$	$2(-3)-3\cdot3=-15$	← Minimum

The maximum value is 5 when $x=3$ and $y=7$. The minimum value is -15 when $x=3$ and $y=-3$.

13. ***Familiarize.*** Let $x =$ the number of gumbo orders and $y =$ the number of sandwiches sold each day.

Translate. The profit P is given by

$$P = \$1.65x + \$1.05y.$$

We wish to maximize P subject to these constraints:

$$10 \leq x \leq 40$$
$$30 \leq y \leq 70$$
$$x + y \leq 90.$$

Carry out. We graph the system of inequalities, determine the vertices and evaluate P at each vertex.

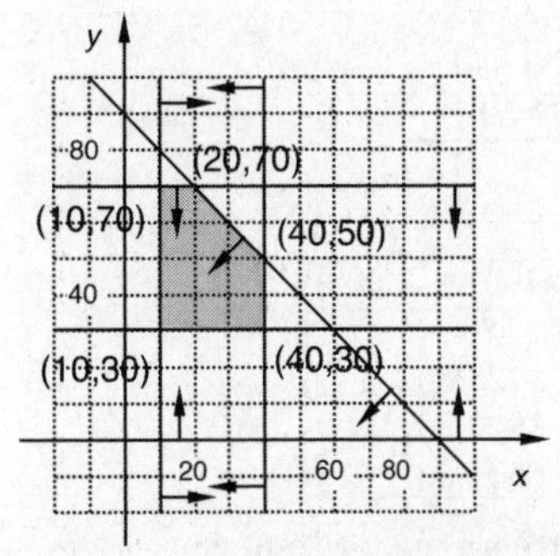

Vertex	$P = \$1.65x + \$1.05y$
(10, 30)	$\$1.65(10) + \$1.05(30) = \$48$
(40, 30)	$\$1.65(40) + \$1.05(30) = \$97.50$
(40, 50)	$\$1.65(40) + \$1.05(50) = \$118.50$
(20, 70)	$\$1.65(20) + \$1.05(70) = \$106.50$
(10, 70)	$\$1.65(10) + \$1.05(70) = \$90$

The largest profit in the table is \$118.50 obtained when 40 orders of gumbo and 50 sandwiches are sold.

Check. Go over the algebra and arithmetic.

State. The maximum profit occurs when 40 orders of gumbo and 50 sandwiches are sold.

15. ***Familiarize.*** Let $x =$ the number of 4 photo pages, and $y =$ the number of 6 photo pages.

Translate. The number of photos N is given by

$$N = 4x + 6y.$$

We wish to maximize N subject to these constraints.

$$x + y \leq 20$$
$$3x + 5y \leq 90$$
$$x \geq 0$$
$$y > 0.$$

Carry out. We graph the system of inequalities, determine the vertices, and evaluate N at each vertex.

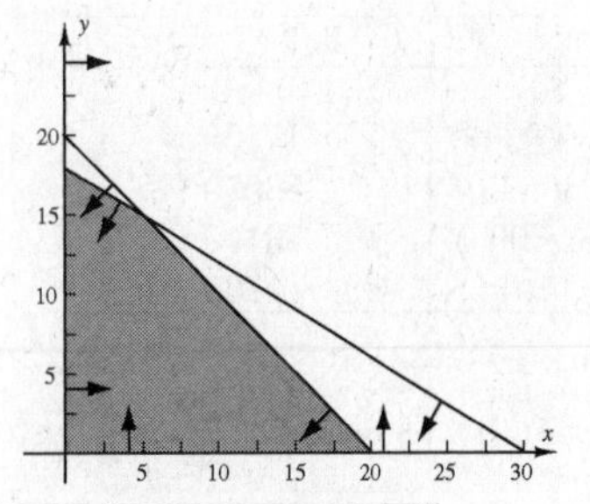

Vertex	$N = 4x + 6y$
(0, 0)	$4 \cdot 0 + 6 \cdot 0 = 0$
(0, 18)	$4 \cdot 0 + 6 \cdot 18 = 108$
(20, 0)	$4 \cdot 20 + 6 \cdot 0 = 80$
(5, 15)	$4 \cdot 5 + 6 \cdot 15 = 110$

The greatest number of photos is 110, obtained when 5 pages of 4-photos and 15 pages of 6-photos are used.

Check. Go over the algebra and arithmetic.

State. The maximum number of photos is achieved by using 5 pages or 4-photos and 15 pages of 6-photos.

17. In order to earn the most interest Rosa should invest the entire \$40,000. She should also invest as much as possible in the type of investment that has the higher interest rate. Thus, she should invest \$22,000 in corporate bonds and the remaining \$18,000 in municipal bonds. The maximum income is

$0.08(\$22{,}000) + 0.075(\$18{,}000) = \$3110$.

We can also solve this problem as follows.

Let $x =$ the amount invested in corporate bonds and $y =$ the amount invested in municipal bonds. Find the maximum value of

$$I = 0.08x + 0.075y$$

subject to

$$x + y \leq \$40{,}000,$$
$$\$6000 \leq x \leq \$22{,}000$$
$$0 \leq y \leq \$30{,}000.$$

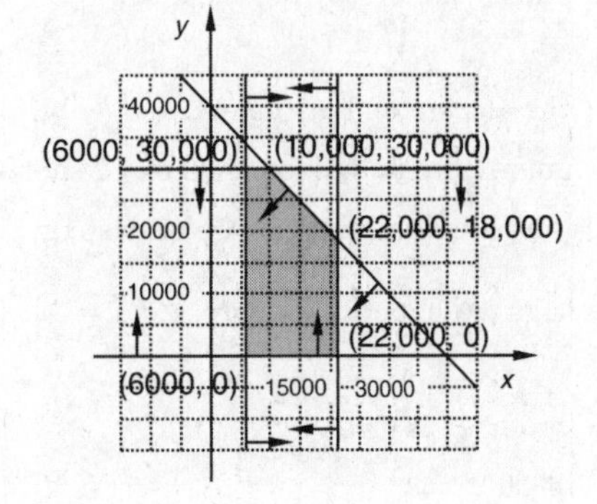

Vertex	$I = 0.08x + 0.075y$
(\$6000, \$0)	\$480
(\$6000, \$30,000)	\$2730
(\$10,000, \$30,000)	\$3050
(\$22,000, \$18,000)	\$3110
(\$22,000, \$0)	\$1760

The maximum income of \$3110 occurs when \$22,000 is invested in corporate bonds and \$18,000 is invested in municipal bonds.

19. ***Familiarize.*** Let $x =$ the number of short-answer questions and $y =$ the number of essay questions answered.

Translate. The score S is given by

$S = 10x + 15y.$

We wish to maximize S subject to these constraints:

$$x + y \leq 16$$
$$3x + 6y \leq 60$$
$$x \geq 0$$
$$y \geq 0.$$

Carry out. We graph the system of inequalities, determine the vertices, and evaluate S at each vertex.

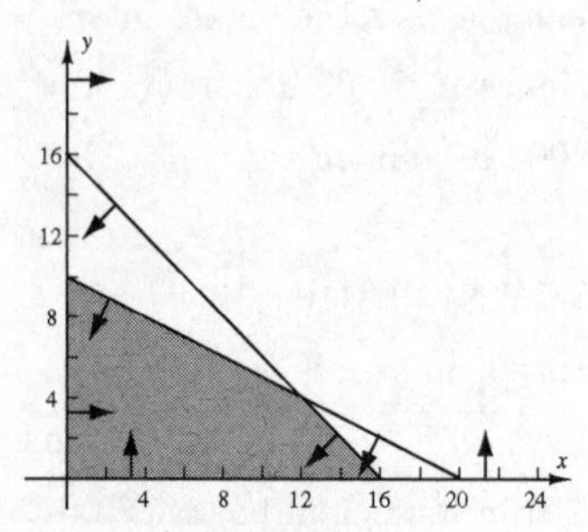

Vertex	$S = 10x + 15y$
(0, 0)	$10 \cdot 0 + 15 \cdot 0 = 0$
(0, 10)	$10 \cdot 0 + 15 \cdot 10 = 150$
(16, 0)	$10 \cdot 16 + 15 \cdot 0 = 160$
(12, 4)	$10 \cdot 12 + 15 \cdot 4 = 180$

The greatest score in the table is 180, obtained when 12 short-answer questions and 4 essay questions are answered.

Check. Go over the algebra and arithmetic.

State. The maximum score is 180 points when 12 short-answer questions and 4 essay questions are answered.

21. ***Familiarize.*** Let $x =$ the Merlot acreage and $y =$ the Cabernet acreage.

Translate. The profit P is given by

$P = \$400x + \$300y$.

We wish to maximize P subject to these constraints:

$$x + y \leq 240,$$
$$2x + y \leq 320,$$
$$x \geq 0,$$
$$y \geq 0.$$

Carry out. We graph the system of inequalities, determine the vertices, and evaluate P at each vertex.

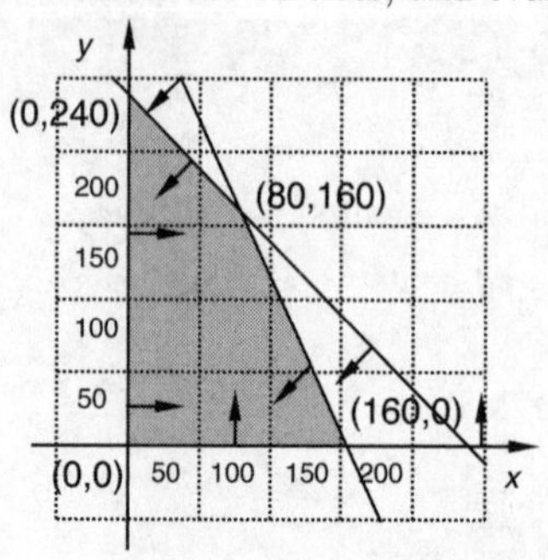

Vertex	$P = \$400x + \$300y$
(0, 0)	\$0
(0, 240)	\$72,000
(80, 160)	\$80,000
(160, 0)	\$64,000

Check. Go over the algebra and arithmetic.

State. The maximum profit occurs by planting 80 acres of Merlot grapes and 160 acres of Cabernet grapes.

23. ***Familiarize.*** Let $x =$ the number of servings of goat cheese and $y =$ the number of servings of hazelnuts.

Translate. The total number of calories is given by

$C = 264x + 628y.$

We wish to minimize C subject to these constraints.

$$15 \leq x + 5y$$
$$x + 5y \leq 45$$
$$1500 \leq 500x + 100y$$
$$500x + 100y \leq 2500$$

Carry out. We graph the system of inequalities, determine the vertices and evaluate C at each vertex.

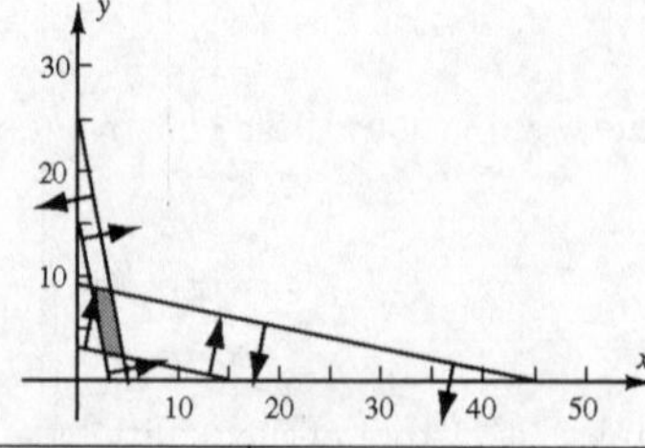

Vertex	$C = 264x + 628y$
(2.5, 2.5)	$264 \cdot 2.5 + 628 \cdot 2.5 = 2230$
(1.25, 8.75)	$264 \cdot 1.25 + 628 \cdot 8.75 = 5825$
$\left(\frac{55}{12}, \frac{25}{12}\right)$	$264 \cdot \frac{55}{12} + 628 \cdot \frac{25}{12} = 2518.3 = \frac{30,220}{12}$
$\left(\frac{10}{3}, \frac{25}{3}\right)$	$264 \cdot \frac{10}{3} + 628 \cdot \frac{25}{3} = 6113.3 = \frac{18,340}{3}$

The least number of calories in the table is 2230, obtained with 2.5 servings of each.

Check. Go over the algebra and arithmetic.

State. The minimum calories consumed is 2230 with 2.5 servings of each.

25. *Writing Exercise.* Answers may vary. Four sections that might be considered are sections on graphing linear equations, graphing linear inequalities, solving systems of equations, and evaluating algebraic expressions.

27.
$$f(x)=4x-7$$
$$f(a)=4a-7$$
$$f(a)+h=4a-7+h$$

29. $f(x)=\dfrac{x-5}{2x+1}$

We exclude the values from the domain that make the denominator zero.

$$2x+1=0$$
$$x=-\frac{1}{2}$$

The domain is $\left\{x \middle| x \text{ is a real number } and\ x\neq-\frac{1}{2}\right\}$, or $\left(-\infty,-\frac{1}{2}\right)\cup\left(-\frac{1}{2},\infty\right)$.

31. $f(x)=\sqrt{2x+8}$

The value inside the radical must be greater than or equal to zero.

$$2x+8\geq0$$
$$2x\geq-8$$
$$x\geq-4$$

The domain is $\{x|x\geq-4\}$, or $[-4,\infty)$.

33. *Writing Exercise.* In regard to Exercise 17, see the solution for that exercise. In Exercise 18, it is logical to invest the maximum amount possible in the investment that generates the most income.

35. ***Familiarize.*** Let x represent the number of T3 planes and y represent the number of S5 planes. Organize the information in a table.

Plane	Number of planes	Passengers		
		First	Tourist	Economy
T3	x	$40x$	$40x$	$120x$
S5	y	$80y$	$30y$	$40y$

Plane	Cost per mile
T3	$30x$
S5	$25y$

Translate. Suppose C is the total cost per mile. Then $C=30x+25y$. We wish to minimize C subject to these facts (constraints) about x and y.

$$40x+80y\geq2000,$$
$$40x+30y\geq1500,$$
$$120x+40y\geq2400,$$
$$x\geq0,$$
$$y\geq0$$

Carry out. Graph the system of inequalities, determine the vertices, and evaluate C at each vertex.

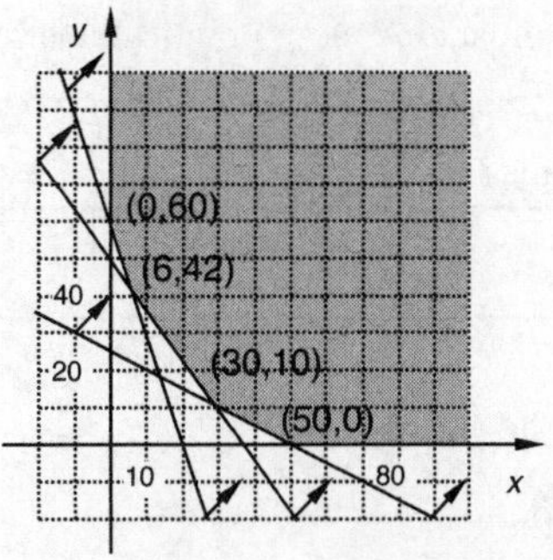

Vertex	$C=30x+25y$
(0, 60)	$30(0)+25(60)=1500$
(6, 42)	$30(6)+25(42)=1230$
(30, 10)	$30(30)+25(10)=1150$
(50, 0)	$30(50)+25(0)=1500$

Check. Go over the algebra and arithmetic.

State. In order to minimize the operating cost, 30 T3 planes and 10 S5 planes should be used.

37. ***Familiarize.*** Let $x=$ the number of chairs and $y=$ the number of sofas produced.

Translate. Find the maximum value of

$I=\$80x+\$1200y$

subject to

$$20x+100y\leq1900,$$
$$x+50y\leq500,$$
$$2x+20y\leq240,$$
$$x\geq0,$$
$$y\geq0.$$

Carry out. Graph the system of inequalities, determine the vertices, and evaluate I at each vertex.

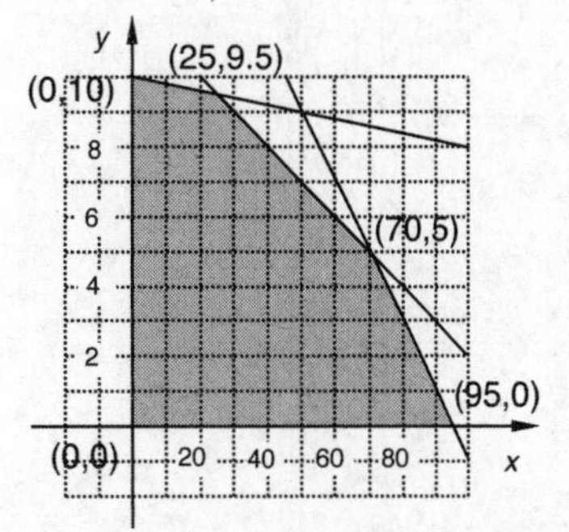

Vertex	$I = \$80x + \$1200y$
(0, 0)	\$0
(0, 10)	\$12,000
(25, 9.5)	\$13,400
(70, 5)	\$11,600
(95, 0)	\$7600

Check. Go over the algebra and arithmetic.

State. The maximum income of \$13,400 occurs when 25 chairs and 9.5 sofas are made. A more practical answer is that the maximum income of \$12,800 is achieved when 25 chairs and 9 sofas are made.

Chapter 9 Review

1. True; see page 582 in the text.

3. True; see page 600 in the text.

5. True; see page 590 in the text.

7. True; see page 600 in the text.

9. False; see page 611 in the text.

11. $-6x-5<4$

$-6x<9$

$x>-\frac{3}{2}$ Reversing the inequality symbol

$\left\{x\middle|x>-\frac{3}{2}\right\}$, or $\left(-\frac{3}{2},\ \infty\right)$

13. $0.3y-7<2.6y+15$

$-2.3y-7<15$

$-2.3y<22$

$y>-\frac{220}{23}$ Reversing the inequality symbol

$\left\{y\middle|y>-\frac{220}{23}\right\}$, or $\left(-\frac{220}{23},\ \infty\right)$

15. Solve graphically: $x-2<3$.

We graph and solve the system of equations

$y=x-2,$

$y=3$

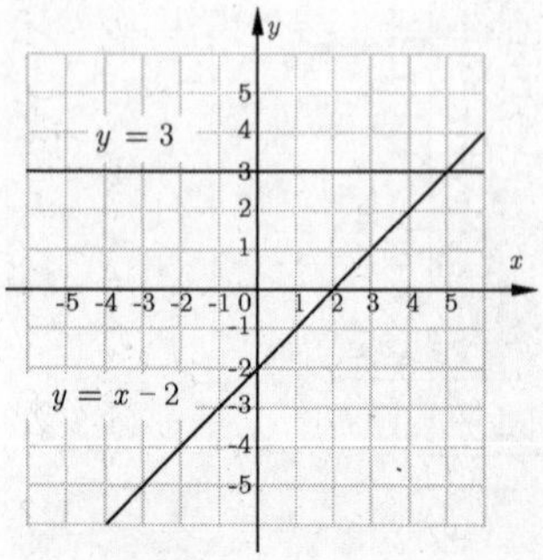

By substitution, we have

$x-2=3$

$x=5$

Then $y = 5 - 2 = 3$.

Thus, the point of intersection is (5, 3). Since the inequality symbol is $<$, we do not include this point in the solution.

We see from the graph, $x-2<3$ is to the left of the point of intersection.

Thus, the solution set is $\{x|x<5\}$, or $(-\infty,\ 5)$.

17. Solve graphically: $x-1\le 2x+3$.

We graph and solve the system of equations

$y=x-1,$

$y=2x+3$

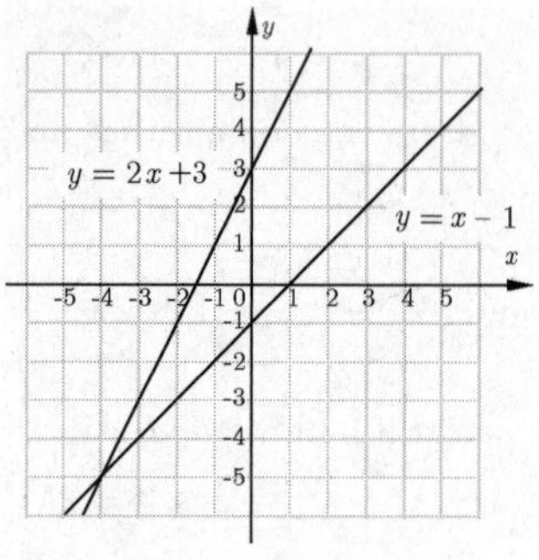

By substitution, we have

$x-1=2x+3$

$-4=x$

Then $y = -4 - 1 = -5$.

Thus, the point of intersection is $(-4, -5)$. Since the inequality symbol is $\le$, we include this point in the solution.

We see from the graph, $x-1\le 2x+3$ is to the right of the point of intersection.

Thus, the solution set is $\{x|x\le -4\}$, or $(-\infty,-4]$.

19. $f(x)\le g(x)$

$3x+2\le 10-x$

$4x\le 8$

$x\le 2$

$\{x|x\le 2\}$ or $(-\infty,\ 2]$

21. Let x = the amount invested at 3% and $9000 - x$ = the amount invested at 3.5%. The interest from the first investment is $0.03x$ and the interest from the second investment is $0.035(9000 - x)$. We solve the inequality.

$$0.03x+0.035(9000-x)\ge 300$$
$$0.03x+315-0.035x\ge 300$$
$$-0.005x+315\ge 300$$
$$-0.005x\ge -15$$
$$x\le 3000$$

Clay should invest at most \$3000 at 3%.

23. $\{a, b, c, d\}\cup\{a, c, e, f, g\}=\{a, b, c, d, e, f, g\}$

25. Graph: $x \le 3$ *or* $x > -5$

$(-\infty, \infty)$

27. $-15 < -4x - 5 < 0$
$-10 < -4x < 5$
$\frac{5}{2} > x > -\frac{5}{4}$, or $-\frac{5}{4} < x < \frac{5}{2}$

$\left\{x \middle| -\frac{5}{2} < x < \frac{5}{2}\right\}$

$\left(-\frac{5}{4}, \frac{5}{2}\right)$

29. $2x + 5 < -17$ *or* $-4x + 10 \le 34$
$2x < -22$ *or* $-4x \le 24$
$x < -11$ *or* $x \ge -6$

$\{x | x < -11 \text{ or } x \ge -6\}$ or $(-\infty, -11) \cup [-6, \infty)$

31. $f(x) < -5$ *or* $f(x) > 5$
$3 - 5x < -5$ *or* $3 - 5x > 5$
$-5x < -8$ *or* $-5x > 2$
$x > \frac{8}{5}$ *or* $x < -\frac{2}{5}$

$\left\{x \middle| x < -\frac{2}{5} \text{ or } x > \frac{8}{5}\right\}$ or $\left(-\infty, -\frac{2}{5}\right) \cup \left(\frac{8}{5}, \infty\right)$

33. $f(x) = \sqrt{5x - 10}$
$5x - 10 \ge 0$
$5x \ge 10$
$x \ge 2$
The domain of f is $[2, \infty)$.

35. $|x| = 11$

$x = -11$ *or* $x = 11$

$\{-11, 11\}$

37. $|x - 8| = 3$

$x - 8 = -3$ *or* $x - 8 = 3$
$x = 5$ *or* $x = 11$

$\{5, 11\}$

39. $|3x - 4| \ge 15$

$3x - 4 \le -15$ *or* $15 \le 3x - 4$
$3x \le -11$ *or* $19 \le 3x$
$x \le -\frac{11}{3}$ *or* $\frac{19}{3} \le x$

$\left\{x \middle| x \le -\frac{11}{3} \text{ or } x \ge \frac{19}{3}\right\}$ or $\left(-\infty, -\frac{11}{3}\right] \cup \left[\frac{19}{3}, \infty\right)$

41. $|5n + 6| = -11$

Absolute value is never negative.

The solution is $\varnothing$.

43. $2|x - 5| - 7 > 3$
$2|x - 5| > 10$
$|x - 5| > 5$
$x - 5 < -5$ *or* $5 < x - 5$
$x < 0$ *or* $10 < x$

$\{x | x < 0 \text{ or } x > 10\}$ or $(-\infty, 0) \cup (10, \infty)$

45. $|8x - 3| < 0$

Absolute value is never negative.

The solution is $\varnothing$.

47. Graph $x + 3y > -1$,
$x + 3y < 4$

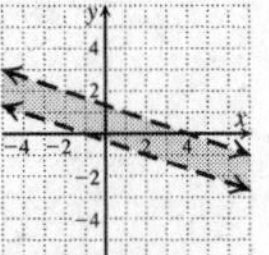

The lines are parallel, there are no vertices.

49. For $F = 3x + y + 4$, subject to
$y \le 2x + 1$,
$x \le 7$,
$y \ge 3$.

Vertices	$F = 3x + y + 4$
(7, 3)	$3 \cdot 7 + 3 + 4 = 28$
(1, 3)	$3 \cdot 1 + 3 + 4 = 10$
(7, 15)	$3 \cdot 7 + 15 + 4 = 40$

The maximum value of F is 40 at $x = 7$, $y = 15$.

The minimum value of F is 10 at $x = 1$, $y = 3$.

51. *Writing Exercise.* The equation $|X| = p$ has two solutions when p is positive because X can be either p or $-p$. The same equation has no solution when p is negative because no number has a negative absolute value.

53. $|2x+5| \le |x+3|$

$2x+5 \le x+3$ *or* $2x+5 \ge -(x+3)$
$x \le -2$ *or* $2x+5 \ge -x-3$
$3x \ge -8$
$x \ge -\frac{8}{3}$

$\left\{x \middle| -\frac{8}{3} \le x \le -2\right\}$

55. $|d - 2.5| \le 0.003$

Chapter 9 Test

1. $-4y-3 \ge 5$
$-4y \ge 8$
$y \le -2$ Reversing the inequality symbol

$\{y|y \le -2\}$, or $(-\infty, -2]$

3. $-2(3x-1)-5 \ge 6x-4(3-x)$
$-6x+2-5 \ge 6x-12+4x$
$-6x-3 \ge 10x-12$
$-16x-3 \ge -12$
$-16x \ge -9$
$x \le \frac{9}{16}$ Reversing the inequality symbol

$\left\{x \middle| x \le \frac{9}{16}\right\}$, or $\left(-\infty, \frac{9}{16}\right]$

5. Solve graphically: $2x-3 \ge x+1$.

We graph and solve the system of equations
$y = 2x-3,$
$y = x+1$

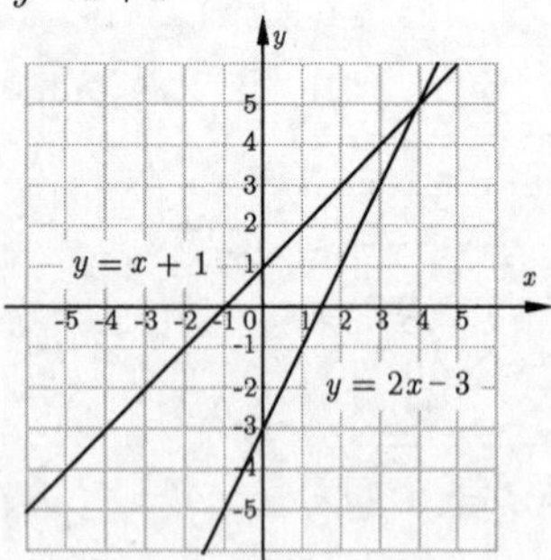

By substitution, we have
$2x-3 = x+1$
$x = 4$

Then $y = 4 + 1 = 5$.

Thus, the point of intersection is (4, 5). Since the inequality symbol is $\ge$, we include this point in the solution.

We see from the graph, $2x-3 \ge x+1$ is to the right of the point of intersection.

Thus, the solution set is $\{x|x \ge 4\}$, or $[4, \infty)$.

7. Let x = the number of miles driven. The cost for unlimited is \$80 and the cost for the other plan is $\$45+\$0.40(x-100)$. We solve the inequality.

$80 < 45+0.40(x-100)$
$80 < 45+0.4x-40$
$80 < 5+0.4x$
$75 < 0.4x$
$187.5 < x$

The unlimited mileage plan is less expensive for more than 187.5 miles.

9. $\{a, e, i, o, u\} \cap \{a, b, c, d, e\} = \{a, e\}$

11. $f(x) = \sqrt{6-3x}$
$6-3x \ge 0$
$6 \ge 3x$
$2 \ge x$

The domain of f is $(-\infty, 2]$.

13. $-5 < 4x+1 \le 3$
$-6 < 4x \le 2$
$-\frac{3}{2} < x \le \frac{1}{2}$

$\left\{x \middle| -\frac{3}{2} < x \le \frac{1}{2}\right\}$ or $\left(-\frac{3}{2}, \frac{1}{2}\right]$

$-\frac{3}{2}$ 0 $\frac{1}{2}$

15. $-3x > 12$ *or* $4x \ge -10$
$x < -4$ *or* $x \ge -\frac{5}{2}$

The solution set is $\left\{x \middle| x < -4 \text{ or } x \ge -\frac{5}{2}\right\}$ or $(-\infty, -4) \cup \left[-\frac{5}{2}, \infty\right)$.

−4 $-\frac{5}{2}$ 0

17. $|n| = 15$

$n = -15$ *or* $n = 15$

$\{-15, 15\}$

−15 0 15

19. $|3x - 1| < 7$
$-7 < 3x-1 < 7$
$-6 < 3x < 8$
$-2 < x < \frac{8}{3}$

$\left\{x \middle| -2 < x < \frac{8}{3}\right\}$, or $\left(-2, \frac{8}{3}\right)$

$\frac{8}{3}$

−2 0 2

21. $|2-5x|=-12$

Absolute value is never negative.

The solution is $\varnothing$.

23. $f(x)=g(x)$

$|2x-1|=|2x+7|$

$2x-1=2x+7$ *or* $2x-1=-(2x+7)$

$-1=7$ $\quad 2x-1=-2x-7$

FALSE $\quad 4x=-6$

$x=-\frac{3}{2}$

$\left\{-\frac{3}{2}\right\}$

25. $x+y\geq 3,$
$x-y\geq 5$

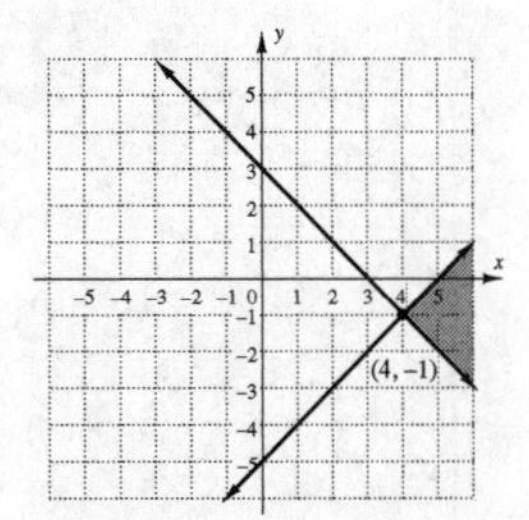

Vertex: (4, –1)

27.

Vertices	$F=5x+3y$
(1, 0)	$5\cdot 1+3\cdot 0=5$
(6, 0)	$5\cdot 6+3\cdot 0=30$
(1, 12)	$5\cdot 1+3\cdot 12=41$
(3, 12)	$5\cdot 3+3\cdot 12=51$
(6, 9)	$5\cdot 6+3\cdot 9=57$

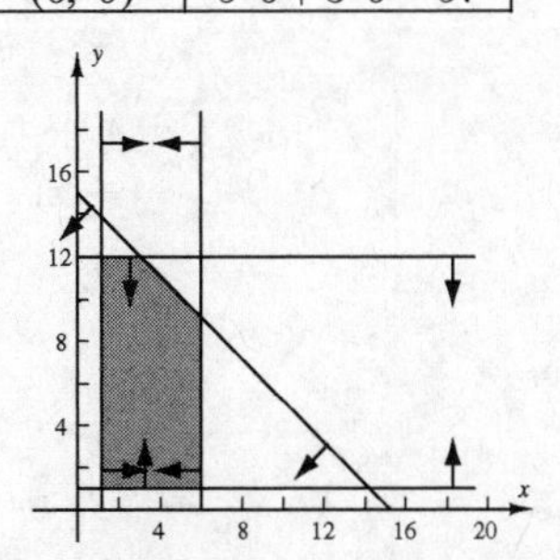

The maximum is 57 when $x=6$, $y=9$.

The minimum is 5 when $x=1$, $y=0$.

29. $|2x-5|\leq 7$ and $|x-2|\geq 2$

$-7\leq 2x-5\leq 7$ *and* $x-2\leq -2$ *or* $2\leq x-2$

$-2\leq 2x\leq 12$ *and* $x\leq 0$ *or* $4\leq x$

$-1\leq x\leq 6$

$\{x|-1\leq x\leq 0 \text{ or } 4\leq x\leq 6\}$ or $[-1, 0]\cup[4, 6]$

31. $\frac{-8+2}{2}=-3$

$\frac{2-(-8)}{2}=5$

$|x-(-3)|\leq 5$ or $|x+3|\leq 5$

Chapter 10

Exponents and Radicals

Exercise Set 10.1

1. two; see page 634 in the text.

3. positive; see Example 4.

5. irrational; see page 635 in the text.

7. nonnegative; see page 638 in the text.

9. The square roots of 64 are 8 and –8, because $8^2 = 64$ and $(-8)^2 = 64$.

11. The square roots of 100 are 10 and –10 because $10^2 = 100$ and $(-10)^2 = 100$.

13. The square roots of 400 are 20 and –20 because $20^2 = 400$ and $(-20)^2 = 400$.

15. The square roots of 625 are 25 and –25 because $25^2 = 625$ and $(-25)^2 = 625$.

17. $\sqrt{49} = 7$ Remember, $\sqrt{\ }$ indicates the principle square root.

19. $-\sqrt{16} = -4$ Since, $\sqrt{16} = 4$, $-\sqrt{16} = -4$

21. $\sqrt{\frac{36}{49}} = \frac{6}{7}$

23. $-\sqrt{169} = -13$ Since, $\sqrt{169} = 13$, $-\sqrt{169} = -13$

25. $-\sqrt{\frac{16}{81}} = -\frac{4}{9}$ Since, $\sqrt{\frac{16}{81}} = \frac{4}{9}$, $-\sqrt{\frac{16}{81}} = -\frac{4}{9}$

27. $\sqrt{0.04} = 0.2$

29. $\sqrt{0.0081} = 0.09$

31. $5\sqrt{p^2+4}$

The radicand is the expression written under the radical sign, p^2+4.

Since the index is not written, we know it is 2.

33. $x^2y^3\sqrt[5]{\frac{x}{y+4}}$

The radicand is the expression written under the radical sign, $\frac{x}{y+4}$.

The index is 5.

35. $f(t) = \sqrt{5t-10}$

$f(3) = \sqrt{5(3)-10} = \sqrt{5}$

$f(2) = \sqrt{5(2)-10} = \sqrt{0} = 0$

$f(1) = \sqrt{5(1)-10} = \sqrt{-5}$

Since negative numbers do not have real-number square roots, $f(1)$ does not exist.

$f(-1) = \sqrt{5(-1)-10} = \sqrt{-15}$

Since negative numbers do not have real-number square roots, $f(1)$ does not exist.

37. $t(x) = -\sqrt{2x^2-1}$

$t(5) = -\sqrt{2\cdot 5^2-1} = -\sqrt{49} = -7$

$t(0) = -\sqrt{2\cdot 0^2-1} = \sqrt{-1}$

$t(0)$ does not exist

$t(-1) = -\sqrt{2(-1)^2-1} = -\sqrt{1} = -1$

$t\left(-\frac{1}{2}\right) = -\sqrt{2\left(-\frac{1}{2}\right)^2-1} = -\sqrt{-\frac{1}{2}}$ does not exist

39. $f(t) = \sqrt{t^2+1}$

$f(0) = \sqrt{0^2+1} = \sqrt{1} = 1$

$f(-1) = \sqrt{(-1)^2+1} = \sqrt{2}$

$f(-10) = \sqrt{(-10)^2+1} = \sqrt{101}$

41. $\sqrt{100x^2} = \sqrt{(10x)^2} = |10x| = 10|x|$

Since x might be negative, absolute-value notation is necessary.

43. $\sqrt{(8-t)^2} = |8-t|$

Since $8 - t$ might be negative, absolute-value notation is necessary.

45. $\sqrt{y^2+16y+64} = \sqrt{(y+8)^2} = |y+8|$

Since $y+8$ might be negative, absolute-value notation is necessary.

47. $\sqrt{4x^2+28x+49}=\sqrt{(2x+7)^2}=|2x+7|$

Since $2x+7$ might be negative, absolute-value notation is necessary.

49. $-\sqrt[4]{256}=-4$ Since $4^4=256$

51. $\sqrt[3]{-1}=-1$ Since $(-1)^3=-1$

53. $-\sqrt[5]{-\frac{32}{243}}=\frac{2}{3}$ Since $\left(-\frac{2}{3}\right)^5=-\frac{32}{243}$

55. $\sqrt[6]{x^6}=|x|$

The index is even. Use absolute-value notation since x could have a negative value.

57. $\sqrt[9]{t^9}=t$

The index is odd. Absolute-value signs are not necessary.

59. $\sqrt[4]{(6a)^4}=|6a|=6|a|$

The index is even. Use absolute-value notation since a could have a negative value.

61. $\sqrt[10]{(-6)^{10}}=|-6|=6$

63. $\sqrt[414]{(a+b)^{414}}=|a+b|$

The index is even. Use absolute-value notation since $a+b$ could have a negative value.

65. $\sqrt{a^{22}}=|a^{11}|$ Note that $(a^{11})^2=a^{22}$; could have a negative value.

67. $\sqrt{-25}$ is not a real number, so $\sqrt{-25}$ cannot be simplified.

69. $\sqrt{16x^2}=\sqrt{(4x)^2}=4x$ Assuming x is nonnegative

71. $-\sqrt{(3t)^2}=-3t$ Assuming t is nonnegative

73. $\sqrt{(-5b)^2}=5b$

75. $\sqrt{a^2+2a+1}=\sqrt{(a+1)^2}=a+1$

77. $\sqrt[3]{27}=3$ $(3^3=27)$

79. $\sqrt[4]{16x^4}=\sqrt[4]{(2x)^4}=2x$

81. $\sqrt[5]{(x-1)^5}=x-1$

83. $-\sqrt[3]{-125y^3}=-(-5y)$ $\left[(-5y^3)=-125y^3\right]$
$=5y$

85. $\sqrt{t^{18}}=\sqrt{(t^9)^2}=t^9$

87. $\sqrt{(x-2)^8}=\sqrt{[(x-2)^4]^2}=(x-2)^4$

89.

$$\begin{aligned}f(x)&=\sqrt[3]{x+1}\\ f(7)&=\sqrt[3]{7+1}=\sqrt[3]{8}=2\\ f(26)&=\sqrt[3]{26+1}=\sqrt[3]{27}=3\\ f(-9)&=\sqrt[3]{-9+1}=\sqrt[3]{-8}=-2\\ f(-65)&=\sqrt[3]{-65+1}=\sqrt[3]{-64}=-4\end{aligned}$$

91.

$$\begin{aligned}g(t)&=\sqrt[4]{t-3}\\ g(19)&=\sqrt[4]{19-3}=\sqrt[4]{16}=2\\ g(-13)&=\sqrt[4]{-13-3}=\sqrt[4]{-16}\\ &g(-13)\text{ does not exist}\\ g(1)&=\sqrt[4]{1-3}=\sqrt[4]{-2}\\ &g(1)\text{ does not exist}\\ g(84)&=\sqrt[4]{84-3}=\sqrt[4]{81}=3\end{aligned}$$

93. $f(x)=\sqrt{x-6}$

Since the index is even, the radicand, $x-6$, must be nonnegative. We solve the inequality:

$$\begin{aligned}x-6&\geq 0\\ x&\geq 6\end{aligned}$$

Domain of $f=\{x|x\geq 6\}$, or $[6,\infty)$

95. $g(t)=\sqrt[4]{t+8}$

Since the index is even, the radicand, $t+8$, must be nonnegative. We solve the inequality:

$$\begin{aligned}t+8&\geq 0\\ t&\geq -8\end{aligned}$$

Domain of $g=\{t|t\geq -8\}$, or $[-8,\infty)$

97. $g(x)=\sqrt[4]{10-2x}$

Since the index is even, the radicand, $10-2x$, must be nonnegative. We solve the inequality:

$$\begin{aligned}10-2x&\geq 0\\ -2x&\geq -10\\ x&\leq 5\end{aligned}$$

Domain of $g=\{x|x\leq 5\}$, or $(-\infty,5]$

99. $f(t)=\sqrt[5]{2t+7}$

Since the index is odd, the radicand can be any real number.

Domain of $f=\{t|t \text{ is a real number}\}$, or $(-\infty,\infty)$

101. $h(z) = -\sqrt[6]{5z+2}$

Since the index is even, the radicand, $5z + 2$, must be nonnegative. We solve the inequality:

$$5z+2 \geq 0$$
$$5z \geq -2$$
$$z \geq -\frac{2}{5}$$

Domain of $h = \left\{z \middle| z \geq -\frac{2}{5}\right\}$, or $\left[-\frac{2}{5},\ \infty\right)$

103. $f(t) = 7 + \sqrt[8]{t^8}$

Since we can compute $7 + \sqrt[8]{t^8}$ for any real number t, the domain is the set of real numbers, or

$\{t|t \text{ is a real number}\}$ or $(-\infty,\ \infty)$.

105. *Writing Exercise.* Write a minus sign in front of the square root notation. For example, for a number n, we can write the negative square root of n as $-\sqrt{n}$.

107. $(a^2b)(a^4b) = a^{2+4}b^{1+1} = a^6b^2$

109. $\left(5x^2y^{-3}\right)^3 = 5^3x^{2\cdot3}y^{-3\cdot3} = 125x^6y^{-9} = \dfrac{125x^6}{y^9}$

111. $\left(\dfrac{10x^{-1}y^5}{5x^2y^{-1}}\right)^{-1} = \left(2x^{-3}y^6\right)^{-1} = 2^{-1}x^3y^{-6} = \dfrac{x^3}{2y^6}$

113. *Writing Exercise.* If n is an odd number, $\sqrt[n]{x^3}$ exists for all real values of x, because every real number has a real root when n is odd. If n is an even number, then $\sqrt[n]{x^3}$ exists for all nonnegative values of x. This is true because when x is nonnegative, then x^3 is nonnegative and even roots of nonnegative numbers exist. If x is negative, then x^3 is also negative and negative numbers do not have real nth roots when n is even.

115. $S = 88.63\ \sqrt[4]{A}$

Substitute 63,000 for A.

$S = 88.63\ \sqrt[4]{63,000}$

$S \approx 1404$

There are about 1404 species of plants.

117. $f(x) = \sqrt{x+5}$

Since the index is even, the radicand, $x + 5$, must be nonnegative. We solve the inequality:

$$x+5 \geq 0$$
$$x \geq -5$$

Domain of $f = \{x|x \geq -5\}$, or $[-5,\ \infty)$

Make a table of values, keeping in mind that x must be –5 or greater. Plot these points and draw the graph.

x	$f(x)$
-5	0
-4	1
-1	2
1	2.4
3	2.8
4	3

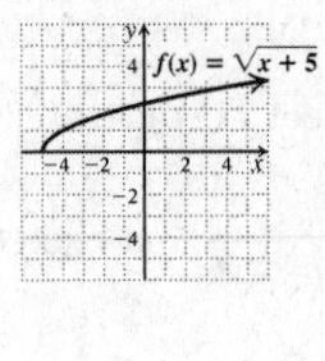

119. $g(x) = \sqrt{x} - 2$

Since the index is even, the radicand, x, must be nonnegative, so we have $x \geq 0$.

Domain of $g = \{x|x \geq 0\}$, or $[0,\ \infty)$

Make a table of values, keeping in mind that x must be 0 or greater. Plot these points and draw the graph.

x	$g(x)$
0	-2
1	-1
4	0
6	0.4
8	0.8

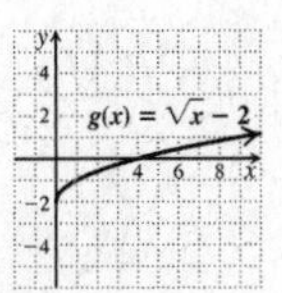

121. $f(x) = \dfrac{\sqrt{x+3}}{\sqrt[4]{2-x}}$

In the numerator we must have $x + 3 \geq 0$, or $x \geq -3$, and in the denominator we must have $2 - x > 0$, or $x < 2$, so

Domain of $f = \{x|-3 \leq x < 2\}$, or $[-3,\ 2)$.

123. $F(x) = \dfrac{x}{\sqrt{x^2-5x-6}}$

Since the radical expression in the denominator has an even index, so the radicand, $x^2 - 5x - 6$, must be nonnegative in order for $\sqrt{x^2-5x-6}$ to exist. In addition, the denominator cannot be zero, so the radicand must be positive. We solve the inequality:

$$x^2 - 5x - 6 > 0$$
$$(x+1)(x-6) > 0$$

We have $x < -1$ *and* $x > 6$, so

Domain of $F = \{x|x < -1 \text{ or } x > 6\}$, or $(-\infty, -1) \cup (6,\ \infty)$.

125. *Graphing Calculator Exercise.*

Exercise Set 10.2

1. Choice (g) is correct because $a^{m/n} = \sqrt[n]{a^m}$.

3. $x^{-5/2} = \frac{1}{x^{5/2}} = \frac{1}{(\sqrt{x})^5}$, so choice (e) is correct.

5. $x^{1/5} \cdot x^{2/5} = x^{1/5+2/5} = x^{3/5}$, so choice (a) is correct.

7. Choice (b) is correct because $\sqrt[n]{a^m}$ and $(\sqrt[n]{a})^m$ are equivalent.

9. $y^{1/3} = \sqrt[3]{y}$

11. $36^{1/2} = \sqrt{36} = 6$

13. $32^{1/5} = \sqrt[5]{32} = 2$

15. $64^{1/2} = \sqrt{64} = 8$

17. $(xyz)^{1/2} = \sqrt{xyz}$

19. $(a^2b^2)^{1/5} = \sqrt[5]{a^2b^2}$

21. $t^{5/6} = \sqrt[6]{t^5}$

23. $16^{3/4} = \sqrt[4]{16^3} = (\sqrt[4]{16})^3 = 2^3 = 8$

25. $125^{4/3} = \sqrt[3]{125^4} = (\sqrt[3]{125})^4 = 5^4 = 625$

27. $(81x)^{3/4} = \sqrt[4]{(81x)^3} = \sqrt[4]{81^3x^3}$, or

$\sqrt[4]{81^3} \cdot \sqrt[4]{x^3} = (\sqrt[4]{81})^3 \cdot (\sqrt[4]{x^3}) = 3^3\sqrt[4]{x^3} = 27\sqrt[4]{x^3}$

29. $(25x^4)^{3/2} = \sqrt{(25x^4)^3} = \sqrt{25^3 \cdot x^{12}} = \sqrt{25^3} \cdot \sqrt{x^{12}}$

$= (\sqrt{25})^3 x^6 = 5^3x^6 = 125x^6$

31. $\sqrt[3]{18} = 18^{1/3}$

33. $\sqrt{30} = 30^{1/2}$

35. $\sqrt{x^7} = x^{7/2}$

37. $\sqrt[5]{m^2} = m^{2/5}$

39. $\sqrt[4]{pq} = (pq)^{1/4}$

41. $\sqrt[5]{xy^2z} = (xy^2z)^{1/5}$

43. $(\sqrt{3mn})^3 = (3mn)^{3/2}$

45. $(\sqrt[7]{8x^2y})^5 = (8x^2y)^{5/7}$

47. $\frac{2x}{\sqrt[3]{z^2}} = \frac{2x}{z^{2/3}}$

49. $a^{-1/4} = \frac{1}{a^{1/4}}$

51. $(2rs)^{-3/4} = \frac{1}{(2rs)^{3/4}}$

53. $\left(\frac{1}{16}\right)^{-3/4} = \left(\frac{16}{1}\right)^{3/4} = (2^4)^{3/4} = 2^{4(3/4)} = 2^3 = 8$

55. $\frac{8c}{a^{-3/5}} = 8a^{3/5}c$

57. $2a^{3/4}b^{-1/2}c^{2/3} = 2 \cdot a^{3/4} \cdot \frac{1}{b^{1/2}} \cdot c^{2/3} = \frac{2a^{3/4}c^{2/3}}{b^{1/2}}$

59. $3^{-5/2}a^3b^{-7/3} = \frac{1}{3^{5/2}} \cdot a^3 \cdot \frac{1}{b^{7/3}} = \frac{a^3}{3^{5/2}b^{7/3}}$

61. $\left(\frac{2ab}{3c}\right)^{-5/6} = \left(\frac{3c}{2ab}\right)^{5/6}$ Finding the reciprocal of the base and changing the sign of the exponent

63. $\frac{6a}{\sqrt[4]{b}} = \frac{6a}{b^{1/4}}$

65. $11^{1/2} \cdot 11^{1/3} = 11^{1/2+1/3} = 11^{3/6+2/6} = 11^{5/6}$

We added exponents after finding a common denominator.

67. $\frac{3^{5/8}}{3^{-1/8}} = 3^{5/8-(-1/8)} = 3^{5/8+1/8} = 3^{6/8} = 3^{3/4}$

We subtracted exponents and simplified.

69. $\frac{4.3^{-1/5}}{4.3^{-7/10}} = 4.3^{-1/5-(-7/10)} = 4.3^{-1/5+7/10}$

$= 4.3^{-2/10+7/10} = 4.3^{5/10} = 4.3^{1/2}$

We subtracted exponents after finding a common denominator. Then we simplified.

71. $(10^{3/5})^{2/5} = 10^{3/5 \cdot 2/5} = 10^{6/25}$

We multiplied exponents.

73. $a^{2/3} \cdot a^{5/4} = a^{2/3+5/4} = a^{8/12+15/12} = a^{23/12}$

We added exponents after finding a common denominator.

75. $(64^{3/4})^{4/3} = 64^{\frac{3}{4}\cdot\frac{4}{3}} = 64^1 = 64$

77. $(m^{2/3}n^{-1/4})^{1/2} = m^{2/3 \cdot 1/2}n^{-1/4 \cdot 1/2} = m^{1/3}n^{-1/8}$

$= m^{1/3} \cdot \frac{1}{n^{1/8}} = \frac{m^{1/3}}{n^{1/8}}$

79. $\sqrt[9]{x^3} = x^{3/9}$ Converting to exponential notation

$= x^{1/3}$ Simplifying the exponent

$= \sqrt[3]{x}$ Returning to radical notation

81. $\sqrt[3]{y^{15}} = y^{15/3}$ Converting to exponential notation
$= y^5$ Simplifying

83. $\sqrt[12]{a^6} = a^{6/12}$ Converting to exponential notation
$= a^{1/2}$ Simplifying the exponent
$= \sqrt{a}$ Returning to radical notation

85. $\left(\sqrt[7]{xy}\right)^{14} = (xy)^{14/7}$ Converting to exponential notation
$= (xy)^2$ Simplifying the exponent
$= x^2y^2$ Using the laws of exponents

87. $\sqrt[4]{(7a)^2} = (7a)^{2/4}$ Converting to exponential notation
$= (7a)^{1/2}$ Simplifying the exponent
$= \sqrt{7a}$ Returning to radical notation

89. $\sqrt[8]{(2x)^6} = (2x)^{6/8}$ Converting to exponential notation
$= (2x)^{3/4}$ Simplifying the exponent
$= \sqrt[4]{(2x)^3}$ Returning to radical notation
$= \sqrt[4]{8x^3}$ Using the laws of exponents

91. $\sqrt{\sqrt[5]{m}} = \sqrt{m^{1/5}}$ Converting to
$= \left(m^{1/5}\right)^{1/2}$ exponential notation
$= m^{1/10}$ Using the laws of exponents
$= \sqrt[10]{m}$ Returning to radical notation

93. $\sqrt[4]{(xy)^{12}} = (xy)^{12/4}$ Converting to exponential notation
$= (xy)^3$ Simplifying the exponent
$= x^3y^3$ Using the laws of exponents

95. $\left(\sqrt[5]{a^2b^4}\right)^{15} = \left(a^2b^4\right)^{15/5}$ Converting to exponential notation
$= \left(a^2b^4\right)^3$ Simplifying the exponent
$= a^6b^{12}$ Using the laws of exponents

97. $\sqrt[3]{\sqrt[4]{xy}} = \sqrt[3]{(xy)^{1/4}}$ Converting to
$= \left[(xy)^{1/4}\right]^{1/3}$ exponential notation
$= (xy)^{1/12}$ Using the laws of exponents
$= \sqrt[12]{xy}$ Returning to radical notation

99. *Writing Exercise.* $f(x) = (x+5)^{1/2}(x+7)^{-1/2}$
Consider $(x+5)^{1/2}$. Since the index is $\frac{1}{2}$, $x+5$ must be nonnegative. Then $x+5 \geq 0$, or $x \geq -5$.
Consider $(x+7)^{-1/2}$. Since the index is $-\frac{1}{2}$, $x+7$ must be positive. Then $x+7>0$, or $x>-7$.
Then the domain of $f = \{x \mid x \geq -5 \text{ and } x > -7\}$, or $f = \{x \mid x \geq -5\}$.

101. $(x+5)(x-5) = x^2 - 25$ Difference of squares

103. $4x^2 + 20x + 25 = (2x+5)(2x+5) = (2x+5)^2$

105. $5t^2 - 10t + 5 = 5\left(t^2 - 2t + 1\right) = 5(t-1)(t-1)$
$= 5(t-1)^2$

107. *Writing Exercise.* For any value of x, $x^6 \geq 0$, and $x^2 \geq 0$, so $\sqrt[3]{x^6} = x^2$. For $x \geq 0$, we have $x^6 \geq 0$ and $x^3 \geq 0$, so $\sqrt[2]{x^6} = x^3$; but for $x < 0$, we have $x^6 > 0$ and $x^3 < 0$. Thus $\sqrt[2]{x^6} \neq x^3$ because $\sqrt[2]{x^6}$ must be nonnegative. Then $\sqrt[2]{x^6} = x^3$ only when $x \geq 0$.

109. $\sqrt{x\sqrt[3]{x^2}} = \sqrt{x \cdot x^{2/3}} = \left(x^{5/3}\right)^{1/2} = x^{5/6} = \sqrt[6]{x^5}$

111. $\sqrt[14]{c^2 - 2cd + d^2} = \sqrt[14]{(c-d)^2} = \left[(c-d)^2\right]^{1/14}$
$= (c-d)^{2/14} = (c-d)^{1/7} = \sqrt[7]{c-d}$

113. $2^{7/12} \approx 1.498 \approx 1.5$ so the G that is 7 half steps above middle C has a frequency that is about 1.5 times that of middle C.

115. a) $L = \dfrac{(0.000169)60^{2.27}}{1} \approx 1.8$ m

b) $L = \dfrac{(0.000169)75^{2.27}}{0.9906} \approx 3.1$ m

c) $L = \dfrac{(0.000169)80^{2.27}}{2.4} \approx 1.5$ m

d) $L = \dfrac{(0.000169)100^{2.27}}{1.1} \approx 5.3$ m

117. $T = 0.936d^{1.97}h^{0.85}$
$= 0.936(3)^{1.97}(80)^{0.85}$
≈ 338 cubic feet

119. *Graphing Calculator Exercise*

Exercise Set 10.3

1. True; see page 650 in the text.

3. False; see page 651 in the text.

5. True; see page 651 in the text.

7. $\sqrt{3}\sqrt{10} = \sqrt{3 \cdot 10} = \sqrt{30}$

9. $\sqrt[3]{7}\,\sqrt[3]{5} = \sqrt[3]{7 \cdot 5} = \sqrt[3]{35}$

11. $\sqrt[4]{6}\,\sqrt[4]{9} = \sqrt[4]{6 \cdot 9} = \sqrt[4]{54}$

13. $\sqrt{2x}\sqrt{13y} = \sqrt{2x \cdot 13y} = \sqrt{26xy}$

15. $\sqrt[5]{8y^3}\,\sqrt[5]{10y} = \sqrt[5]{8y^3 \cdot 10y} = \sqrt[5]{80y^4}$

17. $\sqrt{y-b}\sqrt{y+b} = \sqrt{(y-b)(y+b)} = \sqrt{y^2-b^2}$

19. $\sqrt[3]{0.7y}\,\sqrt[3]{0.3y} = \sqrt[3]{0.7y \cdot 0.3y} = \sqrt[3]{0.21y^2}$

21. $\sqrt[5]{x-2}\,\sqrt[5]{(x-2)^2} = \sqrt[5]{(x-2)(x-2)^2} = \sqrt[5]{(x-2)^3}$

23. $\sqrt{\frac{2}{t}}\sqrt{\frac{3s}{11}} = \sqrt{\frac{2}{t} \cdot \frac{3s}{11}} = \sqrt{\frac{6s}{11t}}$

25. $\sqrt[7]{\frac{x-3}{4}}\,\sqrt[7]{\frac{5}{x+2}} = \sqrt[7]{\frac{x-3}{4} \cdot \frac{5}{x+2}} = \sqrt[7]{\frac{5x-15}{4x+8}}$

27. $\sqrt{12}$
$= \sqrt{4 \cdot 3}$ 4 is the largest perfect square factor of 12
$= \sqrt{4} \cdot \sqrt{3}$
$= 2\sqrt{3}$

29. $\sqrt{45}$
$= \sqrt{9 \cdot 5}$ 9 is the largest perfect square factor of 45
$= \sqrt{9} \cdot \sqrt{5}$
$= 3\sqrt{5}$

31. $\sqrt{8x^9}$
$= \sqrt{4x^8 \cdot 2x}$ $4x^8$ is a perfect square
$= \sqrt{4x^8} \cdot \sqrt{2x}$ Factoring into two radicals
$= 2x^4\sqrt{2x}$ Taking the square root of $4x^8$

33. $\sqrt{120} = \sqrt{4 \cdot 30} = \sqrt{4} \cdot \sqrt{30} = 2\sqrt{30}$

35. $\sqrt{36a^4b}$
$= \sqrt{36a^4 \cdot b}$ $36a^4$ is a perfect square
$= \sqrt{36a^4} \cdot \sqrt{b}$ Factoring into two radicals
$= 6a^2\sqrt{b}$ Taking the square root of $36a^4$

37. $\sqrt[3]{8x^3y^2}$
$= \sqrt[3]{8x^3 \cdot y^2}$ $8x^3$ is a perfect cube
$= \sqrt[3]{8x^3} \cdot \sqrt[3]{y^2}$ Factoring into two radicals
$= 2x\sqrt[3]{y^2}$ Taking the cube root of $8x^3$

39. $\sqrt[3]{-16x^6}$
$= \sqrt[3]{-8x^6 \cdot 2}$ $-8x^6$ is a perfect cube
$= \sqrt[3]{-8x^6} \cdot \sqrt[3]{2}$ Factoring into two radicals
$= -2x^2\,\sqrt[3]{2}$ Taking the cube root of $-8x^6$

41. $f(x) = \sqrt[3]{40x^6}$
$= \sqrt[3]{8x^6 \cdot 5}$
$= \sqrt[3]{8x^6} \cdot \sqrt[3]{5}$
$= 2x^2\,\sqrt[3]{5}$

43. $f(x) = \sqrt{49(x-3)^2}$ $49(x-3)^2$ is a perfect square.
$= |7(x-3)|$, or $7|x-3|$

45. $f(x) = \sqrt{5x^2-10x+5}$
$= \sqrt{5(x^2-2x+1)}$
$= \sqrt{5(x-1)^2}$
$= \sqrt{(x-1)^2} \cdot \sqrt{5}$
$= |x-1|\sqrt{5}$

47. $\sqrt{a^{10}b^{11}}$
$= \sqrt{a^{10} \cdot b^{10} \cdot b}$ Identifying the largest even powers of a and b
$= \sqrt{a^{10}}\sqrt{b^{10}}\sqrt{b}$ Factoring into several radicals
$= a^5b^5\sqrt{b}$

49. $\sqrt[3]{x^5y^6z^{10}}$
$= \sqrt[3]{x^3 \cdot x^2 \cdot y^6 \cdot z^9 \cdot z}$ Identifying the largest perfect-cube powers of x, y and z
$= \sqrt[3]{x^3} \cdot \sqrt[3]{y^6} \cdot \sqrt[3]{z^9} \cdot \sqrt[3]{x^2z}$ Factoring into several radicals
$= xy^2z^3\,\sqrt[3]{x^2z}$

51. $\sqrt[4]{16x^5y^{11}} = \sqrt[4]{2^4 \cdot x^4 \cdot x \cdot y^8 \cdot y^3}$
$= \sqrt[4]{2^4} \cdot \sqrt[4]{x^4} \cdot \sqrt[4]{y^8} \cdot \sqrt[4]{xy^3}$
$= 2xy^2\,\sqrt[4]{xy^3}$

53. $\sqrt[5]{x^{13}y^8z^{17}} = \sqrt[5]{x^{10} \cdot x^3 \cdot y^5 \cdot y^3 \cdot z^{15} \cdot z^2}$
$= \sqrt[5]{x^{10}} \cdot \sqrt[5]{y^5} \cdot \sqrt[5]{z^{15}} \cdot \sqrt[5]{x^3y^3z^2}$
$= x^2yz^3\,\sqrt[5]{x^3y^3z^2}$

55. $\sqrt[3]{-80a^{14}} = \sqrt[3]{-8 \cdot 10 \cdot a^{12} \cdot a^2}$
$= \sqrt[3]{-8} \cdot \sqrt[3]{a^{12}} \cdot \sqrt[3]{10a^2}$
$= -2a^4\,\sqrt[3]{10a^2}$

57. $\sqrt{5}\sqrt{10} = \sqrt{5 \cdot 10} = \sqrt{50} = \sqrt{25 \cdot 2} = 5\sqrt{2}$

59. $\sqrt{6}\sqrt{33} = \sqrt{6 \cdot 33} = \sqrt{198} = \sqrt{9 \cdot 22} = 3\sqrt{22}$

61. $\sqrt[3]{9}\,\sqrt[3]{3} = \sqrt[3]{9 \cdot 3} = \sqrt[3]{27} = 3$

63. $\sqrt{24y^5}\sqrt{24y^5} = \sqrt{\left(24y^5\right)^2} = 24y^5$

65. $\sqrt[3]{5a^2}\,\sqrt[3]{2a} = \sqrt[3]{5a^2 \cdot 2a} = \sqrt[3]{10a^3} = \sqrt[3]{a^3 \cdot 10} = a\sqrt[3]{10}$

67. $\sqrt{2x^5}\sqrt{10x^2} = \sqrt{20x^7} = \sqrt{4x^6 \cdot 5x} = 2x^3\sqrt{5x}$

69. $\sqrt[3]{s^2t^4}\sqrt[3]{s^4t^6} = \sqrt[3]{s^6t^{10}} = \sqrt[3]{s^6t^9 \cdot t} = s^2t^3\sqrt[3]{t}$

71. $\sqrt[3]{(x-y)^2}\sqrt[3]{(x-y)^{10}} = \sqrt[3]{(x-y)^{12}} = (x-y)^4$

73. $\sqrt[4]{20a^3b^7}\sqrt[4]{4a^2b^5} = \sqrt[4]{80a^5b^{12}} = \sqrt[4]{16a^4b^{12} \cdot 5a}$
$= 2ab^3\sqrt[4]{5a}$

75. $\sqrt[5]{x^3(y+z)^6}\sqrt[5]{x^3(y+z)^4} = \sqrt[5]{x^6(y+z)^{10}}$
$= \sqrt[5]{x^5(y+z)^{10} \cdot x} = x(y+z)^2\sqrt[5]{x}$

77. *Writing Exercise.* Suppose that we let $x = 5$. Then we have $\sqrt{x^2-16} = \sqrt{5^2-16} = \sqrt{25-16} = \sqrt{9} = 3$, and $\sqrt{x^2} - \sqrt{16} = \sqrt{5^2} - \sqrt{16} = 5 - 4 = 1$. Thus, $\sqrt{x^2-16} \neq \sqrt{x^2} - \sqrt{16}$.

79. $\frac{15a^2x}{8b} \cdot \frac{24b^2x}{5a} = \frac{(\cancel{5a} \cdot 3ax)(\cancel{8b} \cdot 3bx)}{\cancel{8b} \cdot \cancel{5a}} = 9abx^2$

81.
$$\frac{x-3}{2x-10} - \frac{3x-5}{x^2-25} = \frac{x-3}{2(x-5)} - \frac{3x-5}{(x+5)(x-5)}$$
$$= \frac{x-3}{2(x-5)} \cdot \frac{x+5}{x+5} - \frac{3x-5}{(x+5)(x-5)} \cdot \frac{2}{2}$$
$$= \frac{x^2+2x-15-(6x-10)}{2(x+5)(x-5)}$$
$$= \frac{x^2-4x-5}{2(x+5)(x-5)}$$
$$= \frac{(x-5)(x+1)}{2(x+5)(x-5)}$$
$$= \frac{x+1}{2(x+5)}$$

83. $\frac{a^{-1}+b^{-1}}{ab} = \frac{\frac{1}{a}+\frac{1}{b}}{ab} = \frac{\frac{1}{a}+\frac{1}{b}}{ab} \cdot \frac{ab}{ab} = \frac{b+a}{a^2b^2}$

85. *Writing Exercise.* $\sqrt[n]{ab} = (ab)^{1/n} = a^{1/n}b^{1/n} = \sqrt[n]{a}\sqrt[n]{b}$

87.
$$R(x) = \frac{1}{2}\sqrt[4]{\frac{x \cdot 3.0 \times 10^6}{\pi^2}}$$
$$R(5 \times 10^4) = \frac{1}{2}\sqrt[4]{\frac{5 \times 10^4 \cdot 3.0 \times 10^6}{\pi^2}}$$
$$= \frac{1}{2}\sqrt[4]{\frac{15 \times 10^{10}}{\pi^2}}$$
$$\approx 175.6 \text{ mi}$$

89. a) $T_w = 33 - \frac{(10.45+10\sqrt{8}-8)(33-7)}{22}$
≈ -3.3 °C

b) $T_w = 33 - \frac{(10.45+10\sqrt{12}-12)(33-0)}{22}$
≈ -16.6 °C

c) $T_w = 33 - \frac{(10.45+10\sqrt{14}-14)(33-(-5))}{22}$
≈ -25.5 °C

d) $T_w = 33 - \frac{(10.45+10\sqrt{15}-15)(33-(-23))}{22}$
≈ -54.0 °C

91. $\left(\sqrt[3]{25x^4}\right)^4 = \sqrt[3]{(25x^4)^4} = \sqrt[3]{25^4x^{16}}$
$= \sqrt[3]{25^3 \cdot 25 \cdot x^{15} \cdot x} = \sqrt[3]{25^3}\sqrt[3]{x^{15}}\sqrt[3]{25x}$
$= 25x^5\sqrt[3]{25x}$

93. $\left(\sqrt{a^3b^5}\right)^7 = \sqrt{(a^3b^5)^7} = \sqrt{a^{21}b^{35}}$
$= \sqrt{a^{20} \cdot a \cdot b^{34} \cdot b} = \sqrt{a^{20}}\sqrt{b^{34}}\sqrt{ab} = a^{10}b^{17}\sqrt{ab}$

95.

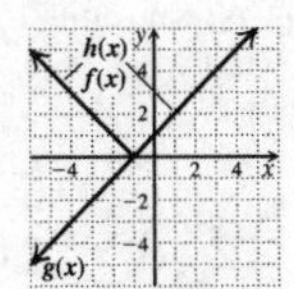

We see that $f(x) = h(x)$ and $f(x) \neq g(x)$.

97. $g(x) = x^2 - 6x + 8$

We must have $x^2 - 6x + 8 \geq 0$, or $(x-2)(x-4) \geq 0$. We graph $y = x^2 - 6x + 8$.

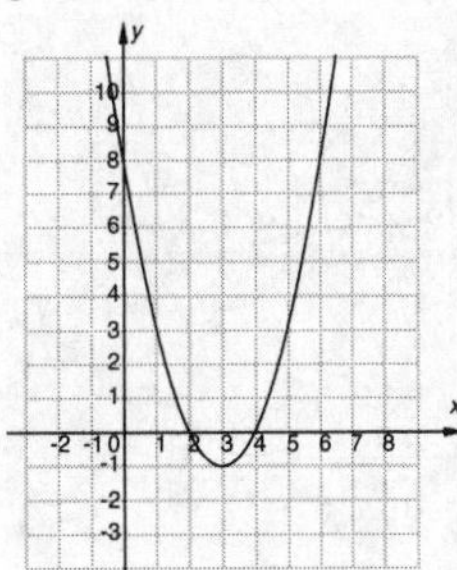

From the graph we see that $y \geq 0$ for $x \leq 2$ or $x \geq 4$, so the domain of g is $\{x \mid x \leq 2 \text{ or } x \geq 4\}$, or $(-\infty, 2] \cup [4, \infty)$.

99.
$$\sqrt[5]{4a^{3k+2}}\sqrt[5]{8a^{6-k}} = 2a^4$$
$$\sqrt[5]{32a^{2k+8}} = 2a^4$$
$$2\sqrt[5]{a^{2k+8}} = 2a^4$$
$$\sqrt[5]{a^{2k+8}} = a^4$$
$$a^{\frac{2k+8}{5}} = a^4$$

Since the base is the same, the exponents must be equal. We have:

$$\frac{2k+8}{5} = 4$$
$$2k+8 = 20$$
$$2k = 12$$
$$k = 6$$

101. *Writing Exercise.* Since $\sqrt{x}$ exists only for $\{x|x \geq 0\}$, this is the domain of $y = \sqrt{x} \cdot \sqrt{x}$.

Exercise Set 10.4

1. $\sqrt[4]{\frac{16a^6}{a^2}} = \sqrt[4]{16a^4} = 2a$, so choice (g) is correct.

3. $\sqrt[5]{\frac{a^6}{b^4}} = \sqrt[5]{\frac{a^6}{b^4} \cdot \frac{b}{b}} = \sqrt[5]{\frac{a^6 b}{b^4 \cdot b}}$, so choice (f) is correct.

5. $\frac{\sqrt[5]{a^2}}{\sqrt[5]{b^2}} = \frac{\sqrt[5]{a^2}}{\sqrt[5]{b^2}} \cdot \frac{\sqrt[5]{b^3}}{\sqrt[5]{b^3}} = \frac{\sqrt[5]{a^2 b^3}}{\sqrt[5]{b^5}}$, so choice (h) is correct.

7. $\frac{\sqrt[5]{a^2}}{\sqrt[5]{b^3}} = \frac{\sqrt[5]{a^2}}{\sqrt[5]{b^3}} \cdot \frac{\sqrt[5]{b^2}}{\sqrt[5]{b^2}} = \frac{\sqrt[5]{a^2 b^2}}{\sqrt[5]{b^5}}$, so choice (a) is correct.

9. $\sqrt{\frac{49}{100}} = \frac{\sqrt{49}}{\sqrt{100}} = \frac{7}{10}$

11. $\sqrt[3]{\frac{125}{8}} = \frac{\sqrt[3]{125}}{\sqrt[3]{8}} = \frac{5}{2}$

13. $\sqrt{\frac{121}{t^2}} = \frac{\sqrt{121}}{\sqrt{t^2}} = \frac{11}{t}$

15. $\sqrt{\frac{36y^3}{x^4}} = \frac{\sqrt{36y^3}}{\sqrt{x^4}} = \frac{\sqrt{36y^2 \cdot y}}{\sqrt{x^4}} = \frac{\sqrt{36y^2}\sqrt{y}}{\sqrt{x^4}} = \frac{6y\sqrt{y}}{x^2}$

17. $\sqrt[3]{\frac{27a^4}{8b^3}} = \frac{\sqrt[3]{27a^4}}{\sqrt[3]{8b^3}} = \frac{\sqrt[3]{27a^3 \cdot a}}{\sqrt[3]{8b^3}} = \frac{\sqrt[3]{27a^3}\sqrt[3]{a}}{\sqrt[3]{8b^3}} = \frac{3a\sqrt[3]{a}}{2b}$

19. $\sqrt[4]{\frac{32a^4}{2b^4c^8}} = \sqrt[4]{\frac{16a^4}{b^4c^8}} = \frac{\sqrt[4]{16a^4}}{\sqrt[4]{b^4c^8}} = \frac{2a}{bc^2}$

21. $\sqrt[4]{\frac{a^5b^8}{c^{10}}} = \frac{\sqrt[4]{a^5b^8}}{\sqrt[4]{c^{10}}} = \frac{\sqrt[4]{a^4b^8 \cdot a}}{\sqrt[4]{c^8 \cdot c^2}} = \frac{\sqrt[4]{a^4b^8}\sqrt[4]{a}}{\sqrt[4]{c^8}\sqrt[4]{c^2}} = \frac{ab^2\sqrt[4]{a}}{c^2\sqrt[4]{c^2}}$,

or $\frac{ab^2}{c^2}\sqrt[4]{\frac{a}{c^2}}$

23. $\sqrt[5]{\frac{32x^6}{y^{11}}} = \frac{\sqrt[5]{32x^6}}{\sqrt[5]{y^{11}}} = \frac{\sqrt[5]{32x^5 \cdot x}}{\sqrt[5]{y^{10} \cdot y}}$

$= \frac{\sqrt[5]{32x^5} \cdot \sqrt[5]{x}}{\sqrt[5]{y^{10}}\sqrt[5]{y}} = \frac{2x\sqrt[5]{x}}{y^2\sqrt[5]{y}}$, or $\frac{2x}{y^2}\sqrt[5]{\frac{x}{y}}$

25. $\sqrt[6]{\frac{x^6y^8}{z^{15}}} = \frac{\sqrt[6]{x^6y^8}}{\sqrt[6]{z^{15}}} = \frac{\sqrt[6]{x^6y^6 \cdot y^2}}{\sqrt[6]{z^{12} \cdot z^3}} = \frac{\sqrt[6]{x^6y^6}\sqrt[6]{y^2}}{\sqrt[6]{z^{12}}\sqrt[6]{z^3}} = \frac{xy\sqrt[6]{y^2}}{z^2\sqrt[6]{z^3}}$,

or $\frac{xy}{z^2}\sqrt[6]{\frac{y^2}{z^3}}$

27. $\frac{\sqrt{18y}}{\sqrt{2y}} = \sqrt{\frac{18y}{2y}} = \sqrt{9} = 3$

29. $\frac{\sqrt[3]{26}}{\sqrt[3]{13}} = \sqrt[3]{\frac{26}{13}} = \sqrt[3]{2}$

31. $\frac{\sqrt{40xy^3}}{\sqrt{8x}} = \sqrt{\frac{40xy^3}{8x}} = \sqrt{5y^3} = \sqrt{y^2 \cdot 5y}$

$= \sqrt{y^2}\sqrt{5y} = y\sqrt{5y}$

33. $\frac{\sqrt[3]{96a^4b^2}}{\sqrt[3]{12a^2b}} = \sqrt[3]{\frac{96a^4b^2}{12a^2b}} = \sqrt[3]{8a^2b} = \sqrt[3]{8}\sqrt[3]{a^2b} = 2\sqrt[3]{a^2b}$

35. $\frac{\sqrt{100ab}}{5\sqrt{2}} = \frac{1}{5}\frac{\sqrt{100ab}}{\sqrt{2}} = \frac{1}{5}\sqrt{\frac{100ab}{2}} = \frac{1}{5}\sqrt{50ab}$

$= \frac{1}{5}\sqrt{25 \cdot 2ab} = \frac{1}{5} \cdot 5\sqrt{2ab} = \sqrt{2ab}$

37. $\frac{\sqrt[4]{48x^9y^{13}}}{\sqrt[4]{3xy^{-2}}} = \sqrt[4]{\frac{48x^9y^{13}}{3xy^{-2}}} = \sqrt[4]{16x^8y^{15}} = \sqrt[4]{16x^8y^{12}}\sqrt[4]{y^3}$

$= 2x^2y^3\sqrt[4]{y^3}$

39. $\frac{\sqrt[3]{x^3-y^3}}{\sqrt[3]{x-y}} = \sqrt[3]{\frac{x^3-y^3}{x-y}} = \sqrt[3]{\frac{(x-y)(x^2+xy+y^2)}{x-y}}$

$= \sqrt[3]{\frac{(x-y)(x^2+xy+y^2)}{x-y}} = \sqrt[3]{x^2+xy+y^2}$

41. $\sqrt{\frac{2}{5}} = \sqrt{\frac{2}{5} \cdot \frac{5}{5}} = \sqrt{\frac{10}{25}} = \frac{\sqrt{10}}{\sqrt{25}} = \frac{\sqrt{10}}{5}$

43. $\frac{2\sqrt{5}}{7\sqrt{3}} = \frac{2\sqrt{5}}{7\sqrt{3}} \cdot \frac{\sqrt{3}}{\sqrt{3}} = \frac{2\sqrt{15}}{21}$

45. $\sqrt[3]{\frac{5}{4}} = \sqrt[3]{\frac{5}{4} \cdot \frac{2}{2}} = \sqrt[3]{\frac{10}{8}} = \frac{\sqrt[3]{10}}{\sqrt[3]{8}} = \frac{\sqrt[3]{10}}{2}$

47. $\frac{\sqrt[3]{3a}}{\sqrt[3]{5c}} = \frac{\sqrt[3]{3a}}{\sqrt[3]{5c}} \cdot \frac{\sqrt[3]{5^2c^2}}{\sqrt[3]{5^2c^2}} = \frac{\sqrt[3]{75ac^2}}{\sqrt[3]{5^3c^3}} = \frac{\sqrt[3]{75ac^2}}{5c}$

49. $\frac{\sqrt[4]{5y^6}}{\sqrt[4]{9x}} = \frac{\sqrt[4]{5y^6}}{\sqrt[4]{9x}} \cdot \frac{\sqrt[4]{9x^3}}{\sqrt[4]{9x^3}} = \frac{\sqrt[4]{5y^6 \cdot 9x^3}}{\sqrt[4]{81x^4}} = \frac{\sqrt[4]{45x^3y^2}}{3x}$

51. $\sqrt[3]{\frac{2}{x^2y}} = \sqrt[3]{\frac{2}{x^2y} \cdot \frac{xy^2}{xy^2}} = \sqrt[3]{\frac{2xy^2}{x^3y^3}} = \frac{\sqrt[3]{2xy^2}}{\sqrt[3]{x^3y^3}} = \frac{\sqrt[3]{2xy^2}}{xy}$

53. $\sqrt{\frac{7a}{18}} = \sqrt{\frac{7a}{18} \cdot \frac{2}{2}} = \sqrt{\frac{14a}{36}} = \frac{\sqrt{14a}}{\sqrt{36}} = \frac{\sqrt{14a}}{6}$

55. $\sqrt[5]{\frac{9}{32x^5y}} = \sqrt[5]{\frac{9}{32x^5y} \cdot \frac{y^4}{y^4}} = \frac{\sqrt[5]{9y^4}}{\sqrt[5]{32x^5y^5}} = \frac{\sqrt[5]{9y^4}}{2xy}$

57. $\sqrt{\frac{10ab^2}{72a^3b}} = \sqrt{\frac{5b}{36a^2}} = \frac{\sqrt{5b}}{6a}$

59. $\sqrt{\frac{5}{11}}=\sqrt{\frac{5}{11}\cdot\frac{5}{5}}=\sqrt{\frac{25}{55}}=\frac{\sqrt{25}}{\sqrt{55}}=\frac{5}{\sqrt{55}}$

61. $\frac{2\sqrt{6}}{5\sqrt{7}}=\frac{2\sqrt{6}}{5\sqrt{7}}\cdot\frac{\sqrt{6}}{\sqrt{6}}=\frac{2\sqrt{36}}{5\sqrt{42}}=\frac{2\cdot 6}{5\sqrt{42}}=\frac{12}{5\sqrt{42}}$

63. $\frac{\sqrt{8}}{2\sqrt{3x}}=\frac{\sqrt{8}}{2\sqrt{3x}}\cdot\frac{\sqrt{8}}{\sqrt{8}}=\frac{\sqrt{64}}{2\sqrt{24x}}=\frac{8}{2\sqrt{24x}}=\frac{8}{4\sqrt{6x}}=\frac{2}{\sqrt{6x}}$

65. $\frac{\sqrt[3]{7}}{\sqrt[3]{2}}=\frac{\sqrt[3]{7}}{\sqrt[3]{2}}\cdot\frac{\sqrt[3]{7^2}}{\sqrt[3]{7^2}}=\frac{\sqrt[3]{7^3}}{\sqrt[3]{98}}=\frac{7}{\sqrt[3]{98}}$

67. $\sqrt{\frac{7x}{3y}}=\sqrt{\frac{7x}{3y}\cdot\frac{7x}{7x}}=\sqrt{\frac{(7x)^2}{21xy}}=\frac{7x}{\sqrt{21xy}}$

69. $\sqrt[3]{\frac{2a^5}{5b}}=\sqrt[3]{\frac{2a^5}{5b}\cdot\frac{4a}{4a}}=\sqrt[3]{\frac{8a^6}{20ab}}=\frac{2a^2}{\sqrt[3]{20ab}}$

71. $\sqrt{\frac{x^3y}{2}}=\sqrt{\frac{x^3y}{2}\cdot\frac{xy}{xy}}=\sqrt{\frac{x^4y^2}{2xy}}=\frac{\sqrt{x^4y^2}}{\sqrt{2xy}}=\frac{x^2y}{\sqrt{2xy}}$

73. *Writing Exercise.* Assuming that a calculator is not available, the calculation is 1.414213562/2 is easier to perform than 1/1.414213562.

75. $3x-8xy+2xz=x(3-8y+2z)$

77. $(a+b)(a-b)=a^2-b^2$ Difference of squares

79. $(8+3x)(7-4x)=56-32x+21x-12x^2$ FOIL
$=56-11x-12x^2$

81. *Writing Exercise.* No; $\frac{\sqrt{8}}{\sqrt{2}}=\sqrt{\frac{8}{2}}=\sqrt{4}=2$.

83. a) $T=2\pi\sqrt{\frac{65}{980}}\approx 1.62$ sec

b) $T=2\pi\sqrt{\frac{98}{980}}\approx 1.99$ sec

c) $T=2\pi\sqrt{\frac{120}{980}}\approx 2.20$ sec

85.
$$\begin{aligned}\frac{\left(\sqrt[3]{81mn^2}\right)^2}{\left(\sqrt[3]{mn}\right)^2}&=\frac{\sqrt[3]{(81mn^2)^2}}{\sqrt[3]{(mn)^2}}\\&=\frac{\sqrt[3]{6561m^2n^4}}{\sqrt[3]{m^2n^2}}\\&=\sqrt[3]{\frac{6561m^2n^4}{m^2n^2}}\\&=\sqrt[3]{6561n^2}\\&=\sqrt[3]{729\cdot 9n^2}\\&=\sqrt[3]{729}\sqrt[3]{9n^2}\\&=9\sqrt[3]{9n^2}\end{aligned}$$

87.
$$\begin{aligned}\sqrt{a^2-3}-\frac{a^2}{\sqrt{a^2-3}}&=\sqrt{a^2-3}-\frac{a^2}{\sqrt{a^2-3}}\cdot\frac{\sqrt{a^2-3}}{\sqrt{a^2-3}}\\&=\sqrt{a^2-3}-\frac{a^2\sqrt{a^2-3}}{a^2-3}\\&=\sqrt{a^2-3}\cdot\frac{a^2-3}{a^2-3}-\frac{a^2\sqrt{a^2-3}}{a^2-3}\\&=\frac{a^2\sqrt{a^2-3}-3\sqrt{a^2-3}-a^2\sqrt{a^2-3}}{a^2-3}\\&=\frac{-3\sqrt{a^2-3}}{a^2-3},\text{ or }\frac{-3}{\sqrt{a^2-3}}\end{aligned}$$

89. Step 1: $\sqrt[n]{x}=x^{1/n}$, by definition;

Step 2: $\left(\frac{x}{y}\right)^n=\frac{x^n}{y^n}$, raising a quotient to a power;

Step 3: $x^{1/n}=\sqrt[n]{x}$, by definition

91. $f(x)=\sqrt{18x^3},\ g(x)=\sqrt{2x}$

$(f/g)(x)=\frac{f(x)}{g(x)}=\frac{\sqrt{18x^3}}{\sqrt{2x}}=\sqrt{\frac{18x^3}{2x}}=\sqrt{9x^2}=3x$

$\sqrt{2x}$ is defined for $2x\geq 0$, or $x\geq 0$. To avoid division by 0, we must exclude 0 from the domain. Thus, the domain of $f/g=\{x|x \text{ is a real number and } x>0\}$, or $(0,\infty)$.

93. $f(x)=\sqrt{x^2-9},\ g(x)=\sqrt{x-3}$

$$\begin{aligned}(f/g)(x)&=\frac{f(x)}{g(x)}=\frac{\sqrt{x^2-9}}{\sqrt{x-3}}=\sqrt{\frac{x^2-9}{x-3}}\\&=\sqrt{\frac{(x+3)(x-3)}{x-3}}=\sqrt{x+3}\end{aligned}$$

$\sqrt{x-3}$ is defined for $x-3\geq 0$, or $x\geq 3$. To avoid division by 0, we must exclude 0 from the domain. Thus, the domain of $f/g=\{x|x \text{ is a real number and } x>3\}$, or $(3,\infty)$.

Exercise Set 10.5

1. To add radical expressions, the indices and the radicands must be the same.

3. To find a product by adding exponents, the bases must be the same.

5. To rationalize the numerator of $\frac{\sqrt{c}-\sqrt{a}}{5}$, we multiply by a form of 1, using the conjugate of $\sqrt{c}-\sqrt{a}$, or $\sqrt{c}+\sqrt{a}$, to write 1.

7. $4\sqrt{3}+7\sqrt{3}=(4+7)\sqrt{3}=11\sqrt{3}$

9. $7\sqrt[3]{4}-5\sqrt[3]{4}=(7-5)\sqrt[3]{4}=2\sqrt[3]{4}$

11. $\sqrt[3]{y}+9\sqrt[3]{y}=(1+9)\sqrt[3]{y}=10\sqrt[3]{y}$

13. $8\sqrt{2}-\sqrt{2}+5\sqrt{2}=(8-1+5)\sqrt{2}=12\sqrt{2}$

15. $9\sqrt[3]{7}-\sqrt{3}+4\sqrt[3]{7}+2\sqrt{3}$
$=(9+4)\sqrt[3]{7}+(-1+2)\sqrt{3}=13\sqrt[3]{7}+\sqrt{3}$

17. $4\sqrt{27}-3\sqrt{3}$
$=4\sqrt{9\cdot 3}-3\sqrt{3}$ Factoring the first radical
$=4\sqrt{9}\cdot\sqrt{3}-3\sqrt{3}$
$=4\cdot 3\sqrt{3}-3\sqrt{3}$ Taking the square root of 9
$=12\sqrt{3}-3\sqrt{3}$
$=9\sqrt{3}$ Combining the radicals

19. $3\sqrt{45}-8\sqrt{20}$
$=3\sqrt{9\cdot 5}-8\sqrt{4\cdot 5}$ Factoring the radicals
$=3\sqrt{9}\cdot\sqrt{5}-8\sqrt{4}\cdot\sqrt{5}$
$=3\cdot 3\sqrt{5}-8\cdot 2\sqrt{5}$ Taking the square roots
$=9\sqrt{5}-16\sqrt{5}$
$=-7\sqrt{5}$ Combining like radicals

21. $3\sqrt[3]{16}+\sqrt[3]{54}=3\sqrt[3]{8\cdot 2}+\sqrt[3]{27\cdot 2}$
$=3\sqrt[3]{8}\cdot\sqrt[3]{2}+\sqrt[3]{27}\cdot\sqrt[3]{2}=3\cdot 2\sqrt[3]{2}+3\sqrt[3]{2}$
$=6\sqrt[3]{2}+3\sqrt[3]{2}=9\sqrt[3]{2}$

23. $\sqrt{a}+3\sqrt{16a^3}=\sqrt{a}+3\sqrt{16a^2\cdot a}=\sqrt{a}+3\sqrt{16a^2}\cdot\sqrt{a}$
$=\sqrt{a}+3\cdot 4a\sqrt{a}=\sqrt{a}+12a\sqrt{a}$
$=(1+12a)\sqrt{a}$

25. $\sqrt[3]{6x^4}-\sqrt[3]{48x}=\sqrt[3]{x^3\cdot 6x}-\sqrt[3]{8\cdot 6x}$
$=\sqrt[3]{x^3}\cdot\sqrt[3]{6x}-\sqrt[3]{8}\cdot\sqrt[3]{6x}=x\sqrt[3]{6x}-2\sqrt[3]{6x}$
$=(x-2)\sqrt[3]{6x}$

27. $\sqrt{4a-4}+\sqrt{a-1}=\sqrt{4(a-1)}+\sqrt{a-1}$
$=\sqrt{4}\sqrt{a-1}+\sqrt{a-1}=2\sqrt{a-1}+\sqrt{a-1}=3\sqrt{a-1}$

29. $\sqrt{x^3-x^2}+\sqrt{9x-9}=\sqrt{x^2(x-1)}+\sqrt{9(x-1)}$
$=\sqrt{x^2}\cdot\sqrt{x-1}+\sqrt{9}\cdot\sqrt{x-1}$
$=x\sqrt{x-1}+3\sqrt{x-1}=(x+3)\sqrt{x-1}$

31. $\sqrt{2}(5+\sqrt{2})=\sqrt{2}\cdot 5+\sqrt{2}\cdot\sqrt{2}=5\sqrt{2}+2$

33. $3\sqrt{5}(\sqrt{6}-\sqrt{7})=3\sqrt{5}\cdot\sqrt{6}-3\sqrt{5}\cdot\sqrt{7}=3\sqrt{30}-3\sqrt{35}$

35. $\sqrt{2}(3\sqrt{10}-\sqrt{8})=\sqrt{2}\cdot 3\sqrt{10}-\sqrt{2}\cdot\sqrt{8}=3\sqrt{20}-\sqrt{16}$
$=3\sqrt{4\cdot 5}-\sqrt{16}=3\cdot 2\sqrt{5}-4$
$=6\sqrt{5}-4$

37. $\sqrt[3]{3}(\sqrt[3]{9}-4\sqrt[3]{21})=\sqrt[3]{3}\cdot\sqrt[3]{9}-\sqrt[3]{3}\cdot 4\sqrt[3]{21}$
$=\sqrt[3]{27}-4\sqrt[3]{63}$
$=3-4\sqrt[3]{63}$

39. $\sqrt[3]{a}(\sqrt[3]{a^2}+\sqrt[3]{24a^2})=\sqrt[3]{a}\cdot\sqrt[3]{a^2}+\sqrt[3]{a}\sqrt[3]{24a^2}$
$=\sqrt[3]{a^3}+\sqrt[3]{24a^3}$
$=\sqrt[3]{a^3}+\sqrt[3]{8a^3\cdot 3}$
$=a+2a\sqrt[3]{3}$

41. $(2+\sqrt{6})(5-\sqrt{6})=2\cdot 5-2\sqrt{6}+5\sqrt{6}-\sqrt{6}\cdot\sqrt{6}$
$=10+3\sqrt{6}-6=4+3\sqrt{6}$

43. $(\sqrt{2}+\sqrt{7})(\sqrt{3}-\sqrt{7})=\sqrt{2}\cdot\sqrt{3}-\sqrt{2}\cdot\sqrt{7}+\sqrt{7}\cdot\sqrt{3}-\sqrt{7}\cdot\sqrt{7}$
$=\sqrt{6}-\sqrt{14}+\sqrt{21}-7$

45. $(2-\sqrt{3})(2+\sqrt{3})=2^2-(\sqrt{3})^2=4-3=1$

47. $(\sqrt{10}-\sqrt{15})(\sqrt{10}+\sqrt{15})=(\sqrt{10})^2-(\sqrt{15})^2$
$=10-15=-5$

49. $(3\sqrt{7}+2\sqrt{5})(2\sqrt{7}-4\sqrt{5})$
$=3\sqrt{7}\cdot 2\sqrt{7}-3\sqrt{7}\cdot 4\sqrt{5}+2\sqrt{5}\cdot 2\sqrt{7}-2\sqrt{5}\cdot 4\sqrt{5}$
$=6\cdot 7-12\sqrt{35}+4\sqrt{35}-8\cdot 5=42-8\sqrt{35}-40$
$=2-8\sqrt{35}$

51. $(4+\sqrt{7})^2=4^2+2\cdot 4\cdot\sqrt{7}+(\sqrt{7})^2=16+8\sqrt{7}+7$
$=23+8\sqrt{7}$

53. $(\sqrt{3}-\sqrt{2})^2=(\sqrt{3})^2-2\cdot\sqrt{3}\cdot\sqrt{2}+(\sqrt{2})^2$
$=3-2\sqrt{6}+2=5-2\sqrt{6}$

55. $(\sqrt{2t}+\sqrt{5})^2=(\sqrt{2t})^2+2\cdot\sqrt{2t}\cdot\sqrt{5}+(\sqrt{5})^2$
$=2t+2\sqrt{10t}+5$

57. $(3-\sqrt{x+5})^2=3^2-2\cdot 3\cdot\sqrt{x+5}+(\sqrt{x+5})^2$
$=9-6\sqrt{x+5}+x+5$
$=14-6\sqrt{x+5}+x$

59. $(2\sqrt[4]{7}-\sqrt[4]{6})(3\sqrt[4]{9}+2\sqrt[4]{5})$
$=2\sqrt[4]{7}\cdot 3\sqrt[4]{9}+2\sqrt[4]{7}\cdot 2\sqrt[4]{5}-\sqrt[4]{6}\cdot 3\sqrt[4]{9}-\sqrt[4]{6}\cdot 2\sqrt[4]{5}$
$=6\sqrt[4]{63}+4\sqrt[4]{35}-3\sqrt[4]{54}-2\sqrt[4]{30}$

61. $\frac{6}{3-\sqrt{2}}=\frac{6}{3-\sqrt{2}}\cdot\frac{3+\sqrt{2}}{3+\sqrt{2}}=\frac{6(3+\sqrt{2})}{(3-\sqrt{2})(3+\sqrt{2})}$
$=\frac{18+6\sqrt{2}}{3^2-(\sqrt{2})^2}=\frac{18+6\sqrt{2}}{9-2}=\frac{18+6\sqrt{2}}{7}$

63. $\dfrac{2+\sqrt{5}}{6+\sqrt{3}}=\dfrac{2+\sqrt{5}}{6+\sqrt{3}}\cdot\dfrac{6-\sqrt{3}}{6-\sqrt{3}}$
$=\dfrac{(2+\sqrt{5})(6-\sqrt{3})}{(6+\sqrt{3})(6-\sqrt{3})}=\dfrac{12-2\sqrt{3}+6\sqrt{5}-\sqrt{15}}{36-3}$
$=\dfrac{12-2\sqrt{3}+6\sqrt{5}-\sqrt{15}}{33}$

65. $\dfrac{\sqrt{a}}{\sqrt{a}+\sqrt{b}}=\dfrac{\sqrt{a}}{\sqrt{a}+\sqrt{b}}\cdot\dfrac{\sqrt{a}-\sqrt{b}}{\sqrt{a}-\sqrt{b}}$
$=\dfrac{\sqrt{a}(\sqrt{a}-\sqrt{b})}{(\sqrt{a}+\sqrt{b})(\sqrt{a}-\sqrt{b})}=\dfrac{a-\sqrt{ab}}{a-b}$

67. $\dfrac{\sqrt{7}-\sqrt{3}}{\sqrt{3}-\sqrt{7}}=\dfrac{-1(\sqrt{3}-\sqrt{7})}{\sqrt{3}-\sqrt{7}}=-1\cdot\dfrac{\sqrt{3}-\sqrt{7}}{\sqrt{3}-\sqrt{7}}=-1\cdot 1=-1$

69. $\dfrac{3\sqrt{2}-\sqrt{7}}{4\sqrt{2}+2\sqrt{5}}=\dfrac{3\sqrt{2}-\sqrt{7}}{4\sqrt{2}+2\sqrt{5}}\cdot\dfrac{4\sqrt{2}-2\sqrt{5}}{4\sqrt{2}-2\sqrt{5}}$
$=\dfrac{(3\sqrt{2}-\sqrt{7})(4\sqrt{2}-2\sqrt{5})}{(4\sqrt{2}+2\sqrt{5})(4\sqrt{2}-2\sqrt{5})}$
$=\dfrac{12\cdot 2-6\sqrt{10}-4\sqrt{14}+2\sqrt{35}}{16\cdot 2-4\cdot 5}$
$=\dfrac{24-6\sqrt{10}-4\sqrt{14}+2\sqrt{35}}{32-20}=\dfrac{24-6\sqrt{10}-4\sqrt{14}+2\sqrt{35}}{12}$
$=\dfrac{2(12-3\sqrt{10}-2\sqrt{14}+\sqrt{35})}{2\cdot 6}=\dfrac{12-3\sqrt{10}-2\sqrt{14}+\sqrt{35}}{6}$

71. $\dfrac{\sqrt{5}+1}{4}=\dfrac{\sqrt{5}+1}{4}\cdot\dfrac{\sqrt{5}-1}{\sqrt{5}-1}=\dfrac{(\sqrt{5}+1)(\sqrt{5}-1)}{4(\sqrt{5}-1)}$
$=\dfrac{(\sqrt{5})^2-1}{4(\sqrt{5}-1)}=\dfrac{5-1}{4(\sqrt{5}-1)}=\dfrac{4}{4(\sqrt{5}-1)}$
$=\dfrac{1}{\sqrt{5}-1}$

73. $\dfrac{\sqrt{6}-2}{\sqrt{3}+7}=\dfrac{\sqrt{6}-2}{\sqrt{3}+7}\cdot\dfrac{\sqrt{6}+2}{\sqrt{6}+2}=\dfrac{(\sqrt{6}-2)(\sqrt{6}+2)}{(\sqrt{3}+7)(\sqrt{6}+2)}$
$=\dfrac{6-4}{\sqrt{18}+2\sqrt{3}+7\sqrt{6}+14}=\dfrac{2}{3\sqrt{2}+2\sqrt{3}+7\sqrt{6}+14}$

75. $\dfrac{\sqrt{x}-\sqrt{y}}{\sqrt{x}+\sqrt{y}}=\dfrac{\sqrt{x}-\sqrt{y}}{\sqrt{x}+\sqrt{y}}\cdot\dfrac{\sqrt{x}+\sqrt{y}}{\sqrt{x}+\sqrt{y}}$
$=\dfrac{(\sqrt{x}-\sqrt{y})(\sqrt{x}+\sqrt{y})}{(\sqrt{x}+\sqrt{y})(\sqrt{x}+\sqrt{y})}=\dfrac{x-y}{x+2\sqrt{xy}+y}$

77. $\dfrac{\sqrt{a+h}-\sqrt{a}}{h}=\dfrac{\sqrt{a+h}-\sqrt{a}}{h}\cdot\dfrac{\sqrt{a+h}+\sqrt{a}}{\sqrt{a+h}+\sqrt{a}}$
$=\dfrac{(\sqrt{a+h}-\sqrt{a})(\sqrt{a+h}+\sqrt{a})}{h(\sqrt{a+h}+\sqrt{a})}=\dfrac{a+h-a}{h(\sqrt{a+h}+\sqrt{a})}$
$=\dfrac{h}{h(\sqrt{a+h}+\sqrt{a})}=\dfrac{1}{\sqrt{a+h}+\sqrt{a}}$

79. $\sqrt[3]{a}\,\sqrt[6]{a}$
$=a^{1/3}\cdot a^{1/6}$ Converting to exponential notation
$=a^{1/2}$ Adding exponents
$=\sqrt{a}$ Returning to radical notation

81. $\sqrt{b^3}\,\sqrt[5]{b^4}$
$=b^{3/2}\cdot b^{4/5}$ Converting to exponential notation
$=b^{23/10}$ Adding exponents
$=b^{2+3/10}$ Writing $\frac{23}{10}$ as a mixed number
$=b^2b^{3/10}$ Factoring
$=b^2\,\sqrt[10]{b^3}$ Returning to radical notation

83. $\sqrt{xy^3}\,\sqrt[3]{x^2y}=(xy^3)^{1/2}(x^2y)^{1/3}$
$=(xy^3)^{3/6}(x^2y)^{2/6}$
$=\left[(xy^3)^3(x^2y)^2\right]^{1/6}$
$=\sqrt[6]{x^3y^9\cdot x^4y^2}$
$=\sqrt[6]{x^7y^{11}}$
$=\sqrt[6]{x^6y^6\cdot xy^5}$
$=xy\sqrt[6]{xy^5}$

85. $\sqrt[4]{9ab^3}\,\sqrt{3a^4b}=(9ab^3)^{1/4}(3a^4b)^{1/2}$
$=(9ab^3)^{1/4}(3a^4b)^{2/4}$
$=\left[(9ab^3)(3a^4b)^2\right]^{1/4}$
$=\sqrt[4]{9ab^3\cdot 9a^8b^2}$
$=\sqrt[4]{81a^9b^5}$
$=\sqrt[4]{81a^8b^4\cdot ab}$
$=3a^2b\sqrt[4]{ab}$

87. $\sqrt{a^4b^3c^4}\,\sqrt[3]{ab^2c}=(a^4b^3c^4)^{1/2}(ab^2c)^{1/3}$
$=(a^4b^3c^4)^{3/6}(ab^2c)^{2/6}$
$=\left[(a^4b^3c^4)^3(ab^2c)^2\right]^{1/6}$
$=\sqrt[6]{a^{12}b^9c^{12}\cdot a^2b^4c^2}$
$=\sqrt[6]{a^{14}b^{13}c^{14}}$
$=\sqrt[6]{a^{12}b^{12}c^{12}\cdot a^2bc^2}$
$=a^2b^2c^2\sqrt[6]{a^2bc^2}$

89. $\dfrac{\sqrt[3]{a^2}}{\sqrt[4]{a}}$
$=\dfrac{a^{2/3}}{a^{1/4}}$ Converting to exponential notation
$=a^{2/3-1/4}$ Subtracting exponents
$=a^{5/12}$ Converting back
$=\sqrt[12]{a^5}$ to radical notation

91. $\dfrac{\sqrt[4]{x^2y^3}}{\sqrt[3]{xy}}$

$=\dfrac{(x^2y^3)^{1/4}}{(xy)^{1/3}}$ Converting to exponential notation

$=\dfrac{x^{2/4}y^{3/4}}{x^{1/3}y^{1/3}}$ Using the power and product rules

$=x^{2/4-1/3}y^{3/4-1/3}$ Subtracting exponents

$=x^{2/12}y^{5/12}$

$=(x^2y^5)^{1/12}$ Converting back to radical notation

$=\sqrt[12]{x^2y^5}$

93. $\dfrac{\sqrt{ab^3}}{\sqrt[5]{a^2b^3}}$

$=\dfrac{(ab^3)^{1/2}}{(a^2b^3)^{1/5}}$ Converting to exponential notation

$=\dfrac{a^{1/2}b^{3/2}}{a^{2/5}b^{3/5}}$

$=a^{1/10}b^{9/10}$ Subtracting exponents

$=(ab^9)^{1/10}$ Converting back to radical notation

$=\sqrt[10]{ab^9}$

95. $\dfrac{\sqrt{(7-y)^3}}{\sqrt[3]{(7-y)^2}}$

$=\dfrac{(7-y)^{3/2}}{(7-y)^{2/3}}$ Converting to exponential notation

$=(7-y)^{3/2-2/3}$ Subtracting exponents

$=(7-y)^{5/6}$

$=\sqrt[6]{(7-y)^5}$ Returning to radical notation

97. $\dfrac{\sqrt[4]{(5+3x)^3}}{\sqrt[3]{(5+3x)^2}}$

$=\dfrac{(5+3x)^{3/4}}{(5+3x)^{2/3}}$ Converting to exponential notation

$=(5+3x)^{3/4-2/3}$ Subtracting exponents

$=(5+3x)^{1/12}$ Converting back to radical notation

$=\sqrt[12]{5+3x}$

99. $\sqrt[3]{x^2y}\left(\sqrt{xy}-\sqrt[5]{xy^3}\right)$

$=(x^2y)^{1/3}\left[(xy)^{1/2}-(xy^3)^{1/5}\right]$

$=x^{2/3}y^{1/3}\left(x^{1/2}y^{1/2}-x^{1/5}y^{3/5}\right)$

$=x^{2/3}y^{1/3}x^{1/2}y^{1/2}-x^{2/3}y^{1/3}x^{1/5}y^{3/5}$

$=x^{2/3+1/2}y^{1/3+1/2}-x^{2/3+1/5}y^{1/3+3/5}$

$=x^{7/6}y^{5/6}-x^{13/15}y^{14/15}$ Writing as a mixed numeral

$=x\cdot x^{1/6}y^{5/6}-x^{13/15}y^{14/15}$

$=x(xy^5)^{1/6}-(x^{13}y^{14})^{1/15}$

$=x\sqrt[6]{xy^5}-\sqrt[15]{x^{13}y^{14}}$

101. $\left(m+\sqrt[3]{n^2}\right)\left(2m+\sqrt[4]{n}\right)$

$=\left(m+n^{2/3}\right)\left(2m+n^{1/4}\right)$ Converting to exponential notation

$=2m^2+mn^{1/4}+2mn^{2/3}+n^{2/3}n^{1/4}$ Using FOIL

$=2m^2+mn^{1/4}+2mn^{2/3}+n^{2/3+1/4}$ Adding exponents

$=2m^2+mn^{1/4}+2mn^{2/3}+n^{11/12}$

$=2m^2+m\sqrt[4]{n}+2m\sqrt[3]{n^2}+\sqrt[12]{n^{11}}$ Converting back to radical notation

103. $f(x)=\sqrt[4]{x},\ g(x)=2\sqrt{x}-\sqrt[3]{x^2}$

$(f\cdot g)(x)=\sqrt[4]{x}\left(2\sqrt{x}-\sqrt[3]{x^2}\right)$

$=x^{1/4}\cdot 2x^{1/2}-x^{1/4}\cdot x^{2/3}$

$=2x^{1/4+1/2}-x^{1/4+2/3}$

$=2x^{1/4+2/4}-x^{3/12+8/12}$

$=2x^{3/4}-x^{11/12}$

$=2\sqrt[4]{x^3}-\sqrt[12]{x^{11}}$

105. $f(x)=x+\sqrt{7},\ g(x)=x-\sqrt{7}$

$(f\cdot g)(x)=\left(x+\sqrt{7}\right)\left(x-\sqrt{7}\right)$

$=x^2-\left(\sqrt{7}\right)^2$

$=x^2-7$

107. $f(x)=x^2$

$f\left(3-\sqrt{2}\right)=\left(3-\sqrt{2}\right)^2=3^2-2\cdot3\cdot\sqrt{2}+\left(\sqrt{2}\right)^2$

$=9-6\sqrt{2}+2=11-6\sqrt{2}$

109. $f(x)=x^2$

$f\left(\sqrt{6}+\sqrt{21}\right)=\left(\sqrt{6}+\sqrt{21}\right)^2$

$=\left(\sqrt{6}\right)^2+2\cdot\sqrt{6}\cdot\sqrt{21}+\left(\sqrt{21}\right)^2$

$=6+2\sqrt{126}+21=27+2\sqrt{9\cdot14}$

$=27+6\sqrt{14}$

111. *Writing Exercise.* The distributive law is used to combine radical expressions with the same indices and radicands just as it is used to combine monomials with the same variables and exponents.

113. $3x-1=125$

$3x=126$

$x=42$

The solution is 42.

115. $x^2+2x+1=22-2x$

$x^2+4x-21=0$

$(x+7)(x-3)=0$

$x+7=0$ *or* $x-3=0$

$x=-7$ *or* $x=3$

The solutions are –7 and 3.

117. $\frac{1}{x}+\frac{1}{2}=\frac{1}{6}$

$\frac{1}{x}\cdot 6x+\frac{1}{2}\cdot 6x=\frac{1}{6}\cdot 6x$

$6+3x=x$

$2x=-6$

$x=-3$

The solution is –3.

119. *Writing Exercise.* It appears that Ramon multiplied the exponents rather than adding them.

121. $f(x)=\sqrt{x^3-x^2}+\sqrt{9x^3-9x^2}-\sqrt{4x^3-4x^2}$
$=\sqrt{x^2(x-1)}+\sqrt{9x^2(x-1)}-\sqrt{4x^2(x-1)}$
$=x\sqrt{x-1}+3x\sqrt{x-1}-2x\sqrt{x-1}$
$=2x\sqrt{x-1}$

123. $f(x)=\sqrt[4]{x^5-x^4}+3\sqrt[4]{x^9-x^8}$
$=\sqrt[4]{x^4(x-1)}+3\sqrt[4]{x^8(x-1)}$
$=\sqrt[4]{x^4}\cdot\sqrt[4]{x-1}+3\sqrt[4]{x^8}\sqrt[4]{x-1}$
$=x\sqrt[4]{x-1}+3x^2\sqrt[4]{x-1}$
$=(x+3x^2)\sqrt[4]{x-1}$

125. $7x\sqrt{(x+y)^3}-5xy\sqrt{x+y}-2y\sqrt{(x+y)^3}$
$=7x\sqrt{(x+y)^2(x+y)}-5xy\sqrt{x+y}-2y\sqrt{(x+y)^2(x+y)}$
$=7x(x+y)\sqrt{x+y}-5xy\sqrt{x+y}-2y(x+y)\sqrt{x+y}$
$=[7x(x+y)-5xy-2y(x+y)]\sqrt{x+y}$
$=(7x^2+7xy-5xy-2xy-2y^2)\sqrt{x+y}$
$=(7x^2-2y^2)\sqrt{x+y}$

127. $\sqrt{8x(y+z)^5}\sqrt[3]{4x^2(y+z)^2}$
$=[8x(y+z)^5]^{1/2}[4x^2(y+z)^2]^{1/3}$
$=[8x(y+z)^5]^{3/6}[4x^2(y+z)^2]^{2/6}$
$=\{[2^3x(y+z)^5]^3[2^2x^2(y+z)^2]^2\}^{1/6}$
$=\sqrt[6]{2^9x^3(y+z)^{15}\cdot 2^4x^4(y+z)^4}$
$=\sqrt[6]{2^{13}x^7(y+z)^{19}}$
$=\sqrt[6]{2^{12}x^6(y+z)^{18}\cdot 2x(y+z)}$
$=2^2x(y+z)^3\sqrt[6]{2x(y+z)}$, or
$4x(y+z)^3\sqrt[6]{2x(y+z)}$

129. $\frac{\frac{1}{\sqrt{w}}-\sqrt{w}}{\frac{\sqrt{w}+1}{\sqrt{w}}}=\frac{\frac{1}{\sqrt{w}}-\sqrt{w}}{\frac{\sqrt{w}+1}{\sqrt{w}}}\cdot\frac{\sqrt{w}}{\sqrt{w}}=\frac{1-w}{\sqrt{w}+1}$
$=\frac{1-w}{\sqrt{w}+1}\cdot\frac{\sqrt{w}-1}{\sqrt{w}-1}=\frac{\sqrt{w}-1-w\sqrt{w}+w}{w-1}$
$=\frac{(w-1)-\sqrt{w}(w-1)}{w-1}=\frac{(w-1)(1-\sqrt{w})}{w-1}$
$=1-\sqrt{w}$

131. $x-5=(\sqrt{x})^2-(\sqrt{5})^2=(\sqrt{x}+\sqrt{5})(\sqrt{x}-\sqrt{5})$

133. $x-a=(\sqrt{x})^2-(\sqrt{a})^2=(\sqrt{x}+\sqrt{a})(\sqrt{x}-\sqrt{a})$

135. $(\sqrt{x+2}-\sqrt{x-2})^2=x+2-2\sqrt{(x+2)(x-2)}+x-2$
$=x+2-2\sqrt{x^2-4}+x-2$
$=2x-2\sqrt{x^2-4}$

Connecting the Concepts

1. $\sqrt{(t+5)^2}=t+5$

3. $\sqrt{6x}\sqrt{15x}=\sqrt{6x\cdot 15x}=\sqrt{90x^2}=\sqrt{9x^2\cdot 10}=3x\sqrt{10}$

5. $\sqrt{15t}+4\sqrt{15t}=(1+4)\sqrt{15t}=5\sqrt{15t}$

7. $\sqrt{6}(\sqrt{10}-\sqrt{33})=\sqrt{6}\sqrt{10}-\sqrt{6}\sqrt{33}=\sqrt{6\cdot 10}-\sqrt{6\cdot 33}$
$=\sqrt{60}-\sqrt{198}=\sqrt{4\cdot 15}-\sqrt{9\cdot 22}$
$=2\sqrt{15}-3\sqrt{22}$

9. $\frac{\sqrt{t}}{\sqrt[8]{t^3}}=\frac{t^{1/2}}{t^{3/8}}=t^{1/2-3/8}=t^{1/8}=\sqrt[8]{t}$

11. $2\sqrt{3}-5\sqrt{12}=2\sqrt{3}-5\sqrt{4\cdot 3}=2\sqrt{3}-5\cdot 2\sqrt{3}$
$=2\sqrt{3}-10\sqrt{3}=-8\sqrt{3}$

13. $(\sqrt{15}+\sqrt{10})^2=(\sqrt{15})^2+2\cdot\sqrt{10}\sqrt{15}+(\sqrt{10})^2$
$=15+2\sqrt{25\cdot 6}+10=25+2\cdot 5\sqrt{6}$
$=25+10\sqrt{6}$

15. $\sqrt{x^3y}\sqrt[5]{xy^4}=x^{3/2}y^{1/2}x^{1/5}y^{4/5}=x^{3/2+1/5}y^{1/2+4/5}$
$=x^{17/10}y^{13/10}=x^{1+7/10}y^{1+3/10}$
$=xy\sqrt[10]{x^7y^3}$

17. $\sqrt{\sqrt[5]{x^2}}=\sqrt{x^{2/5}}=(x^{2/5})^{1/2}=x^{\frac{2}{5}\cdot\frac{1}{2}}=x^{1/5}=\sqrt[5]{x}$

19. $(\sqrt[4]{a^2b^3})^2=(a^{2/4}b^{3/4})^2=a^{\frac{2}{4}\cdot 2}b^{\frac{3}{4}\cdot 2}$
$=ab^{3/2}=ab^1b^{1/2}=ab\sqrt{b}$

Exercise Set 7.6

1. False; if $x^2=25$, then $x=5$, or $x=-5$.

3. True by the principle of powers

5. If we add 8 to both sides of $\sqrt{x}-8=7$, we get $\sqrt{x}=15$, so the statement is true.

7. $\sqrt{5x+1} = 4$

$(\sqrt{5x+1})^2 = 4^2$ Principle of powers (squaring)

$5x+1 = 16$

$5x = 15$

$x = 3$

Check: $\sqrt{5x+1} = 4$

$\sqrt{5 \cdot 3+1} \mid 4$

$\sqrt{16}$

$4 \stackrel{?}{=} 4$ TRUE

The solution is 3.

9. $\sqrt{3x}+1 = 5$

$\sqrt{3x} = 4$ Adding to isolate the radical

$(\sqrt{3x})^2 = 4^2$ Principle of powers (squaring)

$3x = 16$

$x = \frac{16}{3}$

Check: $\sqrt{3x}+1 = 5$

$\sqrt{3 \cdot \frac{16}{3}}+1 \mid 5$

$\sqrt{16}+1$

$4+1$

$5 \stackrel{?}{=} 5$ TRUE

The solution is $\frac{16}{3}$.

11. $\sqrt{y+5}-4 = 1$

$\sqrt{y+5} = 5$ Adding to isolate the radical

$(\sqrt{y+5})^2 = 5^2$ Principle of powers (squaring)

$y+5 = 25$

$y = 20$

Check: $\sqrt{y+5}-4 = 1$

$\sqrt{20+5}-4 \mid 1$

$\sqrt{25}-4$

$5-4$

$1 \stackrel{?}{=} 1$ TRUE

The solution is 20.

13. $\sqrt{8-x}+7 = 10$

$\sqrt{8-x} = 3$ Adding to isolate the radical

$(\sqrt{8-x})^2 = 3^2$ Principle of powers (squaring)

$8-x = 9$

$x = -1$

Check: $\sqrt{8-x}+7 = 10$

$\sqrt{8-(-1)}+7 \mid 10$

$\sqrt{9}+7$

$3+7$

$10 \stackrel{?}{=} 10$ TRUE

The solution is -1.

15. $\sqrt[3]{y+3} = 2$

$(\sqrt[3]{y+3})^3 = 2^3$ Principle of powers (cubing)

$y+3 = 8$

$y = 5$

Check: $\sqrt[3]{y+3} = 2$

$\sqrt[3]{5+3} \mid 2$

$\sqrt[3]{8}$

$2 \stackrel{?}{=} 2$ TRUE

The solution is 5.

17. $\sqrt[4]{t-10} = 3$

$(\sqrt[4]{t-10})^4 = 3^4$

$t-10 = 81$

$t = 91$

Check: $\sqrt[4]{t-10} = 3$

$\sqrt[4]{91-10} \mid 3$

$\sqrt[4]{81}$

$3 \stackrel{?}{=} 3$ TRUE

The solution is 91.

19. $6\sqrt{x} = x$

$(6\sqrt{x})^2 = x^2$

$36x = x^2$

$0 = x^2 - 36x$

$0 = x(x-36)$

$x = 0$ *or* $x = 36$

Check:

For $x = 0$: $6\sqrt{x} = x$

$6\sqrt{0} \mid 0$

$0 \stackrel{?}{=} 0$ TRUE

For $x = 36$: $6\sqrt{x} = x$

$6\sqrt{36} \mid 36$

$6 \cdot 6$

$36 \stackrel{?}{=} 36$ TRUE

The solutions are 0 and 36.

21. $2y^{1/2} - 13 = 7$

$2\sqrt{y} - 13 = 7$

$2\sqrt{y} = 20$

$\sqrt{y} = 10$

$(\sqrt{y})^2 = 10^2$

$y = 100$

Check: $2y^{1/2} - 13 = 7$

$2 \cdot 100^{1/2} - 13 \mid 7$

$2 \cdot 10 - 13$

$20 - 13$

$7 \stackrel{?}{=} 7$ TRUE

The solution is 100.

23. $\sqrt[3]{x} = -5$

$(\sqrt[3]{x})^3 = (-5)^3$

$x = -125$

Check: $\sqrt[3]{x} = -5$

$\sqrt[3]{-125} \mid -5$

$\sqrt[3]{(-5)^3}$

$-5 \stackrel{?}{=} -5$ TRUE

The solution is –125.

25. $z^{1/4} + 8 = 10$

$z^{1/4} = 2$

$(z^{1/4})^4 = 2^4$

$z = 16$

Check: $z^{1/4} + 8 = 10$

$16^{1/4} + 8 \mid 10$

$2 + 8$

$10 \stackrel{?}{=} 10$ TRUE

The solution is 16.

27. $\sqrt{n} = -2$

This equation has no solution, since the principal square root is never negative.

29. $\sqrt[4]{3x+1} - 4 = -1$

$\sqrt[4]{3x+1} = 3$

$(\sqrt[4]{3x+1})^4 = 3^4$

$3x + 1 = 81$

$3x = 80$

$x = \frac{80}{3}$

Check: $\sqrt[4]{3x+1} - 4 = -1$

$\sqrt[4]{3 \cdot \frac{80}{3} + 1} - 4 \mid -1$

$\sqrt[4]{81} - 4$

$3 - 4$

$-1 \stackrel{?}{=} -1$ TRUE

The solution is $\frac{80}{3}$.

31. $(21x + 55)^{1/3} = 10$

$\left[(21x + 55)^{1/3}\right]^3 = 10^3$

$21x + 55 = 1000$

$21x = 945$

$x = 45$

Check: $(21x + 55)^{1/3} = 10$

$(21 \cdot 45 + 55)^{1/3} \mid 10$

$(945 + 55)^{1/3}$

$10 \stackrel{?}{=} 10$ TRUE

The solution is 45.

33. $\sqrt[3]{3y+6} + 7 = 8$

$\sqrt[3]{3y+6} = 1$

$\left(\sqrt[3]{3y+6}\right)^3 = 1^3$

$3y + 6 = 1$

$3y = -5$

$y = -\frac{5}{3}$

Check: $\sqrt[3]{3y+6} + 7 = 8$

$\sqrt[3]{3\left(-\frac{5}{3}\right) + 6} + 7 \mid 8$

$\sqrt[3]{1} + 7$

$1 + 7$

$8 \stackrel{?}{=} 8$ TRUE

The solution is $-\frac{5}{3}$.

35. $\sqrt{3t+4} = \sqrt{4t+3}$

$\left(\sqrt{3t+4}\right)^2 = \left(\sqrt{4t+3}\right)^2$

$3t + 4 = 4t + 3$

$4 = t + 3$

$1 = t$

Check:

$\sqrt{3t+4} = \sqrt{4t+3}$

$\sqrt{3 \cdot 1 + 4} \mid \sqrt{4 \cdot 1 + 3}$

$\sqrt{7} \stackrel{?}{=} \sqrt{7}$ TRUE

The solution is 1.

37. $3(4-t)^{1/4} = 6^{1/4}$

$\left[3(4-t)^{1/4}\right]^4 = \left(6^{1/4}\right)^4$

$81(4-t) = 6$

$324 - 81t = 6$

$-81t = -318$

$t = \frac{106}{27}$

The number $\frac{106}{27}$ checks and is the solution.

39. $3 + \sqrt{5-x} = x$

$\sqrt{5-x} = x - 3$

$\left(\sqrt{5-x}\right)^2 = (x-3)^2$

$5 - x = x^2 - 6x + 9$

$0 = x^2 - 5x + 4$

$0 = (x-1)(x-4)$

$x - 1 = 0$ or $x - 4 = 0$

$x = 1$ or $x = 4$

Check:

For 1: $3 + \sqrt{5-x} = x$

$3 + \sqrt{5-1} \mid 1$

$3 + \sqrt{4}$

$3 + 2$

$5 \stackrel{?}{=} 1$ FALSE

For 4: $3+\sqrt{5-x}=x$

$$\begin{array}{c|c} 3+\sqrt{5-4} & 4 \\ 3+\sqrt{1} & \\ 3+1 & \\ \hline 4 \stackrel{?}{=} 4 & \text{TRUE} \end{array}$$

Since 4 checks but 1 does not, the solution is 4.

41. $\sqrt{4x-3}=2+\sqrt{2x-5}$ One radical is already isolated.

$\left(\sqrt{4x-3}\right)^2=\left(2+\sqrt{2x-5}\right)^2$ Squaring both sides

$4x-3=4+4\sqrt{2x-5}+2x-5$

$2x-2=4\sqrt{2x-5}$

$x-1=2\sqrt{2x-5}$

$x^2-2x+1=8x-20$

$x^2-10x+21=0$

$(x-7)(x-3)=0$

$x-7=0$ or $x-3=0$

$x=7$ or $x=3$

Both numbers check. The solutions are 7 and 3.

43. $\sqrt{20-x}+8=\sqrt{9-x}+11$

$\sqrt{20-x}=\sqrt{9-x}+3$ Isolating one radical

$\left(\sqrt{20-x}\right)^2=\left(\sqrt{9-x}+3\right)^2$ Squaring both sides

$20-x=9-x+6\sqrt{9-x}+9$

$2=6\sqrt{9-x}$ Isolating the remaining radical

$1=3\sqrt{9-x}$ Multiplying by $\frac{1}{2}$

$1^2=\left(3\sqrt{9-x}\right)^2$ Squaring both sides

$1=9(9-x)$

$1=81-9x$

$-80=-9x$

$\frac{80}{9}=x$

The number $\frac{80}{9}$ checks and is the solution.

45. $\sqrt{x+2}+\sqrt{3x+4}=2$

$\sqrt{x+2}=2-\sqrt{3x+4}$ Isolating one radical

$\left(\sqrt{x+2}\right)^2=\left(2-\sqrt{3x+4}\right)^2$

$x+2=4-4\sqrt{3x+4}+3x+4$

$-2x-6=-4\sqrt{3x+4}$ Isolating the remaining radical

$x+3=2\sqrt{3x+4}$ Multiplying by $-\frac{1}{2}$

$(x+3)^2=\left(2\sqrt{3x+4}\right)^2$

$x^2+6x+9=4(3x+4)$

$x^2+6x+9=12x+16$

$x^2-6x-7=0$

$(x-7)(x+1)=0$

$x-7=0$ or $x+1=0$

$x=7$ or $x=-1$

Check:

For 7:

$$\begin{array}{c|c} \multicolumn{2}{c}{\sqrt{x+2}+\sqrt{3x+4}=2} \\ \hline \sqrt{7+2}+\sqrt{3\cdot 7+4} & 2 \\ \sqrt{9}+\sqrt{25} & \\ \hline 8 \stackrel{?}{=} 2 & \text{FALSE} \end{array}$$

For -1:

$$\begin{array}{c|c} \multicolumn{2}{c}{\sqrt{x+2}+\sqrt{3x+4}=2} \\ \hline \sqrt{-1+2}+\sqrt{3\cdot(-1)+4} & \\ \sqrt{1}+\sqrt{1} & \\ \hline 2 \stackrel{?}{=} 2 & \text{TRUE} \end{array}$$

Since -1 checks but 7 does not, the solution is -1.

47. We must have $f(x)=1$, or $\sqrt{x}+\sqrt{x-9}=1$.

$\sqrt{x}+\sqrt{x-9}=1$

$\sqrt{x-9}=1-\sqrt{x}$ Isolating one radical term

$\left(\sqrt{x-9}\right)^2=\left(1-\sqrt{x}\right)^2$

$x-9=1-2\sqrt{x}+x$

$-10=-2\sqrt{x}$ Isolating the remaining radical term

$5=\sqrt{x}$

$25=x$

This value does not check. There is no solution, so there is no value of x for which $f(x)=1$.

49. $\sqrt{t-2}-\sqrt{4t+1}=-3$

$$\begin{aligned}\sqrt{t-2}&=\sqrt{4t+1}-3\\(\sqrt{t-2})^2&=(\sqrt{4t+1}-3)^2\\t-2&=4t+1-6\sqrt{4t+1}+9\\-3t-12&=-6\sqrt{4t+1}\\t+4&=2\sqrt{4t+1}\\(t+4)^2&=(2\sqrt{4t+1})^2\\t^2+8t+16&=4(4t+1)\\t^2+8t+16&=16t+4\\t^2-8t+12&=0\\(t-2)(t-6)&=0\end{aligned}$$

$t-2=0$ or $t-6=0$

$t=2$ or $t=6$

Both numbers check, so we have $f(t)=-3$ when $t=2$ and when $t=6$.

51. We must have $\sqrt{2x-3}=\sqrt{x+7}-2$.

$$\begin{aligned}\sqrt{2x-3}&=\sqrt{x+7}-2\\(\sqrt{2x-3})^2&=(\sqrt{x+7}-2)^2\\2x-3&=x+7-4\sqrt{x+7}+4\\x-14&=-4\sqrt{x+7}\\(x-14)^2&=(-4\sqrt{x+7})^2\\x^2-28x+196&=16(x+7)\\x^2-28x+196&=16x+112\\x^2-44x+84&=0\\(x-2)(x-42)&=0\end{aligned}$$

$x=2$ or $x=42$

Since 2 checks but 42 does not, we have $f(x)=g(x)$ when $x=2$.

53. We must have $4-\sqrt{t-3}=(t+5)^{1/2}$.

$$\begin{aligned}4-\sqrt{t-3}&=(t+5)^{1/2}\\(4-\sqrt{t-3})^2&=\left[(t+5)^{1/2}\right]^2\\16-8\sqrt{t-3}+t-3&=t+5\\-8\sqrt{t-3}&=-8\\\sqrt{t-3}&=1\\(\sqrt{t-3})^2&=1^2\\t-3&=1\\t&=4\end{aligned}$$

The number 4 checks, so we have $f(t)=g(t)$ when $t=4$.

55. *Writing Exercise.* For an even power n, $a^n=(-a)^n$, but $a\neq -a$ (for $a\neq 0$). Thus, we must check the possible solutions found when using the principle of powers.

57. ***Familiarize.*** Let w = the width of the rectangle, in feet. Then $13w + 5$ is the length. Recall the formula for the perimeter of a rectangle with width w and length l is $P = 2w + 2l$.

Translate. Substitute in the formula.

$$430 = 2w + 2(13w + 5)$$

Carry out. We solve the equation.

$$\begin{aligned}430 &= 2w + 2(13w + 5)\\430 &= 2w + 26w + 10\\430 &= 28w + 10\\420 &= 28w\\15 &= w\end{aligned}$$

Check. If the width is 15 ft, then the length is $13\cdot 15+5$, or 200 ft, and the perimeter is $2\cdot 15+2\cdot 200$, or 430 ft. The answer checks.

State. The width is 15 ft and the length is 200 ft.

59. ***Familiarize.*** Let w represent the width of the photo, in inches. Then $w + 4$ represents the length. Recall the formula for the area of a rectangle with width w and length l is $A = \text{width}\times\text{length}$.

Translate. The area is 140 in^2.

$$w(w + 4) = 140$$

Carry out. We solve the equation.

$$\begin{aligned}w(w + 4) &= 140\\w^2 + 4w &= 140\\w^2 + 4w - 140 &= 0\\(w - 10)(w + 14) &= 0\end{aligned}$$

$w - 10 = 0$ *or* $w + 14 = 0$

$w = 10$ *or* $w = -14$

Check. We check only 10, since width cannot be negative. If the width is 10 in., the length is 10 + 4, or 14 in., and the area is $10\cdot 14$, or 140 in^2. We have a solution.

State. The width is 10 in. and the length is 14 in.

61. ***Familiarize.*** Let x represent the first integer, $x + 2$ the second, and $x + 4$ the third.

Translate. We use the Pythagorean theorem.

$$x^2 + (x + 2)^2 = (x + 4)^2$$

Carry out. We solve the equation.

$$\begin{aligned}x^2 + (x + 2)^2 &= (x + 4)^2\\x^2 + x^2 + 4x + 4 &= x^2 + 8x + 16\\2x^2 + 4x + 4 &= x^2 + 8x + 16\\x^2 - 4x - 12 &= 0\\(x - 6)(x + 2) &= 0\end{aligned}$$

$x - 6 = 0$ *or* $x + 2 = 0$

$x = 6$ *or* $x = -2$

Check. We check only 6, since the side cannot be negative. The integers are 6, 8, and 10. The square of 6, or 36 plus the square of 8, or 64, is the square of 10, or 100.

State. The sides are 6, 8, and 10.

63. *Writing Exercise.* Answers may vary. One procedure is to set a radical expression with an even index equal to a negative number This is equivalent to writing an equation similar to those in Exercises 27 and 28.

65. Substitute 100 for $v(p)$ and solve for p.

$$v(p) = 12.1\sqrt{p}$$
$$100 = 12.1\sqrt{p}$$
$$8.2645 \approx \sqrt{p}$$
$$(8.2645)^2 \approx (\sqrt{p})^2$$
$$68.3013 \approx p$$

The nozzle pressure is about 68 psi.

67. Let f be the frequency of the string and t be the tension of the string. Substitute 260 for f, 28 for t, and solve for k, the constant of variation.

$$f = k\sqrt{t}$$
$$260 = k\sqrt{28}$$
$$k = \frac{260}{\sqrt{28}} \approx 49.135$$

Then substitute 32 for t and solve for f.

$$f = 49.135\sqrt{t}$$
$$f = 49.135\sqrt{32} \approx 277.952$$

The frequency is about 278 Hz.

69. Substitute 1880 for $S(t)$ and solve for t.

$$1880 = 1087.7\sqrt{\frac{9t+2617}{2457}}$$
$$1.7284 \approx \sqrt{\frac{9t+2617}{2457}} \quad \text{Dividing by 1087.7}$$
$$(1.7284)^2 \approx \left(\sqrt{\frac{9t+2617}{2457}}\right)^2$$
$$2.9874 \approx \frac{9t+2617}{2457}$$
$$7340.0418 \approx 9t+2617$$
$$4723.0418 \approx 9t$$
$$524.7824 \approx t$$

The temperature is about 524.8°C.

71.
$$S = 1087.7\sqrt{\frac{9t+2617}{2457}}$$
$$\frac{S}{1087.7} = \sqrt{\frac{9t+2617}{2457}}$$
$$\left(\frac{S}{1087.7}\right)^2 = \left(\sqrt{\frac{9t+2617}{2457}}\right)^2$$
$$\frac{S^2}{1087.7^2} = \frac{9t+2617}{2457}$$
$$\frac{2457S^2}{1087.7^2} = 9t+2617$$
$$\frac{2457S^2}{1087.7^2} - 2617 = 9t$$
$$\frac{1}{9}\left(\frac{2457S^2}{1087.7^2} - 2617\right) = t$$

73. $d(n) = 0.75\sqrt{2.8n}$

Substitute 84 for $d(n)$ and solve for n.

$$84 = 0.75\sqrt{2.8n}$$
$$112 = \sqrt{2.8n}$$
$$(112)^2 = (\sqrt{2.8n})^2$$
$$12{,}544 = 2.8n$$
$$4480 = n$$

About 4480 rpm will produce peak performance.

75.
$$v = \sqrt{2gr}\sqrt{\frac{h}{r+h}}$$
$$v^2 = 2gr\cdot\frac{h}{r+h} \quad \text{Squaring both sides}$$
$$v^2(r+h) = 2grh \quad \text{Multiplying by } r+h$$
$$v^2r + v^2h = 2grh$$
$$v^2h = 2grh - v^2r$$
$$v^2h = r(2gh - v^2)$$
$$\frac{v^2h}{2gh - v^2} = r$$

77.
$$\frac{x+\sqrt{x+1}}{x-\sqrt{x+1}} = \frac{5}{11}$$
$$11(x+\sqrt{x+1}) = 5(x-\sqrt{x+1})$$
$$11x + 11\sqrt{x+1} = 5x - 5\sqrt{x+1}$$
$$16\sqrt{x+1} = -6x$$
$$8\sqrt{x+1} = -3x$$
$$(8\sqrt{x+1})^2 = (-3x)^2$$
$$64(x+1) = 9x^2$$
$$64x + 64 = 9x^2$$
$$0 = 9x^2 - 64x - 64$$
$$0 = (9x+8)(x-8)$$
$$9x+8 = 0 \quad \text{or} \quad x-8 = 0$$
$$9x = -8 \quad \text{or} \quad x = 8$$
$$x = -\tfrac{8}{9} \quad \text{or} \quad x = 8$$

Since $-\frac{8}{9}$ checks but 8 does not, the solution is $-\frac{8}{9}$.

79.
$$\left(z^2+17\right)^{3/4}=27$$
$$\left[\left(z^2+17\right)^{3/4}\right]^{4/3}=\left(3^3\right)^{4/3}$$
$$z^2+17=3^4$$
$$z^2+17=81$$
$$z^2-64=0$$
$$(z+8)(z-8)=0$$
$$z=-8 \text{ or } z=8$$

Both -8 and 8 check. They are the solutions.

81.
$$\sqrt{8-b}=b\sqrt{8-b}$$
$$\left(\sqrt{8-b}\right)^2=\left(b\sqrt{8-b}\right)^2$$
$$(8-b)=b^2(8-b)$$
$$0=b^2(8-b)-(8-b)$$
$$0=(8-b)(b^2-1)$$
$$0=(8-b)(b+1)(b-1)$$
$$8-b=0 \quad\text{or}\quad b+1=0 \quad\text{or}\quad b-1=0$$
$$8=b \quad\text{or}\quad b=-1 \quad\text{or}\quad b=1$$

Since the numbers 8 and 1 check but -1 does not, 8 and 1 are the solutions.

83. We find the values of x for which $g(x)=0$.
$$6x^{1/2}+6x^{-1/2}-37=0$$
$$6\sqrt{x}+\frac{6}{\sqrt{x}}=37$$
$$\left(6\sqrt{x}+\frac{6}{\sqrt{x}}\right)^2=37^2$$
$$36x+72+\frac{36}{x}=1369$$
$$36x^2+72x+36=1369x \quad \text{Multiplying by } x$$
$$36x^2-1297x+36=0$$
$$(36x-1)(x-36)=0$$
$$36x-1=0 \quad\text{or}\quad x-36=0$$
$$36x=1 \quad\text{or}\quad x=36$$
$$x=\frac{1}{36} \quad\text{or}\quad x=36$$

Both numbers check. The x-intercepts are $\left(\frac{1}{36}, 0\right)$ and $(36, 0)$.

85. *Graphing Calculator Exercise*

87. *Graphing Calculator Exercise*

Exercise Set 10.7

1. The correct choice is (d) Right; see page 677 in the text.

3. The correct choice is (e) Square roots; see page 677 in the text.

5. The correct choice is (f) 30°-60°-90°; see page 680 in the text.

7. $a=5$, $b=3$

Find c.
$$c^2=a^2+b^2 \quad \text{Pythagorean theorem}$$
$$c^2=5^2+3^2 \quad \text{Substituting}$$
$$c^2=25+9$$
$$c^2=34$$
$$c=\sqrt{34} \quad \text{Exact answer}$$
$$c\approx 5.831 \quad \text{Approximation}$$

9. $a=9$, $b=9$

Observe that the legs have the same length, so this is an isosceles right triangle. Then we know that the length of the hypotenuse is the length of a leg times $\sqrt{2}$, or $9\sqrt{2}$, or approximately 12.728.

11. $b=15$, $c=17$

Find a.
$$a^2+b^2=c^2 \quad \text{Pythagorean theorem}$$
$$a^2+15^2=17^2 \quad \text{Substituting}$$
$$a^2+225=289$$
$$a^2=64$$
$$a=8$$

13.
$$a^2+b^2=c^2 \quad \text{Pythagorean theorem}$$
$$\left(4\sqrt{3}\right)^2+b^2=8^2$$
$$16\cdot 3+b^2=64$$
$$48+b^2=64$$
$$b^2=16$$
$$b=4$$

The other leg is 4 m long.

15.
$$a^2+b^2=c^2 \quad \text{Pythagorean theorem}$$
$$1^2+b^2=\left(\sqrt{20}\right)^2 \quad \text{Substituting}$$
$$1+b^2=20$$
$$b^2=19$$
$$b=\sqrt{19}$$
$$b\approx 4.359$$

The length of the other leg is $\sqrt{19}$ in., or about 4.359 in.

17. Observe that the length of the hypotenuse, $\sqrt{2}$, is $\sqrt{2}$ times the length of the given leg, 1 m. Thus, we have an isosceles right triangle and the length of the other leg is also 1 m.

19. From the drawing in the text we see that we have a right triangle with legs of 150 ft and 200 ft. Let $d =$ the length of the diagonal, in feet. We use the Pythagorean theorem to find d.

$$150^2 + 200^2 = d^2$$
$$22{,}500 + 40{,}000 = d^2$$
$$62{,}500 = d^2$$
$$250 = d$$

Clare travels 250 ft across the parking lot.

21. We make a drawing and let $d =$ the distance from home plate to second base.

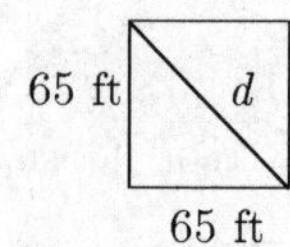

Note that we have an isosceles right triangle. Then the length of the hypotenuse is the length of a leg times $\sqrt{2}$, or $65\sqrt{2}$ ft. This is about 91.924 ft.

(We could also have used the Pythagorean theorem, solving $65^2 + 65^2 = d^2$.)

23. We make a drawing similar to the one in the text.

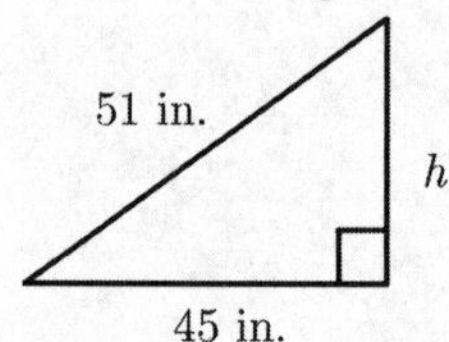

We use the Pythagorean theorem to find h.

$$45^2 + h^2 = 51^2$$
$$2051 + h^2 = 2601$$
$$h^2 = 576$$
$$h = 24$$

The height of the screen is 24 in.

25. First we will find the diagonal distance, d, in feet, across the room. We make a drawing.

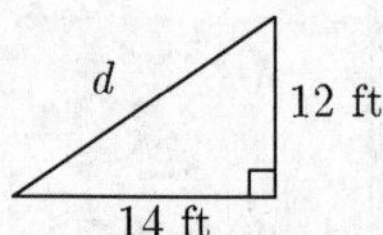

Now we use the Pythagorean theorem.

$$12^2 + 14^2 = d^2$$
$$144 + 196 = d^2$$
$$340 = d^2$$
$$\sqrt{340} = d$$
$$18.439 \approx d$$

Recall that 4 ft of slack is required on each end. Thus, $\sqrt{340} + 2 \cdot 4$, or $(\sqrt{340} + 8)$ ft, of wire should be purchased. This is about 26.439 ft.

27. The diagonal is the hypotenuse of a right triangle with legs of 70 paces and 40 paces. First we use the Pythagorean theorem to find the length d of the diagonal, in paces.

$$70^2 + 40^2 = d^2$$
$$4900 + 1600 = d^2$$
$$6500 = d^2$$
$$\sqrt{6500} = d$$
$$80.623 \approx d$$

If Marissa walks along two sides of the quad she takes $70 + 40$, or 110 paces. Then by using the diagonal she saves $(110 - \sqrt{6500})$ paces. This is approximately $110 - 80.623$, or 29.377 paces.

29. Since one acute angle is $45°$, this is an isosceles right triangle with one leg = 5. Then the other leg = 5 also. And the hypotenuse is the length of the a leg times $\sqrt{2}$, or $5\sqrt{2}$.

Exact answer: Leg$=5$, hypotenuse$=5\sqrt{2}$

Approximation: hypotenuse ≈ 7.071

31. This is a 30-60-90 right triangle with hypotenuse 14. We find the legs:

$2a = 14$, so $a = 7$ and $a\sqrt{3} = 7\sqrt{3}$

Exact answer: shorter leg$=7$; longer leg$=7\sqrt{3}$

Approximation: longer leg ≈ 12.124

33. This is a 30-60-90 right triangle with one leg = 15. We substitute to find the length of the other leg, a, and the hypotenuse, c.

$$b = a\sqrt{3}$$
$$15 = a\sqrt{3}$$
$$\frac{15}{\sqrt{3}} = a$$
$$\frac{15\sqrt{3}}{3} = a \quad \text{Rationalizing the denominator}$$
$$5\sqrt{3} = a \quad \text{Simplifying}$$
$$c = 2a$$
$$c = 2 \cdot 5\sqrt{3}$$
$$c = 10\sqrt{3}$$

Exact answer: $a = 5\sqrt{3}, c = 10\sqrt{3}$

Approximations: $a \approx 8.660, c \approx 17.321$

35. This is an isosceles right triangle with hypotenuse 13. The two legs have the same length, a.

$$a\sqrt{2} = 13$$
$$a = \frac{13}{\sqrt{2}} = \frac{13\sqrt{2}}{2}$$

Exact answer: $\frac{13\sqrt{2}}{2}$

Approximation: 9.192

37. This is a 30-60-90 triangle with the shorter leg = 14. We find the longer leg and the hypotenuse.

$a\sqrt{3} = 14\sqrt{3}$, and $2a = 2 \cdot 14 = 28$.

Exact answer: longer leg $= 14\sqrt{3}$, hypotenuse = 28

Approximation: longer leg ≈ 24.249

39. h is the longer leg of a 30-60-90 right triangle with shorter leg $= 5$. Then $h = 5\sqrt{3} \approx 8.660$.

41. We make a drawing.

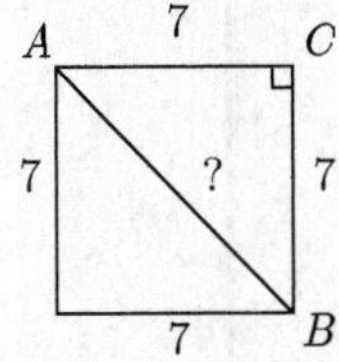

Triangle ABC is an isosceles right triangle with legs of length 7. Then the hypotenuse$=7\sqrt{2} \approx 9.899$.

43. We make a drawing.

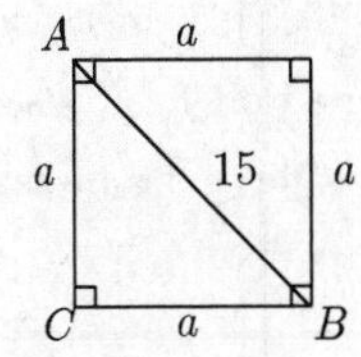

Triangle ABC is an isosceles right triangle with hypotenuse $= 15$. Then $a = \frac{15\sqrt{2}}{2} \approx 10.607$.

45. We will express all distances in feet. Recall that $1 \text{ mi} = 5280 \text{ ft}$.

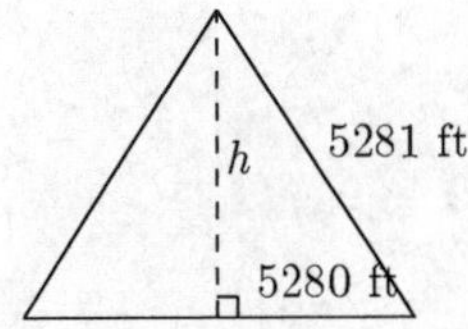

We use the Pythagorean theorem to find h.

$$\begin{aligned} h^2 + (5280)^2 &= (5281)^2 \\ h^2 + 27{,}878{,}400 &= 27{,}888{,}961 \\ h^2 &= 10{,}561 \\ h &= \sqrt{10{,}561} \\ h &\approx 102.767 \end{aligned}$$

The height of the bulge is $\sqrt{10{,}561}$ ft, or about 102.767 ft.

47. We make a drawing.

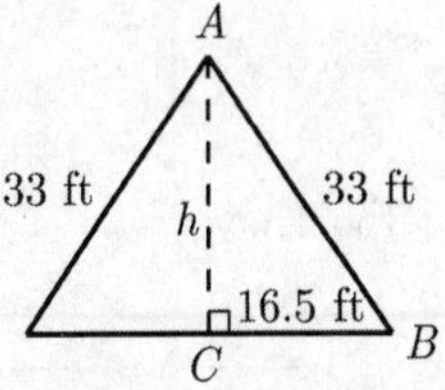

The base of the lodge is an equilateral triangle, so all the angles are 60°. The altitude bisects one angle and one side. Then the triangle ABC is a 30°-60°-90° right triangle with the shorter leg of length $\frac{33}{2}$, or 16.5 ft, and hypotenuse of length 33. Then the height is the length of the shorter leg times $\sqrt{3}$.

Exact answer: $h = \frac{33\sqrt{3}}{2}$ ft

Approximation: $h \approx 28.579$ ft

If the height of triangle ABC is $\frac{33\sqrt{3}}{2}$ and the base is 33 ft, the area is $\frac{1}{2} \cdot 33 \cdot \frac{33\sqrt{3}}{2} = \frac{1089}{4}\sqrt{3}$ ft^2, or about 471.551 ft^2.

49. We make a drawing.

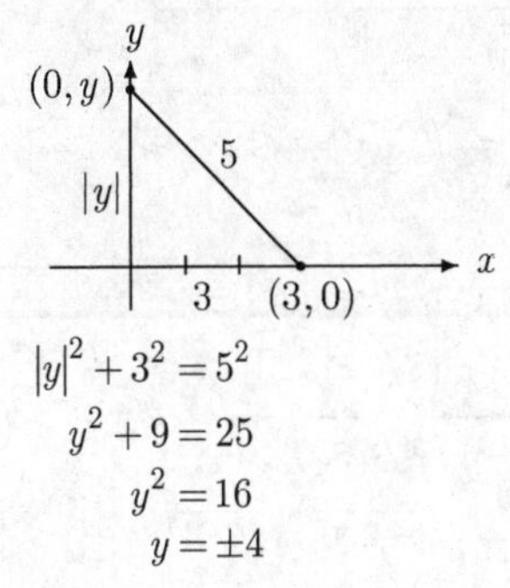

$$\begin{aligned} |y|^2 + 3^2 &= 5^2 \\ y^2 + 9 &= 25 \\ y^2 &= 16 \\ y &= \pm 4 \end{aligned}$$

The points are $(0, -4)$ and $(0, 4)$.

51. Using the distance formula $d = \sqrt{(x_2 - x_1)^2 + (y_2 - y_1)^2}$ for the points (4, 5) and (7, 1),

$$\begin{aligned} d &= \sqrt{(7-4)^2 + (1-5)^2} \\ &= \sqrt{3^2 + (-4)^2} = \sqrt{9+16} = \sqrt{25} \\ &= 5 \end{aligned}$$

53. Using the distance formula $d = \sqrt{(x_2 - x_1)^2 + (y_2 - y_1)^2}$ for the points (1, –2) and (0, –5),

$$\begin{aligned} d &= \sqrt{(1-0)^2 + (-2-(-5))^2} \\ &= \sqrt{1^2 + 3^2} = \sqrt{1+9} = \sqrt{10} \\ &\approx 3.162 \end{aligned}$$

55. Using the distance formula $d=\sqrt{(x_2-x_1)^2+(y_2-y_1)^2}$ for the points (6, –6) and (–4, 4),

$$\begin{aligned} d&=\sqrt{(6-(-4))^2+(-6-4)^2}\\ &=\sqrt{10^2+(-10)^2}=\sqrt{100+100}=\sqrt{200}\\ &\approx 14.142 \end{aligned}$$

57. Using the distance formula $d=\sqrt{(x_2-x_1)^2+(y_2-y_1)^2}$ for the points (–9.2, –3.4) and (8.6, –3.4),

$$\begin{aligned} d&=\sqrt{(-9.8-8.6)^2+(-3.4-(-3.4))^2}\\ &=\sqrt{(-17.8)^2+0^2}=\sqrt{316.84}\\ &\approx 17.8 \end{aligned}$$

59. Using the distance formula $d=\sqrt{(x_2-x_1)^2+(y_2-y_1)^2}$ for the points $\left(\frac{5}{6},-\frac{1}{6}\right)$ and $\left(\frac{1}{2},\ \frac{1}{3}\right)$,

$$\begin{aligned} d&=\sqrt{\left(\frac{5}{6}-\frac{1}{2}\right)^2+\left(-\frac{1}{6}-\frac{1}{3}\right)^2}\\ &=\sqrt{\left(\frac{2}{6}\right)^2+\left(-\frac{3}{6}\right)^2}=\sqrt{\frac{4}{36}+\frac{9}{36}}=\sqrt{\frac{13}{36}}=\frac{\sqrt{13}}{6}\\ &\approx 0.601 \end{aligned}$$

61. Using the distance formula $d=\sqrt{(x_2-x_1)^2+(y_2-y_1)^2}$ for the points (0, 0) and $\left(-\sqrt{6},\ \sqrt{6}\right)$,

$$\begin{aligned} d&=\sqrt{\left(0-\left(-\sqrt{6}\right)\right)^2+\left(0-\sqrt{6}\right)^2}\\ &=\sqrt{\left(\sqrt{6}\right)^2+\left(-\sqrt{6}\right)^2}=\sqrt{6+6}=\sqrt{12}\\ &\approx 3.464 \end{aligned}$$

63. Using the distance formula $d=\sqrt{(x_2-x_1)^2+(y_2-y_1)^2}$ for the points (–2, –40) and (–1, –30),

$$\begin{aligned} d&=\sqrt{(-2-(-1))^2+(-40-(-30))^2}\\ &=\sqrt{(-1)^2+(-10)^2}=\sqrt{1+100}=\sqrt{101}\\ &\approx 10.050 \end{aligned}$$

65. Using the midpoint formula $\left(\frac{x_1+x_2}{2},\ \frac{y_1+y_2}{2}\right)$ for the points (–2, 5) and (8, 3),

$$\left(\frac{-2+8}{2},\ \frac{5+3}{2}\right),\text{ or }\left(\frac{6}{2},\ \frac{8}{2}\right),\text{ or }(3,\ 4)$$

67. Using the midpoint formula $\left(\frac{x_1+x_2}{2},\ \frac{y_1+y_2}{2}\right)$ for the points (2, –1) and (5, 8),

$$\left(\frac{2+5}{2},\ \frac{-1+8}{2}\right),\text{ or }\left(\frac{7}{2},\ \frac{7}{2}\right)$$

69. Using the midpoint formula $\left(\frac{x_1+x_2}{2},\ \frac{y_1+y_2}{2}\right)$ for the points (–8, –5) and (6, –1),

$$\left(\frac{-8+6}{2},\ \frac{-5+(-1)}{2}\right),\text{ or }\left(-\frac{2}{2},\ \frac{-6}{2}\right),\text{ or }(-1,-3)$$

71. Using the midpoint formula $\left(\frac{x_1+x_2}{2},\ \frac{y_1+y_2}{2}\right)$ for the points (–3.4, 8.1) and (4.8, –8.1),

$$\left(\frac{-3.4+4.8}{2},\ \frac{8.1+(-8.1)}{2}\right),\text{ or }\left(\frac{1.4}{2},\ \frac{0}{2}\right),\text{ or }(0.7,\ 0)$$

73. Using the midpoint formula $\left(\frac{x_1+x_2}{2},\ \frac{y_1+y_2}{2}\right)$ for the points $\left(\frac{1}{6},-\frac{3}{4}\right)$ and $\left(-\frac{1}{3},\ \frac{5}{6}\right)$,

$$\left(\frac{\frac{1}{6}+\left(-\frac{1}{3}\right)}{2},\ \frac{-\frac{3}{4}+\frac{5}{6}}{2}\right),\text{ or }\left(\frac{-\frac{1}{6}}{2},\ \frac{\frac{1}{12}}{2}\right),\text{ or }\left(-\frac{1}{12},\ \frac{1}{24}\right)$$

75. Using the midpoint formula $\left(\frac{x_1+x_2}{2},\ \frac{y_1+y_2}{2}\right)$ for the points $\left(\sqrt{2},-1\right)$ and $\left(\sqrt{3},\ 4\right)$,

$$\left(\frac{\sqrt{2}+\sqrt{3}}{2},\ \frac{-1+4}{2}\right),\text{ or }\left(\frac{\sqrt{2}+\sqrt{3}}{2},\ \frac{3}{2}\right)$$

77. *Writing Exercise.* If a right triangle has consecutive numbers for the lengths of its sides, then a, $a+1$, and $a+2$ represent the lengths of the sides and these numbers must satisfy the equation $a^2+(a+1)^2=(a+2)^2$. The solutions of this equation are –1 and 3. Since the length of a side cannot be negative, the only possible sides measure 3, 4, and 5. Thus, the only right triangle that has consecutive numbers for the lengths of its sides has sides measuring 3, 4, and 5.

79. $y=2x-3$

Slope is 2; y-intercept is (0, –3).

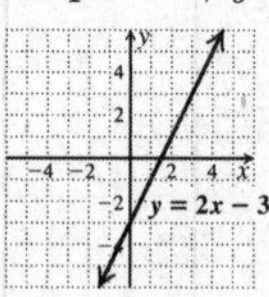

81. $8x-4y=8$

To find the y-intercept, let $x=0$ and solve for y.

$$\begin{aligned} 8\cdot 0-4y&=8\\ y&=-2 \end{aligned}$$

The y-intercept is (0, –2).

To find the x-intercept, let $y=0$ and solve for x.

$$\begin{aligned} 8x-4\cdot 0&=8\\ x&=1 \end{aligned}$$

The x-intercept is (1, 0).

Plot these points and draw the line. A third point could be used as a check.

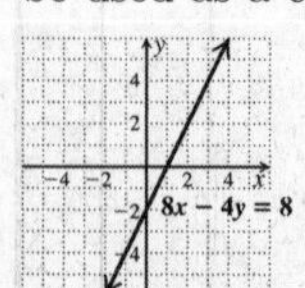

83. $x \geq 1$

Graph the line $x = 1$. Draw the line solid since the inequality symbol is $\geq$. Test the point (0, 0) to determine if it is a solution.

$$\begin{array}{c|c} x \geq 1 \\ \hline 0 & 1 \end{array}$$
$$0 \overset{?}{\geq} 1 \quad \text{FALSE}$$

Since $0 \geq 1$ is false, we shade the half plane that does not contains (0, 0) and obtain the graph.

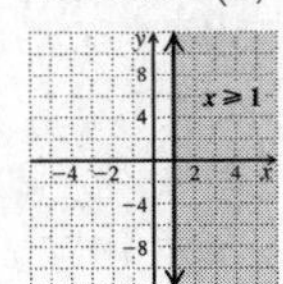

85. *Writing Exercise.* If $P_1(x_1, y_1)$, $P_2(x_2, y_2)$, and $P_3(x_3, y_3)$ are vertices of a right triangle, the distance formula could be used to find the lengths of the legs and the length of the hypotenuse. Then the right triangle can be verified by substituting the lengths for a, b, and c, in the Pythagorean theorem.

87. The length of a side of the hexagon is 72/6, or 12 cm. Then the shaded region is a triangle with base 12 cm. To find the height of the triangle, note that it is the longer leg of a 30°-60°-90° right triangle. Thus its length is the length of the length of the shorter leg times $\sqrt{3}$. The length of the shorter leg is half the length of the base, $\frac{1}{2} \cdot 12$ cm, or 6 cm, so the length of the longer leg is $6\sqrt{3}$ cm. Now we find the area of the triangle.

$$\begin{aligned} A &= \frac{1}{2}bh \\ &= \frac{1}{2}(12 \text{ cm})(6\sqrt{3} \text{ cm}) \\ &= 36\sqrt{3} \text{ cm}^2 \\ &\approx 62.354 \text{ cm}^2 \end{aligned}$$

89. We make a drawing.

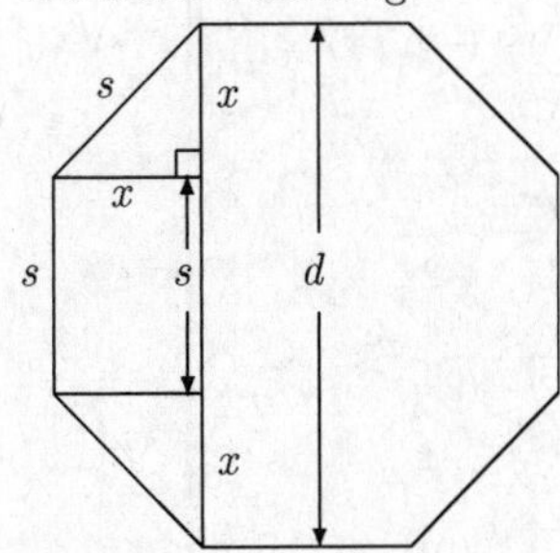

$d = s + 2x$

Use the Pythagoran theorem to find x.

$$\begin{aligned} x^2 + x^2 &= s^2 \\ 2x^2 &= s^2 \\ x^2 &= \frac{s^2}{2} \\ x &= \frac{s}{\sqrt{2}} = \frac{s}{\sqrt{2}} \cdot \frac{\sqrt{2}}{2} = \frac{s\sqrt{2}}{2} \end{aligned}$$

Then $d = s + 2x = s + 2\left(\frac{s\sqrt{2}}{2}\right) = s + s\sqrt{2}$.

91. We make a drawing.

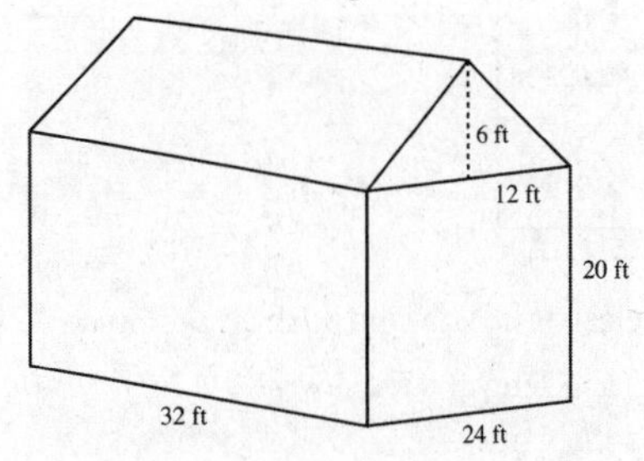

The area to be painted consists of two 20 ft by 24 ft rectangles, two 20 ft by 32 ft rectangles, and two triangles with height 6 ft and base 24 ft. The area of the two 20 ft by 24 ft rectangles is $2 \cdot 20 \text{ ft} \cdot 24 \text{ ft} = 960 \text{ ft}^2$. The area of the two 20 ft by 32 ft rectangles is $2 \cdot 20 \text{ ft} \cdot 32 \text{ ft} = 1280 \text{ ft}^2$. The area of the two triangles is $2 \cdot \frac{1}{2} \cdot 24 \text{ ft} \cdot 6 \text{ ft} = 144 \text{ ft}^2$. Thus, the total area to be painted is $960 \text{ ft}^2 + 1280 \text{ ft}^2 + 144 \text{ ft}^2 = 2384 \text{ ft}^2$.

One gallon of paint covers a minimum of 450 ft^2, so we divide to determine how many gallons of paint are required: $\frac{2384}{450} \approx 5.3$. Thus, 5 gallons of paint should be bought to paint the house. This answer assumes that the total area of the doors and windows is 134 ft^2 or more. ($5 \cdot 450 = 2250$ and $2384 = 2250 + 134$)

93. First we find the radius of a circle with an area of 6160 ft^2. This is the length of the hose.

$$\begin{aligned} A &= \pi r^2 \\ 6160 &= \pi r^2 \\ \frac{6160}{\pi} &= r^2 \\ \sqrt{\frac{6160}{\pi}} &= r \\ 44.28 &\approx r \end{aligned}$$

Now we make a drawing of the room.

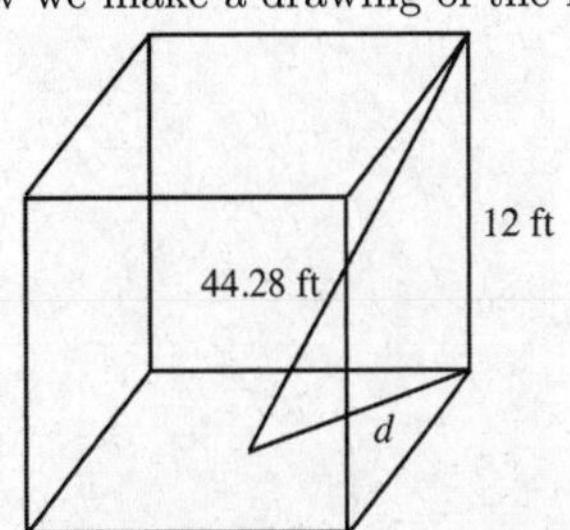

We use the Pythagorean theorem to find d.

$$d^2+12^2=44.28^2$$
$$d^2+144=1960.7184$$
$$d^2=1816.7184$$
$$d\approx 42.623$$

Now we make a drawing of the floor of the room.

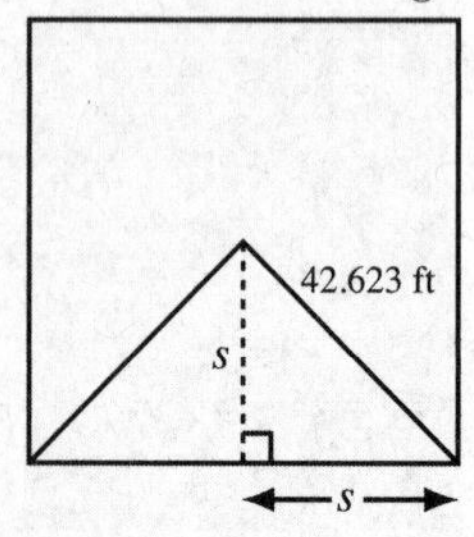

We have an isosceles right triangle with hypotenuse 42.623 ft. We find the length of a side s.

$$s\sqrt{2}=42.623$$
$$s=\frac{42.623}{\sqrt{2}}\approx 30.14\text{ ft}$$

Then the length of a side of the room is $2s=2(30.14\text{ ft})=60.28$ ft; so the dimensions of the largest square room that meets the given conditions is 60.28 ft by 60.28 ft.

95. We make a drawing.

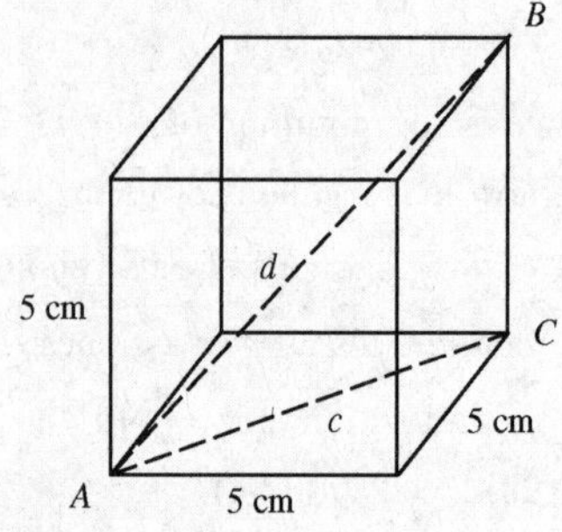

First find the length of a diagonal of the base of the cube. It is the hypotenuse of an isosceles right triangle with legs 5 cm. Then $c=5\sqrt{2}$ cm.

Triangle ABC is a right triangle with legs of $5\sqrt{2}$ cm and 5 cm and hypotenuse d. Use the Pythagorean theorem to find d, the length of the diagonal that connects two opposite corners of the cube.

$$d^2=(5\sqrt{2})^2+5^2$$
$$d^2=25\cdot 2+25$$
$$d^2=50+25$$
$$d^2=75$$
$$d=\sqrt{75}$$

Exact answer: $d=\sqrt{75}$ cm

Exercise Set 10.8

1. False; see page 688 in the text.

3. True; see page 689 in the text.

5. True; see page 690 in the text.

7. False; see Exercises 59-64.

9. $\sqrt{-100}=\sqrt{-1\cdot 100}=\sqrt{-1}\cdot\sqrt{100}=i\cdot 10=10i$

11. $\sqrt{-5}=\sqrt{-1\cdot 5}=\sqrt{-1}\cdot\sqrt{5}=i\cdot\sqrt{5}$, or $\sqrt{5}\,i$

13. $\sqrt{-8}=\sqrt{-1}\cdot\sqrt{4}\cdot\sqrt{2}=i\cdot 2\cdot\sqrt{2}=2i\sqrt{2}$, or $2\sqrt{2}\,i$

15. $-\sqrt{-11}=-\sqrt{-1}\cdot\sqrt{11}=-i\cdot\sqrt{11}=-i\sqrt{11}$, or $-\sqrt{11}\,i$

17. $-\sqrt{-49}=-\sqrt{-1\cdot 49}=-\sqrt{-1}\cdot\sqrt{49}=-i\cdot 7=-7i$

19. $-\sqrt{-300}=-\sqrt{-1}\cdot\sqrt{100}\cdot\sqrt{3}=-i\cdot 10\cdot\sqrt{3}=-10i\sqrt{3}$, or $-10\sqrt{3}i$

21. $6-\sqrt{-84}=6-\sqrt{-1\cdot 4\cdot 21}=6-i\cdot 2\sqrt{21}=6-2i\sqrt{21}$, or $6-2\sqrt{21}i$

23. $-\sqrt{-76}+\sqrt{-125}=-\sqrt{-1\cdot 4\cdot 19}+\sqrt{-1\cdot 25\cdot 5}$
$=-i\cdot 2\sqrt{19}+i\cdot 5\sqrt{5}=-2i\sqrt{19}+5i\sqrt{5}=(-2\sqrt{19}+5\sqrt{5})i$

25. $\sqrt{-18}-\sqrt{-64}=\sqrt{-1\cdot 9\cdot 2}-\sqrt{-1\cdot 64}$
$=i\cdot 3\cdot\sqrt{2}-i\cdot 8=3i\sqrt{2}-8i$, or $(3\sqrt{2}-8)i$

27. $(3+4i)+(2-7i)$
$=(3+2)+(4-7)i$ Combining the real and the imaginary parts
$=5-3i$

29. $(9+5i)-(2+3i)=(9-2)+(5-3)i$
$=7+2i$

31. $(7-4i)-(5-3i)=(7-5)+[-4-(-3)]i=2-i$

33. $(-5-i)-(7+4i)=(-5-7)+(-1-4)i=-12-5i$

35. $5i\cdot 8i=40\cdot i^2=40(-1)=-40$

37. $(-4i)(-6i)=24\cdot i^2=24(-1)=-24$

39. $\sqrt{-36}\sqrt{-9}=\sqrt{-1}\cdot\sqrt{36}\cdot\sqrt{-1}\cdot\sqrt{9}=i\cdot 6\cdot i\cdot 3$
$=i^2\cdot 18=-1\cdot 18=-18$

41. $\sqrt{-3}\sqrt{-10}=\sqrt{-1}\cdot\sqrt{3}\cdot\sqrt{-1}\cdot\sqrt{10}$
$=i\cdot\sqrt{3}\cdot i\cdot\sqrt{10}$
$=\sqrt{30}\cdot i^2=\sqrt{30}\cdot(-1)$
$=-\sqrt{30}$

43. $\sqrt{-6}\sqrt{-21}=\sqrt{-1}\cdot\sqrt{6}\cdot\sqrt{-1}\cdot\sqrt{21}$
$=i\cdot\sqrt{6}\cdot i\cdot\sqrt{21}=i^2\sqrt{126}=-1\cdot\sqrt{9\cdot14}$
$=-3\sqrt{14}$

45. $5i(2+6i)=5i\cdot2+5i\cdot6i=10i+30i^2$
$=10i-30=-30+10i$

47. $-7i(3+4i)=-7i\cdot3-7i\cdot4i$
$=-21i-28i^2$
$=-21i+28=28-21i$

49. $(1+i)(3+2i)=3+2i+3i+2i^2$
$=3+2i+3i-2=1+5i$

51. $(6-5i)(3+4i)=18+24i-15i-20i^2$
$=18+24i-15i+20=38+9i$

53. $(7-2i)(2-6i)=14-42i-4i+12i^2$
$=14-42i-4i-12=2-46i$

55. $(3+8i)(3-8i)=3^2-(8i)^2$ Difference of squares
$=9-64i^2=9-64(-1)$
$=9+64=73$

57. $(-7+i)(-7-i)=(-7)^2-i^2$ Difference of squares
$=49-i^2=49-(-1)$
$=49+1=50$

59. $(4-2i)^2=4^2-2\cdot4\cdot2i+(2i)^2=16-16i+4i^2$
$=16-16i-4=12-16i$

61. $(2+3i)^2=2^2+2\cdot2\cdot3i+(3i)^2=4+12i+9i^2$
$=4+12i-9=-5+12i$

63. $(-2+3i)^2=(-2)^2+2(-2)(3i)+(3i)^2$
$=4-12i+9i^2=4-12i-9=-5-12i$

65. $\frac{10}{3+i}=\frac{10}{3+i}\cdot\frac{3-i}{3-i}$ Multiplying by 1, using the conjugate
$=\frac{30-10i}{9-i^2}$
$=\frac{30-10i}{9-(-1)}$
$=\frac{30-10i}{10}$
$=3-i$

67. $\frac{2}{3-2i}=\frac{2}{3-2i}\cdot\frac{3+2i}{3+2i}$ Multiplying by 1, using the conjugate
$=\frac{6+4i}{9-4i^2}$
$=\frac{6+4i}{9-4(-1)}$
$=\frac{6+4i}{13}$
$=\frac{6}{13}+\frac{4}{13}i$

69. $\frac{2i}{5+3i}=\frac{2i}{5+3i}\cdot\frac{5-3i}{5-3i}=\frac{10i-6i^2}{25-9i^2}=\frac{10i+6}{25+9}$
$=\frac{10i+6}{34}=\frac{6}{34}+\frac{10}{34}i=\frac{3}{17}+\frac{5}{17}i$

71. $\frac{5}{6i}=\frac{5}{6i}\cdot\frac{i}{i}=\frac{5i}{6i^2}=\frac{5i}{-6}=-\frac{5}{6}i$

73. $\frac{5-3i}{4i}=\frac{5-3i}{4i}\cdot\frac{i}{i}=\frac{5i-3i^2}{4i^2}=\frac{5i+3}{-4}$
$=-\frac{3}{4}-\frac{5}{4}i$

75. $\frac{7i+14}{7i}=\frac{7i}{7i}+\frac{14}{7i}=1+\frac{2}{i}=1+\frac{2}{i}\cdot\frac{i}{i}$
$=1+\frac{2i}{i^2}=1+\frac{2i}{-1}=1-2i$

77. $\frac{4+5i}{3-7i}=\frac{4+5i}{3-7i}\cdot\frac{3+7i}{3+7i}=\frac{12+28i+15i+35i^2}{9-49i^2}$
$=\frac{12+28i+15i-35}{9+49}=\frac{-23+43i}{58}$
$=-\frac{23}{58}+\frac{43}{58}i$

79. $\frac{2+3i}{2+5i}=\frac{2+3i}{2+5i}\cdot\frac{2-5i}{2-5i}=\frac{4-10i+6i-15i^2}{4-25i^2}$
$=\frac{4-10i+6i+15}{4+25}=\frac{19-4i}{29}$
$=\frac{19}{29}-\frac{4}{29}i$

81. $\frac{3-2i}{4+3i}=\frac{3-2i}{4+3i}\cdot\frac{4-3i}{4-3i}=\frac{12-9i-8i+6i^2}{16-9i^2}$
$=\frac{12-9i-8i-6}{16+9}=\frac{6-17i}{25}$
$=\frac{6}{25}-\frac{17}{25}i$

83. $i^{32}=\left(i^2\right)^{16}=(-1)^{16}=1$

85. $i^{15}=i^{14}\cdot i=\left(i^2\right)^7\cdot i=(-1)^7\cdot i=-i$

87. $i^{42}=\left(i^2\right)^{21}=(-1)^{21}=-1$

89. $i^9=\left(i^2\right)^4\cdot i=(-1)^4\cdot i=1\cdot i=i$

91. $(-i)^6=(-1\cdot i)^6=(-1)^6\cdot i^6=1\cdot i^6=(i^2)^3=(-1)^3=-1$

93. $(5i)^3=5^3\cdot i^3=125\cdot i^2\cdot i=125(-1)(i)=-125i$

95. $i^2+i^4=-1+(i^2)^2=-1+(-1)^2=-1+1=0$

97. *Writing Exercise.* No; the product of two pure imaginary numbers is a real number.

99. $x^2 - x - 6 = 0$
$(x+2)(x-3) = 0$
$x+2 = 0 \quad or \quad x-3 = 0$
$x = -2 \quad or \quad x = 3$

The solutions are –2 and 3.

101. $t^2 = 100$
$t^2 - 100 = 0$
$(t+10)(t-10) = 0$
$t+10 = 0 \quad or \quad t-10 = 0$
$t = -10 \quad or \quad t = 10$

The solutions are –10 and 10.

103. $15x^2 = 14x + 8$
$15x^2 - 14x - 8 = 0$
$(5x+2)(3x-4) = 0$
$5x+2 = 0 \quad or \quad 3x-4 = 0$
$x = -\frac{2}{5} \quad or \quad x = \frac{4}{3}$

The solutions are $-\frac{2}{5}$ and $\frac{4}{3}$.

105. *Writing Exercise.* Yes; every real number a is a complex number $a + bi$ with $b = 0$.

107.

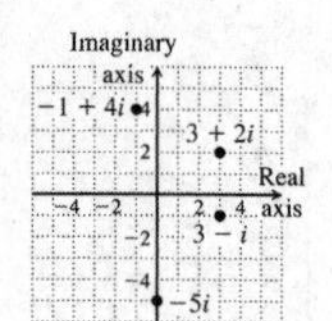

109. $|3+4i| = \sqrt{3^2+4^2} = \sqrt{9+16} = \sqrt{25} = 5$

111. $|-1+i| = \sqrt{(-1)^2+1^2} = \sqrt{1+1} = \sqrt{2}$

113. $g(3i) = \frac{(3i)^4 - (3i)^2}{3i-1} = \frac{81i^4 - 9i^2}{-1+3i} = \frac{81+9}{-1+3i}$
$= \frac{90}{-1+3i} = \frac{90}{-1+3i} \cdot \frac{-1-3i}{-1-3i} = \frac{90(-1-3i)}{1-9i^2}$
$= \frac{90(-1-3i)}{1+9} = \frac{90(-1-3i)}{10} = \frac{9 \cdot 10(-1-3i)}{10}$
$= 9(-1-3i) = -9-27i$

115. First we simplify $g(z)$.

$g(z) = \frac{z^4 - z^2}{z-1} = \frac{z^2(z^2-1)}{z-1} = \frac{z^2(z+1)(z-1)}{z-1}$
$= \frac{z^2(z+1)(z-1)}{z-1} = z^2(z+1)$

Now we substitute.

$g(5i-1) = (5i-1)^2(5i-1+1)$
$= (25i^2 - 10i + 1)(5i)$
$= (-25-10i+1)(5i) = (-24-10i)(5i)$
$= -120i - 50i^2 = 50 - 120i$

117. $\frac{1}{w-w^2} = \frac{1}{\frac{1-i}{10} - \left(\frac{1-i}{10}\right)^2}$
$= \frac{1}{\frac{1-i}{10} - \frac{1-2i+i^2}{100}} = \frac{1}{\frac{10-10i-(1-2i-1)}{100}}$
$= \frac{1}{\frac{10-8i}{100}} = \frac{100}{10-8i} = \frac{50}{5-4i}$
$= \frac{50}{5-4i} \cdot \frac{5+4i}{5+4i} = \frac{250+200i}{25+16}$
$= \frac{250+200i}{41} = \frac{250}{41} + \frac{200}{41}i$

119. $(1-i)^3(1+i)^3$
$= (1-i)(1+i) \cdot (1-i)(1+i) \cdot (1-i)(1+i)$
$= (1-i^2)(1-i^2)(1-i^2) = (1+1)(1+1)(1+1)$
$= 2 \cdot 2 \cdot 2 = 8$

121. $\frac{6}{1+\frac{3}{i}} = \frac{6}{\frac{i+3}{i}} = \frac{6i}{i+3} = \frac{6i}{i+3} \cdot \frac{-i+3}{-i+3}$
$= \frac{-6i^2+18i}{-i^2+9} = \frac{6+18i}{10} = \frac{6}{10} + \frac{18}{10}i$
$= \frac{3}{5} + \frac{9}{5}i$

123. $\frac{i-i^{38}}{1+i} = \frac{i-(i^2)^{19}}{1+i} = \frac{i-(-1)^{19}}{1+i} = \frac{i-(-1)}{1+i} = \frac{i+1}{1+i} = 1$

Chapter 10 Review

1. True, see page 650 of the text.

3. False, see page 636 of the text.

5. True, see page 644 of the text.

7. True, see page 670 of the text.

9. $\sqrt{\frac{100}{121}} = \frac{\sqrt{100}}{\sqrt{121}} = \frac{10}{11}$

11. $f(x) = \sqrt{x+10}$
$f(15) = \sqrt{15+10} = \sqrt{25} = 5$

13. $\sqrt{64t^2} = \sqrt{(8t)^2} = |8t| = 8|t|$

15. $\sqrt{4x^2+4x+1} = \sqrt{(2x+1)^2} = |2x+1|$

17. $(\sqrt[3]{5ab})^4 = (5ab)^{4/3}$

19. $\sqrt{x^6y^{10}} = (x^6y^{10})^{1/2} = x^{6/2}y^{10/2} = x^3y^5$

21. $(x^{-2/3})^{3/5} = x^{-\frac{2}{3}\frac{3}{5}} = x^{-2/5} = \frac{1}{x^{2/5}}$

23. $f(x)=\sqrt{25}\sqrt{(x-6)^2}=5|x-6|$

25. $\sqrt{250x^3y^2}=\sqrt{25x^2y^2\cdot 10x}=5xy\sqrt{10x}$

27. $\sqrt[3]{3x^4b}\sqrt[3]{9xb^2}=\sqrt[3]{3x^4b\cdot 9xb^2}=\sqrt[3]{27x^3b^3\cdot x^2}=3xb\sqrt[3]{x^2}$

29. $\sqrt[3]{\dfrac{-27y^{12}}{64}}=\dfrac{\sqrt[3]{(-3y^4)^3}}{\sqrt[3]{4^3}}=\dfrac{-3y^4}{4}$

31. $\dfrac{\sqrt{75x}}{2\sqrt{3}}=\dfrac{1}{2}\sqrt{\dfrac{75x}{3}}=\dfrac{1}{2}\sqrt{25x}=\dfrac{5\sqrt{x}}{2}$

33. $5\sqrt[3]{4y}+2\sqrt[3]{4y}=(5+2)\sqrt[3]{4y}=7\sqrt[3]{4y}$

35. $\sqrt[3]{8x^4}+\sqrt[3]{xy^6}=\sqrt[3]{8x^3\cdot x}+\sqrt[3]{y^6\cdot x}=2x\sqrt[3]{x}+y^2\sqrt[3]{x}$
$=(2x+y^2)\sqrt[3]{x}$

37. $(3+\sqrt{10})(3-\sqrt{10})=3^2-(\sqrt{10})^2=9-10=-1$

39. $\sqrt[4]{x}\sqrt{x}=x^{1/4}\cdot x^{1/2}=x^{1/4+2/4}=x^{3/4}=\sqrt[4]{x^3}$

41. $f(2-\sqrt{a})=(2-\sqrt{a})^2=2^2-2\cdot 2\sqrt{a}+(\sqrt{a})^2$
$=4-4\sqrt{a}+a$

43. $\dfrac{4\sqrt{5}}{\sqrt{2}+\sqrt{3}}=\dfrac{4\sqrt{5}}{\sqrt{2}+\sqrt{3}}\cdot\dfrac{\sqrt{5}}{\sqrt{5}}=\dfrac{4\sqrt{25}}{\sqrt{10}+\sqrt{15}}=\dfrac{20}{\sqrt{10}+\sqrt{15}}$

45.
$$(x+1)^{1/3}=-5$$
$$\sqrt[3]{x+1}=-5$$
$$(\sqrt[3]{x+1})^3=(-5)^3$$
$$x+1=-125$$
$$x=-126$$

Check: $(x+1)^{1/3}=-5$
$(-126+1)^{1/3} \mid -5$
(-125)
$-5\overset{?}{=}-5$ TRUE

The solution is –126.

47. $f(a)=\sqrt{a+2}+a=4$
$$\sqrt{a+2}+a=4$$
$$\sqrt{a+2}=4-a$$
$$(\sqrt{a+2})^2=(4-a)^2$$
$$a+2=16-8a+a^2$$
$$0=a^2-9a+14$$
$$0=(a-2)(a-7)$$
$$a=2 \quad or \quad a=7$$

Check:

For $a=2$: $\sqrt{a+2}+a=4$
$\sqrt{2+2}+2 \mid 4$
$2+2$
$4\overset{?}{=}4$ TRUE

For $a=7$: $\sqrt{a+2}+a=4$
$\sqrt{7+2}+7 \mid 4$
$3+7$
$10\overset{?}{=}4$ FALSE

The solution is 2.

49. Let b represent the base. We use the Pythagorean theorem to find b.
$$b^2+2^2=6^2$$
$$b^2+4=36$$
$$b^2=32$$
$$b=\sqrt{32}\approx 5.657$$
The base is $\sqrt{32}$ ft or about 5.657 ft.

51. Using the distance formula $d=\sqrt{(x_2-x_1)^2+(y_2-y_1)^2}$ for the points (–6, 4) and (–1, 5),
$$d=\sqrt{(-1-(-6))^2+(5-4)^2}$$
$$=\sqrt{5^2+1^2}=\sqrt{26+1}=\sqrt{26}$$
$$\approx 5.099$$

53. $\sqrt{-45}=\sqrt{-1}\cdot\sqrt{9}\cdot\sqrt{5}=3i\sqrt{5}$ or $3\sqrt{5}\,i$

55. $(9-7i)-(3-8i)=(9-3)+(-7+8)i=6+i$

57. $i^{34}=(i^2)^{17}=(-1)^{17}=-1$

59. $\dfrac{7-2i}{3+4i}=\dfrac{7-2i}{3+4i}\cdot\dfrac{3-4i}{3-4i}=\dfrac{21-28i-6i+8i^2}{9-16i^2}$
$=\dfrac{21-34i-8}{9+16}=\dfrac{13-34i}{25}=\dfrac{13}{25}-\dfrac{34}{25}i$

61. *Writing Exercise.* An absolute-value sign must be used to simplify $\sqrt[n]{x^n}$ when n is even, since x may be negative. If x is negative while n is even, the radical expression cannot be simplified to x, since $\sqrt[n]{x^n}$ represents the principal, or nonnegative, root. When n is odd, there is only one root, and it will be positive or negative depending on the sign of x. Thus, there is no absolute-value sign when n is odd.

63.
$$\sqrt{11x+\sqrt{6+x}}=6$$
$$\left(\sqrt{11x+\sqrt{6+x}}\right)^2=(6)^2$$
$$11x+\sqrt{6+x}=36$$
$$\sqrt{6+x}=36-11x$$
$$(\sqrt{6+x})^2=(36-11x)^2$$
$$6+x=1296-792x+121x^2$$
$$0=121x^2-793x+1290$$
$$0=(x-3)(121x-430)$$
$$x=3 \quad or \quad x=\frac{430}{121}$$

Check:

For $x = 3$: $\sqrt{11x + \sqrt{6+x}} = 6$

$\sqrt{11 \cdot 3 + \sqrt{6+3}}$ | 6

$\sqrt{33+3}$

$6 \stackrel{?}{=} 6$ TRUE

For $x = \frac{430}{121}$: $\sqrt{11x + \sqrt{6+x}} = 6$

$\sqrt{11 \cdot \frac{430}{121} + \sqrt{6 + \frac{430}{121}}}$ | 6

$\sqrt{\frac{430}{11} + \frac{34}{11}}$

$\sqrt{\frac{464}{11}} \stackrel{?}{=} 6$ FALSE

The solution is 3.

65. The isosceles right triangle has hypotenuse 6. Let x represent the leg of the triangle. Using the Pythagorean theorem,

$$x^2 + x^2 = 6^2$$
$$2x^2 = 36$$
$$x^2 = 18$$
$$x = \sqrt{18} \text{ ft}$$

The area of the isosceles right triangle is

$$A = \frac{1}{2}x^2 = \frac{1}{2}(\sqrt{18})^2 = 9 \text{ ft}^2$$

Then in the 30°-60°-90° triangle, a is the shorter leg and we have

$$6 = 2a$$
$$3 = a$$
$$b = a\sqrt{3}$$
$$b = 3\sqrt{3}$$

The area of the 30-60-90 triangle is

$$A = \frac{1}{2}(3\sqrt{3})(3) = \frac{9\sqrt{3}}{2} \text{ ft}^2 \approx 7.794 \text{ ft}^2$$

The area of the isosceles right triangle is larger by about 1.206 ft^2.

Chapter 10 Test

1. $\sqrt{50} = \sqrt{25 \cdot 2} = \sqrt{25} \cdot \sqrt{2} = 5\sqrt{2}$

3. $\sqrt{81a^2} = \sqrt{(9a)^2} = |9a| = 9|a|$

5. $\sqrt{7xy} = (7xy)^{1/2}$

7. $f(x) = \sqrt{2x - 10}$

Since the index is even, the radicand, $2x - 10$, must be non-negative. We solve the inequality:

$$2x - 10 \geq 0$$
$$x \geq 5$$

Domain of $f = \{x | x \geq 5\}$, or $[5, \infty)$

9. $\sqrt[5]{32x^{16}y^{10}} = \sqrt[5]{2^5x^{15}y^{10} \cdot x} = \sqrt[5]{2^5x^{15}y^{10}} \cdot \sqrt[5]{x} = 2x^3y^2\sqrt[5]{x}$

11. $\sqrt{\frac{100a^4}{9b^6}} = \frac{\sqrt{100a^4}}{\sqrt{9b^6}} = \frac{10a^2}{3b^3}$

13. $\sqrt[4]{x^3}\sqrt{x} = x^{3/4}x^{1/2} = x^{3/4+1/2}$

$= x^{5/4} = x^{1+1/4} = x\sqrt[4]{x}$

15. $8\sqrt{2} - 2\sqrt{2} = (8-2)\sqrt{2} = 6\sqrt{2}$

17. $(7 + \sqrt{x})(2 - 3\sqrt{x}) = 14 - 7 \cdot 3\sqrt{x} + 2\sqrt{x} - \sqrt{x} \cdot 3\sqrt{x}$

$= 14 - 21\sqrt{x} + 2\sqrt{x} - 3(\sqrt{x})^2$

$= 14 - 19\sqrt{x} - 3x$

19.
$$6 = \sqrt{x-3} + 5$$
$$1 = \sqrt{x-3}$$
$$1^2 = (\sqrt{x-3})^2$$
$$1 = x - 3$$
$$4 = x$$

Check: $6 = \sqrt{x-3} + 5$

6 | $\sqrt{4-3} + 5$

$\sqrt{1} + 5$

$6 \stackrel{?}{=} 6$ TRUE

The solution is 4.

21.
$$\sqrt{2x} = \sqrt{x+1} + 1$$
$$\sqrt{2x} - 1 = \sqrt{x+1}$$
$$(\sqrt{2x} - 1)^2 = (\sqrt{x+1})^2$$
$$2x - 2\sqrt{2x} + 1 = x + 1$$
$$-2\sqrt{2x} = -x$$
$$(-2\sqrt{2x})^2 = (-x)^2$$
$$4(2x) = x^2$$
$$0 = x^2 - 8x$$
$$0 = x(x-8)$$

$x = 0$ *or* $x = 8$

Check:

For $x = 0$: $\sqrt{2x} = \sqrt{x+1} + 1$

$\sqrt{2 \cdot 0}$ | $\sqrt{0+1} + 1$

$1 + 1$

$0 \stackrel{?}{=} 2$ FALSE

For $x = 8$: $\sqrt{2x} = \sqrt{x+1} + 1$

$\sqrt{2 \cdot 8}$ | $\sqrt{8+1} + 1$

$\sqrt{16}$ | $3 + 1$

$4 \stackrel{?}{=} 4$ TRUE

The solution is 8.

23. This is a 30°-60°-90° right triangle with hypotenuse 10.
Let $a =$ the shorter leg and $b =$ the longer leg.
$2a = 10$, so $a = 5$ cm, and
$b = a\sqrt{3} = 5\sqrt{3} \approx 8.660$ cm.

25. Using the midpoint formula $\left(\frac{x_1 + x_2}{2}, \frac{y_1 + y_2}{2}\right)$ for the points (2, –5) and (1, –7),

$$\left(\frac{2+1}{2}, \frac{-5+(-7)}{2}\right), \text{ or } \left(\frac{3}{2}, \frac{-12}{2}\right), \text{ or } \left(\frac{3}{2}, -6\right)$$

27. $(9+8i)-(-3+6i) = (9+3)+(8-6)i = 12+2i$

29.
$$\begin{aligned} \frac{-2+i}{3-5i} &= \frac{-2+i}{3-5i}\cdot\frac{3+5i}{3+5i} = \frac{-6-10i+3i+5i^2}{9-25i^2} \\ &= \frac{-6-7i-5}{9+25} = \frac{-11-7i}{34} \\ &= -\frac{11}{34} - \frac{7}{34}i \end{aligned}$$

31.
$$\begin{aligned} \sqrt{2x-2}+\sqrt{7x+4} &= \sqrt{13x+10} \\ \left(\sqrt{2x-2}+\sqrt{7x+4}\right)^2 &= \left(\sqrt{13x+10}\right)^2 \\ 2x-2+2\sqrt{2x-2}\sqrt{7x+4}+7x+4 &= 13x+10 \\ 2\sqrt{2x-2}\sqrt{7x+4} &= 4x+8 \\ \left(\sqrt{2x-2}\sqrt{7x+4}\right)^2 &= (2x+4)^2 \\ (2x-2)(7x+4) &= 4x^2+16x+16 \\ 14x^2-6x-8 &= 4x^2+16x+16 \\ 10x^2-22x-24 &= 0 \\ 2\left(5x^2-11x-12\right) &= 0 \\ 2(5x+4)(x-3) &= 0 \\ x = -\frac{4}{5} \quad &or \quad x = 3 \end{aligned}$$

Check:

For $x = -\frac{4}{5}$:

$$\begin{array}{c|c} \multicolumn{2}{c}{\sqrt{2x-2}+\sqrt{7x+4} = \sqrt{13x+10}} \\ \hline \sqrt{2\cdot\left(-\frac{4}{5}\right)+2}+\sqrt{7\left(-\frac{4}{5}\right)+4} & \sqrt{13\left(-\frac{4}{5}\right)+10} \\ \sqrt{\frac{2}{5}}+\sqrt{-\frac{8}{5}} & \sqrt{-\frac{2}{5}} \end{array}$$

Since the values in the check are not real, $-\frac{4}{5}$ is not a solution.

For $x = 3$:

$$\begin{array}{c|c} \multicolumn{2}{c}{\sqrt{2x-2}+\sqrt{7x+4} = \sqrt{13x+10}} \\ \hline \sqrt{2\cdot 3-2}+\sqrt{7\cdot 3+4} & \sqrt{13\cdot 3+10} \\ \sqrt{4}+\sqrt{25} & \sqrt{49} \\ 2+5 & \\ \multicolumn{2}{c}{7 \stackrel{?}{=} 7 \quad \text{TRUE}} \end{array}$$

The solution is 3.

33. Substitute 180 for $D(h)$, and solve for h.

$$\begin{aligned} D(h) &= 1.2\sqrt{h} \\ 180 &= 1.2\sqrt{h} \\ 150 &= \sqrt{h} \\ (150)^2 &= \left(\sqrt{h}\right)^2 \\ 22,500 &= h \end{aligned}$$

The pilot must be above 22,500 ft.

Chapter 11

Quadratic Functions and Equations

Exercise Set 11.1

1. $\sqrt{k};\ -\sqrt{k}$

3. $t+3;\ t+3$

5. $25;\ 5$

7. $x^2=100$

$x=10$ *or* $x=-10$ Using the principle of square roots

The solutions are -10 and 10, or ± 10.

9. $p^2-50=0$

$p^2=50$ Isolating p^2

$p=\sqrt{50}$ *or* $p=-\sqrt{50}$ Principle of square roots

$p=5\sqrt{2}$ *or* $p=-5\sqrt{2}$

The solutions are $5\sqrt{2}$ and $-5\sqrt{2}$ or $\pm 5\sqrt{2}$.

11. $5y^2=30$

$y^2=6$ Isolating y^2

$y=\sqrt{6}$ *or* $y=-\sqrt{6}$ Principle of square roots

The solutions are $\sqrt{6}$ and $-\sqrt{6}$ or $\pm\sqrt{6}$.

13. $9x^2-49=0$

$x^2=\frac{49}{9}$ Isolating x^2

$x=\sqrt{\frac{49}{9}}$ *or* $x=-\sqrt{\frac{49}{9}}$ Principle of square roots

$x=\frac{7}{3}$ *or* $x=-\frac{7}{3}$

The solutions are $\frac{7}{3}$ and $-\frac{7}{3}$ or $\pm\frac{7}{3}$.

15. $6t^2-5=0$

$t^2=\frac{5}{6}$

$t=\sqrt{\frac{5}{6}}$ *or* $t=-\sqrt{\frac{5}{6}}$ Principle of square roots

$t=\sqrt{\frac{5}{6}\cdot\frac{6}{6}}$ *or* $t=-\sqrt{\frac{5}{6}\cdot\frac{6}{6}}$ Rationalizing denominators

$t=\frac{\sqrt{30}}{6}$ *or* $t=-\frac{\sqrt{30}}{6}$

The solutions are $\sqrt{\frac{5}{6}}$ and $-\sqrt{\frac{5}{6}}$. This can also be written as $\pm\sqrt{\frac{5}{6}}$ or, if we rationalize the denominator, $\pm\frac{\sqrt{30}}{6}$.

17. $a^2+1=0$

$a^2=-1$

$a=\sqrt{-1}$ *or* $a=-\sqrt{-1}$

$a=i$ *or* $a=-i$

The solutions are i and $-i$ or $\pm i$.

19. $4d^2+81=0$

$d^2=-\frac{81}{4}$

$d=\sqrt{-\frac{81}{4}}$ *or* $d=-\sqrt{-\frac{81}{4}}$

$d=\frac{9}{2}i$ *or* $d=-\frac{9}{2}i$

The solutions are $\frac{9}{2}i$ and $-\frac{9}{2}i$ or $\pm\frac{9}{2}i$.

21. $(x-3)^2=16$

$x-3=\sqrt{16}$ *or* $x-3=-\sqrt{16}$

$x-3=4$ *or* $x-3=-4$

$x=7$ *or* $x=-1$

The solutions are -1 and 7.

23. $(t+5)^2=12$

$t+5=\sqrt{12}$ *or* $t+5=-\sqrt{12}$

$t+5=2\sqrt{3}$ *or* $t+5=-2\sqrt{3}$

$t=-5+2\sqrt{3}$ *or* $t=-5-2\sqrt{3}$

The solutions are $-5+2\sqrt{3}$ and $-5-2\sqrt{3}$, or $-5\pm 2\sqrt{3}$.

25. $(x+1)^2=-9$

$x+1=\sqrt{-9}$ *or* $x+1=-\sqrt{-9}$

$x+1=3i$ *or* $x+1=-3i$

$x=-1+3i$ *or* $x=-1-3i$

The solutions are $-1+3i$ and $-1-3i$, or $-1\pm 3i$.

27. $\left(y+\frac{3}{4}\right)^2=\frac{17}{16}$

$y+\frac{3}{4}=\pm\frac{\sqrt{17}}{4}$

$y=-\frac{3}{4}\pm\frac{\sqrt{17}}{4}$, or $\frac{-3\pm\sqrt{17}}{4}$

The solutions are $-\frac{3}{4}\pm\frac{\sqrt{17}}{4}$, or $\frac{-3\pm\sqrt{17}}{4}$.

29. $x^2-10x+25=64$
$$(x-5)^2=64$$
$$x-5=\pm 8$$
$$x=5\pm 8$$
$$x=13 \text{ or } x=-3$$
The solutions are 13 and –3.

31. $f(x)=x^2$
$19=x^2$ Substituting
$\sqrt{19}=x$ *or* $-\sqrt{19}=x$
The solutions are $\sqrt{19}$ and $-\sqrt{19}$ or $\pm\sqrt{19}$.

33. $f(x)=16$
$(x-5)^2=16$ Substituting
$x-5=4$ *or* $x-5=-4$
$x=9$ *or* $x=1$
The solutions are 9 and 1.

35. $F(t)=13$
$(t+4)^2=13$ Substituting
$t+4=\sqrt{13}$ *or* $t+4=-\sqrt{13}$
$t=-4+\sqrt{13}$ *or* $t=-4-\sqrt{13}$
The solutions are $-4+\sqrt{13}$ and $-4-\sqrt{13}$, or $-4\pm\sqrt{13}$.

37. $g(x)=x^2+14x+49$
Observe first that $g(0)=49$. Also observe that when $x=-14$, then
$x^2+14x=(-14)^2-(14)(14)=(14)^2-(14)^2=0$, so
$g(-14)=49$ as well. Thus, we have $x=0$ or $x=14$.
We can also do this problem as follows.
$g(x)=49$
$x^2+14x+49=49$ Substituting
$(x+7)^2=49$
$x+7=7$ *or* $x+7=-7$
$x=0$ *or* $x=-14$
The solutions are 0 and -14.

39. x^2+16x
We take half the coefficient of x and square it: Half of 16 is 8, and $8^2=64$. We add 64.
$x^2+16x+64=(x+8)^2$

41. t^2-10t
We take half the coefficient of t and square it:
Half of –10 is –5, and $(-5)^2=25$. We add 25.
$t^2-10t+25=(t-5)^2$

43. t^2-2t
We take half the coefficient of t and square it:
$\frac{1}{2}(-2)=-1$, and $(-1)^2=1$. We add 1.
$t^2-2t+1=(t-1)^2$

45. x^2+3x
We take half the coefficient of t and square it:
$\frac{1}{2}(3)=\frac{3}{2}$, and $\left(\frac{3}{2}\right)^2=\frac{9}{4}$. We add $\frac{9}{4}$.
$x^2+3x+\frac{9}{4}=\left(x+\frac{3}{2}\right)^2$

47. $x^2+\frac{2}{5}x$
$\frac{1}{2}\cdot\frac{2}{5}=\frac{1}{5}$, and $\left(\frac{1}{5}\right)^2=\frac{1}{25}$. We add $\frac{1}{25}$.
$x^2+\frac{2}{5}x+\frac{1}{25}=\left(x+\frac{1}{5}\right)^2$

49. $t^2-\frac{5}{6}t$
$\frac{1}{2}\left(-\frac{5}{6}\right)=-\frac{5}{12}$, and $\left(-\frac{5}{12}\right)^2=\frac{25}{144}$. We add $\frac{25}{144}$.
$t^2-\frac{5}{6}t+\frac{25}{144}=\left(t-\frac{5}{12}\right)^2$

51. $x^2+6x=7$
$x^2+6x+9=7+9$ Adding 9 to both sides to complete the square
$(x+3)^2=16$ Factoring
$x+3=\pm 4$ Principle of square roots
$x=-3\pm 4$
$x=-3+4$ *or* $x=-3-4$
$x=1$ *or* $x=-7$
The solutions are 1 and –7.

53. $t^2-10t=-23$
$t^2-10t+25=-23+25$ Adding 25 to both sides to complete the square
$(t-5)^2=2$ Factoring
$t-5=\pm\sqrt{2}$ Principle of square roots
$t=5\pm\sqrt{2}$
The solutions are $5\pm\sqrt{2}$.

55. $x^2+12x+32=0$
$x^2+12x=-32$
$x^2+12x+36=-32+36$
$(x+6)^2=4$
$x+6=\pm 2$
$x=-6\pm 2$
$x=-6+2$ *or* $x=-6-2$
$x=-4$ *or* $x=-8$
The solutions are –8 and –4.

57. $t^2+8t-3=0$

$$t^2+8t=3$$
$$t^2+8t+16=3+16$$
$$(t+4)^2=19$$
$$t+4=\pm\sqrt{19}$$
$$t=-4\pm\sqrt{19}$$

The solutions are $-4\pm\sqrt{19}$.

59. The value of $f(x)$ must be 0 at any x-intercepts.

$$f(x)=0$$
$$x^2+6x+7=0$$
$$x^2+6x=-7$$
$$x^2+6x+9=-7+9$$
$$(x+3)^2=2$$
$$x+3=\pm\sqrt{2}$$
$$x=-3\pm\sqrt{2}$$

The x-intercepts are $(-3-\sqrt{2},\ 0)$ and $(-3+\sqrt{2},\ 0)$.

61. The value of $g(x)$ must be 0 at any x-intercepts.

$$g(x)=0$$
$$x^2+9x-25=0$$
$$x^2+9x=25$$
$$x^2+9x+\frac{81}{4}=25+\frac{81}{4}$$
$$\left(x+\frac{9}{2}\right)^2=\frac{181}{4}$$
$$x+\frac{9}{2}=\pm\frac{\sqrt{181}}{2}$$
$$x=-\frac{9}{2}\pm\frac{\sqrt{181}}{2}$$

The x-intercepts are $\left(-\frac{9}{2}-\frac{\sqrt{181}}{2},\ 0\right)$ and $\left(-\frac{9}{2}+\frac{\sqrt{181}}{2},\ 0\right)$.

63. The value of $f(x)$ must be 0 at any x-intercepts.

$$f(x)=0$$
$$x^2-10x-22=0$$
$$x^2-10x=22$$
$$x^2-10x+25=22+25$$
$$(x-5)^2=47$$
$$x-5=\pm\sqrt{47}$$
$$x=5\pm\sqrt{47}$$

The x-intercepts are $(5-\sqrt{47},\ 0)$ and $(5+\sqrt{47},\ 0)$.

65. $9x^2+18x=-8$

$$x^2+2x=-\frac{8}{9} \quad \text{Dividing both sides by 9}$$
$$x^2+2x+1=-\frac{8}{9}+1$$
$$(x+1)^2=\frac{1}{9}$$
$$x+1=\pm\frac{1}{3}$$
$$x=-1\pm\frac{1}{3}$$
$$x=-1-\frac{1}{3} \quad or \quad x=-1+\frac{1}{3}$$
$$x=-\frac{4}{3} \quad or \quad x=-\frac{2}{3}$$

The solutions are $-\frac{4}{3}$ and $-\frac{2}{3}$.

67. $3x^2-5x-2=0$

$$3x^2-5x=2$$
$$x^2-\frac{5}{3}x=\frac{2}{3} \quad \text{Dividing both sides by 3}$$
$$x^2-\frac{5}{3}x+\frac{25}{36}=\frac{2}{3}+\frac{25}{36}$$
$$\left(x-\frac{5}{6}\right)^2=\frac{49}{36}$$
$$x-\frac{5}{6}=\pm\frac{7}{6}$$
$$x=\frac{5}{6}\pm\frac{7}{6}$$
$$x=\frac{5}{6}-\frac{7}{6} \quad or \quad x=\frac{5}{6}+\frac{7}{6}$$
$$x=-\frac{1}{3} \quad or \quad x=2$$

The solutions are $-\frac{1}{3}$ and 2.

69. $5x^2+4x-3=0$

$$5x^2+4x=3$$
$$x^2+\frac{4}{5}x=\frac{3}{5} \quad \text{Dividing both sides by 5}$$
$$x^2+\frac{4}{5}x+\frac{4}{25}=\frac{3}{5}+\frac{4}{25}$$
$$\left(x+\frac{2}{5}\right)^2=\frac{19}{25}$$
$$x+\frac{2}{5}=\pm\frac{\sqrt{19}}{5}$$
$$x=-\frac{2}{5}\pm\frac{\sqrt{19}}{5},\ \text{or}\ \frac{-2\pm\sqrt{19}}{5}$$

The solutions are $-\frac{2}{5}\pm\frac{\sqrt{19}}{5}$, or $\frac{-2\pm\sqrt{19}}{5}$.

71. The value of $f(x)$ must be 0 at any x-intercepts.

$$f(x)=0$$
$$4x^2+2x-3=0$$
$$4x^2+2x=3$$
$$x^2+\frac{1}{2}x=\frac{3}{4} \quad \text{Dividing both sides by 4}$$
$$x^2+\frac{1}{2}x+\frac{1}{16}=\frac{3}{4}+\frac{1}{16}$$
$$\left(x+\frac{1}{4}\right)^2=\frac{13}{16}$$
$$x+\frac{1}{4}=\pm\frac{\sqrt{13}}{4}$$
$$x=-\frac{1}{4}\pm\frac{\sqrt{13}}{4}, \text{ or } \frac{-1\pm\sqrt{13}}{4}$$

The x-intercepts are $\left(-\frac{1}{4}-\frac{\sqrt{13}}{4}, 0\right)$ and $\left(-\frac{1}{4}+\frac{\sqrt{13}}{4}, 0\right)$, or $\left(\frac{-1-\sqrt{13}}{4}, 0\right)$ and $\left(\frac{-1+\sqrt{13}}{4}, 0\right)$.

73. The value of $g(x)$ must be 0 at any x-intercepts.

$$g(x)=0$$
$$2x^2-3x-1=0$$
$$2x^2-3x=1$$
$$x^2-\frac{3}{2}x=\frac{1}{2} \quad \text{Dividing both sides by 2}$$
$$x^2-\frac{3}{2}x+\frac{9}{16}=\frac{1}{2}+\frac{9}{16}$$
$$\left(x-\frac{3}{4}\right)^2=\frac{17}{16}$$
$$x-\frac{3}{4}=\pm\frac{\sqrt{17}}{4}$$
$$x=\frac{3}{4}\pm\frac{\sqrt{17}}{4}, \text{ or } \frac{3\pm\sqrt{17}}{4}$$

The x-intercepts are $\left(\frac{3}{4}-\frac{\sqrt{17}}{4}, 0\right)$ and $\left(\frac{3}{4}+\frac{\sqrt{17}}{4}, 0\right)$, or $\left(\frac{3-\sqrt{17}}{4}, 0\right)$ and $\left(\frac{3+\sqrt{17}}{4}, 0\right)$.

75. ***Familiarize.*** We are already familiar with the compound-interest formula.

Translate. We substitute into the formula.

$$A=P(1+r)^t$$
$$2420=2000(1+r)^2$$

Carry out. We solve for r.

$$2420=2000(1+r)^2$$
$$\frac{2420}{2000}=(1+r)^2$$
$$\frac{121}{100}=(1+r)^2$$
$$\pm\sqrt{\frac{121}{100}}=1+r$$
$$\pm\frac{11}{10}=1+r$$
$$-\frac{10}{10}\pm\frac{11}{10}=r$$
$$\frac{1}{10}=r \text{ or } -\frac{21}{10}=r$$

Check. Since the interest rate cannot be negative, we need only check $\frac{1}{10}$, or 10%. If \$2000 were invested at 10% interest, compounded annually, then in 2 years it would grow to $\$2000(1.1)^2$, or \$2420. The number 10% checks.

State. The interest rate is 10%.

77. ***Familiarize.*** We are already familiar with the compound-interest formula.

Translate. We substitute into the formula.

$$A=P(1+r)^t$$
$$6760=6250(1+r)^2$$

Carry out. We solve for r.

$$\frac{6760}{6250}=(1+r)^2$$
$$\frac{676}{625}=(1+r)^2$$
$$\pm\frac{26}{25}=1+r$$
$$-\frac{25}{25}\pm\frac{26}{25}=r$$
$$\frac{1}{25}=r \text{ or } -\frac{51}{25}=r$$

Check. Since the interest rate cannot be negative, we need only check $\frac{1}{25}$, or 4%. If \$6250 were invested at 4% interest, compounded annually, then in 2 years it would grow to $\$6250(1.04)^2$, or \$6760. The number 4% checks.

State. The interest rate is 4%.

79. ***Familiarize.*** We will use the formula $s=16t^2$.

Translate. We substitute into the formula.

$$s=16t^2$$
$$290=16t^2$$

Carry out. We solve for t.

$$290=16t^2$$
$$\frac{290}{16}=t^2$$
$$\sqrt{\frac{290}{16}}=t \quad \text{Principle of square roots;}$$
$$4.3\approx t \quad \text{rejecting the negative square root}$$

Check. Since $16(4.3)^2=295.84\approx290$, our answer checks.

State. It would take an object about 4.3 sec to fall freely from the bridge.

81. ***Familiarize.*** We will use the formula $s=16t^2$.

Translate. We substitute into the formula.

$$s=16t^2$$
$$2063=16t^2$$

Carry out. We solve for t.

$$2063 = 16t^2$$
$$\frac{2063}{16} = t^2$$
$$\sqrt{\frac{2063}{16}} = t \quad \text{Principle of square roots; rejecting the negative square root}$$
$$11.4 \approx t$$

Check. Since $16(11.4)^2 = 2079.63 \approx 2063$, our answer checks.

State. It would take an object about 11.4 sec to fall freely from the top.

83. *Writing Exercise.*

1) If the quadratic equation is of the type $x^2 = k$, use the principle of square roots.

2) If the quadratic equation is of the type $ax^2 + bx + c = 0$, $b \neq 0$, use the principle of zero products, if possible.

3) If the quadratic equation is of the type $ax^2 + bx + c = 0$, $b \neq 0$, and factoring is difficult or impossible, solve by completing the square.

85. $b^2 - 4ac = 2^2 - 4 \cdot 3 \cdot (-5)$
$= 4 + 60$
$= 64$

87. $\sqrt{200} = \sqrt{100} \cdot \sqrt{2} = 10\sqrt{2}$

89. $\sqrt{-4} = \sqrt{4} \cdot \sqrt{-1} = 2i$

91. $\sqrt{-8} = \sqrt{4} \cdot \sqrt{-1} \cdot \sqrt{2} = 2i\sqrt{2}$, or $2\sqrt{2}i$

93. *Writing Exercise.* It would be better to receive 3% interest every 6 months, because the interest compounds faster in this situation.

95. In order for $x^2 + bx + 81$ to be a square, the following must be true:

$$\left(\frac{b}{2}\right)^2 = 81$$
$$\frac{b^2}{4} = 81$$
$$b^2 = 324$$
$$b = 18 \text{ or } b = -18$$

97. We see that x is a factor of each term, so x is also a factor of $f(x)$. We have $f(x) = x(2x^4 - 9x^3 - 66x^2 + 45x + 280)$. Since $x^2 - 5$ is a factor of $f(x)$ it is also a factor of $2x^4 - 9x^3 - 66x^2 + 45x + 280$. We divide to find another factor.

$$\begin{array}{r} 2x^2 - 9x - 56 \\ x^2 - 5 \overline{)\, 2x^4 - 9x^3 - 66x^2 + 45x + 280} \\ \underline{2x^4 \qquad\quad - 10x^2 \qquad\qquad\quad} \\ -9x^3 - 56x^2 + 45x \qquad \\ \underline{-9x^3 \qquad\quad + 45x \qquad} \\ -56x^2 \qquad + 280 \\ \underline{-56x^2 \qquad + 280} \\ 0 \end{array}$$

Then we have $f(x) = x(x^2 - 5)(2x^2 - 9x - 56)$, or $f(x) = x(x^2 - 5)(2x + 7)(x - 8)$. Now we find the values of a for which $f(a) = 0$.

$$f(a) = 0$$
$$a(a^2 - 5)(2a + 7)(a - 8) = 0$$

$a = 0$ *or* $a^2 - 5 = 0$ *or* $2a + 7 = 0$ *or* $a - 8 = 0$
$a = 0$ *or* $a^2 = 5$ *or* $2a = -7$ *or* $a = 8$
$a = 0$ *or* $a = \pm\sqrt{5}$ *or* $a = -\frac{7}{2}$ *or* $a = 8$

The solutions are 0, $\sqrt{5}$, $-\sqrt{5}$, $-\frac{7}{2}$, and 8.

99. ***Familiarize.*** It is helpful to list information in a chart and make a drawing. Let r represent the speed of the fishing boat. Then $r - 7$ represents the speed of the barge.

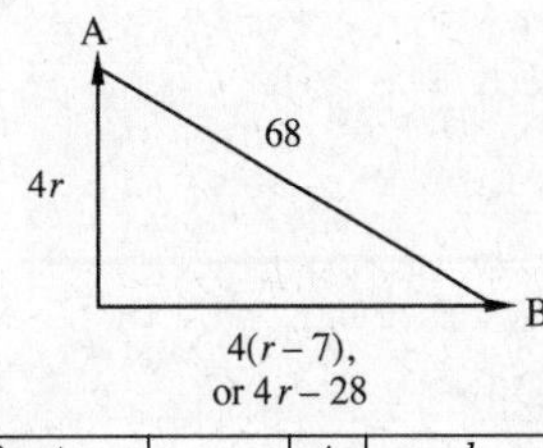

Boat	r	t	d
Fishing	r	4	$4r$
Barge	$r-7$	4	$4(r-7)$

Translate. We use the Pythagorean equation:

$$a^2 + b^2 = c^2$$
$$(4r - 28)^2 + (4r)^2 = 68^2$$

Carry out.

$$(4r - 28)^2 + (4r)^2 = 68^2$$
$$16r^2 - 224r + 784 + 16r^2 = 4624$$
$$32r^2 - 224r - 3840 = 0$$
$$r^2 - 7r - 120 = 0$$
$$(r + 8)(r - 15) = 0$$

$r + 8 = 0$ *or* $r - 15 = 0$
$r = -8$ *or* $r = 15$

Check. We check only 15 since the speeds of the boats cannot be negative. If the speed of the fishing boat is 15 km/h, then the speed of the barge is $15 - 7$, or 8 km/h, and the distances they travel are $4 \cdot 15$ (or 60) and $4 \cdot 8$ (or 32).

$60^2+32^2=3600+1024=4624=68^2$ The values check.

State. The speed of the fishing boat is 15 km/h, and the speed of the barge is 8 km/h.

101. *Graphing Calculator Exercise*

103. *Writing Exercise.* From a reading of the problem we know that we are interested only in positive values of r and it is safe to assume $r \le 1$. We also know that we want to find the value of r for which $4410=4000(1+r)^2$, so the window must include the y-value 4410. A suitable viewing window might be [0, 1, 4000, 4500], Xscl = 0.1, Yscl = 100.

Exercise Set 11.2

1. True; see page 716 in the text.

3. False; see Example 3.

5. False; the quadratic formula yields at most two solutions.

7.
$$2x^2+3x-5=0$$
$$(2x+5)(x-1)=0 \quad \text{Factoring}$$
$$2x+5=0 \quad or \quad x-1=0$$
$$x=-\frac{5}{2} \quad or \quad x=1$$
The solutions are $-\frac{5}{2}$ and 1.

9.
$$u^2+2u-4=0$$
$$u^2+2u=4$$
$$u^2+2u+1=4+1 \quad \text{Completing the square}$$
$$(u+1)^2=5$$
$$u+1=\pm\sqrt{5} \quad \text{Principle of square roots}$$
$$u=-1\pm\sqrt{5}$$
The solutions are $-1+\sqrt{5}$ and $-1-\sqrt{5}$.

11.
$$t^2+3=6t$$
$$t^2-6t=-3$$
$$t^2-6t+9=-3+9$$
$$(t-3)^2=6$$
$$t-3=\pm\sqrt{6}$$
$$t=3\pm\sqrt{6}$$
The solutions are $3+\sqrt{6}$ and $3-\sqrt{6}$.

13.
$$x^2=3x+5$$
$$x^2-3x-5=0$$
$$a=1,\ b=-3,\ c=-5$$
$$x=\frac{-b\pm\sqrt{b^2-4ac}}{2a}$$
$$x=\frac{-(-3)\pm\sqrt{(-3)^2-4\cdot 1\cdot(-5)}}{2\cdot 1}=\frac{3\pm\sqrt{9+20}}{2}$$
$$x=\frac{3\pm\sqrt{29}}{2}=\frac{3}{2}\pm\frac{\sqrt{29}}{2}$$
The solutions are $\frac{3}{2}+\frac{\sqrt{29}}{2}$ and $\frac{3}{2}-\frac{\sqrt{29}}{2}$.

15.
$$3t(t+2)=1$$
$$3t^2+6t=1$$
$$3t^2+6t-1=0$$
$$a=3,\ b=6,\ c=-1$$
$$t=\frac{-b\pm\sqrt{b^2-4ac}}{2a}$$
$$t=\frac{-6\pm\sqrt{6^2-4\cdot 3\cdot(-1)}}{2\cdot 3}=\frac{-6\pm\sqrt{36+12}}{6}$$
$$t=\frac{-6\pm\sqrt{48}}{6}=\frac{-6\pm 4\sqrt{3}}{6}$$
$$t=-\frac{6}{6}\pm\frac{4\sqrt{3}}{6}=-1\pm\frac{2\sqrt{3}}{3}$$
The solutions are $-1+\frac{2\sqrt{3}}{3}$ and $-1-\frac{2\sqrt{3}}{3}$.

17.
$$\frac{1}{x^2}-3=\frac{8}{x}, \quad \text{LCD is } x^2$$
$$x^2\left(\frac{1}{x^2}-3\right)=x^2\cdot\frac{8}{x}$$
$$x^2\cdot\frac{1}{x^2}-x^2\cdot 3=8x$$
$$1-3x^2=8x$$
$$0=3x^2+8x-1$$
a = 3, b = 8, c = -1
$$x=\frac{-8\pm\sqrt{8^2-4\cdot 3\cdot(-1)}}{2\cdot 3}=\frac{-8\pm\sqrt{64+12}}{6}$$
$$x=\frac{-8\pm\sqrt{76}}{6}=\frac{-8\pm\sqrt{4\cdot 19}}{6}=\frac{-8\pm 2\sqrt{19}}{6}$$
$$x=\frac{-4\pm\sqrt{19}}{3}=-\frac{4}{3}\pm\frac{\sqrt{19}}{3}$$
The solutions are $-\frac{4}{3}-\frac{\sqrt{19}}{3}$ and $-\frac{4}{3}+\frac{\sqrt{19}}{3}$.

19. $t^2+10=6t$

$t^2-6t+10=0$

$a=1,\ b=-6,\ c=10$

$t=\dfrac{-b\pm\sqrt{b^2-4ac}}{2a}$

$t=\dfrac{-(-6)\pm\sqrt{(-6)^2-4\cdot1\cdot10}}{2\cdot1}=\dfrac{6\pm\sqrt{36-40}}{6}$

$t=\dfrac{6\pm\sqrt{-4}}{2}=\dfrac{6\pm2i}{2}$

$t=\dfrac{6}{2}\pm\dfrac{2i}{2}=3\pm i$

The solutions are $3+i$ and $3-i$.

21. $p^2-p+1=0$

$a=1,\ b=-1,\ c=1$

$p=\dfrac{-b\pm\sqrt{b^2-4ac}}{2a}$

$p=\dfrac{-(-1)\pm\sqrt{(-1)^2-4\cdot1\cdot1}}{2\cdot1}=\dfrac{1\pm\sqrt{1-4}}{2}$

$p=\dfrac{1\pm\sqrt{-3}}{2}=\dfrac{1}{2}\pm\dfrac{\sqrt{3}}{2}i$

The solutions are $\dfrac{1}{2}+\dfrac{\sqrt{3}}{2}i$ and $\dfrac{1}{2}-\dfrac{\sqrt{3}}{2}i$.

23. $x^2+4x+6=0$

$x=\dfrac{-b\pm\sqrt{b^2-4ac}}{2a}$

$x=\dfrac{-4\pm\sqrt{4^2-4\cdot1\cdot6}}{2\cdot1}=\dfrac{-4\pm\sqrt{16-24}}{2}$

$x=\dfrac{-4\pm\sqrt{-8}}{2}=-\dfrac{4}{2}\pm\dfrac{2\sqrt{2}}{2}i$

$x=-2\pm\sqrt{2}i$

The solutions are $-2+\sqrt{2}i$ and $-2-\sqrt{2}i$.

25. $12t^2+17t=40$

$12t^2+17t-40=0$

$(3t+8)(4t-5)=0$

$3t+8=0$ *or* $4t-5=0$

$t=-\dfrac{8}{3}$ *or* $t=\dfrac{5}{4}$

The solutions are $-\dfrac{8}{3}$ and $\dfrac{5}{4}$.

27. $25x^2-20x+4=0$

$(5x-2)(5x-2)=0$

$5x-2=0$ *or* $5x-2=0$

$5x=2$ *or* $5x=2$

$x=\dfrac{2}{5}$ *or* $x=\dfrac{2}{5}$

The solution is $\dfrac{2}{5}$.

29. $7x(x+2)+5=3x(x+1)$

$7x^2+14x+5=3x^2+3x$

$4x^2+11x+5=0$

$a=4,\ b=11,\ c=5$

$x=\dfrac{-11\pm\sqrt{11^2-4\cdot4\cdot5}}{2\cdot4}=\dfrac{-11\pm\sqrt{121-80}}{8}$

$x=\dfrac{-11\pm\sqrt{41}}{8}=-\dfrac{11}{8}\pm\dfrac{\sqrt{41}}{8}$

The solutions are $-\dfrac{11}{8}-\dfrac{\sqrt{41}}{8}$ and $-\dfrac{11}{8}+\dfrac{\sqrt{41}}{8}$.

31. $14(x-4)-(x+2)=(x+2)(x-4)$

$14x-56-x-2=x^2-2x-8$ Removing parentheses

$13x-58=x^2-2x-8$

$0=x^2-15x+50$

$0=(x-10)(x-5)$

$x-10=0$ *or* $x-5=0$

$x=10$ *or* $x=5$

The solutions are 10 and 5.

33. $51p=2p^2+72$

$0=2p^2-51p+72$

$0=(2p-3)(p-24)$

$2p-3=0$ *or* $p-24=0$

$p=\dfrac{3}{2}$ *or* $p=24$

The solutions are $\dfrac{3}{2}$ and 24.

35. $x(x-3)=x-9$

$x^2-3x=x-9$ Removing parentheses

$x^2-4x=-9$

$x^2-4x+4=-9+4$ Completing the square

$(x-2)^2=-5$

$x-2=\pm\sqrt{-5}$

$x=2\pm\sqrt{5}i$

The solutions are $2+\sqrt{5}i$ and $2-\sqrt{5}i$.

37. $x^3-8=0$

$x^3-2^3=0$

$(x-2)(x^2+2x+4)=0$

$x-2=0$ *or* $x^2+2x+4=0$

$x=2$ *or* $x=\dfrac{-2\pm\sqrt{2^2-4\cdot1\cdot4}}{2\cdot1}$

$x=2$ *or* $x=\dfrac{-2\pm\sqrt{-12}}{2}=\dfrac{-2\pm2i\sqrt{3}}{2}$

$x=2$ *or* $x=-\dfrac{2}{2}\pm\dfrac{2\sqrt{3}}{2}i$

$x=2$ *or* $x=-1\pm\sqrt{3}i$

The solutions are 2, $-1+\sqrt{3}i$, and $-1-\sqrt{3}i$.

39. $f(x)=0$

$$6x^2-7x-20=0$$
$$(3x+4)(2x-5)=0$$
$$3x+4=0 \quad or \quad 2x-5=0$$
$$x=-\frac{4}{3} \quad or \quad x=\frac{5}{2}$$

$f(x)=0$ for $x=-\frac{4}{3}$ and $x=\frac{5}{2}$.

41. $f(x)=1$ Substituting

$$\frac{7}{x}+\frac{7}{x+4}=1$$
$$x(x+4)\left(\frac{7}{x}+\frac{7}{x+4}\right)=x(x+4)\cdot 1$$

Multiplying by the LCD

$$7(x+4)+7x=x^2+4x$$
$$7x+28+7x=x^2+4x$$
$$14x+28=x^2+4x$$
$$0=x^2-10x-28$$

$a=1,\ b=-10,\ c=-28$

$$x=\frac{-(-10)\pm\sqrt{(-10)^2-4\cdot 1\cdot(-28)}}{2\cdot 1}$$
$$x=\frac{10\pm\sqrt{100+112}}{2}=\frac{10\pm\sqrt{212}}{2}$$
$$x=\frac{10\pm\sqrt{4\cdot 53}}{2}=\frac{10\pm 2\sqrt{53}}{2}$$
$$x=5\pm\sqrt{53}$$

$f(x)=1$ for $x=5+\sqrt{53}$ and $x=5-\sqrt{53}$.

43. $F(x)=G(x)$

$$\frac{3-x}{4}=\frac{1}{4x}$$
$$4x\cdot\frac{3-x}{4}=4x\cdot\frac{1}{4x}$$
$$3x-x^2=1$$
$$0=x^2-3x+1$$
$$x=\frac{-(-3)\pm\sqrt{(-3)^2-4\cdot 1\cdot 1}}{2\cdot 1}=\frac{3\pm\sqrt{5}}{2}$$
$$x=\frac{3}{2}\pm\frac{\sqrt{5}}{2}$$

45. $x^2+4x-7=0$

$a=1,\ b=4,\ c=-7$

$$x=\frac{-4\pm\sqrt{4^2-4\cdot 1\cdot(-7)}}{2\cdot 1}=\frac{-4\pm\sqrt{16+28}}{2}$$
$$x=\frac{-4\pm\sqrt{44}}{2}$$

Using a calculator we find that $\frac{-4+\sqrt{44}}{2}\approx 1.317$ and $\frac{-4-\sqrt{44}}{2}\approx -5.317$.

The solutions are approximately 1.317 and –5.317.

47. $x^2-6x+4=0$

$a=1,\ b=-6,\ c=4$

$$x=\frac{-(-6)\pm\sqrt{(-6)^2-4\cdot 1\cdot 4}}{2\cdot 1}=\frac{6\pm\sqrt{36-16}}{2}$$
$$x=\frac{6\pm\sqrt{20}}{2}$$

Using a calculator we find that $\frac{6+\sqrt{20}}{2}\approx 5.236$ and $\frac{6-\sqrt{20}}{2}\approx 0.764$.

The solutions are approximately 5.236 and 0.764.

49. $2x^2-3x-7=0$

$a=2,\ b=-3,\ c=-7$

$$x=\frac{-(-3)\pm\sqrt{(-3)^2-4\cdot 2\cdot(-7)}}{2\cdot 2}$$
$$x=\frac{3\pm\sqrt{9+56}}{4}=\frac{3\pm\sqrt{65}}{4}$$

Using a calculator we find that $\frac{3+\sqrt{65}}{4}\approx 2.766$ and $\frac{3-\sqrt{65}}{4}\approx -1.266$.

The solutions are approximately 2.766 and –1.266.

51. *Writing Exercise.* No; the quadratic formula is derived by solving the equation $ax^2+bx+c=0$ by completing the square, so, any equation that can be solved using the quadratic formula can also be solved by completing the square.

53. $(x-2i)(x+2i)=x^2-(2i)^2=x^2-4i^2$
$$=x^2+4$$

55. $\left[x-(2-\sqrt{7})\right]\left[x-(2+\sqrt{7})\right]$

$=\left[(x-2)+\sqrt{7}\right]\left[(x-2)-\sqrt{7}\right]$ Regrouping

$=(x-2)^2-(\sqrt{7})^2$ Difference of squares

$=x^2-4x+4-7$

$=x^2-4x-3$

57. $\frac{-6\pm\sqrt{(-4)^2-4(2)(2)}}{2(2)}=\frac{-6\pm\sqrt{16-16}}{4}=\frac{-6\pm 0}{4}=-\frac{3}{2}$

59. *Writing Exercise.* A quadratic polynomial in the form ax^2+bx+c can be written in factored form
$$\left(x-\frac{-b+\sqrt{b^2-4ac}}{2a}\right)\left(x+\frac{-b-\sqrt{b^2-4ac}}{2a}\right).$$

61. $f(x)=\frac{x^2}{x-2}+1$

To find the x-coordinates of the x-intercepts of the graph of f, we solve $f(x)=0$.

$$\frac{x^2}{x-2}+1=0$$

$x^2+x-2=0$ Multiplying by $x-2$

$$(x+2)(x-1)=0$$

$x = -2 \ or \ x = 1$

The x-intercepts are (–2, 0) and (1, 0).

63. $f(x) = g(x)$

$\frac{x^2}{x-2} + 1 = \frac{4x-2}{x-2} + \frac{x+4}{2}$ Substituting

$2(x-2)\left(\frac{x^2}{x-2} + 1\right) = 2(x-2)\left(\frac{4x-2}{x-2} + \frac{x+4}{2}\right)$

Multiplying by the LCD

$2x^2 + 2(x-2) = 2(4x-2) + (x-2)(x+4)$

$2x^2 + 2x - 4 = 8x - 4 + x^2 + 2x - 8$

$2x^2 + 2x - 4 = x^2 + 10x - 12$

$x^2 - 8x + 8 = 0$

$a = 1,\ b = -8,\ c = 8$

$x = \frac{-(-8) \pm \sqrt{(-8)^2 - 4\cdot1\cdot8}}{2\cdot1} = \frac{8 \pm \sqrt{64-32}}{2}$

$x = \frac{8 \pm \sqrt{32}}{2} = \frac{8 \pm \sqrt{16\cdot2}}{2} = \frac{8 \pm 4\sqrt{2}}{2}$

$x = \frac{8}{2} \pm \frac{4\sqrt{2}}{2} = 4 \pm 2\sqrt{2}$

The solutions are $4 + 2\sqrt{2}$ and $4 - 2\sqrt{2}$.

65. $z^2 + 0.84z - 0.4 = 0$

$a = 1,\ b = 0.84,\ c = -0.4$

$z = \frac{-0.84 \pm \sqrt{(0.84)^2 - 4\cdot1\cdot(-0.4)}}{2\cdot1}$

$z = \frac{-0.84 \pm \sqrt{2.3056}}{2}$

$z = \frac{-0.84 + \sqrt{2.3056}}{2} \approx 0.339$

$z = \frac{-0.84 - \sqrt{2.3056}}{2} \approx -1.179$

The solutions are approximately 0.339 and –1.179.

67. $\sqrt{2}x^2 + 5x + \sqrt{2} = 0$

$x = \frac{-5 \pm \sqrt{5^2 - 4\cdot\sqrt{2}\cdot\sqrt{2}}}{2\sqrt{2}} = \frac{-5 \pm \sqrt{17}}{2\sqrt{2}}$, or

$x = \frac{-5 \pm \sqrt{17}}{2\sqrt{2}} \cdot \frac{\sqrt{2}}{\sqrt{2}} = \frac{-5\sqrt{2} \pm \sqrt{34}}{4}$

The solutions are $\frac{-5\sqrt{2} \pm \sqrt{34}}{4}$.

69. $kx^2 + 3x - k = 0$

$k(-2)^2 + 3(-2) - k = 0$ Substituting -2 for x

$4k - 6 - k = 0$

$3k = 6$

$k = 2$

$2x^2 + 3x - 2 = 0$ Substituting 2 for k

$(2x-1)(x+2) = 0$

$2x - 1 = 0 \ or \ x + 2 = 0$

$x = \frac{1}{2} \ or \ x = -2$

The other solution is $\frac{1}{2}$.

71. *Graphing Calculator Exercise*

Exercise Set 11.3

1. Since the discriminant is 9, a perfect square, choice (b) two different rational solutions, is correct.

3. Since the discriminant is –1, a negative number, choice (d) two different imaginary-number solutions, is correct.

5. Since the discriminant is 8, a positive number that is not a perfect square, choice (c) two different irrational solutions, is correct.

7. $x^2 - 7x + 5 = 0$

$a = 1,\ b = -7,\ c = 5$

We substitute and compute the discriminant.

$b^2 - 4ac = (-7)^2 - 4\cdot1\cdot5$
$= 49 - 20$
$= 29$

Since the discriminant is a positive number that is not a perfect square, there are two irrational solutions.

9. $x^2 + 11 = 0$

$a = 1,\ b = 0,\ c = 11$

We substitute and compute the discriminant.

$b^2 - 4ac = 0^2 - 4\cdot1\cdot11 = -44$

Since the discriminant is negative, there are two imaginary-number solutions.

11. $x^2 - 11 = 0$

$a = 1,\ b = 0,\ c = -11$

We substitute and compute the discriminant.

$b^2 - 4ac = 0^2 - 4\cdot1\cdot(-11) = 44$

Since the discriminant is a positive number that is not a perfect square, there are two irrational solutions.

13. $4x^2 + 8x - 5 = 0$

$a = 4,\ b = 8,\ c = -5$

We substitute and compute the discriminant.

$b^2 - 4ac = 8^2 - 4\cdot4\cdot(-5)$
$= 64 + 80$
$= 144$

Since the discriminant is a positive number and a perfect square, there are two rational solutions.

15. $x^2 + 4x + 6 = 0$

$a = 1,\ b = 4,\ c = 6$

We substitute and compute the discriminant.

$$b^2-4ac=4^2-4\cdot1\cdot6$$
$$=16-24$$
$$=-8$$

Since the discriminant is negative, there are two imaginary-number solutions.

17. $9t^2-48t+64=0$

$a=9,\ b=-48,\ c=64$

We substitute and compute the discriminant.

$$b^2-4ac=(-48)^2-4\cdot9\cdot64$$
$$=2304-2304$$
$$=0$$

Since the discriminant is 0, there is just one solution and it is a rational number.

19. $9t^2+3t=0$

Observe that we can factor $9t^2+3t$. This tells us that there are two rational solutions. We could also do this problem as follows.

$$b^2-4ac=3^2-4\cdot9\cdot0=9$$

Since the discriminant is a positive number and a perfect square, there are two rational solutions.

21. $x^2+4x=8$

$x^2+4x-8=0$ Standard form

$a=1,\ b=4,\ c=-8$

We substitute and compute the discriminant.

$$b^2-4ac=4^2-4\cdot1\cdot(-8)$$
$$=16+32=48$$

Since the discriminant is a positive number that is not a perfect square, there are two irrational solutions.

23. $2a^2-3a=-5$

$2a^2-3a+5=0$ Standard form

$a=2,\ b=-3,\ c=5$

We substitute and compute the discriminant.

$$b^2-4ac=(-3)^2-4\cdot2\cdot5$$
$$=9-40$$
$$=-31$$

Since the discriminant is negative, there are two imaginary-number solutions.

25. $7x^2=19x$

$7x^2-19x=0$ Standard form

$a=7,\ b=-19,\ c=0$

We substitute and compute the discriminant.

$$b^2-4ac=(-19)^2-4\cdot7\cdot0=361$$

Since the discriminant is a positive number and a perfect square, there are two different rational solutions.

27. $y^2+\frac{9}{4}=4y$

$y^2-4y+\frac{9}{4}=0$ Standard form

$a=1,\ b=-4,\ c=\frac{9}{4}$

We substitute and compute the discriminant.

$$b^2-4ac=(-4)^2-4\cdot1\cdot\frac{9}{4}$$
$$=16-9$$
$$=7$$

The discriminant is a positive number that is not a perfect square. There are two irrational solutions.

29. The solutions are –5 and 4.

$x=-5$ *or* $x=4$

$x+5=0$ *or* $x-4=0$

$(x+5)(x-4)=0$ Principle of zero products

$x^2+x-20=0$ FOIL

31. The only solution is 3. It must be a repeated solution.

$x=3$ *or* $x=3$

$x-3=0$ *or* $x-3=0$

$(x-3)(x-3)=0$ Principle of zero products

$x^2-6x+9=0$ FOIL

33. The solutions are –1 and –3.

$x=-1$ *or* $x=-3$

$x+1=0$ *or* $x+3=0$

$(x+1)(x+3)=0$

$x^2+4x+3=0$

35. The solutions are 5 and $\frac{3}{4}$.

$x=5$ *or* $x=\frac{3}{4}$

$x-5=0$ *or* $x-\frac{3}{4}=0$

$$(x-5)\left(x-\frac{3}{4}\right)=0$$
$$x^2-\frac{3}{4}x-5x+\frac{15}{4}=0$$
$$x^2-\frac{23}{4}x+\frac{15}{4}=0$$

$4x^2-23x+15=0$ Multiplying by 4

37. The solutions are $-\frac{1}{4}$ and $-\frac{1}{2}$.

$x=-\frac{1}{4}$ *or* $x=-\frac{1}{2}$

$x+\frac{1}{4}=0$ *or* $x+\frac{1}{2}=0$

$$\left(x+\frac{1}{4}\right)\left(x+\frac{1}{2}\right)=0$$
$$x^2+\frac{1}{2}x+\frac{1}{4}x+\frac{1}{8}=0$$
$$x^2+\frac{3}{4}x+\frac{1}{8}=0$$

$8x^2+6x+1=0$ Multiplying by 8

39. The solutions are 2.4 and –0.4.

$$x = 2.4 \quad or \quad x = -0.4$$
$$x - 2.4 = 0 \quad or \quad x + 0.4 = 0$$
$$(x-2.4)(x+0.4) = 0$$
$$x^2 + 0.4x - 2.4x - 0.96 = 0$$
$$x^2 - 2x - 0.96 = 0$$

41. The solutions are $-\sqrt{3}$ and $\sqrt{3}$.

$$x = -\sqrt{3} \quad or \quad x = \sqrt{3}$$
$$x + \sqrt{3} = 0 \quad or \quad x - \sqrt{3} = 0$$
$$(x+\sqrt{3})(x-\sqrt{3}) = 0$$
$$x^2 - 3 = 0$$

43. The solutions are $2\sqrt{5}$ and $-2\sqrt{5}$.

$$x = 2\sqrt{5} \quad or \quad x = -2\sqrt{5}$$
$$x - 2\sqrt{5} = 0 \quad or \quad x + 2\sqrt{5} = 0$$
$$(x-2\sqrt{5})(x+2\sqrt{5}) = 0$$
$$x^2 - (2\sqrt{5})^2 = 0$$
$$x^2 - 4 \cdot 5 = 0$$
$$x^2 - 20 = 0$$

45. The solutions are $4i$ and $-4i$.

$$x = 4i \quad or \quad x = -4i$$
$$x - 4i = 0 \quad or \quad x + 4i = 0$$
$$(x-4i)(x+4i) = 0$$
$$x^2 - (4i)^2 = 0$$
$$x^2 + 16 = 0$$

47. The solutions are $2-7i$ and $2+7i$.

$$x = 2-7i \quad or \quad x = 2+7i$$
$$x - 2 + 7i = 0 \quad or \quad x - 2 - 7i = 0$$
$$(x-2) + 7i = 0 \quad or \quad (x-2) - 7i = 0$$
$$[(x-2)+7i][(x-2)-7i] = 0$$
$$(x-2)^2 - (7i)^2 = 0$$
$$x^2 - 4x + 4 - 49i^2 = 0$$
$$x^2 - 4x + 4 + 49 = 0$$
$$x^2 - 4x + 53 = 0$$

49. The solutions are $3-\sqrt{14}$ and $3+\sqrt{14}$.

$$x = 3-\sqrt{14} \quad or \quad x = 3+\sqrt{14}$$
$$x - 3 + \sqrt{14} = 0 \quad or \quad x - 3 - \sqrt{14} = 0$$
$$(x-3) + \sqrt{14} = 0 \quad or \quad (x-3) - \sqrt{14} = 0$$
$$[(x-3)+\sqrt{14}][(x-3)-\sqrt{14}] = 0$$
$$(x-3)^2 - (\sqrt{14})^2 = 0$$
$$x^2 - 6x + 9 - 14 = 0$$
$$x^2 - 6x - 5 = 0$$

51. The solutions are $1-\frac{\sqrt{21}}{3}$ and $1+\frac{\sqrt{21}}{3}$.

$$x = 1-\frac{\sqrt{21}}{3} \quad or \quad x = 1+\frac{\sqrt{21}}{3}$$
$$x - 1 + \frac{\sqrt{21}}{3} = 0 \quad or \quad x - 1 - \frac{\sqrt{21}}{3} = 0$$
$$(x-1) + \frac{\sqrt{21}}{3} = 0 \quad or \quad (x-1) - \frac{\sqrt{21}}{3} = 0$$
$$\left[(x-1)+\frac{\sqrt{21}}{3}\right]\left[(x-1)-\frac{\sqrt{21}}{3}\right] = 0$$
$$(x-1)^2 - \left(\frac{\sqrt{21}}{3}\right)^2 = 0$$
$$x^2 - 2x + 1 - \frac{21}{9} = 0$$
$$x^2 - 2x + 1 - \frac{7}{3} = 0$$
$$x^2 - 2x - \frac{4}{3} = 0$$
$$3x^2 - 6x - 4 = 0 \quad \text{Multiplying by 3}$$

53. The solutions are –2, 1, and 5.

$$x = -2 \quad or \quad x = 1 \quad or \quad x = 5$$
$$x + 2 = 0 \quad or \quad x - 1 = 0 \quad or \quad x - 5 = 0$$
$$(x+2)(x-1)(x-5) = 0$$
$$(x^2 + x - 2)(x-5) = 0$$
$$x^3 + x^2 - 2x - 5x^2 - 5x + 10 = 0$$
$$x^3 - 4x^2 - 7x + 10 = 0$$

55. The solutions are –1, 0, and 3.

$$x = -1 \quad or \quad x = 0 \quad or \quad x = 3$$
$$x + 1 = 0 \quad or \quad x = 0 \quad or \quad x - 3 = 0$$
$$(x+1)(x)(x-3) = 0$$
$$(x^2 + x)(x-3) = 0$$
$$x^3 - 3x^2 + x^2 - 3x = 0$$
$$x^3 - 2x^2 - 3x = 0$$

57. *Writing Exercise.* When the discriminant, $b^2 - 4ac$ is 0, then the quadratic formula can be simplified to

$$x = \frac{-b \pm \sqrt{b^2 - 4ac}}{2a} = \frac{-b \pm \sqrt{0}}{2a} = -\frac{b}{2a}.$$

There is only one solution, $-\frac{b}{2a}$.

59.

$$\frac{c}{d} = c + d$$
$$d \cdot \frac{c}{d} = d(c+d)$$
$$c = cd + d^2$$
$$c - cd = d^2$$
$$c(1-d) = d^2$$
$$c = \frac{d^2}{1-d}$$

61.
$$x=\frac{3}{1-y}$$
$$(1-y)x=(1-y)\frac{3}{1-y}$$
$$x-xy=3$$
$$x-3=xy$$
$$\frac{x-3}{x}=y \text{ or } y=1-\frac{3}{x}$$

63. ***Familiarize.*** Let r = Jamal's speed in mph. Then r – 1.5 = Kade's speed in mph. We organize the information in a table. The time is the same for both so we use t for each time.

	Distance	Speed	Time
Jamal	7	r	t
Kade	4	$r-1.5$	t

Translate. Using the formula Time = Distance/Rate in each row of the table and the fact that the times are the same, we can write an equation.

$$\frac{7}{r}=\frac{4}{r-1.5}$$

Carry out. We solve the equation.

$$\frac{7}{r}=\frac{4}{r-1.5}, \quad \text{LCD is } r(r-1.5)$$
$$r(r-1.5)\cdot\frac{7}{r}=r(r-1.5)\cdot\frac{4}{r-1.5}$$
$$7(r-1.5)=4r$$
$$7r-10.5=4r$$
$$-10.5=-3r$$
$$3.5=r$$

Check. If the Jamal's speed is 3.5 mph, then Kade's speed is $3.5-1.5$, or 2 mph. Traveling 7 mi at 3.5 mph takes $\frac{7}{3.5}=2\text{ hr}$. Traveling 4 mi at 2 mph takes $\frac{4}{2}=2\text{ hr}$. Since the times are the same, the answer checks.

State. The Jamal's speed is 3.5 mph and Kade's speed is 2 mph.

65. *Writing Exercise.* Consider a quadratic equation in standard form, $ax^2+bx+c=0$. The solutions are

$$\frac{-b\pm\sqrt{b^2-4ac}}{2a}.$$

The product of the solutions is

$$\left(\frac{-b+\sqrt{b^2-4ac}}{2a}\right)\left(\frac{-b-\sqrt{b^2-4ac}}{2a}\right)$$
$$=\frac{(-b)^2-\left(\sqrt{b^2-4ac}\right)^2}{(2a)^2}=\frac{b^2-(b^2-4ac)}{4a^2}=\frac{4ac}{4a^2}=\frac{c}{a}$$

For integers c and a, this is a real number.

67. The graph includes the points (–3, 0), (0, –3), and (1, 0). Substituting in $y=ax^2+bx+c$, we have three equations.

$$0 = 9a - 3b + c,$$
$$-3 = c,$$
$$0 = a + b + c$$

The solution of this system of equations is a = 1, b = 2, $c=-3$.

69. a) $kx^2-2x+k=0$; one solution is –3

We first find k by substituting –3 for x.

$$k(-3)^2-2(-3)+k=0$$
$$9k+6+k=0$$
$$10k=-6$$
$$k=-\frac{6}{10}$$
$$k=-\frac{3}{5}$$

b) Now substitute $-\frac{3}{5}$ for k in the original equation.

$$-\frac{3}{5}x^2-2x+\left(-\frac{3}{5}\right)=0$$
$$3x^2+10x+3=0 \quad \text{Multiplying by} -5$$
$$(3x+1)(x+3)=0$$
$$x=-\frac{1}{3} \text{ or } x=-3$$

The other solution is $-\frac{1}{3}$.

71. a) $x^2-(6+3i)x+k=0$; one solution is 3.

We first find k by substituting 3 for x.

$$3^2-(6+3i)3+k=0$$
$$9-18-9i+k=0$$
$$-9-9i+k=0$$
$$k=9+9i$$

b) Now we substitute $9+9i$ for k in the original equation.

$$x^2-(6+3i)x+(9+9i)=0$$
$$x^2-(6+3i)x+3(3+3i)=0$$
$$[x-(3+3i)][x-3]=0$$

The other solution is $3+3i$.

73. The solutions of $ax^2+bx+c=0$ are $x=\frac{-b\pm\sqrt{b^2-4ac}}{2a}$.

When there is just one solution, $b^2-4ac=0$, so $x=\frac{-b\pm0}{2a}=-\frac{b}{2a}$.

75. We substitute (–3, 0), $\left(\frac{1}{2}, 0\right)$, and (0,–12) in $f(x)=ax^2+bx+c$ and get three equations.

$$0=9a-3b+c,$$
$$0=\frac{1}{4}a+\frac{1}{2}b+c,$$
$$-12=c$$

The solution of this system of equations is $a=8$, $b=20$, $c=-12$.

77. If $-\sqrt{2}$ is one solution then $\sqrt{2}$ is another solution. Then

$$x=-\sqrt{2} \quad or \quad x=\sqrt{2}$$
$$x=\pm\sqrt{2}$$
$$x^2=2 \quad \text{Principle of square roots}$$
$$x^2-2=0$$

79. If $1-\sqrt{5}$ and $3+2i$ are two solutions, then $1+\sqrt{5}$ and $3-2i$ are also solutions. The equation of lowest degree that has these solutions is found as follows.

$$[x-(1-\sqrt{5})][x-(1+\sqrt{5})][x-(3+2i)][x-(3-2i)]=0$$
$$(x^2-2x-4)(x^2-6x+13)=0$$
$$x^4-8x^3+21x^2-2x-52=0$$

81. *Writing Exercise.* If the discriminant indicates that there are two imaginary-number solutions, or only one rational solution, Keisha would know that her graph is not correct.

Exercise Set 11.4

1. ***Familiarize.*** We first make a drawing, labeling it with the known and unknown information. We can also organize the information in a table. We let r represent the speed and t the time for the first part of the trip.

r mph t hr $r-10$ mph $4-t$ hr

120 mi 100 mi

Trip	Distance	Speed	Time
1st part	120	r	t
2nd part	100	$r-10$	$4-t$

Translate. Using $r=\frac{d}{t}$, we get two equations from the table, $r=\frac{120}{t}$ and $r-10=\frac{100}{4-t}$.

Carry out. We substitute $\frac{120}{t}$ for r in the second equation and solve for t.

$$\frac{120}{t}-10=\frac{100}{4-t}, \quad \text{LCD is } t(4-t)$$
$$t(4-t)\left(\frac{120}{t}-10\right)=t(4-t)\cdot\frac{100}{4-t}$$
$$120(4-t)-10t(4-t)=100t$$
$$480-120t-40t+10t^2=100t$$
$$10t^2-260t+480=0 \quad \text{Standard form}$$
$$t^2-26t+48=0 \quad \text{Multiplying by } \frac{1}{10}$$
$$(t-2)(t-24)=0$$
$$t=2 \quad or \quad t=24$$

Check. Since the time cannot be negative (If $t=24$, $4-t=-20$.), we check only 2 hr. If $t=2$, then $4-t=2$. The speed of the first part is $\frac{120}{2}$, or 60 mph. The speed of the second part is $\frac{100}{2}$, or 50 mph. The speed of the second part is 10 mph slower than the first part. The value checks.

State. The speed of the first part was 60 mph, and the speed of the second part was 50 mph.

3. ***Familiarize.*** We first make a drawing. We also organize the information in a table. We let $r=$ the speed and $t=$ the time of the slower trip.

200 mi r mph t hr

200 mi $r+10$ mph $t-1$ hr

Trip	Distance	Speed	Time
Slower	200	r	t
Faster	200	$r+10$	$t-1$

Translate. Using $t=d/r$, we get two equations from the table:

$$t=\frac{200}{r} \text{ and } t-1=\frac{200}{r+10}$$

Carry out. We substitute $\frac{200}{r}$ for t in the second equation and solve for r.

$$\frac{200}{r}-1=\frac{200}{r+10}, \quad \text{LCD is } r(r+10)$$
$$r(r+10)\left(\frac{200}{r}-1\right)=r(r+10)\cdot\frac{200}{r+10}$$
$$200(r+10)-r(r+10)=200r$$
$$200r+2000-r^2-10r=200r$$
$$0=r^2+10r-2000$$
$$0=(r+50)(r-40)$$
$$r=-50 \quad or \quad r=40$$

Check. Since negative speed has no meaning in this problem, we check only 40. If $r=40$, then the time for the slower trip is $\frac{200}{40}$, or 5 hours. If $r=40$, then $r+10=50$ and the time for the faster trip is $\frac{200}{50}$, or 4 hours. This is 1 hour less time than the slower trip took, so we have an answer to the problem.

State. The speed is 40 mph.

5. ***Familiarize.*** We make a drawing and then organize the information in a table. We let $r=$ the speed and $t=$ the time of the Cessna.

600 mi r mph t hr

1000 mi $r+50$ mph $t+1$ hr

Plane	Distance	Speed	Time
Cessna	600	r	t
Beechcraft	1000	$r+50$	$t+1$

Translate. Using $t = d/r$, we get two equations from the table:

$$t = \frac{600}{r} \text{ and } t+1 = \frac{1000}{r+50}$$

Carry out. We substitute $\frac{600}{r}$ for t in the second equation and solve for r.

$$\frac{600}{r}+1 = \frac{1000}{r+50}, \quad \text{LCD is } r(r+50)$$
$$r(r+50)\left(\frac{600}{r}+1\right) = r(r+50)\cdot\frac{1000}{r+50}$$
$$600(r+50)+r(r+50) = 1000r$$
$$600r+30{,}000+r^2+50r = 1000r$$
$$r^2-350r+30{,}000 = 0$$
$$(r-150)(r-200) = 0$$

$r = 150$ *or* $r = 200$

Check. If $r = 150$, then the Cessna's time is $\frac{600}{150}$, or 4 hr and the Beechcraft's time is $\frac{1000}{150+50}$, or $\frac{1000}{200}$, or 5 hr. If $r = 200$, then the Cessna's time is $\frac{600}{200}$, or 3 hr and the Beechcraft's time is $\frac{1000}{200+50}$, or $\frac{1000}{250}$, or 4 hr. Since the Beechcraft's time is 1 hr longer in each case, both values check. There are two solutions.

State. The speed of the Cessna is 150 mph and the speed of the Beechcraft is 200 mph; or the speed of the Cessna is 200 mph and the speed of the Beechcraft is 250 mph.

7. ***Familiarize.*** We make a drawing and then organize the information in a table. We let r represent the speed and t the time of the trip to Hillsboro.

Hillsboro

36 mi — r mph — t hr →

← 36 mi — $r-3$ mph — $7-t$ hr

Trip	Distance	Speed	Time
To Hillsboro	36	r	t
Return	36	$r-3$	$7-t$

Translate. Using $t = \frac{d}{r}$, we get two equations from the table,

$$t = \frac{36}{r} \text{ and } 7-t = \frac{36}{r-3}.$$

Carry out. We substitute $\frac{36}{r}$ for t in the second equation and solve for r.

$$7-\frac{36}{r} = \frac{36}{r-3}, \quad \text{LCD is } r(r-3)$$
$$r(r-3)\left(7-\frac{36}{r}\right) = r(r-3)\cdot\frac{36}{r-3}$$
$$7r(r-3)-36(r-3) = 36r$$
$$7r^2-21r-36r+108 = 36r$$
$$7r^2-93r+108 = 0$$
$$(7r-9)(r-12) = 0$$
$$r = \frac{9}{7} \text{ or } r = 12$$

Check. Since negative speed has no meaning in this problem (If $r = \frac{9}{7}$, then $r-3 = -\frac{12}{7}$.), we check only 12 mph. If $r = 12$, then the time of the trip to Hillsboro is $\frac{36}{12}$, or 3 hr. The speed of the return trip is 12 – 3, or 9 mph, and the time is $\frac{36}{9}$, or 4 hr. The total time for the round trip is $3\text{ hr}+4\text{ hr}$, or 7 hr. The value checks.

State. Naoki's speed on the trip to Hillsboro was 12 mph and it was 9 mph on the return trip.

9. ***Familiarize.*** We make a drawing and organize the information in a table. Let r represent the speed of the boat in still water, and let t represent the time of the trip upriver.

60 mi — $r-3$ mph — t hr → Upriver

Downriver ← 60 mi — $r+3$ mph — $9-t$ hr

Trip	Distance	Speed	Time
Upriver	60	$r-3$	t
Downriver	60	$r+3$	$9-t$

Translate. Using $t = \frac{d}{r}$, we get two equations from the table,

$$t = \frac{60}{r-3} \text{ and } 9-t = \frac{60}{r+3}.$$

Carry out. We substitute $\frac{60}{r-3}$ for t in the second equation and solve for r.

$$9-\frac{60}{r-3} = \frac{60}{r+3}$$
$$(r-3)(r+3)\left(9-\frac{60}{r-3}\right) = (r-3)(r+3)\cdot\frac{60}{r+3}$$
$$9(r-3)(r+3)-60(r+3) = 60(r-3)$$
$$9r^2-81-60r-180 = 60r-180$$
$$9r^2-120r-81 = 0$$
$$3r^2-40r-27 = 0 \quad \text{Dividing by 3}$$

We use the quadratic formula.

$$r = \frac{-(-40)\pm\sqrt{(-40)^2-4\cdot3\cdot(-27)}}{2\cdot3}$$
$$r = \frac{40\pm\sqrt{1924}}{6}$$
$$r \approx 14 \text{ or } r \approx -0.6$$

Check. Since negative speed has no meaning in this problem, we check only 14 mph. If $r \approx 14$, then the speed upriver is about $14-3$, or 11 mph, and the time is about $\frac{60}{11}$, or 5.5 hr. The speed downriver is about $14+3$, or 17 mph, and the time is about $\frac{60}{17}$, or 3.5 hr. The total time of the round trip is $5.5+3.5$, or 9 hr. The value checks.

State. The speed of the boat in still water is about 14 mph.

11. ***Familiarize.*** Let x represent the time it take the spring to fill the pool. Then $x-8$ represents the time it takes the well to fill the pool. It takes them 3 hr to fill the pool working together, so they can fill $\frac{1}{3}$ of the pool in 1 hr. The spring will fill $\frac{1}{x}$ of the pool in 1 hr, and the well will fill $\frac{1}{x-8}$ of the pool in 1 hr.

Translate. We have an equation.

$$\frac{1}{x}+\frac{1}{x-8}=\frac{1}{3}$$

Carry out. We solve the equation.

We multiply by the LCD, $3x(x-8)$.

$$3x(x-8)\left(\frac{1}{x}+\frac{1}{x-8}\right)=3x(x-8)\cdot\frac{1}{3}$$
$$3(x-8)+3x=x(x-8)$$
$$3x-24+3x=x^2-8x$$
$$0=x^2-14x+24$$
$$0=(x-2)(x-12)$$

Check. Since negative time has no meaning in this problem, 2 is not a solution $(2-8=-6)$. We check only 12 hr. This is the time it would take the spring working alone. Then the well would take $12-8$, or 4 hr working alone. The well would fill $3\left(\frac{1}{4}\right)$, or $\frac{3}{4}$ of the pool in 3 hr, and the spring would fill $3\left(\frac{1}{12}\right)$, or $\frac{1}{4}$ of the pool in 3 hr. Thus, in 3 hr they would fill $\frac{3}{4}+\frac{1}{4}$ of the pool. This is all of it, so the numbers check.

State. It takes the spring, working alone, 12 hr to fill the pool.

13. We make a drawing and then organize the information in a table. We let r represent Kofi's speed in still water. Then $r-2$ is the speed upstream and $r+2$ is the speed downstream. Using $t=\frac{d}{r}$, we let $\frac{1}{r-2}$ represent the time upstream and $\frac{1}{r+2}$ represent the time downstream.

1 mi $r-2$ mph → Upstream

Downstream ← 1 mi $r+2$ mph

Trip	Distance	Speed	Time
Upstream	1	$r-2$	$\frac{1}{r-2}$
Downstream	1	$r+2$	$\frac{1}{r+2}$

Translate. The time for the round trip is 1 hour. We now have an equation.

$$\frac{1}{r-2}+\frac{1}{r+2}=1$$

Carry out. We solve the equation. We multiply by the LCD, $(r-2)(r+2)$.

$$(r-2)(r+2)\left(\frac{1}{r-2}+\frac{1}{r+2}\right)=(r-2)(r+2)\cdot 1$$
$$(r+2)+(r-2)=(r-2)(r+2)$$
$$2r=r^2-4$$
$$0=r^2-2r-4$$

$a=1,\ b=-2,\ c=-4$

$$r=\frac{-(-2)\pm\sqrt{(-2)^2-4\cdot 1(-4)}}{2\cdot 1}$$
$$r=\frac{2\pm\sqrt{4+16}}{2}=\frac{2\pm\sqrt{20}}{2}$$
$$r=\frac{2\pm 2\sqrt{5}}{2}=1\pm\sqrt{5}$$

$1+\sqrt{5}\approx 1+2.236\approx 3.24$

$1-\sqrt{5}\approx 1-2.236\approx -1.24$

Check. Since negative speed has no meaning in this problem, we check only 3.24 mph. If $r\approx 3.24$, then $r-2\approx 1.24$ and $r+2\approx 5.24$. The time it takes to travel upstream is approximately $\frac{1}{1.24}$, or 0.806 hr, and the time it takes to travel downstream is approximately $\frac{1}{5.24}$, or 0.191 hr. The total time is 0.997 which is approximately 1 hour. The value checks.

State. Kofi's speed in still water is approximately 3.24 mph.

15.
$$A=4\pi r^2$$
$$\frac{A}{4\pi}=r^2 \quad \text{Dividing by } 4\pi$$
$$\frac{1}{2}\sqrt{\frac{A}{\pi}}=r \quad \text{Taking the positive square root}$$

17. $A = 2\pi r^2 + 2\pi rh$

$0 = 2\pi r^2 + 2\pi rh - A$ Standard form

$a = 2\pi,\ b = 2\pi h,\ c = -A$

$r = \frac{-2\pi h \pm \sqrt{(2\pi h)^2 - 4 \cdot 2\pi \cdot (-A)}}{2 \cdot 2\pi}$ Using the quadratic formula

$r = \frac{-2\pi h \pm \sqrt{4\pi^2 h^2 + 8\pi A}}{4\pi}$

$r = \frac{-2\pi h \pm 2\sqrt{\pi^2 h^2 + 2\pi A}}{4\pi}$

$r = \frac{-\pi h \pm \sqrt{\pi^2 h^2 + 2\pi A}}{2\pi}$

Since taking the negative square root would result in a negative answer, we take the positive one.

$r = \frac{-\pi h + \sqrt{\pi^2 h^2 + 2\pi A}}{2\pi}$

19. $F = \frac{Gm_1 m_2}{r^2}$

$Fr^2 = Gm_1 m_2$

$r^2 = \frac{Gm_1 m_2}{F}$

$r = \sqrt{\frac{Gm_1 m_2}{F}}$

21. $c = \sqrt{gH}$

$c^2 = gH$ Squaring

$\frac{c^2}{g} = H$

23. $a^2 + b^2 = c^2$

$b^2 = c^2 - a^2$

$b = \sqrt{c^2 - a^2}$

25. $s = v_0 t + \frac{gt^2}{2}$

$0 = \frac{gt^2}{2} + v_0 t - s$ Standard form

$a = \frac{g}{2},\ b = v_0,\ c = -s$

$t = \frac{-v_0 \pm \sqrt{v_0^2 - 4\left(\frac{g}{2}\right)(-s)}}{2\left(\frac{g}{2}\right)}$

$t = \frac{-v_0 \pm \sqrt{v_0^2 + 2gs}}{g}$

Since taking the negative square root would result in a negative answer, we take the positive one.

$t = \frac{-v_0 + \sqrt{v_0^2 + 2gs}}{g}$

27. $N = \frac{1}{2}(n^2 - n)$

$N = \frac{1}{2}n^2 - \frac{1}{2}n$

$0 = \frac{1}{2}n^2 - \frac{1}{2}n - N$

$a = \frac{1}{2},\ b = -\frac{1}{2},\ c = -N$

$n = \frac{-\left(-\frac{1}{2}\right) \pm \sqrt{\left(-\frac{1}{2}\right)^2 - 4 \cdot \frac{1}{2} \cdot (-N)}}{2\left(\frac{1}{2}\right)}$

$n = \frac{1}{2} \pm \sqrt{\frac{1}{4} + 2N}$

$n = \frac{1}{2} \pm \sqrt{\frac{1 + 8N}{4}}$

$n = \frac{1}{2} \pm \frac{1}{2}\sqrt{1 + 8N}$

Since taking the negative square root would result in a negative answer, we take the positive one.

$n = \frac{1}{2} + \frac{1}{2}\sqrt{1 + 8N}$, or $\frac{1 + \sqrt{1 + 8N}}{2}$

29. $T = 2\pi\sqrt{\frac{l}{g}}$

$\frac{T}{2\pi} = \sqrt{\frac{l}{g}}$ Multiplying by $\frac{1}{2\pi}$

$\frac{T^2}{4\pi^2} = \frac{l}{g}$ Squaring

$gT^2 = 4\pi^2 l$ Multiplying by $4\pi^2 g$

$g = \frac{4\pi^2 l}{T^2}$ Multiplying by $\frac{1}{T^2}$

31. $at^2 + bt + c = 0$

The quadratic formula gives the result.

$t = \frac{-b \pm \sqrt{b^2 - 4ac}}{2a}$

33. a) ***Familiarize and Translate***. From Example 4, we know

$t = \frac{-v_0 + \sqrt{v_0^2 + 19.6s}}{9.8}$.

Carry out. Substituting 500 for s and 0 for v_0, we have

$t = \frac{0 + \sqrt{0^2 + 19.6(500)}}{9.8}$

$t \approx 10.1$

Check. Substitute 10.1 for t and 0 for v_0 in the original formula. (See Example 4.)

$s = 4.9t^2 + v_0 t = 4.9(10.1)^2 + 0 \cdot (10.1)^2 \approx 500$

The answer checks.

State. It takes the bolt about 10.1 sec to reach the ground.

b) ***Familiarize and Translate.*** From Example 4, we know

$$t=\frac{-v_0+\sqrt{v_0^2+19.6s}}{9.8}$$

Carry out. Substitute 500 for s and 30 for v_0.

$$t=\frac{-30+\sqrt{30^2+19.6(500)}}{9.8}$$
$$t\approx 7.49$$

Check. Substitute 30 for v_0 and 7.49 for t in the original formula. (See Example 4.)

$$s=4.9t^2+v_0t=4.9(7.49)^2+(30)(7.49)\approx 500$$

The answer checks.

State. It takes the ball about 7.49 sec to reach the ground.

c) ***Familiarize and Translate.*** We will use the formula in Example 4, $s=4.9t^2+v_0t$.

Carry out. Substitute 5 for t and 30 for v_0.

$$s=4.9(5)^2+30(5)=272.5$$

Check. We can substitute 30 for v_0 and 272.5 for s in the form of the formula we used in part (b).

$$t=\frac{-v_0+\sqrt{v_0^2+19.6s}}{9.8}$$
$$=\frac{-30+\sqrt{(30)^2+19.6(272.5)}}{9.8}=5$$

The answer checks.

State. The object will fall 272.5 m.

35. ***Familiarize.*** We will use the formula $4.9t^2=s$.

Translate. Substitute 40 for s.

$$4.9t^2=40$$

Carry out. We solve the equation.

$$4.9t^2=40$$
$$t^2=\frac{40}{4.9}$$
$$t=\sqrt{\frac{40}{4.9}}$$
$$t\approx 2.9$$

Check. Substitute 2.9 for t in the formula.

$$s=4.9(2.9)^2=41.209\approx 40$$

The answer checks.

State. Chad will fall for about 2.9 sec before the cord begins to stretch.

37. ***Familiarize.*** We will use the formula $V=48T^2$.

Translate. Substitute 38 for V.

$$38=48T^2$$

Carry out. We solve the equation.

$$38=48T^2$$
$$\frac{38}{48}=T^2$$
$$T=\sqrt{\frac{38}{48}}$$
$$T\approx 0.890$$

Check. Substitute 0.890 for T in the formula.

$$V=48(0.890)^2\approx 38$$

The answer checks.

State. Dwight Howard's hang time is about 0.890 sec.

39. ***Familiarize and Translate.*** We will use the formula in Example 4, $s=4.9t^2+v_0t$.

Carry out. Solve the formula for v_0.

$$s-4.9t^2=v_0t$$
$$\frac{s-4.9t^2}{t}=v_0$$

Now substitute 51.6 for s and 3 for t.

$$\frac{51.6-4.9(3)^2}{3}=v_0$$
$$2.5=v_0$$

Check. Substitute 3 for t and 2.5 for v_0 in the original formula.

$$s=4.9(3)^2+2.5(3)=51.6$$

The solution checks.

State. The initial velocity is 2.5 m/sec.

41. ***Familiarize and Translate.*** From Exercise 32 we know that $r=-1+\frac{-P_2+\sqrt{P_2^2+4AP_1}}{2P_1}$

where A is the total amount in the account after two years, P_1 is the amount of the original deposit, P_2 is deposited at the beginning of the second year, and r is the annual interest rate.

Carry out. Substitute 3200 for P_1, 1800 for P_2, and 5375.48 for A.

$$r=-1+\frac{-1800+\sqrt{(1800)^2+4(5375.48)(3200)}}{2(3200)}$$

Using a calculator we have $r=0.045$.

Check. Substitute in the original formula in Exercise 32.

$$A=P_1(1+r)^2+P_2(1+r)$$
$$A=3200(1.045)^2+1800(1.045)=5375.48$$

The solution checks.

State. The annual interest rate is 0.045 or 4.5%.

43. *Writing Exercise.* Let the length of Rafe's cord be s, in meters. Then the length of Marti's cord is $2s$. Then, using the formula in Example 4, we find that the time t_1 that Rafe will fall before his cord begins to stretch is given by

$t_1 = \dfrac{0 \pm \sqrt{0^2 + 19.6s}}{9.8} \approx 0.5\sqrt{s}$ sec. Similarly, the time t_2 that Marti will fall before her cord begins to stretch is given by $t_2 = \dfrac{0 \pm \sqrt{0^2 + 19.6(2s)}}{9.8} \approx 0.6\sqrt{s}$ sec. Since $t_2 \neq 2t_1$, Marti's fall will not take twice as long as Rafe's.

45. $(m^{-1})^2 = m^{-1\cdot 2} = m^{-2}$ or $\dfrac{1}{m^2}$

47. $(y^{1/6})^2 = y^{\frac{1}{6}\cdot 2} = y^{2/6} = y^{1/3}$

49.
$$t^{-1} = \frac{1}{2}$$
$$\frac{1}{t} = \frac{1}{2}$$
$$2t \cdot \frac{1}{t} = 2t \cdot \frac{1}{2}$$
$$2 = t$$

51. *Writing Exercise.* Answers may vary.
An express train travels 30 mph faster than a local train. The express train travels 450 mi in 2 hr less time than it takes the local train to travel 420 mi. Find the speed of the express train.

53.
$$A = 6.5 - \frac{20.4t}{t^2+36}$$
$$(t^2+36)A = (t^2+36)\left(6.5 - \frac{20.4t}{t^2+36}\right)$$
$$At^2 + 36A = (t^2+36)(6.5) - (t^2+36)\left(\frac{20.4t}{t^2+36}\right)$$
$$At^2 + 36A = 6.5t^2 + 234 - 20.4t$$
$$At^2 - 6.5t^2 + 20.4t + 36A - 234 = 0$$
$$(A-6.5)t^2 + 20.4t + (36A-234) = 0$$
$$a = A - 6.5,\ b = 20.4,\ c = 36A - 234$$
$$t = \frac{-20.4 + \sqrt{(20.4)^2 - 4(A-6.5)(36A-234)}}{2(A-6.5)}$$
$$t = \frac{-20.4 + \sqrt{416.16 - 144A^2 + 1872A - 6084}}{2(A-6.5)}$$
$$t = \frac{-20.4 + \sqrt{-144A^2 + 1872A - 5667.84}}{2(A-6.5)}$$
$$t = \frac{-20.4 + \sqrt{144(-A^2 + 13A - 39.36)}}{2(A-6.5)}$$
$$t = \frac{-20.4 + 12\sqrt{-A^2 + 13A - 39.36}}{2(A-6.5)}$$
$$t = \frac{2(-10.2 + 6\sqrt{-A^2 + 13A - 39.36})}{2(A-6.5)}$$
$$t = \frac{-10.2 + 6\sqrt{-A^2 + 13A - 39.36}}{A-6.5}$$

55. ***Familiarize.*** Let a = the number. Then $a - 1$ is 1 less than a and the reciprocal of that number is $\dfrac{1}{a-1}$. Also, 1 more than the number is $a + 1$.

Translate.

The reciprocal of 1 less than a number	is	1 more than the number.
↓	↓	↓
$\dfrac{1}{(a-1)}$	$=$	$a+1$

Carry out. We solve the equation.
$$\frac{1}{a-1} = a+1, \quad \text{LCD is } a-1$$
$$(a-1)\cdot\frac{1}{a-1} = (a-1)(a+1)$$
$$1 = a^2 - 1$$
$$2 = a^2$$
$$\pm\sqrt{2} = a$$

Check. $\dfrac{1}{\sqrt{2}-1} \approx 2.4142 \approx \sqrt{2}+1$ and $\dfrac{1}{-\sqrt{2}-1} \approx -0.4142 \approx -\sqrt{2}+1$. The answers check.

State. The numbers are $\sqrt{2}$ and $-\sqrt{2}$, or $\pm\sqrt{2}$.

57.
$$\frac{w}{l} = \frac{l}{w+l}$$
$$l(w+l)\cdot\frac{w}{l} = l(w+l)\cdot\frac{l}{w+l}$$
$$w(w+l) = l^2$$
$$w^2 + lw = l^2$$
$$0 = l^2 - lw - w^2$$
Use the quadratic formula with $a = 1$, $b = -w$, and $c = -w^2$.
$$l = \frac{-(-w) \pm \sqrt{(-w)^2 - 4\cdot 1(-w^2)}}{2\cdot 1}$$
$$l = \frac{w \pm \sqrt{w^2 + 4w^2}}{2} = \frac{w \pm \sqrt{5w^2}}{2}$$
$$l = \frac{w \pm w\sqrt{5}}{2}$$
Since $\dfrac{w - w\sqrt{5}}{2}$ is negative we use the positive square root: $l = \dfrac{w + w\sqrt{5}}{2}$

59. $mn^4 - r^2pm^3 - r^2n^2 + p = 0$
Let $u = n^2$. Substitute and rearrange.
$$mu^2 - r^2u - r^2pm^3 + p = 0$$
$$a = m,\ b = -r^2,\ c = -r^2pm^3 + p$$

$$u = \frac{-(-r^2) \pm \sqrt{(-r^2)^2 - 4 \cdot m(-r^2pm^3 + p)}}{2 \cdot m}$$
$$u = \frac{r^2 \pm \sqrt{r^4 + 4m^4r^2p - 4mp}}{2m}$$
$$n^2 = \frac{r^2 \pm \sqrt{r^4 + 4m^4r^2p - 4mp}}{2m}$$
$$n = \pm\sqrt{\frac{r^2 \pm \sqrt{r^4 + 4m^4r^2p - 4mp}}{2m}}$$

61. Let s represent a length of a side of the cube, let S represent the surface area of the cube, and let A represent the surface area of the sphere. Then the diameter of the sphere is s, so the radius r is $s/2$. From Exercise 15, we know, $A = 4\pi r^2$, so when $r = s/2$ we have

$$A = 4\pi\left(\frac{s}{2}\right)^2 = 4\pi \cdot \frac{s^2}{4} = \pi s^2.$$

From the formula for the surface area of a cube (See Exercise 16.) we know that $S = 6s^2$, so $\frac{S}{6} = s^2$ and then

$$A = \pi \cdot \frac{S}{6}, \text{ or } A(S) = \frac{\pi S}{6}.$$

Exercise Set 11.5

1. $x^6 = (x^3)^2$, so (f) is an appropriate choice.

3. $x^8 = (x^4)^2$, so (h) is an appropriate choice.

5. $x^{4/3} = (x^{2/3})^2$, so (g) is an appropriate choice.

7. $x^{-4/3} = (x^{-2/3})^2$, so (e) is an appropriate choice.

9. Since $\left(\sqrt{p}\right)^2 = p$, use $u = \sqrt{p}$.

11. Since $\left(x^2+3\right)^2 = \left(x^2+3\right)^2$, use $u = x^2 + 3$.

13. Since $\left[(1+t)^2\right]^2 = (1+t)^2$, use $u = (1+t)^2$.

15. $x^4 - 13x^2 + 36 = 0$

Let $u = x^2$ and $u^2 = x^4$.

$u^2 - 13u + 36 = 0$ Substituting u for x^2
$(u-4)(u-9) = 0$
$u - 4 = 0$ *or* $u - 9 = 0$
$u = 4$ *or* $u = 9$

Now replace u with x^2 and solve these equations.

$x^2 = 4$ *or* $x^2 = 9$
$x = \pm 2$ *or* $x = \pm 3$

The numbers 2, –2, 3, and –3 check. They are the solutions.

17. $t^4 - 7t^2 + 12 = 0$

Let $u = t^2$ and $u^2 = t^4$.

$u^2 - 7u + 12 = 0$ Substituting u for t^2
$(u-3)(u-4) = 0$
$u - 3 = 0$ *or* $u - 4 = 0$
$u = 3$ *or* $u = 4$

Now replace u with t^2 and solve these equations.

$t^2 = 3$ *or* $t^2 = 4$
$t = \pm\sqrt{3}$ *or* $t = \pm 2$

The numbers $\sqrt{3}, -\sqrt{3}$, 2, and –2 check. They are the solutions.

19. $4x^4 - 9x^2 + 5 = 0$

Let $u = x^2$ and $u^2 = x^4$.

$4u^2 - 9u + 5 = 0$ Substituting u for x^2
$(4u-5)(u-1) = 0$
$4u - 5 = 0$ *or* $u - 1 = 0$
$u = \frac{5}{4}$ *or* $u = 1$

Now replace u with x^2 and solve these equations.

$x^2 = \frac{5}{4}$ *or* $x^2 = 1$
$x = \pm\sqrt{\frac{5}{4}} = \pm\frac{\sqrt{5}}{2}$ *or* $x = \pm 1$

The numbers $\frac{\sqrt{5}}{2}, -\frac{\sqrt{5}}{2}$, 1, and –1 check. They are the solutions.

21. $w + 4\sqrt{w} - 12 = 0$

Let $u = \sqrt{w}$ and $u^2 = \left(\sqrt{w}\right)^2 = w$.

$u^2 + 4u - 12 = 0$
$(u+6)(u-2) = 0$
$u + 6 = 0$ *or* $u - 2 = 0$
$u = -6$ *or* $u = 2$

Now replace u with $\sqrt{w}$ and solve these equations.

$\sqrt{w} = -6$ *or* $\sqrt{w} = 2$
or $w = 4$

Since the principal square root cannot be negative, only 4 checks as a solution.

23. $(x^2 - 7)^2 - 3(x^2 - 7) + 2 = 0$

Let $u = x^2 - 7$ and $u^2 = (x^2 - 7)^2$.

$u^2 - 3u + 2 = 0$ Substituting
$(u-1)(u-2) = 0$
$u = 1$ *or* $u = 2$
$x^2 - 7 = 1$ *or* $x^2 - 7 = 2$ Replacing u with $x^2 - 7$
$x^2 = 8$ *or* $x^2 = 9$
$x = \pm\sqrt{8} = \pm 2\sqrt{2}$ *or* $x = \pm 3$

The numbers $2\sqrt{2}$, $-2\sqrt{3}$, 3, and –3 check. They are the solutions.

25. $r-2\sqrt{r}-6=0$

Let $u=\sqrt{r}$ and $u^2=r$.

$u^2-2u-6=0$

$$u=\frac{-(-2)\pm\sqrt{(-2)^2-4\cdot1\cdot(-6)}}{2\cdot1}$$
$$u=\frac{2\pm\sqrt{28}}{2}=\frac{2+2\sqrt{7}}{2}$$
$$u=1\pm\sqrt{7}$$

Replace u with $\sqrt{r}$ and solve these equations:

$\sqrt{r}=1+\sqrt{7}$ *or* $\sqrt{r}=1-\sqrt{7}$

$(\sqrt{r})^2=(1+\sqrt{7})^2$

$r=1+2\sqrt{7}+7$

$r=8+2\sqrt{7}$

The number $1-\sqrt{7}$ is not a solution since it is negative.

The number $8+2\sqrt{7}$ checks. It is the solution.

27. $(1+\sqrt{x})^2+5(1+\sqrt{x})+6=0$

Let $u=1+\sqrt{x}$ and $u^2=(1+\sqrt{x})^2$.

$u^2+5u+6=0$ Substituting

$(u+3)(u+2)=0$

$u=-3$ *or* $u=-2$

$1+\sqrt{x}=-3$ *or* $1+\sqrt{x}=-2$ Replacing u with $1+\sqrt{x}$

$\sqrt{x}=-4$ *or* $\sqrt{x}=-3$

Since the principal square root cannot be negative, this equation has no solution.

29. $x^{-2}-x^{-1}-6=0$

Let $u=x^{-1}$ and $u^2=x^{-2}$.

$u^2-u-6=0$ Substituting

$(u-3)(u+2)=0$

$u=3$ *or* $u=-2$

Now we replace u with x^{-1} and solve these equations:

$x^{-1}=3$ *or* $x^{-1}=-2$

$\frac{1}{x}=3$ *or* $\frac{1}{x}=-2$

$\frac{1}{3}=x$ *or* $-\frac{1}{2}=x$

Both $\frac{1}{3}$ and $-\frac{1}{2}$ check. They are the solutions.

31. $4t^{-2}-3t^{-1}-1=0$

Let $u=t^{-1}$ and $u^2=t^{-2}$.

$4u^2-3u-1=0$ Substituting

$(4u+1)(u-1)=0$

$4u+1=0$ *or* $u-1=0$

$u=-\frac{1}{4}$ *or* $u=1$

Now replace u with t^{-1} and solve these equations.

$t^{-1}=-\frac{1}{4}$ *or* $t^{-1}=1$

$\frac{1}{t}=-\frac{1}{4}$ *or* $\frac{1}{t}=1$

$-4=t$ *or* $1=t$

Both –4 and 1 check. They are the solutions.

33. $t^{2/3}+t^{1/3}-6=0$

Let $u=t^{1/3}$ and $u^2=t^{2/3}$.

$u^2+u-6=0$ Substituting

$(u+3)(u-2)=0$

$u=-3$ *or* $u=2$

Now we replace u with $t^{1/3}$ and solve these equations:

$t^{1/3}=-3$ or $t^{1/3}=2$

$t=(-3)^3$ or $t=2^3$ Raising to the third power

$t=-27$ or $t=8$

Both –27 and 8 check. They are the solutions.

35. $y^{1/3}-y^{1/6}-6=0$

Let $u=y^{1/6}$ and $u^2=y^{2/3}$.

$u^2-u-6=0$ Substituting

$(u-3)(u+2)=0$

$u=3$ or $u=-2$

Now we replace u with $y^{1/6}$ and solve these equations:

$y^{1/6}=3$ *or* $y^{1/6}=-2$

$\sqrt[6]{y}=3$ *or* $\sqrt[6]{y}=-2$

$y=3^6$

$y=729$

The equation $\sqrt[6]{y}=-2$ has no solution since principal sixth roots are never negative. The number 729 checks and is the solution.

37. $t^{1/3}+2t^{1/6}=3$

$t^{1/3}+2t^{1/6}-3=0$

Let $u=t^{1/6}$ and $u^2=t^{2/6}=t^{1/3}$.

$u^2+2u-3=0$ Substituting

$(u+3)(u-1)=0$

$u=-3$ *or* $u=1$

$t^{1/6}=-3$ *or* $t^{1/6}=1$ Substituting $t^{1/6}$ for u

$t=1$

Since principal sixth roots are never negative, the equation $t^{1/6}=-3$ has no solution.

The number 1 checks and is the solution.

39. $(3-\sqrt{x})^2-10(3-\sqrt{x})+23=0$

Let $u=3-\sqrt{x}$ and $u^2=(3-\sqrt{x})^2$.

$u^2-10u+23=0$ Substituting

$u=\dfrac{-(-10)\pm\sqrt{(-10)^2-4\cdot1\cdot23}}{2\cdot1}$

$u=\dfrac{10\pm\sqrt{8}}{2}=\dfrac{2\cdot5\pm2\sqrt{2}}{2}$

$u=5\pm\sqrt{2}$

$u=5+\sqrt{2}$ *or* $u=5-\sqrt{2}$

Now we replace u with $3-\sqrt{x}$ and solve these equations:

$3-\sqrt{x}=5+\sqrt{2}$ *or* $3-\sqrt{x}=5-\sqrt{2}$

$-\sqrt{x}=2+\sqrt{2}$ *or* $-\sqrt{x}=2-\sqrt{2}$

$\sqrt{x}=-2-\sqrt{2}$ *or* $\sqrt{x}=-2+\sqrt{2}$

Since both $-2-\sqrt{2}$ and $-2+\sqrt{2}$ are negative and principal square roots are never negative, the equation has no solution.

41. $16\left(\dfrac{x-1}{x-8}\right)^2+8\left(\dfrac{x-1}{x-8}\right)+1=0$

Let $u=\dfrac{x-1}{x-8}$ and $u^2=\left(\dfrac{x-1}{x-8}\right)^2$.

$16u^2+8u+1=0$ Substituting

$(4u+1)(4u+1)=0$

$u=-\dfrac{1}{4}$

Now we replace u with $\dfrac{x-1}{x-8}$ and solve this equation:

$\dfrac{x-1}{x-8}=-\dfrac{1}{4}$

$4x-4=-x+8$ Multiplying by $4(x-8)$

$5x=12$

$x=\dfrac{12}{5}$

The number $\dfrac{12}{5}$ checks and is the solution.

43. $x^4+5x^2-36=0$

Let $u=x^2$ and $u^2=x^4$.

$u^2+5u-36=0$

$(u-4)(u+9)=0$

$u-4=0$ *or* $u+9=0$

$u=4$ *or* $u=-9$

Now replace u with x^2 and solve these equations.

$x^2=4$ *or* $x^2=-9$

$x=\pm2$ *or* $x=\pm\sqrt{-9}=\pm3i$

The numbers 2, –2, $3i$, and $-3i$ check. They are the solutions.

45. $\left(n^2+6\right)^2-7\left(n^2+6\right)+10=0$

Let $u=n^2+6$ and $u^2=\left(n^2+6\right)^2$.

$u^2-7u+10=0$

$(u-2)(u-5)=0$

$u-2=0$ *or* $u-5=0$

$u=2$ *or* $u=5$

Now replace u with n^2+6 and solve these equations.

$n^2+6=2$ *or* $n^2+6=5$

$n^2=-4$ *or* $n^2=-1$

$n=\pm\sqrt{-4}=\pm2i$ *or* $n=\pm\sqrt{-1}=\pm i$

$2i$, $-2i$, i, and $-i$ check. They are the solutions.

47. The x-intercepts occur where $f(x)=0$. Thus, we must have $5x+13\sqrt{x}-6=0$.

Let $u=\sqrt{x}$ and $u^2=x$.

$5u^2+13u-6=0$ Substituting

$(5u-2)(u+3)=0$

$u=\dfrac{2}{5}$ *or* $u=-3$

Now replace u with $\sqrt{x}$ and solve these equations:

$\sqrt{x}=\dfrac{2}{5}$ *or* $\sqrt{x}=-3$ has no solution

$x=\dfrac{4}{25}$

The number $\dfrac{4}{25}$ checks. Thus, the x-intercept is $\left(\dfrac{4}{25},\ 0\right)$.

49. The x-intercepts occur where $f(x)=0$. Thus, we must have $(x^2-3x)^2-10(x^2-3x)+24=0$.

Let $u=x^2-3x$ and $u^2=(x^2-3x)^2$.

$u^2-10u+24=0$ Substituting

$(u-6)(u-4)=0$

$u=6$ *or* $u=4$

Now replace u with x^2-3x and solve these equations:

$x^2-3x=6$ *or* $x^2-3x=4$

$x^2-3x-6=0$ *or* $x^2-3x-4=0$

$x=\dfrac{-(-3)\pm\sqrt{(-3)^2-4(1)(-6)}}{2\cdot1}$ *or* $(x-4)(x+1)=0$

$x=\dfrac{3}{2}\pm\dfrac{\sqrt{33}}{2}$ *or* $x=4$ *or* $x=-1$

All four numbers check. Thus, the x-intercepts are $\left(\dfrac{3}{2}+\dfrac{\sqrt{33}}{2},\ 0\right)$, $\left(\dfrac{3}{2}-\dfrac{\sqrt{33}}{2},\ 0\right)$, (4, 0), and (–1, 0).

51. The x-intercepts occur where $f(x)=0$. Thus, we must have $x^{2/5}+x^{1/5}-6=0$.

Let $u=x^{1/5}$ and $u^2=x^{2/5}$.

$u^2+u-6=0$ Substituting

$(u+3)(u-2)=0$

$u=-3$ *or* $u=2$

$x^{1/5}=-3$ *or* $x^{1/5}=2$ Replacing u with $x^{1/5}$

$x=-243$ $x=32$ Raising to the fifth power

Both –243 and 32 check. Thus, the x-intercepts are (–243, 0) and (32, 0).

53. $f(x)=\left(\frac{x^2+2}{x}\right)^4+7\left(\frac{x^2+2}{x}\right)^2+5$

Observe that, for all real numbers x, each term is positive. Thus, there are no real-number values of x for which $f(x)=0$ and hence no x-intercepts.

55. *Writing Exercise.* They can both be correct. Jose's substitution reduces the original equation to the quadratic equation $u^2-2u+1=0$ while Robin's produces the equation $25u^2-10u+1=0$. Both equations can be solved and they yield the same result.

57. Graph $f(x)=x$.

We find some ordered pairs, plot points, and draw the graph.

x	y
-2	-2
-1	-1
0	0
1	1
2	2

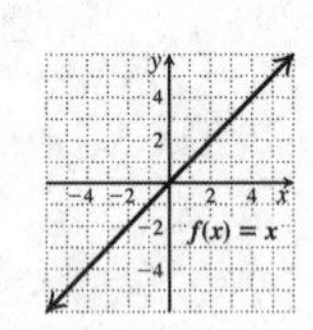

59. Graph $h(x)=x-2$.

We find some ordered pairs, plot points, and draw the graph.

x	y
-1	-3
0	-2
1	-1
2	0
3	1

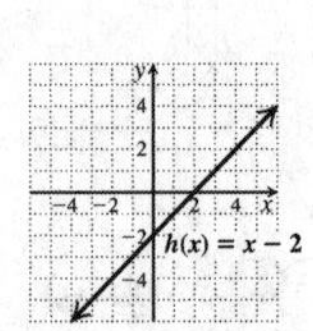

61. Graph $g(x)=x^2+2$.

We find some ordered pairs, plot points, and draw the graph.

x	y
-2	6
-1	3
0	2
1	3
2	6

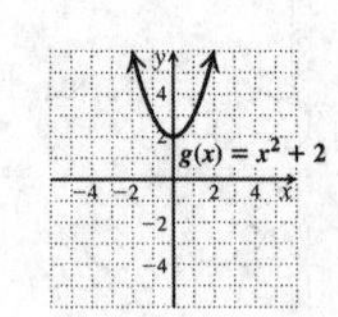

63. *Writing Exercise.* Substitute u for x^2 and solve $au^2+bu+c=0$. Then replace u with x^2 and solve for x.

65. $3x^4+5x^2-1=0$

Let $u=x^2$ and $u^2=x^4$.

$3u^2+5u-1=0$ Substituting

$$u=\frac{-5\pm\sqrt{5^2-4\cdot3\cdot(-1)}}{2\cdot3}$$

$$u=\frac{-5\pm\sqrt{37}}{6}$$

$$x^2=\frac{-5\pm\sqrt{37}}{6}\quad\text{Replacing } u \text{ with } x^2$$

$$x=\pm\sqrt{\frac{-5\pm\sqrt{37}}{6}}$$

All four numbers check and are the solutions.

67. $(x^2-5x-1)^2-18(x^2-5x-1)+65=0$

Let $u=x^2-5x-1$ and $u^2=(x^2-5x-1)^2$.

$u^2-18u+65=0$ Substituting

$(u-5)(u-13)=0$

$u=5$ *or* $u=13$

$x^2-5x-1=5$ *or* $x^2-5x-1=13$

Replacing u with x^2-4x-2

$x^2-5x-6=0$ *or* $x^2-5x-14=0$

$(x-6)(x+1)=0$ *or* $(x-7)(x+2)=0$

$x=6$ *or* $x=-1$ *or* $x=7$ *or* $x=-2$

The numbers 6, –1, 7 and –2 check and are the solutions.

69. $\frac{x}{x-1}-6\sqrt{\frac{x}{x-1}}-40=0$

Let $u=\sqrt{\frac{x}{x-1}}$ and $u^2=\frac{x}{x-1}$.

$u^2-6u-40=0$ Substituting

$(u-10)(u+4)=0$

$u=10$ *or* $u=-4$

$\sqrt{\frac{x}{x-1}}=10$ *or* $\sqrt{\frac{x}{x-1}}=-4$ has no solution

$\frac{x}{x-1}=100$

$x=100x-100$ Multiplying by $(x-1)$

$100=99x$

$\frac{100}{99}=x$

The number $\frac{100}{99}$ checks. It is the solution.

71. $a^5(a^2-25)+13a^3(25-a^2)+36a(a^2-25)=0$

$a^5(a^2-25)-13a^3(a^2-25)+36a(a^2-25)=0$

$a(a^2-25)(a^4-13a^2+36)=0$

$a(a^2-25)(a^2-4)(a^2-9)=0$

$a=0$ *or* $a^2-25=0$ *or* $a^2-4=0$ *or* $a^2-9=0$

$a=0$ *or* $a^2=25$ *or* $a^2=4$ *or* $a^2=9$

$a=0$ *or* $a=\pm5$ *or* $a=\pm2$ *or* $a=\pm3$

All seven numbers check. The solutions are 0, 5, –5, 2, –2 , 3, and –3.

73. $x^6 - 28x^3 + 27 = 0$

Let $u = x^3$.

$u^2 - 28u + 27 = 0$
$(u-27)(u-1) = 0$
$u = 27 \quad or \quad u = 1$
$x^3 = 27 \quad or \quad x^3 = 1$
$x^3 - 27 = 0 \quad or \quad x^3 - 1 = 0$

First we solve $x^3 - 27 = 0$.

$x^3 - 27 = 0$
$(x-3)(x^2+3x+9) = 0$
$x - 3 = 0 \quad or \quad x^2 + 3x + 9 = 0$

$x = 3 \quad or \quad x = \dfrac{-3 \pm \sqrt{3^2 - 4\cdot1\cdot9}}{2\cdot1}$

$x = 3 \quad or \quad x = \dfrac{-3 \pm \sqrt{-27}}{2}$

$x = 3 \quad or \quad x = -\dfrac{3}{2} \pm \dfrac{3\sqrt{3}}{2}i$

Next we solve $x^3 - 1 = 0$.

$x^3 - 1 = 0$
$(x-1)(x^2+x+1) = 0$
$x - 1 = 0 \quad or \quad x^2 + x + 1 = 0$

$x = 1 \quad or \quad x = \dfrac{-1 \pm \sqrt{1^2 - 4\cdot1\cdot1}}{2\cdot1}$

$x = 1 \quad or \quad x = \dfrac{-1 \pm \sqrt{-3}}{2}$

$x = 1 \quad or \quad x = -\dfrac{1}{2} \pm \dfrac{\sqrt{3}}{2}i$

All six numbers check.

75. *Graphing Calculator Exercise*

77. *Writing Exercise.* Salam can find only two real-number solutions using a graphing calculator. There are also two imaginary-number solutions which must be found algebraically.

Connecting the Concepts

1. $x^2 - 3x - 10 = 0$
$(x+2)(x-5) = 0$
$x + 2 = 0 \quad or \quad x - 5 = 0$
$x = -2 \quad or \quad x = 5$

3. $x^2 + 6x = 10$
$x^2 + 6x + 9 = 10 + 9$ Completing the square
$(x+3)^2 = 19$
$x + 3 = \pm\sqrt{19}$
$x = -3 \pm \sqrt{19}$

5. $(x+1)^2 = 2$
$x + 1 = \pm\sqrt{2}$
$x = -1 \pm \sqrt{2}$

7. $x^2 - x - 1 = 0$

$a = 1,\ b = -1,\ c = -1$

$x = \dfrac{-(-1) \pm \sqrt{(-1)^2 - 4\cdot1\cdot(-1)}}{2\cdot1} = \dfrac{1 \pm \sqrt{1+4}}{2}$

$= \dfrac{1 \pm \sqrt{5}}{2} = \dfrac{1}{2} \pm \dfrac{\sqrt{5}}{2}$

9. $4t^2 = 11$

$t^2 = \dfrac{11}{4}$

$t = \pm\sqrt{\dfrac{11}{4}} = \pm\dfrac{\sqrt{11}}{2}$

11. $c^2 + c + 1 = 0$

$a = 1,\ b = 1,\ c = 1$

$c = \dfrac{-1 \pm \sqrt{1^2 - 4\cdot1\cdot1}}{2\cdot1} = \dfrac{-1 \pm \sqrt{1-4}}{2}$

$= \dfrac{-1 \pm \sqrt{-3}}{2} = -\dfrac{1}{2} \pm \dfrac{\sqrt{3}}{2}i$

13. $6y^2 - 7y - 10 = 0$
$(6y+5)(y-2) = 0$
$6y + 5 = 0 \quad or \quad y - 2 = 0$
$y = -\dfrac{5}{6} \quad or \quad y = 2$

15. $x^4 - 10x^2 + 9 = 0$

Let $u = x^2$ and $u^2 = x^4$.

$u^2 - 10u + 9 = 0$
$(u-1)(u-9) = 0$
$u - 1 = 0 \quad or \quad u - 9 = 0$
$u = 1 \quad or \quad u = 9$

Replace u with x^2.

$x^2 = 1 \quad or \quad x^2 = 9$
$x = \pm1 \quad or \quad x = \pm3$

17. $t(t-3) = 2t(t+1)$
$t^2 - 3t = 2t^2 + 2t$
$0 = t^2 + 5t$
$0 = t(t+5)$
$t = 0 \quad or \quad t + 5 = 0$
$t = 0 \quad or \quad t = -5$

19. $(m^2+3)^2 - 4(m^2+3) - 5 = 0$

Let $u = m^2 + 3$ and $u^2 = (m^2+3)^2$.

$u^2 - 4u - 5 = 0$
$(u-5)(u+1) = 0$
$u - 5 = 0 \quad or \quad u + 1 = 0$
$u = 5 \quad or \quad u = -1$

Replace u with m^2+3.

$m^2+3=5$ *or* $m^2+3=-1$
$m^2=2$ *or* $m^2=-4$
$m=\pm\sqrt{2}$ *or* $m=\pm\sqrt{-4}=\pm 2i$

Exercise Set 11.6

1. The graph of $f(x)=2(x-1)^2+3$ has vertex (1, 3) and opens up. Choice (h) is correct.

3. The graph of $f(x)=2(x+1)^2+3$ has vertex (–1, 3) and opens up. Choice (f) is correct.

5. The graph of $f(x)=-2(x+1)^2+3$ has vertex (–1, 3) and opens down. Choice (b) is correct.

7. The graph of $f(x)=2(x+1)^2-3$ has vertex (–1, –3) and opens up. Choice (e) is correct.

9. $f(x)=x^2$

See Example 1 in the text.

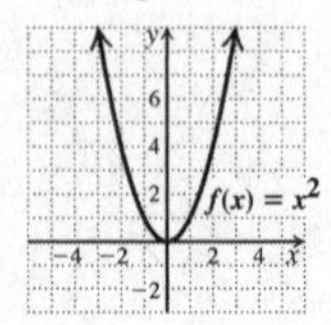

11. $f(x)=-2x^2$

We choose some numbers for x and compute $f(x)$ for each one. Then we plot the ordered pairs $(x, f(x))$ and connect them with a smooth curve.

x	$f(x)=-2x^2$
0	0
1	-2
2	-8
-1	-2
-2	-8

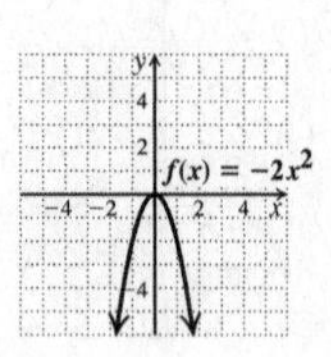

13. $g(x)=\frac{1}{3}x^2$

x	$g(x)=\frac{1}{3}x^2$
0	0
1	$\frac{1}{3}$
2	$\frac{4}{3}$
3	3
-1	$\frac{1}{3}$
-2	$\frac{4}{3}$
-3	3

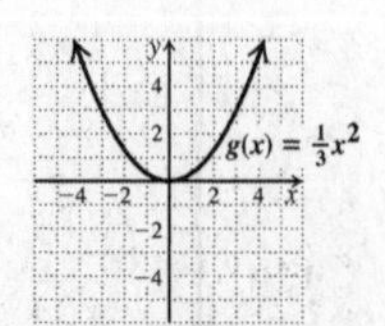

15. $h(x)=-\frac{1}{3}x^2$

Observe that the graph of $h(x)=-\frac{1}{3}x^2$ is the reflection of the graph of $g(x)=\frac{1}{3}x^2$ across the x-axis. We graphed $g(x)$ in Exercise 13, so we can use it to graph $h(x)$. If we did not make this observation we could find some ordered pairs, plot points, and connect them with a smooth curve.

x	$h(x)=-\frac{1}{3}x^2$
0	0
1	$-\frac{1}{3}$
2	$-\frac{4}{3}$
3	-3
-1	$-\frac{1}{3}$
-2	$-\frac{4}{3}$
-3	-3

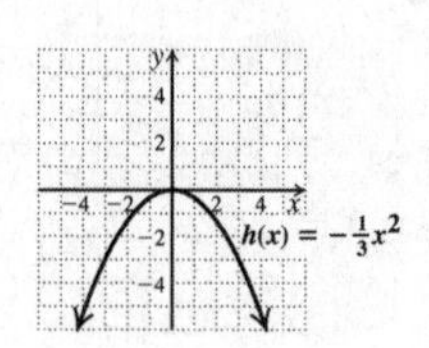

17. $f(x)=\frac{5}{2}x^2$

x	$f(x)=\frac{5}{2}x^2$
0	0
1	$\frac{5}{2}$
2	10
-1	$\frac{5}{2}$
-2	10

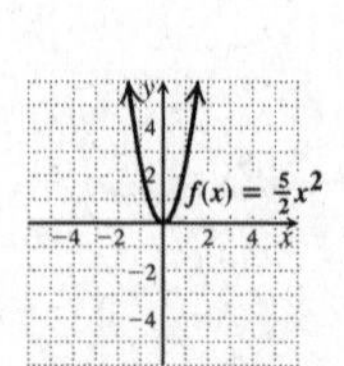

19. $g(x)=(x+1)^2=[x-(-1)]^2$

We know that the graph of $g(x)=(x+1)^2$ looks like the graph of $f(x)=x^2$ (see Exercise 9) but moved to the left 1 unit.

Vertex: (–1, 0), axis of symmetry: $x=-1$

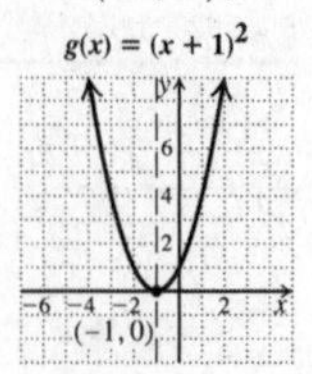

21. $f(x)=(x-2)^2$

The graph of $f(x)=(x-2)^2$ looks like the graph of $f(x)=x^2$ (see Exercise 9) but moved to the right 2 units.

Vertex: (2, 0), axis of symmetry: $x = 2$

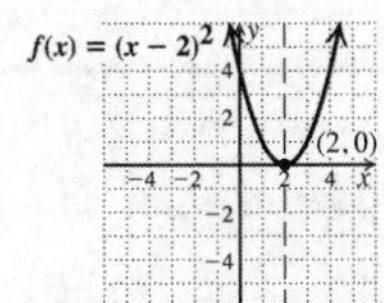

23. $g(x)=-(x+1)^2$

The graph of $g(x)=-(x+1)^2$ looks like the graph of $f(x)=x^2$ (see Exercise 9) but moved to the left 1 unit. It will also open downward because of the negative coefficient, –1.

Vertex: (–1, 0), axis of symmetry: $x = -1$

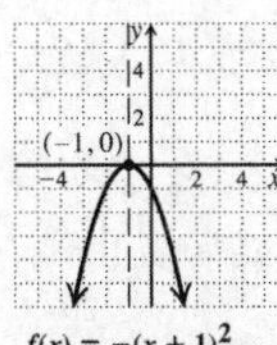

25. $g(x)=-(x-2)^2$

The graph of $g(x)=-(x-2)^2$ looks like the graph of $f(x)=x^2$ (see Exercise 9) but moved to the right 2 units. It will also open downward because of the negative coefficient, –1.

Vertex: (2, 0), axis of symmetry: $x = 2$

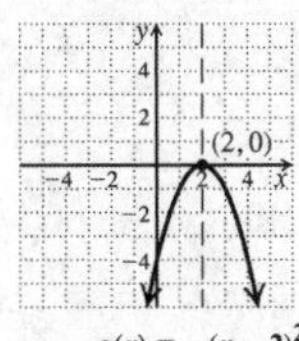

27. $f(x)=2(x+1)^2$

The graph of $f(x)=2(x+1)^2$ looks like the graph of $h(x)=2x^2$ (see graph following Example 1) but moved to the left 1 unit.

Vertex: (–1, 0), axis of symmetry: $x = -1$

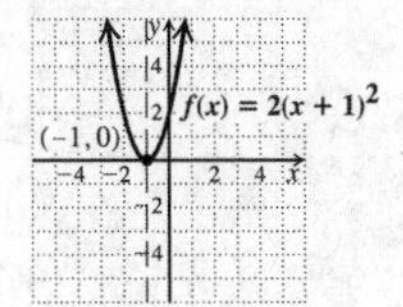

29. $g(x)=3(x-4)^2$

The graph of $g(x)=3(x-4)^2$ looks like the graph of $g(x)=3x^2$ but moved to the right 4 units.

Vertex: (4, 0), axis of symmetry: $x = 4$

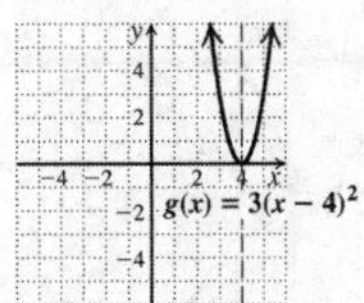

31. $h(x)=-\frac{1}{2}(x-4)^2$

The graph of $h(x)=-\frac{1}{2}(x-4)^2$ looks like the graph of $g(x)=\frac{1}{2}x^2$ (see graph following Example 1) but moved to the right 4 units. It will also open downward because of the negative coefficient, $-\frac{1}{2}$.

Vertex: (4,0) , axis of symmetry: $x=4$

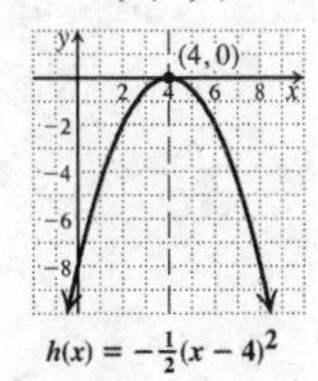

33. $f(x)=\dfrac{1}{2}(x-1)^2$

The graph of $f(x)=\dfrac{1}{2}(x-1)^2$ looks like the graph of $g(x)=\dfrac{1}{2}x^2$ (see graph following Example 1) but moved to the right 1 unit.

Vertex: (1, 0), axis of symmetry: $x = 1$

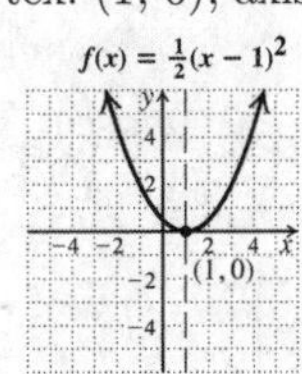

35. $f(x)=-2(x+5)^2=-2[x-(-5)]^2$

The graph of $f(x)=-2(x+5)^2$ looks like the graph of $h(x)=2x^2$ (see the graph following Example 1) but moved to the left 5 units. It will also open downward because of the negative coefficient, –2.

Vertex: (–5, 0), axis of symmetry: $x = -5$

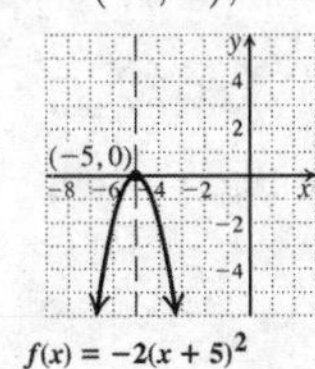

37. $h(x)=-3\left(x-\frac{1}{2}\right)^2$

The graph of $h(x)=-3\left(x-\frac{1}{2}\right)^2$ looks like the graph of $f(x)=-3x^2$ (see Exercise 12) but moved to the right $\frac{1}{2}$ unit.

Vertex: $\left(\frac{1}{2},\ 0\right)$, axis of symmetry: $x=\frac{1}{2}$

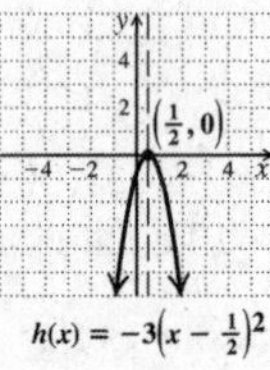

39. $f(x)=(x-5)^2+2$

We know that the graph looks like the graph of $f(x)=x^2$ (see Example 1) but moved to the right 5 units and up 2 units. The vertex is (5, 2), and the axis of symmetry is $x=5$. Since the coefficient of $(x-5)^2$ is positive $(1>0)$, there is a minimum function value, 2.

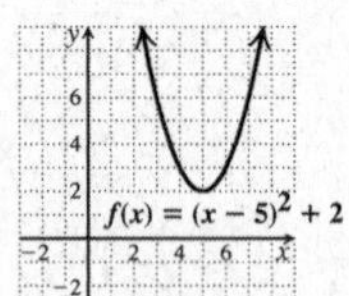

41. $f(x)=(x+1)^2-3$

We know that the graph looks like the graph of $f(x)=x^2$ (see Example 1) but moved to the left 1 unit and down 3 units. The vertex is (–1, –3), and the axis of symmetry is $x=-1$. Since the coefficient of $(x+1)^2$ is positive $(1>0)$, there is a minimum function value, –3.

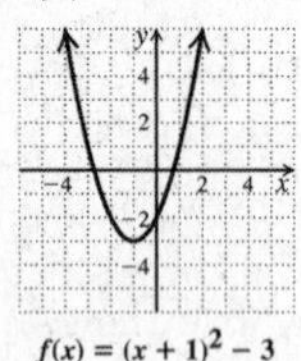

43. $g(x)=\frac{1}{2}(x+4)^2+1$

We know that the graph looks like the graph of $f(x)=\frac{1}{2}x^2$ (see graph following Example 1) but moved to the left 4 units and up 1 unit. The vertex is (–4, 1), and the axis of symmetry is $x = -4$, and the minimum function value is 1.

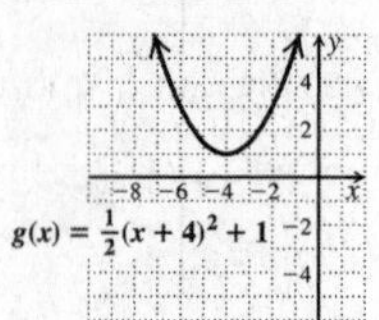

45. $h(x)=-2(x-1)^2-3$

We know that the graph looks like the graph of $h(x)=2x^2$ (see graph following Example 1) but moved to the right 1 unit and down 3 units and turned upside down. The vertex is (1, –3), and the axis of symmetry is $x=1$. The maximum function value is –3.

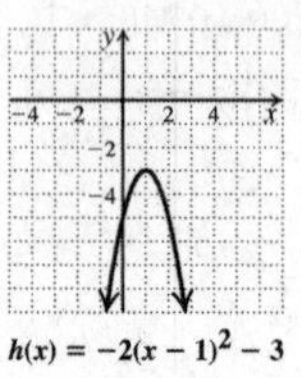

47. $f(x)=2(x+3)^2+1$

We know that the graph looks like the graph of $f(x)=2x^2$ (see graph following Example 1) but moved to the left 3 units and up 1 unit. The vertex is (–3, 1), and the axis of symmetry is $x = -3$. The minimum function value is 1.

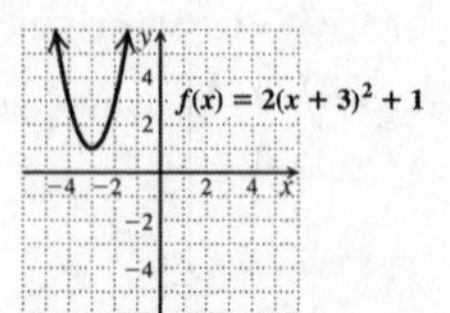

49. $g(x)=-\frac{3}{2}(x-2)^2+4$

We know that the graph looks like the graph of $f(x)=\frac{3}{2}x^2$ (see Exercise 18) but moved to the right 2 units and up 4 units and turned upside down. The vertex is (2, 4), and the axis of symmetry is $x = 2$, and the maximum function value is 4.

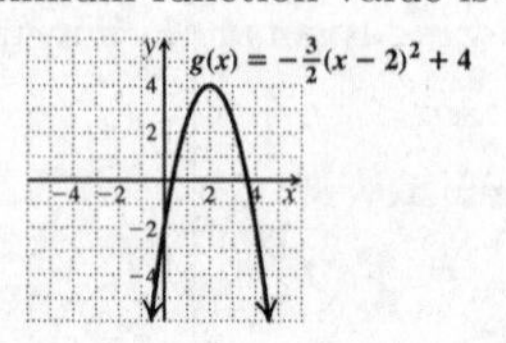

51. $f(x)=5(x-3)^2+9$

The function is of the form $f(x)=a(x-h)^2+k$ with $a = 5$, $h = 3$, and $k = 9$. The vertex is (h, k), or (3, 9). The axis of symmetry is $x = h$, or $x = 3$. Since $a > 0$, then k, or 9, is the minimum function value.

53. $f(x)=-\frac{3}{7}(x+8)^2+2$

The function is of the form $f(x)=a(x-h)^2+k$ with $a=-\frac{3}{7}$, $h=-8$, and $k=2$. The vertex is (h, k), or $(-8, 2)$. The axis of symmetry is $x=h$, or $x=-8$. Since $a<0$, then k, or 2, is the maximum function value.

55. $f(x)=\left(x-\frac{7}{2}\right)^2-\frac{29}{4}$

The function is of the form $f(x)=a(x-h)^2+k$ with $a=1$, $h=\frac{7}{2}$, and $k=-\frac{29}{4}$. The vertex is (h, k), or $\left(\frac{7}{2},-\frac{29}{4}\right)$. The axis of symmetry is $x=h$, or $x=\frac{7}{2}$. Since $a>0$, then k, or $-\frac{29}{4}$, is the minimum function value.

57. $f(x)=-\sqrt{2}(x+2.25)^2-\pi$

The function is of the form $f(x)=a(x-h)^2+k$ with $a=-\sqrt{2}$, $h=-2.25$, and $k=-\pi$. The vertex is (h, k), or $(-2.25,-\pi)$. The axis of symmetry is $x=h$, or $x=-2.25$. Since $a<0$, then k, or $-\pi$, is the maximum function value.

59. *Writing Exercise.* For any input, the output of $y=x^2-4$ is 4 less than (or 4 units down from) the output of $y=x^2$.

61. $8x-6y=24$

Find the x-intercept.

$$8x-6\cdot 0=24$$
$$8x=24$$
$$x=3$$

The x-intercept is $(3, 0)$.

Find the y-intercept.

$$8\cdot 0-6y=24$$
$$-6y=24$$
$$y=-4$$

The y-intercept is $(0, -4)$.

63. $f(x)=x^2+8x+15$

Find the x-intercepts.

$$x^2+8x+15=0$$
$$(x+5)(x+3)=0$$
$$x+5=0 \quad or \quad x+3=0$$
$$x=-5 \quad or \quad x=-3$$

The x-intercepts are $(-5, 0)$ and $(-3, 0)$.

65. x^2-14x

We take half the coefficient of x and square it.

$\frac{1}{2}(-14)=-7,\quad (-7)^2=49$

Then we have $x^2-14x+49=(x-7)^2$.

67. *Writing Exercise.* She uses symmetry to find the mirror images of the two ordered pairs she calculates after plotting the vertex.

69. The equation will be of the form $f(x)=\frac{3}{5}(x-h)^2+k$ with $h=1$ and $k=3$:

$$f(x)=\frac{3}{5}(x-1)^2+3$$

71. The equation will be of the form $f(x)=\frac{3}{5}(x-h)^2+k$ with $h=4$ and $k=-7$:

$$f(x)=\frac{3}{5}(x-4)^2-7$$

73. The equation will be of the form $f(x)=\frac{3}{5}(x-h)^2+k$ with $h=-2$ and $k=-5$:

$$f(x)=\frac{3}{5}[x-(-2)]^2+(-5), \text{or}$$
$$f(x)=\frac{3}{5}(x+2)^2-5$$

75. Since there is a minimum at $(2, 0)$, the parabola will have the same shape as $f(x)=2x^2$. It will be of the form $f(x)=2(x-h)^2+k$ with $h=2$ and $k=0$:

$$f(x)=2(x-2)^2$$

77. Since there is a maximum at $(0, -5)$, the parabola will have the same shape as $g(x)=-2x^2$. It will be of the form $g(x)=-2(x-h)^2+k$ with $h=0$ and $k=-5$:

$g(x)=-2(x-0)^2-5$, or $g(x)=-2x^2-5$.

79. If h is increased, the graph will move to the right.

81. If a is replaced with $-a$, the graph will be reflected across the x-axis.

83. The maximum value of $g(x)$ is 1 and occurs at the point $(5, 1)$, so for $F(x)$ we have $h=5$ and $k=1$. $F(x)$ has the same shape as $f(x)$ and has a minimum, so $a=3$. Thus, $F(x)=3(x-5)^2+1$.

85. The graph of $y = f(x-1)$ looks like the graph of $y = f(x)$ moved 1 unit to the right.

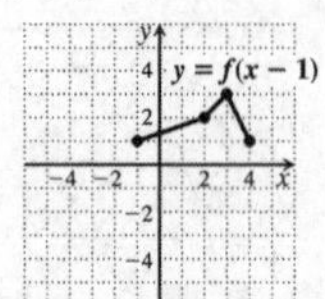

87. The graph of $y = f(x)+2$ looks like the graph of $y = f(x)$ moved up 2 units.

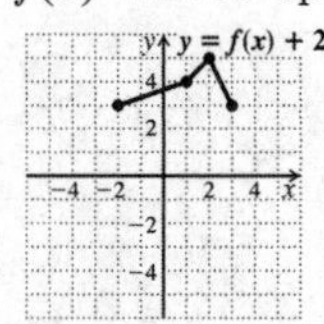

89. The graph of $y = f(x+3)-2$ looks like the graph of $y = f(x)$ moved 3 units to the left and also moved down 2 units.

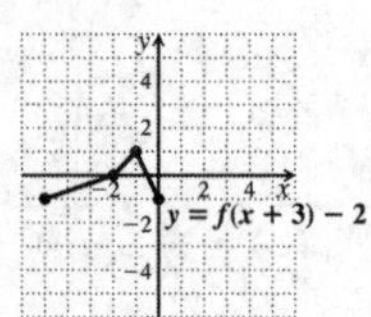

91. *Graphing Calculator Exercise*

93. *Writing Exercise.* The coefficient of x^2 is negative, so the parabola should open down.

Exercise Set 11.7

1. True; since $a = 3 > 0$, the graph opens upward.

3. True

5. False; the axis of symmetry is $x = \frac{3}{2}$.

7. False; the y-intercept is (0, 7).

9. $\frac{1}{2}\cdot(-8) = -4;\ \ (-4)^2 = 16$

$$\begin{aligned} f(x) &= x^2 - 8x + 2 \\ &= \left(x^2 - 8x + 16\right) - 16 + 2 \\ &= (x-4)^2 + (-14) \end{aligned}$$

11. $\frac{1}{2}\cdot 3 = \frac{3}{2};\ \ \left(\frac{3}{2}\right)^2 = \frac{9}{4}$

$$\begin{aligned} f(x) &= x^2 + 3x - 5 \\ &= \left(x^2 + 3x + \frac{9}{4}\right) - \frac{9}{4} - 5 \\ &= \left[x - \left(-\frac{3}{2}\right)\right]^2 + \left(-\frac{29}{4}\right) \end{aligned}$$

13. $\frac{1}{2}\cdot 2 = 1;\ \ 1^2 = 1$

$$\begin{aligned} f(x) &= 3x^2 + 6x - 2 \\ &= 3\left(x^2 + 2x\right) - 2 \\ &= 3\left(x^2 + 2x + 1\right) + 3(-1) - 2 \\ &= 3[x - (-1)]^2 + (-5) \end{aligned}$$

15. $\frac{1}{2}\cdot 4 = 2;\ \ 2^2 = 4$

$$\begin{aligned} f(x) &= -x^2 - 4x - 7 \\ &= -1\left(x^2 + 4x\right) - 7 \\ &= -\left(x^2 + 4x + 4\right) + -1(-4) - 7 \\ &= -[x - (-2)]^2 + (-3) \end{aligned}$$

17. $\frac{1}{2}\cdot\left(-\frac{5}{2}\right) = -\frac{5}{4};\ \ \left(-\frac{5}{4}\right)^2 = \frac{25}{16}$

$$\begin{aligned} f(x) &= 2x^2 - 5x + 10 \\ &= 2\left(x^2 - \frac{5}{2}x\right) + 10 \\ &= 2\left(x^2 - \frac{5}{2}x + \frac{25}{16}\right) + 2\left(-\frac{25}{16}\right) + 10 \\ &= 2\left(x - \frac{5}{4}\right)^2 + \frac{55}{8} \end{aligned}$$

19. a) $f(x) = x^2 + 4x + 5$

$$\begin{aligned} &= (x^2 + 4x + 4 - 4) + 5 \quad \text{Adding } 4-4 \\ &= (x^2 + 4x + 4) - 4 + 5 \quad \text{Regrouping} \\ &= (x+2)^2 + 1 \end{aligned}$$

The vertex is (–2, 1), the axis of symmetry is $x = -2$, and the graph opens upward since the coefficient 1 is positive. We plot a few points as a check and draw the curve.

b)

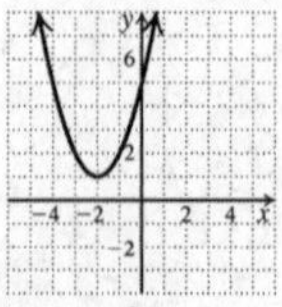

$f(x) = x^2 + 4x + 5$

21. a) $f(x) = x^2 + 8x + 20$

$$\begin{aligned} &= (x^2 + 8x + 16 - 16) + 20 \quad \text{Adding 16–16} \\ &= (x^2 + 8x + 16) - 16 + 20 \quad \text{Regrouping} \\ &= (x+4)^2 + 4 \end{aligned}$$

The vertex is (–4, 4), the axis of symmetry is

$x = -4$, and the graph opens upward since the coefficient 1 is positive.

b)

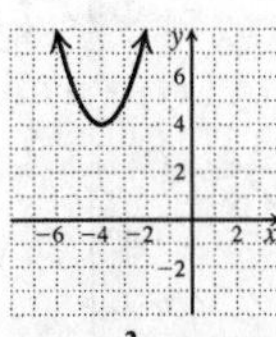

$f(x) = x^2 + 8x + 20$

23. a) $h(x) = 2x^2 - 16x + 25$

$= 2(x^2 - 8x) + 25$ Factoring 2 from the first two terms

$= 2(x^2 - 8x + 16 - 16) + 25$ Adding $16 - 16$ inside the parentheses

$= 2(x^2 - 8x + 16) + 2(-16) + 25$ Distributing to obtain a trinomial square

$= 2(x-4)^2 - 7$

The vertex is (4, –7), the axis of symmetry is $x = 4$, and the graph opens upward since the coefficient 2 is positive.

b)

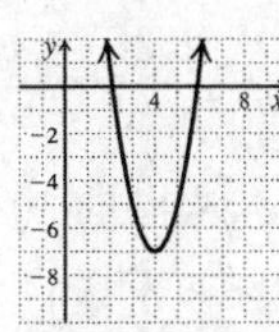

$h(x) = 2x^2 - 16x + 25$

25. a) $f(x) = -x^2 + 2x + 5$

$= -(x^2 - 2x) + 5$ Factoring -1 from the first two terms

$= -(x^2 - 2x + 1 - 1) + 5$ Adding $1 - 1$ inside the parentheses

$= -(x^2 - 2x + 1) - (-1) + 5$

$= -(x-1)^2 + 6$

The vertex is (1, 6), the axis of symmetry is $x = 1$, and the graph opens downward since the coefficient –1 is negative.

b)

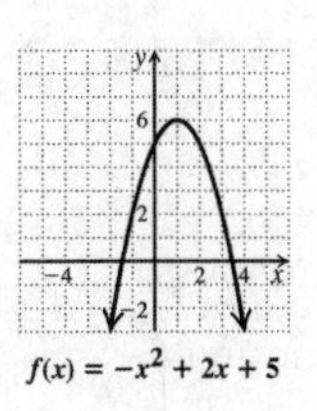

$f(x) = -x^2 + 2x + 5$

27. a) $g(x) = x^2 + 3x - 10$

$= \left(x^2 + 3x + \frac{9}{4} - \frac{9}{4}\right) - 10$

$= \left(x^2 + 3x + \frac{9}{4}\right) - \frac{9}{4} - 10$

$= \left(x + \frac{3}{2}\right)^2 - \frac{49}{4}$

The vertex is $\left(-\frac{3}{2}, -\frac{49}{4}\right)$, the axis of symmetry is $x = -\frac{3}{2}$, and the graph opens upward since the coefficient 1 is positive.

b)

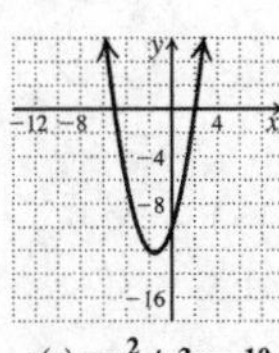

$g(x) = x^2 + 3x - 10$

29. a) $h(x) = x^2 + 7x$

$= \left(x^2 + 7x + \frac{49}{4}\right) - \frac{49}{4}$

$= \left(x + \frac{7}{2}\right)^2 - \frac{49}{4}$

The vertex is $\left(-\frac{7}{2}, -\frac{49}{4}\right)$, the axis of symmetry is $x = -\frac{7}{2}$, and the graph opens upward since the coefficient 1 is positive.

b)

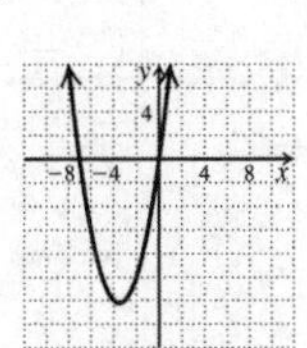

$h(x) = x^2 + 7x$

31. a) $f(x) = -2x^2 - 4x - 6$

$= -2(x^2 + 2x) - 6$ Factoring

$= -2(x^2 + 2x + 1 - 1) - 6$ Adding $1 - 1$ inside the parentheses

$= -2(x^2 + 2x + 1) - 2(-1) - 6$

$= -2(x+1)^2 - 4$

The vertex is (–1, –4), the axis of symmetry is $x = -1$, and the graph opens downward since the coefficient –2 is negative.

b)

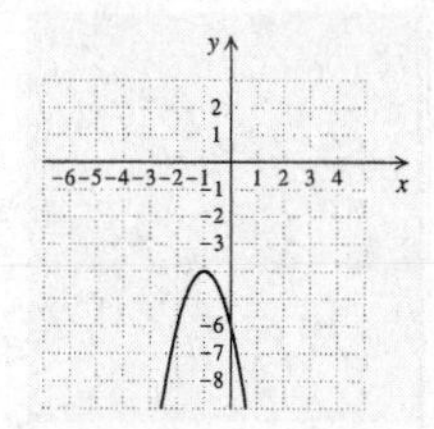

33. a)

$$\begin{aligned} g(x) &= x^2 - 6x + 13 \\ &= (x^2 - 6x + 9 - 9) + 13 \quad \text{Adding } 9 - 9 \\ &= (x^2 - 6x + 9) - 9 + 13 \quad \text{Regrouping} \\ &= (x - 3)^2 + 4 \end{aligned}$$

The vertex is (3, 4), the axis of symmetry is $x = 3$, and the graph opens upward since the coefficient 1 is positive. The minimum is 4.

b)

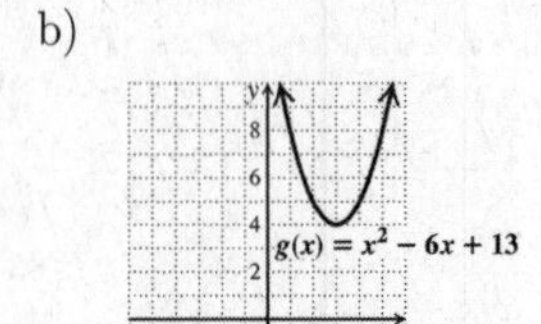

35. a)

$$\begin{aligned} g(x) &= 2x^2 - 8x + 3 \\ &= 2(x^2 - 4x) + 3 \quad \text{Factoring} \\ &= 2(x^2 - 4x + 4 - 4) + 3 \quad \text{Adding } 4 - 4 \text{ inside the parentheses} \\ &= 2(x^2 - 4x + 4) + 2(-4) + 3 \\ &= 2(x - 2)^2 - 5 \end{aligned}$$

The vertex is (2, –5), the axis of symmetry is $x = 2$, and the graph opens upward since the coefficient 2 is positive. The minimum is –5.

b)

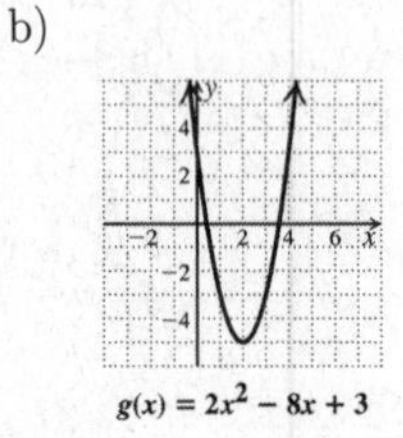
g(x) = 2x² − 8x + 3

37. a)

$$\begin{aligned} f(x) &= 3x^2 - 24x + 50 \\ &= 3(x^2 - 8x) + 50 \quad \text{Factoring} \\ &= 3(x^2 - 8x + 16 - 16) + 50 \quad \text{Adding } 16 - 16 \text{ inside the parentheses} \\ &= 3(x^2 - 8x + 16) - 3\cdot 16 + 50 \\ &= 3(x - 4)^2 + 2 \end{aligned}$$

The vertex is (4, 2), the axis of symmetry is $x = 4$, and the graph opens upward since the coefficient 3 is positive. The minimum is 2.

b)

f(x) = 3x² − 24x + 50

39. a)

$$\begin{aligned} f(x) &= -3x^2 + 5x - 2 \\ &= -3\left(x^2 - \frac{5}{3}x\right) - 2 \quad \text{Factoring} \\ &= -3\left(x^2 - \frac{5}{3}x + \frac{25}{36} - \frac{25}{36}\right) - 2 \quad \text{Adding } \frac{25}{36} - \frac{25}{36} \text{ inside the parentheses} \\ &= -3\left(x^2 - \frac{5}{3}x + \frac{25}{36}\right) - 3\left(-\frac{25}{36}\right) - 2 \\ &= -3\left(x - \frac{5}{6}\right)^2 + \frac{1}{12} \end{aligned}$$

The vertex is $\left(\frac{5}{6}, \frac{1}{12}\right)$, the axis of symmetry is $x = \frac{5}{6}$, and the graph opens downward since the coefficient –3 is negative. The maximum is $\frac{1}{12}$.

b)

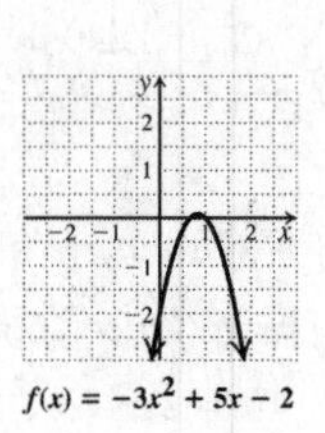
f(x) = −3x² + 5x − 2

41. a)

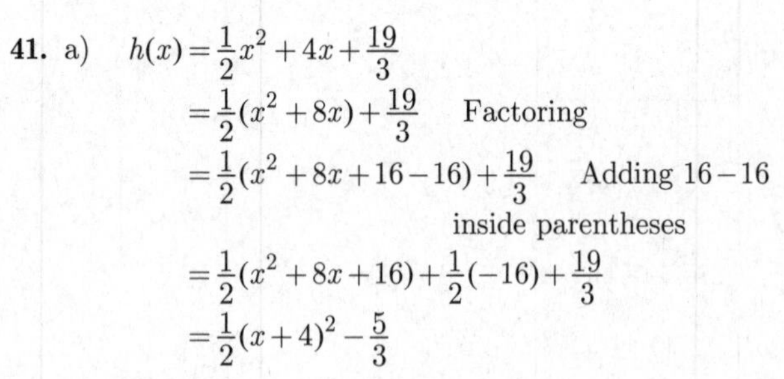

$$\begin{aligned} h(x) &= \frac{1}{2}x^2 + 4x + \frac{19}{3} \\ &= \frac{1}{2}(x^2 + 8x) + \frac{19}{3} \quad \text{Factoring} \\ &= \frac{1}{2}(x^2 + 8x + 16 - 16) + \frac{19}{3} \quad \text{Adding } 16 - 16 \text{ inside parentheses} \\ &= \frac{1}{2}(x^2 + 8x + 16) + \frac{1}{2}(-16) + \frac{19}{3} \\ &= \frac{1}{2}(x + 4)^2 - \frac{5}{3} \end{aligned}$$

The vertex is $\left(-4, -\frac{5}{3}\right)$, the axis of symmetry is $x = -4$, and the graph opens upward since the coefficient $\frac{1}{2}$ is positive. The minimum is $-\frac{5}{3}$.

b)

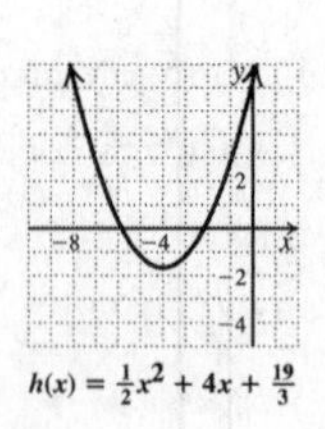
$h(x) = \frac{1}{2}x^2 + 4x + \frac{19}{3}$

43. $f(x) = x^2 - 6x + 3$

To find the x-intercepts, solve the equation $0 = x^2 - 6x + 3$. Use the quadratic formula.

$$x = \frac{-(-6) \pm \sqrt{(-6)^2 - 4\cdot 1\cdot 3}}{2\cdot 1}$$

$$x = \frac{6 \pm \sqrt{24}}{2} = \frac{6 \pm 2\sqrt{6}}{2} = 3 \pm \sqrt{6}$$

The x-intercepts are $(3 - \sqrt{6}, 0)$ and $(3 + \sqrt{6}, 0)$. The y-intercept is $(0, f(0))$, or (0, 3).

45. $g(x)=-x^2+2x+3$

To find the x-intercepts, solve the equation

$0=-x^2+2x+3$. We factor.

$$\begin{aligned} 0&=-x^2+2x+3 \\ 0&=x^2-2x-3 \quad \text{Multiplying by } -1 \\ 0&=(x-3)(x+1) \\ x&=3 \text{ or } x=-1 \end{aligned}$$

The x-intercepts are (–1, 0) and (3, 0).

The y-intercept is $(0,\ g(0))$, or (0, 3).

47. $f(x)=x^2-9x$

To find the x-intercepts, solve the equation $0=x^2-9x$.

We factor.

$$\begin{aligned} 0&=x^2-9x \\ 0&=x(x-9) \\ x&=0 \text{ or } x=9 \end{aligned}$$

The x-intercepts are (0, 0) and (9, 0).

Since (0, 0) is an x-intercept, we observe that (0, 0) is also the y-intercept.

49. $h(x)=-x^2+4x-4$

To find the x-intercepts, solve the equation

$0=-x^2+4x-4$. We factor.

$$\begin{aligned} 0&=-x^2+4x-4 \\ 0&=x^2-4x+4 \quad \text{Multiplying by} -1 \\ 0&=(x-2)(x-2) \\ x&=2 \text{ or } x=2 \end{aligned}$$

The x-intercept is (2, 0).

The y-intercept is $(0,\ h(0))$, or (0, –4).

51. $g(x)=x^2+x-5$

To find the x-intercepts, solve the equation

$0=x^2+x-5$. Use the quadratic formula.

$$x=\frac{-1\pm\sqrt{1^2-4\cdot 1\cdot(-5)}}{2\cdot 1}$$

$$x=\frac{-1\pm\sqrt{21}}{2}=-\frac{1}{2}\pm\frac{\sqrt{21}}{2}$$

The x-intercepts are $\left(-\frac{1}{2}-\frac{\sqrt{21}}{2},\ 0\right)$ and $\left(-\frac{1}{2}+\frac{\sqrt{21}}{2},\ 0\right)$.

The y-intercept is $(0,\ g(0))$, or (0, –5).

53. $f(x)=2x^2-4x+6$

To find the x-intercepts, solve the equation

$0=2x^2-4x+6$. We use the quadratic formula.

$$x=\frac{-(-4)\pm\sqrt{(-4)^2-4\cdot 2\cdot 6}}{2\cdot 2}$$

$$x=\frac{4\pm\sqrt{-32}}{4}=\frac{4\pm 4i\sqrt{2}}{2}=2\pm 2i\sqrt{2}$$

There are no real-number solutions, so there is no x-intercept.

The y-intercept is $(0,\ f(0))$, or (0, 6).

55. *Writing Exercise.* If the quadratic function opens downward and has no x-intercepts it must lie either in quadrant III or IV.

57.

$$\begin{aligned} x+y+z&=3, &(1) \\ x-y+z&=1, &(2) \\ -x-y+z&=-1 &(3) \end{aligned}$$

We eliminate y from two different pairs of equations.

$$\begin{aligned} x+y+z&=3 &(1) \\ x-y+z&=1 &(2) \\ \hline 2x\quad +2z&=4 & \\ x+z&=2 &(4) \end{aligned}$$

$$\begin{aligned} x+y+z&=3 &(1) \\ -x-y+z&=-1 &(3) \\ \hline 2z&=2 & \\ z&=1 & \end{aligned}$$

We eliminate not only y, but also x and found z = 1

Substitute 1 for z in Equation (4) to find x.

$$\begin{aligned} x+1&=2 \quad \text{Substituting 1 for } z \text{ in (4)} \\ x&=1 \end{aligned}$$

Substitute in one of the original equations to find y.

$$\begin{aligned} 1+y+1&=3 \quad \text{Substituting 1 for } z \text{ and 1 for } x \text{ in (1)} \\ y&=1 \end{aligned}$$

The solution is (1, 1, 1).

59.

$$\begin{aligned} z&=8, &(1) \\ x+y+z&=23, &(2) \\ 2x+y-z&=17 &(3) \end{aligned}$$

We eliminate y from equations (2) and (3).

$$\begin{aligned} x+y+z&=23 &(2) \\ -2x-y+z&=-17 &\text{Multiply (3) by } -1 \\ \hline -x\quad +2z&=6 &(4) \end{aligned}$$

Substitute 8 for z in Equation (4) to find x.

$$\begin{aligned} -x+2(8)&=6 \quad \text{Substituting 8 for } z \text{ in (4)} \\ -x&=-10 \\ x&=10 \end{aligned}$$

Substitute in one of the original equations to find y.

$$\begin{aligned} 10+y+8&=23 \quad \text{Substituting 8 for } z \text{ and 10 for } x \text{ in (2)} \\ y&=5 \end{aligned}$$

The solution is (10, 5, 8).

61.

$$\begin{aligned} 1.5&=c, &(1) \\ 52.5&=25a+5b+c, &(2) \\ 7.5&=4a+2b+c &(3) \end{aligned}$$

We eliminate b from equations (2) and (3).

$$\begin{aligned} 50a+10b+2c&=105 &\text{Multiply (2) by 2} \\ -20a-10b-5c&=-37.5 &\text{Multiply (3) } -5 \\ \hline 30a\qquad -3c&=67.5 &(4) \end{aligned}$$

Substitute 1.5 for c in Equation (4) to find a.

$30a - 3(1.5) = 67.5$ Substituting 1.5 for c in (4)
$30a = 72$
$a = 2.4$

Substitute in one of the original equations to find b.

$7.5 = 4(2.4) + 2b + 1.5$ Substituting 2.4 for a and 1.5 for c in (3)
$-1.8 = b$

The solution is (2.4, –1.8, 1.5).

63. *Writing Exercise.* No; the graphs could open in different directions and have different vertices. Consider the graphs of $f(x) = x^2 - 4$ and $g(x) = -x^2 + 4$, for example. Both have x-intercepts (–2, 0) and (2, 0), but the vertex of $f(x) = (0, -4)$ while the vertex of $g(x)$ is (0, 4).

65. a) $f(x) = 2.31x^2 - 3.135x - 5.89$
$= 2.31(x^2 - 1.357142857x) - 5.89$
$= 2.31(x^2 - 1.357142857x + 0.460459183 - 0.460459183) - 5.89$
$= 2.31(x^2 - 1.357142857x + 0.460459183) + 2.31(-0.460459183) - 5.89$
$= 2.31(x - 0.678571428)^2 - 6.953660714$

Since the coefficient 2.31 is positive, the function has a minimum value. It is –6.953660714.

b) To find the x-intercepts, solve $0 = 2.31x^2 - 3.135x - 5.89$.

$$x = \frac{-(-3.135) \pm \sqrt{(-3.135)^2 - 4(2.31)(-5.89)}}{2(2.31)}$$
$$x \approx \frac{3.135 \pm 8.015723611}{4.62}$$
$x \approx -1.056433682$ *or* $x \approx 2.413576539$

The x-intercepts are (–1.056433682, 0) and (2.413576539, 0).

The y-intercept is $(0, f(0))$, or (0, –5.89).

67. $f(x) = x^2 - x - 6$

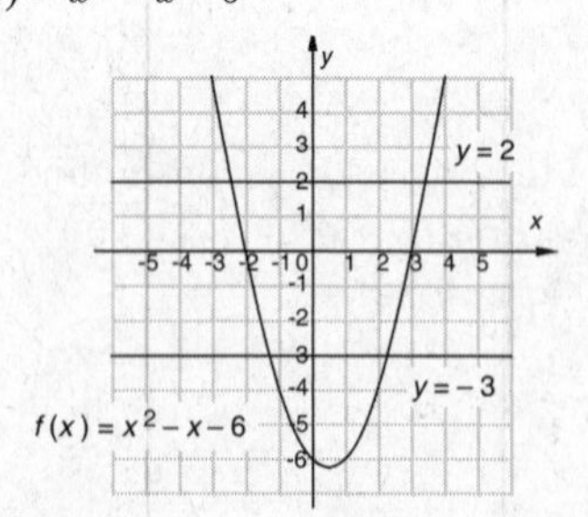

a) The solutions of $x^2 - x - 6 = 2$ are the first coordinates of the points of intersection of the graphs of $f(x) = x^2 - x - 6$ and $y = 2$. From the graph we see that the solutions are approximately –2.4 and 3.4.

b) The solutions of $x^2 - x - 6 = -3$ are the first coordinates of the points of intersection of the graphs of $f(x) = x^2 - x - 6$ and $y = -3$. From the graph we see that the solutions are approximately –1.3 and 2.3.

69. $f(x) = mx^2 - nx + p$
$$= m\left(x^2 - \frac{n}{m}x\right) + p$$
$$= m\left(x^2 - \frac{n}{m}x + \frac{n^2}{4m^2} - \frac{n^2}{4m^2}\right) + p$$
$$= m\left(x - \frac{n}{2m}\right)^2 - \frac{n^2}{4m} + p$$
$$= m\left(x - \frac{n}{2m}\right)^2 + \frac{-n^2 + 4mp}{4m}, \text{or}$$
$$m\left(x - \frac{n}{2m}\right)^2 + \frac{4mp - n^2}{4m}$$

71. The horizontal distance from (–1, 0) to (3, –5) is $|3 - (-1)|$, or 4, so by symmetry the other x-intercept is $(3 + 4, 0)$, or (7, 0). Substituting the three ordered pairs (–1, 0), (3,–5), and (7, 0) in the equation $f(x) = ax^2 + bx + c$ yields a system of equations:

$0 = a - b + c,$
$-5 = 9a + 3b + c,$
$0 = 49a + 7b + c$

The solution of this system of equations is $\left(\frac{5}{16}, -\frac{15}{8}, -\frac{35}{16}\right)$, so $f(x) = \frac{5}{16}x^2 - \frac{15}{8}x - \frac{35}{16}$.

If we complete the square we find that this function can also be expressed as $f(x) = \frac{5}{16}(x - 3)^2 - 5$.

73. $f(x) = |x^2 - 1|$

We plot some points and draw the curve. Note that it will lie entirely on or above the x-axis since absolute value is never negative.

x	$f(x)$
–3	8
–2	3
–1	0
0	1
1	0
2	3
3	8

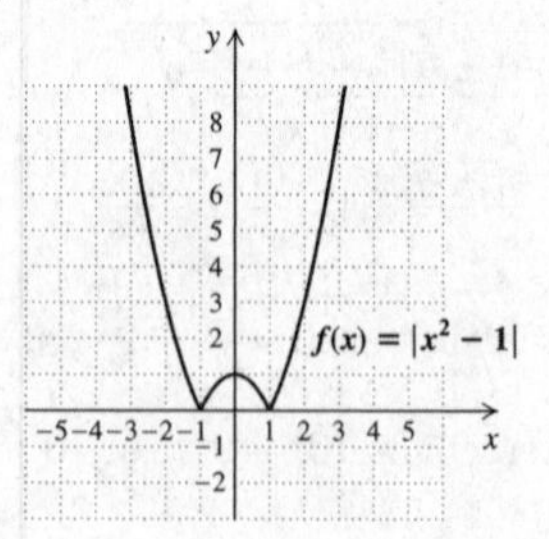

75. $f(x)=\left|2(x-3)^2-5\right|$

We plot some points and draw the curve. Note that it will lie entirely on or above the $x-$axis since absolute value is never negative.

x	$f(x)$
−1	27
0	13
1	3
2	3
3	5
4	3
5	3
6	13

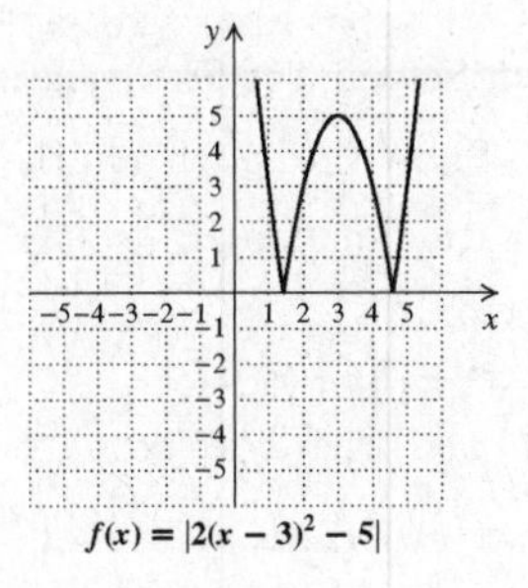

$f(x) = |2(x-3)^2 - 5|$

Section 11.8

1. e

3. c

5. d

7. ***Familiarize and Translate.*** We are given the formula

$p(x)=-0.2x^2+1.3x+6.2$.

Carry out. To find the value of x for which $p(x)$ is a maximum, we first find $-\dfrac{b}{2a}$:

$$-\frac{b}{2a}=-\frac{1.3}{2(-0.2)}=3.25, \text{ or } 3\frac{1}{4}$$

Now we find the maximum value of the function $p(3.25)$:

$$p(3.25)=-0.2(3.25)^2+1.3(3.25)+6.2=8.3125$$

The minimum function value of about 8.3 occurs when $x=3.25$.

Check. We can go over the calculations again. We could also solve the problem again by completing the square. The answer checks.

State. A calf's daily milk consumption is greatest at 3.25 weeks at about 8.3 lb of milk per day.

9. ***Familiarize and Translate.*** We want to find the value of x for which $C(x)=0.1x^2-0.7x+2.425$ is a minimum.

Carry out. We complete the square.

$$C(x)=0.1(x^2-7x+12.25)+2.425-1.225$$
$$C(x)=0.1(x-3.5)^2+1.2$$

The minimum function value of 1.2 occurs when $x=3.5$.

Check. Check a function value for x less than 3.5 and for x greater than 3.5.

$$C(3)=0.1(3)^2-0.7(3)+2.425=1.225$$
$$C(4)=0.1(4)^2-0.7(4)+2.425=1.225$$

Since 1.2 is less than these numbers, it looks as though we have a minimum.

State. The minimum average cost is \$1.2 hundred, or \$120. To achieve the minimum cost, 3.5 hundred, or 350 dulcimers should be built.

11. ***Familiarize.*** We make a drawing and label it.

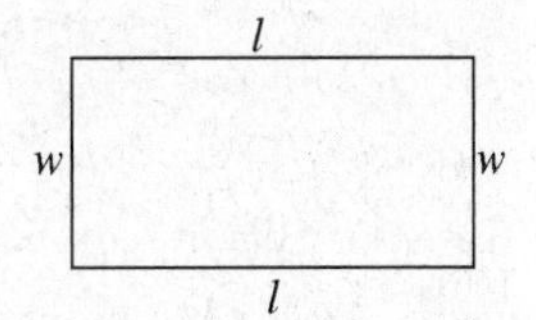

Perimeter: $2l + 2w = 720$ ft

Area: $A=l\cdot w$

Translate. We have a system of equations.

$$2l+2w=720,$$
$$A=lw$$

Carry out. Solving the first equation for l, we get $l = 360 - w$. Substituting for l in the second equation we get a quadratic function A:

$$A=(360-w)w$$
$$A=-w^2+360w$$

Completing the square, we get

$$A=-(w-180)^2+32{,}400$$

The maximum function value is 32,400. It occurs when w is 180. When $w = 180$, $l = 360 - 180$, or 180.

Check. We check a function value for w less than 180 and for w greater than 180.

$$A(179)=-179^2+360\cdot 179=32{,}399$$
$$A(181)=-181^2+360\cdot 181=32{,}399$$

Since 32,400 is greater than these numbers, it looks as though we have a maximum.

State. The maximum area occurs when the dimensions are 180 ft by 180 ft.

13. ***Familiarize.*** We make a drawing and label it.

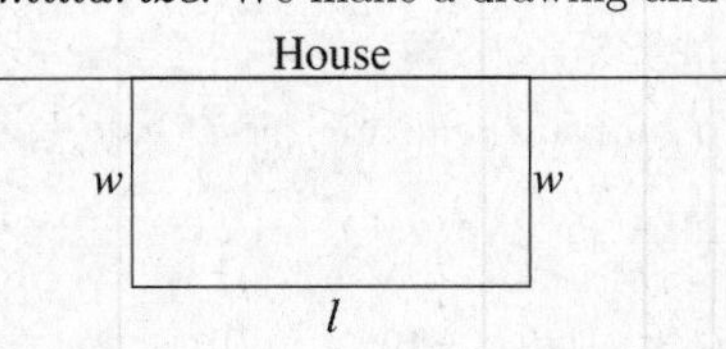

Translate. We have two equations.

$$l+2w=60,$$
$$A=lw$$

Carry out. Solve the first equation for l.

$$l=60-2w$$

Substitute for l in the second equation.

$A = (60 - 2w)w$
$A = -2w^2 + 60w$

Completing the square, we get

$A = -2(w - 15)^2 + 450$.

The maximum function value of 450 occurs when $w = 15$.

When $w = 15$, $l = 60 - 2 \cdot 15 = 30$.

Check. Check a function value for w less than 15 and for w greater than 15.

$A(14) = -2 \cdot 14^2 + 60 \cdot 14 = 448$
$A(16) = -2 \cdot 16^2 + 60 \cdot 16 = 448$

Since 450 is greater than these numbers, it looks as though we have a maximum.

State. The maximum area of 450 ft ^{2} will occur when the dimensions are 15 ft by 30 ft.

15. ***Familiarize.*** Let x represent the height of the file and y represent the width. We make a drawing.

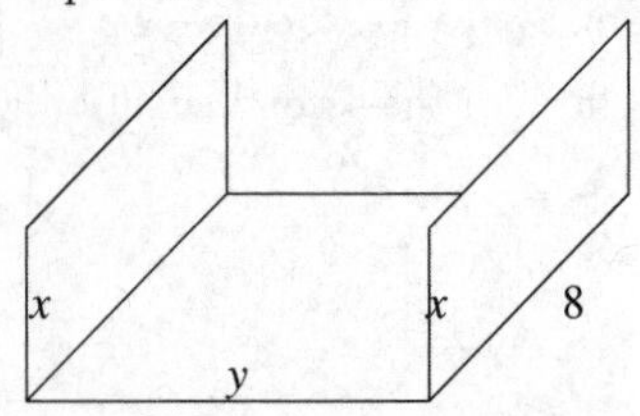

Translate. We have two equations.

$2x + y = 14$
$V = 8xy$

Carry out. Solve the first equation for y.

$y = 14 - 2x$

Substitute for y in the second equation.

$V = 8x(14 - 2x)$
$V = -16x^2 + 112x$

Completing the square, we get

$V = -16\left(x - \frac{7}{2}\right)^2 + 196$.

The maximum function value of 196 occurs when $x = \frac{7}{2}$.

When $x = \frac{7}{2}$, $y = 14 - 2 \cdot \frac{7}{2} = 7$.

Check. Check a function value for x less than $\frac{7}{2}$ and for x greater than $\frac{7}{2}$.

$V(3) = -16 \cdot 3^2 + 112 \cdot 3 = 192$
$V(4) = -16 \cdot 4^2 + 112 \cdot 4 = 192$

Since 196 is greater than these numbers, it looks as though we have a maximum.

State. The file should be $\frac{7}{2}$ in., or 3.5 in. tall.

17. ***Familiarize.*** We let x and y represent the numbers, and we let P represent their product.

Translate. We have two equations.

$x + y = 18,$
$P = xy$

Carry out. Solving the first equation for y, we get $y = 18 - x$. Substituting for y in the second equation we get a quadratic function P:

$P = x(18 - x)$
$P = -x^2 + 18x$

Completing the square, we get

$P = -(x - 9)^2 + 81$.

The maximum function value is 81. It occurs when $x = 9$.

When $x = 9$, $y = 18 - 9$, or 9.

Check. We can check a function value for x less than 9 and for x greater than 9.

$P(10) = -10^2 + 18 \cdot 10 = 80$
$P(8) = -8^2 + 18 \cdot 8 = 80$

Since 81 is greater than these numbers, it looks as though we have a maximum.

State. The maximum product of 81 occurs for the numbers 9 and 9.

19. ***Familiarize.*** We let x and y represent the two numbers, and we let P represent their product.

Translate. We have two equations.

$x - y = 8,$
$P = xy$

Carry out. Solve the first equation for x.

$x = 8 + y$

Substitute for x in the second equation.

$P = (8 + y)y$
$P = y^2 + 8y$

Completing the square, we get

$P = (y + 4)^2 - 16$.

The minimum function value is –16. It occurs when $y = -4$. When $y = -4$, $x = 8 + (-4)$, or 4.

Check. Check a function value for y less than –4 and for y greater than –4.

$P(-5) = (-5)^2 + 8(-5) = -15$
$P(-3) = (-3)^2 + 8(-3) = -15$

Since –16 is less than these numbers, it looks as though we have a minimum.

State. The minimum product of –16 occurs for the numbers 4 and –4.

21. From the results of Exercises 17 and 18, we might observe that the numbers are –5 and –5 and that the maximum product is 25. We could also solve this problem as follows.

Familiarize. We let x and y represent the two numbers, and we let P represent their product.

Translate. We have two equations.

$$x+y=-10,$$
$$P=xy$$

Carry out. Solve the first equation for y.

$$y=-10-x$$

Substitute for y in the second equation.

$$P=x(-10-x)$$
$$P=-x^2-10x$$

Completing the square, we get

$$P=-(x+5)^2+25$$

The maximum function value is 25. It occurs when $x=-5$. When $x=-5$, $y=-10-(-5)$, or -5.

Check. Check a function value for x less than –5 and for x greater than –5.

$$P(-6)=-(-6)^2-10(-6)=24$$
$$P(-4)=-(-4)^2-10(-4)=24$$

Since 25 is greater than these numbers, it looks as though we have a maximum.

State. The maximum product of 25 occurs for the numbers –5 and –5.

23. The data points rise and then fall. The graph appears to represent a quadratic function that opens downward. Thus a quadratic function $f(x)=ax^2+bx+c,\ a<0,$ might be used to model the data.

25. The data points rise. The graph does not appear to represent a quadratic function in which the data points would rise and then fall or vice versa. Thus a linear function $f(x)=mx+b$ might be used to model the data.

27. The data points do not represent a linear or quadratic pattern. Thus, it does not appear that the data can be modeled with either a quadratic or a linear function.

29. The data points fall and then rise. The graph appears to represent a quadratic function that opens upward. Thus a quadratic function $f(x)=ax^2+bx+c,\ a>0,$ might be used to model the data.

31. The data points appear to represent the right half of a quadratic function that opens upward. Thus a quadratic function $f(x)=ax^2+bx+c,\ a>0,$ might be used to model the data.

33. The data points fall. The graph does not appear to represent a quadratic function in which the data points would rise and then fall or vice versa. Thus a linear function $f(x)=mx+b$ might be used to model the data.

35. We look for a function of the form $f(x)=ax^2+bx+c$. Substituting the data points, we get

$$4=a(1)^2+b(1)+c,$$
$$-2=a(-1)^2+b(-1)+c,$$
$$13=a(2)^2+b(2)+c,$$

or

$$4=a+b+c,$$
$$-2=a-b+c,$$
$$13=4a+2b+c.$$

Solving this system, we get

$a=2$, $b=3$, and $c=-1$.

Therefore the function we are looking for is

$$f(x)=2x^2+3x-1.$$

37. We look for a function of the form $f(x)=ax^2+bx+c$. Substituting the data points, we get

$$0=a(2)^2+b(2)+c,$$
$$3=a(4)^2+b(4)+c,$$
$$-5=a(12)^2+b(12)+c,$$

or

$$0=4a+2b+c,$$
$$3=16a+4b+c,$$
$$-5=144a+12b+c.$$

Solving this system, we get

$a=-\frac{1}{4}$, $b=3$, $c=-5$.

Therefore the function we are looking for is

$$f(x)=-\frac{1}{4}x^2+3x-5.$$

39. a) ***Familiarize.*** We look for a function of the form $A(s)=as^2+bs+c$, where $A(s)$ represents the number of nighttime accidents (for every 200 million km) and s represents the travel speed (in km/h).

Translate. We substitute the given values of s and $A(s)$.

$$400 = a(60)^2 + b(60) + c,$$
$$250 = a(80)^2 + b(80) + c,$$
$$250 = a(100)^2 + b(100) + c,$$

or

$$400 = 3600a + 60b + c,$$
$$250 = 6400a + 80b + c,$$
$$250 = 10{,}000a + 100b + c.$$

Carry out. Solving the system of equations, we get

$a = \frac{3}{16}$, $b = -\frac{135}{4}$, $c = 1750$.

Check. Recheck the calculations.

State. The function

$A(s) = \frac{3}{16}s^2 - \frac{135}{4}s + 1750$ fits the data.

b) Find $A(50)$.

$$A(50) = \frac{3}{16}(50)^2 - \frac{135}{4}(50) + 1750 = 531.25$$

About 531 accidents occur at 50 km/h.

41. ***Familiarize.*** Think of a coordinate system placed on the drawing in the text with the origin at the point where the arrow is released. Then three points on the arrow's parabolic path are (0, 0), (63, 27), and (126, 0). We look for a function of the form $h(d) = ad^2 + bd + c$, where $h(d)$ represents the arrow's height and d represents the distance the arrow has traveled horizontally.

Translate. We substitute the values given above for d and $h(d)$.

$$0 = a \cdot 0^2 + b \cdot 0 + c,$$
$$27 = a \cdot 63^2 + b \cdot 63 + c,$$
$$0 = a \cdot 126^2 + b \cdot 126 + c$$

or

$$0 = c,$$
$$27 = 3969a + 63b + c,$$
$$0 = 15{,}876a + 126b + c$$

Carry out. Solving the system of equations, we get $a \approx -0.0068$, $b \approx 0.8571$, and $c = 0$.

Check. Recheck the calculations.

State. The function $h(d) = -0.0068d^2 + 0.8571d$ expresses the arrow's height as a function of the distance it has traveled horizontally.

43. *Writing Exercise.* The graph of a nonlinear function could extend without bound in both the positive and negative directions and thus have neither a minimum nor a maximum value.

45.
$$2x - 3 > 5$$
$$2x > 8$$
$$x > 4$$

The solution set is $\{x | x > 4\}$, or $(4, \infty)$.

47. $|9 - x| \geq 2$

$$9 - x \leq -2 \quad or \quad 2 \leq 9 - x$$
$$-x \leq -11 \quad or \quad -7 \leq -x$$
$$x \geq 11 \quad or \quad 7 \geq x$$

The solution set is $\{x | x \leq 7 \text{ or } x \geq 11\}$, or $(-\infty, 7] \cup [11, \infty)$.

49. $f(x) = \frac{x-3}{x+4} - 5$

$$= \frac{x-3}{x+4} - 5$$

Note that $x \neq -4$.

$$= \frac{x-3}{x+4} - 5 \cdot \frac{x+4}{x+4}$$
$$= \frac{x-3-5(x+4)}{x+4}$$
$$= \frac{x-3-5x-20}{x+4}$$
$$= \frac{-4x-23}{x+4},\ x \neq -4$$

51. Note the restriction that $x \neq -4$.

$$\frac{x-3}{x+4} = 5$$
$$(x+4) \cdot \frac{x-3}{x+4} = (x+4) \cdot 5$$
$$x - 3 = 5x + 20$$
$$-23 = 4x$$
$$-\frac{23}{4} = x$$

The solution is $-\frac{23}{4}$.

53. Note the restriction that $x \neq -7, 3$.

$$\frac{x}{(x-3)(x+7)} = 0$$
$$(x-3)(x+7) \cdot \frac{x}{(x-3)(x+7)} = (x-3)(x+7) \cdot 0$$
$$x = 0$$

The solution is 0.

55. *Writing Exercise.* The graph for the other pitchers appears to be quadratic. It starts at 3.5 at age 20, seems to have a minimum of 3.3 around the age of 30, and a maximum of 4.5 at age 46. For Clemens, the graph appears to be more linear, having a minimum of 3.1 at 23 and continuing to a maximum of 3.3 at age 46. Clemens has no fall and rise with age, just a small rise.

57. ***Familiarize.*** Position the bridge on a coordinate system as shown with the vertex of the parabola at (0, 30).

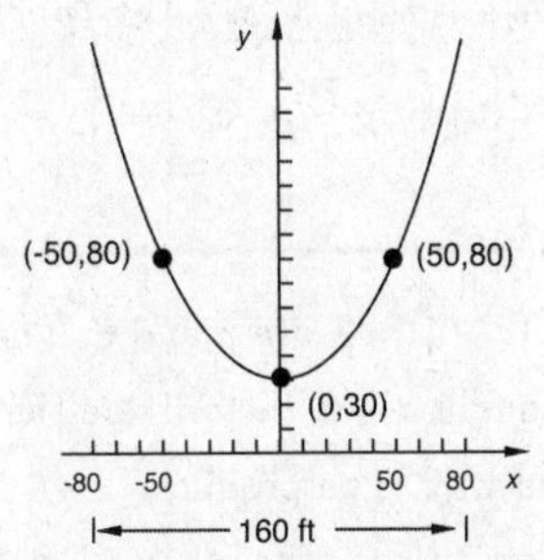

We find a function of the form $y = ax^2 + bx + c$ which represents the parabola containing the points (0, 30), (–50, 80), and (50, 80).

Translate. Substitute for x and y.

$$30 = a \cdot 0^2 + b \cdot 0 + c,$$
$$80 = a(-50)^2 + b(-50) + c,$$
$$80 = a(50)^2 + b(50) + c,$$

or

$$30 = c,$$
$$80 = 2500a - 50b + c,$$
$$80 = 2500a + 50b + c.$$

Carry out. Solving the system of equations, we get

$a = 0.02$, $b = 0$, $c = 30$.

The function $y = 0.02x^2 + 30$ represents the parabola. Because the cable supports are 160 ft apart, the tallest supports are positioned 160/2, or 80 ft, to the left and right of the midpoint. This means that the longest vertical cables occur at $x = -80$ and $x = 80$. For $x = \pm 80$,

$$y = 0.02(\pm 80)^2 + 30$$
$$= 128 + 30$$
$$= 158 \text{ ft}$$

Check. We go over the calculations.

State. The longest vertical cables are 158 ft long.

59. ***Familiarize.*** Let x represent the number of 25 increases in the admission price. Then $10 + 0.25x$ represents the admission price, and $80 - x$ represents the corresponding average attendance. Let R represent the total revenue.

Translate. Since the total revenue is the product of the cover charge and the number attending a show, we have the following function for the amount of money the owner makes.

$$R(x) = (10 + 0.25x)(80 - x), \text{ or}$$
$$R(x) = -0.25x^2 + 10x + 800$$

Carry out. Completing the square, we get

$$R(x) = -0.25(x - 20)^2 + 900$$

The maximum function value of 900 occurs when $x = 20$. The owner should charge \$10 + \$0.25(20), or \$15.

Check. We check a function value for x less than 20 and for x greater than 20.

$$R(19) = -0.25(19)^2 + 10 \cdot 19 + 800 = 899.75$$
$$R(21) = -0.25(21)^2 + 10 \cdot 21 + 800 = 899.75$$

Since 900 is greater than these numbers, it looks as though we have a maximum.

State. The owner should charge \$15.

61. ***Familiarize.*** We add labels to the drawing in the text.

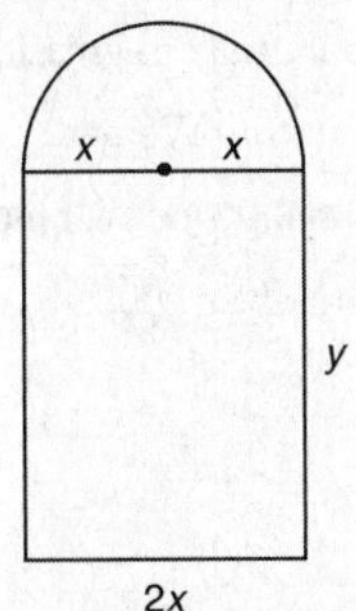

The perimeter of the semicircular portion of the window is $\frac{1}{2} \cdot 2\pi x$, or πx. The perimeter of the rectangular portion is $y + 2x + y$, or $2x + 2y$. The area of the semicircular portion of the window is $\frac{1}{2} \cdot \pi x^2$, or $\frac{\pi}{2} x^2$. The area of the rectangular portion is $2xy$.

Translate. We have two equations, one giving the perimeter of the window and the other giving the area.

$$\pi x + 2x + 2y = 24,$$
$$A = \frac{\pi}{2} x^2 + 2xy$$

Carry out. Solve the first equation for y.

$$\pi x + 2x + 2y = 24$$
$$2y = 24 - \pi x - 2x$$
$$y = 12 - \frac{\pi x}{2} - x$$

Substitute for y in the second equation.

$$A = \frac{\pi}{2} x^2 + 2x\left(12 - \frac{\pi x}{2} - x\right)$$
$$A = \frac{\pi}{2} x^2 + 24x - \pi x^2 - 2x^2$$
$$A = -2x^2 - \frac{\pi}{2} x^2 + 24x$$
$$A = -\left(2x + \frac{\pi}{2}\right) x^2 + 24x$$

Completing the square, we get

$$A = -\left(2 + \frac{\pi}{2}\right)\left(x^2 + \frac{24}{-\left(2 + \frac{\pi}{2}\right)} x\right)$$
$$A = -\left(2 + \frac{\pi}{2}\right)\left(x^2 - \frac{48}{4 + \pi} x\right)$$
$$A = -\left(2 + \frac{\pi}{2}\right)\left(x - \frac{24}{4 + \pi}\right)^2 + \left(\frac{24}{4 + \pi}\right)^2$$

The maximum function value occurs when

$x=\frac{24}{4+\pi}$. When $x=\frac{24}{4+\pi}$,

$y=12-\frac{\pi}{2}\left(\frac{24}{4+\pi}\right)-\frac{24}{4+\pi}$

$=\frac{48+12\pi}{4+\pi}-\frac{12\pi}{4+\pi}-\frac{24}{4+\pi}=\frac{24}{4+\pi}$

Check. Recheck the calculations.

State. The radius of the circular portion of the window and the height of the rectangular portion should each be $\frac{24}{4+\pi}$ ft.

63. a) Enter the data and use the quadratic regression operation on a graphing calculator. We get $h(x)=11{,}090.60714x^2-29{,}069.62143x+39{,}983.8$, where x is the number of years after 2000.

b) In 2010, $x = 2010 - 2000 = 10$.

$h(10)\approx 858{,}348$ vehicles

Exercise Set 11.9

1. The solutions of $(x-3)(x+2)=0$ are 3 and –2 and for a test value in [–2, 3], say 0, $(x-3)(x+2)$ is negative so the statement is true. (Note that the endpoints must be included in the solution set because the inequality symbol is $\leq$.)

3. The solutions of $(x-1)(x-6)=0$ are 1 and 6. For a value of x less than 1, say 0, $(x-1)(x-6)$ is positive; for a value of x greater than 6, say 7, $(x-1)(x-6)$ is also positive. Thus, the statement is true. (Note that the endpoints of the intervals are not included because the inequality symbol is >.)

5. Since $x + 2 = 0$ when $x = -2$ and $x - 3 = 0$ when $x = 3$, the statement is false.

7. $p(x)\leq 0$ when $-4\leq x\leq \frac{3}{2}$,

$\left[-4, \frac{3}{2}\right]$ or $\left\{x\middle|-4\leq x\leq \frac{3}{2}\right\}$

9. $x^4+12x>3x^2+4x^2$ is equivalent to $x^4-3x^2-4x^2+12x>0$, which is the graph in the text.

$p(x)>0$ when $(-\infty,-2)\cup(0, 2)\cup(3, \infty)$ or $\{x|x<-2 \text{ or } 0<x<2 \text{ or } x>3\}$

11. $\frac{x-1}{x+2}<3$ is equivalent to finding the values of x for which the graph $r(x)$ is less than 3, or below $g(x)$.

$\left(-\infty,-\frac{7}{2}\right)\cup(-2, \infty)$ or $\left\{x\middle|x<-\frac{7}{2} \text{ or } x>-2\right\}$

13. $(x-6)(x-5)<0$

The solutions of $(x-6)(x-5)=0$ are 5 and 6. They are not solutions of the inequality, but they divide the real number line in a natural way. The product $(x-6)(x-5)$ is positive or negative, for values other than 5 and 6., depending on the signs of the factors $x - 6$ and $x - 5$.

$x - 6 > 0$ when $x > 6$ and $x - 6 < 0$ when $x < 6$.

$x - 5 > 0$ when $x > 5$ and $x - 5 < 0$ when $x < 5$

We make a diagram.

		5		6	
Sign of $x-6$	–		–		+
Sign of $x-5$	–		+		+
Sign of product	+		–		+

For the product $(x-6)(x-5)$ to be negative, one factor must be positive and the other negative. We see from the diagram that numbers satisfying $5 < x < 6$ are solutions. The solution set of the inequality is (5, 6) or $\{x|5<x<6\}$.

15. $(x+7)(x-2)\geq 0$

The solutions of $(x+7)(x-2)=0$ are –7 and 2. They divide the number line into three intervals as shown:

A B C
–7 2

We try test numbers in each interval.

A: Test –8, $f(-8)=(-8+7)(-8-2)=10$

B: Test 0, $f(0)=(0+7)(0-2)=-14$

C: Test 3, $f(3)=(3+7)(3-2)=10$

Since $f(-8)$ and $f(3)$ are positive, the function value will be positive for all numbers in the intervals containing –8 and 3. The inequality symbol is $\leq$, so we need to include the endpoints. The solution set is $(-\infty,-7]\cup[2, \infty)$, or $\{x\,|\,x\leq -7 \text{ or } x\geq 2\}$.

17. $x^2-x-2>0$

$(x+1)(x-2)>0$ Factoring

The solutions of $(x+1)(x-2)=0$ are –1 and 2. They divide the number line into three intervals as shown:

A B C
–1 2

We try test numbers in each interval.

A: Test –2, $f(-2)=(-2+1)(-2-2)=4$

B: Test 0, $f(0)=(0+1)(0-2)=-2$

C: Test 3, $f(3)=(3+1)(3-2)=4$

Since $f(-2)$ and $f(3)$ are positive, the function value will be positive for all numbers in the intervals containing –2 and 3. The solution set is $(-\infty,-1)\cup(2,\ \infty)$, or $\{x\,|\,x<-1 \text{ or } x>2\}$.

19. $x^2+4x+4<0$

$(x+2)^2<0$

Observe that $(x+2)^2\geq 0$ for all values of x. Thus, the solution set is $\varnothing$.

21.
$$x^2-4x\leq 3$$
$$x^2-4x+4\leq 3+4$$
$$(x-2)^2\leq 7$$
$$x-2\leq\pm\sqrt{7}$$
$$x\leq 2\pm\sqrt{7}$$

The solutions of $x^2-4x-3\leq 0$ are $2\pm\sqrt{7}$. They divide the number line into three intervals as shown:

A B C

$2-\sqrt{7}$ $2+\sqrt{7}$

We try test numbers in each interval.

A: Test –1, $f(-1)=(-1)^2-4(-1)-3=2$

B: Test 0, $f(0)=0^2-4(0)-3=-3$

C: Test 5, $f(5)=5^2-4(5)-3=2$

Since $f(0)$ is negative, the function value will be negative for all numbers in the interval containing 0. The solution set is $[2-\sqrt{7},\ 2+\sqrt{7}]$, or $\{x\,|\,2-\sqrt{7}\leq x\leq 2+\sqrt{7}\}$.

23. $3x(x+2)(x-2)<0$

The solutions of $3x(x+2)(x-2)=0$ are 0, –2, and 2. They divide the real-number line into four intervals as shown:

A B C D

–2 0 2

We try test numbers in each interval.

A: Test –3, $f(-3)=3(-3)(-3+2)(-3-2)=-45$

B: Test –1, $f(-1)=3(-1)(-1+2)(-1-2)=9$

C: Test 1, $f(1)=3(1)(1+2)(1-2)=-9$

D: Test 3, $f(3)=3(3)(3+2)(3-2)=45$

Since $f(-3)$ and $f(1)$ are negative, the function value will be negative for all numbers in the intervals containing –3 and 1. The solution set is $(-\infty,-2)\cup(0,\ 2)$, or $\{x\,|\,x<-2 \text{ or } 0<x<2\}$.

25. $(x-1)(x+2)(x-4)\geq 0$

The solutions of $(x-1)(x+2)(x-4)=0$ are 1, –2, and 4. They divide the real-number line in a natural way. The product $(x-1)(x+2)(x-4)$ is positive or negative depending on the signs of $x-1$, $x+2$, and $x-4$.

Sign of $x-1$	–	–	+	+
Sign of $x+2$	–	+	+	+
Sign of $x-4$	–	–	–	+
Sign of product	–	+	–	+

–2 1 4

A product of three numbers is positive when all three factors are positive or when two are negative and one is positive. Since the $\geq$ symbol allows for equality, the endpoints –2, 1, and 4 are solutions. From the chart we see that the solution set is $[-2,\ 1]\cup[4,\ \infty)$, or $\{x\,|\,-2\leq x\leq 1 \text{ or } x\geq 4\}$.

27.
$$f(x)\geq 3$$
$$7-x^2\geq 3$$
$$-x^2+4\geq 0$$
$$x^2-4\leq 0$$
$$(x-2)(x+2)\leq 0$$

The solutions of $(x-2)(x+2)=0$ are 2 and –2. They divide the real-number line as shown below.

Sign of $x-2$	–	–	+
Sign of $x+2$	–	+	+
Sign of product	+	–	+

–2 2

Because the inequality symbol is $\leq$, we must include the endpoints in the solution set. From the chart, we see that the solution set is $[-2,\ 2]$, or $\{x\,|\,-2\leq x\leq 2\}$.

29.
$$g(x)>0$$
$$(x-2)(x-3)(x+1)>0$$

The solutions of $(x-2)(x-3)(x+1)=0$ are 2, 3, and -1. They divide the real-number line into four intervals as shown below.

A B C D

–1 2 3

We try test numbers in each interval.

A: Test –2, $f(-2)=(-2-2)(-2-3)(-2+1)=-20$

B: Test 0, $f(0)=(0-2)(0-3)(0+1)=6$

C: Test $\frac{5}{2}$, $f\left(\frac{5}{2}\right)=\left(\frac{5}{2}-2\right)\left(\frac{5}{2}-3\right)\left(\frac{5}{2}+1\right)=-\frac{7}{8}$

D: Test 4, $f(4)=(4-2)(4-3)(4+1)=10$

The function value will be positive for all numbers in intervals B and D. The solution set is $(-1,\ 2)\cup(3,\ \infty)$, or $\{x\,|\,-1<x<2 \text{ or } x>3\}$.

31. $F(x) \le 0$

$x^3 - 7x^2 + 10x \le 0$

$x(x^2 - 7x + 10) \le 0$

$x(x-2)(x-5) \le 0$

The solutions of $x(x-2)(x-5) = 0$ are 0, 2, and 5. They divide the real-number line as shown below.

Sign of x	−	+	+	+
Sign of $x-2$	−	−	+	+
Sign of $x-5$	−	−	−	+
Sign of product	−	+	−	+

0 2 5

Because the inequality symbol is $\le$ we must include the endpoints in the solution set. From the chart we see that the solution set is $(-\infty, 0] \cup [2, 5]$ or $\{x \mid x \le 0 \text{ or } 2 \le x \le 5\}$.

33. $\frac{1}{x-5} < 0$

We write the related equation by changing the < symbol to =:.

$\frac{1}{x-5} = 0$

We solve the related equation.

$(x-5) \cdot \frac{1}{x-5} = (x-5) \cdot 0$

$1 = 0$

The related equation has no solution.

Next we find the values that make the denominator 0 by setting the denominator equation to 0 and solving:

$x - 5 = 0$

$x = 5$

We use 5 to divide the number line into two intervals as shown:

A B
5

A: Test 0, $\frac{1}{0-5} = \frac{1}{-5} = -\frac{1}{5} < 0$

The number 0 is a solution of the inequality, so the interval A is part of the solution set.

B: Test 6, $\frac{1}{6-5} = 1 \not< 0$

The number 6 is not a solution of the inequality, so the interval B is part of the solution set.

The solution set is $(-\infty, 5)$, or $\{x \mid x < 5\}$.

35. $\frac{x+1}{x-3} \ge 0$

Solve the related equation.

$\frac{x+1}{x-3} = 0$

$x + 1 = 0$

$x = -1$

Find the values that make the denominator 0.

$x - 3 = 0$

$x = 3$

Use the numbers –1 and 3 to divide the number line into intervals as shown:

A B C
−1 3

Try test numbers in each interval.

A: Test –2, $\frac{-2+1}{-2-3} = \frac{-1}{-5} = \frac{1}{5} > 0$

The number –2 is a solution of the inequality, so the interval A is part of the solution set.

B: Test 0, $\frac{0+1}{0-3} = \frac{1}{-3} = -\frac{1}{3} \not> 0$

The number 0 is not a solution of the inequality, so the interval B is not part of the solution set.

C: Test 4, $\frac{4+1}{4-3} = \frac{5}{1} = 5 > 0$

The number 4 is a solution of the inequality, so the interval C is part of the solution set.

The solution set includes intervals A and C. The number –1 is also included since the inequality symbol is $\ge$ and –1 is the solution of the related equation. The number 3 is not included since $\frac{x+1}{x-3}$ is undefined for $x = 3$. The solution set is $(-\infty, -1] \cup (3, \infty)$, or $\{x \mid x \le -1 \text{ or } x > 3\}$.

37. $\frac{x+1}{x+6} \ge 1$

Solve the related equation.

$\frac{x+1}{x+6} = 1$

$x + 1 = x + 6$

$1 = 6$

The related equation has no solution.

Find the values that make the denominator 0.

$x + 6 = 0$

$x = -6$

Use the number –6 to divide the number line into two intervals.

A B
−6

Try test numbers in each interval.

A: Test –7, $\frac{-7+1}{-7+6} = \frac{-6}{-1} = 6 > 1$.

The number –7 is a solution of the inequality, so the interval A is part of the solution set.

B: Test 0, $\frac{0+1}{0+6} = \frac{1}{6} \not> 1$

The number 0 is not a solution of the inequality, so the interval B is not part of the solution set. The number –6 is not included in the solution set since $\frac{x+1}{x+6}$ is undefined for $x=-6$. The solution set is $(-\infty,-6)$, or $\{x|x<-6\}$.

39. $\frac{(x-2)(x+1)}{x-5}\le 0$

Solve the related equation.

$$\frac{(x-2)(x+1)}{x-5}=0$$
$$(x-2)(x+1)=0$$
$$x=2 \text{ or } x=-1$$

Find the values that make the denominator 0.

$$x-5=0$$
$$x=5$$

Use the numbers 2, –1, and 5 to divide the number line into intervals as shown:

A B C D
–1 2 5

Try test numbers in each interval.

A: Test –2, $\frac{(-2-2)(-2+1)}{-2-5}=\frac{-4(-1)}{-7}=-\frac{4}{7}\le 0$

Interval A is part of the solution set.

B: Test 0, $\frac{(0-2)(0+1)}{0-5}=\frac{-2\cdot 1}{-5}=\frac{2}{5}\not\le 0$

Interval B is not part of the solution set.

C: Test 3, $\frac{(3-2)(3+1)}{3-5}=\frac{1\cdot 4}{-2}=-2\le 0$

Interval C is part of the solution set.

D: Test 6, $\frac{(6-2)(6+1)}{6-5}=\frac{4\cdot 7}{1}=28\not\le 0$

Interval D is not part of the solution set.

The solution set includes intervals A and C. The numbers –1 and 2 are also included since the inequality symbol is $\le$ and –1 and 2 are the solutions of the related equation. The number 5 is not included since $\frac{(x-2)(x+1)}{x-5}$ is undefined for $x=5$. The solution set is $(-\infty,-1]\cup[2, 5)$, or $\{x|x\le -1 \text{ or } 2\le x<5\}$.

41. $\frac{x}{x+3}\ge 0$

Solve the related equation.

$$\frac{x}{x+3}=0$$
$$x=0$$

Find the values that make the denominator 0.

$$x+3=0$$
$$x=-3$$

Use the numbers 0 and –3 to divide the number line into intervals as shown.

A B C
–3 0

Try test numbers in each interval.

A: Test –4, $\frac{-4}{-4+3}=\frac{-4}{-1}=4\ge 0$

Interval A is part of the solution set.

B: Test –1, $\frac{-1}{-1+3}=\frac{-1}{2}=-\frac{1}{2}\not\ge 0$

Interval B is not part of the solution set.

C: Test 1, $\frac{1}{1+3}=\frac{1}{4}\ge 0$

The interval C is part of the solution set.

The solution set includes intervals A and C. The number 0 is also included since the inequality symbol is $\ge$ and 0 is the solution of the related equation. The number –3 is not included since $\frac{x}{x+3}$ is undefined for $x=-3$. The solution set is $(-\infty,-3)\cup[0, \infty)$, or $\{x|x<-3 \text{ or } x\ge 0\}$.

43. $\frac{x-5}{x}<1$

Solve the related equation.

$$\frac{x-5}{x}=1$$
$$x-5=x$$
$$-5=0$$

The related equation has no solution.

Find the values that make the denominator 0.

$$x=0$$

Use the number 0 to divide the number line into two intervals as shown.

A B
0

Try test numbers in each interval.

A: Test -1, $\frac{-1-5}{-1}=\frac{-6}{-1}=6\not<1$

Interval A is not part of the solution set.

B: Test 1, $\frac{1-5}{1}=\frac{-4}{1}=-4<1$

Interval B is part of the solution set.

The solution set is $(0, \infty)$ or $\{x|x>0\}$.

45. $\frac{x-1}{(x-3)(x+4)}\le 0$

Solve the related equation.

$$\frac{x-1}{(x-3)(x+4)}=0$$
$$x-1=0$$
$$x=1$$

Find the values that make the denominator 0.

$$(x-3)(x+4)=0$$

$x=3$ *or* $x=-4$

Use the numbers 1, 3, and –4 to divide the number line into intervals as shown:

A B C D

–4 1 3

Try test numbers in each interval.

A: Test –5, $\frac{-5-1}{(-5-3)(-5+4)}=\frac{-6}{-8(-1)}=-\frac{3}{4}<0$

Interval A is part of the solution set.

B: Test 0, $\frac{0-1}{(0-3)(0+4)}=\frac{-1}{-3\cdot 4}=\frac{1}{12}\not<0$

Interval B is not part of the solution set.

C: Test 2, $\frac{2-1}{(2-3)(2+4)}=\frac{1}{-1\cdot 6}=-\frac{1}{6}<0$

Interval C is part of the solution set.

D: Test 4, $\frac{4-1}{(4-3)(4+4)}=\frac{3}{1\cdot 8}=\frac{3}{8}\not<0$

Interval D is not part of the solution set.

The solution set includes intervals A and C. The number 1 is also included since the inequality symbol is $\leq$ and 1 is the solution of the related equation. The numbers –4 and 3 are not included since $\frac{x-1}{(x-3)(x+4)}$ is undefined for $x=-4$ and for $x=3$.

The solution set is $(-\infty,-4)\cup[1,\ 3)$, or $\{x\,|\,x<-4 \text{ or } 1\leq x<3\}$.

47. $f(x)\geq 0$

$\frac{5-2x}{4x+3}\geq 0$

Solve the related equation.

$\frac{5-2x}{4x+3}=0$
$5-2x=0$
$5=2x$
$\frac{5}{2}=x$

Find the values that make the denominator 0.

$4x+3=0$
$4x=-3$
$x=-\frac{3}{4}$

Use the numbers $\frac{5}{2}$ and $-\frac{3}{4}$ to divide the number line as shown:

A B C

$-\frac{3}{4}$ $\frac{5}{2}$

Try test numbers in each interval.

A: Test –1, $\frac{5-2(-1)}{4(-1)+3}=-7\not\geq 0$

Interval A is not part of the solution set.

B: Test 0, $\frac{5-2\cdot 0}{4\cdot 0+3}=\frac{5}{3}>0$

Interval B is part of the solution set.

C: Test 3, $\frac{5-2\cdot 3}{4\cdot 3+3}=-\frac{1}{15}\not\geq 0$

Interval C is not part of the solution set.

The solution set includes interval B. The number $\frac{5}{2}$ is also included since the inequality symbol is $\geq$ and $\frac{5}{2}$ is the solution of the related equation. The number $-\frac{3}{4}$ is not included since $\frac{5-2x}{4x+3}$ is undefined for $x=-\frac{3}{4}$. The solution set is $\left(-\frac{3}{4},\ \frac{5}{2}\right]$, or $\left\{x\middle|-\frac{3}{4}<x\leq\frac{5}{2}\right\}$.

49. $G(x)\leq 1$

$\frac{1}{x-2}\leq 1$

Solve the related equation.

$\frac{1}{x-2}=1$
$1=x-2$
$3=x$

Find the values of x that make the denominator 0.

$x-2=0$
$x=2$

Use the numbers 2 and 3 to divide the number line as shown.

A B C

2 3

Try a test number in each interval.

A: Test 0, $\frac{1}{0-2}=-\frac{1}{2}\leq 1$

Interval A is part of the solution set.

B: Test $\frac{5}{2}$, $\frac{1}{\frac{5}{2}-2}=\frac{1}{\frac{1}{2}}=2\not\leq 1$

Interval B is not part of the solution set.

C: Test 4, $\frac{1}{4-2}=\frac{1}{2}\leq 1$

Interval C is part of the solution set.

The solution set includes intervals A and B. The number 3 is also included since the inequality symbol is $\leq$ and 3 is the solution of the related equation. The number 2 is not included since $\frac{1}{x-2}$ is undefined for $x=2$. The solution set is $(-\infty,2)\cup[3,\ \infty)$, or $\{x\,|\,x<2 \text{ or } x\geq 3\}$.

51. *Writing Exercise.* Consider the quadratic portion of the inequality, ax^2+bx+c. The graph of $f(x)=ax^2+bx+c$ is a parabola and we can solve the inequality as in Example 1.

53. Graph $f(x)=x^3-2$.

x	y
-2	-10
-1	-3
0	-2
1	-1
2	6

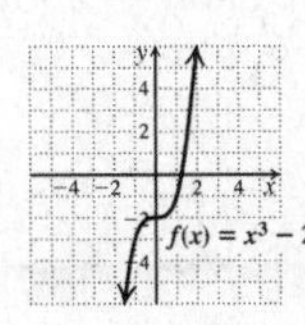

54. Graph $g(x)=\frac{2}{x}$. Note $x \neq 0$.

x	y
-2	-1
-1	-2
0	undefined
1	2
2	1

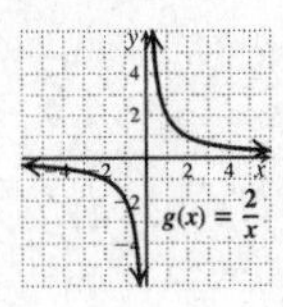

55. $f(x)=x+7$

$$f\left(\frac{1}{a^2}\right)=\frac{1}{a^2}+7$$

57.

$$g(x)=x^2+2$$
$$g(2a+5)=(2a+5)^2+2$$
$$=4a^2+20a+25+2$$
$$=4a^2+20a+27$$

59. *Writing Exercise.* If any solutions from step (1) are also replacements for which the rational expression is undefined, they must be excluded from the solution set.

61.

$$x^2+2x<5$$
$$x^2+2x-5<0$$

Using the quadratic formula, we find that the solutions of the related equation are $x=-1\pm\sqrt{6}$. These numbers divide the real-number line into three intervals as shown:

A B C

$-1-\sqrt{6}$ $-1+\sqrt{6}$

We try test numbers in each interval.

A: Test -4, $f(-4)=(-4)^2+2(-4)-5=3$

B: Test 0, $f(0)=0^2+2\cdot0-5=-5$

C: Test 2, $f(2)=2^2+2\cdot2-5=3$

The function value will be negative for all numbers in interval B. The solution set is $\left(-1-\sqrt{6},-1+\sqrt{6}\right)$, or $\left\{x\middle|-1-\sqrt{6}<x<-1+\sqrt{6}\right\}$.

63.

$$x^4+3x^2\le 0$$
$$x^2(x^2+3)\le 0$$

$x^2=0$ for $x=0$, $x^2>0$ for $x\neq0$, $x^2+3>0$ for all x

The solution set is $\{0\}$.

65. a)

$$-3x^2+630x-6000>0$$
$$x^2-210x+2000<0 \quad \text{Multiplying by } -\frac{1}{3}$$
$$(x-200)(x-10)<0$$

The solutions of $f(x)=(x-200)(x-10)=0$ are 200 and 10. They divide the number line as shown:

A B C

10 200

A: Test 0, $f(0)=0^2-210\cdot0+2000=2000$

B: Test 20, $f(20)=20^2-210\cdot20+2000=-1800$

C: Test 300, $f(300)=300^2-210\cdot300+2000=29{,}000$

The company makes a profit for values of x such that $10<x<200$, or for values of x in the interval $(10,\ 200)$.

b) See part (a). Keep in mind that x must be nonnegative since negative numbers have no meaning in this application.

The company loses money for values of x such that $0\le x<10$ or $x>200$, or for values of x in the interval $[0,\ 10)\cup(200,\ \infty)$.

67. We find values of n such that $N\ge66$ *and* $N\le300$.

For $N\ge66$:

$$\frac{n(n-1)}{2}\ge66$$
$$n(n-1)\ge132$$
$$n^2-n-132\ge0$$
$$(n-12)(n+11)\ge0$$

The solutions of $f(n)=(n-12)(n+11)=0$ are 12 and -11. They divide the number line as shown:

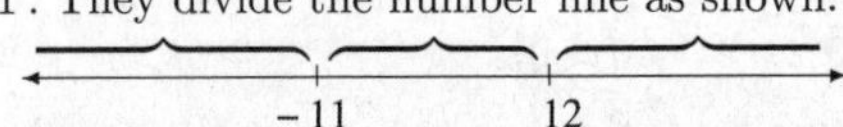

However, only positive values of n have meaning in this exercise so we need only consider the intervals shown below:

A B

0 12

A: Test 1, $f(1)=1^2-1-132=-132$

B: Test 20, $f(20)=20^2-20-132=248$

Thus, $N\ge66$ for $\{n|n\ge12\}$.

For $N\le300$:

$$\frac{n(n-1)}{2}\le300$$
$$n(n-1)\le600$$
$$n^2-n-600\le0$$
$$(n-25)(n+24)\le0$$

The solutions of $f(n)=(n-25)(n+24)=0$ are 25 and

-24. They divide the number line as shown:

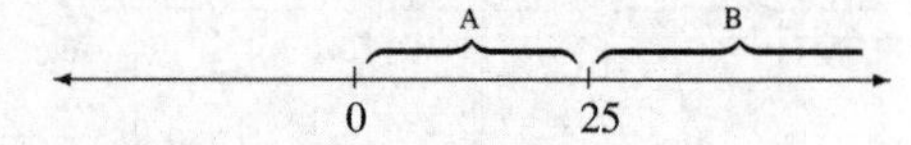

However, only positive values of n have meaning in this exercise so we need only consider the intervals shown below:

A B

0 25

A: Test 1, $f(1)=1^2-1-600=-600$

B: Test 30, $f(30)=30^2-30-600=270$

Thus, $N \le 300$ (and $n>0$) for $\{n \mid 0<n\le 25\}$.

Then $66 \le N \le 300$ for

$\{n \mid n \text{ is an integer } \textit{and } 12 \le n \le 25\}$.

69. From the graph we determine the following:

The solutions of $f(x)=0$ are –2, 1, and 3.

The solution of $f(x)<0$ is $(-\infty,-2)\cup(1,\ 3)$, or

$\{x|x<-2 \textit{ or } 1<x<3\}$.

The solution of $f(x)>0$ is $(-2,\ 1)\cup(3,\ \infty)$, or

$\{x|-2<x<1 \textit{ or } x>3\}$.

71. From the graph we determine the following:

$f(x)$ has no zeros.

The solutions of $f(x)<0$ are $(-\infty,\ 0)$, or $\{x|x<0\}$;

The solutions of $f(x)>0$ are $(0,\ \infty)$, or $\{x|x>0\}$.

73. From the graph we determine the following:

The solutions of $f(x)=0$ are –1 and 0.

The solution of $f(x)<0$ is $(-\infty,-3)\cup(-1,\ 0)$, or

$\{x|x<-3 \textit{ or } -1<x<0\}$.

The solution of $f(x)>0$ is $(-3,\ -1)\cup(0,\ 2)\cup(2,\ \infty)$, or

$\{x|-3<x<-1 \textit{ or } 0<x<2 \textit{ or } x>2\}$.

75. For $f(x)=\sqrt{x^2-4x-45}$, we find the domain:

$$x^2-4x-45\ge 0$$
$$(x+5)(x-9)\ge 0$$

The quadratic is nonnegative when $(-\infty,-5]\cup[9,\ \infty)$, or

$\{x|x\le -5 \textit{ or } x\ge 9\}$.

77. For $f(x)=\sqrt{x^2+8x}$, we find the domain:

$$x^2+8x\ge 0$$
$$x(x+8)\ge 0$$

The quadratic is nonnegative when $(-\infty,-8]\cup[0,\ \infty)$, or

$\{x|x\le -8 \textit{ or } x\ge 0\}$.

79. *Writing Exercise.* Answers may vary.

For $a<b$, write the inequality $(x-a)(x-b)\ge 0$, or

$x^2-(a+b)x+ab\ge 0$.

Chapter 11 Review

1. False; see page 722 in the text.

3. True

5. False; the vertex is (–3, –4).

7. True; since the coefficient of x^2 is –2, the graph opens down and therefore has no minimum.

9. False; see page 722 in the text.

11. $9x^2-2=0$

$$9x^2=2$$
$$x^2=\frac{2}{9}$$
$$x=\pm\sqrt{\frac{2}{9}}=\pm\frac{\sqrt{2}}{3}$$

The solutions are $-\frac{\sqrt{2}}{3}$ and $\frac{\sqrt{2}}{3}$.

13. $x^2-12x+36=9$

$$(x-6)^2=9$$
$$x-6=\pm 3$$
$$x=6\pm 3$$

The solutions are 3 and 9.

15. $x(3x+4)=4x(x-1)+15$

$$3x^2+4x=4x^2-4x+15$$
$$0=x^2-8x+15$$
$$0=(x-3)(x-5)$$
$$x-3=0 \quad \textit{or} \quad x-5=0$$
$$x=3 \quad \textit{or} \quad x=5$$

The solutions are 3 and 5.

17. $x^2-5x-2=0$

$a=1,\ b=-5,\ c=-2$

$$x=\frac{-(-5)\pm\sqrt{(-5)^2-4\cdot 1\cdot(-2)}}{2\cdot 1}=\frac{5\pm\sqrt{25+8}}{2}$$
$$x=\frac{5\pm\sqrt{33}}{2}$$

$x\approx -0.372,\ 5.372$

19. $\frac{1}{2}\cdot(-18)=-9;\quad (-9)^2=81$

$x^2-18x+81=(x-9)^2$

21. $x^2-6x+1=0$
$$x^2-6x=-1$$
$$x^2-6x+9=9-1$$
$$(x-3)^2=8$$
$$x-3=\pm\sqrt{8}$$
$$x=3\pm\sqrt{8}$$
$$x=3\pm2\sqrt{2}$$

23. $s=16t^2$
$$1018=16t^2$$
$$\frac{509}{8}=t^2$$
$$\sqrt{\frac{509}{8}}=t \quad \text{Principle of square roots; rejecting the negative square root.}$$
$$8.0\approx t$$
It will take an object about 8.0 sec to fall.

25. $x^2+2x+5=0$
$b^2-4ac=2^2-4\cdot1\cdot5=-16$
There are two imaginary numbers.

27. The only solution is –5. It must be a repeated solution.
$$x=-5 \quad or \quad x=-5$$
$$x+5=0 \quad or \quad x+5=0$$
$$(x+5)(x+5)=0$$
$$x^2+10x+25=0$$

29. ***Familiarize***. Let x represent the time it takes Cheri to reply. Then $x+6$ represents the time it takes Dani to reply. It takes them 4 hr to reply working together, so they can reply to $\frac{1}{4}$ of the emails in 1 hr. Cheri will reply to $\frac{1}{x}$ of the emails in 1 hr, and Dani will reply to $\frac{1}{x+6}$ of the emails in 1 hr.

Translate. We have an equation.
$$\frac{1}{x}+\frac{1}{x+6}=\frac{1}{4}$$
Carry out. We solve the equation.
We multiply by the LCD, $4x(x+6)$.
$$4x(x+6)\left(\frac{1}{x}+\frac{1}{x+6}\right)=4x(x+6)\cdot\frac{1}{4}$$
$$4(x+6)+4x=x(x+6)$$
$$4x+24+4x=x^2+6x$$
$$0=x^2-2x-24$$
$$0=(x-6)(x+4)$$
Check. Since negative time has no meaning in this problem, –4 is not a solution. We check only 6 hr. This is the time it would take Cheri working alone. Then Dani would take 6 + 6, or 12 hr working alone. Cheri would reply to $4\left(\frac{1}{6}\right)$, or $\frac{2}{3}$ of the emails in 4 hr, and Dani would reply to $4\left(\frac{1}{12}\right)$, or $\frac{1}{3}$ of the emails in 4 hr. Thus, in 4 hr they would reply to $\frac{2}{3}+\frac{1}{3}$ of the emails. This is all of it, so the numbers check.

State. It Cheri, working alone, 6 hr to reply to the emails.

31. $15x^{-2}-2x^{-1}-1=0$
Let $u=x^{-1}$ and $u^2=x^{-2}$.
$$15u^2-2u-1=0$$
$$(5u+1)(3u-1)=0$$
$$5u+1=0 \quad or \quad 3u-1=0$$
$$u=-\frac{1}{5} \quad or \quad u=\frac{1}{3}$$
Replace u with x^{-1}.
$$x^{-1}=-\frac{1}{5} \quad or \quad x^{-1}=\frac{1}{3}$$
$$x=-5 \quad or \quad x=3$$
The numbers –5 and 3 check. They are the solutions.

33. $f(x)=-3(x+2)^2+4$
We know that the graph looks like the graph of $h(x)=3x^2$ but moved to the left 2 units and up 4 units and turned upside down. The vertex is (–2, 4), and the axis of symmetry is $x=-2$. The maximum function value is 4.

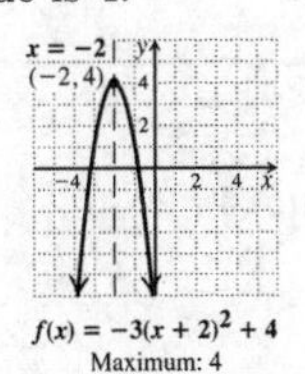

35. $f(x)=x^2-9x+14$
To find the x-intercepts, solve the equation $0=x^2-9x+14$. We factor.
$$0=x^2-9x+14$$
$$0=(x-2)(x-7)$$
$$x=2 \text{ or } x=7$$
The x-intercepts are (2, 0) and (7, 0).
The y-intercept is $(0, f(0))$, or (0, 14).

37. $2A+T=3T^2$
$$3T^2-T-2A=0$$
$a=3$, $b=-1$, $c=-2A$
$$T=\frac{-(-1)\pm\sqrt{(-1)^2-4\cdot3\cdot(-2A)}}{2\cdot3}$$
$$T=\frac{1\pm\sqrt{1+24A}}{6}$$

39. The data points fall. The graph does not appear to represent a quadratic function in which the data points would rise and then fall or vice versa. Thus a linear function $f(x)=mx+b$ might be used to model the data.

41. a) ***Familiarize.*** We look for a function of the form $f(x)=ax^2+bx+c$, where $f(x)$ represents the percent increase in premiums and x represents the years after 2000.

Translate. We substitute the given values of x and $f(x)$.

$$\begin{aligned} 8 &= a(0)^2+b(0)+c, \\ 11 &= a(2)^2+b(2)+c, \\ 8 &= a(6)^2+b(6)+c, \end{aligned}$$

or

$$\begin{aligned} 8 &= 0a+0b+c, \\ 11 &= 4a+2b+c, \\ 8 &= 36a+6b+c. \end{aligned}$$

Carry out. Solving the system of equations, we get

$a=-\frac{3}{8},\ b=\frac{9}{4},\ c=8.$

Check. Recheck the calculations.

State. The function

$f(x)=-\frac{3}{8}x^2+\frac{9}{4}x+8$ fits the data.

b) Find $f(5)$.

$f(5)=-\frac{3}{8}(5)^2+\frac{9}{4}(5)+8=9.875\approx 10$

About 10% increase for premiums in 2005 is estimated.

43. $\frac{x-5}{x+3}\le 0$

Solve the related equation.

$$\begin{aligned} \frac{x-5}{x+3} &= 0 \\ x-5 &= 0 \\ x &= 5 \end{aligned}$$

Find the values that make the denominator 0.

$$\begin{aligned} x+3 &= 0 \\ x &= -3 \end{aligned}$$

Use the numbers 5 and –3 to divide the number line into intervals as shown:

A B C

–3 5

Try test numbers in each interval.

A: Test –4, $\frac{-4-5}{-5+3}=\frac{-9}{-2}=\frac{9}{2}\not<0$

The number –4 is a not solution of the inequality, so the interval A is not part of the solution set.

B: Test 0, $\frac{0-5}{0+3}=\frac{-5}{3}=-\frac{5}{3}<0$

The number 0 is a solution of the inequality, so the interval B is part of the solution set.

C: Test 6, $\frac{6-5}{6+3}=\frac{1}{9}\not<0$

The number 6 is a not solution of the inequality, so the interval C is not part of the solution set.

The solution set includes interval B. The number 5 is also included since the inequality symbol is $\le$ and 5 is the solution of the related equation. The number –3 is not included since $\frac{x-5}{x+3}$ is undefined for $x=-3$. The solution set is $(-3,-5]$, or $\{x\mid -3<x\le 5\}$.

45. *Writing Exercise.* Yes; if the discriminant is a perfect square, then the solutions are rational numbers, p/q and r/s. (Note that if the discriminant is 0, then $p/q=r/s$.) Then the equation can be written in factored form, $(qx-p)(sx-r)=0$.

47. *Writing Exercise.* Completing the square was used to solve quadratic equations and to graph quadratic functions by rewriting the function in the form $f(x)=a(x-h)^2+k$.

49. From Section 8.4, we know the sum of the solutions of $ax^2+bx+c=0$ is $-\frac{b}{a}$, and the product is $\frac{c}{a}$.

$$\begin{aligned} 3x^2-hx+4k &= 0 \\ a=3,\ b=-h,\ c&=4k \end{aligned}$$

Substituting

For $-\frac{b}{a}$: $-\frac{-h}{3}=20$

$\frac{h}{3}=20$

$h=60$

For $\frac{c}{a}$: $\frac{4k}{3}=80$

$4k=240$

$k=60$

Chapter 11 Test

1. $25x^2-7=0$

$25x^2=7$

$x^2=\frac{7}{25}$

$x=\pm\sqrt{\frac{7}{25}}=\pm\frac{\sqrt{7}}{5}$

The solutions are $-\frac{\sqrt{7}}{5}$ and $\frac{\sqrt{7}}{5}$.

3. $x^2+2x+3=0$

$a=1,\ b=2,\ c=3$

$x=\frac{-2\pm\sqrt{2^2-4(1)(3)}}{2(1)}=\frac{-2\pm\sqrt{4-12}}{2}$

$=\frac{-2\pm\sqrt{-8}}{2}=\frac{-2\pm 2i\sqrt{2}}{2}=\frac{-2}{2}\pm\frac{2i\sqrt{2}}{2}$

$=-1\pm i\sqrt{2}$ or $-1\pm\sqrt{2}i$

The solutions are $-1+\sqrt{2}i$ and $-1-\sqrt{2}i$.

5. $x^{-2}-x^{-1}=\frac{3}{4}$

$x^{-2}-x^{-1}-\frac{3}{4}=0$

$4x^{-2}-4x^{-1}-3=0$ Clearing fractions

Let $u=x^{-1}$ and $u^2=x^{-2}$.

$4u^2-4u-3=0$

$(2u-3)(2u+1)=0$

$2u-3=0$ *or* $2u+1=0$

$u=\frac{3}{2}$ *or* $u=-\frac{1}{2}$

Now we replace u with x^{-1} and solve these equations:

$x^{-1}=\frac{3}{2}$ *or* $x^{-1}=-\frac{1}{2}$

$\frac{1}{x}=\frac{3}{2}$ *or* $\frac{1}{x}=-\frac{1}{2}$

$2=3x$ *or* $2=-x$

$\frac{2}{3}=x$ $\quad -2=x$

The solutions are -2 and $\frac{2}{3}$.

7. Let $f(x)=0$ and solve for x.

$0=12x^2-19x-21$

$0=(4x+3)(3x-7)$

$x=-\frac{3}{4}$ *or* $x=\frac{7}{3}$

The solutions are $-\frac{3}{4}$ and $\frac{7}{3}$.

9. $\frac{1}{2}\cdot\frac{2}{7}=\frac{1}{7};\ \left(\frac{1}{7}\right)^2=\frac{1}{49}$

$x^2+\frac{2}{7}x+\frac{1}{49}=\left(x+\frac{1}{7}\right)^2$

11. $x^2+2x+5=0$

$b^2-4ac=2^2-4(1)(5)=-16$

Two imaginary numbers

13. ***Familiarize.*** Let r represent the cruiser's speed in still water. Then $r-4$ is the speed upriver and $r+4$ is the speed downriver. Using $t=\frac{d}{r}$, we let $\frac{60}{r-4}$ represent the time upriver and $\frac{60}{r+4}$ represent the time downriver.

Trip	Distance	Speed	Time
Upriver	60	$r-4$	$\frac{60}{r-4}$
Downriver	60	$r+4$	$\frac{60}{r+4}$

Translate. We have an equation.

$\frac{60}{r-4}+\frac{60}{r+4}=8$

Carry out. We solve the equation.

We multiply by the LCD, $(r-4)(r+4)$.

$(r-4)(r+4)\cdot\left(\frac{60}{r-4}+\frac{60}{r+4}\right)=(r-4)(r+4)\cdot 8$

$60(r+4)+60(r-4)=8(r-4)(r+4)$

$60r+240+60r-240=8r^2-128$

$0=8r^2-120r-128$

$0=8(r^2-15r-16)$

$0=8(r-16)(r+1)$

$r=16$ *or* $r=-1$

Check. Since negative time has no meaning in this problem, -1 is not a solution. We check only 16 hr. If $r=16$, then the speed upriver is $16-4$, or 12 km/h, and the time is $\frac{60}{12}$, or 5 hr. The speed downriver is $16+4$, or 20 km/h, and the time is $\frac{60}{20}$, or 3 hr. The total time of the round trip is $5+3$, or 8 hr. The value checks.

State. The speed of the cruiser in still water is 16 km/h.

15. $f(x)=x^4-15x^2-16$

To find the x-intercepts, solve the equation

$0=x^4-15x^2-16$.

Let $u=x^2$ and $u^2=x^4$. .

$0=u^2-15u-16$

$0=(u-16)(u+1)$

$u=16$ *or* $u=-1$

Replace u with x^2.

$x^2=16$ *or* $x^2=-1$ Has no real solutions

$x=\pm 4$

The x-intercepts are $(-4,0)$ and $(4,0)$.

17. $f(x)=2x^2+4x-6$
$=2(x^2+2x)-6$
$=2(x^2+2x+1)-6-2$
$=2(x+1)^2-8$

We know that the graph looks like the graph of $h(x)=2x^2$ but moved to the left 1 unit and down 8 units. The vertex is (–1, –8), and the axis of symmetry is $x=-1$.

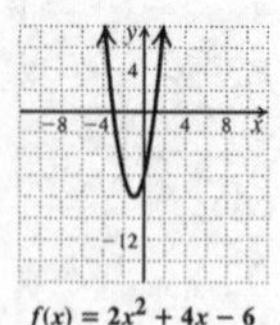

$f(x) = 2x^2 + 4x - 6$

19.
$$V=\frac{1}{3}\pi(R^2+r^2)$$
$$\frac{3V}{\pi}=R^2+r^2$$
$$\frac{3V}{\pi}-R^2=r^2$$
$$\sqrt{\frac{3V}{\pi}-R^2}=r$$

We only consider the positive square root as instructed.

21. $C(x)=0.2x^2-1.3x+3.4025$
$C(x)=0.2(x^2-6.5x)+3.4025$
$C(x)=0.2(x-6.5x+10.5625)-0.2(10.5625)+3.4025$
$C(x)=0.2(x-3.25)^2+1.29$

3.25 hundred or 324 cabinets should be built to have a minimum at \$1.29 hundred, or \$129 per cabinet.

23.
$$x^2+5x<6$$
$$x^2+5x-6<0$$
$$(x+6)(x-1)<0$$

The solutions of $(x+6)(x-1)=0$ are –6 and 1. They divide the number line into three intervals as shown:

A B C

–6 1

We try test numbers in each interval.

A: Test –7, $f(-7)=(-7+6)(-7-1)=8$

B: Test 0, $f(0)=(0+6)(0-1)=-6$

C: Test 2, $f(2)=(2+6)(2-1)=8$

Since $f(0)$ is negative, the function value will be negative for all numbers in the interval containing 0. Because the symbol is $<$, we do not include the endpoints in the solution. The solution set is $(-6, 1)$, or $\{x|-6<x<1\}$.

25.
$$kx^2+3x-k=0$$
$$k(-2)^2+3(-2)-k=0 \quad \text{Substitute } -2 \text{ for } x$$
$$4k-6-k=0$$
$$3k=6$$
$$k=2$$

Then $2x^2+3x-2=0$
$(x+2)(2x-1)=0$

The other solution is $\frac{1}{2}$.

27. $x^4-4x^2-1=0$

Let $u=x^2$ and $u^2=x^4$.
$$u^2-4u-1=0$$
$$u^2-4u=1$$
$$u^2-4u+4=1+4$$
$$(u-2)^2=5$$
$$u-2=\pm\sqrt{5}$$
$$u=2\pm\sqrt{5}$$

Replace u with x^2.
$$x^2=2+\sqrt{5} \quad \text{or} \quad x^2=2-\sqrt{5}$$
$$x=\pm\sqrt{2+\sqrt{5}} \quad \text{or} \quad x=\pm\sqrt{2-\sqrt{5}}$$

Since $2-\sqrt{5}$ is negative, we can rewrite it as follows:
$$\pm\sqrt{2-\sqrt{5}}=\pm\sqrt{\sqrt{5}-2}\,i$$

The solutions are $\pm\sqrt{\sqrt{5}+2}$ and $\pm\sqrt{\sqrt{5}-2}\,i$.

Chapter 12

Exponential and Logarithmic Functions

Exercise Set 12.1

1. True; see page 785 in the text.

3. $(g\circ f)=g(f(x))=x^2+3\neq(x+3)^2$, so the statement is false.

5. False; see page 787 in the text.

7. True; see page 788 in the text.

9. a) $(f\circ g)(1)=f(g(1))=f(1-3)$
$=f(-2)=(-2)^2+1$
$=4+1=5$

b) $(g\circ f)(1)=g(f(1))=g(1^2+1)$
$=g(2)=2-3=-1$

c) $(f\circ g)=f(g(x))=f(x-3)$
$=(x-3)^2+1=x^2-6x+9+1$
$=x^2-6x+10$

d) $(g\circ f)(x)=g(f(x))=g(x^2+1)$
$=x^2+1-3=x^2-2$

11. a) $(f\circ g)(1)=f(g(1))=f(2\cdot 1^2-7)$
$=f(-5)=5(-5)+1=-24$

b) $(g\circ f)(1)=g(f(1))=g(5\cdot 1+1)$
$=g(6)=2\cdot 6^2-7=65$

c) $(f\circ g)(x)=f(g(x))=f(2x^2-7)$
$=5(2x^2-7)+1=10x^2-34$

d) $(g\circ f)(x)=g(f(x))=g(5x+1)$
$=2(5x+1)^2-7=2(25x^2+10x+1)-7$
$=50x^2+20x-5$

13. a) $(f\circ g)(1)=f(g(1))=f\left(\frac{1}{1^2}\right)$
$=f(1)=1+7=8$

b) $(g\circ f)(1)=g(f(1))=g(1+7)=g(8)=\frac{1}{8^2}=\frac{1}{64}$

c) $(f\circ g)(x)=f(g(x))=f\left(\frac{1}{x^2}\right)=\frac{1}{x^2}+7$

d) $(g\circ f)(x)=g(f(x))=g(x+7)=\frac{1}{(x+7)^2}$

15. a) $(f\circ g)(1)=f(g(1))=f(1+3)$
$=f(4)=\sqrt{4}=2$

b) $(g\circ f)(1)=g(f(1))=g(\sqrt{1})$
$=g(1)=1+3=4$

c) $(f\circ g)(x)=f(g(x))=f(x+3)=\sqrt{x+3}$

d) $(g\circ f)(x)=g(f(x))=g(\sqrt{x})=\sqrt{x}+3$

17. a) $(f\circ g)(1)=f(g(1))=f\left(\frac{1}{1}\right)=f(1)=\sqrt{4\cdot 1}=\sqrt{4}=2$

b) $(g\circ f)(1)=g(f(1))=g(\sqrt{4\cdot 1})=g(\sqrt{4})=g(2)=\frac{1}{2}$

c) $(f\circ g)(x)=f(g(x))=f\left(\frac{1}{x}\right)=\sqrt{4\cdot\frac{1}{x}}=\sqrt{\frac{4}{x}}$

d) $(g\circ f)(x)=g(f(x))=g(\sqrt{4x})=\frac{1}{\sqrt{4x}}$

19. a) $(f\circ g)(1)=f(g(1))=f(\sqrt{1-1})$
$=f(\sqrt{0})=f(0)=0^2+4=4$

b) $(g\circ f)(1)=g(f(1))=g(1^2+4)$
$=g(5)=\sqrt{5-1}=\sqrt{4}=2$

c) $(f\circ g)(x)=f(g(x))=f(\sqrt{x-1})$
$=(\sqrt{x-1})^2+4=x-1+4=x+3$

d) $(g\circ f)(x)=g(f(x))=g(x^2+4)$
$=\sqrt{x^2+4-1}=\sqrt{x^2+3}$

21. $h(x)=(3x-5)^4$

This is $3x-5$ raise to the fourth power, so the two most obvious functions are $f(x)=x^4$ and $g(x)=3x-5$.

23. $h(x)=\sqrt{9x+1}$

We have $9x+1$ and take the square root of their expression, so the two most obvious functions are $f(x)=\sqrt{x}$ and $g(x)=9x+1$.

25. $h(x)=\frac{6}{5x-2}$

This is 6 divided by $5x-2$, so two functions that can be used are $f(x)=\frac{6}{x}$ and $g(x)=5x-2$.

27. The graph of $f(x) = -x$ is shown below.

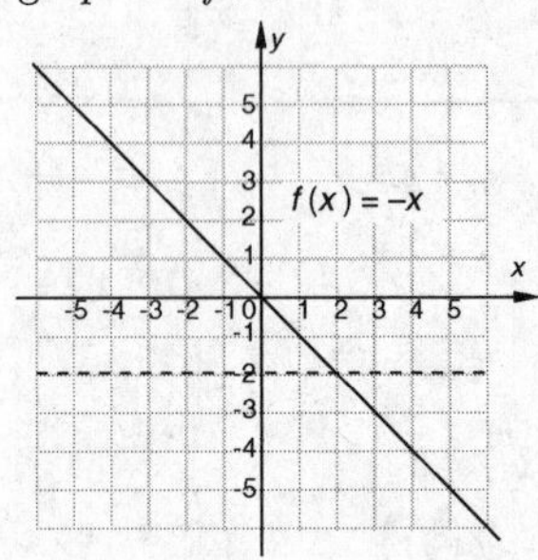

Since there is no horizontal line that crosses the graph more than once, the function is one-to-one.

29. $f(x) = x^2 + 3$

Observe that the graph of this function is a parabola that opens up. Thus, there are many horizontal lines that cross the graph more than once, so the function is not one-to-one. We can also draw the graph as shown below.

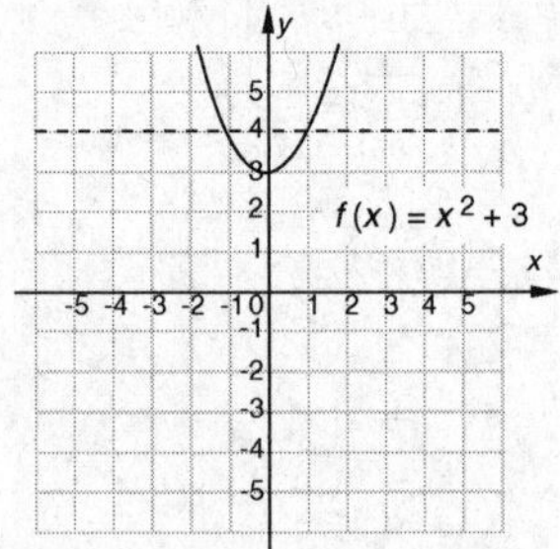

There are many horizontal lines that cross the graph more than once. In particular, the line $y = 4$ crosses the graph more than once. The function is not one-to-one.

31. Since there is no horizontal line that crosses the graph more than once, the function is one-to-one.

33. There are many horizontal lines that cross the graph more than once, the function is not one-to-one.

35. a) The function $f(x) = x + 3$ is a linear function that is not constant, so it passes the horizontal-line test. Thus, f is one-to-one.

b) Replace $f(x)$ by y: $y = x + 3$
Interchange x and y: $x = y + 3$
Solve for y: $x - 3 = y$
Replace y by $f^{-1}(x)$: $f^{-1}(x) = x - 3$

37. a) The function $f(x) = 2x$ is a linear function that is not constant, so it passes the horizontal-line test. Thus, f is one-to-one.

b) Replace $f(x)$ by y: $y = 2x$
Interchange x and y: $x = 2y$
Solve for y: $\frac{x}{2} = y$
Replace y by $f^{-1}(x)$: $f^{-1}(x) = \frac{x}{2}$

39. a) The function $g(x) = 3x - 1$ is a linear function that is not constant, so it passes the horizontal-line test. Thus, g is one-to-one.

b) Replace $g(x)$ by y: $y = 3x - 1$
Interchange x and y: $x = 3y - 1$
Solve for y: $x + 1 = 3y$
$\frac{x+1}{3} = y$
Replace y by $g^{-1}(x)$: $f^{-1}(x) = \frac{x+1}{3}$

41. a) The function $f(x) = \frac{1}{2}x + 1$ is a linear function that is not constant, so it passes the horizontal-line test. Thus, f is one-to-one.

b) Replace $f(x)$ by y: $y = \frac{1}{2}x + 1$
Interchange variables: $x = \frac{1}{2}y + 1$
Solve for y: $x - 1 = \frac{1}{2}y$
$2x - 2 = y$
Replace y by $f^{-1}(x)$: $f^{-1}(x) = 2x - 2$

43. a) The graph of $g(x) = x^2 + 5$ is shown below. There are many horizontal lines that cross the graph more than once. For example, the line $y = 8$ crosses the graph more than once. The function is not one-to-one.

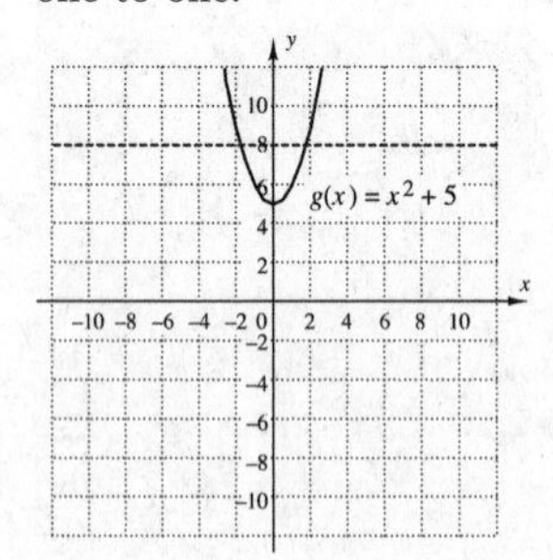

45. a) The function $h(x) = -10 - x$ is a linear function that is not constant, so it passes the horizontal-line test. Thus, h is one-to-one.

b) Replace $h(x)$ by y: $y = -10 - x$
Interchange variables: $x = -10 - y$
Solve for y: $x + 10 = -y$
$-x - 10 = y$
Replace y by $h^{-1}(x)$: $h^{-1}(x) = -x - 10$

47. a) The graph of $f(x)=\dfrac{1}{x}$ is shown below. It passes the horizontal-line test, so the function is one-to-one.

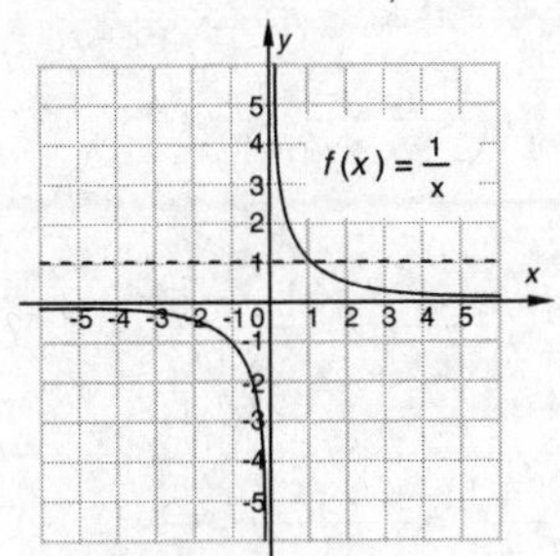

b) Replace $f(x)$ by y: $y=\dfrac{1}{x}$

Interchange x and y: $x=\dfrac{1}{y}$

Solve for y: $xy=1$

$y=\dfrac{1}{x}$

Replace y by $f^{-1}(x)$: $f^{-1}(x)=\dfrac{1}{x}$

49. a) The graph of $g(x)=1$ is shown below. The horizontal line $y=1$ crosses the graph more than once, so the function is not one-to-one.

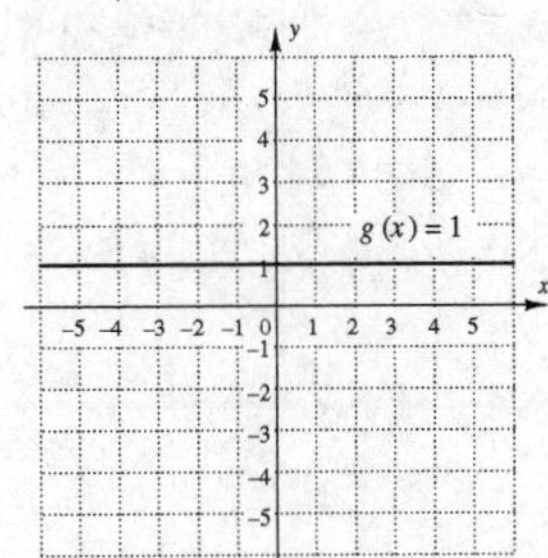

51. a) The function $f(x)=\dfrac{2x+1}{3}=\dfrac{2}{3}x+\dfrac{1}{3}$ is a linear function that is not constant, so it passes the horizontal-line test. Thus, f is one-to-one.

b) Replace $f(x)$ by y: $y=\dfrac{2x+1}{3}$

Interchange x and y: $x=\dfrac{2y+1}{3}$

Solve for y: $3x=2y+1$

$3x-1=2y$

$\dfrac{3x-1}{2}=y$

Replace y by $f^{-1}(x)$: $f^{-1}(x)=\dfrac{3x-1}{2}$

53. a) The graph of $f(x)=x^3+5$ is shown below. It passes the horizontal-line test, so the function is one-to-one.

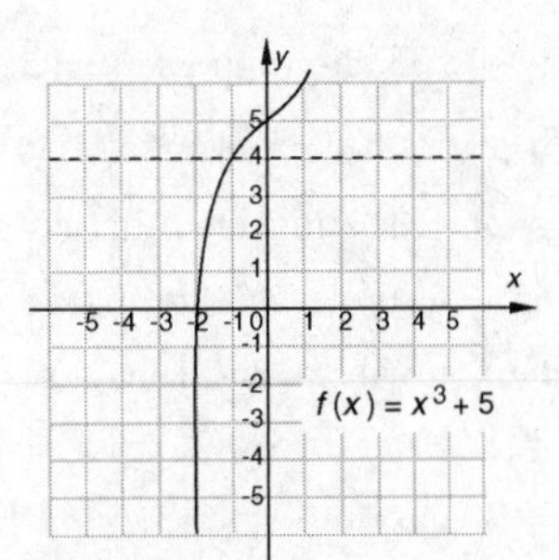

b) Replace $f(x)$ by y: $y=x^3+5$

Interchange x and y: $x=y^3+5$

Solve for y: $x-5=y^3$

$\sqrt[3]{x-5}=y$

Replace y by $f^{-1}(x)$: $f^{-1}(x)=\sqrt[3]{x-5}$

55. a) The graph of $g(x)=(x-2)^3$ is shown below. It passes the horizontal-line test, so the function is one-to-one.

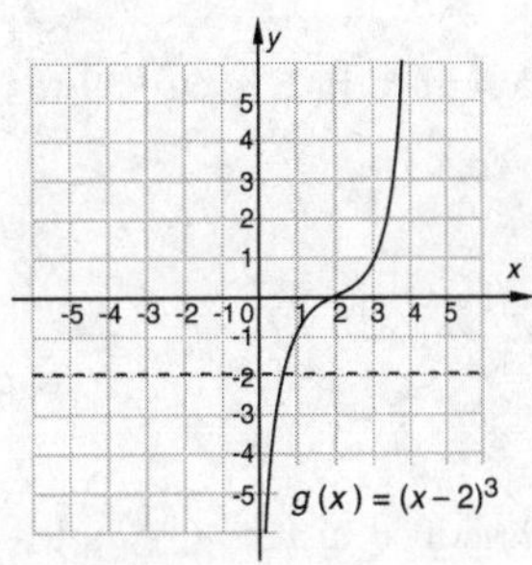

b) Replace $g(x)$ by y: $y=(x-2)^3$

Interchange x and y: $x=(y-2)^3$

Solve for y: $\sqrt[3]{x}=y-2$

$\sqrt[3]{x}+2=y$

Replace y by $g^{-1}(x)$: $g^{-1}(x)=\sqrt[3]{x}+2$

57. a) The graph of $f(x)=\sqrt{x}$ is shown below. It passes the horizontal-line test, so the function is one-to-one.

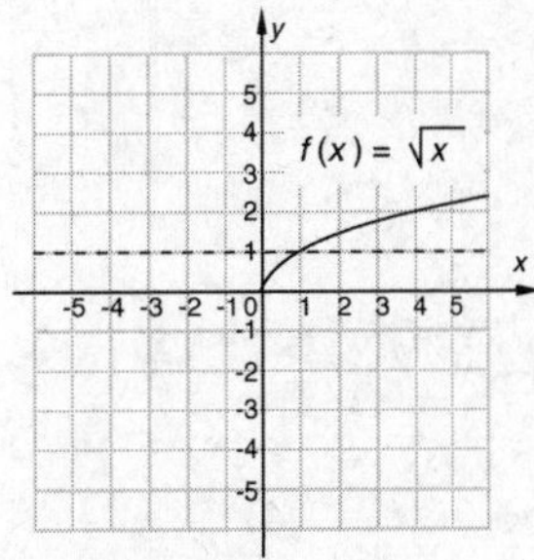

b) Replace $f(x)$ by y: $y=\sqrt{x}$ (Note that $f(x)\geq 0$)

Interchange x and y: $x=\sqrt{y}$

Solve for y: $x^2=y$

Replace y by $f^{-1}(x)$: $f^{-1}(x)=x^2; x\geq 0$

59. First graph $f(x)=\frac{2}{3}x+4$. Then graph the inverse function by reflecting the graph of $f(x)=\frac{2}{3}x+4$ across the line $y=x$. The graph of the inverse function can also be found by first finding a formula for the inverse, substituting to find function values, and then plotting points.

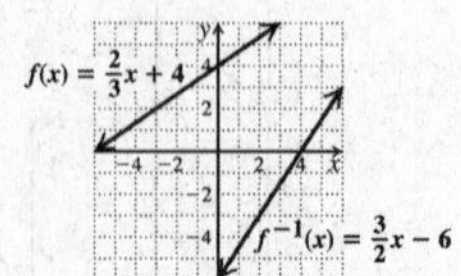

61. Follow the procedure described in Exercise 59 to graph the function and its inverse.

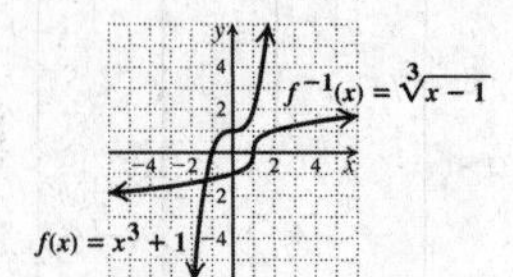

63. Follow the procedure described in Exercise 59 to graph the function and its inverse.

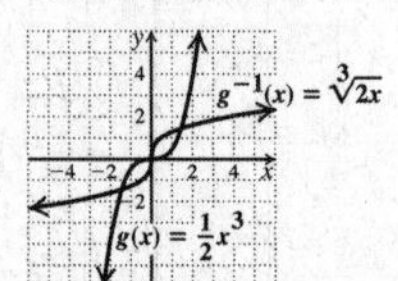

65. Follow the procedure described in Exercise 59 to graph the function and its inverse.

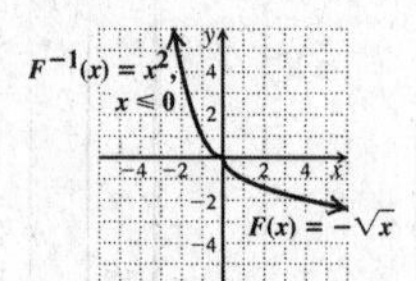

67. Follow the procedure described in Exercise 59 to graph the function and its inverse.

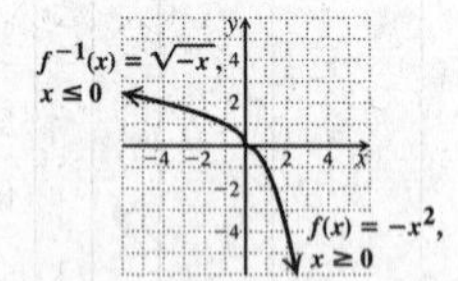

69. We check to see that $\left(f^{-1}\circ f\right)(x)=x$ and $\left(f\circ f^{-1}\right)(x)=x$.

$$(f^{-1}\circ f)(x)=f^{-1}(f(x))=f^{-1}(\sqrt[3]{x-4})$$
$$=\left(\sqrt[3]{x-4}\right)^3+4=x-4+4=x$$
$$(f\circ f^{-1})(x)=f(f^{-1}(x))=f(x^3+4)$$
$$=\sqrt[3]{x^3+4-4}=\sqrt[3]{x^3}=x$$

71. We check to see that $f^{-1}\circ f(x)=x$ and $f\circ f^{-1}(x)=x$.

$$f^{-1}\circ f(x)=f^{-1}(f(x))=f^{-1}\left(\frac{1-x}{x}\right)=\frac{1}{\frac{1-x}{x}+1}$$
$$=\frac{1}{\frac{1-x}{x}+1}\cdot\frac{x}{x}=\frac{x}{1-x+x}=\frac{x}{1}=x$$
$$f\circ f^{-1}(x)=f(f^{-1}(x))=f\left(\frac{1}{x+1}\right)=\frac{1-\frac{1}{x+1}}{\frac{1}{x+1}}$$
$$=\frac{1-\frac{1}{x+1}}{\frac{1}{x+1}}\cdot\frac{x+1}{x+1}=\frac{x+1-1}{1}=\frac{x}{1}=x$$

73. a) $f(8)=2(8+12)=2\cdot 20=40$

Size 40 in Italy corresponds to size 8 in the U.S.

$f(10)=2(10+12)=2\cdot 22=44$

Size 44 in Italy corresponds to size 10 in the U.S.

$f(14)=2(14+12)=2\cdot 26=52$

Size 52 in Italy corresponds to size 14 in the U.S.

$f(18)=2(18+12)=2\cdot 30=60$

Size 60 in Italy corresponds to size 18 in the U.S.

b) The function $f(x)=2(x+12)$ is a linear function that is not constant, so it passes the horizontal-line test and has an inverse that is a function.

Replace $f(x)$ by y: $\quad y=2(x+12)$

Interchange x and y: $\quad x=2(y+12)$

Solve for y: $\quad x=2y+24$

$$x-24=2y$$
$$\frac{x-24}{2}=y$$

Replace y by $f^{-1}(x)$: $\quad f^{-1}(x)=\frac{x-24}{2}$ or $\frac{x}{2}-12$

c) $f^{-1}(40)=\frac{40-24}{2}=\frac{16}{2}=8$

Size 8 in the U.S. corresponds to size 40 in Italy.

$f^{-1}(44)=\frac{44-24}{2}=\frac{20}{2}=10$

Size 10 in the U.S. corresponds to size 44 in Italy.

$f^{-1}(52)=\frac{52-24}{2}=\frac{28}{2}=14$

Size 14 in the U.S. corresponds to size 52 in Italy.

$f^{-1}(60)=\frac{60-24}{2}=\frac{36}{2}=18$

Size 18 in the U.S. corresponds to size 60 in Italy.

75. *Writing Exercise.* No; several items can have the same price.

77. $2^{-3}=\frac{1}{2^3}=\frac{1}{8}$

79. $4^{5/2}=\left(\sqrt{4}\right)^5=2^5=32$

81. Graph $y = x^3$.

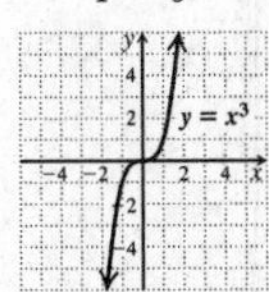

83. *Writing Exercise.* $V^{-1}(t)$ could be used to determine the number of years from 2008 where t is the value of the stamp.

85. Reflect the graph of f across the line $y = x$.

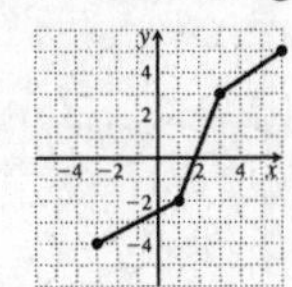

87. From Exercise 73(b), we know that a function that converts dress sizes in Italy to those in the United States is $g(x) = \frac{x-24}{2}$. From Exercise 74, we know that a function that converts dress sizes in the United States to those in France is $f(x) = x + 32$. Then a function that converts dress sizes in Italy to those in France is

$$h(x) = (f \circ g)(x)$$
$$h(x) = f\left(\frac{x-24}{2}\right)$$
$$h(x) = \frac{x-24}{2} + 32$$
$$h(x) = \frac{x}{2} - 12 + 32$$
$$h(x) = \frac{x}{2} + 20.$$

89. *Writing Exercise.* The functions found in Exercises 87 and 88 are inverses. We can show that $(h \circ d)(x) = x$ and $(d \circ h)(x) = x$.

91. Suppose that $h(x) = (f \circ g)(x)$. First note that for $I(x) = x$, $(f \circ I)(x) = f(I(x))$ for any function f.

i) $$((g^{-1} \circ f^{-1}) \circ h)(x) = ((g^{-1} \circ f^{-1}) \circ (f \circ g))(x)$$
$$= ((g^{-1} \circ (f^{-1} \circ f)) \circ g)(x)$$
$$= ((g^{-1} \circ I) \circ g)(x)$$
$$= (g^{-1} \circ g)(x) = x$$

ii) $$(h \circ (g^{-1} \circ f^{-1}))(x) = ((f \circ g) \circ (g^{-1} \circ f^{-1}))(x)$$
$$= ((f \circ (g \circ g^{-1})) \circ f^{-1})(x)$$
$$= ((f \circ I) \circ f^{-1})(x)$$
$$= (f \circ f^{-1})(x) = x$$

Therefore, $(g^{-1} \circ f^{-1})(x) = h^{-1}(x)$.

93. $(f \circ g)(x) = x$ and $(g \circ f)(x) = x$, so the functions are inverses.

95. $(f \circ g)(x) \neq x$, so the functions are not inverses. (It is also true that $(g \circ f)(x) \neq x$.)

97. (1) C; (2) A; (3) B; (4) D

99. *Writing Exercise.*

a) For $x = 2$ and $x = 4$, $y_2(y_1(x)) = x$ and $y_1(y_2(x)) = x$.

b) $Y_1 = 0.5x + 1$; $Y_2 = 2x - 2$

c) Following the procedure on page 793 in the text, we find that $Y_1^{-1} = 2x - 2$, so the functions are inverses.

Exercise Set 9.2

1. True; see page 797 in the text.

3. True; the graph of $y = f(x-3)$ is a translation of the graph of $y = f(x)$, 3 units to the right.

5. False; the graph of $y = 3^x$ crosses the y-axis at (0, 1).

7. Graph: $y = f(x) = 3^x$

We compute some function values, thinking of y as $f(x)$, and keep the results in a table.

$f(0) = 3^0 = 1$

$f(1) = 3^1 = 3$

$f(2) = 3^2 = 9$

$f(-1) = 3^{-1} = \frac{1}{3^1} = \frac{1}{3}$

$f(-2) = 3^{-2} = \frac{1}{3^2} = \frac{1}{9}$

x	y, or $f(x)$
0	1
1	3
2	9
−1	$\frac{1}{3}$
−2	$\frac{1}{9}$

Next we plot these points and connect them with a smooth curve.

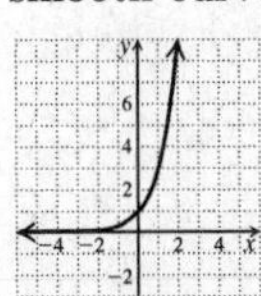

9. Graph: $y = 6^x$

We compute some function values, thinking of y as $f(x)$, and keep the results in a table.

$f(0) = 6^0 = 1$

$f(1) = 6^1 = 6$

$f(2) = 6^2 = 36$

$f(-1) = 6^{-1} = \frac{1}{6^1} = \frac{1}{6}$

$f(-2) = 6^{-2} = \frac{1}{6^2} = \frac{1}{36}$

x	y, or $f(x)$
0	1
1	6
2	36
-1	$\frac{1}{6}$
-2	$\frac{1}{36}$

Next we plot these points and connect them with a smooth curve.

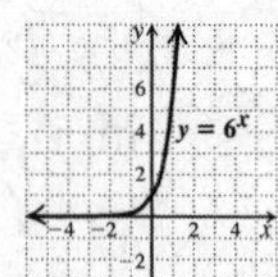

11. Graph: $y = 2^x + 1$

We compute some function values, thinking of y as $f(x)$, and keep the results in a table.

$f(-4) = 2^{-4} + 1 = \frac{1}{2^4} + 1 = \frac{1}{16} + 1 = 1\frac{1}{16}$

$f(-2) = 2^{-2} + 1 = \frac{1}{2^2} + 1 = \frac{1}{4} + 1 = 1\frac{1}{4}$

$f(0) = 2^0 + 1 = 1 + 1 = 2$

$f(1) = 2^1 + 1 = 2 + 1 = 3$

$f(2) = 2^2 + 1 = 4 + 1 = 5$

x	y, or $f(x)$
-4	$1\frac{1}{16}$
-2	$1\frac{1}{4}$
0	2
1	3
2	5

Next we plot these points and connect them with a smooth curve.

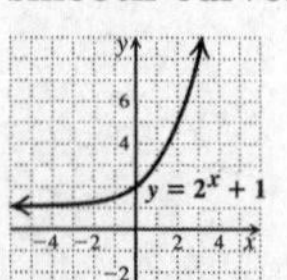

13. Graph: $y = 3^x - 2$

We compute some function values, thinking of y as $f(x)$, and keep the results in a table.

$f(-3) = 3^{-3} - 2 = \frac{1}{3^3} - 2 = \frac{1}{27} - 2 = -\frac{53}{27}$

$f(-1) = 3^{-1} - 2 = \frac{1}{3} - 2 = -\frac{5}{3}$

$f(0) = 3^0 - 2 = 1 - 2 = -1$

$f(1) = 3^1 - 2 = 3 - 2 = 1$

$f(2) = 3^2 - 2 = 9 - 2 = 7$

x	y, or $f(x)$
-3	$-\frac{53}{27}$
-1	$-\frac{5}{3}$
0	-1
1	1
2	7

Next we plot these points and connect them with a smooth curve.

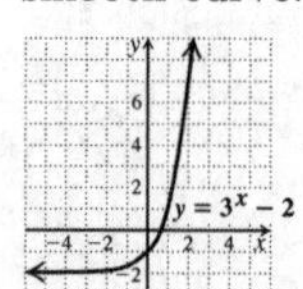

15. Graph: $y = 2^x - 5$

We construct a table of values, thinking of y as $f(x)$. Then we plot the points and connect them with a smooth curve.

$f(0) = 2^0 - 5 = 1 - 5 = -4$

$f(1) = 2^1 - 5 = 2 - 5 = -3$

$f(2) = 2^2 - 5 = 4 - 5 = -1$

$f(3) = 2^3 - 5 = 8 - 5 = 3$

$f(-1) = 2^{-1} - 5 = \frac{1}{2} - 5 = -\frac{9}{2}$

$f(-2) = 2^{-2} - 5 = \frac{1}{4} - 5 = -\frac{19}{4}$

$f(-4) = 2^{-4} - 5 = \frac{1}{16} - 5 = -\frac{79}{16}$

x	y, or $f(x)$
0	-4
1	-3
2	-1
3	3
-1	$-\frac{9}{2}$
-2	$-\frac{19}{4}$
-4	$-\frac{79}{16}$

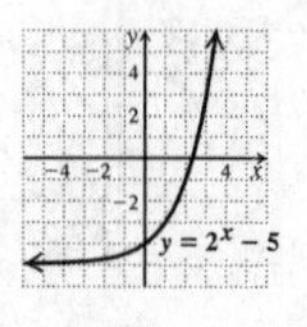

17. Graph: $y = 2^{x-3}$

We construct a table of values, thinking of y as $f(x)$. Then we plot the points and connect them with a smooth curve.

$f(0) = 2^{0-3} = 2^{-3} = \frac{1}{8}$

$f(-1) = 2^{-1-3} = 2^{-4} = \frac{1}{16}$

$f(1) = 2^{1-3} = 2^{-2} = \frac{1}{4}$

$f(2) = 2^{2-3} = 2^{-1} = \frac{1}{2}$

$f(3) = 2^{3-3} = 2^{0} = 1$

$f(4) = 2^{4-3} = 2^{1} = 2$

$f(5) = 2^{5-3} = 2^{2} = 4$

x	y, or $f(x)$
0	$\frac{1}{8}$
-1	$\frac{1}{16}$
1	$\frac{1}{4}$
2	$\frac{1}{2}$
3	1
4	2
5	3

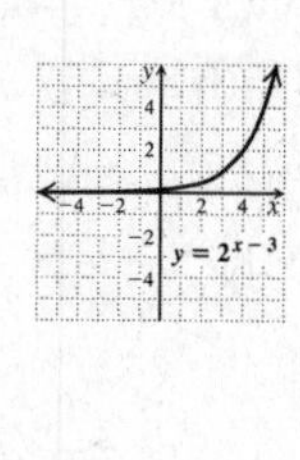

19. Graph: $y = 2^{x+1}$

We construct a table of values, thinking of y as $f(x)$. Then we plot the points and connect them with a smooth curve.

$f(-3) = 2^{-3+1} = 2^{-2} = \frac{1}{4}$

$f(-1) = 2^{-1+1} = 2^{0} = 1$

$f(0) = 2^{0+1} = 2^{1} = 2$

$f(1) = 2^{1+1} = 2^{2} = 4$

x	y, or $f(x)$
-3	$\frac{1}{4}$
-1	1
0	2
1	4

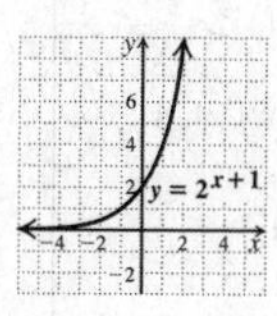

21. Graph: $y = \left(\frac{1}{4}\right)^x$

We construct a table of values, thinking of y as $f(x)$. Then we plot the points and connect them with a smooth curve.

$f(0) = \left(\frac{1}{4}\right)^0 = 1$

$f(1) = \left(\frac{1}{4}\right)^1 = \frac{1}{4}$

$f(2) = \left(\frac{1}{4}\right)^2 = \frac{1}{16}$

$f(-1) = \left(\frac{1}{4}\right)^{-1} = \frac{1}{\frac{1}{4}} = 4$

$f(-2) = \left(\frac{1}{4}\right)^{-2} = \frac{1}{\frac{1}{16}} = 16$

x	y, or $f(x)$
0	1
1	$\frac{1}{4}$
2	$\frac{1}{16}$
-1	4
-2	16

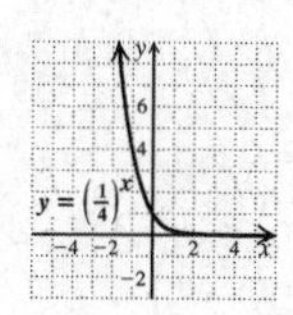

23. Graph: $y = \left(\frac{1}{3}\right)^x$

We construct a table of values, thinking of y as $f(x)$. Then we plot the points and connect them with a smooth curve.

$f(0) = \left(\frac{1}{3}\right)^0 = 1$

$f(1) = \left(\frac{1}{3}\right)^1 = \frac{1}{3}$

$f(2) = \left(\frac{1}{3}\right)^2 = \frac{1}{9}$

$f(3) = \left(\frac{1}{3}\right)^3 = \frac{1}{27}$

$f(-1) = \left(\frac{1}{3}\right)^{-1} = \frac{1}{\left(\frac{1}{3}\right)^1} = \frac{1}{\frac{1}{3}} = 3$

$f(-2) = \left(\frac{1}{3}\right)^{-2} = \frac{1}{\left(\frac{1}{3}\right)^2} = \frac{1}{\frac{1}{9}} = 9$

$f(-3) = \left(\frac{1}{3}\right)^{-3} = \frac{1}{\left(\frac{1}{3}\right)^3} = \frac{1}{\frac{1}{27}} = 27$

x	y, or $f(x)$
0	1
1	$\frac{1}{3}$
2	$\frac{1}{9}$
3	$\frac{1}{27}$
-1	3
-2	9
-3	27

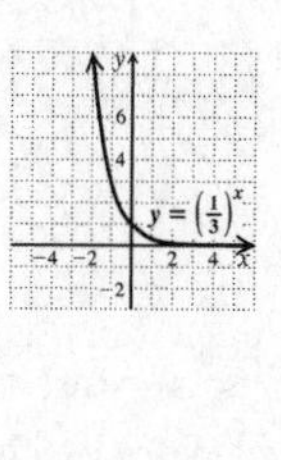

25. Graph: $y = 2^{x+1} - 3$

We construct a table of values, thinking of y as $f(x)$. Then we plot the points and connect them with a smooth curve.

$f(0)=2^{0+1}-3=2-3=-1$
$f(1)=2^{1+1}-3=4-3=1$
$f(2)=2^{2+1}-3=8-3=5$
$f(-1)=2^{-1+1}-3=1-3=-2$
$f(-2)=2^{-2+1}-3=\frac{1}{2}-3=-\frac{5}{2}$
$f(-3)=2^{-3+1}-3=\frac{1}{4}-3=-\frac{11}{4}$

x	y, or $f(x)$
0	-1
1	1
2	5
-1	-2
-2	$-\frac{5}{2}$
-3	$-\frac{11}{4}$

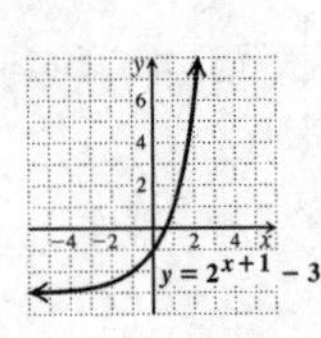

27. Graph: $x = 6^y$

We can find ordered pairs by choosing values for y and then computing values for x.

For $y=0, x=6^0=1$.

For $y=1, x=6^1=6$.

For $y=-1, x=6^{-1}=\frac{1}{6^1}=\frac{1}{6}$.

For $y=-2, x=6^{-2}=\frac{1}{6^2}=\frac{1}{36}$.

x	y
1	0
6	1
$\frac{1}{6}$	-1
$\frac{1}{36}$	-2

↑ ↑ (1) Choose values for y.

(2) Compute values for x.

We plot the points and connect them with a smooth curve.

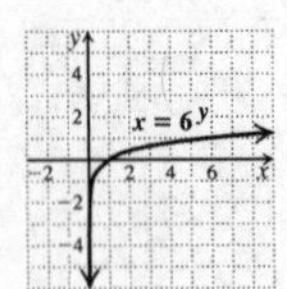

29. Graph: $x = 3^{-y} = (\frac{1}{3})^y$

We can find ordered pairs by choosing values for y and then computing values for x. Then we plot these points and connect them with a smooth curve.

For $y=0,\ x=\left(\frac{1}{3}\right)^0=1$.

For $y=1,\ x=\left(\frac{1}{3}\right)^1=\frac{1}{3}$.

For $y=2,\ x=\left(\frac{1}{3}\right)^2=\frac{1}{9}$.

For $y=-1,\ x=\left(\frac{1}{3}\right)^{-1}=\frac{1}{\frac{1}{3}}=3$.

For $y=-2,\ x=\left(\frac{1}{3}\right)^{-2}=\frac{1}{\frac{1}{9}}=9$.

x	y
1	0
$\frac{1}{3}$	1
$\frac{1}{9}$	2
3	-1
9	-2

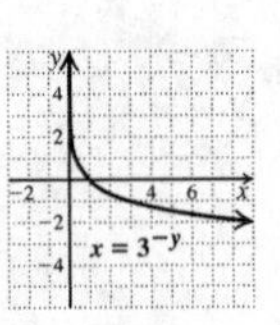

31. Graph: $x = 4^y$

We can find ordered pairs by choosing values for y and then computing values for x. Then we plot these points and connect them with a smooth curve.

For $y=0,\ x=4^0=1$.

For $y=1,\ x=4^1=4$.

For $y=2,\ x=4^2=16$.

For $y=-1,\ x=4^{-1}=\frac{1}{4}$.

For $y=-2,\ x=4^{-2}=\frac{1}{16}$.

x	y
1	0
4	1
16	2
$\frac{1}{4}$	-1
$\frac{1}{16}$	-2

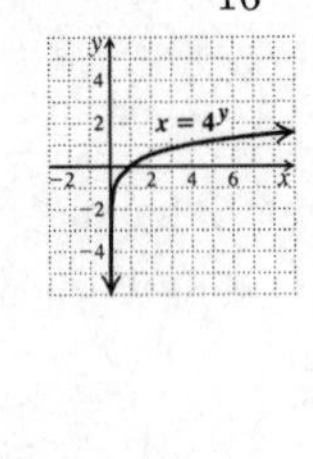

33. Graph: $x = (\frac{4}{3})^y$

We can find ordered pairs by choosing values for y and then computing values for x. Then we plot these points and connect them with a smooth curve.

For $y=0,\ x=\left(\frac{4}{3}\right)^0=1$.

For $y=1,\ x=\left(\frac{4}{3}\right)^1=\frac{4}{3}$.

For $y=2,\ x=\left(\frac{4}{3}\right)^2=\frac{16}{9}$.

For $y=3,\ x=\left(\frac{4}{3}\right)^3=\frac{64}{27}$.

For $y=-1,\ x=\left(\frac{4}{3}\right)^{-1}=\frac{3}{4}$.

For $y=-2,\ x=\left(\frac{4}{3}\right)^{-2}=\left(\frac{3}{4}\right)^2=\frac{9}{16}$.

For $y=-3,\ x=\left(\frac{4}{3}\right)^{-3}=\left(\frac{3}{4}\right)^3=\frac{27}{64}$.

x	y
1	0
$\frac{4}{3}$	1
$\frac{16}{9}$	2
$\frac{64}{27}$	3
$\frac{3}{4}$	-1
$\frac{9}{16}$	-2
$\frac{27}{64}$	-3

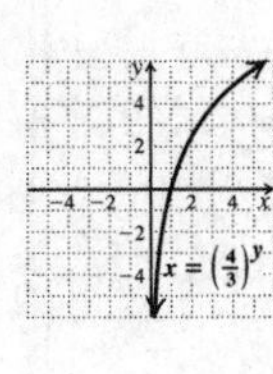

35. Graph $y = 3^x$ (see Exercise 8) and $x = 3^y$ (see Exercise 28) using the same set of axes.

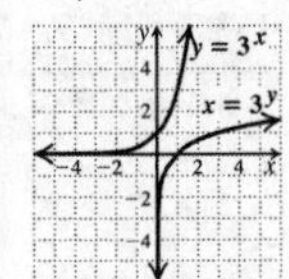

37. Graph $y = \left(\frac{1}{2}\right)^x$ (see Exercise 24) and $x = \left(\frac{1}{2}\right)^y$ (see Exercise 30) using the same set of axes.

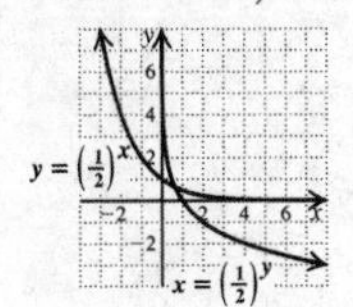

39. a) In 2006, $t = 2006 - 2003 = 3$

$M(3) = 0.353(1.244)^3 \approx 0.680$ billion tracks

In 2008, $t = 2008 - 2003 = 5$

$M(5) = 0.353(1.244)^5 \approx 1.052$ billion tracks

In 2012, $t = 2012 - 2003 = 9$

$M(9) = 0.353(1.244)^9 \approx 2.519$ billion tracks

b) Use the function values computed in part (a) and others, if desired, and draw the graph.

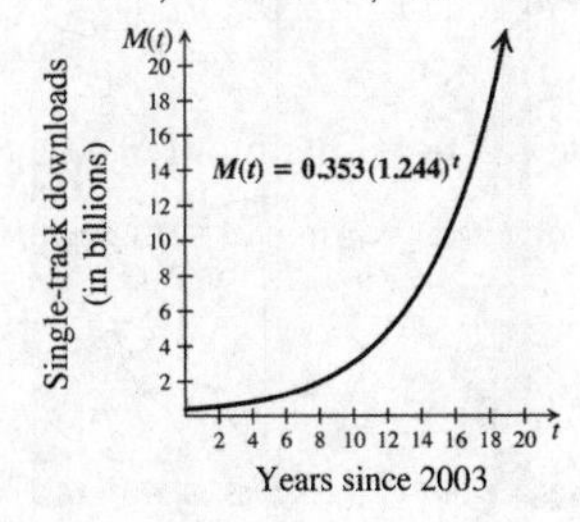

41. a) $P(1) = 21.4(0.914)^1 \approx 19.6\%$

$P(3) = 21.4(0.914)^3 \approx 16.3\%$

1 yr = 12 months; $P(12) = 21.4(0.914)^{12} \approx 7.3\%$

b)

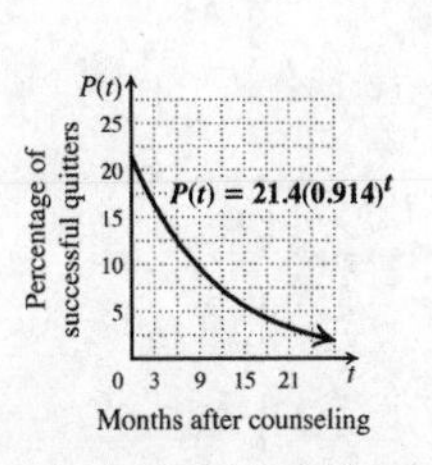

43. a) In 1930, $t = 1930 - 1900 = 30$.

$$P(t) = 150(0.960)^t$$
$$P(30) = 150(0.960)^{30}$$
$$\approx 44.079$$

In 1930, about 44.079 thousand, or 44,079, humpback whales were alive.

In 1960, $t = 1960 - 1900 = 60$.

$$P(t) = 150(0.960)^t$$
$$P(60) = 150(0.960)^{60}$$
$$\approx 12.953$$

In 1960, about 12.953 thousand, or 12,953, humpback whales were alive.

b) Plot the points found in part (a), (30, 44,079) and (60, 12,953) and additional points as needed and graph the function.

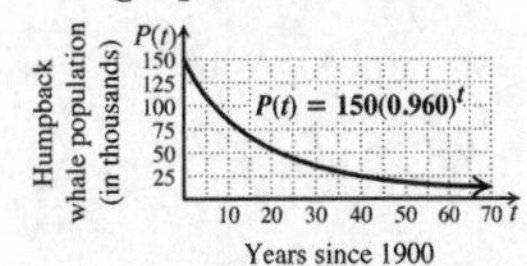

45. a) In 1992, $t = 1992 - 1982 = 10$.

$P(10) = 5.5(1.047)^{10} \approx 8.706$

In 1992, about 8.706 thousand, or 8706, humpback whales were alive.

In 2004, $t = 2004 - 1982 = 22$.

$P(22) = 5.5(1.047)^{22} \approx 15.107$

In 2004, about 15.107 thousand, or 15,107, humpback whales were alive.

b) Use the function values computed in part (a) and others, if desired, and draw the graph.

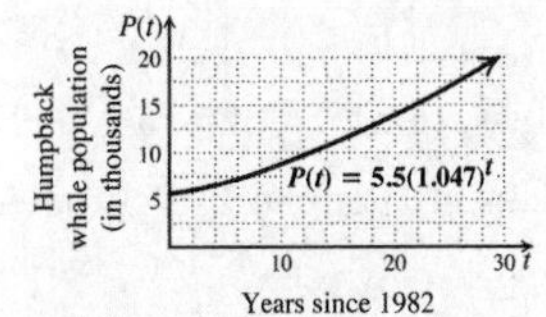

47. a) $A(5) = 10 \cdot 34^5 = 454{,}354{,}240 \text{ cm}^2$

$A(7) = 10 \cdot 34^7 = 525{,}233{,}501{,}400 \text{ cm}^2$

b) Use the function values computed in part (a) and others, if desired, and draw the graph.

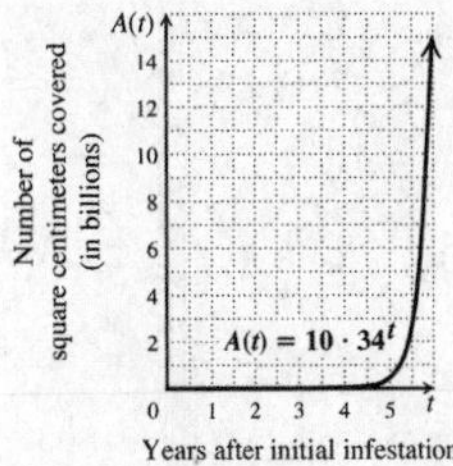

49. *Writing Exercise.* Since $3<\pi<4$, $2^3<2^\pi<2^4$; that is, 2^π is greater than 2^3, or 8, but less than 2^4, or 16.

51. $3x^2-48=3(x^2-16)=3(x+4)(x-4)$

53. $6x^2+x-12=(2x+3)(3x-4)$

55. t^2-y^2+2y-1
$=t^2-(y^2-2y+1)$
$=t^2-(y-1)^2$ Difference of squares
$=[t-(y-1)][t+(y-1)]$
$=(t-y+1)(t+y-1)$

57. *Writing Exercise.* No; the number computed for 2010 far exceeds the U.S. population, so the number that would be computed for 20 years from now would not be realistic.

59. Since the bases are the same, the one with the larger exponent is the larger number. Thus $\pi^{2.4}$ is larger.

61. Graph: $f(x)=2.5^x$

Use a calculator with a power key to construct a table of values. (We will round values of $f(x)$ to the nearest hundredth.) Then plot these points and connect them with a smooth curve.

x	y
0	1
1	2.5
2	6.25
3	15.63
−1	0.4
−2	0.16

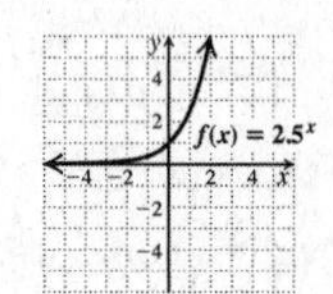

63. Graph: $y=2^x+2^{-x}$

Construct a table of values, thinking of y as $f(x)$. Then plot these points and connect them with a curve.

$f(0)=2^0+2^{-0}=1+1=2$

$f(1)=2^1+2^{-1}=2+\frac{1}{2}=2\frac{1}{2}$

$f(2)=2^2+2^{-2}=4+\frac{1}{4}=4\frac{1}{4}$

$f(3)=2^3+2^{-3}=8+\frac{1}{8}=8\frac{1}{8}$

$f(-1)=2^{-1}+2^{-(-1)}=\frac{1}{2}+2=2\frac{1}{2}$

$f(-2)=2^{-2}+2^{-(-2)}=\frac{1}{4}+4=4\frac{1}{4}$

$f(-3)=2^{-3}+2^{-(-3)}=\frac{1}{8}+8=8\frac{1}{8}$

x	y, or $f(x)$
0	2
1	$2\frac{1}{2}$
2	$4\frac{1}{4}$
3	$8\frac{1}{8}$
−1	$2\frac{1}{2}$
−2	$4\frac{1}{4}$
−3	$8\frac{1}{8}$

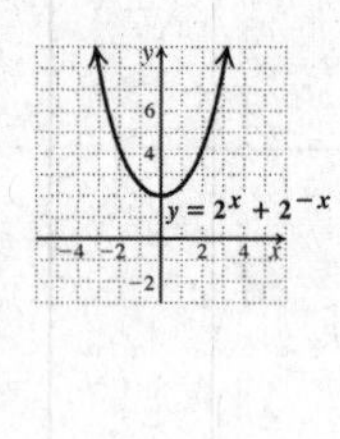

65. Graph: $y=|2^x-2|$

We construct a table of values, thinking of y as $f(x)$. Then plot these points and connect them with a curve.

$f(0)=|2^0-2|=|1-2|=|-1|=1$
$f(1)=|2^1-2|=|2-2|=|0|=0$
$f(2)=|2^2-2|=|4-2|=|2|=2$
$f(3)=|2^3-2|=|8-2|=|6|=6$
$f(-1)=|2^{-1}-2|=\left|\frac{1}{2}-2\right|=\left|-\frac{3}{2}\right|=\frac{3}{2}$
$f(-3)=|2^{-3}-2|=\left|\frac{1}{8}-2\right|=\left|-\frac{15}{8}\right|=\frac{15}{8}$
$f(-5)=|2^{-5}-2|=\left|\frac{1}{32}-2\right|=\left|-\frac{63}{32}\right|=\frac{63}{32}$

x	y, or $f(x)$
0	1
1	0
2	2
3	6
−1	$\frac{3}{2}$
−3	$\frac{15}{8}$
−5	$\frac{63}{32}$

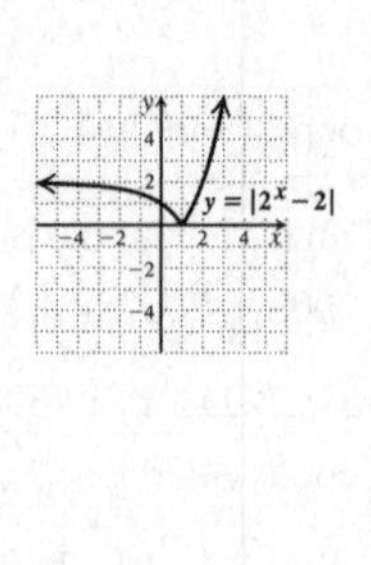

67. Graph: $y=|2^{x^2}-1|$

We construct a table of values, thinking of y as $f(x)$. Then we plot these points and connect them with a curve.

$f(0)=|2^{0^2}-1|=|1-1|=0$
$f(1)=|2^{1^2}-1|=|2-1|=1$
$f(2)=|2^{2^2}-1|=|16-1|=15$
$f(-1)=|2^{(-1)^2}-1|=|2-1|=1$
$f(-2)=|2^{(-2)^2}-1|=|16-1|=15$

x	y, or $f(x)$
0	0
1	1
2	15
−1	1
−2	15

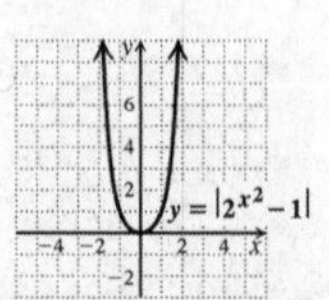

69. $y = 3^{-(x-1)}$ $x = 3^{-(y-1)}$

x	y
0	3
1	1
2	$\frac{1}{3}$
3	$\frac{1}{9}$
−1	9

x	y
3	0
1	1
$\frac{1}{3}$	2
$\frac{1}{9}$	3
9	−1

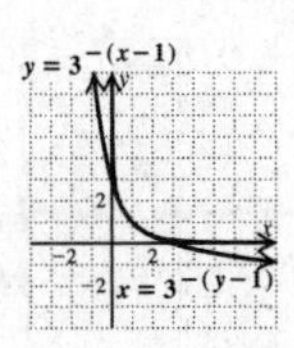

71. Enter the data points (0, 0.5), (4, 4) and (8, 50) and then use the ExpReg option from the STAT CALC menu of a graphing calculator to find an exponential function that models the data:

$N(t) = 0.464(1.778)^t$, where $N(t)$ is the number of navigational devices in use, in millions, t years after 2000.

In 2012, $t = 2012 - 2000 = 12$

$N(12) \approx 463$ million devices.

73. *Writing Exercise.*

a) Section A appears to grow at a Linear rate.
Section B appears to increase at a Exponential rate.
Section C appears to have nearly stopped growth so it should be labeled Saturation.

b) Each of the forces affect wave height in different ways. Gravity is a constant force, while wind shows linear growth, and wave height can grow exponentially due to these forces and surface roughness.

75. *Graphing Calculator Exercise*

Exercise Set 12.3

1. $5^2 = 25$, so choice (g) is correct.

3. $5^1 = 5$, so choice (a) is correct.

5. The exponent to which we raise 5 to get 5^x is x, so choice (b) is correct.

7. $5 = 2^x$ is equivalent to $\log_2 5 = x$, so choice (e) is correct.

9. $\log_{10} 1000$ is the exponent to which we raise 10 to get 1000. Since $10^3 = 1000$, $\log_{10} 1000 = 3$.

11. $\log_7 49$ is the exponent to which we raise 7 to get 49. Since $7^2 = 49$, $\log_7 49 = 2$.

13. $\log_3 81$ is the exponent to which we raise 3 to get 81. Since $3^4 = 81$, $\log_3 81 = 4$.

15. $\log_5 \frac{1}{25}$ is the exponent to which we raise 5 to get $\frac{1}{25}$. Since $5^{-2} = \frac{1}{25}$, $\log_5 \frac{1}{25} = -2$.

17. Since $8^{-1} = \frac{1}{8}$, $\log_8 \frac{1}{8} = -1$.

19. Since $5^4 = 625$, $\log_5 625 = 4$.

21. Since $7^1 = 7$, $\log_7 7 = 1$.

23. Since $3^0 = 1$, $\log_3 1 = 0$.

25. $\log_6 6^5$ is the exponent to which we raise 6 to get 6^5. Clearly, this power is 5, so $\log_6 6^5 = 5$.

27. Since $10^{-2} = \frac{1}{100} = 0.01$, $\log_{10} 0.01 = -2$.

29. Since $16^{1/2} = 4$, $\log_{16} 4 = \frac{1}{2}$.

31. Since $9 = 3^2$ and $(3^2)^{3/2} = 3^3 = 27$, $\log_9 27 = \frac{3}{2}$.

33. Since $1000 = 10^3$ and $(10^3)^{2/3} = 10^2 = 100$, $\log_{1000} 100 = \frac{2}{3}$.

35. Since $\log_3 29$ is the power to which we raise 3 to get 29, then 3 raised to this power is 29. That is $3^{\log_3 29} = 29$.

37. Graph: $y = \log_{10} x$

The equation $y = \log_{10} x$ is equivalent to $10^y = x$. We can find ordered pairs by choosing values for y and computing the corresponding x-values.

For $y = 0$, $x = 10^0 = 1$.

For $y = 1$, $x = 10^1 = 10$.

For $y = 2$, $x = 10^2 = 100$.

For $y = -1$, $x = 10^{-1} = \frac{1}{10}$.

For $y = -2$, $x = 10^{-2} = \frac{1}{100}$.

x, or 10^y	y
1	0
10	1
100	2
$\frac{1}{10}$	-1
$\frac{1}{100}$	-2

↑ ↑ (1) Select y.

(2) Compute x.

We plot the set of ordered pairs and connect the points with a smooth curve.

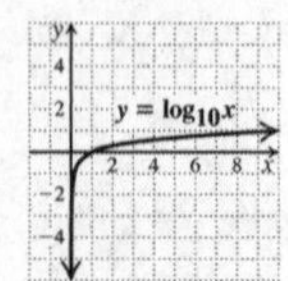

39. Graph: $y = \log_3 x$

The equation $y = \log_3 x$ is equivalent to $3^y = x$. We can find ordered pairs by choosing values for y and computing the corresponding x-values.

For $y = 0$, $x = 3^0 = 1$.
For $y = 1$, $x = 3^1 = 3$.
For $y = 2$, $x = 3^2 = 9$.
For $y = -1$, $x = 3^{-1} = \frac{1}{3}$.
For $y = -2$, $x = 3^{-2} = \frac{1}{9}$.

x, or 3^y	y
1	0
3	1
9	2
$\frac{1}{3}$	-1
$\frac{1}{9}$	-2

We plot the set of ordered pairs and connect the points with a smooth curve.

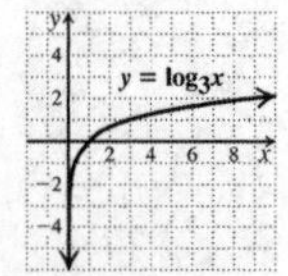

41. Graph: $f(x) = \log_6 x$

Think of $f(x)$ as y. Then $y = \log_6 x$ is equivalent to $6^y = x$. We find ordered pairs by choosing values for y and computing the corresponding x-values. Then we plot the points and connect them with a smooth curve.

For $y = 0$, $x = 6^0 = 1$.
For $y = 1$, $x = 6^1 = 6$.
For $y = 2$, $x = 6^2 = 36$.
For $y = -1$, $x = 6^{-1} = \frac{1}{6}$.
For $y = -2$, $x = 6^{-2} = \frac{1}{36}$.

x, or 6^y	y
1	0
6	1
36	2
$\frac{1}{6}$	-1
$\frac{1}{36}$	-2

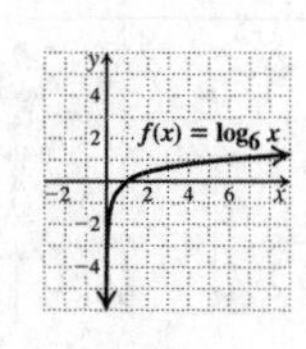

43. Graph: $f(x) = \log_{2.5} x$

Think of $f(x)$ as y. Then $y = \log_{2.5} x$ is equivalent to $2.5^y = x$. We construct a table of values, plot these points and connect them with a smooth curve.

For $y = 0$, $x = 2.5^0 = 1$.
For $y = 1$, $x = 2.5^1 = 2.5$.
For $y = 2$, $x = 2.5^2 = 6.25$.
For $y = 3$, $x = 2.5^3 = 15.625$.
For $y = -1$, $x = 2.5^{-1} = 0.4$.
For $y = -2$, $x = 2.5^{-2} = 0.16$.

x, or 2.5^y	y
1	0
2.5	1
6.25	2
15.625	3
0.4	-1
0.16	-2

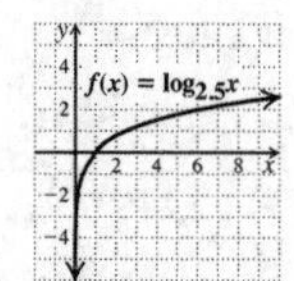

45. Graph $f(x) = 3^x$ (see Exercise Set 12.2, Exercise 7) and $f^{-1}(x) = \log_3 x$ (see Exercise 39 above) on the same set of axes.

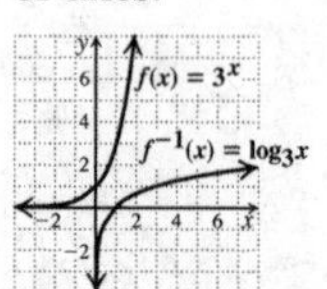

47. $x = \log_{10} 8 \Rightarrow 10^x = 8$

The base remains the same.

The logarithm is the exponent.

49. $\log_9 9 = 1 \Rightarrow 9^1 = 9$

The logarithm is the exponent.

The base remains the same.

51. $\log_{10} 0.1 = -1$ is equivalent to $10^{-1} = 0.1$.

53. $\log_{10} 7 = 0.845$ is equivalent to $10^{0.845} = 7$.

55. $\log_c m = 8$ is equivalent to $c^8 = m$.

57. $\log_r C = t$ is equivalent to $r^t = C$.

59. $\log_e 0.25 = -1.3863$ is equivalent to $e^{-1.3863} = 0.25$.

61. $\log_r T = -x$ is equivalent to $r^{-x} = T$.

63. $10^2 = 100 \Rightarrow 2 = \log_{10} 100$

The exponent is the logarithm.

The base remains the same.

65. $5^{-3} = \frac{1}{125} \Rightarrow -3 = \log_5 \frac{1}{125}$

The logarithm is the exponent.

The base remains the same.

67. $16^{1/4} = 2$ is equivalent to $\frac{1}{4} = \log_{16} 2$.

69. $10^{0.4771} = 3$ is equivalent to $0.4771 = \log_{10} 3$.

71. $z^m = 6$ is equivalent to $m = \log_z 6$.

73. $p^t = q$ is equivalent to $t = \log_p q$.

75. $e^3 = 20.0855$ is equivalent to $3 = \log_e 20.0855$.

77. $e^{-4} = 0.0183$ is equivalent to $-4 = \log_e 0.0183$.

79. $\log_6 x = 2$

$6^2 = x$ Converting to an exponential equation

$36 = x$ Computing 6^2

81. $\log_2 32 = x$

$2^x = 32$ Converting to an exponential equation

$2^x = 2^5$

$x = 5$ The exponents must be the same.

83. $\log_x 9 = 1$

$x^1 = 9$ Converting to an exponential equation

$x = 9$ Simplifying x^1

85. $\log_x 7 = 1$

$x^{1/2} = 7$ Converting to an exponential equation

$x = 49$ Simplifying $x^{1/2}$

87. $\log_3 x = -2$

$3^{-2} = x$ Converting to an exponential equation

$\frac{1}{9} = x$ Simplifying

89. $\log_{32} x = \frac{2}{5}$

$32^{2/5} = x$ Converting to an exponential equation

$(2^5)^{2/5} = x$

$4 = x$

91. *Writing Exercise.* By definition, $m = \log_a x$ is equivalent to $a^m = x$. So the number $\log_a x$ is the exponent, m.

93. $\sqrt{18a^3b}\sqrt{50ab^7} = \sqrt{18a^3b \cdot 50ab^7} = \sqrt{900a^4b^8} = 30a^2b^4$

95. $\sqrt{192x} - \sqrt{75x} = 8\sqrt{3x} - 5\sqrt{3x}$

$= (8-5)\sqrt{3x} = 3\sqrt{3x}$

97. $\dfrac{\frac{3}{x} - \frac{2}{xy}}{\frac{2}{x^2} + \frac{1}{xy}}$

The LCD of all the denominators is x^2y. We multiply numerator and denominator by the LCD.

$$\frac{\frac{3}{x} - \frac{2}{xy}}{\frac{2}{x^2} + \frac{1}{xy}} \cdot \frac{x^2y}{x^2y} = \frac{\left(\frac{3}{x} - \frac{2}{xy}\right)x^2y}{\left(\frac{2}{x^2} + \frac{1}{xy}\right)x^2y}$$

$$= \frac{\frac{3}{x} \cdot x^2y - \frac{2}{xy} \cdot x^2y}{\frac{2}{x^2} \cdot x^2y + \frac{1}{xy} \cdot x^2y}$$

$$= \frac{3xy - 2x}{2y + x}, \text{ or } \frac{x(3y-2)}{2y+x}$$

99. *Writing Exercise.* The graph of a logarithmic function $f(x) = \log_a x$ increases slowly as x increases. Thus, although the manufacturer would be pleased that sales were growing, he or she would probably prefer that they were growing more rapidly.

101. Graph: $y = \left(\frac{3}{2}\right)^x$

x	y, or $\left(\frac{3}{2}\right)^x$
0	1
1	$\frac{3}{2}$
2	$\frac{9}{4}$
3	$\frac{27}{8}$
-1	$\frac{2}{3}$
-2	$\frac{4}{9}$

Graph: $y = \log_{3/2} x$, or $x = \left(\frac{3}{2}\right)^y$

x, or $\left(\frac{3}{2}\right)^y$	y
1	0
$\frac{3}{2}$	1
$\frac{9}{4}$	2
$\frac{27}{8}$	3
$\frac{2}{3}$	-1
$\frac{4}{9}$	-2

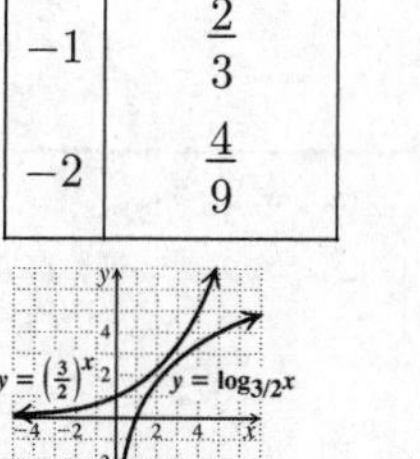

103. Graph: $y = \log_3 |x+1|$

x	y
0	0
2	1
8	2
−2	0
−4	1
−9	2

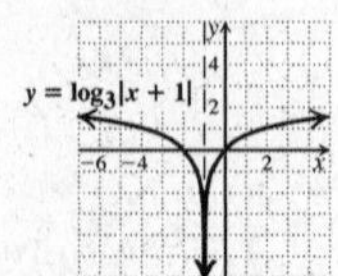

105. $\log_4(3x-2)=2$
$$4^2=3x-2$$
$$16=3x-2$$
$$18=3x$$
$$6=x$$

107. $\log_{10}(x^2+21x)=2$
$$10^2=x^2+21x$$
$$0=x^2+21x-100$$
$$0=(x+25)(x-4)$$

$x=-25$ or $x=4$

109. Let $\log_{1/5} 25 = x$. Then
$$\left(\frac{1}{5}\right)^x=25$$
$$\left(5^{-1}\right)^x=25$$
$$5^{-x}=5^2$$
$$-x=2$$
$$x=-2.$$
Thus, $\log_{1/5} 25=-2$.

111. $\log_{10}\left(\log_4\left(\log_3 81\right)\right)$
$=\log_{10}\left(\log_4 4\right)$ $\quad (\log_3 81=4)$
$=\log_{10} 1$ $\quad (\log_4 4=1)$
$=0$

113. Let $b=0$, $x=1$, and $y=2$. Then $0^1=0^2$, but $1\neq 2$.
Let $b=1$, $x=1$, and $y=2$. Then $1^1=1^2$, but $1\neq 2$.

Exercise Set 12.4

1. Use the product rule for logarithms.
$\log_7 20=\log_7(5\cdot 4)=\log_7 5+\log_7 4$; choice (e) is correct.

3. Use the quotient rule for logarithms.
$\log_7 \frac{5}{4}=\log_7 5-\log_7 4$; choice (a) is correct.

5. The exponent to which we raise 7 to get 1 is 0, so choice (c) is correct.

7. $\log_3(81\cdot 27)=\log_3 81+\log_3 27$ Using the product rule

9. $\log_4(64\cdot 16)=\log_4 64+\log_4 16$ Using the product rule

11. $\log_c rst=\log_c r+\log_c s+\log_c t$ Using the product rule

13. $\log_a 2+\log_a 10=\log_a(2\cdot 10)$ Using the product rule
The result can also be expressed as $\log_a 20$.

15. $\log_c t+\log_c y=\log_c(t\cdot y)$ Using the product rule

17. $\log_a r^8=8\log_a r$ Using the power rule

19. $\log_2 y^{1/3}=\frac{1}{3}\log_2 y$ Using the power rule

21. $\log_b C^{-3}=-3\log_b C$ Using the power rule

23. $\log_2 \frac{5}{11}=\log_2 5-\log_2 11$ Using the quotient rule

25. $\log_b \frac{m}{n}=\log_b m-\log_b n$ Using the quotient rule

27. $\log_a 19-\log_a 2=\log_a \frac{19}{2}$ Using the quotient rule

29. $\log_b 36-\log_b 4=\log_b \frac{36}{4}$, Using the quotient rule
or $\log_b 9$

31. $\log_a x-\log_a y=\log_a \frac{x}{y}$ Using the quotient rule

33. $\log_a(xyz)$
$=\log_a x+\log_a y+\log_a z$ Using the product rule

35. $\log_a\left(x^3z^4\right)$
$=\log_a x^3+\log_a z^4$ Using the product rule
$=3\log_a x+4\log_a z$ Using the power rule

37. $\log_a\left(w^2x^{-2}y\right)$
$=\log_a w^2+\log_a x^{-2}+\log_a y$ Using the product rule
$=2\log_a w-2\log_a x+\log_a y$ Using the power rule

39. $\log_a \frac{x^5}{y^3z}$
$=\log_a x^5-\log_a y^3z$ Using the quotient rule
$=\log_a x^5-\left(\log_a y^3+\log_a z\right)$ Using the product rule
$=\log_a x^5-\log_a y^3-\log_a z$ Removing the parentheses
$=5\log_a x-3\log_a y-\log_a z$ Using the power rule

41. $\log_b \frac{xy^2}{wz^3}$

$= \log_b xy^2 - \log_b wz^3$ Using the quotient rule

$= \log_b x + \log_b y^2 - (\log_b w + \log_b z^3)$

Using the product rule

$= \log_b x + \log_b y^2 - \log_b w - \log_b z^3$

Removing parentheses

$= \log_b x + 2\log_b y - \log_b w - 3\log_b z$

Using the power rule

43. $\log_a \sqrt{\frac{x^7}{y^5 z^8}}$

$= \log_a \left(\frac{x^7}{y^5 z^8}\right)^{1/2}$

$= \frac{1}{2}\log_a \frac{x^7}{y^5 z^8}$ Using the power rule

$= \frac{1}{2}(\log_a x^7 - \log_a y^5 z^8)$ Using the quotient rule

$= \frac{1}{2}[\log_a x^7 - (\log_a y^5 + \log_a z^8)]$ Using the product rule

$= \frac{1}{2}(\log_a x^7 - \log_a y^5 - \log_a z^8)$ Removing parentheses

$= \frac{1}{2}(7\log_a x - 5\log_a y - 8\log_a z)$ Using the power rule

45. $\log_a \sqrt[3]{\frac{x^6 y^3}{a^2 z^7}}$

$= \log_a \left(\frac{x^6 y^3}{a^2 z^7}\right)^{1/3}$

$= \frac{1}{3}\log_a \frac{x^6 y^3}{a^2 z^7}$ Using the power rule

$= \frac{1}{3}(\log_a x^6 y^3 - \log_a a^2 z^7)$ Using the quotient rule

$= \frac{1}{3}[\log_a x^6 + \log_a y^3 - (\log_a a^2 + \log_a z^7)]$ Using the product rule

$= \frac{1}{3}(\log_a x^6 + \log_a y^3 - \log_a a^2 - \log_a z^7)$

Removing parentheses

$= \frac{1}{3}(\log_a x^6 + \log_a y^3 - 2 - \log_a z^7)$

2 is the number to which we raise a to get a^2.

$= \frac{1}{3}(6\log_a x + 3\log_a y - 2 - 7\log_a z)$

Using the power rule

47. $8\log_a x + 3\log_a z$

$= \log_a x^8 + \log_a z^3$ Using the power rule

$= \log_a (x^8 z^3)$ Using the product rule

49. $\log_a x^2 - 2\log_a \sqrt{x}$

$= \log_a x^2 - \log_a (\sqrt{x})^2$ Using the power rule

$= \log_a x^2 - \log_a (\sqrt{x})^2 = x$

$= \log_a \frac{x^2}{x}$ Using the quotient rule

$= \log_a x$ Simplifying

51. $\frac{1}{2}\log_a x + 5\log_a y - 2\log_a x$

$= \log_a x^{1/2} + \log_a y^5 - \log_a x^2$ Using the power rule

$= \log_a x^{1/2} y^5 - \log_a x^2$ Using the product rule

$= \log_a \frac{x^{1/2} y^5}{x^2}$ Using the quotient rule

The result can also be expressed as

$\log_a \frac{\sqrt{x} y^5}{x^2}$ or as $\log_a \frac{y^5}{x^{3/2}}$.

53. $\log_a(x^2 - 9) - \log_a(x + 3)$

$= \log_a \frac{x^2 - 9}{x + 3}$ Using the quotient rule

$= \log_a \frac{(x+3)(x-3)}{x+3}$

$= \log_a \frac{\cancel{(x+3)}(x-3)}{\cancel{x+3}}$ Simplifying

$= \log_a (x - 3)$

55. $\log_b 15 = \log_b (3 \cdot 5)$

$= \log_b 3 + \log_b 5$ Using the product rule

$= 0.792 + 1.161$

$= 1.953$

57. $\log_b \frac{3}{5} = \log_b 3 - \log_b 5$ Using the quotient rule

$= 0.792 - 1.161$

$= -0.369$

59. $\log_b \frac{1}{5} = \log_b 1 - \log_b 5$ Using the quotient rule

$= 0 - 1.161$ $(\log_b 1 = 0)$

$= -1.161$

61. $\log_b \sqrt{b^3} = \log_b b^{3/2} = \frac{3}{2}$ 3/2 is the number to which we raise b to get $b^{3/2}$.

63. $\log_b 8$

Since 8 cannot be expressed using the numbers 1, 3, and 5, we cannot find $\log_b 8$ using the given information.

65. $\log_t t^{10} = 10$ 10 is the exponent to which we raise t to get t^{10}.

67. $\log_e e^m = m$ m is the exponent to which we raise e to get e^m.

69. *Writing Exercise.* The logarithm of a quotient is an expression like $\log_a \frac{x}{y}$ which can be simplified, using the quotient rule, to $\log_a x - \log_a y$. A quotient of logarithms is an expression like $\frac{\log_a x}{\log_a y}$, which cannot be simplified.

71. Graph $f(x)=\sqrt{x}-3$.

We construct a table of values, plot points, and connect them with a smooth curve. Note that we must choose nonnegative values of x in order for $\sqrt{x}$ to be a real number.

x	$f(x)$
0	−3
1	−2
4	−1
9	0

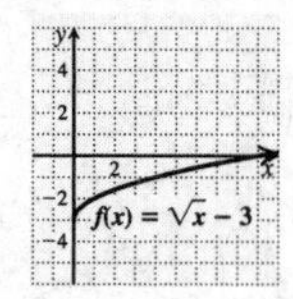

73. Graph: $g(x)=x^3+2$.

We construct a table of values, plot points, and connect them with a smooth curve.

x	$g(x)$
−2	−6
−1	1
0	2
1	3
2	10

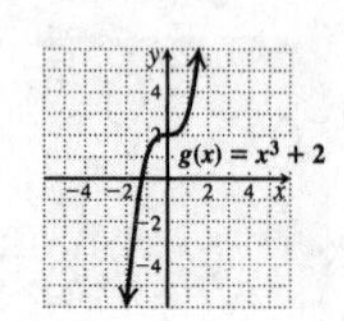

75. For $f(x)=\dfrac{x-3}{x+7}$,

$$x+7=0$$
$$x=-7$$

The domain is $\{x \mid x \text{ is a real number and } x\neq -7\}$ or $(-\infty,-7)\cup(-7,\infty)$.

77. For $g(x)=\sqrt{10-x}$

$$10-x\geq 0$$
$$-x\geq -10$$
$$x\leq 10$$

The domain is $\{x \mid x\leq 10\}$ or $(-\infty, 10]$.

79. *Writing Exercise.* The student didn't subtract the logarithm of the entire denominator after using the quotient rule. The correct procedure is as follows:

$$\log_b \frac{1}{x}=\log_b \frac{x}{xx}$$
$$=\log_b x-\log_b xx$$
$$=\log_b x-(\log_b x+\log_b x)$$
$$=\log_b x-\log_b x-\log_b x$$
$$=-\log_b x$$

(Note that $-\log_b x$ is equivalent to $\log_b 1-\log_b x$.)

81. $\log_a\left(x^8-y^8\right)-\log_a\left(x^2+y^2\right)$

$$=\log_a \frac{x^8-y^8}{x^2+y^2}$$
$$=\log_a \frac{\left(x^4+y^4\right)\left(x^2+y^2\right)(x+y)(x-y)}{x^2+y^2}$$
$$=\log_a\left[\left(x^4+y^4\right)\left(x^2-y^2\right)\right] \qquad \text{Simplifying}$$
$$=\log_a\left(x^6-x^4y^2+x^2y^4-y^6\right)$$

83. $\log_a \sqrt{1-s^2}$

$$=\log_a\left(1-s^2\right)^{1/2}$$
$$=\frac{1}{2}\log_a\left(1-s^2\right)$$
$$=\frac{1}{2}\log_a[(1-s)(1+s)]$$
$$=\frac{1}{2}\log_a(1-s)+\frac{1}{2}\log_a(1+s)$$

85. $\log_a \dfrac{\sqrt[3]{x^2z}}{\sqrt[3]{y^2z^{-2}}}$

$$=\log_a\left(\frac{x^2z^3}{y^2}\right)^{1/3}$$
$$=\frac{1}{3}\left(\log_a x^2z^3-\log_a y^2\right)$$
$$=\frac{1}{3}(2\log_a x+3\log_a z-2\log_a y)$$
$$=\frac{1}{3}[2\cdot 2+3\cdot 4-2\cdot 3]$$
$$=\frac{1}{3}(10)$$
$$=\frac{10}{3}$$

87. $\log_a x=2$, so $a^2=x$.

Let $\log_{1/a} x=n$ and solve for n.

$$\log_{1/a} a^2=n \qquad \text{Substituting } a^2 \text{ for } x$$
$$\left(\frac{1}{a}\right)^n=a^2$$
$$\left(a^{-1}\right)^n=a^2$$
$$a^{-n}=a^2$$
$$-n=2$$
$$n=-2$$

Thus, $\log_{1/a} x=-2$ when $\log_a x=2$.

89. $\log_2 80+\log_2 x=5$

$$\log_2 80x=5$$
$$2^5=80x$$
$$\frac{32}{80}=x$$
$$\frac{2}{5}=x$$

91. True; $\log_a\left(Q+Q^2\right)=\log_a[Q(1+Q)]$
$$=\log_a Q+\log_a(1+Q)=\log_a Q+\log_a(Q+1)$$

Exercise Set 12.5

1. True; see page 817 in the text.

3. True; see page 818 in the text.

5. True; $\log 18 - \log 2 = \log \frac{18}{2} = \log 9$

7. True; $\ln 81 = \ln 9^2 = 2\ln 9$

9. True; see Example 7.

11. 0.8451

13. 1.1367

15. Since $10^3 = 1000$, $\log 1000 = 3$.

17. -0.1249

19. 13.0014

21. 50.1187

23. 0.0011

25. 2.1972

27. -5.0832

29. 96.7583

31. 15.0293

33. 0.0305

35. We will use common logarithms for the conversion. Let $a = 10$, $b = 3$, and $M = 28$ and substitute in the change-of-base formula.

$$\log_b M = \frac{\log_a M}{\log_a b}$$
$$\log_3 28 = \frac{\log_{10} 28}{\log_{10} 3} \approx \frac{1.447158031}{0.477121254} \approx 3.0331$$

37. We will use common logarithms for the conversion. Let $a = 10$, $b = 2$, and $M = 100$ and substitute in the change-of-base formula.

$$\log_2 100 = \frac{\log_{10} 100}{\log_{10} 2} \approx \frac{2}{0.3010} \approx 6.6439$$

39. We will use natural logarithms for the conversion. Let $a = e$, $b = 4$, and $M = 5$ and substitute in the change-of-base formula.

$$\log_4 5 = \frac{\ln 5}{\ln 4} \approx \frac{1.6094}{1.3863} \approx 1.1610$$

41. We will use natural logarithms for the conversion. Let $a = e$, $b = 0.1$, and $M = 2$ and substitute in the change-of-base formula.

$$\log_{0.1} 2 = \frac{\ln 2}{\ln 0.1} \approx \frac{0.6931}{-2.3026} \approx -0.3010$$

43. We will use common logarithms for the conversion. Let $a = 10$, $b = 2$, and $M = 0.1$ and substitute in the change-of-base formula.

$$\log_2 0.1 = \frac{\log_{10} 0.1}{\log_{10} 2} \approx \frac{-1}{0.3010} \approx -3.3220$$

45. We will use natural logarithms for the conversion. Let $a = e$, $b = \pi$, and $M = 10$ and substitute in the change-of-base formula.

$$\log_\pi 10 = \frac{\ln 10}{\ln \pi} \approx \frac{2.3026}{1.1447} \approx 2.0115$$

47. Graph: $f(x) = e^x$

We find some function values with a calculator. We use these values to plot points and draw the graph.

x	e^x
1	2.7
2	7.4
3	20.1
−1	0.4
−2	0.1

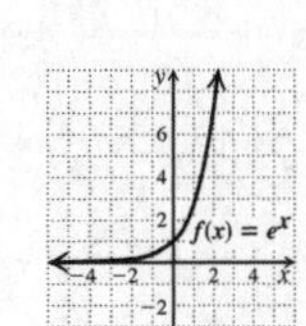

The domain is the set of real numbers and the range is $(0, \infty)$.

49. Graph: $f(x) = e^x + 3$

We find some function values, plot points, and draw the graph.

x	$e^x + 3$
0	4
1	5.72
2	10.39
−1	3.37
−2	3.14

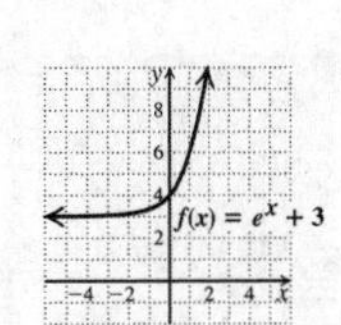

The domain is the set of real numbers and the range is $(3, \infty)$.

51. Graph: $f(x) = e^x - 2$

We find some function values, plot points, and draw the graph.

x	$e^x - 2$
0	-1
1	0.72
2	5.4
-1	-1.6
-2	-1.9

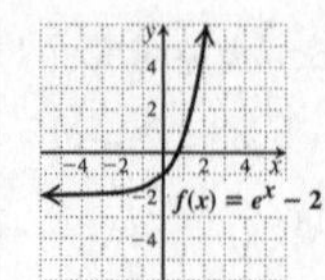

The domain is the set of real numbers and the range is $(-2, \infty)$.

53. Graph: $f(x) = 0.5e^x$

We find some function values, plot points, and draw the graph.

x	$0.5e^x$
0	0.5
1	1.36
2	3.69
-1	0.18
-2	0.07

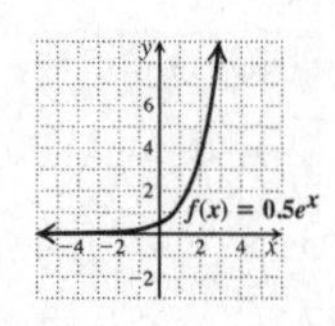

The domain is the set of real numbers and the range is $(0, \infty)$.

55. Graph: $f(x) = 0.5e^{2x}$

We find some function values, plot points, and draw the graph.

x	$0.5e^{2x}$
0	0.5
1	3.69
2	27.30
-1	0.07
-2	0.01

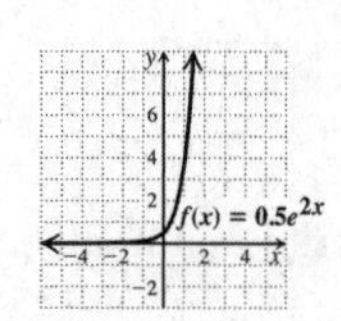

The domain is the set of real numbers and the range is $(0, \infty)$.

57. Graph: $f(x) = e^{x-3}$

We find some function values, plot points, and draw the graph.

x	e^{x-3}
2	0.37
3	1
4	2.72
-2	0.01

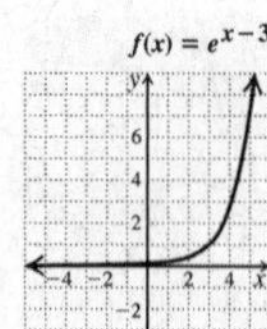

The domain is the set of real numbers and the range is $(0, \infty)$.

59. Graph: $f(x) = e^{x+2}$

We find some function values, plot points, and draw the graph.

x	e^{x+2}
-1	2.72
-2	1
-3	0.37
-4	0.14

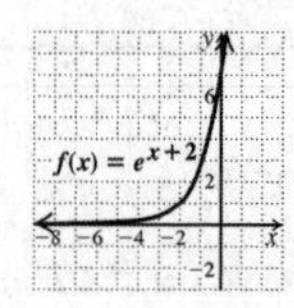

The domain is the set of real numbers and the range is $(0, \infty)$.

61. Graph: $f(x) = -e^x$

We find some function values, plot points, and draw the graph.

x	$-e^x$
0	-1
1	-2.72
2	-7.39
-1	-0.37
-3	-0.05

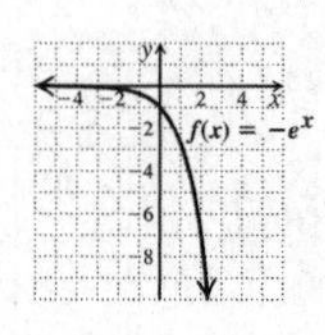

The domain is the set of real numbers and the range is $(-\infty, 0)$.

63. Graph: $g(x) = \ln x + 1$

We find some function values, plot points, and draw the graph.

x	$\ln x + 1$
1	1
3	2.10
5	2.61
7	2.95

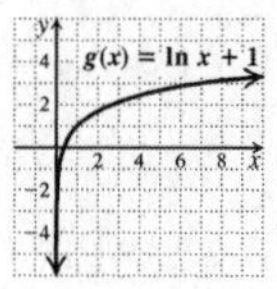

The domain is $(0, \infty)$ and the range is the set of real numbers.

65. Graph: $g(x) = \ln x - 2$

x	$\ln x + 1$
1	1
3	-0.90
5	-0.39
7	-0.05

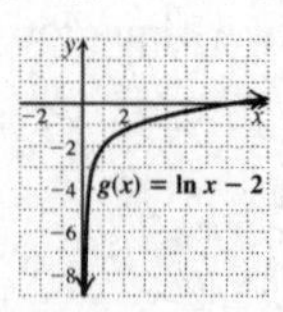

The domain is $(0, \infty)$ and the range is the set of real numbers.

67. Graph: $f(x) = 2 \ln x$

x	$2 \ln x$
0.5	-1.4
1	0
2	1.4
3	2.2
4	2.8
5	3.2
6	3.6

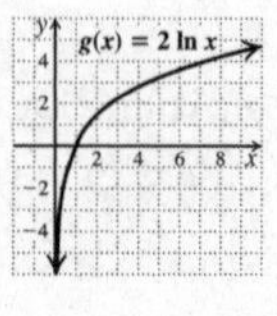

The domain is $(0, \infty)$ and the range is the set of real numbers.

69. Graph: $g(x) = -2 \ln x$

x	$-2\ln x$
0.5	1.4
1	0
2	−1.4
3	−2.2
4	−2.8
5	−3.2
6	−3.6

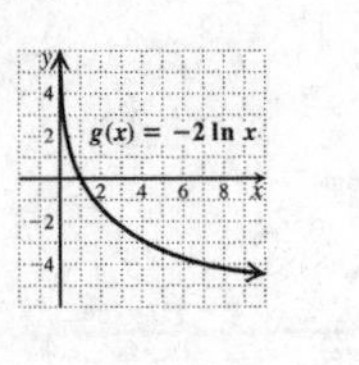

The domain is $(0, \infty)$ and the range is the set of real numbers.

71. Graph: $f(x) = \ln(x+2)$

We find some function values, plot points, and draw the graph.

x	$\ln(x+2)$
1	1.10
3	1.61
5	1.95
−1	0
−2	Undefined

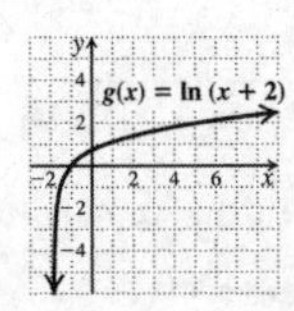

The domain is $(-2, \infty)$ and the range is the set of real numbers.

73. Graph: $g(x) = \ln(x-1)$

We find some function values, plot points, and draw the graph.

x	$\ln(x-1)$
2	0
3	0.69
4	1.10
6	1.61

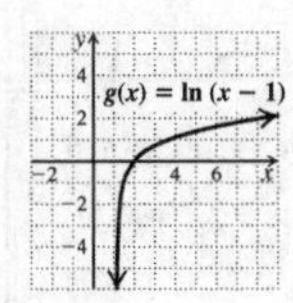

The domain is $(1, \infty)$ and the range is the set of real numbers.

75. *Writing Exercise.* $\log 10 < \log 79 < \log 100$, or $1 < \log 79 < 2$

77.
$$x^2 - 3x - 28 = 0$$
$$(x+4)(x-7) = 0$$
$$x+4=0 \quad or \quad x-7=0$$
$$x=-4 \quad or \quad x=7$$

The solutions are –4 and 7.

79.
$$17x - 15 = 0$$
$$17x = 15$$
$$x = \frac{15}{17}$$

The solution is $\frac{15}{17}$.

81.
$$(x-5)\cdot 9 = 11$$
$$x-5 = \frac{11}{9}$$
$$x = \frac{56}{9}$$

The solution is $\frac{56}{9}$.

83. $x^{1/2} - 6x^{1/4} + 8 = 0$

Let $u = x^{1/4}$.

$$u^2 - 6u + 8 = 0 \quad \text{Substituting}$$
$$(u-4)(u-2) = 0$$
$$u = 4 \quad or \quad u = 2$$
$$x^{1/4} = 4 \quad or \quad x^{1/4} = 2$$
$$x = 256 \quad or \quad x = 16 \quad \text{Raising both sides to the fourth power}$$

Both numbers check. The solutions are 256 and 16.

85. *Writing Exercise.* Reflect the graph of $f(x) = e^x$ across the line $y = x$ and then translate it up one unit.

87. We use the change-of-base formula.

$$\log_6 81 = \frac{\log 81}{\log 6}$$
$$= \frac{\log 3^4}{\log(2\cdot 3)}$$
$$= \frac{4\log 3}{\log 2 + \log 3}$$
$$\approx \frac{4(0.477)}{0.301 + 0.477}$$
$$\approx 2.452$$

89. We use the change-of-base formula.

$$\log_{12} 36 = \frac{\log 36}{\log 12}$$
$$= \frac{\log(2\cdot 3)^2}{\log(2^2\cdot 3)}$$
$$= \frac{2\log(2\cdot 3)}{\log 2^2 + \log 3}$$
$$= \frac{2(\log 2 + \log 3)}{2\log 2 + \log 3}$$
$$\approx \frac{2(0.301 + 0.477)}{2(0.301) + 0.477}$$
$$\approx 1.442$$

91. Use the change-of-base formula with $a = e$ and $b = 10$. We obtain

$$\log M = \frac{\ln M}{\ln 10}.$$

93.
$$\log(492x) = 5.728$$
$$10^{5.728} = 492x$$
$$\frac{10^{5.728}}{492} = x$$
$$1086.5129 \approx x$$

95. $\log 692 + \log x = \log 3450$
$$\log x = \log 3450 - \log 692$$
$$\log x = \log \frac{3450}{692}$$
$$x = \frac{3450}{692}$$
$$x \approx 4.9855$$

97. a) Domain: $\{x \mid x > 0\}$, or $(0, \infty)$; range: $\{y \mid y < 0.5135\}$, or $(-\infty, 0.5135)$;

b) $[-1, 5, -10, 5]$;

c)

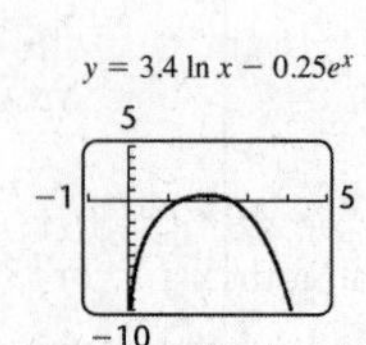

99. a) Domain $\{x \mid x > 0\}$, or $(0, \infty)$;

range: $\{y \mid y > -0.2453\}$, or $(-0.2453, \infty)$

b) $[-1, 5, -1, 10]$;

c)

y = 2x³ ln x
10
−1
5
−1

101. *Graphing Calculator Exercise*

Connecting the Concepts

1. $\log_4 16 = \log_4 4^2$
$$= 2\log_4 4 \quad \text{Using the power rule}$$
$$= 2 \quad \log_4 4 = 1$$

3. $\log_{100} 10 = \log_{100} 100^{1/2}$
$$= \frac{1}{2}\log_{100} 100$$
$$= \frac{1}{2}$$

5. $\log 10 = \log_{10} 10 = 1$

7. $\log 10^4 = 4\log_{10} 10 = 4$

9. $e^{\ln 7} = 7$

11. $\log_x 3 = m$
$$x^m = 3$$

13. $e^t = x$
$$\ln x = t$$

15. $\log_x 64 = 3$
$$x^3 = 64$$
$$x^3 = 4^3$$
$$x = 4$$

17. $\log \sqrt{\frac{x^2}{yz^3}} = \log\left(\frac{x^2}{yz^3}\right)^{1/2} = \frac{1}{2}\log\frac{x^2}{yz^3}$
$$= \frac{1}{2}\left(\log x^2 - \log y - \log z^3\right)$$
$$= \frac{1}{2}(2\log x - \log y - 3\log z)$$
$$= \log x - \frac{1}{2}\log y - \frac{3}{2}\log z$$

19. $\log_4 8 = \frac{\log 8}{\log 4} \approx \frac{0.9031}{0.6021} \approx 1.5$

Exercise Set 12.6

1. If we take the common logarithm on both sides, we see that choice (e) is correct.

3. $\ln x = 3$ means that 3 is the exponent to which we raise e to get x, so choice (f) is correct.

5. By the product rule for logarithms, $\log_5 x + \log_5 (x-2)$ $= \log_5 [x(x-2)] = \log_5 (x^2 - 2x)$, so choice (b) is correct.

7. By the quotient rule for logarithms,

$\ln x - \ln(x-2) = \ln\frac{x}{x-2}$, so choice (g) is correct.

9. $3^{2x} = 81$
$$3^{2x} = 3^4$$
$$2x = 4 \quad \text{Equating the exponents}$$
$$x = 2$$

11. $4^x = 32$
$$2^{2x} = 2^5$$
$$2x = 5 \quad \text{Equating the exponents}$$
$$x = \frac{5}{2}$$

13. $2^x = 10$
$$\log 2^x = \log 10 \quad \text{Taking log on both sides}$$
$$\log 2^x = 1 \quad \log 10 = 1$$
$$x\log 2 = 1$$
$$x = \frac{1}{\log 2}$$
$$x \approx 3.322 \quad \text{Using a calculator}$$

15. $2^{x+5} = 16$
$$2^{x+5} = 2^4$$
$$x + 5 = 4 \quad \text{Equating the exponents}$$
$$x = -1$$

17.
$$\begin{aligned} 8^{x-3} &= 19 \\ \log 8^{x-3} &= \log 19 \quad \text{Taking log on both sides} \\ (x-3)\log 8 &= \log 19 \\ x-3 &= \frac{\log 19}{\log 8} \\ x &= \frac{\log 19}{\log 8}+3 \\ x &\approx 4.416 \quad \text{Using a calculator} \end{aligned}$$

19.
$$\begin{aligned} e^t &= 50 \\ \ln e^t &= \ln 50 \quad \text{Taking ln on both sides} \\ t &= \ln 50 \quad \ln e^t = t\ln e = t\cdot 1 = t \\ t &\approx 3.912 \end{aligned}$$

21.
$$\begin{aligned} e^{-0.02t} &= 8 \\ \ln e^{-0.02t} &= \ln 8 \quad \text{Taking ln on both sides} \\ -0.02t &= \ln 8 \\ t &= \frac{\ln 8}{-0.02} \\ t &\approx -103.972 \end{aligned}$$

23.
$$\begin{aligned} 4.9^x - 87 &= 0 \\ 4.9^x &= 87 \\ \log 4.9^x &= \log 87 \\ x\log 4.9 &= \log 87 \\ x &= \frac{\log 87}{\log 4.9} \\ x &\approx 2.810 \end{aligned}$$

25.
$$\begin{aligned} 19 &= 2e^{4x} \\ \frac{19}{2} &= e^{4x} \\ \ln\left(\frac{19}{2}\right) &= \ln e^{4x} \\ \ln\left(\frac{19}{2}\right) &= 4x \\ \frac{\ln\left(\frac{19}{2}\right)}{4} &= x \\ 0.563 &\approx x \end{aligned}$$

27.
$$\begin{aligned} 7+3e^{-x} &= 13 \\ 3e^{-x} &= 6 \\ e^{-x} &= 2 \\ \ln e^{-x} &= \ln 2 \\ -x &= \ln 2 \\ x &= -\ln 2 \\ x &\approx -0.693 \end{aligned}$$

29.
$$\begin{aligned} \log_3 x &= 4 \\ x &= 3^4 \quad \text{Writing an equivalent exponential equation} \\ x &= 81 \end{aligned}$$

31.
$$\begin{aligned} \log_4 x &= -2 \\ x &= 4^{-2} \quad \text{Writing an equivalent exponential equation} \\ x &= \frac{1}{4^2}, \text{ or } \frac{1}{16} \end{aligned}$$

33.
$$\begin{aligned} \ln x &= 5 \\ x &= e^5 \quad \text{Writing an equivalent exponential equation} \\ x &\approx 148.413 \end{aligned}$$

35.
$$\begin{aligned} \ln 4x &= 3 \\ 4x &= e^3 \\ x &= \frac{e^3}{4} \approx 5.021 \end{aligned}$$

37.
$$\begin{aligned} \log x &= 1.2 \quad \text{The base is 10.} \\ x &= 10^{1.2} \\ x &\approx 15.849 \end{aligned}$$

39.
$$\begin{aligned} \ln(2x+1) &= 4 \\ 2x+1 &= e^4 \\ 2x &= e^4 - 1 \\ x &= \frac{e^4-1}{2} \approx 26.799 \end{aligned}$$

41.
$$\begin{aligned} \ln x &= 1 \\ x &= e \approx 2.718 \end{aligned}$$

43.
$$\begin{aligned} 5\ln x &= -15 \\ \ln x &= -3 \\ x &= e^{-3} \approx 0.050 \end{aligned}$$

45.
$$\begin{aligned} \log_2(8-6x) &= 5 \\ 8-6x &= 2^5 \\ 8-6x &= 32 \\ -6x &= 24 \\ x &= -4 \end{aligned}$$

The answer checks. The solution is –4.

47.
$$\begin{aligned} \log(x-9)+\log x &= 1 \quad \text{The base is 10.} \\ \log_{10}[(x-9)(x)] &= 1 \quad \text{Using the product rule} \\ x(x-9) &= 10^1 \\ x^2 - 9x &= 10 \\ x^2 - 9x - 10 &= 0 \\ (x+1)(x-10) &= 0 \end{aligned}$$

$x = -1$ *or* $x = 10$

Check: For -1:

$$\begin{array}{c|c} \log(x-9)+\log x & 1 \\ \hline \log(-1-9)+\log(-1) \overset{?}{=} 1 & \text{FALSE} \end{array}$$

For 10:

$$\begin{array}{r|l} \log(x-9)+\log x & 1 \\ \hline \log(10-9)+\log(10) & 1 \\ \log 1 + \log 10 & \\ 0+1 & \end{array}$$

$1 \overset{?}{=} 1$ TRUE

The number –1 does not check, because negative numbers do not have logarithms. The solution is 10.

49. $\log x - \log(x+3) = 1$ The base is 10.

$$\log_{10}\frac{x}{x+3} = 1 \quad \text{Using the quotient rule}$$
$$\frac{x}{x+3} = 10^1$$
$$x = 10(x+3)$$
$$x = 10x + 30$$
$$-9x = 30$$
$$x = -\frac{10}{3}$$

The number $-\frac{10}{3}$ does not check. The equation has no solution.

51. We observe that since $\log(2x+1) = \log 5$, then

$$2x+1 = 5$$
$$2x = 4$$
$$x = 2$$

53.
$$\log_4(x+3) = 2 + \log_4(x-5)$$
$$\log_4(x+3) - \log_4(x-5) = 2$$
$$\log_4\frac{x+3}{x-5} = 2 \quad \text{Using the quotient rule}$$
$$\frac{x+3}{x-5} = 4^2$$
$$\frac{x+3}{x-5} = 16$$
$$x+3 = 16(x-5)$$
$$x+3 = 16x - 80$$
$$83 = 15x$$
$$\frac{83}{15} = x$$

The number $\frac{83}{15}$ checks. It is the solution.

55. $\log_7(x+1) + \log_7(x+2) = \log_7 6$

$$\log_7[(x+1)(x+2)] = \log_7 6 \quad \text{Using the product rule}$$
$$\log_7(x^2+3x+2) = \log_7 6$$
$$x^2+3x+2 = 6 \quad \text{Using the property of logarithmic equality}$$
$$x^2+3x-4 = 0$$
$$(x+4)(x-1) = 0$$

$x = -4$ *or* $x = 1$

The number 1 checks, but –4 does not.

The solution is 1.

57. $\log_5(x+4) + \log_5(x-4) = \log_5 20$

$$\log_5[(x+4)(x-4)] = \log_5 20 \quad \text{Using the product rule}$$
$$\log_5(x^2-16) = \log_5 20$$
$$x^2-16 = 20 \quad \text{Using the property of logarithmic equality}$$
$$x^2 = 36$$
$$x = \pm 6$$

The number 6 checks, but –6 does not.

The solution is 6.

59. $\ln(x+5) + \ln(x+1) = \ln 12$

$$\ln[(x+5)(x+1)] = \ln 12$$
$$\ln(x^2+6x+5) = \ln 12$$
$$x^2+6x+5 = 12$$
$$x^2+6x-7 = 0$$
$$(x+7)(x-1) = 0$$

$x = -7$ *or* $x = 1$

The number –7 does not check, but 1 does.

The solution is 1.

61. $\log_2(x-3) + \log_2(x+3) = 4$

$$\log_2[(x-3)(x+3)] = 4$$
$$(x-3)(x+3) = 2^4$$
$$x^2 - 9 = 16$$
$$x^2 = 25$$
$$x = \pm 5$$

The number 5 checks, but –5 does not.

The solution is 5.

63. $\log_{12}(x+5) - \log_{12}(x-4) = \log_{12} 3$

$$\log_{12}\frac{x+5}{x-4} = \log_{12} 3$$
$$\frac{x+5}{x-4} = 3 \quad \text{Using the property of logarithmic equality}$$
$$x+5 = 3(x-4)$$
$$x+5 = 3x - 12$$
$$17 = 2x$$
$$\frac{17}{2} = x$$

The number $\frac{17}{2}$ checks and is the solution.

65. $\log_2(x-2) + \log_2 x = 3$

$$\log_2[(x-2)(x)] = 3$$
$$x(x-2) = 2^3$$
$$x^2 - 2x = 8$$
$$x^2 - 2x - 8 = 0$$
$$(x-4)(x+2) = 0$$

$x = 4$ *or* $x = -2$

The number 4 checks, but –2 does not.

The solution is 4.

67. *Writing Exercise.* Madison mistakenly thinks that, because negative numbers do not have logarithms, negative numbers cannot be solutions of logarithmic equations.

69. ***Familiarize.*** Let w represent the width of the rectangle, in ft. Then $w + 6$ represents the length. Recall that the formula for the perimeter P of a rectangle with length l and width w is $P = 2l + 2w$.

Translate.

$$\underbrace{\text{Perimeter}} \quad \text{is} \quad \underbrace{26 \text{ ft.}}$$
$$2\cdot(w+6)+2\cdot w \quad = \quad 26$$

Carry out. We solve the equation.

$$\begin{aligned} 2\cdot(w+6)+2\cdot w &= 26 \\ 2w+12+2w &= 26 \\ 4w+12 &= 26 \\ 4w &= 14 \\ w &= 3.5 \end{aligned}$$

When $w=3.5$, then $w+6=3.5+6=9.5$.

Check. If the length is 9.5 ft and the width is 3.5 ft, then the length is 6 ft longer than the width. Also $P=2\cdot 9.5+2\cdot 3.5=19+7=26$ ft. The answer checks.

State. The length of the rectangle is 9.5 ft, and the width is 3.5 ft.

71. ***Familiarize.*** Let $x=$ the number of pounds of Golden Days and $y=$ the number of pounds of Snowy Friends to be used in the mixture. The amount of sunflower seeds in the mixture is 33%(50 lb), or 16.5.

Translate. We organize the information in a table.

	Golden Days	Snowy Friends	Mixture
Number of pounds	x	y	50
Percent of sunflower seeds	25%	40%	33%
Amount of sunflower seeds	$0.25x$	$0.40y$	16.5 lb

We get one equation from the "Number of pounds" row of the table:

$$x+y=50$$

The last row of the table yields a second equation:

$$0.25x+0.4y=16.5$$

After clearing decimals, we have the problem translated to a system of equations:

$$\begin{aligned} x+y&=50, \quad (1) \\ 25x+40y&=1650 \quad (2) \end{aligned}$$

Carry out. We use the elimination method to solve the system of equations.

$$\begin{aligned} -25x-25y&=-1250 \quad \text{Multiplying (1) by } -25 \\ 25x+40y&=1650 \\ \hline 15y&=400 \\ y&=26\frac{2}{3} \end{aligned}$$

Substitute $26\frac{2}{3}$ for y in (1) and solve for x.

$$\begin{aligned} x+26\frac{2}{3}&=50 \\ x&=23\frac{1}{3} \end{aligned}$$

Check. The amount of the mixture is $26\frac{2}{3}$ lb $+23\frac{1}{3}$ lb or 50 lb. The amount of sunflower seeds in the mixture is $0.25\left(23\frac{1}{3}\text{ lb}\right)+0.4\left(26\frac{2}{3}\text{ lb}\right)=5\frac{5}{6}\text{ lb}+10\frac{2}{3}\text{ lb}=16.5\text{ lb}$. The answer checks.

State. $23\frac{1}{3}$ lb of Golden Days and $26\frac{2}{3}$ lb of Snowy Friends should be mixed.

73. ***Familiarize.*** Let $t=$ the time, in hours, it takes Max and Miles to key in the score, working together. Then in t hours Max does $\frac{t}{2}$ of the job, Miles does $\frac{t}{3}$, and together they do 1 entire job.

Translate.

$$\frac{t}{2}+\frac{t}{3}=1$$

Carry out. We solve the equation. First we multiply by the LCD, 6.

$$\begin{aligned} 6\left(\frac{t}{2}+\frac{t}{3}\right)&=6\cdot 1 \\ 6\cdot\frac{t}{2}+6\cdot\frac{t}{3}&=6 \\ 3t+2t&=6 \\ 5t&=6 \\ t&=\frac{6}{5} \end{aligned}$$

Check. In $\frac{6}{5}$ hr Max does $\frac{6/5}{2}$, or $\frac{3}{5}$ of the job, and Miles does $\frac{6/5}{2}$, or $\frac{2}{5}$ of the job. Together they do $\frac{3}{5}+\frac{2}{5}$ or 1 entire job. The answer checks.

State. It takes Max and Miles $\frac{6}{5}$ hr, or $1\frac{1}{5}$ hr, to do the job, working together.

75. *Writing Exercise.* No; let $m=2$, $n=-2$, and $f(x)=x^2$. Then $f(2)=f(-2)$, but $2\neq -2$.

77.
$$\begin{aligned} 8^x &= 16^{3x+9} \\ (2^3)^x &= (2^4)^{3x+9} \\ 2^{3x} &= 2^{12x+36} \\ 3x &= 12x+36 \\ -36 &= 9x \\ -4 &= x \end{aligned}$$

The solution is –4.

79.
$$\begin{aligned} \log_6(\log_2 x) &= 0 \\ \log_2 x &= 6^0 \\ \log_2 x &= 1 \\ x &= 2^1 \\ x &= 2 \end{aligned}$$

The solution is 2.

81. $\log_5 \sqrt{x^2-9}=1$

$$\sqrt{x^2-9}=5^1$$
$$\left(\sqrt{x^2-9}\right)^2=5^2$$
$$x^2-9=25$$
$$x^2=34$$
$$x=\pm\sqrt{34}$$

The solutions are $\pm\sqrt{34}$.

83. $2^{x^2+4x}=\frac{1}{8}$

$$2^{x^2+4x}=\frac{1}{2^3}$$
$$2^{x^2+4x}=2^{-3}$$
$$x^2+4x=-3$$
$$x^2+4x+3=0$$
$$(x+3)(x+1)=0$$
$$x=-3 \quad or \quad x=-1$$

The solutions are –3 and –1.

85. $\log_5 |x|=4$

$$|x|=5^4$$
$$|x|=625$$
$$x=625 \quad or \quad x=-625$$

The solutions are 625 and –625.

87. $\log\sqrt{2x}=\sqrt{\log 2x}$

$$\log(2x)^{1/2}=\sqrt{\log 2x}$$
$$\frac{1}{2}\log 2x=\sqrt{\log 2x}$$
$$\frac{1}{4}(\log 2x)^2=\log 2x \quad \text{Squaring both sides}$$
$$\frac{1}{4}(\log 2x)^2-\log 2x=0$$

Let $u=\log 2x$.

$$\frac{1}{4}u^2-u=0$$
$$u\left(\frac{1}{4}u-1\right)=0$$
$$u=0 \quad or \quad \frac{1}{4}u-1=0$$
$$u=0 \quad or \quad \frac{1}{4}u=1$$
$$u=0 \quad or \quad u=4$$
$$\log 2x=0 \quad or \quad \log 2x=4 \quad \text{Replacing } u \text{ with } \log 2$$
$$2x=10^0 \quad or \quad 2x=10^4$$
$$2x=1 \quad or \quad 2x=10{,}000$$
$$x=\frac{1}{2} \quad or \quad x=5000$$

Both numbers check. The solutions are $\frac{1}{2}$ and 5000.

89. $3^{x^2}\cdot 3^{4x}=\frac{1}{27}$

$$3^{x^2+4x}=3^{-3}$$
$$x^2+4x=-3 \quad \text{The exponents must be equal.}$$
$$x^2+4x+3=0$$
$$(x+1)(x+3)=0$$
$$x=-1 \quad or \quad x=-3$$

Both numbers check. The solutions are –1 and –3.

91. $\log x^{\log x}=25$

$$\log x(\log x)=25 \quad \text{Using the power rule}$$
$$(\log x)^2=25$$
$$\log x=\pm 5$$
$$x=10^5 \quad or \quad x=10^{-5}$$
$$x=100{,}000 \quad or \quad x=\frac{1}{100{,}000}$$

Both numbers check. The solutions are 100,000 and $\frac{1}{100{,}000}$.

93. $(81^{x-2})(27^{x+1})=9^{2x-3}$

$$\left[(3^4)^{x-2}\right]\left[(3^3)^{x+1}\right]=(3^2)^{2x-3}$$
$$(3^{4x-8})(3^{3x+3})=3^{4x-6}$$
$$3^{7x-5}=3^{4x-6}$$
$$7x-5=4x-6$$
$$3x=-1$$
$$x=-\frac{1}{3}$$

The solution is $-\frac{1}{3}$.

95. $2^y=16^{x-3} \quad and \quad 3^{y+2}=27^x$

$$2^y=(2^4)^{x-3} \quad and \quad 3^{y+2}=(3^3)^x$$
$$y=4x-12 \quad and \quad y+2=3x$$
$$12=4x-y \quad and \quad 2=3x-y$$

Solving this system of equations we get $x=10$ and $y=28$. Then $x+y=10+28=38$.

97. Find the first coordinate of the point of intersection of $y_1=\ln x$ and $y_2=\log x$. The value of x for which the natural logarithm of x is the same as the common logarithm of x is 1.

Exercise Set 12.7

1. a) Replace $A(t)$ with 4000 and solve for t.

$$\begin{aligned} A(t) &= 77(1.283)^t \\ 4000 &= 77(1.283)^t \\ \frac{4000}{77} &= 1.283^t \\ \ln\frac{4000}{77} &= \ln 1.283^t \\ \ln\frac{4000}{77} &= t\ln 1.283 \\ \frac{\ln\frac{4000}{77}}{\ln 1.283} &= t \\ 16 &\approx t \end{aligned}$$

The number of known asteroids first reached 4000 about 16 yr after 1990, or in 2006.

b) $A(0)=77(1.283)^0=77\cdot 1=77$, so to find the doubling time, we replace $A(t)$ with 2(77), or 154 and solve for t.

$$\begin{aligned} 154 &= 77(1.283)^t \\ 2 &= 1.283^t \\ \ln 2 &= \ln 1.283^t \\ \ln 2 &= t\ln 1.283 \\ \frac{\ln 2}{\ln 1.283} &= t \\ 2.8 &\approx t \end{aligned}$$

The doubling time is about 2.8 years.

3. a) Replace $S(t)$ with 100 and solve for t.

$$\begin{aligned} S(t) &= 180(0.97)^t \\ 100 &= 180(0.97)^t \\ \frac{100}{180} &= (0.97)^t \\ \ln\frac{100}{180} &= \ln(0.97)^t \\ \ln\frac{100}{180} &= t\ln 0.97 \\ \frac{\ln\frac{100}{180}}{\ln 0.97} &= t \\ 19 &\approx t \end{aligned}$$

The death rate reached 100 about 19 yr after 1960, or in 1979.

b) Replace $S(t)$ with 25 and solve for t.

$$\begin{aligned} S(t) &= 180(0.97)^t \\ 25 &= 180(0.97)^t \\ \frac{25}{180} &= (0.97)^t \\ \ln\frac{25}{180} &= \ln(0.97)^t \\ \ln\frac{25}{180} &= t\ln 0.97 \\ \frac{\ln\frac{25}{180}}{\ln 0.97} &= t \\ 65 &\approx t \end{aligned}$$

The death rate will reach 25 about 65 yr after 1960, or in 2025.

5. a) Replace $A(t)$ with 35,000 and solve for t.

$$\begin{aligned} A(t) &= 29{,}000(1.03)^t \\ 35{,}000 &= 29{,}000(1.03)^t \\ 1.207 &\approx (1.03)^t \\ \log 1.207 &\approx \log(1.03)^t \\ \log 1.207 &\approx t\log 1.03 \\ \frac{\log 1.207}{\log 1.03} &\approx t \\ 6.4 &\approx t \end{aligned}$$

The amount due will reach \$35,000 after about 6.4 years.

b) Replace $A(t)$ with 2(29,000), or 58,000, and solve for t.

$$\begin{aligned} 58{,}000 &= 29{,}000(1.03)^t \\ 2 &= (1.03)^t \\ \log 2 &= \log(1.03)^t \\ \log 2 &= t\log 1.03 \\ \frac{\log 2}{\log 1.03} &= t \\ 23.4 &\approx t \end{aligned}$$

The doubling time is about 23.4 years.

7. a) Substitute 50 for $W(t)$ and solve for t.

$$\begin{aligned} W(t) &= 89(0.837)^t \\ 50 &= 89(0.837)^t \\ \frac{50}{89} &= (0.837)^t \\ \log\frac{50}{89} &= \log(0.837)^t \\ \log\frac{50}{89} &= t\log 0.837 \\ \frac{\log\frac{50}{89}}{\log 0.837} &= t \\ 3 &\approx t \end{aligned}$$

The number dropped below 50% 3 yr after 1988, or 1991.

b) Substitute 1 for $W(t)$ and solve for t.

$$W(t)=89(0.837)^t$$
$$1=89(0.837)^t$$
$$\frac{1}{89}=(0.837)^t$$
$$\log\frac{1}{89}=\log(0.837)^t$$
$$-\log 89=t\log 0.837 \qquad \log\frac{1}{89}=-\log 89$$
$$\frac{-\log 89}{\log 0.837}=t$$
$$25\approx t$$

The number will drop below 1% 25 yr after 1988, or 2013.

9. a) $P(t)$ is given in thousands, so we substitute 30 for $P(t)$ and solve for t.

$$P(t)=5.5(1.047)^t$$
$$30=5.5(1.047)^t$$
$$5.455\approx 1.047^t$$
$$\log 5.455\approx\log 1.047^t$$
$$\log 5.455\approx t\log 1.047$$
$$\frac{\log 5.455}{\log 1.047}\approx t$$
$$36.9\approx t$$

The humpback whale population will reach 30,000 about 36.9 yr after 1982, or in 2018.

b) $P(0)=5.5(1.047)^0=5.5(1)=5.5$ and $2(5.5)=11$, so we substitute 11 for $P(t)$ and solve for t.

$$11=5.5(1.047)^t$$
$$2=1.047^t$$
$$\log 2=\log 1.047^t$$
$$\log 2=t\log 1.047$$
$$\frac{\log 2}{\log 1.047}=t$$
$$15.1\approx t$$

The doubling time is about 15.1 yr.

11. $\mathrm{pH}=-\log[H^+]$

$$=-\log(1.3\times 10^{-5})$$
$$\approx -(-4.886057) \quad \text{Using a calculator}$$
$$\approx 4.9$$

The pH of fresh-brewed coffee is about 4.9.

13.

$$\mathrm{pH}=-\log[H^+]$$
$$7.0=-\log[H^+]$$
$$-7.0=\log[H^+]$$
$$10^{-7.0}=[H^+] \quad \text{Converting to an exponential equation}$$

The hydrogen ion concentration is 10^{-7} moles per liter.

15. $L=10\cdot\log\frac{I}{I_0}$

$$=10\cdot\log\frac{10}{10^{-12}}$$
$$=10\cdot\log 10^{13}$$
$$=10\cdot 13\cdot\log 10$$
$$=130$$

The sound level is 130 dB.

17.

$$L=10\cdot\log\frac{I}{I_0}$$
$$128.8=10\cdot\log\frac{I}{10^{-12}}$$
$$12.88=\log\frac{I}{10^{-12}}$$
$$12.88=\log I-\log 10^{-12} \quad \text{Using the quotient rule}$$
$$12.88=\log I-(-12)$$
$$12.88=\log I+12$$
$$0.88=\log I$$
$$10^{0.88}=I \quad \text{Converting to an exponential equation}$$
$$7.6\approx I$$

The intensity of the sound is $10^{0.88}$ W/m^2, or about 7.6 W/m^2.

19.

$$M=\log\frac{v}{1.34}$$
$$7.5=\log\frac{v}{1.34}$$
$$10^{7.5}=\frac{v}{1.34}$$
$$1.34(10^{7.5})=v$$
$$42{,}400{,}000\approx v$$

Approximately 42.4 million messages per day are sent by that network.

21. a) Substitute 0.025 for k:

$$P(t)=P_0e^{0.025t}$$

b) To find the balance after one year, replace P_0 with 5000 and t with 1. We find $P(1)$:

$$P(1)=5000e^{0.025(1)}=5000e^{0.025}\approx \$5126.58$$

To find the balance after 2 years, replace P_0 with 5000 and t with 2. We find $P(2)$:

$$P(2)=5000e^{0.025(2)}=5000e^{0.05}\approx \$5256.36$$

c) To find the doubling time, replace P_0 with 5000 and $P(t)$ with 10,000 and solve for t.

$$10{,}000=5000e^{0.025t}$$
$$2=e^{0.025t}$$
$$\ln 2=\ln e^{0.025t} \quad \text{Taking the natural logarithm on both sides}$$
$$\ln 2=0.025t \quad \text{Finding the logarithm of the base to a power}$$
$$\frac{\ln 2}{0.025}=t$$
$$27.7\approx t$$

The investment will double in about 27.7 years.

23. a) $P(t)=304e^{0.009t}$, where $P(t)$ is in millions and t is the number of years after 2008.

b) In 2012, t = 2012 – 2008 = 4. Find $P(4)$.

$$P(4)=304e^{0.009(4)}=304e^{0.036}\approx 315$$

The U.S. population will be about 315 million in 2012.

c) Substitute 325 for $P(t)$ and solve for t.

$$325=304e^{0.009t}$$
$$\frac{325}{304}=e^{0.009t}$$
$$\ln\frac{325}{304}=\ln e^{0.009t}$$
$$\ln\frac{325}{304}=0.009t$$
$$\frac{\ln\frac{325}{304}}{0.009}=t$$
$$7\approx t$$

The U.S. population will reach 325 million about 7 yr after 2008, or in 2015.

25. The exponential growth function is $S(t)=S_0e^{3.40t}$. We replace $S(t)$ with $2S_0$ and solve for t.

$$2S_0=S_0e^{3.40t}$$
$$2=e^{3.40t}$$
$$\ln 2=\ln e^{3.40t}$$
$$\ln 2=3.40t$$
$$\frac{\ln 2}{3.40}=t$$
$$0.2\approx t$$

The doubling time for the zebra mussels is about 0.2 yr.

27. $Y(x)=71.41\ln\frac{x}{4.6}$

a) $Y(10)=71.41\ln\frac{10}{4.6}\approx 55$

The world population will reach 10 billion about 55 yr after 2000, or in 2055.

b) $Y(12)=71.41\ln\frac{12}{4.6}\approx 68$

The world population will reach 12 billion about 68 yr after 2000, or in 2068.

c) Plot the points found in parts (a) and (b) and others as necessary and draw the graph.

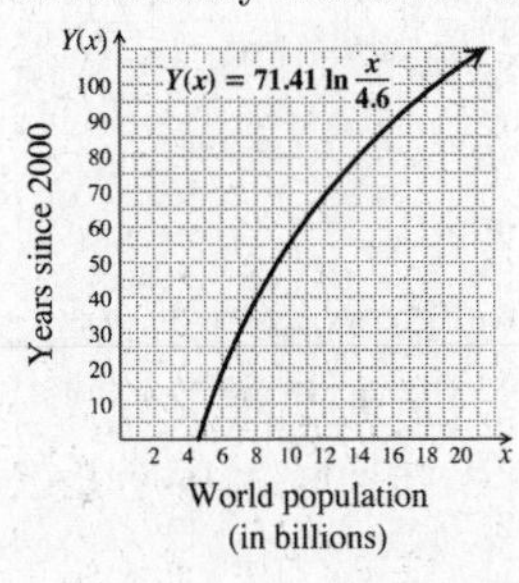

29. a) $S(0)=68-20\log(0+1)=68-20\log 1$
$=68-20(0)=68\%$

b) $S(4)=68-20\log(4+1)=68-20\log 5$
$\approx 68-20(0.69897)\approx 54\%$

$S(24)=68-20\log(24+1)$
$=68-20\log 25\approx 68-20(1.39794)\approx 40\%$

c) Using the values we computed in parts (a) and (b) and any others we wish to calculate, we sketch the graph:

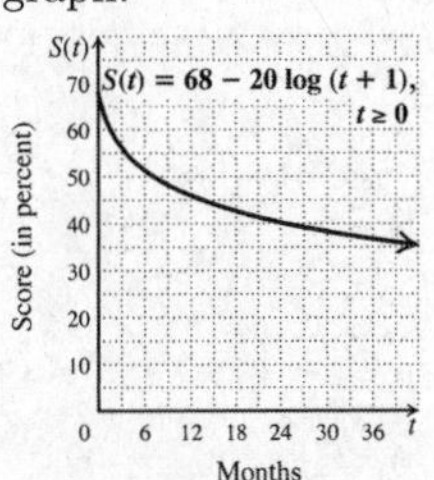

d)
$$50=68-20\log(t+1)$$
$$-18=-20\log(t+1)$$
$$0.9=\log(t+1)$$
$$10^{0.9}=t+1$$
$$7.9\approx t+1$$
$$6.9\approx t$$

After about 6.9 months, the average score was 50.

31. a) We start with the exponential growth equation

$$P(t)=P_0e^{kt}.$$

Substituting 2000 for P_0, we have $P(t)=2000e^{kt}$. To find the exponential growth rate k, observe that the wind-power capacity was 17,000 mW in 2007 or 17 years after 1990. We substitute and solve for k.

$$P(t)=2000e^{k\cdot 17}$$
$$17{,}000=2000e^{17k}$$
$$8.5=e^{17k}$$
$$\ln 8.5=\ln e^{17k}$$
$$\ln 8.5=17k$$
$$\frac{\ln 8.5}{17}=k$$
$$0.126\approx k$$

Thus, the exponential growth function is

$P(t)=2000e^{0.126t}$, where t is the number of years after 1990.

b) Substitute 50,000 for $P(t)$ and solve for t.

$$50{,}000=2000e^{0.126t}$$
$$25=e^{0.126t}$$
$$\ln 25=\ln e^{0.126t}$$
$$\ln 25=0.126t$$
$$\frac{\ln 25}{0.126}=t$$
$$25\approx t$$

The wind-power capacity will reach 50,000 mW about 25 yr after 1990, or in 2015.

33. a) We start with the exponential growth equation

$P(t)=P_0e^{kt}$, where t is the number of years after 1997.

Substituting 8200 for P_0, we have

$P(t)=8200e^{kt}$.

To find the exponential growth rate k, observe that the cost was \$500 in 2007 or 10 years after 1997. We substitute and solve for k.

$$P(t)=8200e^{k\cdot 10}$$
$$500=8200e^{10k}$$
$$\frac{5}{82}=e^{10k}$$
$$\ln\frac{5}{82}=\ln e^{10k}$$
$$\ln\frac{5}{82}=10k$$
$$\frac{\ln\frac{5}{82}}{10}=k$$
$$-0.280\approx k$$

Thus, the exponential growth function is

$P(t)=8200e^{-0.280t}$, where t is the number of years after 1997.

b) In 2010, $t=2010-1997=13$

$P(t)=8200e^{-0.280(15)}\approx \$215$

The cost in 2010 will be about \$215 per gigabit per second per mile.

c) Substitute 1 for $P(t)$ and solve for t.

$$1=8200e^{-0.280t}$$
$$\frac{1}{8200}=e^{-0.280t}$$
$$\ln\frac{1}{8200}=\ln e^{-0.280t}$$
$$\ln\frac{1}{8200}=-0.280t$$
$$\frac{\ln\frac{1}{8200}}{-0.280}=t$$
$$32\approx t$$

The cost will be \$1 per gigabit per second per mile about 32 yr after 1997, or in 2029.

35. We will use the function derived in Example 7:

$P(t)=P_0e^{-0.00012t}$

If the seed had lost 21% of its carbon-14 from the initial amount P_0, then $79\%(P_0)$ is the amount present. To find the age t of the seed, we substitute $79\%(P_0)$, or $0.79P_0$ for $P(t)$ in the function above and solve for t.

$$0.79P_0=P_0e^{-0.00012t}$$
$$0.79=e^{-0.00012t}$$
$$\ln 0.79=\ln e^{-0.00012t}$$
$$\ln 0.79=-0.00012t$$
$$\frac{\ln 0.79}{-0.00012}=t$$
$$1964\approx t$$

The seed is about 1964 yr old.

37. The function $P(t)=P_0\,e^{-kt}$, $k>0$, can be used to model decay. For iodine-131, $k=9.6\%$, or 0.096. To find the half-life we substitute 0.096 for k and $\frac{1}{2}$ P_0 for $P(t)$, and solve for t.

$$\frac{1}{2}P_0=P_0e^{-0.096t},\text{ or }\frac{1}{2}=e^{-0.096t}$$
$$\ln\frac{1}{2}=\ln e^{-0.096t}=-0.096t$$
$$t=\frac{\ln 0.5}{-0.096}\approx\frac{-0.6931}{-0.096}\approx 7.2\text{ days}$$

39. a) The function $P(t)=P_0e^{-kt}$, $k>0$, can be used to model decay. We substitute $\frac{1}{2}P_0$ for $P(t)$ and 5 for t and solve for the decay rate k.

$$\frac{1}{2}P_0=P_0e^{-k\cdot 5}$$
$$\frac{1}{2}=e^{-5k}$$
$$\ln\frac{1}{2}=\ln e^{-5k}$$
$$-\ln 2=-5k$$
$$\frac{\ln 2}{5}=k$$
$$0.139\approx k$$

The decay rate is 0.139, or 13.9% per hour.

b) 95% consumed = 5% remains

$$0.05P_0=P_0e^{-0.139t}$$
$$0.05=e^{-0.139t}$$
$$\ln 0.05=\ln e^{-0.139t}$$
$$\ln 0.05=-0.139t$$
$$\frac{\ln 0.05}{-0.139}=t$$
$$21.6\approx t$$

How will take approximately 21.6 hr for 95% of the caffeine to leave the body.

41. a) We start with the exponential growth equation

$V(t)=V_0e^{kt}$, where t is the number of years after 1991.

Substituting 451,000 for V_0, we have

$V(t)=451{,}000\,e^{kt}$.

To find the exponential growth rate k, observe that the card sold for \$2.8 million, or \$2,800,000 in 2007, or 16 years after 1991. We substitute and solve for k.

$$V(16) = 451{,}000e^{k\cdot 16}$$
$$2{,}800{,}000 = 451{,}000e^{16k}$$
$$\frac{2800}{451} = e^{16k}$$
$$\ln\frac{2800}{451} = \ln e^{16k}$$
$$\ln\frac{2800}{451} = 16k$$
$$\frac{\ln\frac{2800}{451}}{16} = k$$
$$0.114 \approx k$$

Thus, the exponential growth function is

$V(t) = 451{,}000e^{0.114t}$, where t is the number of years after 1991.

b) In 2012, $t = 2012 - 1991 = 21$

$V(21) = 451{,}000e^{0.114(21)} \approx 4{,}900{,}000$

The card's value in 2012 will be about \$4.9 million.

c) Substitute 2(\$451,000), or \$902,000 for $V(t)$ and solve for t.

$$902{,}000 = 451{,}000e^{0.114t}$$
$$2 = e^{0.114t}$$
$$\ln 2 = \ln e^{0.114t}$$
$$\ln 2 = 0.114t$$
$$\frac{\ln 2}{0.114} = t$$
$$6.1 \approx t$$

The doubling time is about 6.1 years.

d) Substitute \$4,000,000 for $V(t)$ and solve for t.

$$4{,}000{,}000 = 451{,}000e^{0.114t}$$
$$\frac{4000}{451} = e^{0.114t}$$
$$\ln\frac{4000}{451} = \ln e^{0.114t}$$
$$\ln\frac{4000}{451} = 0.114t$$
$$\frac{\ln\frac{4000}{451}}{0.114} = t$$
$$19 \approx t$$

The value of the card will first exceed \$4,000,000 about 19 years after 1991, or in 2010.

43. *Writing Exercise.* Answers will vary. The problem could be modeled after Exercises 19-22, 29, 30, 39, or 40.

45. Using the distance formula $d = \sqrt{(x_2 - x_1)^2 + (y_2 - y_1)^2}$ for the points (–3, 7) and (–2, 6),

$$d = \sqrt{[-2-(-3)]^2 + (6-7)^2}$$
$$= \sqrt{1^2 + (-1)^2} = \sqrt{1+1}$$
$$= \sqrt{2}$$

47. Using the midpoint formula $\left(\frac{x_1 + x_2}{2}, \frac{y_1 + y_2}{2}\right)$ for the points (3, –8) and (5, –6),

$\left(\frac{3+5}{2}, \frac{-8+(-6)}{2}\right)$, or $(4, -7)$

49.

$$x^2 + 8x = 1$$
$$x^2 + 8x + 16 = 1 + 16$$
$$(x+4)^2 = 17$$
$$x + 4 = \pm\sqrt{17}$$
$$x = -4 \pm \sqrt{17}$$

51. Graph $y = x^2 - 5x - 6$

First we find the vertex.

$$-\frac{b}{2a} = -\frac{-5}{2\cdot 1} = \frac{5}{2}$$

When $x = \frac{5}{2}$, $y = \left(\frac{5}{2}\right)^2 - 5\left(\frac{5}{2}\right) - 6 = -12\frac{1}{4}$

The vertex is $\left(\frac{5}{2}, -12\frac{1}{4}\right)$ and the axis of symmetry is $x = \frac{5}{2}$.

x	y
−1	0
0	−6
1	−10
4	−10
5	−6
6	0

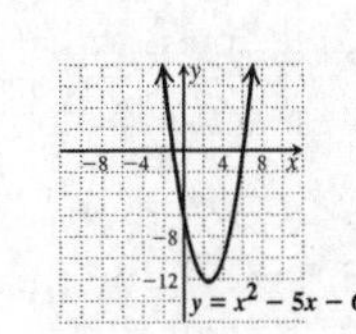

53. *Writing Exercise.* No; the model predicts that the number of text messages will be 1.4×10^{15} in 2030. This is not realistic.

55. We will use the exponential growth function $V(t) = V_0e^{kt}$, where t is the number of years after 2008 and $V(t)$ is in millions of dollars. Substitute 20 for $V(t)$, 0.04 for k, and 8 for t and solve for V_0.

$$V(t) = V_0e^{kt}$$
$$20 = V_0e^{0.04(8)}$$
$$\frac{20}{e^{0.32}} = V_0$$
$$14.5 \approx V_0$$

About \$14.5 million would need to be invested.

57. a) Substitute 1390 for I and solve for m.

$$m(I) = -(19 + 2.5\cdot\log I)$$
$$m = -(19 + 2.5\cdot\log 1390)$$
$$m \approx -26.9$$

The apparent stellar magnitude is about –26.9.

b) Substitute 23 for m and solve for I.

$$\begin{aligned} m(I) &= -(19+2.5\cdot\log I) \\ 23 &= -(19+2.5\cdot\log I) \\ -23 &= 19+2.5\cdot\log I \\ -42 &= 2.5\cdot\log I \\ -16.8 &= \log I \\ 10^{-16.8} &= I \\ 1.58\times10^{-17} &\approx I \end{aligned}$$

The intensity is about 1.58×10^{-17} W/m^2.

59. Consider an exponential growth function $P(t)=P_0e^{kt}$.
Suppose that at time T, $P(T)=2P_0$.
Solve for T:

$$\begin{aligned} 2P_0 &= P_0e^{kT} \\ 2 &= e^{kT} \\ \ln 2 &= \ln e^{kT} \\ \ln 2 &= kT \\ \frac{\ln 2}{k} &= T \end{aligned}$$

61. *Writing Exercise and Graphing Calculator Exercise.*

Answers may vary.

a) Using a graphing calculator to graph the number of applications data, it appears that an exponential function might be a better fit for this data.

b) Using a graphing calculator to graph the number of approvals data, it appears that a linear function might be a better fit for this data.

c) For the first set of data, we find

$f(x)=241(1.24)^x$ as the regression model.

For the second set of data, we find

$g(x)=41x+266$ as the regression model.

d) To find x when $\frac{1}{2}f(x)=g(x)$, use the graphing calculator to make an approximation. The solution is about 7 yr. In approximately 2001+7, or 2008, there will be only half as many approvals as applications.

Chapter 12 Review

1. True.

3. True.

5. False; log, which has base 10, is not the same as ln, which has base e. In addition, the power rule states $\log x^a = a\log x$.

7. False; the domain of $f(x)=3^x$ is all real numbers.

9. True; if $F(-2)=F(-5)$, then the function has the same output for two different inputs, and therefore is not one-to-one.

11.
$$\begin{aligned} (f\circ g)(x) &= f(g(x)) = f(2x-3) \\ &= (2x-3)^2+1 \\ &= 4x^2-12x+9+1 \\ &= 4x^2-12x+10 \end{aligned}$$
$$\begin{aligned} (g\circ f)(x) &= g(f(x)) = g(x^2+1) \\ &= 2(x^2+1)-3 \\ &= 2x^2+2-3 \\ &= 2x^2-1 \end{aligned}$$

13. $f(x)=4-x^2$

The graph of this function is a parabola that opens down. Thus, there are many horizontal lines that cross the graph more than once. In particular, the line $y=-4$ crosses the graph more than once. The function is not one-to-one.

15.

$y=\frac{3x+1}{2}$ Replace $g(x)$.

$x=\frac{3y+1}{2}$ Interchange variables.

$\frac{2x-1}{3}=y$ Solve for y.

$g^{-1}(x)=\frac{2x-1}{3}$ Replace y.

17. Graph $f(x)=3^x+1$

We compute some function values, thinking of y as $f(x)$, and keep the results in a table.

$$\begin{aligned} f(-2) &= 3^{-2}+1=\frac{1}{9}+1=\frac{10}{9} \\ f(-1) &= 3^{-1}+1=\frac{1}{3}+1=\frac{4}{3} \\ f(0) &= 3^0+1=1 \\ f(1) &= 3^1+1=4 \\ f(2) &= 3^2+1=10 \end{aligned}$$

x	y, or $f(x)$
-2	$\frac{10}{9}$
-1	$\frac{4}{3}$
0	1
1	4
2	10

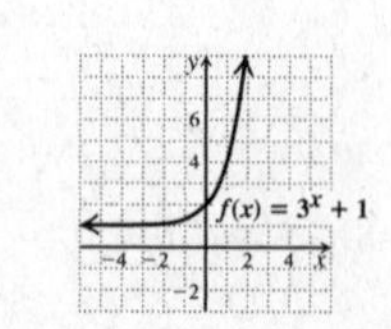

19. Graph: $f(x)=\log_5 x$

Think of $f(x)$ as y. Then $y=\log_5 x$ is equivalent to $5^y=x$. We find ordered pairs by choosing values for y and computing the corresponding x-values. Then we plot the points and connect them with a smooth curve.

For $y=0$, $x=5^0=1$
For $y=1$, $x=5^1=5$
For $y=2$, $x=5^2=25$
For $y=-1$, $x=5^{-1}=\frac{1}{5}$
For $y=-2$, $x=5^{-2}=\frac{1}{25}$

x, or 5^y	y
1	0
5	1
25	2
$\frac{1}{5}$	−1
$\frac{1}{25}$	−2

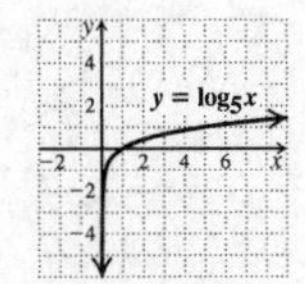

21. $\log_3 \frac{1}{9}=\log_3 3^{-2}=-2\log_3 3=-2$

23. $\log_{16} 4=\log_{16} 16^{1/2}=\frac{1}{2}\log_{16} 16=\frac{1}{2}$

25. $25^{1/2}=5$ is equivalent to $\frac{1}{2}=\log_{25} 5$.

27. $\log_8 1=0$ is equivlant to $1=8^0$.

29. $\log_a \frac{x^5}{yz^2}=\log_a x^5-\log_a y-\log_a z^2$
$=5\log_a x-\log_a y-2\log_a z$

31. $\log_a 5+\log_a 8=\log_a (5\cdot 8)=\log_a 40$

33. $\frac{1}{2}\log a-\log b-2\log c=\log a^{1/2}-\log b-\log c^2$
$=\log \frac{a^{1/2}}{bc^2}$

35. $\log_m m=1$

37. $\log_m m^{17}=17\log_m m=17\cdot 1=17$

39. $\log_a \frac{2}{7}=\log_a 2-\log_a 7$
$=1.8301-5.0999$
$=-3.2698$

41. $\log_a 3.5=\log_a \frac{7}{2}$
$=\log_a 7-\log_a 2$
$=5.0999-1.8301$
$=3.2698$

43. $\log_a \frac{1}{4}=\log_a 1-\log_a 4$
$=0-\log_a 2^2$
$=-2\log_a 2$
$=-2(1.8301)$
$=-3.6602$

45. 61.5177

47. 0.3753

49. We will use common logarithms for the conversion. Let $a=10$, $b=6$, and $M=5$ and substitute in the change-of-base formula.

$$\log_6 5=\frac{\log 5}{\log 6}\approx\frac{0.6890}{0.7782}\approx 0.8982$$

51. Graph $g(x)=0.6\ln x$.

We find some function values, plot points, and draw the graph.

x	$0.6\ln x$
0.5	−0.42
1	0
2	0.42
3	0.66

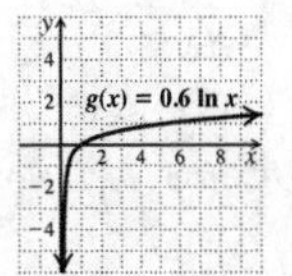

The domain is $(0,\infty)$ and the range is all real numbers.

53. $3^{2x}=\frac{1}{9}$
$3^{2x}=3^{-2}$
$2x=-2$
$x=-1$
The solution is −1.

55. $\log_x 16=4$
$x^4=16$
$x=\pm\sqrt[4]{16}=\pm 2$
Since x cannot be negative, we use only the positive solution. The solution is 2.

57. $6\ln x=18$
$\ln x=3$
$x=e^3\approx 20.0855$
The solution is e^3 or approximately 20.0855.

59. $2^x=12$
$x=\log_2 12=\frac{\log 12}{\log 2}\approx 3.5850$

61. $2\ln x=-6$
$\ln x=-3$
$x=e^{-3}\approx 0.0498$

63. $\log_4 x-\log_4 (x-15)=2$
$\log_4 \frac{x}{x-15}=2$
$\frac{x}{x-15}=4^2$
$\frac{x}{x-15}=16$
$x=16(x-15)$
$x=16x-240$
$240=15x$
$16=x$

65. $S(t)=82-18\log(t+1)$

a) $S(0)=82-18\log(0+1)=82-18\log 1=82$

b) $S(6)=82-18\log(6+1)=82-18\log 7\approx 66.8$

c) Substitute 54 for $S(t)$ and solve for t.

$$54=82-18\log(t+1)$$
$$-28=-18\log(t+1)$$
$$\frac{14}{9}=\log(t+1)$$
$$t+1=10^{14/9}$$
$$t=10^{14/9}-1\approx 35$$

The average score will be 54 after about 35 months.

67. a) We start with the exponential growth equation

$A(t)=A_0e^{kt}$, where t is the number of years after 2005.

Substituting \$885 for A_0, we have

$$A(t)=885e^{kt}.$$

Substitute 5 for t, \$1100 for $A(t)$ and solve for k.

$$A(5)=885e^{k(5)}$$
$$1100=885e^{5k}$$
$$\frac{1100}{885}=e^{5k}$$
$$\ln\frac{1100}{885}=\ln e^{5k}$$
$$\ln\frac{1100}{885}=5k$$
$$\frac{\ln\frac{1100}{885}}{5}=k$$
$$0.043\approx k$$

Thus, the exponential growth function is

$A(t)=885e^{0.043t}$, where t is the number of years after 2005.

b) In 2008, $t=2008-2005=3$

$$A(3)=885e^{0.043(3)}\approx 1000$$

The amount spent in 2008 will be about \$1.0 billion.

c) Substitute \$2000 for $A(t)$ and solve for t.

$$2000=885e^{0.043t}$$
$$\frac{2000}{885}=e^{0.043t}$$
$$\ln\frac{2000}{885}=\ln e^{0.043t}$$
$$\ln\frac{2000}{885}=0.043t$$
$$\frac{\ln\frac{2000}{885}}{0.043}=t$$
$$18\approx t$$

The \$2 billion will be spent on advertising about 18 years after 2005, or in 2023.

d) Substitute 2(\$885), or \$1170 for $A(t)$ and solve for t.

$$1770=885e^{0.043t}$$
$$2=e^{0.043t}$$
$$\ln 2=\ln e^{0.043t}$$
$$\ln 2=0.043t$$
$$\frac{\ln 2}{0.043}=t$$
$$16.1\approx t$$

The doubling time is about 16.1 years.

69. The doubling time of the initial investment P_0 would be $2P_0$ and $t=6$ years. Substitute this information into the exponential growth formula and solve for k.

$$2P_0=P_0e^{k(6)}$$
$$2=e^{6k}$$
$$\ln 2=\ln e^{6k}$$
$$\ln 2=6k$$
$$\frac{\ln 2}{6}=k$$
$$0.11553\approx k$$

The rate is 11.553% per year.

71. We will use the function $P(t)=P_0e^{-0.00012t}$

If the skull had lost 34% of its carbon-14 from the initial amount P_0, then $66\%(P_0)$ is the amount present. To find the age t of the skull, we substitute $66\%(P_0)$, or $0.66P_0$ for $P(t)$ in the function above and solve for t.

$$0.66P_0=P_0e^{-0.00012t}$$
$$0.66=e^{-0.00012t}$$
$$\ln 0.66=\ln e^{-0.00012t}$$
$$\ln 0.66=-0.00012t$$
$$\frac{\ln 0.66}{-0.00012}=t$$
$$3463\approx t$$

The skull is about 3463 yr old.

73. $L=10\cdot\log\frac{I}{I_0}$

$$=10\cdot\log\frac{2.5\times 10^{-1}}{10^{-12}}$$
$$=10\log(2.5\times 10^{11})$$
$$\approx 114$$

The sound is about 114 dB.

75. *Writing Exercise.* If $f(x)=e^x$, then to find the inverse function, we let $y=e^x$ and interchange x and y: $x=e^y$. If $x=e^y$, then $\log_e x=y$ by the definition of logarithms. Since $\log_e x=\ln x$, we have $y=\ln x$ or $f^{-1}(x)=\ln x$. Thus, $g(x)=\ln x$ is the inverse of $f(x)=e^x$. Another approach is to find $(f\circ g)(x)$ and $(g\circ f)(x)$:

$(f\circ g)(x)=e^{\ln x}=x$, and

$(g\circ f)(x)=\ln e^{x}=x$.

Thus, g and f are inverse functions.

77. $2^{x^2+4x}=\frac{1}{8}$ can be written as $2^{x^2+4x}=2^{-3}$.

So the exponents must be equal.

$$\begin{aligned} x^2+4x&=-3\\ x^2+4x+3&=0\\ (x+3)(x+1)&=0 \end{aligned}$$

$$x+3=0 \quad or \quad x+1=0$$
$$x=-3 \quad or \quad x=-1$$

The solutions are –3 and –1.

Chapter 12 Test

1. $$\begin{aligned}(f\circ g)(x)&=f(g(x))=f(2x+1)\\ &=(2x+1)+(2x+1)^2\\ &=2x+1+4x^2+4x+1\\ &=4x^2+6x+2\end{aligned}$$

$$\begin{aligned}(g\circ f)(x)&=g(f(x))=g(x+x^2)\\ &=2(x+x^2)+1\\ &=2x+2x^2+1\\ &=2x^2+2x+1\end{aligned}$$

3. $f(x)=x^2+3$

Observe that the graph of this function is a parabola that opens up. Thus, there are many horizontal lines that cross the graph more than once. In particular, the line $y=4$ crosses the graph more than once. The function is not one-to-one.

5.

$y=(x+1)^3$	Replace $g(x)$.
$x=(y+1)^3$	Interchange variables.
$\sqrt[3]{x}-1=y$	Solve for y.
$g^{-1}(x)=\sqrt[3]{x}-1$	Replace y.

7. Graph: $f(x)=\log_7 x$

Think of $f(x)$ as y. Then $y=\log_7 x$ is equivalent to $7^y=x$. We find ordered pairs by choosing values for y and computing the corresponding x-values. Then we plot the points and connect them with a smooth curve.

For $y=0$, $x=7^0=1$
For $y=1$, $x=7^1=7$
For $y=2$, $x=7^2=49$
For $y=-1$, $x=7^{-1}=\frac{1}{7}$
For $y=-2$, $x=7^{-2}=\frac{1}{49}$

x, or 7^y	y
1	0
7	1
49	2
$\frac{1}{7}$	-1
$\frac{1}{49}$	-2

9. $\log_{100}10=\log_{100}100^{1/2}=\frac{1}{2}\log_{100}100=\frac{1}{2}\cdot 1=\frac{1}{2}$

11. $\log_n n=1$

13. $\log_a a^{19}=19\log_a a=19\cdot 1=19$

15. $m=\log_2\frac{1}{2}$ is equivalent to $2^m=\frac{1}{2}$.

17. $$\begin{aligned}\frac{1}{3}\log_a x+2\log_a z&=\log_a x^{1/3}+\log_a z^2\\ &=\log_a\sqrt[3]{x}+\log_a z^2\\ &=\log_a\left(z^2\sqrt[3]{x}\right)\end{aligned}$$

19. $$\begin{aligned}\log_a 3&=\log_a\frac{6}{2}\\ &=\log_a 6-\log_a 2\\ &=0.778-0.301\\ &=0.477\end{aligned}$$

21. 1.3979

23. –0.9163

25. We will use common logarithms for the conversion. Let $a=10$, $b=3$, and $M=14$ and substitute in the change-of-base formula.

$$\log_3 14=\frac{\log 14}{\log 3}\approx\frac{1.1461}{0.4771}\approx 2.4022$$

27. Graph $g(x)=\ln(x-4)$.

We find some function values, plot points, and draw the graph.

x	$\ln(x-4)$
4.5	-0.69
5	0
6	0.69
7	1.10

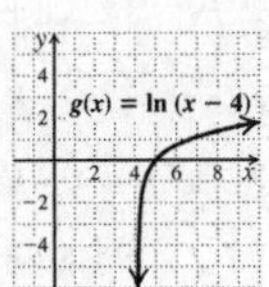

The domain is $(4,\ \infty)$ and the range is all real numbers.

29. $$\begin{aligned}\log_4 x&=\frac{1}{2}\\ x&=4^{1/2}\\ x&=2\end{aligned}$$

31. $5^{4-3x}=87$

$$4-3x=\log_5 87$$
$$-3x=\log_5 87-4$$
$$x=-\frac{1}{3}\left(\frac{\log 87}{\log 5}-4\right)\approx 0.4084$$

33. $\ln x=3$

$$x=e^3\approx 20.0855$$

35. $R=0.37\ln P+0.05$

a) Substitute 383 for P and solve for R.

$$R=0.37\ln 383+0.05=2.25$$

The average walking speed is about 2.25 ft/sec.

b) Substitute 3 for R and solve for P.

$$3=0.37\ln P+0.05$$
$$2.95=0.37\ln P$$
$$\frac{2.95}{0.37}=\ln P$$
$$P=e^{2.95/0.37}\approx 2{,}901$$

The population is approximately 2,901,000.

37. a) We start with the exponential growth equation

$C(t)=C_0e^{kt}$, where t is the number of years after 2001.

Substituting \$21,855 for C_0, we have

$$C(t)=21{,}855e^{kt}.$$

Substitute 5 for t, \$27,317 for $C(t)$ and solve for k.

$$27{,}317=21{,}855e^{k(5)}$$
$$\frac{27{,}317}{21{,}855}=e^{5k}$$
$$\ln\frac{27{,}317}{21{,}855}=\ln e^{5k}$$
$$\ln\frac{27{,}317}{21{,}855}=5k$$
$$\frac{\ln\frac{27{,}317}{21{,}855}}{5}=k$$
$$0.045\approx k$$

Thus, the exponential growth function is

$C(t)=21{,}855e^{0.045t}$, where t is the number of years after 2001.

b) In 2012, $t=2012-2001=11$

$$C(11)=21{,}855e^{0.045(11)}\approx 35{,}853$$

The cost in 2012 will be about \$35,853.

c) Substitute \$50,000 for $C(t)$ and solve for t.

$$50{,}000=21{,}855e^{0.045t}$$
$$\frac{50{,}000}{21{,}855}=e^{0.045t}$$
$$\ln\frac{50{,}000}{21{,}855}=\ln e^{0.045t}$$
$$\ln\frac{50{,}000}{21{,}855}=0.045t$$
$$\frac{\ln\frac{50{,}000}{21{,}855}}{0.045}=t$$
$$18\approx t$$

Cost for college will be \$50,000 about 18 years after 2001, or in 2019.

39. We will use the function $P(t)=P_0e^{-0.00012t}$

If the bone had lost 43% of its carbon-14 from the initial amount P_0, then $57\%(P_0)$ is the amount present. To find the age t of the bone, we substitute $57\%(P_0)$, or $0.57P_0$ for $P(t)$ in the function above and solve for t.

$$0.57P_0=P_0e^{-0.00012t}$$
$$0.57=e^{-0.00012t}$$
$$\ln 0.57=\ln e^{-0.00012t}$$
$$\ln 0.57=-0.00012t$$
$$\frac{\ln 0.57}{-0.00012}=t$$
$$4684\approx t$$

The bone is about 4684 yr old.

41. $\text{pH}=-\log[H^+]$

$$=-\log(1.0\times 10^{-7})$$
$$=7.0$$

The pH of water is 7.0.

43. $\log_a\frac{\sqrt[3]{x^2z}}{\sqrt[3]{y^2z^{-1}}}$

$$=\log_a\sqrt[3]{\frac{x^2z}{y^2z^{-1}}}$$
$$=\log_a\sqrt[3]{\frac{x^2z^2}{y^2}}$$
$$=\log_a\left(\frac{x^2z^2}{y^2}\right)^{1/3}$$
$$=\frac{1}{3}\log_a\frac{x^2z^2}{y^2}$$
$$=\frac{1}{3}(\log_a x^2+\log_a z^2-\log_a y^2)$$
$$=\frac{1}{3}(2\log_a x+2\log_a z-2\log_a y)$$
$$=\frac{1}{3}(2\cdot 2+2\cdot 4-2\cdot 3)$$
$$=2$$

Chapter 13

Conic Sections

Exercise Set 13.1

1. $(x-2)^2+(y+5)^2=9$, or $(x-2)^2+[y-(-5)]^2=3^2$, is the equation of a circle with center $(2,-5)$ and radius 3, so choice (f) is correct.

3. $(x-5)^2+(y+2)^2=9$, or $(x-5)^2+[y-(-2)]^2=3^2$, is the equation of a circle with center $(5,-2)$ and radius 3, so choice (g) is correct.

5. $y=(x-2)^2-5$ is the equation of a parabola with vertex $(2,-5)$ that opens upward, so choice (c) is correct.

7. $x=(y-2)^2-5$ is the equation of a parabola with vertex $(-5, 2)$ that opens to the right, so choice (d) is correct.

9. $y=-x^2$

This is equivalent to $y=-(x-0)^2+0$. The vertex is $(0, 0)$.

We choose some x-values on both sides of the vertex and compute the corresponding values of y. The graph opens down, because the coefficient of x^2, -1, is negative.

x	y
0	0
1	-1
2	-4
-1	-1
-2	-4

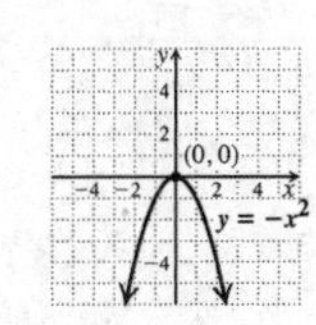

11. $y=-x^2+4x-5$

We can find the vertex by computing the first coordinate, $x=-b/2a$, and then substituting to find the second coordinate:

$$x=-\frac{b}{2a}=-\frac{4}{2(-1)}=2$$

$$y=-x^2+4x-5=-(2)^2+4(2)-5=-1$$

The vertex is $(2,-1)$.

We choose some x-values and compute the corresponding values for y. The graph opens downward because the coefficient of x^2, -1, is negative.

x	y
2	-1
3	-2
4	-5
1	-2
0	-5

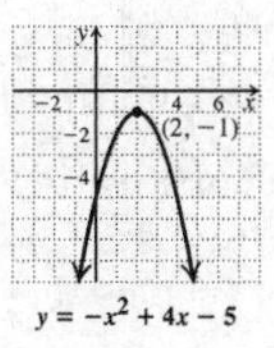

13. $x=y^2-4y+2$

We find the vertex by completing the square.

$$x=(y^2-4y+4)+2-4$$

$$x=(y-2)^2-2$$

The vertex is $(-2,2)$.

To find ordered pairs, we choose values for y and compute the corresponding values of x. The graph opens to the right, because the coefficient of y^2, 1, is positive.

x	y
7	-1
2	0
-1	1
-2	2
-1	3

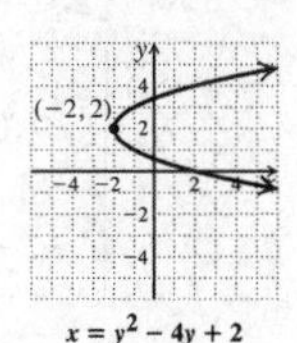

15. $x=y^2+3$

$x=(y-0)^2+3$

The vertex is (3, 0).

To find the ordered pairs, we choose y-values and compute the corresponding values for x. The graph opens to the right, because the coefficient of y^2, 1, is positive.

x	y
3	0
4	1
7	2
4	-1
7	-2

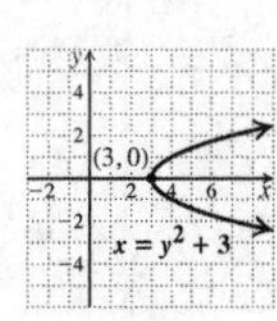

17. $x=2y^2$

$x=2(y-0)^2+0$

The vertex is (0, 0).

We choose y-values and compute the corresponding values for x. The graph opens to the right, because the coefficient of y^2, 2, is positive.

x	y
0	0
2	1
2	−1
8	2
8	−2

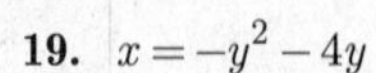

19. $x=-y^2-4y$

We find the vertex by computing the second coordinate, $y=-b/2a$, and then substituting to find the first coordinate:

$$y=-\frac{b}{2a}=-\frac{-4}{2(-1)}=-2$$
$$x=-y^2-4y=-(-2)^2-4(-2)=4$$

The vertex is $(4,-2)$.

We choose y-values and compute the corresponding values for x. The graph opens to the left, because the coefficient of y^2, –1, is negative.

x	y
4	−2
−5	1
0	0
3	−1
3	−3

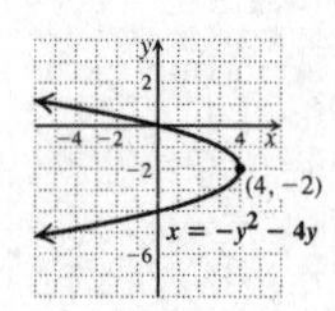

21. $y=x^2-2x+1$

$y=(x-1)^2+0$

The vertex is (1, 0).

We choose x-values and compute the corresponding values for y. The graph opens upward, because the coefficient of x^2, 1, is positive.

x	y
1	0
0	1
−1	4
2	1
3	4

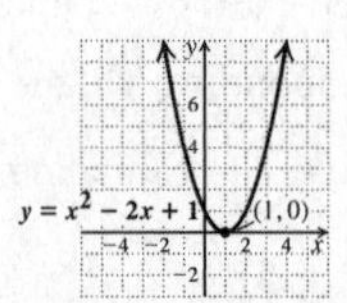

23. $x=-\frac{1}{2}y^2$

$x=-\frac{1}{2}(y-0)^2+0$

The vertex is (0, 0).

We choose y-values and compute the corresponding values for x. The graph opens to the left, because the coefficient of y^2, $-\frac{1}{2}$, is negative.

x	y
0	0
−2	2
−8	4
−2	−2
−8	−4

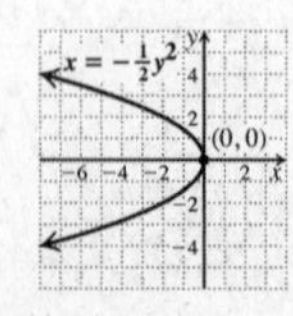

25. $x=-y^2+2y-1$

We find the vertex by computing the second coordinate, $y=-b/2a$, and then substituting to find the first coordinate.

$$y=-\frac{b}{2a}=-\frac{2}{2(-1)}=1$$
$$x=-y^2+2y-1=-(1)^2+2(1)-1=0$$

The vertex is (0, 1).

We choose y-values and compute the corresponding values for x. The graph opens to the left, because the coefficient of y^2, –1, is negative.

x	y
−4	3
−1	2
−1	0
−4	−1
−4	3

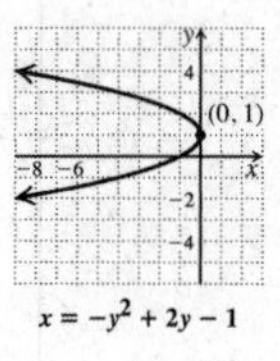

27. $x=-2y^2-4y+1$

We find the vertex by completing the square.

$$x=-2(y^2+2y)+1$$
$$x=-2(y^2+2y+1)+1+2$$
$$x=-2(y+1)^2+3$$

The vertex is $(3,-1)$.

We choose y-values and compute the corresponding values for x. The graph opens to the left, because the coefficient of y^2, –2, is negative.

x	y
3	−1
1	−2
−5	−3
1	0
−5	1

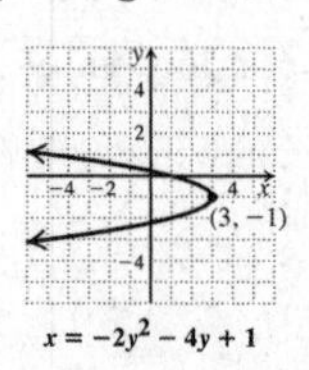

29. $(x-h)^2+(y-k)^2=r^2$ Standard form
$(x-0)^2+(y-0)^2=8^2$ Substituting
$x^2+y^2=64$ Simplifying

31. $(x-h)^2+(y-k)^2=r^2$ Standard form
$(x-7)^2+(y-3)^2=(\sqrt{6})^2$ Substituting
$(x-7)^2+(y-3)^2=6$

33.
$$(x-h)^2+(y-k)^2=r^2$$
$$[x-(-4)]^2+(y-3)^2=(3\sqrt{2})^2$$
$$(x+4)^2+(y-3)^2=18$$

35.
$$(x-h)^2+(y-k)^2=r^2$$
$$[x-(-5)]^2+[y-(-8)]^2=(10\sqrt{3})^2$$
$$(x+5)^2+(y+8)^2=300$$

37. Since the center is (0, 0), we have

$(x-0)^2+(y-0)^2=r^2$ or $x^2+y^2=r^2$

The circle passes through (–3, 4). We find r^2 by substituting –3 for x and 4 for y.

$$(-3)^2+4^2=r^2$$
$$9+16=r^2$$
$$25=r^2$$

Then $x^2+y^2=25$ is an equation of the circle.

39. Since the center is (–4, 1), we have

$$[x-(-4)]^2+(y-1)^2=r^2, \text{ or}$$
$$(x+4)^2+(y-1)^2=r^2.$$

The circle passes through (–2, 5). We find r^2 by substituting –2 for x and 5 for y.

$$(-2+4)^2+(5-1)^2=r^2$$
$$4+16=r^2$$
$$20=r^2$$

Then $(x+4)^2+(y-1)^2=20$ is an equation of the circle.

41. We write standard form.

$(x-0)^2+(y-0)^2=1^2$

The center is (0, 0), and the radius is 1.

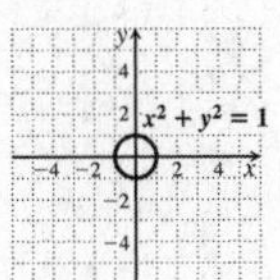

43. $(x+1)^2+(y+3)^2=49$

$[x-(-1)]^2+[y-(-3)]^2=7^2$ Standard form

The center is (–1, –3), and the radius is 7.

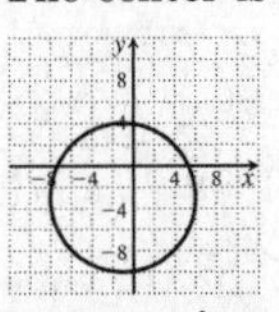

45. $(x-4)^2+(y+3)^2=10$

$(x-4)^2+[y-(-3)]^2=(\sqrt{10})^2$

The center is (4, –3), and the radius is $\sqrt{10}$.

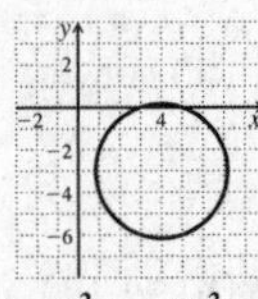

47. $x^2+y^2=8$

$(x-0)^2+(y-0)^2=(\sqrt{8})^2$ Standard form

The center is (0, 0), and the radius is $\sqrt{8}$, or $2\sqrt{2}$.

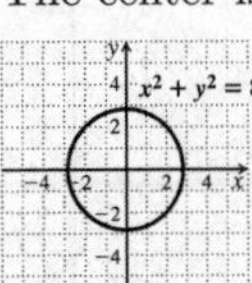

49. $(x-5)^2+y^2=\frac{1}{4}$

$(x-5)^2+(y-0)^2=\left(\frac{1}{2}\right)^2$ Standard form

The center is (5, 0), and the radius is $\frac{1}{2}$.

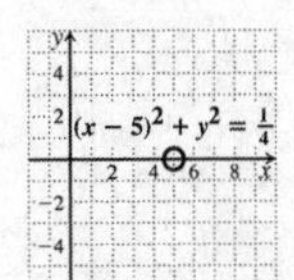

51.

$$x^2+y^2+8x-6y-15=0$$
$$x^2+8x+y^2-6y=15$$
$$(x^2+8x+16)+(y^2-6y+9)=15+16+9 \quad \text{Completing the square twice}$$
$$(x+4)^2+(y-3)^2=40$$
$$[x-(-4)]^2+(y-3)^2=(\sqrt{40})^2 \quad \text{Standard form}$$

The center is (–4, 3), and the radius is $\sqrt{40}$, or $2\sqrt{10}$.

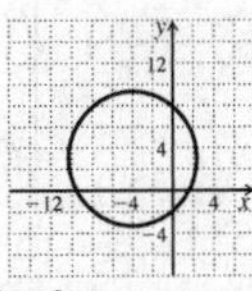

53.

$$x^2+y^2-8x+2y+13=0$$
$$x^2-8x+y^2+2y=-13$$
$$(x^2-8x+16)+(y^2+2y+1)=-13+16+1 \quad \text{Completing the square twice}$$
$$(x-4)^2+(y+1)^2=4$$
$$(x-4)^2+[y-(-1)]^2=2^2 \quad \text{Standard form}$$

The center is (4, –1), and the radius is 2.

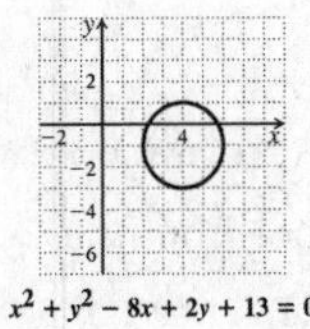

55.
$$x^2+y^2+10y-75=0$$
$$x^2+y^2+10y=75$$
$$x^2+(y^2+10y+25)=75+25$$
$$(x-0)^2+(y+5)^2=100$$
$$(x-0)^2+[y-(-5)]^2=10^2$$

The center is (0, –5), and the radius is 10.

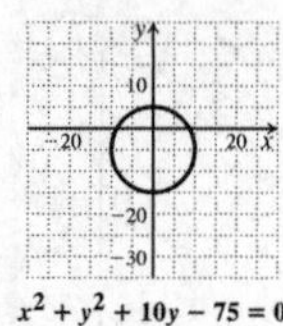

$x^2+y^2+10y-75=0$

57.
$$x^2+y^2+7x-3y-10=0$$
$$x^2+7x+y^2-3y=10$$
$$\left(x^2+7x+\frac{49}{4}\right)+\left(y^2-3y+\frac{9}{4}\right)=10+\frac{49}{4}+\frac{9}{4}$$
$$\left(x+\frac{7}{2}\right)^2+\left(y-\frac{3}{2}\right)^2=\frac{98}{4}$$
$$\left[x-\left(-\frac{7}{2}\right)\right]^2+\left(y-\frac{3}{2}\right)^2=\left(\sqrt{\frac{98}{4}}\right)^2$$

The center is $\left(-\frac{7}{2}, \frac{3}{2}\right)$, and the radius is $\sqrt{\frac{98}{4}}$, or $\frac{\sqrt{98}}{2}$, or $\frac{7\sqrt{2}}{2}$.

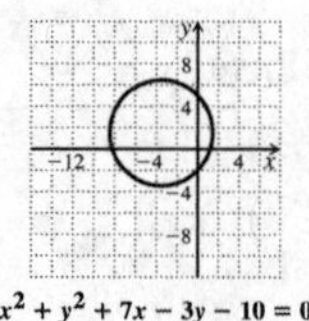

$x^2+y^2+7x-3y-10=0$

59.
$$36x^2+36y^2=1$$
$$x^2+y^2=\frac{1}{36} \quad \text{Multiplying by } \frac{1}{36} \text{ on both sides}$$
$$(x-0)^2+(y-0)^2=\left(\frac{1}{6}\right)^2$$

The center is (0, 0), and the radius is $\frac{1}{6}$.

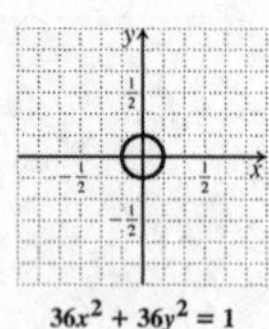

$36x^2+36y^2=1$

61. *Writing Exercise.* No; a circle is defined to be the set of points in a plane that are a fixed distance from the center. Thus, unless $r=0$ and the circle is one point, the center is not part of the circle.

63.
$$\frac{y^2}{16}=1$$
$$y^2=16$$
$y=4$ or $y=-4$ Using the principle of square roots

The solutions are ± 4.

65.
$$\frac{(x-1)^2}{25}=1$$
$$(x-1)^2=25$$
$$x-1=5 \quad or \quad x-1=-5$$
$$x=6 \quad or \quad x=-4$$

The solutions are –4 and 6.

67.
$$\frac{1}{4}+\frac{(y+3)^2}{36}=1$$
$$\frac{(y+3)^2}{36}=\frac{3}{4}$$
$$(y+3)^2=27$$
$$y+3=\sqrt{27} \quad or \quad y+3=-\sqrt{27}$$
$$y=-3+3\sqrt{3} \quad or \quad y=-3-3\sqrt{3}$$

The solutions are $-3\pm 3\sqrt{3}$.

69. *Writing Exercise.* The points appear to form a parabola. A circle is formed from a fixed distance, the radius, from a point. The result is a circle. This set of points was formed from a fixed distance from a point *and* a line. The result is a U-shaped figure.

71. We make a drawing of the circle with center (3, –5) and tangent to the y-axis.

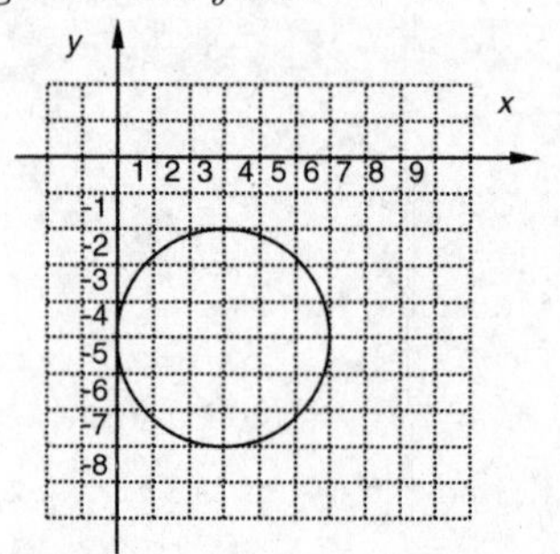

We see that the circle touches the y-axis at (0, –5). Hence the radius is the distance between (0, –5) and (3, –5), or $\sqrt{(3-0)^2+[-5-(-5)]^2}$, or 3. Now we write the equation of the circle.

$$(x-h)^2+(y-k)^2=r^2$$
$$(x-3)^2+[y-(-5)]^2=3^2$$
$$(x-3)^2+(y+5)^2=9$$

73. First we use the midpoint formula to find the center:

$\left(\frac{7+(-1)}{2}, \frac{3+(-3)}{2}\right)$, or $\left(\frac{6}{2}, \frac{0}{2}\right)$, or (3, 0)

The length of the radius is the distance between the center (3, 0) and either endpoint of a diameter. We will use endpoint (7, 3) in the distance formula:

$$r=\sqrt{(7-3)^2+(3-0)^2}=\sqrt{25}=5$$

Now we write the equation of the circle:

$$(x-h)^2+(y-k)^2=r^2$$
$$(x-3)^2+(y-0)^2=5^2$$
$$(x-3)^2+y^2=25$$

75. Let $(0, y)$ be the point on the y-axis that is equidistant from (2, 10) and (6, 2). Then the distance between (2, 10) and $(0, y)$ is the same as the distance between (6, 2) and $(0, y)$.

$$\sqrt{(0-2)^2+(y-10)^2}=\sqrt{(0-6)^2+(y-2)^2}$$
$$(-2)^2+(y-10)^2=(-6)^2+(y-2)^2 \quad \text{Squaring both sides}$$
$$4+y^2-20y+100=36+y^2-4y+4$$
$$64=16y$$
$$4=y$$

This number checks. The point is (0, 4).

77. For the outer circle, $r^2=\frac{81}{4}$. For the inner circle, $r^2=16$. The area of the red zone is the difference between the areas of the outer and inner circles. Recall that the area A of a circle with radius r is given by the formula $A=\pi r^2$.

$$\pi\cdot\frac{81}{4}-\pi\cdot 16=\frac{81}{4}\pi-\frac{64}{4}\pi=\frac{17}{4}\pi$$

The area of the red zone is $\frac{17}{4}\pi$ m^2, or about 13.4 m^2.

79. Superimposing a coordinate system on the snowboard as in Exercise 78, and observing that 1160/2 = 580, we know that three points on the circle are (–580, 0), (0, 23.5) and (580, 0). Let $(0, k)$ represent the center of the circle. Use the fact that $(0, k)$ is equidistant from (–580, 0) and (0, 23.5).

$$\sqrt{(-580-0)^2+(0-k)^2}=\sqrt{(0-0)^2+(23.5-k)^2}$$
$$\sqrt{336{,}400+k^2}=\sqrt{552.25-47k+k^2}$$
$$336{,}400+k^2=552.25-47k+k^2$$
$$335{,}847.75=-47k$$
$$-7145.7\approx k$$

Then to find the radius, we find the distance from the center (0, –7145.7) to any one of the three known points on the circle. We use (0, 23.5).

$$r=\sqrt{(0-0)^2+(-7145.7-23.5)^2}\approx 7169 \text{ mm}$$

81. a) When the circle is positioned on a coordinate system as shown in the text, the center lies on the y-axis. To find the center, we will find the point on the y-axis that is equidistant from (–4, 0) and (0, 2). Let $(0, y)$ be this point.

$$\sqrt{[0-(-4)]^2+(y-0)^2}=\sqrt{(0-0)^2+(y-2)^2}$$
$$4^2+y^2=0^2+(y-2)^2 \quad \text{Squaring both sides}$$
$$16+y^2=y^2-4y+4$$
$$12=-4y$$
$$-3=y$$

The center of the circle is (0, –3).

b) We find the radius of the circle.

$$(x-0)^2+[y-(-3)]^2=r^2 \quad \text{Standard form}$$
$$x^2+(y+3)^2=r^2$$
$$(-4)^2+(0+3)^2=r^2 \quad \text{Substituting}$$
$$16+9=r^2 \quad (-4,\ 0) \text{ for } (x,\ y)$$
$$25=r^2$$
$$5=r$$

The radius is 5 ft.

83. We write the equation of a circle with center (0, 30.6) and radius 24.3:

$$x^2+(y-30.6)^2=590.49$$

85. Substitute 6 for N.

$$H=\frac{D^2N}{2.5}=\frac{D^2\cdot 6}{2.5}=2.4D^2$$

Find some ordered pairs for $2.5\le D\le 8$ and draw the graph.

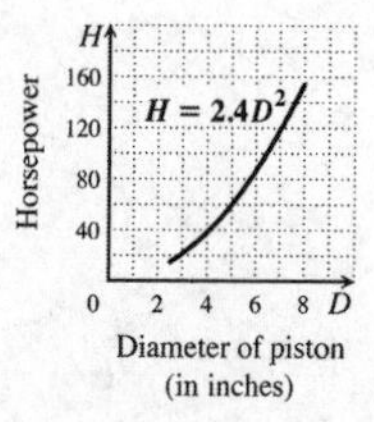

Using the graph, a horse power of 120, on the vertical axis, relates to a diameter of 7 in., on the horizontal axis.

87. *Writing Exercise.* One method is to enter $y_1=ax^2+bx+c$, deselect it, and then use the Draw Inverse operation to graph the inverse of y_1. This inverse is the graph of $x_1=ay^2+by+c$.

Exercise Set 13.2

1. True; see page 864 in the text.

3. False; see page 864 in the text.

5. True; see page 864 in the text.

7. True; see page 866 in the text.

9. $\frac{x^2}{1}+\frac{y^2}{4}=1$

$\frac{x^2}{1^2}+\frac{y^2}{2^2}=1$

The x-intercepts are (1, 0) and (–1, 0), and the y-intercepts are (0, 2) and (0, –2). We plot these points and connect them with an oval-shaped curve.

$\frac{x^2}{1}+\frac{y^2}{4}=1$

11. $\frac{x^2}{25}+\frac{y^2}{9}=1$

$\frac{x^2}{5^2}+\frac{y^2}{3^2}=1$

The x-intercepts are (5, 0) and (–5, 0), and the y-intercepts are (0, 3) and (0, –3). We plot these points and connect them with an oval-shaped curve.

$\frac{x^2}{25}+\frac{y^2}{9}=1$

13. $4x^2+9y^2=36$

$\frac{1}{36}(4x^2+9y^2)=\frac{1}{36}(36)$ Multiplying by $\frac{1}{36}$

$\frac{x^2}{9}+\frac{y^2}{4}=1$

$\frac{x^2}{3^2}+\frac{y^2}{2^2}=1$

The x-intercepts are (–3, 0) and (3, 0), and the y-intercepts are (0, –2) and (0, 2). We plot these points and connect them with an oval-shaped curve.

$4x^2+9y^2=36$

15. $16x^2+9y^2=144$

$\frac{x^2}{9}+\frac{y^2}{16}=1$ Multiplying by $\frac{1}{144}$

$\frac{x^2}{3^2}+\frac{y^2}{4^2}=1$

The x-intercepts are (3, 0) and (–3, 0), and the y-intercepts are (0, 4) and (0, –4). We plot these points and connect them with an oval-shaped curve.

$16x^2+9y^2=144$

17. $2x^2+3y^2=6$

$\frac{x^2}{3}+\frac{y^2}{2}=1$ Multiplying by $\frac{1}{6}$

$\frac{x^2}{(\sqrt{3})^2}+\frac{y^2}{(\sqrt{2})^2}=1$

The x-intercepts are $(\sqrt{3},\ 0)$ and $(-\sqrt{3},\ 0)$, and the y-intercepts are $(0,\ \sqrt{2})$ and $(0, -\sqrt{2})$. We plot these points and connect them with an oval-shaped curve.

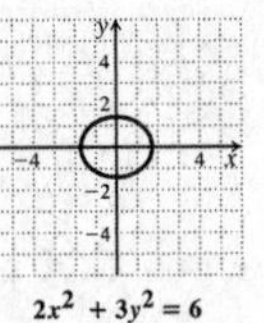

$2x^2+3y^2=6$

19. $5x^2+5y^2=125$

Observe that the x^2- and y^2-terms have the same coefficient. We divide both sides of the equation by 5 to obtain $x^2+y^2=25$. This is the equation of a circle with center (0, 0) and radius 5.

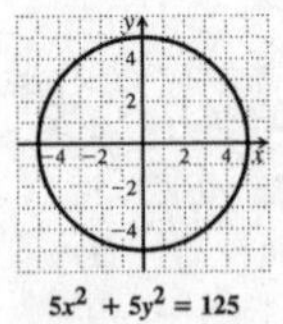

$5x^2+5y^2=125$

21. $3x^2+7y^2-63=0$

$3x^2+7y^2=63$

$\frac{x^2}{21}+\frac{y^2}{9}=1$ Multiplying by $\frac{1}{63}$

$\frac{x^2}{(\sqrt{21})^2}+\frac{y^2}{3^2}=1$

The x-intercepts are $(\sqrt{21},\ 0)$ and $(-\sqrt{21},\ 0)$, or about $(4.583,\ 0)$ and $(-4.583,\ 0)$. The y-intercepts are (0, 3) and (0, –3). We plot these points and connect them with an oval-shaped curve.

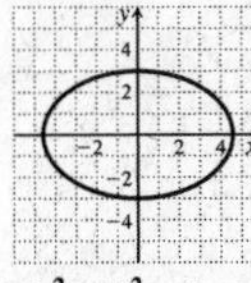

$3x^2 + 7y^2 - 63 = 0$

23. $16x^2 = 16 - y^2$
$$16x^2 + y^2 = 16$$
$$\frac{x^2}{1} + \frac{y^2}{16} = 1$$

The x-intercepts are (1, 0) and (–1, 0), and the y-intercepts are (0, 4) and (0, –4). We plot these points and connect them with an oval-shaped curve.

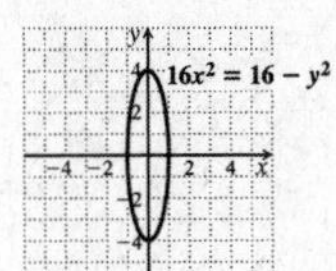

25. $16x^2 + 25y^2 = 1$

Note that $16 = \frac{1}{\frac{1}{16}}$ and $25 = \frac{1}{\frac{1}{25}}$. Thus, we can rewrite the equation:

$$\frac{x^2}{\frac{1}{16}} + \frac{y^2}{\frac{1}{25}} = 1$$
$$\frac{x^2}{\left(\frac{1}{4}\right)^2} + \frac{y^2}{\left(\frac{1}{5}\right)^2} = 1$$

The x-intercepts are $\left(\frac{1}{4}, 0\right)$ and $\left(-\frac{1}{4}, 0\right)$, and the y-intercepts are $\left(0, \frac{1}{5}\right)$ and $\left(0, -\frac{1}{5}\right)$. We plot these points and connect them with an oval-shaped curve.

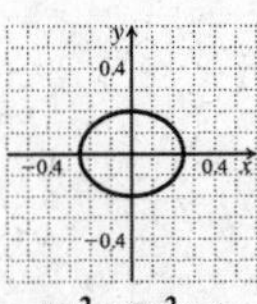

$16x^2 + 25y^2 = 1$

27. $\frac{(x-3)^2}{9} + \frac{(y-2)^2}{25} = 1$
$$\frac{(x-3)^2}{3^2} + \frac{(y-2)^2}{5^2} = 1$$

The center of the ellipse is (3, 2). Note that $a = 3$ and $b = 5$. We locate the center and then plot the points $(3+3,\ 2)$ (3 – 3, 2), (3, 2 + 5), and (3, 2 – 5), or (6, 2), (0, 2), (3, 7), and (3, –3). Connect these points with an oval-shaped curve.

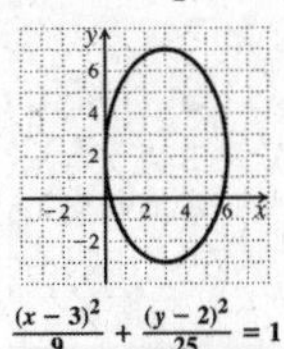

$\frac{(x-3)^2}{9} + \frac{(y-2)^2}{25} = 1$

29. $\frac{(x+4)^2}{16} + \frac{(y-3)^2}{49} = 1$
$$\frac{(x-(-4))^2}{4^2} + \frac{(y-3)^2}{7^2} = 1$$

The center of the ellipse is (–4, 3). Note that $a = 4$ and $b = 7$. We locate the center and then plot the points $(-4+4,\ 3)$, (–4 – 4, 3), (–4, 3 + 7), and (–4, 3 – 7), or (0, 3), (–8, 3), (–4, 10), and (–4,–4). Connect these points with an oval-shaped curve.

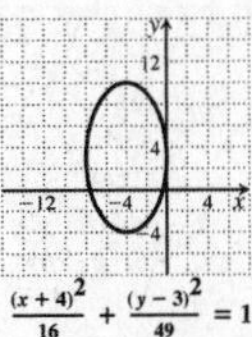

$\frac{(x+4)^2}{16} + \frac{(y-3)^2}{49} = 1$

31. $12(x-1)^2 + 3(y+4)^2 = 48$
$$\frac{(x-1)^2}{4} + \frac{(y+4)^2}{16} = 1$$
$$\frac{(x-1)^2}{2^2} + \frac{(y-(-4))^2}{4^2} = 1$$

The center of the ellipse is (1, –4). Note that $a = 2$ and $b = 4$. We locate the center and then plot the points $(1+2,\ -4)$, (1 – 2, –4), (1, –4 + 4), and (1, –4 – 4), or (3,–4), (–1, –4), (1, 0), and (1, –8). Connect these points with an oval-shaped curve.

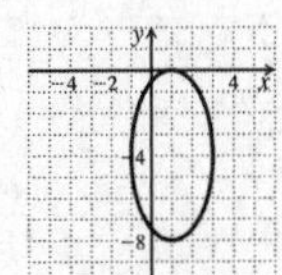

$12(x-1)^2 + 3(y+4)^2 = 48$

33. $4(x+3)^2 + 4(y+1)^2 - 10 = 90$
$$4(x+3)^2 + 4(y+1)^2 = 100$$

Observe that the x^2- and y^2-terms have the same coefficient. Dividing both sides by 4, we have
$$(x+3)^2 + (y+1)^2 = 25.$$

This is the equation of a circle with center (–3, –1) and radius 5.

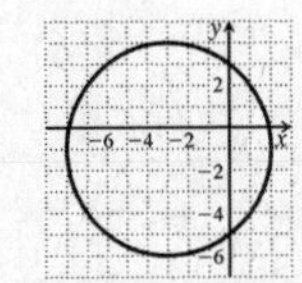

$4(x+3)^2 + 4(y+1)^2 - 10 = 90$

35. *Writing Exercise.* Write the equation of an ellipse in standard form, $\frac{x^2}{a^2}+\frac{y^2}{b^2}=1$. If $a^2>b^2$, then the ellipse is horizontal. If $b^2>a^2$, then the ellipse is vertical.

37. $x^2-5x+3=0$

$$x=\frac{5\pm\sqrt{25-4\cdot1\cdot3}}{2\cdot1} \quad \text{Using the quadratic formula}$$
$$=\frac{5\pm\sqrt{13}}{2}$$

39.
$$\frac{4}{x+2}+\frac{3}{2x-1}=2$$
$$4(2x-1)+3(x+2)=2(x+2)(2x-1)$$
$$8x-4+3x+6=4x^2+6x-4$$
$$4x^2-5x-6=0$$
$$(4x+3)(x-2)=0$$
$$4x+3=0 \quad or \quad x-2=0$$
$$x=-\tfrac{3}{4} \quad or \quad x=2$$

41. $x^2=11$

$x=\pm\sqrt{11}$

43. *Writing Exercise.* Since the coefficients of the squared terms $9x^2$ and y^2 have the same sign, then the equation is an ellipse.

45. Plot the given points.

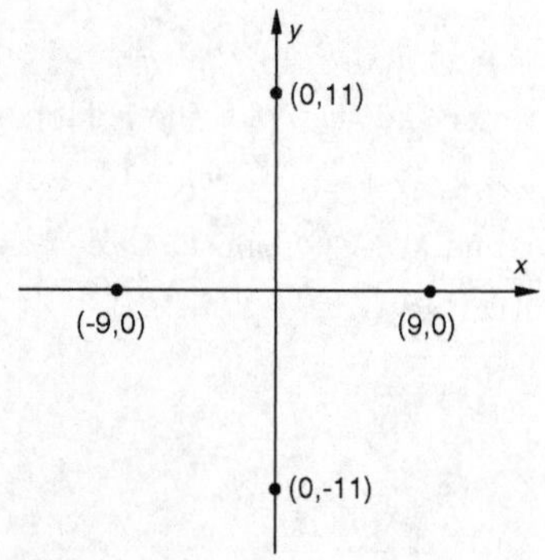

From the location of these points, we see that the ellipse that contains them is centered at the origin with $a=9$ and $b=11$. We write the equation of the ellipse:

$$\frac{x^2}{9^2}+\frac{y^2}{11^2}=1$$
$$\frac{x^2}{81}+\frac{y^2}{121}=1$$

47. Plot the given points.

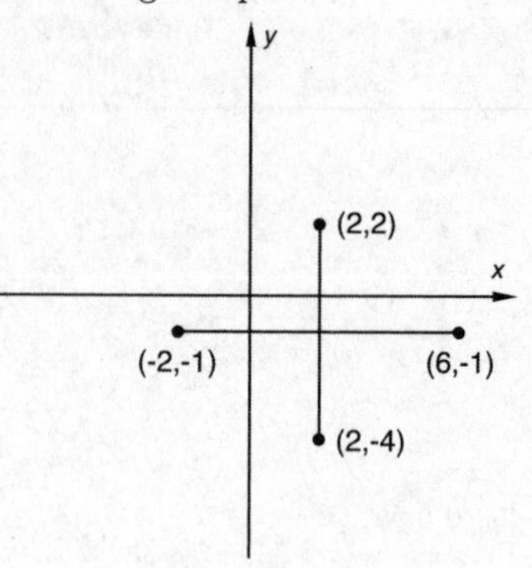

The midpoint of the segment from (–2, –1) to (6, –1) is $\left(\frac{-2+6}{2}, \frac{-1-1}{2}\right)$, or (2, –1). The midpoint of the segment from (2, –4) to (2, 2) is $\left(\frac{2+2}{2}, \frac{-4+2}{2}\right)$, or $(2,-1)$. Thus, we can conclude that (2, –1) is the center of the ellipse. The distance from (–2, –1) to (2, –1) is $\sqrt{[2-(-2)]^2+[-1-(-1)]^2}=\sqrt{16}=4$, so $a=4$. The distance from (2, 2) to (2, –1) is $\sqrt{(2-2)^2+(-1-2)^2}$ $=\sqrt{9}=3$, so $b=3$. We write the equation of the ellipse.

$$\frac{(x-2)^2}{4^2}+\frac{(y-(-1))^2}{3^2}=1$$
$$\frac{(x-2)^2}{16}+\frac{(y+1)^2}{9}=1$$

49. We make a drawing.

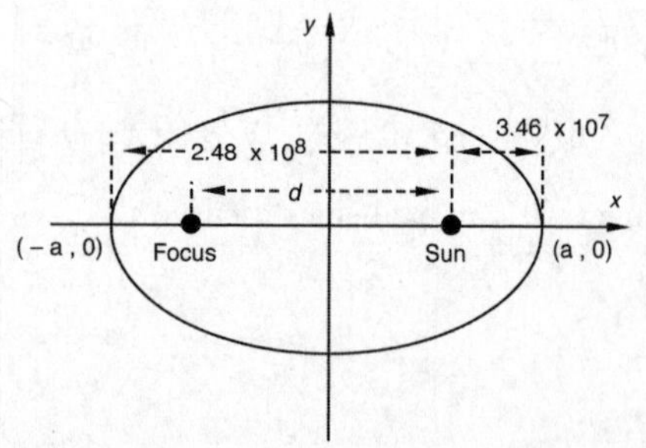

The distance between vertex $(a, 0)$ and the sun is the same as the distance between vertex $(-a, 0)$ and the other focus. Then

$$d=2.48\times10^8-3.46\times10^7$$
$$=2.48\times10^8-0.346\times10^8=2.134\times10^8 \text{ mi.}$$

51. a) Let $F_1=(-c,\ 0)$ and $F_2=(c,\ 0)$. Then the sum of the distances from the foci to P is $2a$. By the distance formula,

$$\sqrt{(x+c)^2+y^2}+\sqrt{(x-c)^2+y^2}=2a \text{, or}$$
$$\sqrt{(x+c)^2+y^2}=2a-\sqrt{(x-c)^2+y^2}.$$

Squaring, we get

$$(x+c)^2+y^2=4a^2-4a\sqrt{(x-c)^2+y^2}+(x-c)^2+y^2,$$

or $x^2+2cx+c^2+y^2$

$$=4a^2-4a\sqrt{(x-c)^2+y^2}+x^2-2cx+c^2+y^2.$$

Thus

$$-4a^2+4cx=-4a\sqrt{(x-c)^2+y^2}$$
$$a^2-cx=a\sqrt{(x-c)^2+y^2}.$$

Squaring again, we get

$$a^4-2a^2cx+c^2x^2=a^2(x^2-2cx+c^2+y^2)$$
$$a^4-2a^2cx+c^2x^2=a^2x^2-2a^2cx+a^2c^2+a^2y^2,$$

or

$$x^2(a^2-c^2)+a^2y^2=a^2(a^2-c^2)$$
$$\frac{x^2}{a^2}+\frac{y^2}{a^2-c^2}=1.$$

b) When P is at (0, b), it follows that $b^2=a^2-c^2$.

Substituting, we have $\frac{x^2}{a^2}+\frac{y^2}{b^2}=1$.

53. For the given ellipse, $a=6/2$, or 3, and $b=2/2$, or 1. The patient's mouth should be at a distance 2c from the light source, where the coordinates of the foci of the ellipse are $(-c, 0)$ and $(c, 0)$. From Exercise 51(b), we know $b^2=a^2-c^2$. We use this to find c.

$$b^2=a^2-c^2$$
$$1^2=3^2-c^2 \quad \text{Substituting}$$
$$c^2=8$$
$$c=\sqrt{8}$$

Then $2c=2\sqrt{8}\approx 5.66$. The patient's mouth should be about 5.66 ft from the light source.

55.

$$x^2-4x+4y^2+8y-8=0$$
$$x^2-4x+4y^2+8y=8$$
$$x^2-4x+4(y^2+2y)=8$$
$$(x^2-4x+4-4)+4(y^2+2y+1-1)=8$$
$$(x^2-4x+4)+4(y^2+2y+1)=8+4+4\cdot 1$$
$$(x-2)^2+4(y+1)^2=16$$
$$\frac{(x-2)^2}{16}+\frac{(y+1)^2}{4}=1$$
$$\frac{(x-2)^2}{4^2}+\frac{(y-(-1))^2}{2^2}=1$$

The center of the ellipse is (2, –1). Note that a = 4 and $b=2$. We locate the center and then plot the points $(2+4,-1)$, $(2-4,-1)$, $(2,-1+2)$, $(2,-1-2)$, or $(6,-1)$, $(-2,-1)$, (2, 1), and $(2,-3)$. Connect these points with an oval-shaped curve.

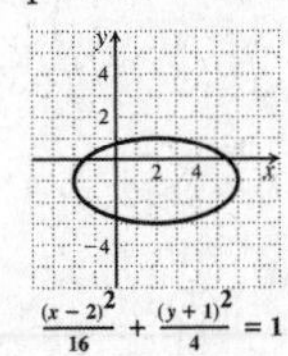

57. *Graphing Calculator Exercise*

Exercise Set 13.3

1. d; see page 870 in the text.

3. h; see page 874 in the text.

5. g; see page 874 in the text.

7. c; see page 874 in the text.

9.
$$\frac{y^2}{16}-\frac{x^2}{16}=1$$
$$\frac{y^2}{4^2}-\frac{x^2}{4^2}=1$$

a = 4 and b = 4, so the asymptotes are $y=\frac{4}{4}x$ and $y=-\frac{4}{4}x$, or $y=x$ and $y=-x$. We sketch them.

Replacing x with 0 and solving for y, we get $y=\pm 4$, so the intercepts are (0, 4) and (0, –4).

We plot the intercepts and draw smooth curves through them that approach the asymptotes.

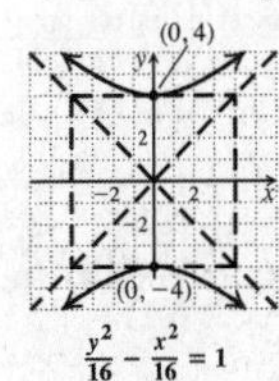

11.
$$\frac{x^2}{4}-\frac{y^2}{25}=1$$
$$\frac{x^2}{2^2}-\frac{y^2}{5^2}=1$$

a = 2 and b = 5, so the asymptotes are $y=\frac{5}{2}x$ and $y=-\frac{5}{2}x$. We sketch them.

Replacing y with 0 and solving for x, we get $x=\pm 2$, so the intercepts are (2, 0) and (–2, 0).

We plot the intercepts and draw smooth curves through

them that approach the asymptotes.

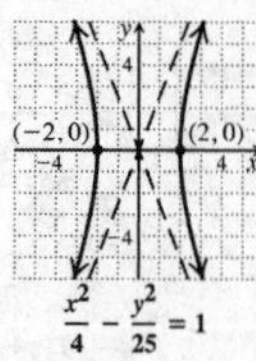

13. $\frac{y^2}{36}-\frac{x^2}{9}=1$

$\frac{y^2}{6^2}-\frac{x^2}{3^2}=1$

$a=3$ and $b=6$, so the asymptotes are $y=\frac{6}{3}x$ and $y=-\frac{6}{3}x$, or $y=2x$ and $y=-2x$. We sketch them.

Replacing x with 0 and solving for y, we get $y=\pm 6$, so the intercepts are (0, 6) and (0, –6).

We plot the intercepts and draw smooth curves through them that approach the asymptotes.

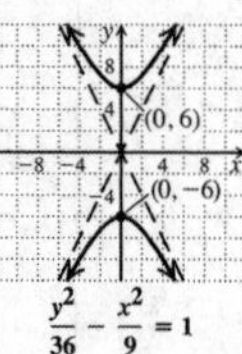

15. $y^2-x^2=25$

$\frac{y^2}{25}-\frac{x^2}{25}=1$

$\frac{y^2}{5^2}-\frac{x^2}{5^2}=1$

$a=5$ and $b=5$, so the asymptotes are $y=\frac{5}{5}x$ and $y=-\frac{5}{5}x$, or $y=x$ and $y=-x$. We sketch them.

Replacing x with 0 and solving for y, we get $y=\pm 5$, so the intercepts are (0, 5) and (0, –5).

We plot the intercepts and draw smooth curves through them that approach the asymptotes.

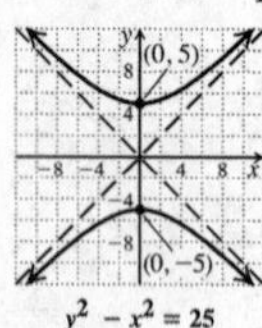

17. $25x^2-16y^2=400$

$\frac{x^2}{16}-\frac{y^2}{25}=1$ Multiplying by $\frac{1}{400}$

$\frac{x^2}{4^2}-\frac{y^2}{5^2}=1$

$a=4$ and $b=5$, so the asymptotes are $y=\frac{5}{4}x$ and $y=-\frac{5}{4}x$. We sketch them.

Replacing y with 0 and solving for x, we get $x=\pm 4$, so the intercepts are (4, 0) and (–4, 0).

We plot the intercepts and draw smooth curves through them that approach the asymptotes.

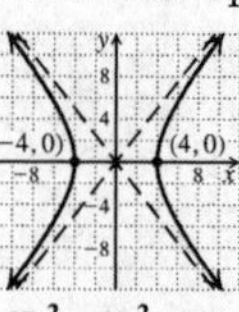

19. $xy=-6$

$y=-\frac{6}{x}$ Solving for y

We find some solutions, keeping the results in a table.

x	y
$\frac{1}{6}$	36
1	−6
6	−1
12	$-\frac{1}{2}$
$-\frac{1}{6}$	36
−1	6
−6	1
−12	$\frac{1}{2}$

Note that we cannot use 0 for x. The x-axis and the y-axis are the asymptotes.

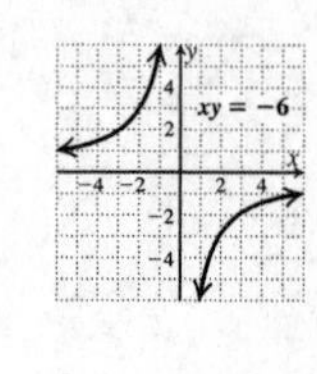

21. $xy=4$

$y=\frac{4}{x}$ Solving for y

We find some solutions, keeping the results in a table.

x	y
$\frac{1}{2}$	8
1	4
4	1
8	$\frac{1}{2}$
$-\frac{1}{2}$	−8
−1	−4
−2	−2
−4	−1

Note that we cannot use 0 for x. The x-axis and the y-axis are the asymptotes.

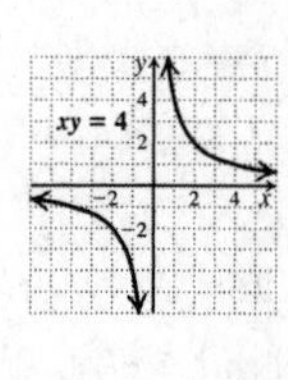

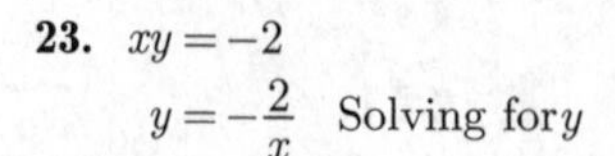

23. $xy=-2$

$y=-\frac{2}{x}$ Solving for y

x	y
$\frac{1}{2}$	−4
1	−2
2	−1
4	$-\frac{1}{2}$
$-\frac{1}{2}$	4
−1	2
−2	1
−4	$\frac{1}{2}$

Note that we cannot use 0 for x. The x-axis and the y-axis are the asymptotes.

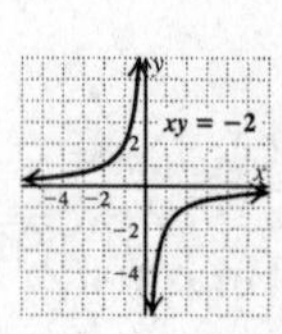

25. $xy = 1$

$y = \frac{1}{x}$ Solving for y

x	y
$\frac{1}{4}$	4
$\frac{1}{2}$	2
1	1
2	$\frac{1}{2}$
4	$\frac{1}{4}$
$-\frac{1}{4}$	-4
$-\frac{1}{2}$	-2
-1	-1
-2	$-\frac{1}{2}$
-4	$-\frac{1}{4}$

Note that we cannot use 0 for x. The x-axis and the y-axis are the asymptotes.

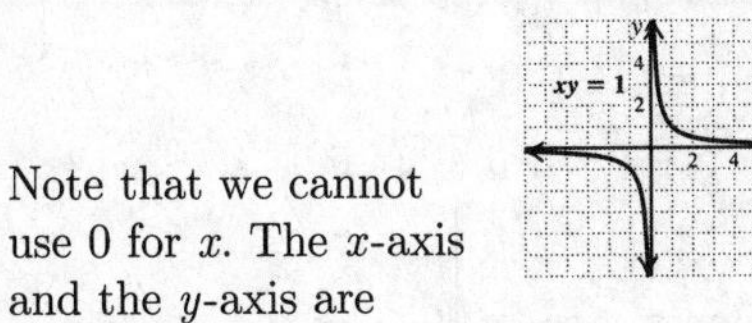

27. $x^2 + y^2 - 6x + 10y - 40 = 0$

Completing the square twice, we obtain an equivalent equation:

$$(x^2 - 6x) + (y^2 + 10y) = 40$$
$$(x^2 - 6x + 9) + (y^2 + 10y + 25) = 40 + 9 + 25$$
$$(x - 3)^2 + (y + 5)^2 = 74$$

The graph is a circle.

29. $9x^2 + 4y^2 - 36 = 0$

$$9x^2 + 4y^2 = 36$$
$$\frac{x^2}{4} + \frac{y^2}{9} = 1$$

The graph is an ellipse.

31. $4x^2 - 9y^2 - 72 = 0$

$$4x^2 - 9y^2 = 72$$
$$\frac{x^2}{18} - \frac{y^2}{8} = 1$$

The graph is a hyperbola.

33. $y^2 = 20 - x^2$

$$x^2 + y^2 = 20$$

The graph is a circle.

35. $x - 10 = y^2 - 6y$

$$x - 10 + 9 = y^2 - 6y + 9$$
$$x - 1 = (y - 3)^2$$

The graph is a parabola.

37. $x - \frac{3}{y} = 0$

$$x = \frac{3}{y}$$
$$xy = 3$$

The graph is a hyperbola.

39. $y + 6x = x^2 + 5$

$$y = x^2 - 6x + 5$$

The graph is a parabola

41. $25y^2 = 100 + 4x^2$

$$25y^2 - 4x^2 = 100$$
$$\frac{y^2}{4} - \frac{x^2}{25} = 1$$

The graph is a hyperbola.

43. $3x^2 + y^2 - x = 2x^2 - 9x + 10y + 40$

$$x^2 + y^2 + 8x - 10y = 40$$

Both variables are squared, so the graph is not a parabola. The plus sign between x^2 and y^2 indicates that we have either a circle or an ellipse. Since the coefficients of x^2 and y^2 are the same, the graph is a circle.

45. $16x^2 + 5y^2 - 12x^2 + 8y^2 - 3x + 4y = 568$

$$4x^2 + 13y^2 - 3x + 4y = 568$$

Both variables are squared, so the graph is not a parabola. The plus sign between x^2 and y^2 indicates that we have either a circle or an ellipse. Since the coefficients of x^2 and y^2 are different, the graph is an ellipse.

47. *Writing Exercise.* The equation of the ellipse in standard form is the sum of the squared terms. The equation of the hyperbola in standard form is the difference of the squared terms.

49. $5x + 2y = -3$ (1)

$2x + 3y = 12$ (2)

We multiply twice to make two terms become additive inverses.

From (1) $15x + 6y = -9$ Multiplying by 3

From (2) $-4x - 6y = -24$ Multiplying by -2

$$11x = -33$$
$$x = -3$$

Substitute -3 for x in Equation (2) and solve for y.

$$2x + 3y = 12$$
$$2(-3) + 3y = 12$$
$$-6 + 3y = 12$$
$$3y = 18$$
$$y = 6$$

The solution is $(-3, 6)$.

51.
$$\frac{3}{4}x^2 + x^2 = 7$$
$$\frac{7}{4}x^2 = 7$$
$$x^2 = 4$$
$$x^2 - 4 = 0$$
$$(x+2)(x-2) = 0$$
$$x+2=0 \quad or \quad x-2=0$$
$$x=-2 \quad or \quad x=2$$

53. $x^2 - 3x - 1 = 0$
$$x = \frac{3 \pm \sqrt{(-3)^2 - 4\cdot 1\cdot(-1)}}{2\cdot 1}$$
$$x = \frac{3 \pm \sqrt{9+4}}{2}$$
$$x = \frac{3 \pm \sqrt{13}}{2}$$

55. *Writing Exercise.* The ratio b/a controls how wide open the branches of a hyperbola are. The larger this ratio, the steeper the slant of the asymptotes. For a hyperbola with a horizontal axis, the steeper the slant the more wide open the branches. For a hyperbola with a vertical axis, the steeper the slant the less wide open the branches.

57. Since the intercepts are (0, 6) and (0, –6), we know that the hyperbola is of the form $\frac{y^2}{b^2} - \frac{x^2}{a^2} = 1$ and that $b = 6$. The equations of the asymptotes tell us that $b/a = 3$, so
$$\frac{6}{a} = 3$$
$$a = 2.$$

The equation is $\frac{y^2}{6^2} - \frac{x^2}{2^2} = 1$, or $\frac{y^2}{36} - \frac{x^2}{4} = 1$.

59.
$$\frac{(x-5)^2}{36} - \frac{(y-2)^2}{25} = 1$$
$$\frac{(x-5)^2}{6^2} - \frac{(y-2)^2}{5^2} = 1$$

$h = 5,\ k = 2,\ a = 6,\ b = 5$

Center: (5, 2)

Vertices: (5 – 6, 2) and (5 + 6, 2), or (–1, 2) and (11, 2)

Asymptotes: $y-2 = \frac{5}{6}(x-5)$ and $y-2 = -\frac{5}{6}(x-5)$

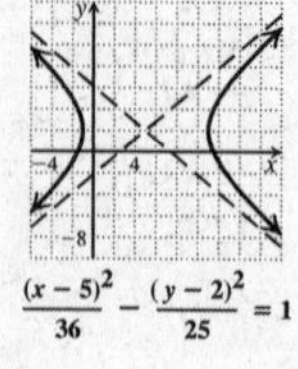

$\frac{(x-5)^2}{36} - \frac{(y-2)^2}{25} = 1$

61.
$$8(y+3)^2 - 2(x-4)^2 = 32$$
$$\frac{(y+3)^2}{4} - \frac{(x-4)^2}{16} = 1$$
$$\frac{(y-(-3))^2}{2^2} - \frac{(x-4)^2}{4^2} = 1$$

$h = 4,\ k = -3,\ a = 4,\ b = 2$

Center: (4, –3)

Vertices: (4, –3 + 2) and (4, – 3 – 2), or (4, –1) and (4,–5)

Asymptotes: $y-(-3) = \frac{2}{4}(x-4)$ and $y-(-3) = -\frac{2}{4}(x-4)$, or $y+3 = \frac{1}{2}(x-4)$ and $y+3 = -\frac{1}{2}(x-4)$

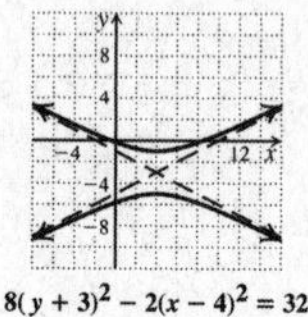

$8(y+3)^2 - 2(x-4)^2 = 32$

63.
$$4x^2 - y^2 + 24x + 4y + 28 = 0$$
$$4(x^2+6x) - (y^2-4y) = -28$$
$$4(x^2+6x+9-9) - (y^2-4y+4-4) = -28$$
$$4(x^2+6x+9) - (y^2-4y+4) = -28 + 4\cdot 9 - 4$$
$$4(x+3)^2 - (y-2)^2 = 4$$
$$\frac{(x+3)^2}{1} - \frac{(y-2)^2}{4} = 1$$
$$\frac{(x-(-3))^2}{1^2} - \frac{(y-2)^2}{2^2} = 1$$

$h = -3,\ k = 2,\ a = 1,\ b = 2$

Center: (–3, 2)

Vertices: (–3 – 1, 2), and (–3 + 1, 2), or (–4, 2) and (–2, 2)

Asymptotes: $y-2 = \frac{2}{1}(x-(-3))$ and $y-2 = -\frac{2}{1}(x-(-3))$, or $y-2 = 2(x+3)$ and $y-2 = -2(x+3)$

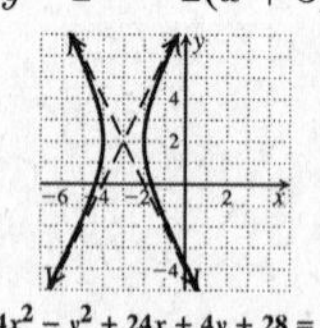

$4x^2 - y^2 + 24x + 4y + 28 = 0$

65. *Graphing Calculator Exercise*

Connecting the Concepts

1. $y = 3(x-4)^2 + 1$ parabola

Vertex: (4, 1)

Axis of symmetry: $x = 4$

3. $(x-3)^2+(y-2)^2=5$ circle

Center: (3, 2)

5. $\frac{x^2}{144}+\frac{y^2}{81}=1$ ellipse

x-intercepts: (–12, 0) and (12, 0)

y-intercepts: (0, 9) and (0, –9)

7. $4y^2-x^2=4$ hyperbola

$\frac{y^2}{1}-\frac{x^2}{4}=1$

Vertices: (0, –1) and (0, 1)

9. $x^2+y^2=36$ is a circle.

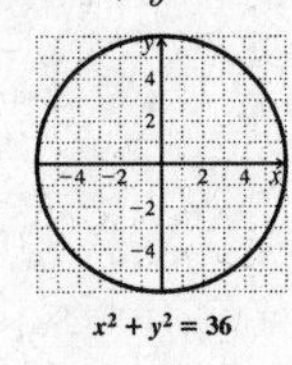

$x^2+y^2=36$

11. $\frac{x^2}{25}+\frac{y^2}{49}=1$ is an ellipse.

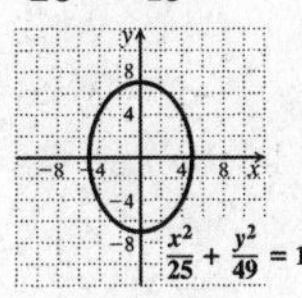

13. $x=(y+3)^2+2$ is a parabola.

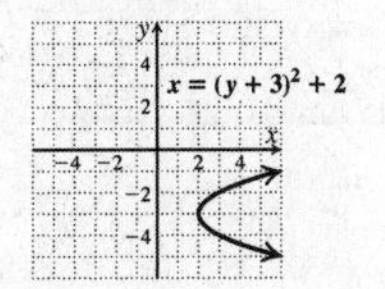

15. $xy=-4$ is a hyperbola.

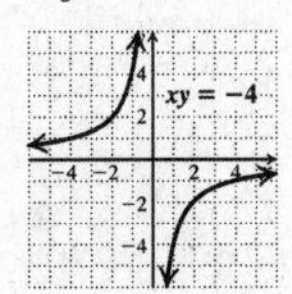

17. $x^2+y^2-8y-20=0$

$x^2+y^2-8y+16=20+16$

$x^2+(y-4)^2=36$ is a circle.

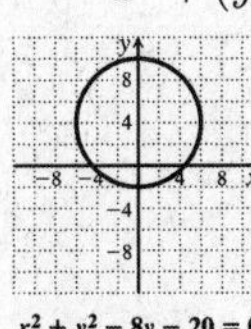

$x^2+y^2-8y-20=0$

19. $16y^2-x^2=16$

$\frac{y^2}{1}-\frac{x^2}{16}=1$ is a hyperbola.

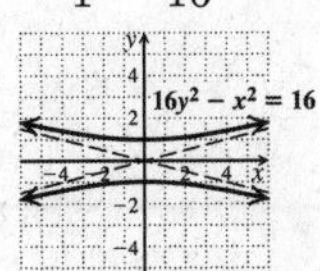

Exercise Set 13.4

1. True.

3. False; see page 881 in the text.

5. True; see page 879 in the text.

7. $x^2+y^2=41,$ (1)
$y-x=1$ (2)

First solve Equation (2) for y.

$y=x+1$ (3)

Then substitute $x+1$ for y in Equation (1) and solve for x.

$$x^2+y^2=41$$
$$x^2+(x+1)^2=41$$
$$x^2+x^2+2x+1=41$$
$$2x^2+2x-40=0$$
$$x^2+x-20=0 \quad \text{Multiplying by } \tfrac{1}{2}$$
$$(x+5)(x-4)=0$$
$$x+5=0 \quad or \quad x-4=0 \quad \text{Principle of zero products}$$
$$x=-5 \quad or \quad x=4$$

Now substitute these numbers in Equation (3) and solve for y.

For $x=-5$, $y=-5+1=-4$
For $x=4$, $y=4+1=5$

The pairs (–5, –4) and (4, 5) check, so they are the solutions.

9. $4x^2+9y^2=36,$ (1)
$3y+2x=6$ (2)

First solve Equation (2) for y.

$3y=-2x+6$
$y=-\frac{2}{3}x+2$ (3)

Then substitute $-\frac{2}{3}x+2$ for y in Equation (1) and solve for x.

$$4x^2+9y^2=36$$
$$4x^2+9\left(-\tfrac{2}{3}x+2\right)^2=36$$
$$4x^2+9\left(\tfrac{4}{9}x^2-\tfrac{8}{3}x+4\right)=36$$
$$4x^2+4x^2-24x+36=36$$
$$8x^2-24x=0$$
$$x^2-3x=0$$
$$x(x-3)=0$$

$x=0$ *or* $x=3$

Now substitute these numbers in Equation (3) and solve for y.

For $x=0$, $y=-\frac{2}{3}\cdot 0+2=2$

For $x=3$, $y=-\frac{2}{3}\cdot 3+2=0$

The pairs (0, 2) and (3, 0) check, so they are the solutions.

11. $y^2=x+3$, (1)
$2y=x+4$ (2)

First solve Equation (2) for x.

$2y-4=x$ (3)

Then substitute $2y-4$ for x in Equation (1) and solve for y.

$$y^2=x+3$$
$$y^2=(2y-4)+3$$
$$y^2=2y-1$$
$$y^2-2y+1=0$$
$$(y-1)(y-1)=0$$

$y-1=0$ *or* $y-1=0$
$y=1$ *or* $y=1$

Now substitute 1 for y in Equation (3) and solve for x.

$2\cdot 1-4=x$
$-2=x$

The pair (–2, 1) checks. It is the solution.

13. $x^2-xy+3y^2=27$, (1)
$x-y=2$ (2)

First solve Equation (2) for y.

$x-2=y$ (3)

Then substitute x - 2 for y in Equation (1) and solve for x.

$$x^2-xy+3y^2=27$$
$$x^2-x(x-2)+3(x-2)^2=27$$
$$x^2-x^2+2x+3x^2-12x+12=27$$
$$3x^2-10x-15=0$$
$$x=\frac{-(-10)\pm\sqrt{(-10)^2-4(3)(-15)}}{2\cdot 3}$$
$$x=\frac{10\pm\sqrt{100+180}}{6}=\frac{10\pm\sqrt{280}}{6}$$
$$x=\frac{10\pm 2\sqrt{70}}{6}=\frac{5\pm\sqrt{70}}{3}$$

Now substitute these numbers in Equation (3) and solve for y.

For $x=\frac{5+\sqrt{70}}{3}$, $y=\frac{5+\sqrt{70}}{3}-2=\frac{-1+\sqrt{70}}{3}$

For $x=\frac{5-\sqrt{70}}{3}$, $y=\frac{5-\sqrt{70}}{3}-2=\frac{-1-\sqrt{70}}{3}$

The pairs $\left(\frac{5+\sqrt{70}}{3}, \frac{-1+\sqrt{70}}{3}\right)$ and $\left(\frac{5-\sqrt{70}}{3}, \frac{-1-\sqrt{70}}{3}\right)$ check, so they are the solutions.

15. $x^2+4y^2=25$, (1)
$x+2y=7$ (2)

First solve Equation (2) for x.

$x=-2y+7$ (3)

Then substitute $-2y+7$ for x in Equation (1) and solve for y.

$$x^2+4y^2=25$$
$$(-2y+7)^2+4y^2=25$$
$$4y^2-28y+49+4y^2=25$$
$$8y^2-28y+24=0$$
$$2y^2-7y+6=0$$
$$(2y-3)(y-2)=0$$

$y=\frac{3}{2}$ *or* $y=2$

Now substitute these numbers in Equation (3) and solve for x.

For $y=\frac{3}{2}$, $x=-2\cdot\frac{3}{2}+7=4$

For $y=2$, $x=-2\cdot 2+7=3$

The pairs $\left(4, \frac{3}{2}\right)$ and (3, 2) check, so they are the solutions.

17. $x^2-xy+3y^2=5$, (1)
$x-y=2$ (2)

First solve Equation (2) for y.

$x-2=y$ (3)

Then substitute $x-2$ for y in Equation (1) and solve for x.

$$x^2-xy+3y^2=5$$
$$x^2-x(x-2)+3(x-2)^2=5$$
$$x^2-x^2+2x+3x^2-12x+12=5$$
$$3x^2-10x+7=0$$
$$(3x-7)(x-1)=0$$

$x=\frac{7}{3}$ *or* $x=1$

Now substitute these numbers in Equation (3) and solve for y.

For $x=\frac{7}{3}$, $y=\frac{7}{3}-2=\frac{1}{3}$

For $x=1$, $y=1-2=-1$

The pairs $\left(\frac{7}{3}, \frac{1}{3}\right)$ and (1, –1) check, so they are the solutions.

19. $3x+y=7,\quad (1)$
$4x^2+5y=24\quad (2)$

First solve Equation (1) for y.

$$y=7-3x\quad (3)$$

Then substitute $7-3x$ for y in Equation (2) and solve for x.

$$\begin{aligned}4x^2+5y&=24\\4x^2+5(7-3x)&=24\\4x^2+35-15x&=24\\4x^2-15x+11&=0\\(4x-11)(x-1)&=0\end{aligned}$$

$$x=\frac{11}{4}\quad or\quad x=1$$

Now substitute these numbers into Equation (3) and solve for y.

For $x=\frac{11}{4}$, $\quad y=7-3\cdot\frac{11}{4}=-\frac{5}{4}$
For $x=1$, $\quad y=7-3\cdot1=4$

The pairs $\left(\frac{11}{4},-\frac{5}{4}\right)$ and $(1, 4)$ check, so they are the solutions.

21. $a+b=6,\quad (1)$
$ab=8\quad (2)$

First solve Equation (1) for a.

$$a=-b+6\quad (3)$$

Then substitute $-b+6$ for a in Equation (2) and solve for b.

$$\begin{aligned}(-b+6)b&=8\\-b^2+6b&=8\\0&=b^2-6b+8\\0&=(b-2)(b-4)\end{aligned}$$

$b-2=0\quad or\quad b-4=0$
$b=2\quad or\quad b=4$

Now substitute these numbers in Equation (3) and solve for a.

For $b=2$, $\quad a=-2+6=4$
For $b=4$, $\quad a=-4+6=2$

The pairs $(4, 2)$ and $(2, 4)$ check, so they are the solutions.

23. $2a+b=1,\quad (1)$
$b=4-a^2\quad (2)$

Equation (2) is already solved for b. Substitute $4-a^2$ for b in Equation (1) and solve for a.

$$\begin{aligned}2a+4-a^2&=1\\0&=a^2-2a-3\\0&=(a-3)(a+1)\end{aligned}$$

$a=3\quad or\quad a=-1$

Substitute these numbers in Equation (2) and solve for b.

For $a=3$, $\quad b=4-3^2=-5$
For $a=-1$, $\quad b=4-(-1)^2=3$

The pairs $(3, -5)$ and $(-1, 3)$ check, so they are the solutions.

25. $a^2+b^2=89,\quad (1)$
$a-b=3\quad (2)$

First solve Equation (2) for a.

$$a=b+3\quad (3)$$

Then substitute $b+3$ for a in Equation (1) and solve for b.

$$\begin{aligned}(b+3)^2+b^2&=89\\b^2+6b+9+b^2&=89\\2b^2+6b-80&=0\\b^2+3b-40&=0\\(b+8)(b-5)&=0\end{aligned}$$

$b=-8\quad or\quad b=5$

Substitute these numbers in Equation (3) and solve for a.

For $b=-8$, $\quad a=-8+3=-5$
For $b=5$, $\quad a=5+3=8$

The pairs $(-5, -8)$ and $(8, 5)$ check, so they are the solutions.

27. $y=x^2,\quad (1)$
$x=y^2\quad (2)$

Equation (1) is already solved for y. Substitute x^2 for y in Equation (2) and solve for x.

$$\begin{aligned}x&=y^2\\x&=\left(x^2\right)^2\\x&=x^4\\0&=x^4-x\\0&=x(x^3-1)\\0&=x(x-1)(x^2+x+1)\end{aligned}$$

$x=0\quad or\quad x=1\quad or\quad x=\frac{-1\pm\sqrt{1^2-4\cdot1\cdot1}}{2}$

$x=0\quad or\quad x=1\quad or\quad x=-\frac{1}{2}\pm\frac{\sqrt{3}}{2}i$

Substitute these numbers in Equation (1) and solve for y.

For $x=0$, $\quad y=0^2=0$
For $x=1$, $\quad y=1^2=1$
For $x=-\frac{1}{2}+\frac{\sqrt{3}}{2}i$, $\quad y=\left(-\frac{1}{2}+\frac{\sqrt{3}}{2}i\right)^2=-\frac{1}{2}-\frac{\sqrt{3}}{2}i$
For $x=-\frac{1}{2}-\frac{\sqrt{3}}{2}i$, $\quad y=\left(-\frac{1}{2}-\frac{\sqrt{3}}{2}i\right)^2=-\frac{1}{2}+\frac{\sqrt{3}}{2}i$

The pairs $(0, 0)$, $(1, 1)$, $\left(-\frac{1}{2}+\frac{\sqrt{3}}{2}i,\ -\frac{1}{2}-\frac{\sqrt{3}}{2}i\right)$, and $\left(-\frac{1}{2}-\frac{\sqrt{3}}{2}i,\ -\frac{1}{2}+\frac{\sqrt{3}}{2}i\right)$ check, so they are the solutions.

29. $x^2+y^2=16$, (1)
$x^2-y^2=16$ (2)

Here we use the elimination method.

$$\begin{aligned} x^2+y^2&=16 \quad (1)\\ x^2-y^2&=16 \quad (2)\\ \hline 2x^2&=32 \quad \text{Adding}\\ x^2&=16\\ x&=\pm 4 \end{aligned}$$

If $x=4$, $x^2=16$, and if $x=-4$, $x^2=16$, so substituting 4 or –4 in Equation (2) gives us

$$\begin{aligned} x^2+y^2&=16\\ 16+y^2&=16\\ y^2&=0\\ y&=0 \end{aligned}$$

The pairs (4, 0) and (–4, 0) check. They are the solutions.

31. $x^2+y^2=25$, (1)
$xy=12$ (2)

First we solve Equation (2) for y.

$$\begin{aligned} xy&=12\\ y&=\frac{12}{x} \end{aligned}$$

Then we substitute $\frac{12}{x}$ for y in Equation (1) and solve for x.

$$\begin{aligned} x^2+y^2&=25\\ x^2+\left(\frac{12}{x}\right)^2&=25\\ x^2+\frac{144}{x^2}&=25\\ x^4+144&=25x^2 \quad \text{Multiplying by } x^2\\ x^4-25x^2+144&=0\\ u^2-25u+144&=0 \quad \text{Letting } u=x^2\\ (u-9)(u-16)&=0\\ u=9 \quad &or \quad u=16 \end{aligned}$$

We now substitute x^2 for u and solve for x.

$$\begin{aligned} x^2=9 \quad &or \quad x^2=16\\ x=\pm 3 \quad &or \quad x=\pm 4 \end{aligned}$$

Since $y=12/x$, if $x=3$, $y=4$; if $x=-3$, $y=-4$; if $x=4$, $y=3$; and if $x=-4$, $y=-3$. The pairs (3, 4), $(-3,-4)$, (4, 3), and $(-4,-3)$ check. They are the solutions.

33. $x^2+y^2=9$, (1)
$25x^2+16y^2=400$ (2)

$$\begin{aligned} -16x^2-16y^2&=-144 \quad \text{Multiplying (1) by } -16\\ 25x^2+16y^2&=400\\ \hline 9x^2&=256 \quad \text{Adding}\\ x&=\pm\frac{16}{3} \end{aligned}$$

$$\begin{aligned} \frac{256}{9}+y^2&=9 \quad \text{Substituting in (1)}\\ y^2&=9-\frac{256}{9}\\ y^2&=-\frac{175}{9}\\ y&=\pm\sqrt{-\frac{175}{9}}=\pm\frac{5\sqrt{7}}{3}i \end{aligned}$$

The pairs $\left(\frac{16}{3}, \frac{5\sqrt{7}}{3}i\right)$, $\left(\frac{16}{3}, -\frac{5\sqrt{7}}{3}i\right)$, $\left(-\frac{16}{3}, \frac{5\sqrt{7}}{3}i\right)$, and $\left(-\frac{16}{3}, -\frac{5\sqrt{7}}{3}i\right)$ check. They are the solutions.

35. $x^2+y^2=14$, (1)

$$\begin{aligned} x^2-y^2&=4 \quad (2)\\ \hline 2x^2&=18 \quad \text{Adding}\\ x^2&=9\\ x&=\pm 3 \end{aligned}$$

$$\begin{aligned} 9+y^2&=14 \quad \text{Substituting in Eq. (1)}\\ y^2&=5\\ y&=\pm\sqrt{5} \end{aligned}$$

The pairs $(-3,-\sqrt{5})$, $(-3, \sqrt{5})$, $(3,-\sqrt{5})$, and $(3, \sqrt{5})$ check. They are the solutions.

37. $x^2+y^2=10$, (1)
$xy=3$ (2)

First we solve Equation (2) for y.

$$\begin{aligned} xy&=3\\ y&=\frac{3}{x} \end{aligned}$$

Then we substitute $\frac{3}{x}$ for y in Equation (1) and solve for x.

$$\begin{aligned} x^2+y^2&=10\\ x^2+\left(\frac{3}{x}\right)^2&=10\\ x^2+\frac{9}{x^2}&=10\\ x^4+9&=10x^2 \quad \text{Multiplying by } x^2\\ x^4-10x^2+9&=0\\ u^2-10u+9&=0 \quad \text{Letting } u=x^2\\ (u-1)(u-9)&=0\\ u=1 \quad &or \quad u=9\\ x^2=1 \quad &or \quad x^2=9 \quad \text{Substitute } x^2 \text{ for } u\\ x=\pm 1 \quad &or \quad x=\pm 3 \quad \text{and solve for } x. \end{aligned}$$

$y=3/x$, so if $x=1$, $y=3$; if $x=-1$, $y=-3$; if $x=3$, $y=1$; if $x=-3$, $y=-1$. The pairs (1, 3), (–1, –3), (3, 1), and (–3, –1) check. They are the solutions.

39. $x^2+4y^2=20$, (1)
$xy=4$ (2)

First we solve Equation (2) for y.

$$y=\frac{4}{x}$$

Then we substitute $\frac{4}{x}$ for y in Equation (1) and solve for x.

$$x^2+4\left(\frac{4}{x}\right)^2=20$$
$$x^2+\frac{64}{x^2}=20$$
$$x^4+64=20x^2$$
$$x^4-20x^2+64=0$$
$$u^2-20u+64=0 \quad \text{Letting } u=x^2$$
$$(u-16)(u-4)=0$$
$$u=16 \quad or \quad u=4$$
$$x^2=16 \quad or \quad x^2=4$$
$$x=\pm 4 \quad or \quad x=\pm 2$$

$y=4/x$, so if $x=4$, $y=1$; if $x=-4$, $y=-1$; if $x=2$, $y=2$; and if $x=-2$, $y=-2$. The pairs (4, 1), (−4,−1), (2, 2), and (−2,−2) check. They are the solutions.

41. $2xy+3y^2=7$, (1)
$3xy-2y^2=4$ (2)

$$6xy+9y^2=21 \quad \text{Multiplying (1) by 3}$$
$$-6xy+4y^2=-8 \quad \text{Multiplying (2) by} -2$$
$$13y^2=13$$
$$y^2=1$$
$$y=\pm 1$$

Substitute for y in Equation (1) and solve for x.

When $y=1$: $2\cdot x\cdot 1+3\cdot 1^2=7$
$2x=4$
$x=2$

When $y=-1$: $2\cdot x\cdot(-1)+3(-1)^2=7$
$-2x=4$
$x=-2$

The pairs (2, 1) and (−2,−1) check. They are the solutions.

43. $4a^2-25b^2=0$, (1)
$2a^2-10b^2=3b+4$ (2)

$$4a^2-25b^2=0$$
$$-4a^2+20b^2=-6b-8 \quad \text{Multiplying (2) by} -2$$
$$-5b^2=-6b-8$$

$$0=5b^2-6b-8$$
$$0=(5b+4)(b-2)$$
$$b=-\frac{4}{5} \quad or \quad b=2$$

Substitute for b in Equation (1) and solve for a.

When $b=-\frac{4}{5}$: $4a^2-25\left(-\frac{4}{5}\right)^2=0$
$4a^2=16$
$a^2=4$
$a=\pm 2$

When $b=2$: $4a^2-25(2)^2=0$
$4a^2=100$
$a^2=25$
$a=\pm 5$

The pairs $\left(2,-\frac{4}{5}\right)$, $\left(-2,-\frac{4}{5}\right)$, (5, 2) and (−5, 2) check. They are the solutions.

45. $ab-b^2=-4$, (1)
$ab-2b^2=-6$ (2)

$$ab-b^2=-4$$
$$-ab+2b^2=6 \quad \text{Multiplying (2) by} -1$$
$$b^2=2$$
$$b=\pm\sqrt{2}$$

Substitute for b in Equation (1) and solve for a.

When $b=\sqrt{2}$: $a(\sqrt{2})-(\sqrt{2})^2=-4$
$a\sqrt{2}=-2$
$a=-\frac{2}{\sqrt{2}}=-\sqrt{2}$

When $b=-\sqrt{2}$: $a(-\sqrt{2})-(-\sqrt{2})^2=-4$
$-a\sqrt{2}=-2$
$a=\frac{-2}{-\sqrt{2}}=\sqrt{2}$

The pairs $(-\sqrt{2}, \sqrt{2})$ and $(\sqrt{2},-\sqrt{2})$ check. They are the solutions.

47. ***Familiarize.*** We first make a drawing. We let l and w represent the length and width, respectively.

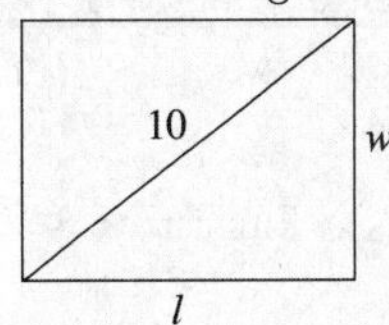

Translate. The perimeter is 28 cm.

$2l+2w=28$, or $l+w=14$

Using the Pythagorean theorem we have another equation.

$l^2+w^2=10^2$, or $l^2+w^2=100$

Carry out. We solve the system:

$l+w=14$, (1)
$l^2+w^2=100$ (2)

First solve Equation (1) for w.

$w=14-l$ (3)

Then substitute $14-l$ for w in Equation (2) and solve for l.

$$l^2+w^2=100$$
$$l^2+(14-l)^2=100$$
$$l^2+196-28l+l^2=100$$
$$2l^2-28l+96=0$$
$$l^2-14l+48=0$$
$$(l-8)(l-6)=0$$
$$l=8 \text{ or } l=6$$

If $l=8$, then $w=14-8$, or 6. If $l=6$, then $w=14-6$, or 8. Since the length is usually considered to be longer than the width, we have the solution $l=8$ and $w=6$, or $(8, 6)$.

Check. If $l=8$ and $w=6$, then the perimeter is $2\cdot 8+2\cdot 6$, or 28. The length of a diagonal is $\sqrt{8^2+6^2}$, or $\sqrt{100}$, or 10. The numbers check.

State. The length is 8 cm, and the width is 6 cm.

49. ***Familiarize.*** Let $l=$ the length and $w=$ the width of the rectangle.

Translate. The perimeter is 6 in., so we have one equation:

$$2l+2w=6\text{, or } l+w=3$$

Using the Pythagorean theorem we have another equation.

$$l^2+w^2=\left(\sqrt{5}\right)^2\text{, or } l^2+w^2=5$$

Carry out. We solve the system of equations:

$$l+w=3, \quad (1)$$
$$l^2+w^2=5 \quad (2)$$

Solve Equation (1) for l: $l=3-w$. Substitute $3-w$ for l in Equation (2) and solve for w.

$$l^2+w^2=5$$
$$(3-w)^2+w^2=5$$
$$9-6w+w^2+w^2=5$$
$$2w^2-6w+4=0$$
$$2\left(w^2-3w+2\right)=0$$
$$2(w-2)(w-1)=0$$
$$w=2 \text{ or } w=1$$

For $w=2$, $l=3-2=1$.

For $w=1$, $l=3-1=2$.

Check. The solutions are (1, 2) and (2, 1). We choose the larger number for the length. $2+1=3$, and $2^2+1^2=5$. The solution checks.

State. The length is 2 in. and the width is 1 in.

51. ***Familiarize.*** We first make a drawing. Let $l=$ the length and $w=$ the width of the cargo area, in feet.

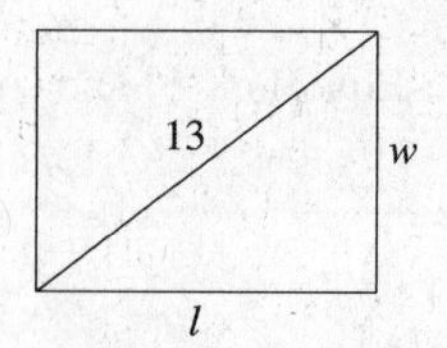

Translate. The cargo area must be 60 ft^2, so we have one equation:

$$lw=60$$

The Pythagorean equation gives us another equation:

$$l^2+w^2=13^2\text{, or } l^2+w^2=169$$

Carry out. We solve the system of equations.

$$lw=60, \quad (1)$$
$$l^2+w^2=169 \quad (2)$$

First solve Equation (1) for w:

$$lw=60$$
$$w=\tfrac{60}{l} \quad (3)$$

Then substitute $60/l$ for w in Equation (2) and solve for l.

$$l^2+w^2=169$$
$$l^2+\left(\frac{60}{l}\right)^2=169$$
$$l^2+\frac{3600}{l^2}=169$$
$$l^4+3600=169l^2$$
$$l^4-169l^2+3600=0$$

Let $u=l^2$ and $u^2=l^4$ and substitute.

$$u^2-169u+3600=0$$
$$(u-144)(u-25)=0$$
$$u=144 \quad \text{or} \quad u=25$$
$$l^2=144 \quad \text{or} \quad l^2=25 \quad \text{Replacing } u \text{ with } l^2$$
$$l=\pm 12 \quad \text{or} \quad l=\pm 5$$

Since the length cannot be negative, we consider only 12 and 5. We substitute in Equation (3) to find w. When $l=12$, $w=60/12=5$; when $l=5$, $w=60/5=12$. Since we usually consider length to be longer than width, we check the pair (12, 5).

Check. If the length is 12 ft and the width is 5 ft, then the area is $12\cdot 5$, or 60 ft^2. Also $12^2+5^2=144+25=169=13^2$. The answer checks.

State. The length is 12 ft and the width is 5 ft.

53. ***Familiarize.*** Let x and y represent the numbers.

Translate. The product of the numbers is 90, so we have

$$xy=90 \quad (1)$$

The sum of the squares of the numbers is 261, so we have

$$x^2+y^2=261 \quad (2)$$

Carry out. We solve the system of equations.

$$xy = 90, \quad (1)$$
$$x^2 + y^2 = 261 \quad (2)$$

First solve Equation (1) for y:

$$xy = 90$$
$$y = \frac{90}{x} \quad (3)$$

Then substitute $90/x$ for y in Equation (2) and solve for x.

$$x^2 + y^2 = 261$$
$$x^2 + \left(\frac{90}{x}\right)^2 = 261$$
$$x^2 + \frac{8100}{x^2} = 261$$
$$x^4 + 8100 = 261x^2$$
$$x^4 - 261x^2 + 8100 = 0$$

Let $u = x^2$ and $u^2 = x^4$ and substitute.

$$u^2 - 261u + 8100 = 0$$
$$(u-36)(u-225) = 0$$
$$u = 36 \quad or \quad u = 225$$
$$x^2 = 36 \quad or \quad x^2 = 225 \quad \text{Replacing } u \text{ with } x^2$$
$$x = \pm 6 \quad or \quad x = \pm 15$$

We use Equation (3) to find y.

When $x = 6$, $y = 90/6 = 15$;

when $x = -6$, $y = 90/(-6) = -15$;

when $x = 15$, $y = 90/15 = 6$;

when $x = -15$, $y = 90/(-15) = -6$. We see that the numbers can be 6 and 15 or -6 and -15.

Check. $6 \cdot 15 = 90$ and $6^2 + 15^2 = 261$; also $-6(-15) - 90$ and $(-6)^2 + (-15)^2 = 261$. The solutions check.

State. The numbers are 6 and 15 or -6 and -15.

55. ***Familiarize.*** We let x = the length of a side of one flower bed, in feet, and y = the length of a side of the other flower bed. Make a drawing.

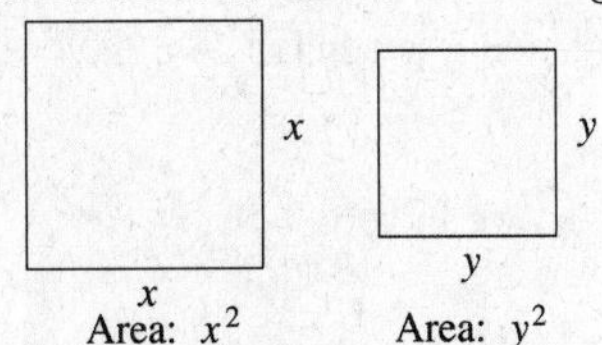

Translate. The sum of the areas is 832 ft^2, so we have

$$x^2 + y^2 = 832. \quad (1)$$

The difference of the areas is 320 ft^2, so we have

$$x^2 - y^2 = 320. \quad (2)$$

Carry out. We solve the system of equations.

$$x^2 + y^2 = 832 \quad (1)$$
$$\underline{x^2 - y^2 = 320} \quad (2)$$
$$2x^2 \quad = 1152 \quad \text{Adding}$$
$$x^2 = 576$$
$$x = \pm 24$$

Since the length cannot be negative we consider only 24. We substitute 24 for x in Equation (1) and solve for y.

$$24^2 + y^2 = 832$$
$$576 + y^2 = 832$$
$$y^2 = 256$$
$$y = \pm 16$$

Again we consider only the positive number.

Check. If the lengths of the sides of the beds are 24 ft and 16 ft, the areas of the beds are 24^2, or 576 ft^2, and 16^2, or 256 ft^2, respectively. Then $576 \text{ ft}^2 + 256 \text{ ft}^2 = 832 \text{ ft}^2$, and $576 \text{ ft}^2 - 256 \text{ ft}^2 = 320 \text{ ft}^2$, so the answer checks.

State. The lengths of the sides of the beds are 24 ft and 16 ft.

57. ***Familiarize.*** Let l = the length and w = the width of the rectangle area, in meters.

Translate. The area must be $\sqrt{3} \text{ m}^2$, so we have one equation: $lw = \sqrt{3}$

The Pythagorean equation gives us another equation:

$$l^2 + w^2 = 2^2, \text{ or } l^2 + w^2 = 4$$

Carry out. We solve the system of equations.

$$lw = \sqrt{3}, \quad (1)$$
$$l^2 + w^2 = 4 \quad (2)$$

First solve Equation (1) for l:

$$l = \frac{\sqrt{3}}{w} \quad (3)$$

Then substitute $\sqrt{3}/w$ for l in Equation (2) and solve for w.

$$l^2 + w^2 = 4$$
$$\left(\frac{\sqrt{3}}{w}\right)^2 + w^2 = 4$$
$$\frac{3}{w^2} + w^2 = 4$$
$$3 + w^4 = 4w^2$$
$$w^4 - 4w^2 + 3 = 0$$

Let $u = w^2$ and $u^2 = w^4$ and substitute.

$$u^2 - 4u + 3 = 0$$
$$(u-1)(u-3) = 0$$
$$u = 1 \quad \text{or} \quad u = 3$$
$$w^2 = 1 \quad \text{or} \quad w^2 = 3 \quad \text{Replacing } u \text{ with } w^2$$
$$w = \pm 1 \quad \text{or} \quad w = \pm\sqrt{3}$$

Since the length cannot be negative, we consider only 1 and $\sqrt{3}$. We substitute in Equation (3) to find l. When $w=\sqrt{3}$, $l=\sqrt{3}/\sqrt{3}=1$; when $w=1$, $l=\sqrt{3}/1=\sqrt{3}$. Since we usually consider length to be longer than width, we check the pair $(\sqrt{3}, 1)$.

Check. If the length is $\sqrt{3}$ m and the width is 1 m, then the area is $\sqrt{3}\cdot 1$, or $\sqrt{3}$ m^2. Also $\sqrt{3}^2+1^2=3+1=4=2^2$. The answer checks.

State. The length is $\sqrt{3}$ m and the width is 1 m.

59. *Writing Exercise.* When we can visualize the graphs of the equations in the system, we can determine how many real number solutions the system has and whether the solutions found algebraically seem reasonable.

61. $(-1)^9(-3)^2=-1\cdot 9=-9$

63. $\frac{(-1)^k}{k-6}=\frac{(-1)^7}{7-6}=\frac{-1}{1}=-1$

65. $\frac{n}{2}(3+n)=\frac{11}{2}(3+11)=\frac{11}{2}(14)=77$

67. *Writing Exercise.* Answers may vary. One possibility is given. A rectangular banner has a diagonal of 2 ft and an area of 5 ft^2. Find its dimensions.

69. Let (h, k) represent the point on the line $5x+8y=-2$ which is the center of a circle that passes through the points (–2, 3) and (–4, 1). The distance between (h, k) and (–2, 3) is the same as the distance between (h, k) and (–4, 1). This gives us one equation:

$$\sqrt{[h-(-2)]^2+(k-3)^2}=\sqrt{[h-(-4)]^2+(k-1)^2}$$
$$(h+2)^2+(k-3)^2=(h+4)^2+(k-1)^2$$
$$h^2+4h+4+k^2-6k+9=h^2+8h+16+k^2-2k+1$$
$$4h-6k+13=8h-2k+17$$
$$-4h-4k=4$$
$$h+k=-1$$

We get a second equation by substituting (h, k) in $5x+8y=-2$.

$$5h+8k=-2$$

We now solve the following system:

$$h+k=-1,$$
$$5h+8k=-2$$

The solution, which is the center of the circle, is (–2, 1). Next we find the length of the radius. We can find the distance between either (–2, 3) or (–4, 1) and the center (–2, 1). We use (–2, 3).

$$r=\sqrt{[-2-(-2)]^2+(1-3)^2}$$
$$r=\sqrt{0^2+(-2)^2}$$
$$r=\sqrt{4}=2$$

We can write the equation of the circle with center (–2, 1) and radius 2.

$$(x-h)^2+(y-k)^2=r^2$$
$$[x-(-2)]^2+(y-1)^2=2^2$$
$$(x+2)^2+(y-1)^2=4$$

71. $p^2+q^2=13,$ (1)

$\frac{1}{pq}=-\frac{1}{6}$ (2)

Solve Equation (2) for p.

$$\frac{1}{q}=-\frac{p}{6}$$
$$-\frac{6}{q}=p$$

Substitute $-6/q$ for p in Equation (1) and solve for q.

$$\left(-\frac{6}{q}\right)^2+q^2=13$$
$$\frac{36}{q^2}+q^2=13$$
$$36+q^4=13q^2$$
$$q^4-13q^2+36=0$$
$$u^2-13u+36=0 \quad \text{Letting } u=q^2$$
$$(u-9)(u-4)=0$$
$$u=9 \quad \text{or} \quad u=4$$
$$x^2=9 \quad \text{or} \quad x^2=4$$
$$x=\pm 3 \quad \text{or} \quad x=\pm 2$$

Since $p=-6/q$, if $q=3$, $p=-2$; if $q=-3$, $p=2$; if $q=2$, $p=-3$; and if $q=-2$, $p=3$. The pairs (–2, 3), (2,–3), (–3, 2), and (3,–2) check. They are the solutions.

73. ***Familiarize.*** Let $l=$ the length of the rectangle, in feet, and let $w=$ the width.

Translate. 100 ft of fencing is used, so we have

$l+w=100$. (1)

The area is 2475 ft^2, so we have

$lw=2475$. (2)

Carry out. Solving the system of equations, we get (55, 45) and (45, 55). Since length is usually considered to be longer than width, we have $l=55$ and $w=45$.

Check. If the length is 55 ft and the width is 45 ft, then $55+45$, or 100 ft, of fencing is used. The area is $55\cdot 45$, or 2475 ft^2. The answer checks.

State. The length of the rectangle is 55 ft, and the width is 45 ft.

75. ***Familiarize.*** We let x and y represent the length and width of the base of the box, in inches, respectively. Make a drawing.

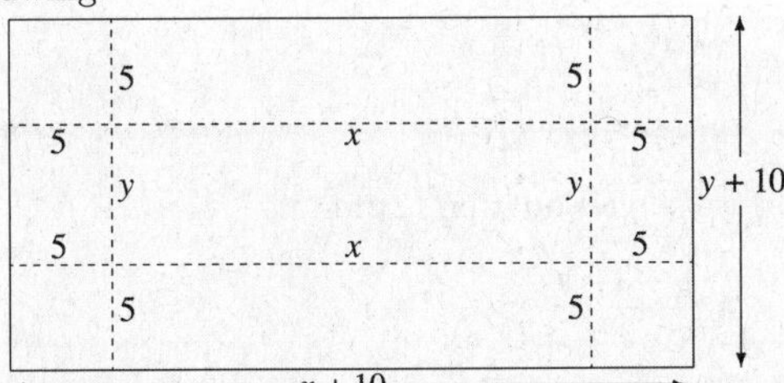

The dimensions of the metal sheet are $x + 10$ and $y + 10$.

Translate. The area of the sheet of metal is 340 in^2, so we have

$(x+10)(y+10)=340$. (1)

The volume of the box is 350 in^3, so we have

$x \cdot y \cdot 5 = 350$. (2)

Carry out. Solving the system of equations, we get (10, 7) and (7, 10). Since length is usually considered to be longer than width, we have $l = 10$ and $w = 7$.

Check. The dimensions of the metal sheet are 10 + 10, or 20, and 7 + 10, or 17, so the area is $20 \cdot 17$, or 340 in^2. The volume of the box is $7 \cdot 10 \cdot 5$, or 350 in^3. The answer checks.

State. The dimensions of the box are 10 in. by 7 in. by 5 in.

77. ***Familiarize.*** Let $l =$ the length, and $h =$ the height, in inches.

Translate. Since the ratio of the length to height is 16 to 9, we have one equation:

$$\frac{l}{h} = \frac{16}{9}$$

The Pythagorean equation gives us a second equation:

$$l^2 + h^2 = 73^2$$

We have a system of equations.

$$\frac{l}{h} = \frac{16}{9}$$
$$l^2 + h^2 = 5329$$

Carry out. Solving the system of equations, we get (35.8, 63.6) and (–35.8, –63.6). Since the dimensions cannot be negative, we consider only (35.8, 63.6).

Check. The ratio of 63.6 to 35.8 is $\frac{63.6}{35.8} \approx 1.777 \approx \frac{16}{9}$.

Also, $(63.6)^2 + (35.8)^2 = 5326.6 \approx 5329$. The answer checks.

State. The length is about 63.6 in., and the height is about 35.8 in.

79. *Graphing Calculator Exercise*

Chapter 13 Review

1. True; see page 856 in the text.

3. False; see page 863 in the text.

5. True; see page 874 in the text.

7. False; see page 881 in the text.

9.
$$(x+3)^2 + (y-2)^2 = 16$$
$$[x-(-3)]^2 + (y-2)^2 = 16 \quad \text{Standard form}$$

The center is (–3, 2) and the radius is 4.

11.
$$x^2 + y^2 - 6x - 2y + 1 = 0$$
$$x^2 - 6x + y^2 - 2y = -1$$
$$(x^2 - 6x + 9) + (y^2 - 2y + 1) = -1 + 9 + 1$$
$$(x-3)^2 + (y-1)^2 = 9$$
$$(x-3)^2 + (y-1)^2 = 3^2$$

The center is (3, 1) and the radius is 3.

13.
$$(x-h)^2 + (y-k)^2 = r^2$$
$$[x-(-4)]^2 + (y-3)^2 = 4^2$$
$$(x+4)^2 + (y-3)^2 = 16$$

15. Circle

$$5x^2 + 5y^2 = 80$$
$$x^2 + y^2 = 16$$

Center: (0, 0); radius: 4

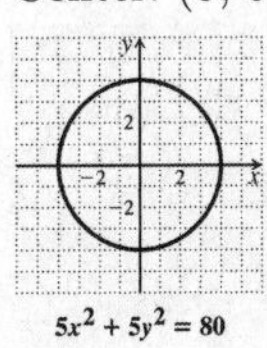

17. Parabola

$$y = -x^2 + 2x - 3$$
$$x = -\frac{b}{2a} = -\frac{2}{2(-1)} = 1$$
$$y = -x^2 + 2x - 3 = -1^2 + 2 \cdot 1 - 3 = -2$$

The vertex is (1, –2). The graph opens downward because the coefficient of x^2, –1 is negative.

x	y
–1	–6
0	–3
1	–2
2	–3
3	–6

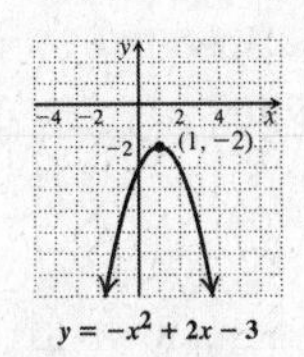

19. Hyperbola

$xy=9$

$y=\frac{9}{x}$ Solving for y

x	y
1	9
3	3
9	1
$\frac{1}{2}$	18
-1	-9
-3	-3
-9	-1
$-\frac{1}{2}$	-18

Note that we cannot use 0 for x. The x-axis and the y-axis are the asymptotes.

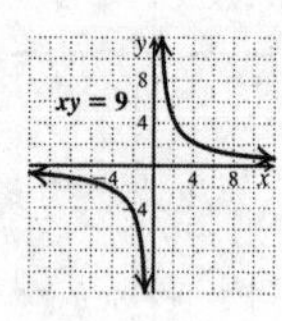

21. Ellipse

$\frac{(x+1)^2}{3}+(y-3)^2=1$ 3>1 ellipse is horizontal

The center is (–1, 3). Note $a=\sqrt{3}$ and $b=1$.

Vertices: (–1, 2), (–1, 4), $\left(-1-\sqrt{3},\ 3\right)$, $\left(-1+\sqrt{3},\ 3\right)$

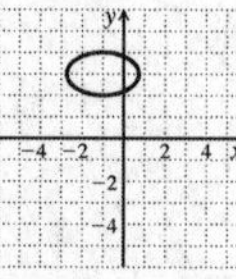

$\frac{(x+1)^2}{3}+(y-3)^2=1$

23. $x^2-y^2=21$, (1)

$x+y=3$ (2)

First we solve Equation (2) for y.

$y=-x+3$ (3)

Then we substitute $-x+3$ for y in Equation (1) and solve for x.

$$x^2-(-x+3)^2=21$$
$$x^2-x^2+6x-9=21$$
$$6x=30$$
$$x=5$$

Substitute 5 for x in Equation (3).

$y=-5+3=-2$

The solution is (5, –2).

25. $x^2-y=5$, (1)

$2x-y=5$ (2)

First we solve Equation (2) for y.

$y=2x-5$ (3)

Here we multiply Equation (2) by –1 and then add.

$$x^2-y=5$$
$$-2x+y=-5 \quad \text{Multiplying by } -1$$
$$x^2-2x=0 \quad \text{Adding}$$
$$x(x-2)=0$$

$x=0$ or $x=2$

Now we substitute these numbers for x in Equation (3).

If $x=0$, $y=2\cdot 0-5=-5$

If $x=2$, $y=2\cdot 2-5=-1$

The solutions are (0, –5) and (2, –1).

27. $x^2-y^2=3$, (1)

$y=x^2-3$ (2)

First we solve Equation (2) for x.

$$y+3=x^2$$
$$\pm\sqrt{y+3}=x \quad (3)$$

Adding Equations (1) and (2).

$$x^2-y^2=3 \quad (1)$$
$$-x^2+y=-3 \quad (2)$$
$$-y^2+y=0 \quad \text{Adding}$$
$$-y(y-1)=0$$

$y=0$ or $y=1$

Now we substitute these numbers for y in Equation (3).

For $y=0$, $x=\pm\sqrt{0+3}=\pm\sqrt{3}$

For $y=1$, $x=\pm\sqrt{1+3}=\pm\sqrt{4}=\pm 2$

The solutions are $\left(\sqrt{3},\ 0\right)$, $\left(-\sqrt{3},\ 0\right)$, (–2, 1) and (2, 1).

29. $x^2+y^2=100$, (1)

$2x^2-3y^2=-120$ (2)

We use the elimination method.

$$-2x^2-2y^2=-200 \quad \text{Multiplying (1) by } -2$$
$$2x^2-3y^2=-120 \quad (2)$$
$$-5y^2=-320 \quad \text{Adding}$$
$$y^2=64$$
$$y=\pm 8$$

Solve Equation (1) for x.

$$x^2+y^2=100$$
$$x^2=100-y^2$$
$$x=\pm\sqrt{100-y^2} \quad (3)$$

Since x is solved in terms of y^2, we need only substitute once in Equation (3).

For $y^2=64$, $x=\pm\sqrt{100-64}=\pm\sqrt{36}=\pm 6$

The solutions are (6, 8), (6, –8), (–6, 8) and (–6, –8).

31. ***Familiarize.*** Let l = the length and w = the width of the bandstand.

Translate.

Perimeter: $2l+2w=38$, or $l+w=19$

Area: $lw=84$

Carry out. We solve the system:

Solve the first equation for l: $l=19-w$.

Substitute $19-w$ for l in the second equation and solve for w.

$$(19-w)w=84$$
$$19w-w^2=84$$
$$0=w^2-19w+84$$
$$0=(w-7)(w-12)$$
$$w=7 \quad or \quad w=12$$

If $w = 7$, then $l = 19 - 7$, or 12. If $w = 12$, then $l = 19 - 12$, or 7. Since length is usually considered to be longer than width, we have the solution $l = 12$ and $w = 7$, or (12, 7)

Check. If $l = 12$ and $w = 7$, the area is $12 \cdot 7$, or 84. The perimeter is $2 \cdot 12 + 2 \cdot 7$, or 38. The numbers check.

State. The length is 12 m and the width is 7 m.

33. ***Familiarize.*** Let x represent the length of a side of one square mirror and y represent the length of a side of the other square mirror.

Translate.

$4x = 4y + 12$, or $x = y + 3$

$x^2 = y^2 + 39$

Carry out. We solve the system of equations.

$$x = y + 3 \quad (1)$$
$$x^2 = y^2 + 39 \quad (2)$$

We substitute $y + 3$ for x into Equation (2).

$$(y+3)^2 = y^2 + 39$$
$$y^2 + 6y + 9 = y^2 + 39$$
$$6y + 9 = 39$$
$$6y = 30$$
$$y = 5$$

Then $x = 5 + 3 = 8$.

Check. If $x = 8$ and $y = 5$, then $4 \cdot 5 + 12 = 32 = 4 \cdot 8$, and $5^2 + 39 = 64 = 8^2$. The numbers check. The perimeter of the first mirror is $4 \cdot 8$, or 32, and the perimeter of the second mirror is $4 \cdot 5$, or 20.

State. The perimeter of each mirror is 32 cm and 20 cm, respectively.

35. *Writing Exercise.* The graph of a parabola has one branch whereas the graph of a hyperbola has two branches. A hyperbola has asymptotes, but a parabola does not.

37. $4x^2 - x - 3y^2 = 9, \quad (1)$

$-x^2 + x + y^2 = 2 \quad (2)$

We use the elimination method.

$$\begin{array}{rl} 4x^2 - x - 3y^2 = 9 & (1) \\ -3x^2 + 3x + 3y^2 = 6 & \text{Multiplying (2) by 3} \\ \hline x^2 + 2x = 15 & \text{Adding} \end{array}$$
$$x^2 + 2x - 15 = 0$$
$$(x+5)(x-3) = 0$$
$$x = -5 \quad or \quad x = 3$$

Solving Equation (2) for y:

$$-x^2 + x + y^2 = 2$$
$$y^2 = x^2 - x + 2$$
$$y = \pm\sqrt{x^2 - x + 2}$$

For $x = -5$, $y = \pm\sqrt{(-5)^2 - (-5) + 2} = \pm\sqrt{32} = \pm 4\sqrt{2}$

For $x = 3$, $y = \pm\sqrt{3^2 - 3 + 2} = \pm\sqrt{8} = \pm 2\sqrt{2}$

The solutions are $(-5, -4\sqrt{2})$, $(-5, 4\sqrt{2})$, $(3, -2\sqrt{2})$, $(3, 2\sqrt{2})$.

39. The three points are equidistant from the center of the circle, (h, k). Using each of the three points in the equation of a circle, we get three different equations.

For (–2, –4), $[x-(-2)]^2 + [y-(-4)]^2 = r^2$

$$(x+2)^2 + (y+4)^2 = r^2$$
$$x^2 + 4x + 4 + y^2 + 8y + 16 = r^2$$
$$x^2 + 4x + y^2 + 8y + 20 = r^2 \quad (1)$$

For (5, –5), $(x-5)^2 + [y-(-5)]^2 = r^2$

$$(x-5)^2 + (y+5)^2 = r^2$$
$$x^2 - 10x + 25 + y^2 + 10y + 25 = r^2$$
$$x^2 - 10x + y^2 + 10y + 50 = r^2 \quad (2)$$

For (6, 2), $(x-6)^2 + (y-2)^2 = r^2$

$$x^2 - 12x + 36 + y^2 - 4y + 4 = r^2$$
$$x^2 - 12x + y^2 - 4y + 40 = r^2 \quad (3)$$

Since the radius is equal, we can set Equation (1) equal to Equation (2) and simplify.

$$x^2 + 4x + y^2 + 8y + 20 = x^2 - 10x + y^2 + 10y + 50$$
$$14x - 2y = 30$$
$$7x - y = 15 \quad (4)$$

Next, we set Equation (1) equal to Equation (3).

$$x^2 + 4x + y^2 + 8y + 20 = x^2 - 12x + y^2 - 4y + 40$$
$$16x + 12y = 20$$
$$4x + 3y = 5 \quad (5)$$

We solve the system of Equations (4) and (5) using the elimination method.

$$\begin{array}{rl} 21x - 3y = 45 & \text{Multiplying (4) by 3} \\ 4x + 3y = 5 & (5) \\ \hline 25x \quad\quad = 50 & \text{Adding} \end{array}$$
$$x = 2$$
$$y = -1$$

The center of the circle is (2, –1). Thus the equation is $(x-2)^2 + (y+1)^2 = r^2$.

We may choose any of the three points on the circle to determine r^2.

$$(6-2)^2+(2+1)^2=r^2$$
$$4^2+3^2=r^2$$
$$25=r^2$$

The equation of the circle is $(x-2)^2+(y+1)^2=25$.

41. Let $(x, 0)$ be the point on the x-axis that is equidistant from (–3, 4) and (5, 6).

$$\sqrt{[x-(-3)]^2+(0-4)^2}=\sqrt{(x-5)^2+(0-6)^2}$$
$$x^2+6x+9+16=x^2-10x+25+36$$
$$16x=36$$
$$x=\frac{9}{4}$$

The point is $\left(\frac{9}{4},\ 0\right)$.

Chapter 13 Test

1. For circle with center (3, –4) and radius $2\sqrt{3}$,

$$(x-3)^2+[y-(-4)]^2=(2\sqrt{3})^2$$
$$(x-3)^2+(y+4)^2=12$$

3.
$$x^2+y^2+4x-6y+4=0$$
$$x^2+4x+y^2-6y=-4$$
$$(x^2+4x+4)+(y^2-6y+9)=-4+4+9$$
$$(x+2)^2+(y-3)^2=3^2$$

The center is (–2, 3) and the radius is 3.

5. Circle

$$x^2+y^2+2x+6y+6=0$$
$$x^2+2x+y^2+6y=-6$$
$$(x^2+2x+1)+(y^2+6y+9)=-6+1+9$$
$$(x+1)^2+(y+3)^2=4$$

The center is (–1, –3) and the radius is 2.

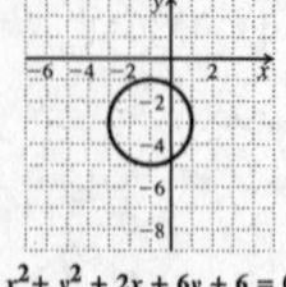

7. Ellipse

$16x^2+4y^2=64$ 4<16 ellipse is vertical

$$\frac{x^2}{4}+\frac{y^2}{16}=1$$

The center is (0, 0). Note $a=2$ and $b=4$.

Vertices: (0, 2), (0, –2), (4, 0), and (–4, 0)

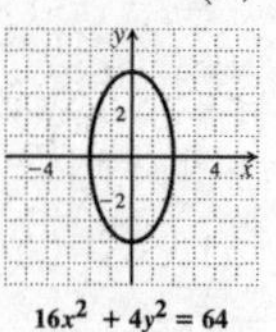

9. Parabola

$$x=-y^2+4y$$
$$y=-\frac{b}{2a}=-\frac{4}{2(-1)}=2$$
$$x=-y^2+4y=-2^2+4\cdot 2=4$$

The vertex is (4, 2). The graph opens to the left because the coefficient of y^2, –1 is negative.

x	y
0	0
3	1
4	2
3	3
0	4

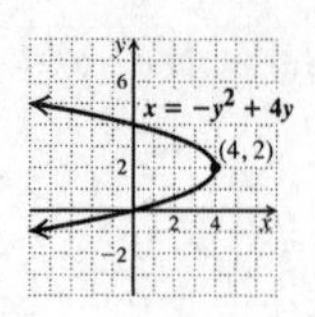

11. $x^2-y=3$, (1)
$2x+y=5$ (2)

Use the elimination method.

$$x^2-y=3 \quad (1)$$
$$2x+y=5 \quad (2)$$
$$x^2+2x=8 \quad \text{Adding}$$
$$x^2+2x-8=0$$
$$(x+4)(x-2)=0$$

$x=-4$ *or* $x=2$

Substitute for x in Equation (2).

$y=5-2x$ Solving Eq. (2) for y

For $x=-4$, $y=5-2(-4)=13$
For $x=2$, $y=5-2(2)=1$

The pairs (–4, 13) and (2, 1) check. They are the solutions.

13. $x^2+y^2=10$, (1)
$x^2=y^2+2$ (2)

Substitute y^2+2 for x^2 in Equation (1) and solve for y.

$$x^2+y^2=10$$
$$y^2+2+y^2=10$$
$$2y^2-8=0$$
$$2(y^2-4)=0$$
$$2(y+2)(y-2)=0$$

$y=-2$ *or* $y=2$

Now substitute these numbers in Equation (2) and solve for x.

For $y=-2$, $x^2=(-2)^2+2=6$, so $x=\pm\sqrt{6}$
For $y=2$, $x^2=2^2+2=6$, so $x=\pm\sqrt{6}$

The pairs $(-\sqrt{6},-2)$, $(-\sqrt{6},\ 2)$, $(-\sqrt{6},\ 2)$, $(\sqrt{6},\ 2)$ check, so they are the solutions.

15. ***Familiarize.*** We let x = the length of a side of one square, in meters, and y = the length of a side of the other square. Make a drawing.

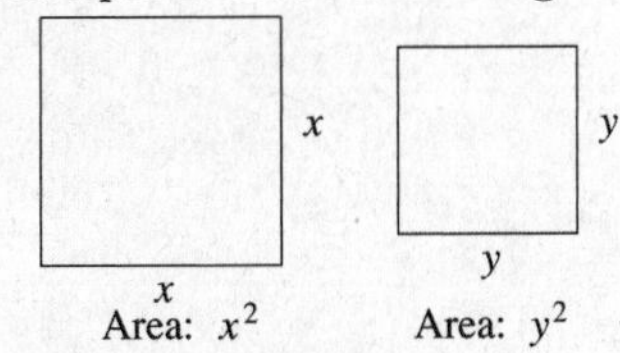

Translate. The sum of the areas is 8 m^2, so we have

$x^2+y^2=8$. (1)

The difference of the areas is 2 m^2, so we have

$x^2-y^2=2$. (2)

Carry out. We solve the system of equations.

$$\begin{aligned} x^2+y^2&=8 && (1)\\ x^2-y^2&=2 && (2)\\ \hline 2x^2\quad&=10 && \text{Adding}\\ x^2&=5\\ x&=\pm\sqrt{5} \end{aligned}$$

Since the length cannot be negative we consider only $\sqrt{5}$. We substitute $\sqrt{5}$ for x in Equation (1) and solve for y.

$$\begin{aligned} (\sqrt{5})^2+y^2&=8\\ 5+y^2&=8\\ y^2&=3\\ y&=\pm\sqrt{3} \end{aligned}$$

Again we consider only the positive number.

Check. If the lengths of the sides of the squares are $\sqrt{5}$ m and $\sqrt{3}$ m, the areas of the squares are $(\sqrt{5})^2$, or 5 m^2, and $(\sqrt{3})^2$, or 3 ft^2, respectively. Then $5\text{ m}^2+3\text{ m}^2=8\text{ m}^2$, and $5\text{ m}^2-3\text{ m}^2=2\text{ m}^2$, so the answer checks.

State. The lengths of the sides of the squares are $\sqrt{5}$ m and $\sqrt{3}$ m.

17. ***Familiarize.*** Let p = the principal and r = the interest rate. We recall the formula $I=prt$. Since the time is one year, t = 1, and the formula simplifies to $I=pr$.

Translate.

Brett invested p dollars at interest rate r with \$72 in interest.

Erin invested $p+240$ dollars at $\frac{5}{6}r$ interest rate with \$72 in interest. We have two equations.

$$\begin{aligned} 72&=pr && (1)\\ 72&=(p+240)\frac{5}{6}r && (2) \end{aligned}$$

Carry out. We solve the system of equations using substitution.

We solve Equation (1) for r: $r=\frac{72}{p}$.

We substitute $\frac{72}{p}$ for r in Equation (2) and solve for p.

$$\begin{aligned} 72&=(p+240)\frac{5}{6}\left(\frac{72}{p}\right)\\ 72&=(p+240)\frac{60}{p}\\ 72p&=60p+14{,}400\\ 12p&=14{,}400\\ p&=1200\\ r&=\frac{72}{1200}=0.06\text{ or }6\% \end{aligned}$$

Check. $1200\cdot 0.06=72$ and $(1200+240)\frac{5}{6}\cdot 0.06=72$, so the numbers check.

State. The principal was \$1200 and the interest rate was 6%.

19. Let $(0,\ y)$ be the point on the y-axis which is equidistant from $(-3,\ -5)$ and $(4,\ -7)$. We equate their distances and solve

$$\begin{aligned} \sqrt{[0-(-3)]^2+[y-(-5)]^2}&=\sqrt{(0-4)^2+[y-(-7)]^2}\\ \sqrt{9+y^2+10y+25}&=\sqrt{16+y^2+14y+49}\\ y^2+10y+34&=y^2+14y+65\\ -4y&=31\\ y&=-\frac{31}{4} \end{aligned}$$

The point on the y-axis is $\left(0,-\frac{31}{4}\right)$.

21. Let the actor be in the center at $(0,\ 0)$. Using the information, we have $(-4,\ 0)$ and $(4,\ 0)$ and $(0,\ -7)$ and $(0,\ 7)$. Thus, a = 4 and b = 7. We write the equation of the ellipse.

$$\begin{aligned} \frac{x^2}{4^2}+\frac{y^2}{7^2}&=1\\ \frac{x^2}{16}+\frac{y^2}{49}&=1 \end{aligned}$$

Chapter 14

Sequences, Series, and the Binomial Theorem

Exercise Set 14.1

1. f

3. d

5. c

7. $a_n = 5n+3$
$a_8 = 5\cdot 8+3 = 40+3 = 43$

9. $a_n = (3n+1)(2n-5)$
$a_9 = (3\cdot 9+1)(2\cdot 9-5) = 28\cdot 13 = 364$

11. $a_n = (-1)^{n-1}(3.4n-17.3)$
$a_{12} = (-1)^{12-1}[3.4(12)-17.3] = -23.5$

13. $a_n = 3n^2(9n-100)$
$a_{11} = 3\cdot 11^2(9\cdot 11-100) = 3\cdot 121(-1) = -363$

15. $a_n = \left(1+\frac{1}{n}\right)^2$

$a_{20} = \left(1+\frac{1}{20}\right)^2 = \left(\frac{21}{20}\right)^2 = \frac{441}{400}$

17. $a_n = 3n-1$
$a_1 = 3\cdot 1-1 = 2$
$a_2 = 3\cdot 2-1 = 5$
$a_3 = 3\cdot 3-1 = 8$
$a_4 = 3\cdot 4-1 = 11$
$a_{10} = 3\cdot 10-1 = 29$
$a_{15} = 3\cdot 15-1 = 44$

19. $a_n = n^2+2$
$a_1 = 1^2+2 = 3$
$a_2 = 2^2+2 = 6$
$a_3 = 3^2+2 = 11$
$a_4 = 4^2+2 = 18$
$a_{10} = 10^2+2 = 102$
$a_{15} = 15^2+2 = 227$

21. $a_n = \frac{n}{n+1}$

$a_1 = \frac{1}{1+1} = \frac{1}{2}$

$a_2 = \frac{2}{2+1} = \frac{2}{3}$

$a_3 = \frac{3}{3+1} = \frac{3}{4}$

$a_4 = \frac{4}{4+1} = \frac{4}{5}$

$a_{10} = \frac{10}{10+1} = \frac{10}{11}$

$a_{15} = \frac{15}{15+1} = \frac{15}{16}$

23. $a_n = \left(-\frac{1}{2}\right)^{n-1}$

$a_1 = \left(-\frac{1}{2}\right)^{1-1} = 1$

$a_2 = \left(-\frac{1}{2}\right)^{2-1} = -\frac{1}{2}$

$a_3 = \left(-\frac{1}{2}\right)^{3-1} = \frac{1}{4}$

$a_4 = \left(-\frac{1}{2}\right)^{4-1} = -\frac{1}{8}$

$a_{10} = \left(-\frac{1}{2}\right)^{10-1} = -\frac{1}{512}$

$a_{15} = \left(-\frac{1}{2}\right)^{15-1} = \frac{1}{16,384}$

25. $a_n = \frac{(-1)^n}{n}$

$a_1 = \frac{(-1)^1}{1} = -1$

$a_2 = \frac{(-1)^2}{2} = \frac{1}{2}$

$a_3 = \frac{(-1)^3}{3} = -\frac{1}{3}$

$a_4 = \frac{(-1)^4}{4} = \frac{1}{4}$

$a_{10} = \frac{(-1)^{10}}{10} = \frac{1}{10}$

$a_{15} = \frac{(-1)^{15}}{15} = -\frac{1}{15}$

27. $a_n = (-1)^n (n^3 - 1)$
$a_1 = (-1)^1 (1^3 - 1) = 0$
$a_2 = (-1)^2 (2^3 - 1) = 7$
$a_3 = (-1)^3 (3^3 - 1) = -26$
$a_4 = (-1)^4 (4^3 - 1) = 63$
$a_{10} = (-1)^{10} (10^3 - 1) = 999$
$a_{15} = (-1)^{15} (15^3 - 1) = -3374$

29. 2, 4, 6, 8, 10,...
These are even integers beginning with 2, so the general term could be $2n$.

31. -1, 1, -1, 1,...
-1 and 1 alternate, beginning with -1, so the general term could be $(-1)^n$.

33. 1, -2, 3, -4,...
These are the first four natural numbers, but with alternating signs, beginning with a positive number. The general term could be $(-1)^{n+1} \cdot n$.

35. 3, 5, 7, 9,...
These are odd integers beginning with 3, so the general term could be $2n + 1$.

37. 0, 3, 8, 15, 24,...
We can see a pattern if we write the sequence $1^2 - 1$, $2^2 - 1$, $3^2 - 1$, $4^2 - 1$, $5^2 - 1$,... The general term could be $n^2 - 1$, or $(n+1)(n-1)$.

39. $\frac{1}{2}, \frac{2}{3}, \frac{3}{4}, \frac{4}{5}, \frac{5}{6}, ...$
These are fractions in which the denominator is 1 greater than the numerator. Also, each numerator is 1 greater than the preceding numerator. The general term could be $\frac{n}{n+1}$.

41. 0.1, 0.01, 0.001, 0.0001,...
This is negative powers of 10, or positive powers of 0.1, so the general term is 10^{-n}, or $(0.1)^n$.

43. -1, 4, -9, 16,...
This is the squares of the first four natural numbers, but with alternating signs, beginning with a negative number. The general terms could be $(-1)^n \cdot n^2$.

45. -1, 2, -3, 4, -5, 6,...
$S_{10} = -1 + 2 - 3 + 4 - 5 + 6 - 7 + 8 - 9 + 10 = 5$

47. $1, \frac{1}{10}, \frac{1}{100}, \frac{1}{1000}, ...$
$S_6 = 1 + \frac{1}{10} + \frac{1}{100} + \frac{1}{1000} + \frac{1}{10,000} + \frac{1}{100,000} = 1.11111$

49.
$$\sum_{k=1}^{5} \frac{1}{2k} = \frac{1}{2\cdot1} + \frac{1}{2\cdot2} + \frac{1}{2\cdot3} + \frac{1}{2\cdot4} + \frac{1}{2\cdot5}$$
$$= \frac{1}{2} + \frac{1}{4} + \frac{1}{6} + \frac{1}{8} + \frac{1}{10}$$
$$= \frac{60}{120} + \frac{30}{120} + \frac{20}{120} + \frac{15}{120} + \frac{12}{120}$$
$$= \frac{137}{120}$$

51.
$$\sum_{k=0}^{4} 10^k = 10^0 + 10^1 + 10^2 + 10^3 + 10^4$$
$$= 1 + 10 + 100 + 1000 + 10,000$$
$$= 11,111$$

53.
$$\sum_{k=2}^{8} \frac{k}{k-1}$$
$$= \frac{2}{2-1} + \frac{3}{3-1} + \frac{4}{4-1} + \frac{5}{5-1} + \frac{6}{6-1} + \frac{7}{7-1} + \frac{8}{8-1}$$
$$= \frac{2}{1} + \frac{3}{2} + \frac{4}{3} + \frac{5}{4} + \frac{6}{5} + \frac{7}{6} + \frac{8}{7}$$
$$= \frac{1343}{140}$$

55.
$$\sum_{k=1}^{8} (-1)^{k+1} 2^k$$
$$= (-1)^{1+1} 2^1 + (-1)^{2+1} 2^2 + (-1)^{3+1} 2^3 + (-1)^{4+1} 2^4$$
$$+ (-1)^{5+1} 2^5 + (-1)^{6+1} 2^6 + (-1)^{7+1} 2^7 + (-1)^{8+1} 2^8$$
$$= 2 - 4 + 8 - 16 + 32 - 64 + 128 - 256$$
$$= -170$$

57.
$$\sum_{k=0}^{5} (k^2 - 2k + 3)$$
$$= (0^2 - 2\cdot0 + 3) + (1^2 - 2\cdot1 + 3) + (2^2 - 2\cdot2 + 3)$$
$$+ (3^2 - 2\cdot3 + 3) + (4^2 - 2\cdot4 + 3) + (5^2 - 2\cdot5 + 3)$$
$$= 3 + 2 + 3 + 6 + 11 + 18$$
$$= 43$$

59.
$$\sum_{k=3}^{5} \frac{(-1)^k}{k(k+1)} = \frac{(-1)^3}{3(3+1)} + \frac{(-1)^4}{4(4+1)} + \frac{(-1)^5}{5(5+1)}$$
$$= \frac{-1}{3\cdot4} + \frac{1}{4\cdot5} + \frac{-1}{5\cdot6}$$
$$= -\frac{1}{12} + \frac{1}{20} - \frac{1}{30}$$
$$= -\frac{4}{60} = -\frac{1}{15}$$

61. $\frac{2}{3} + \frac{3}{4} + \frac{4}{5} + \frac{5}{6} + \frac{6}{7}$
This is a sum of fractions in which the denominator is one greater than the numerator. Also, each numerator is 1 greater than the preceding numerator. Sigma notation is
$$\sum_{k=1}^{5} \frac{k+1}{k+2}.$$

63. $1+4+9+16+25+36$

This is the sum of the squares of the first six natural numbers. Sigma notation is

$$\sum_{k=1}^{6} k^2.$$

65. $4-9+16-25+...+(-1)^n n^2$

This is a sum of terms of the form $(-1)^k k^2$, beginning with $k=2$ and continuing through $k=n$. Sigma notation is

$$\sum_{k=2}^{n} (-1)^k k^2.$$

67. $6+12+18+24+...$

This is the sum of all the positive multiples of 6. It is an infinite series. Sigma notation is

$$\sum_{k=1}^{\infty} 6k.$$

69. $\frac{1}{1\cdot 2}+\frac{1}{2\cdot 3}+\frac{1}{3\cdot 4}+\frac{1}{4\cdot 5}+...$

This is a sum of fractions in which the numerator is 1 and the denominator is a product of two consecutive integers. The larger integer in each product is the smaller integer in the succeeding product. It is an infinite series. Sigma notation is

$$\sum_{k=1}^{\infty} \frac{1}{k(k+1)}.$$

71. *Writing Exercise.* The graph of *f* is a set of points $(x,\ x^2)$ where x is a natural number. The graph of $y=x^2$ for $x>0$ is formed by connecting these points with a smooth curve. The graph of $y=x^2$ also contains the points $(x,\ x^2)$ for $x\le 0$.

73. $\frac{7}{2}(a_1+a_7)=\frac{7}{2}(8+20)=\frac{7}{2}(28)=98$

75. $(a_1+3d)+d=a_1+3d+d=a_1+4d$

77. $(a_1+a_n)+(a_1+a_n)+(a_1+a_n)$
$=3(a_1+a_n)=3a_1+3a_n$

79. *Writing Exercise.*

$$\sum_{k=1}^{n}(a_k+b_k)=(a_1+b_1)+(a_2+b_2)+...+(a_n+b_n)$$
$$=(a_1+a_2+...+a_n)+(b_1+b_2+...+b_n)$$

Using the commutative and associative laws of addition

$$=\sum_{k=1}^{n} a_k+\sum_{k=1}^{n} b_k$$

81. $a_1=1,\ a_{n+1}=5a_n-2$

$a_1=1$
$a_2=5\cdot 1-2=3$
$a_3=5\cdot 3-2=13$
$a_4=5\cdot 13-2=63$
$a_5=5\cdot 63-2=313$
$a_6=5\cdot 313-2=1563$

83. Find each term by multiplying the preceding term by 0.80: \$2500, \$2000, \$1600, \$1280, \$1024, \$819.20, \$655.36, \$524.29, \$419.43, \$335.54.

85. $a_n=(-1)^n$

This sequence is of the form –1, 1, –1, 1, ... Each pair of terms adds to 0. S_{100} has 50 such pairs, so $S_{100}=0$. S_{101} consists of the 50 pairs in S_{100} that add to 0 as well as a_{101}, or –1, so $S_{101}=-1$.

87. $a_n=i^n$

$a_1=i^1=1$
$a_2=i^2=-1$
$a_3=i^3=i^2\cdot i=-1\cdot i=-i$
$a_4=i^4=(i^2)^2=(-1)^2=1$
$a_5=i^5=(i^2)^2\cdot i=(-1)^2\cdot i=1\cdot i=i$
$S_5=i-1-i+1+i=i$

89. Enter $y_1=x^5-14x^4+6x^3+416x^2-655x-1050$. Then scroll through a table of values. We see that $y_1=6144$ when $x=11$, so the 11th term of the sequence is 6144.

Exercise Set 14.2

1. True; see page 903 in the text.

3. False; see page 903 in the text.

5. True; see page 905 in the text.

7. False; $S_5=a_1+a_2+a_3+a_4+a_5$.

9. 8, 13, 18, 23, ...

$a_1 = 8$

$d = 5$ (13 – 8 = 5, 18 – 13 = 5, 23 – 18 = 5)

11. 7, 3, –1, –5, ...

$a_1 = 7$

$d = -4$ (3 – 7 = –4, –1 – 3 = –4, –5 – (–1) = –4)

13. $\frac{3}{2}, \frac{9}{4}, 3, \frac{15}{4}, \ldots$

$a_1 = \frac{3}{2}$

$d = \frac{3}{4}$ $\left(\frac{9}{4} - \frac{3}{2} = \frac{3}{4},\ 3 - \frac{9}{4} = \frac{3}{4}\right)$

15. \$8.16, \$8.46, \$8.76, \$9.06, ...

$a_1 = \$8.16$

$d = \$0.30$ (\$8.46 – \$8.16 = \$0.30,
\$8.76 – \$8.46 = \$0.30,
\$9.06 – \$8.76 = \$0.30)

17. 10, 18, 26, ...

$a_1 = 10,\ d = 8, \text{ and } n = 19$
$a_n = a_1 + (n-1)d$
$a_{19} = 10 + (19-1)8 = 10 + 18 \cdot 8 = 10 + 144 = 154$

19. 8, 2, –4, ...

$a_1 = 8,\ d = -6, \text{ and } n = 18$
$a_n = a_1 + (n-1)d$
$a_{18} = 8 + (18-1)(-6) = 8 + 17(-6) = 8 - 102 = -94$

21. \$1200, \$964.32, \$728.64, ...

$a_1 = \$1200,\ d = \$964.32 - \$1200 = -\$235.68$ and $n = 13$
$a_n = a_1 + (n-1)d$
$a_{13} = \$1200 + (13-1)(-\$235.68)$
$= \$1200 + 12(-\$235.68) = \$1200 - \$2828.16$
$= -\$1628.16$

23. $a_1 = 10,\ d = 8$
$a_n = a_1 + (n-1)d$

Let $a_n = 210$, and solve for n.

$210 = 10 + (n-1)8$
$210 = 10 + 8n - 8$
$210 = 2 + 8n$
$208 = 8n$
$26 = n$

The 26th term is 210.

25. $a_1 = 8,\ d = -6$
$a_n = a_1 + (n-1)d$
$-328 = 8 + (n-1)(-6)$
$-328 = 8 - 6n + 6$
$-328 = 14 - 6n$
$-342 = -6n$
$57 = n$

The 57th term is –328.

27. $a_n = a_1 + (n-1)d$
$a_{18} = 8 + (18-1)10$ Substituting 18 for n, 8 for a_1, and 10 for d
$= 8 + 17 \cdot 10$
$= 8 + 170$
$= 178$

29. $a_n = a_1 + (n-1)d$
$33 = a_1 + (8-1)4$ Substituting 33 for a_8, 8 for n, and 4 for d
$33 = a_1 + 28$
$5 = a_1$

31. $a_n = a_1 + (n-1)d$
$-76 = 5 + (n-1)(-3)$ Substituting -76 for a_n, 5 for a_1, and -3 for d
$-76 = 5 - 3n + 3$
$-76 = 8 - 3n$
$-84 = -3n$
$28 = n$

33. We know that $a_{17} = -40$ and $a_{28} = -73$. We would have to add d eleven times to get from a_{17} to a_{28}. That is,

$-40 + 11d = -73$
$11d = -33$
$d = -3.$

Since $a_{17} = -40$, we subtract d sixteen times to get a_1.

$a_1 = -40 - 16(-3) = -40 + 48 = 8$

We write the first five terms of the sequence:

8, 5, 2, –1, –4.

35. $a_{13} = 13$ and $a_{54} = 54$

Observe that for this to be true, $a_1 = 1$ and $d = 1$.

37. $1 + 5 + 9 + 13 + \ldots$

Note that $a_1 = 1,\ d = 4$, and $n = 20$. Before using the formula for S_n, we find a_{20}:

$a_{20} = 1 + (20-1)4$ Substituting into the formula for a_n
$= 1 + 19 \cdot 4$
$= 77$

Then using the formula for S_n,

$$S_{20} = \frac{20}{2}(1 + 77) = 10(78) = 780.$$

39. The sum is $1+2+3+\ldots+249+250$. This is the sum of the arithmetic sequence for which $a_1=1$, $a_n=250$, and $n=250$. We use the formula for S_n.

$$S_n=\frac{n}{2}(a_1+a_n)$$
$$S_{250}=\frac{250}{2}(1+250)=125(251)=31{,}375$$

41. The sum is $2+4+6+\ldots+98+100$. This is the sum of the arithmetic sequence for which $a_1=2$, $a_n=100$, and $n=50$. We use the formula for S_n.

$$S_n=\frac{n}{2}(a_1+a_n)$$
$$S_{50}=\frac{50}{2}(2+100)=25(102)=2550$$

43. The sum is $6+12+18+\ldots+96+102$. This is the sum of the arithmetic sequence for which $a_1=6$, $a_n=102$, and $n=17$. We use the formula for S_n.

$$S_n=\frac{n}{2}(a_1+a_n)$$
$$S_{17}=\frac{17}{2}(6+102)=\frac{17}{2}(108)=918$$

45. Before using the formula for S_n, we find a_{20}:

$$a_{20}=4+(20-1)5 \quad \text{Substituting into the formula for } a_n$$
$$=4+19\cdot 5$$
$$=99$$

Then using the formula for S_n,

$$S_{20}=\frac{20}{2}(4+99)=10(103)=1030.$$

47. ***Familiarize.*** We want to find the fifteenth term and the sum of an arithmetic sequence with $a_1=7$, $d=2$, and $n=15$. We will first use the formula for a_n to find a_{15}. This result is the number of musicians in the last row. Then we will use the formula for S_n to find S_{15}. This is the total number of musicians .

Translate. Substituting into the formula for a_n, we have

$$a_{15}=7+(15-1)2.$$

Carry out. We first find a_{15}.

$$a_{15}=7+14\cdot 2=35$$

Then use the formula for S_n to find S_{15}.

$$S_{15}=\frac{15}{2}(7+35)=\frac{15}{2}(42)=315$$

Check. We can do the calculations again. We can also do the entire addition.

$$7+9+11+\ldots+35.$$

State. There are 35 musicians in the last row, and there are 315 musicians altogether.

49. ***Familiarize.*** We want to find the sum of the arithmetic sequence $36+32+\ldots+4$. Note that $a_1=36$, and $d=-4$. We will first use the formula for a_n to find n. Then we will use the formula for S_n.

Translate. Substituting into the formula for a_n, we have

$$4=36+(n-1)(-4).$$

Carry out. We solve for n.

$$4=36+(n-1)(-4)$$
$$4=36-4n+4$$
$$4=40-4n$$
$$-36=-4n$$
$$9=n$$

Now we find S_9.

$$S_9=\frac{9}{2}(36+4)=\frac{9}{2}(40)=180$$

Check. We can do the calculations again. We can also do the entire addition.

$$36+32+\ldots+4.$$

State. There are 180 stones in the pyramid.

51. ***Familiarize.*** We want to find the sum of the arithmetic sequence with $a_1=10¢$, $d=10¢$, and $n=31$. First we will find a_{31} and then we will find S_{31}.

Translate. Substituting in the formula for a_n, we have

$$a_{31}=10+(31-1)(10).$$

Carry out. First we find a_{31}.

$$a_{31}=10+30\cdot 10=10+300=310$$

Then we use the formula for S_n to find S_{31}.

$$S_{31}=\frac{31}{2}(10+310)=\frac{31}{2}(320)=4960$$

Check. We can do the calculations again.

State. The amount saved is 4960¢, or \$49.60.

53. ***Familiarize.*** We want to find the sum of an arithmetic sequence with $a_1=20$, $d=2$, and $n=16$. We will use the formula for a_n to find a_{16}, and then we will use the formula for S_n to find S_{16}.

Translate. Substituting into the formula for a_n, we have $a_{16}=20+(16-1)2$.

Carry out. We find a_{16}.

$$a_{16}=20+15\cdot 2=50$$

Then we use the formula for S_n to find S_{16}.

$$S_{16}=\frac{16}{2}(20+50)=560$$

Check. We do the calculations again.

State. There are 560 seats.

55. *Writing Exercise.*

$$\begin{aligned}&1+2+3+...+100\\&=(1+100)+(2+99)+(3+98)+...+(50+51)\\&=\underbrace{101+101+101+...+101}_{50 \text{ addends of } 101}\\&=50\cdot 101\\&=5050\end{aligned}$$

57. Using the slope-intercept form, where $m=\frac{1}{3}$ and $b = 10$, we have $y=\frac{1}{3}x+10$.

59. Rewrite the equation.

$$\begin{aligned}2x+y&=8\\y&=-2x+8\end{aligned}$$

The slope of the parallel line is –2.

Use point-slope form.

$$\begin{aligned}y-y_1&=m(x-x_1)\\y-0&=-2(x-5)\\y&=-2x+10\end{aligned}$$

61. A circle with center (0, 0) and radius 4 is $x^2+y^2=16$.

63. *Writing Exercise.* To explain why S_n is always an integer, we recall how S_n was first developed.

$S_n=a_1+(a_1+d)+(a_1+2d)+...+(a_n-2d)+(a_n-d)+a_n$

Since a_1, d and a_n are all integers, the sum is also an integer.

65. The frog climbs 4 – 1, or 3 ft, with each jump. Then the total distance the frog has jumped with each successive jump is given by the arithmetic sequence 3, 6, 9, ..., 96. When the frog has climbed 96 ft, it will reach the top of the hole on the next jump because it will have climbed $96+4$, or 100 ft with that jump. Then the total number of jumps is the number of terms of the sequence above plus the final jump. We find n for the sequence with $a_1=3$, $d=3$, and $a_n=96$:

$$\begin{aligned}a_n&=a_1+(n-1)d\\96&=3+(n-1)3\\96&=3+3n-3\\96&=3n\\32&=n\end{aligned}$$

The total number of jumps is 32 + 1, or 33 jumps.

67. Let d = the common difference. Since p, m, and q form an arithmetic sequence, $m = p + d$, and $q = p + 2d$. Then

$$\frac{p+q}{2}=\frac{p+(p+2d)}{2}=p+d=m.$$

69. Each integer from 501 through 750 is 500 more than the corresponding integer from 1 through 250. There are 250 integers from 501 through 750, so their sum is the sum of the integers from 1 to 250 plus $250\cdot 500$. From Exercise 39, we know that the sum of the integers from 1 through 250 is 31,375. Thus, we have

$$31{,}375+250\cdot 500, \text{ or } 156{,}375.$$

Exercise Set 14.3

1. $\frac{a_{n+1}}{a_n}=2$, so this is a geometric sequence.

3. $a_{n+1}=a_n-3$, so this is a arithmetic sequence.

5. $\frac{a_{n+1}}{a_n}=5$, so this is a geometric series.

7. $\frac{a_{n+1}}{a_n}=-\frac{1}{2}$, so this is a geometric series.

9. 10, 20, 40, 80, ...

$\frac{20}{10}=2,\ \frac{40}{20}=2,\ \frac{80}{40}=2$

$r = 2$

11. 6, –0.6, 0.06, –0.006, ...

$-\frac{0.6}{6}=-0.1,\ \frac{0.06}{-0.6}=-0.1,\ \frac{-0.006}{0.06}=-0.1$

$r = -0.1$

13. $\frac{1}{2},\ -\frac{1}{4},\ \frac{1}{8},\ -\frac{1}{16},...$

$\frac{-\frac{1}{4}}{\frac{1}{2}}=-\frac{1}{4}\cdot\frac{2}{1}=-\frac{2}{4}=-\frac{1}{2},\ \frac{\frac{1}{8}}{-\frac{1}{4}}=\frac{1}{8}\cdot\left(-\frac{4}{1}\right)=-\frac{4}{8}=-\frac{1}{2},$

$\frac{-\frac{1}{16}}{\frac{1}{8}}=-\frac{1}{16}\cdot\frac{8}{1}=-\frac{8}{16}=-\frac{1}{2}$

$r=-\frac{1}{2}$

15. 75, 15, 3, $\frac{3}{5}$, ...

$\frac{15}{75}=\frac{1}{5},\ \frac{3}{15}=\frac{1}{5},\ \frac{\frac{3}{5}}{3}=\frac{3}{5}\cdot\frac{1}{3}=\frac{1}{5}$

$r=\frac{1}{5}$

17. $\frac{1}{m}, \frac{6}{m^2}, \frac{36}{m^3}, \frac{216}{m^4},...$

$\frac{\frac{6}{m^2}}{\frac{1}{m}} = \frac{6}{m^2}\cdot\frac{m}{1} = \frac{6}{m}, \quad \frac{\frac{36}{m^3}}{\frac{6}{m^2}} = \frac{36}{m^3}\cdot\frac{m^2}{6} = \frac{6}{m}$

$\frac{\frac{216}{m^4}}{\frac{36}{m^3}} = \frac{216}{m^4}\cdot\frac{m^3}{36} = \frac{6}{m}$

$r = \frac{6}{m}$

19. 2, 6, 18, ...

$a_1 = 2,\ n = 7,$ and $r = \frac{6}{2} = 3$

We use the formula $a_n = a_1r^{n-1}$.

$a_7 = 2\cdot 3^{7-1} = 2\cdot 3^6 = 2\cdot 729 = 1458$

21. $\sqrt{3},\ 3,\ 3\sqrt{3},...$

$a_1 = \sqrt{3},\ n = 10,$ and $r = \frac{3\sqrt{3}}{3} = \sqrt{3}$

$a_n = a_1r^{n-1}$

$a_{10} = \sqrt{3}\left(\sqrt{3}\right)^{10-1} = \sqrt{3}\left(\sqrt{3}\right)^9 = \left(\sqrt{3}\right)^{10} = 243$

23. $-\frac{8}{243}, \frac{8}{81}, -\frac{8}{27},...$

$a_1 = -\frac{8}{243},\ n = 14,$ and $r = \frac{\frac{8}{81}}{-\frac{8}{243}} = \frac{8}{81}\left(-\frac{243}{8}\right) = -3$

$a_n = a_1r^{n-1}$

$a_{14} = -\frac{8}{243}(-3)^{14-1} = -\frac{8}{243}(-3)^{13}$
$= -\frac{8}{243}(-1{,}594{,}323) = 52{,}488$

25. \$1000, \$1040, \$1081.60, ...

$a_1 = \$1000,\ n = 10,$ and $r = \frac{1040}{1000} = 1.04$

$a_n = a_1r^{n-1}$

$a_{10} = \$1000(1.04)^{10-1} \approx \$1000(1.423311812) \approx \$1423.31$

27. 1, 5, 25, 125,...

$a_1 = 1,$ and $r = \frac{5}{1} = 5$

$a_n = a_1r^{n-1}$

$a_n = 1\cdot 5^{n-1} = 5^{n-1}$

29. $1,\ -1,\ 1,\ -1,...$

$a_1 = 1,$ and $r = \frac{-1}{1} = -1$

$a_n = a_1r^{n-1}$

$a_n = 1(-1)^{n-1} = (-1)^{n-1}$

31. $\frac{1}{x}, \frac{1}{x^2}, \frac{1}{x^3},...$

$a_1 = \frac{1}{x},$ and $r = \frac{\frac{1}{x^2}}{\frac{1}{x}} = \frac{1}{x^2}\cdot\frac{x}{1} = \frac{1}{x}$

$a_n = a_1r^{n-1}$

$a_n = \frac{1}{x}\left(\frac{1}{x}\right)^{n-1} = \frac{1}{x}\cdot\frac{1}{x^{n-1}} = \frac{1}{x^{1+n-1}} = \frac{1}{x^n},$ or x^{-n}

33. $6 + 12 + 24 + ...$

$a_1 = 6,\ n = 9,$ and $r = \frac{12}{6} = 2$

$S_n = \frac{a_1\left(1 - r^n\right)}{1 - r}$

$S_9 = \frac{6\left(1 - 2^9\right)}{1 - 2} = \frac{6(1 - 512)}{-1} = \frac{6(-511)}{-1} = 3066$

35. $\frac{1}{18} - \frac{1}{6} + \frac{1}{2} - ...$

$a_1 = \frac{1}{18},\ n = 7,$ and $r = \frac{-\frac{1}{6}}{\frac{1}{18}} = -\frac{1}{6}\cdot\frac{18}{1} = -3$

$S_n = \frac{a_1\left(1 - r^n\right)}{1 - r}$

$S_7 = \frac{\frac{1}{18}\left[1 - (-3)^7\right]}{1 - (-3)} = \frac{\frac{1}{18}(1 + 2187)}{4} = \frac{\frac{1}{18}(2188)}{4}$
$= \frac{1}{18}(2188)\left(\frac{1}{4}\right) = \frac{547}{18}$

37. $1 + x + x^2 + x^3 + ...$

$a_1 = 1,\ n = 8,$ and $r = \frac{x}{1},$ or x

$S_n = \frac{a_1\left(1 - r^n\right)}{1 - r}$

$S_8 = \frac{1\left(1 - x^8\right)}{1 - x} = \frac{\left(1 + x^4\right)\left(1 - x^4\right)}{1 - x}$
$= \frac{\left(1 + x^4\right)\left(1 + x^2\right)\left(1 - x^2\right)}{1 - x}$
$= \frac{\left(1 + x^4\right)\left(1 + x^2\right)(1 + x)(1 - x)}{1 - x}$
$= \left(1 + x^4\right)\left(1 + x^2\right)(1 + x)$

39. $\$200 + \$200(1.06) + \$200(1.06)^2 + ...$

$a_1 = \$200,\ n = 16,$ and $r = \frac{\$200(1.06)}{\$200} = 1.06$

$S_n = \frac{a_1\left(1 - r^n\right)}{1 - r}$

$S_{16} = \frac{\$200\left[1 - (1.06)^{16}\right]}{1 - 1.06} \approx \frac{\$200(1 - 2.540351685)}{-0.06}$
$\approx \$5134.51$

41. $18 + 6 + 2 + \ldots$

$|r| = \left|\frac{6}{18}\right| = \left|\frac{1}{3}\right| = \frac{1}{3}$, and since $|r| < 1$, the series does have a limit.

$$S_\infty = \frac{a_1}{1-r} = \frac{18}{1-\frac{1}{3}} = \frac{18}{\frac{2}{3}} = 18 \cdot \frac{3}{2} = 27$$

43. $7 + 3 + \frac{9}{7} + \ldots$

$|r| = \left|\frac{3}{7}\right| = \frac{3}{7}$, and since $|r| < 1$, the series does have a limit.

$$S_\infty = \frac{a_1}{1-r} = \frac{7}{1-\frac{3}{7}} = \frac{7}{\frac{4}{7}} = 7 \cdot \frac{7}{4} = \frac{49}{4}$$

45. $3 + 15 + 75 + \ldots$

$|r| = \left|\frac{15}{3}\right| = |5| = 5$, and since $|r| \not< 1$, the series does not have a limit.

47. $4 - 6 + 9 - \frac{27}{2} + \ldots$

$|r| = \left|\frac{-6}{4}\right| = \left|-\frac{3}{2}\right| = \frac{3}{2}$, and since $|r| \not< 1$, the series does not have a limit.

49. $0.43 + 0.0043 + 0.000043 + \ldots$

$|r| = \left|\frac{0.0043}{0.43}\right| = |0.01| = 0.01$, and since $|r| < 1$, the series does have a limit.

$$S_\infty = \frac{a_1}{1-r} = \frac{0.43}{1-0.01} = \frac{0.43}{0.99} = \frac{43}{99}$$

51. $\$500(1.02)^{-1} + \$500(1.02)^{-2} + \$500(1.02)^{-3} + \ldots$

$|r| = \left|\frac{\$500(1.02)^{-2}}{\$500(1.02)^{-1}}\right| = \left|(1.02)^{-1}\right| = (1.02)^{-1}$, or $\frac{1}{1.02}$, and since $|r| < 1$, the series does have a limit.

$$S_\infty = \frac{a_1}{1-r} = \frac{\$500(1.02)^{-1}}{1-\left(\frac{1}{1.02}\right)} = \frac{\frac{\$500}{1.02}}{\frac{0.02}{1.02}} = \frac{\$500}{1.02} \cdot \frac{1.02}{0.02}$$

$$= \$25{,}000$$

53. $0.5555\ldots = 0.5 + 0.05 + 0.005 + 0.0005 + \ldots$

This is an infinite geometric series with $a_1 = 0.5$.

$|r| = \left|\frac{0.05}{0.5}\right| = |0.1| = 0.1 < 1$, so the series has a limit.

$$S_\infty = \frac{a_1}{1-r} = \frac{0.5}{1-0.1} = \frac{0.5}{0.9} = \frac{5}{9}$$

Fractional notation for $0.5555\ldots$ is $\frac{5}{9}$.

55. $3.4646\ldots = 3 + 0.4646\ldots$

$0.464646\ldots = 0.46 + 0.0046 + 0.000046 + \ldots$

$|r| = \left|\frac{0.0046}{0.46}\right| = |0.01| = 0.01 < 1$, so the series has a limit.

$$S_\infty = \frac{a_1}{1-r} = \frac{0.46}{1-0.01} = \frac{0.46}{0.99} = \frac{46}{99}$$

Fractional notation for $0.4646\ldots$ is $\frac{46}{99}$.

Fractional notation for $3.4646\ldots$ is $3 + \frac{46}{99} = \frac{343}{99}$.

57. $0.15151515\ldots = 0.15 + 0.0015 + 0.000015 + \ldots$

This is an infinite geometric series with $a_1 = 0.15$.

$|r| = \left|\frac{0.0015}{0.15}\right| = |0.01| = 0.01 < 1$, so the series has a limit.

$$S_\infty = \frac{a_1}{1-r} = \frac{0.15}{1-0.01} = \frac{0.15}{0.99} = \frac{15}{99} = \frac{5}{33}$$

Fractional notation for $0.15151515\ldots$ is $\frac{5}{33}$.

59. ***Familiarize.*** The rebound distances form a geometric sequence:

$$\frac{1}{4} \times 20,\ \left(\frac{1}{4}\right)^2 \times 20,\ \left(\frac{1}{4}\right)^3 \times 20, \ldots,$$

$$\text{or } 5,\ \frac{1}{4} \times 5,\ \left(\frac{1}{4}\right)^2 \times 5, \ldots$$

The height of the 6th rebound is the 6th term of the sequence.

Translate. We will use the formula $a_n = a_1 r^{n-1}$, with $a_1 = 5$, $r = \frac{1}{4}$, and $n = 6$:

$$a_6 = 5\left(\frac{1}{4}\right)^{6-1}$$

Carry out. We calculate to obtain $a_6 = \frac{5}{1024}$.

Check. We can do the calculation again.

State. It rebounds $\frac{5}{1024}$ ft the 6th time.

61. ***Familiarize.*** In one year, the population will be $100{,}000 + 0.03(100{,}000)$, or $(1.03)100{,}000$. In two years, the population will be $(1.03)100{,}000 + 0.03(1.03)100{,}000$, or $(1.03)^2 100{,}000$. Thus, the populations form a geometric sequence:

$100{,}000,\ (1.03)100{,}000,\ (1.03)^2 100{,}000, \ldots$

The population in 15 years will be the 16th term of the sequence.

Translate. We will use the formula $a_n = a_1 r^{n-1}$ with $a_1 = 100{,}000$, $r = 1.03$, and $n = 16$:

$$a_{16} = 100{,}000(1.03)^{16-1}$$

Carry out. We calculate to obtain $a_{16} \approx 155{,}797$.

Check. We can do the calculation again.

State. In 15 years the population will be about 155,797.

63. ***Familiarize.*** At the end of each minute the population is 96% of the previous population.

We have a geometric sequence:

$$5000,\ 5000(0.96),\ 5000(0.96)^2, \ldots$$

The number of fruit flies remaining alive after 15 minutes is given by the 16th term of the sequence.

Translate. We use the formula $a_n = a_1 r^{n-1}$ with $a_1 = 5000$, $r = 0.96$, and $n = 16$:

$$a_{16} = 5000(0.96)^{16-1}$$

Carry out. We calculate to obtain $a_{16} \approx 2710$.

Check. We can do the calculation again.

State. About 2710 flies will be alive after 15 min.

65. ***Familiarize.*** Each year the number of espresso-based coffees sold in the U.S. is 104% of the number sold the previous year. These numbers form a geometric sequence:

$$17,\ 17(1.04),\ 17(1.04)^2, \ldots$$

The number of espresso-based coffees sold from 2007 to 2015 is the sum of the first 9 terms of this sequence.

Translate. We use the formula $S_n = \dfrac{a_1(1-r^n)}{1-r}$ with $a_1 = 17$, $r = 1.04$ and $n = 9$.

$$S_9 = \frac{17(1-1.04^9)}{1-1.04}$$

Carry out. We use a calculator to obtain

$$S_9 \approx 179.9 \text{ billion}$$

Check. We can do the calculation again.

State. About 179.9 billion espresso-based coffees were sold from 2007 to 2015.

67. ***Familiarize.*** The lengths of the falls form a geometric sequence:

$$556,\ 556\left(\frac{3}{4}\right),\ 556\left(\frac{3}{4}\right)^2,\ 556\left(\frac{3}{4}\right)^3, \ldots$$

The total length of the first 6 falls is the sum of the first six terms of this sequence. The heights of the rebounds also form a geometric sequence:

$$556\left(\frac{3}{4}\right),\ 556\left(\frac{3}{4}\right)^2,\ 556\left(\frac{3}{4}\right)^3, \ldots \text{ or}$$

$$417,\ 417\left(\frac{3}{4}\right),\ 417\left(\frac{3}{4}\right)^2, \ldots$$

When the ball hits the ground for the 6th time, it will have rebounded 5 times. Thus the total length of the rebounds is the sum of the first five terms of this sequence.

Translate. We use the formula $S_n = \dfrac{a_1(1-r^n)}{1-r}$ twice, once with $a_1 = 556$, $r = \dfrac{3}{4}$, and $n = 6$ and a second time with $a_1 = 417$, $r = \dfrac{3}{4}$, and $n = 5$.

D = Length of falls + length of rebounds

$$= \frac{556\left[1-\left(\frac{3}{4}\right)^6\right]}{1-\frac{3}{4}} + \frac{417\left[1-\left(\frac{3}{4}\right)^5\right]}{1-\frac{3}{4}}$$

Carry out. We use a calculator to obtain $D \approx 3100.35$.

Check. We can do the calculations again.

State. The ball will have traveled about 3100.35 ft.

69. ***Familiarize.*** The heights of the stack form a geometric sequence:

$$0.02,\ 0.02(2),\ 0.02(2)^2, \ldots$$

The height of the stack after it is doubled 10 times is given by the 11th term of this sequence.

Translate. We have a geometric sequence with $a_1 = 0.02$, $r = 2$, and $n = 11$. We use the formula $a_n = a_1 r^{n-1}$.

Carry out. We substitute and calculate.

$$a_{11} = 0.02(2^{11-1})$$
$$a_{11} = 0.02(1024) = 20.48$$

Check. We can do the calculations again.

State. The final stack will be 20.48 in. high.

71. *Writing Exercise.* One circumstance in which this situation occurs is in an alternating sequence, with $a_1 > 0$ and $r > 1$. One example is the sequence

1, –2, 4, –8, 16, –32, 64, ...

73. $(x+y)^2 = (x+y)(x+y) = x^2 + 2xy + y^2$

75.
$$\begin{aligned}(x-y)^3 &= (x-y)(x-y)(x-y)\\ &= (x^2 - 2xy + y^2)(x-y)\\ &= x^3 - 3x^2y + 3xy^2 - y^3\end{aligned}$$

77.
$$\begin{aligned}(2x+y)^3 &= (2x+y)(2x+y)(2x+y)\\ &= (4x^2 + 4xy + y^2)(2x+y)\\ &= 8x^3 + 12x^2y + 6xy^2 + y^3\end{aligned}$$

79. *Writing Exercise.* Answers may vary. One possibility is given.

Casey invests \$900 at 8% interest, compounded annually. How much will be in the account at the end of 40 years?

81. $\sum_{k=1}^{\infty} 6(0.9)^k = 6(0.9)+6(0.9)^2+6(0.9)^3+...$

$|r| = \left|\frac{6(0.9)^2}{6(0.9)}\right| = |0.9| = 0.9 < 1$, so the series has a limit.

$$S_\infty = \frac{a_1}{1-r} = \frac{6(0.9)}{1-0.9} = \frac{5.4}{0.1} = 54$$

83. $x^2 - x^3 + x^4 - x^5 + ...$

This is a geometric series with $a_1 = x^2$ and $r = -x$.

$$S_n = \frac{a_1(1-r^n)}{1-r} = \frac{x^2[1-(-x)^n]}{1-(-x)} = \frac{x^2[1-(-x)^n]}{1+x}$$

85. The length of a side of the first square is 16 cm. The length of a side of the next square is the length of the hypotenuse of a right triangle with legs 8 cm and 8 cm, or $8\sqrt{2}$ cm. The length of a side of the next square is the length of the hypotenuse of a right triangle with legs $4\sqrt{2}$ cm and $4\sqrt{2}$ cm, or 8 cm. The areas of the squares form a sequence:

$(16)^2,\ (8\sqrt{2})^2,\ (8)^2,\ ...,$ or
$256,\ 128,\ 64,\ ...$

This is a geometric series with $a_1 = 256$ and $r = \frac{1}{2}$.

We find the sum of the infinite geometric series $256+128+64+...$

$$S_\infty = \frac{a_1}{1-r} = \frac{256}{1-\frac{1}{2}} = \frac{256}{\frac{1}{2}} = 512 \text{ cm}^2$$

87. *Writing Exercise.* If the graph shows that the points $(n,\ a_n)$ approach a horizontal line as n increases, then the geometric series has a limit. If this does not occur, then the series does not have a limit.

Connecting the Concepts

1. $a_n = n^2 - 5n$

$a_{20} = 20^2 - 5\cdot 20 = 300$

3. 1, 2, 3, 4, ...

Note that $a_1 = 1$, $d = 1$, and $n = 12$. Before using the formula to find S_{12}, we find a_{12}.

$$a_{12} = 1+(12-1)1 = 12$$

Then using the formula for S_n,

$$S_{12} = \frac{12}{2}(1+12) = 6(13) = 78.$$

5. $1-2+3-4+5-6 = \sum_{k=1}^{6}(-1)^{k+1}\cdot k$

7. 10, 15, 20, 25, ...

Note that $a_1 = 10$, $d = 5$, and $n = 21$.

$$a_{21} = 10+(21-1)5 = 10+100 = 110$$

9. $a_n = a_1 + (n-1)d$

$a_{25} = 9+(25-1)(-2) = 9+(24)(-2) = -39$

11. $a_n = a_1 + (n-1)d$

$$0 = 5+(n-1)\left(-\frac{1}{2}\right)$$
$$-5 = -\frac{n}{2}+\frac{1}{2}$$
$$10 = n-1$$
$$11 = n$$

13. $\frac{1}{3},\ -\frac{1}{6},\ \frac{1}{12},\ -\frac{1}{24},\ ...$

$$r = \frac{-\frac{1}{6}}{\frac{1}{3}} = -\frac{1}{6}\cdot\frac{3}{1} = -\frac{1}{2}$$

15. 2, −2, 2, −2, ...

$a_1 = 2$, and $r = \frac{-2}{2} = -1$

$a_n = a_1 r^{n-1}$

$a_n = 2(-1)^{n-1}$ or $2(-1)^{n+1}$

17. $0.9+0.09+0.009+...$

$|r| = \left|\frac{0.09}{0.9}\right| = |0.1| = 0.1 < 1$, so the series has a limit.

$$S_\infty = \frac{0.9}{1-0.1} = \frac{0.9}{0.9} = 1$$

Thus, $0.9+0.09+0.009+... = 1$.

19. \$1+\$2+\$3+\$4+...

This is an arithmetic sequence $a_1 = 1$, $d = 1$, and $n = 30$. Before using the formula to find S_{30}, we find a_{30}.

$$a_{30} = 1+(30-1)1 = 30$$

Then using the formula for S_n,

$$S_{30} = \frac{30}{2}(1+30) = 15(31) = 465.$$

She earns \$465.

Exercise Set 14.4

1. 2^5, or 32

3. 9

5. $\binom{8}{5}$, or $\binom{8}{3}$

7. 1

9. $4! = 4\cdot 3\cdot 2\cdot 1 = 24$

11. $10! = 10\cdot 9\cdot 8\cdot 7\cdot 6\cdot 5\cdot 4\cdot 3\cdot 2\cdot 1 = 3{,}628{,}800$

13. $\frac{10!}{8!} = \frac{10\cdot 9\cdot 8!}{8!} = 10\cdot 9 = 90$

15. $\frac{9!}{4!5!} = \frac{9\cdot 8\cdot 7\cdot 6\cdot 5!}{4!5!} = \frac{9\cdot 8\cdot 7\cdot 6}{4\cdot 3\cdot 2\cdot 1} = 3\cdot 7\cdot 6 = 126$

17. $\binom{10}{4} = \frac{10!}{6!4!} = \frac{10\cdot 9\cdot 8\cdot 7\cdot 6!}{6!4!} = \frac{10\cdot 9\cdot 8\cdot 7}{4\cdot 3\cdot 2\cdot 1} = 10\cdot 3\cdot 7 = 210$

19. $\binom{9}{9} = \frac{9!}{0!9!} = \frac{9!}{1\cdot 9!} = \frac{9!}{9!} = 1$

21. $\binom{30}{2} = \frac{30!}{28!2!} = \frac{30\cdot 29\cdot 28!}{28!2!} = \frac{30\cdot 29}{2\cdot 1} = 15\cdot 29 = 435$

23. $\binom{40}{38} = \frac{40!}{2!38!} = \frac{40\cdot 39\cdot 38!}{2!38!} = \frac{40\cdot 39}{2\cdot 1} = 20\cdot 39 = 780$

25. Expand $(a-b)^4$.

We have $a = a$, $b = -b$, and $n = 4$.

Form 1: We use the fifth row of Pascal's triangle:

1 4 6 4 1

$$(a-b)^4 = 1\cdot a^4 + 4a^3(-b) + 6a^2(-b)^2 + 4a(-b)^3 + 1\cdot(-b)^4$$
$$= a^4 - 4a^3b + 6a^2b^2 - 4ab^3 + b^4$$

Form 2:

$$(a-b)^4 = \binom{4}{0}a^4 + \binom{4}{1}a^3(-b) + \binom{4}{2}a^2(-b)^2 + \binom{4}{3}a(-b)^3 + \binom{4}{4}(-b)^4$$
$$= \frac{4!}{4!0!}a^4 + \frac{4!}{3!1!}a^3(-b) + \frac{4!}{2!2!}a^2(-b)^2 + \frac{4!}{1!3!}a(-b)^3 + \frac{4!}{0!4!}(-b)^4$$
$$= a^4 - 4a^3b + 6a^2b^2 - 4ab^3 + b^4$$

27. Expand $(p+q)^7$.

We have $a = p$, $b = q$, and $n = 7$.

Form 1: We use the 8th row of Pascal's triangle:

1 7 21 35 35 21 7 1

$$(p+q)^7 = p^7 + 7p^6q^1 + 21p^5q^2 + 35p^4q^3 + 35p^3q^4 + 21p^2q^5 + 7pq^6 + q^7$$

Form 2:

$$(p+q)^7 = \binom{7}{0}p^7 + \binom{7}{1}p^6q^1 + \binom{7}{2}p^5q^2 + \binom{7}{3}p^4q^3 + \binom{7}{4}p^3q^4 + \binom{7}{5}p^2q^5 + \binom{7}{6}pq^6 + \binom{7}{7}q^7$$
$$= \frac{7!}{7!0!}p^7 + \frac{7!}{6!1!}p^6q^1 + \frac{7!}{5!2!}p^5q^2 + \frac{7!}{4!3!}p^4q^3 + \frac{7!}{3!4!}p^3q^4 + \frac{7!}{2!5!}p^2q^5 + \frac{7!}{1!6!}pq^6 + \frac{7!}{0!7!}q^7$$
$$= p^7 + 7p^6q^1 + 21p^5q^2 + 35p^4q^3 + 35p^3q^4 + 21p^2q^5 + 7pq^6 + q^7$$

29. Expand $(3c-d)^7$.

We have $a = 3c$, $b = -d$, and $n = 7$.

Form 1: We use the 8th row of Pascal's triangle:

1 7 21 35 35 21 7 1

$$(3c-d)^7 = (3c)^7 + 7(3c)^6(-d)^1 + 21(3c)^5(-d)^2 + 35(3c)^4(-d)^3 + 35(3c)^3(-d)^4 + 21(3c)^2(-d)^5 + 7(3c)(-d)^6 + (-d)^7$$
$$= 2187c^7 - 5103c^6d + 5103c^5d^2 - 2835c^4d^3 + 945c^3d^4 - 189c^2d^5 + 21cd^6 - d^7$$

Form 2:

$$(3c-d)^7 = \binom{7}{0}(3c)^7 + \binom{7}{1}(3c)^6(-d)^1 + \binom{7}{2}(3c)^5(-d)^2 + \binom{7}{3}(3c)^4(-d)^3 + \binom{7}{4}(3c)^3(-d)^4 + \binom{7}{5}(3c)^2(-d)^5 + \binom{7}{6}(3c)(-d)^6 + \binom{7}{7}(-d)^7$$
$$= \frac{7!}{7!0!}(3c)^7 + \frac{7!}{6!1!}(3c)^6(-d)^1 + \frac{7!}{5!2!}(3c)^5(-d)^2 + \frac{7!}{4!3!}(3c)^4(-d)^3 + \frac{7!}{3!4!}(3c)^3(-d)^4 + \frac{7!}{2!5!}(3c)^2(-d)^5 + \frac{7!}{1!6!}(3c)(-d)^6 + \frac{7!}{0!7!}(-d)^7$$
$$= 2187c^7 - 5103c^6d + 5103c^5d^2 - 2835c^4d^3 + 945c^3d^4 - 189c^2d^5 + 21cd^6 - d^7$$

31. Expand $(t^{-2}+2)^6$.

We have $a = t^{-2}$, $b = 2$, and $n = 6$.

Form 1: We use the 7th row of Pascal's triangle:

1 6 15 20 15 6 1

$$(t^{-2}+2)^6 = 1\cdot(t^{-2})^6 + 6(t^{-2})^5(2)^1 + 15(t^{-2})^4(2)^2 + 20(t^{-2})^3(2)^3 + 15(t^{-2})^2(2)^4 + 6(t^{-2})^1(2)^5 + 1\cdot(2)^6$$
$$= t^{-12} + 12t^{-10} + 60t^{-8} + 160t^{-6} + 240t^{-4} + 192t^{-2} + 64$$

Form 2:

$$\begin{aligned}
&\left(t^{-2}+2\right)^6\\
&=\binom{6}{0}\left(t^{-2}\right)^6+\binom{6}{1}\left(t^{-2}\right)^5(2)^1+\binom{6}{2}\left(t^{-2}\right)^4(2)^2\\
&\quad+\binom{6}{3}\left(t^{-2}\right)^3(2)^3+\binom{6}{4}\left(t^{-2}\right)^2(2)^4\\
&\quad+\binom{6}{5}\left(t^{-2}\right)^1(2)^5+\binom{6}{6}(2)^6\\
&=\frac{6!}{6!0!}\left(t^{-2}\right)^6+\frac{6!}{5!1!}\left(t^{-2}\right)^5(2)^1+\frac{6!}{4!2!}\left(t^{-2}\right)^4(2)^2\\
&\quad+\frac{6!}{3!3!}\left(t^{-2}\right)^3(2)^3+\frac{6!}{2!4!}\left(t^{-2}\right)^2(2)^4\\
&\quad+\frac{6!}{1!5!}\left(t^{-2}\right)^1(2)^5+\frac{6!}{0!6!}(2)^6\\
&=t^{-12}+12t^{-10}+60t^{-8}+160t^{-6}+240t^{-4}+192t^{-2}+64
\end{aligned}$$

33. Expand $(x-y)^5$.

We have $a=x$, $b=-y$, and $n=5$.

Form 1: We use the 6th row of Pascal's triangle:

1 5 10 10 5 1

$$\begin{aligned}
&(x-y)^5\\
&=1\cdot x^5+5x^4(-y)^1+10x^3(-y)^2\\
&\quad+10x^2(-y)^3+5x^1(-y)^4+1\cdot(-y)^5\\
&=x^5-5x^4y+10x^3y^2-10x^2y^3+5xy^4-y^5
\end{aligned}$$

Form 2:

$$\begin{aligned}
(x-y)^5&=\binom{5}{0}x^5+\binom{5}{1}x^4(-y)+\binom{5}{2}x^3(-y)^2\\
&\quad+\binom{5}{3}x^2(-y)^3+\binom{5}{4}x(-y)^4+\binom{5}{5}(-y)^5\\
&=\frac{5!}{5!0!}x^5+\frac{5!}{4!1!}x^4(-y)+\frac{5!}{3!2!}x^3(-y)^2\\
&\quad+\frac{5!}{2!3!}x^2(-y)^3+\frac{5!}{1!4!}x(-y)^4+\frac{5!}{0!5!}(-y)^5\\
&=x^5-5x^4y+10x^3y^2-10x^2y^3+5xy^4-y^5
\end{aligned}$$

35. Expand $\left(3s+\frac{1}{t}\right)^9$.

We have $a=3s$, $b=\frac{1}{t}$, and $n=9$.

Form 1: We use the tenth row of Pascal's triangle:

1 9 36 84 126 126 84 36 9 1

$$\begin{aligned}
&\left(3s+\frac{1}{t}\right)^9\\
&=1\cdot(3s)^9+9(3s)^8\left(\frac{1}{t}\right)^1+36(3s)^7\left(\frac{1}{t}\right)^2+84(3s)^6\left(\frac{1}{t}\right)^3\\
&\quad+126(3s)^5\left(\frac{1}{t}\right)^4+126(3s)^4\left(\frac{1}{t}\right)^5+84(3s)^3\left(\frac{1}{t}\right)^6\\
&\quad+36(3s)^2\left(\frac{1}{t}\right)^7+9(3s)^1\left(\frac{1}{t}\right)^8+1\cdot\left(\frac{1}{t}\right)^9\\
&=19{,}683s^9+\frac{59{,}049s^8}{t}+\frac{78{,}732s^7}{t^2}+\frac{61{,}236s^6}{t^3}\\
&\quad+\frac{30{,}618s^5}{t^4}+\frac{10{,}206s^4}{t^5}+\frac{2268s^3}{t^6}+\frac{324s^2}{t^7}+\frac{27s}{t^8}+\frac{1}{t^9}
\end{aligned}$$

Form 2:

$$\begin{aligned}
&\left(3s+\frac{1}{t}\right)^9\\
&=\binom{9}{0}(3s)^9+\binom{9}{1}(3s)^8\left(\frac{1}{t}\right)^1+\binom{9}{2}(3s)^7\left(\frac{1}{t}\right)^2+\binom{9}{3}(3s)^6\left(\frac{1}{t}\right)^3\\
&\quad+\binom{9}{4}(3s)^5\left(\frac{1}{t}\right)^4+\binom{9}{5}(3s)^4\left(\frac{1}{t}\right)^5+\binom{9}{6}(3s)^3\left(\frac{1}{t}\right)^6\\
&\quad+\binom{9}{7}(3s)^2\left(\frac{1}{t}\right)^7+\binom{9}{8}(3s)^1\left(\frac{1}{t}\right)^8+\binom{9}{9}\left(\frac{1}{t}\right)^9\\
&=\frac{9!}{9!0!}(3s)^9+\frac{9!}{8!1!}(3s)^8\left(\frac{1}{t}\right)^1+\frac{9!}{7!2!}(3s)^7\left(\frac{1}{t}\right)^2+\frac{9!}{6!3!}(3s)^6\left(\frac{1}{t}\right)^3\\
&\quad+\frac{9!}{5!4!}(3s)^5\left(\frac{1}{t}\right)^4+\frac{9!}{4!5!}(3s)^4\left(\frac{1}{t}\right)^5+\frac{9!}{3!6!}(3s)^3\left(\frac{1}{t}\right)^6\\
&\quad+\frac{9!}{2!7!}(3s)^2\left(\frac{1}{t}\right)^7+\frac{9!}{1!8!}(3s)^1\left(\frac{1}{t}\right)^8+\frac{9!}{0!9!}\left(\frac{1}{t}\right)^9\\
&=19{,}683s^9+\frac{59{,}049s^8}{t}+\frac{78{,}732s^7}{t^2}+\frac{61{,}236s^6}{t^3}\\
&\quad+\frac{30{,}618s^5}{t^4}+\frac{10{,}206s^4}{t^5}+\frac{2268s^3}{t^6}+\frac{324s^2}{t^7}+\frac{27s}{t^8}+\frac{1}{t^9}
\end{aligned}$$

37. Expand $\left(x^3-2y\right)^5$.

We have $a=x^3$, $b=-2y$, and $n=5$.

Form 1: We use the 6th row of Pascal's triangle:

1 5 10 10 5 1

$$\begin{aligned}
&\left(x^3-2y\right)^5\\
&=1\cdot\left(x^3\right)^5+5\left(x^3\right)^4(-2y)^1+10\left(x^3\right)^3(-2y)^2\\
&\quad+10\left(x^3\right)^2(-2y)^3+5\left(x^3\right)^1(-2y)^4+1\cdot(-2y)^5\\
&=x^{15}-10x^{12}y+40x^9y^2-80x^6y^3+80x^3y^4-32y^5
\end{aligned}$$

Form 2:

$$\begin{aligned}
&\left(x^3-2y\right)^5\\
&=\binom{5}{0}\left(x^3\right)^5+\binom{5}{1}\left(x^3\right)^4(-2y)+\binom{5}{2}\left(x^3\right)^3(-2y)^2\\
&\quad+\binom{5}{3}\left(x^3\right)^2(-2y)^3+\binom{5}{4}\left(x^3\right)(-2y)^4+\binom{5}{5}(-2y)^5\\
&=\frac{5!}{5!0!}\left(x^3\right)^5+\frac{5!}{4!1!}\left(x^3\right)^4(-2y)+\frac{5!}{3!2!}\left(x^3\right)^3(-2y)^2\\
&\quad+\frac{5!}{2!3!}\left(x^3\right)^2(-2y)^3+\frac{5!}{1!4!}\left(x^3\right)(-2y)^4+\frac{5!}{0!5!}(-2y)^5\\
&=x^{15}-10x^{12}y+40x^9y^2-80x^6y^3+80x^3y^4-32y^5
\end{aligned}$$

39. Expand $\left(\sqrt{5}+t\right)^6$.

We have $a=\sqrt{5}$, $b=t$, and $n=6$.

Form 1: We use the 7th row of Pascal's triangle:

1 6 15 20 15 6 1

$$\begin{aligned}
&\left(\sqrt{5}+t\right)^6\\
&=1\cdot\left(\sqrt{5}\right)^6+6\left(\sqrt{5}\right)^5t^1+15\left(\sqrt{5}\right)^4t^2+20\left(\sqrt{5}\right)^3t^3\\
&\quad+15\left(\sqrt{5}\right)^2t^4+6\left(\sqrt{5}\right)^1t^5+1\cdot t^6\\
&=125+125\sqrt{5}t+375t^2+100\sqrt{5}t^3+75t^4+6\sqrt{5}t^5+t^6
\end{aligned}$$

Form 2:

$$\left(\sqrt{5}+t\right)^6$$
$$=\binom{6}{0}\left(\sqrt{5}\right)^6+\binom{6}{1}\left(\sqrt{5}\right)^5t^1+\binom{6}{2}\left(\sqrt{5}\right)^4t^2+\binom{6}{3}\left(\sqrt{5}\right)^3t^3$$
$$+\binom{6}{4}\left(\sqrt{5}\right)^2t^4+\binom{6}{5}\left(\sqrt{5}\right)^1t^5+\binom{6}{6}t^6$$
$$=\frac{6!}{6!0!}\left(\sqrt{5}\right)^6+\frac{6!}{5!1!}\left(\sqrt{5}\right)^5t^1+\frac{6!}{4!2!}\left(\sqrt{5}\right)^4t^2+\frac{6!}{3!3!}\left(\sqrt{5}\right)^3t^3$$
$$+\frac{6!}{2!4!}\left(\sqrt{5}\right)^2t^4+\frac{6!}{1!5!}\left(\sqrt{5}\right)^1t^5+\frac{6!}{0!6!}t^6$$
$$=125+125\sqrt{5}t+375t^2+100\sqrt{5}t^3+75t^4+6\sqrt{5}t^5+t^6$$

41. Expand $\left(\frac{1}{\sqrt{x}}-\sqrt{x}\right)^6$.

We have $a=\frac{1}{\sqrt{x}}$, $b=-\sqrt{x}$, and $n=6$.

Form 1: We use the 7th row of Pascal's triangle:

1 6 15 20 15 6 1

$$\left(\frac{1}{\sqrt{x}}-\sqrt{x}\right)^6$$
$$=1\cdot\left(\frac{1}{\sqrt{x}}\right)^6+6\left(\frac{1}{\sqrt{x}}\right)^5\left(-\sqrt{x}\right)^1+15\left(\frac{1}{\sqrt{x}}\right)^4\left(-\sqrt{x}\right)^2$$
$$+20\left(\frac{1}{\sqrt{x}}\right)^3\left(-\sqrt{x}\right)^3+15\left(\frac{1}{\sqrt{x}}\right)^2\left(-\sqrt{x}\right)^4$$
$$+6\left(\frac{1}{\sqrt{x}}\right)^1\left(-\sqrt{x}\right)^5+1\cdot\left(-\sqrt{x}\right)^6$$
$$=x^{-3}-6x^{-2}+15x^{-1}-20+15x-6x^2+x^3$$

Form 2:

$$\left(\frac{1}{\sqrt{x}}-\sqrt{x}\right)^6$$
$$=\binom{6}{0}\left(\frac{1}{\sqrt{x}}\right)^6+\binom{6}{1}\left(\frac{1}{\sqrt{x}}\right)^5\left(-\sqrt{x}\right)^1+\binom{6}{2}\left(\frac{1}{\sqrt{x}}\right)^4\left(-\sqrt{x}\right)^2$$
$$+\binom{6}{3}\left(\frac{1}{\sqrt{x}}\right)^3\left(-\sqrt{x}\right)^3+\binom{6}{4}\left(\frac{1}{\sqrt{x}}\right)^2\left(-\sqrt{x}\right)^4$$
$$+\binom{6}{5}\left(\frac{1}{\sqrt{x}}\right)^1\left(-\sqrt{x}\right)^5+\binom{6}{6}\left(-\sqrt{x}\right)^6$$
$$=\frac{6!}{6!0!}\left(\frac{1}{\sqrt{x}}\right)^6+\frac{6!}{5!1!}\left(\frac{1}{\sqrt{x}}\right)^5\left(-\sqrt{x}\right)^1+\frac{6!}{4!2!}\left(\frac{1}{\sqrt{x}}\right)^4\left(-\sqrt{x}\right)^2$$
$$+\frac{6!}{3!3!}\left(\frac{1}{\sqrt{x}}\right)^3\left(-\sqrt{x}\right)^3+\frac{6!}{2!4!}\left(\frac{1}{\sqrt{x}}\right)^2\left(-\sqrt{x}\right)^4$$
$$+\frac{6!}{1!5!}\left(\frac{1}{\sqrt{x}}\right)^1\left(-\sqrt{x}\right)^5+\frac{6!}{0!6!}\left(-\sqrt{x}\right)^6$$
$$=x^{-3}-6x^{-2}+15x^{-1}-20+15x-6x^2+x^3$$

43. Find the 3rd term of $(a+b)^6$.

First we note that $3 = 2 + 1$, $a = a$, $b = b$, and $n = 6$.

Then the 3rd term of the expansion of $(a+b)^6$ is

$\binom{6}{2}a^{6-2}b^2$, or $\frac{6!}{4!2!}a^4b^2$, or $15a^4b^2$.

45. Find the 12th term of $(a-3)^{14}$.

First we note that $12 = 11+1$, $a = a$, $b = -3$, and $n = 14$.

Then the 12th term of the expansion of $(a-3)^{14}$ is

$$\binom{14}{11}a^{14-11}\cdot(-3)^{11}=\frac{14!}{3!11!}a^3(-177{,}147)$$
$$=364a^3(-177{,}147)$$
$$=-64{,}481{,}508a^3$$

47. Find the 5th term of $\left(2x^3+\sqrt{y}\right)^8$.

First we note that $5 = 4+1$, $a=2x^3$, $b=\sqrt{y}$, and $n = 8$.

Then the 5th term of the expansion of $\left(2x^3+\sqrt{y}\right)^8$ is

$$\binom{8}{4}\left(2x^3\right)^{8-4}\left(\sqrt{y}\right)^4=\frac{8!}{4!4!}\left(2x^3\right)^4\left(\sqrt{y}\right)$$
$$=70\left(16x^{12}\right)\left(y^2\right)$$
$$=1120x^{12}y^2$$

49. The expansion of $\left(2u+3v^2\right)^{10}$ has 11 terms so the 6th term is the middle term. Note that $6 = 5+1$, $a=2u$, $b=-3v^2$, and $n = 10$. Then the 6th term of the expansion of $\left(2u+3v^2\right)^{10}$ is

$$\binom{10}{5}(2u)^{10-5}\left(3v^2\right)^5=\frac{10!}{5!5!}(2u)^5\left(3v^2\right)^5$$
$$=252\left(32u^5\right)\left(243v^{10}\right)$$
$$=1{,}959{,}552u^5v^{10}$$

51. The 9th term of $(x-y)^8$ is the last term, y^8.

53. *Writing Exercise.* The binomial coefficients of the first and last terms are always 1, so Maya needs only to find a^n and b^n to find these two terms. The binomial coefficients of the second term and the next-to-the-last term are always n, so to find these two terms Maya has only to compute na^{n-1} and nb^{n-1}.

55. Graph $y=x^2-5$.

This is a parabola with vertex (0, –5).

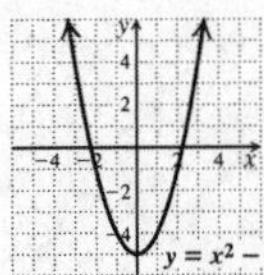

57. Graph $y\geq x-5$.

Use a solid line to form the line $y = x - 5$. Since the test point (0, 0) is a solution, shade this side of the line.

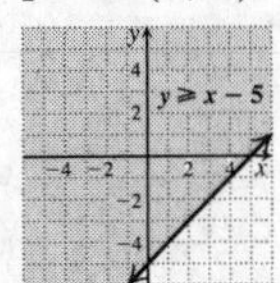

59. Graph $f(x) = \log_5 x$.

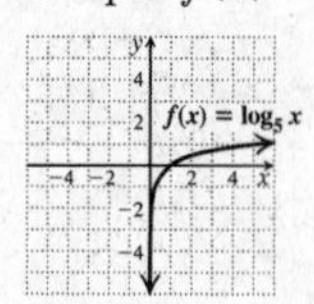

61. *Writing Exercise.* The $(r + 1)$st term of $\left(x - \frac{3}{x}\right)^{10}$ is $\binom{10}{r} x^{10-r}\left(-\frac{3}{x}\right)^r$. In the x^2-term the exponent $10 - r$ is 2 more than the exponent r, so we have:

$$\begin{aligned} 10 - r &= 2 + r \\ 8 &= 2r \\ 4 &= r \end{aligned}$$

Thus, we would find the fifth term of the expansion $\binom{10}{4} x^6 \left(-\frac{3}{x}\right)^4$.

63. Consider the set of 5 elements $\{a, b, c, d, e\}$. List all the subsets of size 3:

$\{a, b, c\}$, $\{a, b, d\}$, $\{a, b, e\}$, $\{a, c, d\}$, $\{a, c, e\}$, $\{a, d, e\}$, $\{b, c, d\}$, $\{b, c, e\}$, $\{b, d, e\}$, $\{c, d, e\}$.

There are exactly 10 subsets of size 3 and $\binom{5}{3} = 10$, so there are exactly $\binom{5}{3}$ ways of forming a subset of size 3 from a set of 5 elements.

65. Find the sixth term of $(0.15 + 0.85)^8$.

$$\binom{8}{5}(0.15)^{8-5}(0.85)^5 = \frac{8!}{3!5!}(0.15)^3(0.85)^5 \approx 0.084$$

67. Find and add the 7th through 9th terms of $(0.15 + 0.85)^9$.

$$\binom{8}{6}(0.15)^2(0.85)^6 + \binom{8}{7}(0.15)(0.85)^7 + \binom{8}{8}(0.85)^8 \approx 0.89$$

69. $\binom{n}{n-r} = \frac{n!}{[n-(n-r)]!(n-r)!} = \frac{n!}{r!(n-r)!} = \binom{n}{r}$

71.
$$\frac{\binom{5}{3}(p^2)^2\left(-\frac{1}{2}p\sqrt[3]{q}\right)^3}{\binom{5}{2}(p^2)^3\left(-\frac{1}{2}p\sqrt[3]{q}\right)^2} = \frac{-\frac{1}{8}p^7q}{\frac{1}{4}p^8\sqrt[3]{q^2}} = -\frac{\frac{1}{8}p^7q}{\frac{1}{4}p^8q^{2/3}}$$
$$= -\frac{1}{8}\cdot\frac{4}{1}\cdot p^{7-8}\cdot q^{1-2/3} = -\frac{1}{2}p^{-1}q^{1/3} = -\frac{\sqrt[3]{q}}{2p}$$

73.
$$\begin{aligned} &(x^2 + 2xy + y^2)(x^2 + 2xy + y^2)^2(x + y) \\ &= (x+y)^2\left[(x+y)^2\right]^2(x+y) = (x+y)^7 \end{aligned}$$

We can find the given product by finding the binomial expansion of $(x + y)^7$. It is (See Exercise 27.)

$$x^7 + 7x^6y + 21x^5y^2 + 35x^4y^3 + 35x^3y^4 + 21x^2y^5 + 7xy^6 + y^7.$$

Chapter 14 Review

1. False; the next term of the arithmetic sequence 10, 15, 20, ... is 20 + 5, or 25.

3. True.

5. False; a geometric sequence has a common *ratio*.

7. False; $n! = n\cdot(n-1)\cdot(n-2)\cdot\ldots\cdot3\cdot2\cdot1$.

9.
$$\begin{aligned} a_n &= 10n - 9 \\ a_1 &= 10\cdot1 - 9 = 1 \\ a_2 &= 10\cdot2 - 9 = 11 \\ a_3 &= 10\cdot3 - 9 = 21 \\ a_4 &= 10\cdot4 - 9 = 31 \\ a_8 &= 10\cdot8 - 9 = 71 \\ a_{12} &= 10\cdot12 - 9 = 111 \end{aligned}$$

11. $-5, -10, -15, -20, \ldots$

These are negative multiples of 5 beginning with –5, so the general term could be $-5n$.

13.
$$\begin{aligned} \sum_{k=1}^{5}(-2)^k &= (-2)^1 + (-2)^2 + (-2)^3 + (-2)^4 + (-2)^5 \\ &= -2 + 4 + (-8) + 16 + (-32) = -22 \end{aligned}$$

15. $7 + 14 + 21 + 28 + 35 + 42$

This is the sum of the first six positive multiples of 7. It is an finite series. Sigma notation is

$$\sum_{k=1}^{6} 7k.$$

17. –3, –7, –11, ...

$$\begin{aligned} a_1 &= -3,\ d = -4,\ \text{and } n = 14 \\ a_n &= a_1 + (n-1)d \\ a_{14} &= -3 + (14-1)(-4) = -3 + 13(-4) = -55 \end{aligned}$$

19. We know that $a_8 = 20$ and $a_{24} = 100$. We would have to add d sixteen times to get from a_8 to a_{24}. That is,

$$20 + 16d = 100$$
$$16d = 80$$
$$d = 5.$$

Since $a_8 = 20$, we subtract d seven times to get a_1.

$$a_1 = 20 - 7(5) = 20 - 35 = -15$$

21. The sum is $5 + 10 + 15 + ... + 495 + 500$. This is the sum of the arithmetic sequence for which $a_1 = 5$, $a_n = 500$, and $n = 100$. We use the formula for S_n.

$$S_n = \frac{n}{2}(a_1 + a_n)$$
$$S_{100} = \frac{100}{2}(5 + 500) = 50(505) = 25{,}250$$

23. $r = \frac{30}{40} = \frac{3}{4}$

25. $3, \frac{3}{4}x, \frac{3}{16}x^2, ...$

$a_1 = 3$, and $r = \frac{\frac{3}{4}x}{3} = \frac{3x}{4} \cdot \frac{1}{3} = \frac{x}{4}$

$$a_n = a_1 r^{n-1}$$
$$a_n = 3\left(\frac{x}{4}\right)^{n-1}$$

27. $3x - 6x + 12x - ...$

$a_1 = 3x$, $n = 12$, and $r = \frac{-6x}{3x} = -2$

$$S_n = \frac{a_1(1 - r^n)}{1 - r}$$
$$S_{12} = \frac{3x\left(1 - (-2)^{12}\right)}{1 - (-2)} = \frac{3x(1 - 4096)}{3} = x(-4095) = -4095x$$

29. $7 - 4 + \frac{16}{7} - ...$

$|r| = \left|\frac{-4}{7}\right| = \frac{4}{7} < 1$, the series has a limit.

$$S_\infty = \frac{a_1}{1 - r} = \frac{7}{1 - \left(-\frac{4}{7}\right)} = \frac{7}{\frac{11}{7}} = 7 \cdot \frac{7}{11} = \frac{49}{11}$$

31. $0.04 + 0.08 + 0.16 + 0.32 + ...$

$|r| = \left|\frac{0.08}{0.04}\right| = |2| = 2$, and since $|r| \not< 1$, the series does not have a limit.

33. $0.5555... = 0.5 + 0.05 + 0.005 + 0.0005 + ...$

This is an infinite geometric series with $a_1 = 0.5$.

$|r| = \left|\frac{0.05}{0.5}\right| = |0.1| = 0.1 < 1$, so the series has a limit.

$$S_\infty = \frac{a_1}{1 - r} = \frac{0.5}{1 - 0.1} = \frac{0.5}{0.9} = \frac{5}{9}$$

Fractional notation for 0.5555... is $\frac{5}{9}$.

35. ***Familiarize.*** A \$0.40 raise every 3 months for 8 years, is 32 raises. We want to find the 33th term of an arithmetic sequence with $a_1 = \$11.50$, $d = \$0.40$, and $n = 33$. We will first use the formula for a_n to find a_{33}. This result is the hourly wage after 8 years.

Translate. Substituting into the formula for a_n, we have

$$a_{33} = \$11.50 + (33 - 1)(\$0.40).$$

Carry out. We first find a_{33}.

$$a_{33} = \$11.50 + 32(\$0.40) = \$24.30$$

Check. We can do the calculations again.

State. After 8 years, Tyrone earns \$24.30 an hour.

37. ***Familiarize.*** At the end of each year, the interest at 4% will be added to the previous year's amount.

We have a geometric sequence:

$\$12{,}000, \$12{,}000(1.04), \$12{,}000(1.04)^2, ...$

We are looking for the amount after 7 years.

Translate. We use the formula $a_n = a_1 r^{n-1}$ with $a_1 = \$12{,}000$, $r = 1.04$, and $n = 8$:

$$a_8 = \$12{,}000(1.04)^{8-1}$$

Carry out. We calculate to obtain $a_8 \approx \$15{,}791.18$.

Check. We can do the calculation again.

State. After 7 years, the amount of the loan will be about \$15,791.18.

39. $7! = 7 \cdot 6 \cdot 5 \cdot 4 \cdot 3 \cdot 2 \cdot 1 = 5040$

41. $\binom{20}{2} a^{20-2} b^2 = 190 a^{18} b^2$

43. *Writing Exercise.* For a geometric sequence with $|r| < 1$, as n gets larger, the absolute value of the terms gets smaller, since $|r^n|$ gets smaller.

45. $1 - x + x^2 - x^3 + ...$

$a_1 = 1$, $n = n$, and $r = \frac{-x}{1} = -x$

$$S_n = \frac{1\left(1 - (-x)^n\right)}{1 - (-x)} = \frac{1 - (-x)^n}{1 + x}$$

Chapter 14 Test

1. $a_n = \dfrac{1}{n^2+1}$

$a_1 = \dfrac{1}{1^2+1} = \dfrac{1}{2}$

$a_2 = \dfrac{1}{2^2+1} = \dfrac{1}{5}$

$a_3 = \dfrac{1}{3^2+1} = \dfrac{1}{10}$

$a_4 = \dfrac{1}{4^2+1} = \dfrac{1}{17}$

$a_5 = \dfrac{1}{5^2+1} = \dfrac{1}{26}$

$a_{12} = \dfrac{1}{12^2+1} = \dfrac{1}{145}$

3. $\sum_{k=2}^{5}(1-2^k) = (1-2^2)+(1-2^3)+(1-2^4)+(1-2^5)$

$= -3+(-7)+(-15)+(-31) = -56$

5. $\frac{1}{2}, 1, \frac{3}{2}, 2, ...$

$a_1 = \frac{1}{2},\ d = \frac{1}{2}$, and $n = 13$

$a_n = a_1 + (n-1)d$

$a_{13} = \frac{1}{2} + (13-1)\left(\frac{1}{2}\right) = \frac{1}{2} + 12\left(\frac{1}{2}\right) = \frac{13}{2}$

7. We know that $a_5 = 16$ and $a_{10} = -3$. We would have to add d five times to get from a_5 to a_{10}. That is,

$16 + 5d = -3$

$5d = -19$

$d = -3.8.$

Since $a_5 = 16$, we subtract d four times to get a_1.

$a_1 = 16 - 4(-3.8) = 16 + 15.2 = 31.2$

9. $-3, 6, -12, ...$

$a_1 = -3,\ n = 10$, and $r = \dfrac{6}{-3} = -2$

$a_n = a_1 r^{n-1}$

$a_{10} = -3(-2)^{10-1} = -3(-2)^9 = -3(-512) = 1536$

11. $3, 9, 27, ...$

$a_1 = 3$, and $r = \dfrac{9}{3} = 3$

$a_n = a_1 r^{n-1}$

$a_n = 3(3)^{n-1} = 3^n$

13. $0.5 + 0.25 + 0.125 + ...$

$|r| = \left|\dfrac{0.25}{0.5}\right| = |0.5| = 0.5 < 1$, the series has a limit.

$S_\infty = \dfrac{a_1}{1-r} = \dfrac{0.5}{1-0.5} = \dfrac{0.5}{0.5} = 1$

15. $\$1000 + \$80 + \$6.40 + ...$

$|r| = \left|\dfrac{\$80}{\$1000}\right| = |0.08| = 0.08 < 1$, the series has a limit.

$S_\infty = \dfrac{a_1}{1-r} = \dfrac{\$1000}{1-0.08} = \dfrac{\$1000}{0.92} = \dfrac{\$25,000}{23} \approx \$1086.96$

17. ***Familiarize.*** We want to find the seventeenth term of an arithmetic sequence with $a_1 = 31$, $d = 2$, and $n = 17$. We will first use the formula for a_n to find a_{17}. This result is the number of seats in the 17th row.

Translate. Substituting into the formula for a_n, we have

$a_{17} = 31 + (17-1)2.$

Carry out. We first find a_{17}.

$a_{17} = 31 + 16 \cdot 2 = 63$

Check. We can do the calculations again.

State. There are 63 seats in the 17th row.

19. ***Familiarize.*** At the end of each week, the price will be 95% of the previous week's price.

We have a geometric sequence:

$\$10,000,\ \$10,000(0.95),\ \$10,000(0.95)^2, ...$

We are looking for the price after 10 weeks.

Translate. We use the formula $a_n = a_1 r^{n-1}$ with $a_1 = \$10,000$, $r = 0.95$, and $n = 11$:

$a_{10} = \$10,000(0.95)^{11-1}$

Carry out. We calculate to obtain $a_{11} \approx \$5987.37$.

Check. We can do the calculation again.

State. After 10 weeks, the price of the boat will be about \$5987.37.

21. $\dbinom{12}{9} = \dfrac{12!}{3!9!} = \dfrac{12 \cdot 11 \cdot 10 \cdot 9!}{3!9!} = \dfrac{12 \cdot 11 \cdot 10}{3 \cdot 2 \cdot 1} = 4 \cdot 11 \cdot 5 = 220$

23. $\dbinom{12}{3} a^{12-3} x^3 = 220a^9x^3$

25. $1 + \dfrac{1}{x} + \dfrac{1}{x^2} + \dfrac{1}{x^3} + ...$

$a_1 = 1,\ n = n$, and $r = \dfrac{\frac{1}{x}}{1} = \dfrac{1}{x}$

$S_n = \dfrac{1\left(1-\left(\frac{1}{x}\right)^n\right)}{1-\frac{1}{x}} = \dfrac{1-\left(\frac{1}{x}\right)^n}{1-\frac{1}{x}}$, or $\dfrac{x^n-1}{x^{n-1}(x-1)}$